Inverse problems lie at the heart of contemporary scientific inquiry and technological development. Applications include a variety of medical and other imaging techniques, which are used for early detection of cancer and pulmonary edema, location of oil and mineral deposits in the Earth's interior, creation of astrophysical images from telescope data, finding cracks and interfaces within materials, shape optimization, model identification in growth processes, and modeling in the life sciences, among others.

The expository survey essays in this book describe recent developments in inverse problems and imaging, including hybrid or couple-physics methods arising in medical imaging, Calderón's problem and electrical impedance tomography, inverse problems arising in global seismology and oil exploration, inverse spectral problems, and the study of asymptotically hyperbolic spaces. The book is suitable for graduate students and researchers interested in inverse problems and their applications.

Mathematical Sciences Research Institute
Publications

60

Inverse Problems and Applications: Inside Out II

Mathematical Sciences Research Institute Publications

1	Freed/Uhlenbeck: *Instantons and Four-Manifolds*, second edition
2	Chern (ed.): *Seminar on Nonlinear Partial Differential Equations*
3	Lepowsky/Mandelstam/Singer (eds.): *Vertex Operators in Mathematics and Physics*
4	Kac (ed.): *Infinite Dimensional Groups with Applications*
5	Blackadar: *K-Theory for Operator Algebras*, second edition
6	Moore (ed.): *Group Representations, Ergodic Theory, Operator Algebras, and Mathematical Physics*
7	Chorin/Majda (eds.): *Wave Motion: Theory, Modelling, and Computation*
8	Gersten (ed.): *Essays in Group Theory*
9	Moore/Schochet: *Global Analysis on Foliated Spaces*, second edition
10–11	Drasin/Earle/Gehring/Kra/Marden (eds.): *Holomorphic Functions and Moduli*
12–13	Ni/Peletier/Serrin (eds.): *Nonlinear Diffusion Equations and Their Equilibrium States*
14	Goodman/de la Harpe/Jones: *Coxeter Graphs and Towers of Algebras*
15	Hochster/Huneke/Sally (eds.): *Commutative Algebra*
16	Ihara/Ribet/Serre (eds.): *Galois Groups over* $\mathbb{Q}$
17	Concus/Finn/Hoffman (eds.): *Geometric Analysis and Computer Graphics*
18	Bryant/Chern/Gardner/Goldschmidt/Griffiths: *Exterior Differential Systems*
19	Alperin (ed.): *Arboreal Group Theory*
20	Dazord/Weinstein (eds.): *Symplectic Geometry, Groupoids, and Integrable Systems*
21	Moschovakis (ed.): *Logic from Computer Science*
22	Ratiu (ed.): *The Geometry of Hamiltonian Systems*
23	Baumslag/Miller (eds.): *Algorithms and Classification in Combinatorial Group Theory*
24	Montgomery/Small (eds.): *Noncommutative Rings*
25	Akbulut/King: *Topology of Real Algebraic Sets*
26	Judah/Just/Woodin (eds.): *Set Theory of the Continuum*
27	Carlsson/Cohen/Hsiang/Jones (eds.): *Algebraic Topology and Its Applications*
28	Clemens/Kollár (eds.): *Current Topics in Complex Algebraic Geometry*
29	Nowakowski (ed.): *Games of No Chance*
30	Grove/Petersen (eds.): *Comparison Geometry*
31	Levy (ed.): *Flavors of Geometry*
32	Cecil/Chern (eds.): *Tight and Taut Submanifolds*
33	Axler/McCarthy/Sarason (eds.): *Holomorphic Spaces*
34	Ball/Milman (eds.): *Convex Geometric Analysis*
35	Levy (ed.): *The Eightfold Way*
36	Gavosto/Krantz/McCallum (eds.): *Contemporary Issues in Mathematics Education*
37	Schneider/Siu (eds.): *Several Complex Variables*
38	Billera/Björner/Green/Simion/Stanley (eds.): *New Perspectives in Geometric Combinatorics*
39	Haskell/Pillay/Steinhorn (eds.): *Model Theory, Algebra, and Geometry*
40	Bleher/Its (eds.): *Random Matrix Models and Their Applications*
41	Schneps (ed.): *Galois Groups and Fundamental Groups*
42	Nowakowski (ed.): *More Games of No Chance*
43	Montgomery/Schneider (eds.): *New Directions in Hopf Algebras*
44	Buhler/Stevenhagen (eds.): *Algorithmic Number Theory: Lattices, Number Fields, Curves and Cryptography*
45	Jensen/Ledet/Yui: *Generic Polynomials: Constructive Aspects of the Inverse Galois Problem*
46	Rockmore/Healy (eds.): *Modern Signal Processing*
47	Uhlmann (ed.): *Inside Out: Inverse Problems and Applications*
48	Gross/Kotiuga: *Electromagnetic Theory and Computation: A Topological Approach*
49	Darmon/Zhang (eds.): *Heegner Points and Rankin L-Series*
50	Bao/Bryant/Chern/Shen (eds.): *A Sampler of Riemann–Finsler Geometry*
51	Avramov/Green/Huneke/Smith/Sturmfels (eds.): *Trends in Commutative Algebra*
52	Goodman/Pach/Welzl (eds.): *Combinatorial and Computational Geometry*
53	Schoenfeld (ed.): *Assessing Mathematical Proficiency*
54	Hasselblatt (ed.): *Dynamics, Ergodic Theory, and Geometry*
55	Pinsky/Birnir (eds.): *Probability, Geometry and Integrable Systems*
56	Albert/Nowakowski (eds.): *Games of No Chance 3*
57	Kirsten/Williams (eds.): *A Window into Zeta and Modular Physics*
58	Friedman/Hunsicker/Libgober/Maxim (eds.): *Topology of Stratified Spaces*
59	Caporaso/McKernan/Mustaţă/Popa (eds.): *Current Developments in Algebraic Geometry*

Volumes 1–4, 6–8, and 10–27 are published by Springer-Verlag

Inverse Problems and Applications: Inside Out II

Edited by

Gunther Uhlmann
University of Washington

Gunther Uhlmann
Department of Mathematics
University of Washington
gunther@math.uwashington.edu

Silvio Levy (*Series Editor*)
Mathematical Sciences Research Institute
Berkeley, CA 94720
levy@msri.org

The Mathematical Sciences Research Institute wishes to acknowledge support by the National Science Foundation and the *Pacific Journal of Mathematics* for the publication of this series.

CAMBRIDGE UNIVERSITY PRESS
Cambridge, New York, Melbourne, Madrid, Cape Town,
Singapore, São Paulo, Delhi, Mexico City

Cambridge University Press
32 Avenue of the Americas, New York, NY 10013-2473, USA

www.cambridge.org
Information on this title: www.cambridge.org/9781107032019

© Mathematical Sciences Research Institute 2013

This publication is in copyright. Subject to statutory exception and to the provisions of relevant collective licensing agreements, no reproduction of any part may take place without the written permission of Cambridge University Press.

First published 2013

Printed in the United States of America

A catalog record for this publication is available from the British Library.

ISBN 978-1-107-03201-9 Hardback

Cambridge University Press has no responsibility for the persistence or accuracy of URLs for external or third-party Internet Web sites referred to in this publication and does not guarantee that any content on such Web sites is, or will remain, accurate or appropriate.

Contents

Preface	ix
Calderón's inverse problem: imaging and invisibility	1
KARI ASTALA, MATTI LASSAS, AND LASSI PÄIVÄRINTA	
Resistor network approaches to electrical impedance tomography	55
LILIANA BORCEA, VLADIMIR DRUSKIN, FERNANDO GUEVARA VASQUEZ, AND ALEXANDER V. MAMONOV	
The Calderón inverse problem in two dimensions	119
COLIN GUILLARMOU AND LEO TZOU	
The Calderón problem on Riemannian manifolds	167
MIKKO SALO	
Enclosure methods for Helmholtz-type equations	249
JENN-NAN WANG AND TING ZHOU	
Multiwave methods via ultrasound	271
PLAMEN STEFANOV AND GUNTHER UHLMANN	
Hybrid inverse problems and internal functionals	325
GUILLAUME BAL	
Inverse problems for connections	369
GABRIEL P. PATERNAIN	
Elastic-wave inverse scattering based on reverse time migration with active and passive source reflection data	411
VALERIY BRYTIK, MAARTEN V. DE HOOP, AND ROBERT D. VAN DER HILST	
Inverse problems in spectral geometry	455
KIRIL DATCHEV AND HAMID HEZARI	
Microlocal analysis of asymptotically hyperbolic spaces and high energy resolvent estimates	487
ANDRÁS VASY	
Transmission eigenvalues in inverse scattering theory	529
FIORALBA CAKONI AND HOUSSEM HADDAR	

Preface

Inverse problems are those where one tries to obtain information about a system (such as a patient's body, an industrial object, or the Earth's interior) without taking it apart: the system is a "black box" whose properties we would like to reconstruct from its response to appropriate stimuli, such as waves or fields of various types.

Such problems lie at the heart of contemporary scientific inquiry and technological development. Their study has led to the development of a vast array of techniques in medical, geophysical, and industrial imaging, radar, sonar, and other areas. Practical applications of these techniques include the early detection of cancer and pulmonary edema, location of oil and mineral deposits in the Earth's interior, creation of astrophysical images from telescope data, detection of cracks and interfaces within materials, shape optimization, model identification in growth processes, and much more.

Most inverse problems arise from a physical situation modeled by a partial differential equation. The inverse problem is to determine parameters of the equation given some information about the solutions. The analysis of such problems brings together diverse areas of mathematics such as complex analysis, differential geometry, harmonic analysis, integral geometry, microlocal analysis, numerical analysis, optimization, partial differential equations, and probability. It is a fertile area for interaction between pure and applied mathematics. This book includes several chapters on some of the topics discussed at the Inverse Problems and Applications Semester held at MSRI in Fall 2010.

A prototypical example of an inverse boundary problem for an elliptic equation is the by now classical Calderón problem, also called electrical impedance tomography (EIT). Calderón proposed the problem in the mathematical literature in 1980. In EIT one attempts to determine the electrical conductivity of a medium by making voltage and current measurements at the boundary of the medium. The information is encoded in the Dirichlet-to-Neumann map associated to the conductivity equation. EIT arises in several applications including geophysical prospection (Calderón's original motivation) and medical imaging. In the last 30 years or so there has been remarkable progress on this problem. In the last few years this includes the two-dimensional problem, the case of partial data, and the discrete problem. The chapter by Astala, Lassas and Päivärinta describes some of this progress in the two-dimensional case. The chapter by Borcea, Druskin, Guevara Vasquez and Mamonov discusses recent progress on discrete network problems and their approximation to the continuous problem, also in the two-dimensional case. The chapter by Guillarmou and Tzou deals with the Calderón problem on two-dimensional manifolds and, more generally, with inverse boundary value problems for the Schrödinger equation on Riemann surfaces, including the case of partial data. The chapter by Salo gives a detailed account of progress achieved for the anisotropic problem in dimensions three and higher. The chapter by Wang and Zhou considers the determination of the location of discontinuous electromagnetic or elastic parameters using the enclosure method.

New inverse problems and methodologies arise all the time because of the applications. For example, in medical imaging, there has been considerable interest in recent years in multiwave methods, also called hybrid methods, which combine a high-resolution modality with a high-contrast one through a physical principle. For example, in breast imaging ultrasound provides high (submillimeter) resolution, but suffers from low contrast. On the other hand, many tumors absorb much more energy from electromagnetic waves than healthy cells. Thus using electromagnetic waves offers very high contrast. Examples of novel hybrid imaging methods are photoacoustic tomography (PAT), thermoacoustic tomography (TAT), ultrasound modulation tomography, transient elastography, and magnetic resonance elastography. We describe briefly PAT and TAT. PAT consists of exposing tissues to relatively harmless optical radiation that causes temperature increases in the millikelvin range, resulting in the generation of propagating ultrasound waves (the photoacoustic effect). Such ultrasound waves are readily measurable. The inverse problem then consists of reconstructing the optical properties of the tissue. In TAT, low-frequency microwaves, with wavelengths on the order of 1 m, are sent into the medium. The rationale for using the latter frequencies is that they are less absorbed than optical frequencies. PAT and TAT offer potential breakthroughs in the clinical application of hybrid imaging modalities to early detection of cancer, functional imaging, and molecular imaging. The inverse problem for these two modalities has two steps. The first step is to solve a well-posed inverse boundary problem for the wave equation. The chapter by Stefanov and Uhlmann surveys recent progress for this step, including the case of discontinuous sound speeds. From the first step one obtains different internal functionals for the elastic, electromagnetic, or optical properties of the medium, depending on the imaging modality. The inverse problem is to recover the parameters of the medium from knowledge of these internal functionals. The chapter by Bal summarizes in detail what is known about this second step.

An outstanding inverse problem in geophysics consists in determining the inner structure of the Earth from measurements of travel times of seismic waves.

From a mathematical point of view, the inner structure of the Earth is modelled by a Riemannian metric, and travel times by the lengths of geodesics between boundary points. This gives rise to a typical geometric inverse problem: is it possible to determine a Riemannian metric from its boundary distance function? The geodesic ray transform, where one integrates a function or a tensor field along geodesics of a Riemannian metric, is closely related to the boundary rigidity problem. The integration of a function along geodesics is the linearization of the boundary rigidity problem in a fixed conformal class. The standard X-ray transform, where one integrates a function along straight lines, corresponds to the case of the Euclidean metric and is the basis of medical imaging techniques such as CT and PET. The case of integration along more general geodesics arises in geophysical imaging in determining the inner structure of the Earth, since the speed of elastic waves generally increases with depth, thus curving the rays back to the surface. It also arises in ultrasound imaging, where the Riemannian metric models the anisotropic index of refraction. In tensor tomography problems one would like to determine a symmetric tensor field up to natural obstruction from its integrals over geodesics. Recently there has been much activity in the study of these transforms. Besides their importance in imaging technology, they arise naturally in various inverse problems in geometry as explained above.

In the case of Euclidean space with the Euclidean metric the attenuated ray transform is the basis of the medical imaging technology of SPECT and has been extensively studied. There are two natural directions in which this transform can be extended: one is to replace Euclidean space by a Riemannian manifold, and the other is to consider the case of systems where the attenuation is given, for example, by a unitary connection. The chapter by Paternain on inverse problems for connections deals with this latter case as well as other geometric inverse problems for connections.

One of the fascinating aspects of inverse problems is the continuous interplay between pure and applied mathematics. This interplay has been particularly noticeable in the applications of microlocal analysis (MA) to inverse problems. MA — which is, roughly speaking, local analysis in phase space — was developed about 40 years ago by Hörmander, Maslov, Sato, and many others in order to understand the propagation of singularities of solutions of partial differential equations. The early roots of MA were in the theory of geometrical optics. Microlocal analysis has been used successfully in determining the singularities of medium parameters in several inverse problems ranging from X-ray tomography to reflection seismology and electrical impedance tomography.

Brytik, De Hoop, and Van der Hilst's chapter considers applications of MA to geophysics. In particular it contains a new analysis of elastic reverse time migration, using MA as a tool, with applications to exploration seismology and global seismology. Another area where MA has been applied successfully in recent years is inverse spectral problems centered around Marc Kac's problem: can one hear the shape of a drum? The chapter by Datchev and Hezari surveys recent progress on this and other inverse spectral questions as well as inverse problems for resonances.

Vasy writes in his chapter about a compelling new way he found of doing scattering theory for the Laplacian on conformally compact spaces, including nontrapping high-energy bounds for the analytic continuation of the resolvent in appropriate circumstances, by appropriately extending the problem to a larger space. The resulting problems are not elliptic, but fit into a very nice microlocal framework on manifolds without boundary. The microlocal machinery being used is also very useful in other settings, including asymptotically hyperbolic spaces and Lorentzian geometries including Kerr–de Sitter spaces. Vasy develops this approach in detail in his chapter for the case of asymptotically hyperbolic spaces.

In scattering theory there has been a flurry of activity recently about transmission eigenvalues — wavenumbers at which the scattering operator has a nontrivial kernel and cokernel. The chapter by Cakoni and Haddar describes the recent developments as well as several open problems.

I would like to express my deep gratitude to the authors of the chapters in this book for their hard work and outstanding contributions. The book would not have been possible without MSRI's support of the inverse problems program in Fall 2010. Many thanks are also due to Silvio Levy for his great help in editing this book.

Gunther Uhlmann

Calderón's inverse problem: imaging and invisibility

KARI ASTALA, MATTI LASSAS AND LASSI PÄIVÄRINTA

We consider the determination of a conductivity function in a two-dimensional domain from the Cauchy data of the solutions of the conductivity equation on the boundary. In the first sections of the paper we consider this inverse problem, posed by Calderón, for conductivities that are in L^∞ and are bounded from below by a positive constant. After this we consider uniqueness results and counterexamples for conductivities that are degenerate, that is, not necessarily bounded from above or below. Elliptic equations with such coefficient functions are essential for physical models used in transformation optics and metamaterial constructions. The present counterexamples for the inverse problem have been related to invisibility cloaking. This means that there are conductivities for which a part of the domain is shielded from detection via boundary measurements. Such conductivities are called invisibility cloaks. At the end of the paper we consider the borderline of the smoothness required for the visible conductivities and the borderline of smoothness needed for invisibility cloaking conductivities.

1. Introduction

In electrical impedance tomography one aims to determine the internal structure of a body from electrical measurements on its surface. To consider the precise mathematical formulation of the electrical impedance tomography problem, suppose that $\Omega \subset \mathbb{R}^n$ is a bounded domain with connected complement and let us start with the case when $\sigma : \Omega \to (0, \infty)$ be a measurable function that is bounded away from zero and infinity.

Then the Dirichlet problem

$$\nabla \cdot \sigma \nabla u = 0 \quad \text{in } \Omega, \tag{1}$$

$$u\big|_{\partial\Omega} = \phi \in W^{1/2,2}(\partial\Omega) \tag{2}$$

The authors were supported by the Academy of Finland, the Finnish Centres of Excellence in Analysis and Dynamics and Inverse Problems, and the Mathematical Sciences Research Institute.

admits a unique solution u in the Sobolev space $W^{1,2}(\Omega)$. Here

$$W^{1/2,2}(\partial\Omega) = H^{1/2}(\partial\Omega) = W^{1,2}(\Omega)/W_0^{1,2}(\Omega)$$

stands for the space of equivalence classes of functions $W^{1,2}(\Omega)$ that are the same up to a function in $W_0^{1,2}(\Omega) = \mathrm{cl}_{W^{1,2}(\Omega)}(C_0^\infty(\Omega))$. This is the most general space of functions that can possibly arise as Dirichlet boundary values or traces of general $W^{1,2}(\Omega)$-functions in a bounded domain Ω.

In terms of physics, if the electric potential of a body Ω at point x is $u(x)$, having the boundary value $\phi = u|_{\partial\Omega}$, and there are no sources inside the body, u satisfies the equations (1)–(2). The electric current J in the body is equal to

$$J = -\sigma\nabla u.$$

In electrical impedance tomography, one measures only the normal component of the current, $\nu \cdot J|_{\partial\Omega} = -\nu \cdot \sigma\nabla u$, where ν is the unit outer normal to the boundary. For smooth σ this quantity is well defined pointwise, while for general bounded measurable σ we need to use the (equivalent) definition of $\nu \cdot \sigma\nabla u|_{\partial\Omega}$,

$$\langle \nu \cdot \sigma\nabla u, \psi \rangle = \int_\Omega \sigma(x)\nabla u(x) \cdot \nabla\psi(x)\, dm(x) \quad \text{for all } \psi \in W^{1,2}(\Omega)(\Omega), \quad (3)$$

as an element of $H^{-1/2}(\partial\Omega)$, the dual of space of $H^{1/2}(\partial\Omega) = W^{1/2,2}(\partial\Omega)$. Here, m is the Lebesgue measure.

Calderón's inverse problem is the question whether an unknown conductivity distribution inside a domain can be determined from the voltage and current measurements made on the boundary. The measurements correspond to the knowledge of the Dirichlet-to-Neumann map Λ_σ (or the equivalent quadratic form) associated to σ, i.e., the map taking the Dirichlet boundary values of the solution of the conductivity equation

$$\nabla \cdot \sigma(x)\nabla u(x) = 0 \qquad (4)$$

to the corresponding Neumann boundary values,

$$\Lambda_\sigma : u|_{\partial\Omega} \mapsto \nu \cdot \sigma\nabla u|_{\partial\Omega}. \qquad (5)$$

For sufficiently regular conductivities the Dirichlet-to-Neumann map Λ_σ can be considered as an operator from $W^{1/2,2}(\partial\Omega)$ to $W^{-1/2,2}(\partial\Omega)$. In the classical theory of the problem, the conductivity σ is bounded uniformly from above and below. The problem was originally proposed by Calderón [1980]. Sylvester and Uhlmann [1987] proved the unique identifiability of the conductivity in dimensions three and higher for isotropic conductivities which are C^∞-smooth, and Nachman [1988] gave a reconstruction method. In three dimensions or higher unique identifiability of the conductivity is proven for conductivities with 3/2 derivatives [Päivärinta et al. 2003; Brown and Torres 2003] and $C^{1,\alpha}$-smooth conductivities which are C^∞ smooth outside surfaces on which they have conormal singularities [Greenleaf et al. 2003b]. Haberman and Tataru [2011] have recently proven uniqueness for the C^1-smooth conductivities. The problems has also been solved with measurements only on a part of the boundary [Kenig et al. 2007].

In two dimensions the first global solution of the inverse conductivity problem is due to Nachman [1996a] for conductivities with two derivatives. In this seminal paper the $\bar{\partial}$ technique was first time used in the study of Calderón's inverse problem. The smoothness requirements were reduced in [Brown and Uhlmann 1997a] to Lipschitz conductivities. Finally, in [Astala and Päivärinta 2006] the uniqueness of the inverse problem was proven in the form that the problem was originally formulated in [Calderón 1980], i.e., for general isotropic conductivities in L^∞ which are bounded from below and above by positive constants.

The Calderón problem with an anisotropic, i.e., matrix-valued, conductivity that is uniformly bounded from above and below has been studied in two dimensions [Sylvester 1990; Nachman 1996a; Lassas and Uhlmann 2001; Astala et al. 2005; Imanuvilov et al. 2010] and in dimensions $n \geq 3$ [Lee and Uhlmann 1989; Lassas and Uhlmann 2001; Ferreira et al. 2009]. For example, for the anisotropic inverse conductivity problem in the two-dimensional case it is known that the Dirichlet-to-Neumann map determines a regular conductivity tensor up to a diffeomorphism $F : \overline{\Omega} \to \overline{\Omega}$, i.e., one can obtain an image of the interior of Ω in deformed coordinates. This implies that the inverse problem is not uniquely solvable, but the nonuniqueness of the problem can be characterized. We note that the problem in higher dimensions is presently solved only in special cases, like when the conductivity is real analytic.

Electrical impedance tomography has a variety of different applications for instance in engineering and medical diagnostics. For a general expository presentations see [Borcea 2002; Cheney et al. 1999], for medical applications see [Dijkstra et al. 1993].

In the last section we will study the inverse conductivity problem in the two-dimensional case with degenerate conductivities. Such conductivities appear in physical models where the medium varies continuously from a perfect conductor to a perfect insulator. As an example, we may consider a case where the conductivity goes to zero or to infinity near ∂D where $D \subset \Omega$ is a smooth open set. We ask what kind of degeneracy prevents solving the inverse problem, that is, we study what is the border of visibility. We also ask what kind of degeneracy makes it even possible to coat of an arbitrary object so that it appears the same as a homogeneous body in all static measurements, that is, we study what is the border of the invisibility cloaking. Surprisingly, these borders are not the same; we identify these borderlines and show that between them there are the electric holograms, that is, the conductivities creating an illusion of a nonexisting body (see Figure 1 on page 43). These conductivities are the counterexamples for the unique solvability of inverse problems for which even the topology of the domain can not be determined using boundary measurements.

In this presentation we concentrate on solving Calderón's inverse problem in two dimensions. The presentation is based on the works [Astala and Päivärinta 2006; Astala et al. 2009; 2005; 2011a], where the problem is considered using quasiconformal techniques. In higher dimensions the usual method is to reduce, by substituting $v = \sigma^{1/2} u$, the conductivity equation (1) to the Schrödinger equation and then to apply the methods of scattering theory. Indeed, after such a substitution v satisfies

$$\Delta v - qv = 0,$$

where $q = \sigma^{-1/2} \Delta \sigma^{1/2}$. This substitution is possible only if σ has some smoothness. In the case $\sigma \in L^\infty$, relevant for practical applications the reduction to the Schrödinger equation fails. In the two-dimensional case we can overcome this by using methods of complex analysis. However, what we adopt from the scattering theory type approaches is the use of exponentially growing solutions, the so-called geometric optics solutions to the conductivity equation (1). These are specified by the condition

$$u(z, \xi) = e^{i\xi z}\left(1 + \mathcal{O}\left(\frac{1}{|z|}\right)\right) \quad \text{as } |z| \to \infty, \tag{6}$$

where $\xi, z \in \mathbb{C}$ and ξz denotes the usual product of these complex numbers. Here we have set $\sigma \equiv 1$ outside Ω to get an equation defined globally. Studying the ξ-dependence of these solutions then gives rise to the basic concept of this presentation, the *nonlinear Fourier transform* $\tau_\sigma(\xi)$. The detailed definition will be given in Section 2F.

Thus to start the study of $\tau_\sigma(\xi)$ we need first to establish the existence of exponential solutions. Already here the quasiconformal techniques are essential. We note that the study of the inverse problems is closely related to the nonlinear Fourier transform: It is not difficult to show that the Dirichlet-to-Neumann boundary operator Λ_σ determines the nonlinear Fourier transforms $\tau_\sigma(\xi)$ for all $\xi \in \mathbb{C}$. Therefore the main difficulty, and our main strategy, is to invert the nonlinear Fourier transform, show that $\tau_\sigma(\xi)$ determines $\sigma(z)$ almost everywhere.

The properties of the nonlinear Fourier transform depend on the underlying differential equation. In one dimension the basic properties of the transform are fairly well understood, while deeper results such as analogs of Carleson's

L^2-convergence theorem remain open. The reader should consult the excellent lecture notes by Tao and Thiele [2003] for an introduction to the one-dimensional theory.

For (1) with nonsmooth σ, many basic questions concerning the nonlinear Fourier transform, even such as finding a right version of the Plancherel formula, remain open. What we are able to show is that for $\sigma^{\pm 1} \in L^\infty$, with $\sigma \equiv 1$ near ∞, we have a Riemann–Lebesgue type result,

$$\tau_\sigma \in C_0(\mathbb{C}).$$

Indeed, this requires the asymptotic estimates of the solutions (6), and these are the key point and main technical part of our argument. For results on related equations, see [Brown 2001]. The nonlinear Fourier transform in the two-dimensional case is also closely related to the Novikov–Veselov (NV) equation, which is a (2+1)-dimensional nonlinear evolution equation that generalizes the (1+1)-dimensional Korteweg-deVries(KdV) equation; see [Boiti et al. 1987; Lassas et al. 2007; Tsai 1993; Veselov and Novikov 1984].

2. Calderón's inverse for isotropic L^∞-conductivity

To avoid some of the technical complications, below we assume that the domain $\Omega = \mathbb{D} = \mathbb{D}(1)$, the unit disk. In fact the reduction of general Ω to this case is not difficult; see [Astala and Päivärinta 2006]. Our main aim in this section is to consider the following uniqueness result and its generalizations:

Theorem 2.1 [Astala and Päivärinta 2006]. *Let $\sigma_j \in L^\infty(\mathbb{D})$, $j = 1, 2$. Suppose that there is a constant $c > 0$ such that $c^{-1} \leq \sigma_j \leq c$. If*

$$\Lambda_{\sigma_1} = \Lambda_{\sigma_2},$$

then $\sigma_1 = \sigma_2$ almost everywhere. Here Λ_{σ_i}, $i = 1, 2$, are defined by (5).

For the first steps in numerical implementation of the solution of the inverse problem based on quasiconformal methods see [Astala et al. 2011b].

Our approach will be based on quasiconformal methods, which also enables the use of tools from complex analysis. These are not available in higher dimensions, at least to the same extent, and this is one of the reasons why the problem is still open for L^∞-coefficients in dimensions three and higher. The complex analytic connection comes as follows: From Theorem 2.3 below we see that if $u \in W^{1,2}(\mathbb{D})$ is a real-valued solution of (1), then it has the σ-harmonic conjugate $v \in W^{1,2}(\mathbb{D})$ such that

$$\partial_x v = -\sigma \partial_y u, \quad \partial_y v = \sigma \partial_x u. \tag{7}$$

Equivalently (see (26)), the function $f = u + iv$ satisfies the $\mathbb{R}$-linear Beltrami equation

$$\frac{\partial f}{\partial \bar{z}} = \mu(z) \overline{\frac{\partial f}{\partial z}}, \qquad (8)$$

where $\frac{\partial f}{\partial \bar{z}} = \partial_{\bar{z}} f = \frac{1}{2}(\partial_x f + i \partial_y f)$, $\frac{\partial f}{\partial z} = \partial_z f = \frac{1}{2}(\partial_x f - i \partial_y f)$, and

$$\mu = \frac{1-\sigma}{1+\sigma}.$$

In particular, note that μ is real-valued and that the assumptions on σ in Theorem 2.1 imply $\|\mu\|_{L^\infty} \leq k < 1$. This reduction to the Beltrami equation and the complex analytic methods it provides will be the main tools in our analysis of the Dirichlet-to-Neumann map and the solutions to (1).

2A. *Linear and nonlinear Beltrami equations.* A powerful tool for finding the exponential growing solutions to the conductivity equation (including degenerate conductivities) are given by the nonlinear Beltrami equation. We therefore first review a few of the basic facts here. For more details and results see [Astala et al. 2009].

We start with general facts on the linear divergence-type equation

$$\text{div } A(z) \nabla u = 0, \qquad z \in \Omega \subset \mathbb{R}^2 \qquad (9)$$

where we assume that $u \in W^{1,2}_{\text{loc}}(\Omega)$ and that the coefficient matrix

$$A = A(z) = \begin{bmatrix} \alpha_{11} & \alpha_{12} \\ \alpha_{21} & \alpha_{22} \end{bmatrix}, \qquad \alpha_{21} = \alpha_{12}, \qquad (10)$$

is symmetric and elliptic:

$$\frac{1}{K(z)} |\xi|^2 \leq \langle A(z)\xi, \xi \rangle \leq K(z)|\xi|^2, \qquad \xi \in \mathbb{R}^2, \qquad (11)$$

almost everywhere in Ω. Here, $\langle \eta, \xi \rangle = \eta_1 \xi_1 + \eta_2 \xi_2$ for $\eta, \xi \in \mathbb{R}^2$. We denote by $K_A(z)$ the smallest number for which (11) is valid. We start with the case when $A(z)$ is assumed to be isotropic, $A(z) = \sigma(z)\mathbf{I}$ with $\sigma(z) \in \mathbb{R}_+$. We also assume that there is $K \in \mathbb{R}_+$ such that $K_A(z) \leq K$ almost everywhere.

For many purposes it is convenient to express the above ellipticity condition (11) in terms of the single inequality

$$|\xi|^2 + |A(z)\xi|^2 \leq \left(K_A(z) + \frac{1}{K_A(z)} \right) \langle A(z)\xi, \xi \rangle \qquad (12)$$

valid for almost every $z \in \Omega$ and all $\xi \in \mathbb{R}^2$. For the symmetric matrix $A(z)$ this is seen by representing the matrix as a diagonal matrix in the coordinates given by the eigenvectors.

Below we will study the divergence equation (9) by reducing it to the complex Beltrami system. For solutions to (9) a conjugate structure, similar to harmonic functions, is provided by the *Hodge star* operator $*$, which here really is nothing more than the (counterclockwise) rotation by 90 degrees,

$$* = \begin{bmatrix} 0 & -1 \\ 1 & 0 \end{bmatrix} : \mathbb{R}^2 \to \mathbb{R}^2, \qquad ** = -\mathbf{I}. \tag{13}$$

There are two vector fields associated with each solution to the homogeneous equation

$$\operatorname{div} A(z)\nabla u = 0, \qquad u \in W^{1,2}_{\text{loc}}(\Omega).$$

The first, $E = \nabla u$, has zero curl (in the sense of distributions, the curl of any gradient field is zero), while the second, $B = A(z)\nabla u$, is divergence-free as a solution to the equation.

It is the Hodge star $*$ operator that transforms curl-free fields into divergence-free fields, and vice versa. In particular, if

$$E = \nabla w = (w_x, w_y), \qquad w \in W^{1,1}_{\text{loc}}(\Omega),$$

then $*E = (-w_y, w_x)$ and hence

$$\operatorname{div}(*E) = \operatorname{div}(*\nabla w) = 0,$$

at least in the distributional sense. We recall here a well-known fact from calculus (the Poincaré lemma):

Lemma 2.2. *Let $E \in L^p(\Omega, \mathbb{R}^2)$, $p \geq 1$, be a vector field defined on a simply connected domain Ω. If $\operatorname{Curl} E = 0$, then E is a gradient field; that is, there exists a real-valued function $u \in W^{1,p}(\Omega)$ such that $\nabla u = E$.*

When u is A-harmonic function in a simply connected domain Ω, that is, u solves the equation $\operatorname{div} A(z)\nabla u = 0$, then the field $*A\nabla u$ is curl-free and may be rewritten as

$$\nabla v = *A(z)\nabla u, \tag{14}$$

where $v \in W^{1,2}_{\text{loc}}(\Omega)$ is some Sobolev function unique up to an additive constant. This function v we call the *A-harmonic conjugate* of u. Sometimes in the literature one also finds the term *stream function* used for v.

The ellipticity conditions for A can be equivalently formulated for the induced complex function $f = u + iv$. We arrive, after a lengthy but quite routine purely algebraic manipulation, at the equivalent complex first-order equation for $f = u + iv$, which we record in the following theorem.

Theorem 2.3. *Let Ω be a simply connected domain and let $u \in W^{1,1}_{\text{loc}}(\Omega)$ be a solution to*

$$\text{div } A \nabla u = 0. \tag{15}$$

If $v \in W^{1,1}(\Omega)$ is a solution to the conjugate A-harmonic equation (14), the function $f = u + iv$ satisfies the homogeneous Beltrami equation

$$\frac{\partial f}{\partial \bar{z}} - \mu(z)\frac{\partial f}{\partial z} - \nu(z)\overline{\frac{\partial f}{\partial z}} = 0. \tag{16}$$

The coefficients are given by

$$\mu = \frac{\alpha_{22} - \alpha_{11} - 2i\alpha_{12}}{1 + \text{trace } A + \det A}, \quad \nu = \frac{1 - \det A}{1 + \text{trace } A + \det A}. \tag{17}$$

Conversely, if $f \in W^{1,1}_{\text{loc}}(\Omega, \mathbb{C})$ is a mapping satisfying (16), then $u = \text{Re}(f)$ and $v = \text{Im}(f)$ satisfy (14) with A given by solving the complex equations in (17):

$$\alpha_{11}(z) = \frac{|1-\mu|^2 - |\nu|^2}{|1+\nu|^2 - |\mu|^2}, \tag{18}$$

$$\alpha_{22}(z) = \frac{|1+\mu|^2 - |\nu|^2}{|1+\nu|^2 - |\mu|^2}, \tag{19}$$

$$\alpha_{12}(z) = \alpha_{21}(z) = \frac{-2\,\text{Im}(\mu)}{|1+\nu|^2 - |\mu|^2}, \tag{20}$$

The ellipticity of A can be explicitly measured in terms of μ and ν. The optimal ellipticity bound in (11) is

$$K_A(z) = \max\{\lambda_1(z), 1/\lambda_2(z)\}, \tag{21}$$

where $0 < \lambda_2(z) \le \lambda_1(z) < \infty$ are the eigenvalues of $A(z)$. With this choice we have pointwise

$$|\mu(z)| + |\nu(z)| = \frac{K_A(z) - 1}{K_A(z) + 1} < 1. \tag{22}$$

We also denote by $K_f(z)$ the smallest number for which the inequality

$$\|Df(z)\|^2 \le K_f(z) J(z, f) \tag{23}$$

is valid. Here, $Df(z) \in \mathbb{R}^2$ is the Jacobian matrix (or the derivative) of f at z and $J(z, f) = \det(Df(z))$ is the Jacobian determinant of f.

Below, let $k \in [0, 1]$ and $K \in [1, \infty]$ be constants satisfying

$$\sup_{z \in \Omega}(|\mu(z)| + |\nu(z)|) \le k \quad \text{and} \quad K := \frac{1+k}{1-k}. \tag{24}$$

Then (16) yields
$$\left|\frac{\partial f}{\partial \bar{z}}\right| \le k \left|\frac{\partial f}{\partial z}\right|.$$

The above ellipticity bounds have then the relation
$$K_f(z) \le K_A(z) \le K \quad \text{for a.e. } z \in \Omega. \tag{25}$$

A mapping $f \in W^{1,2}_{\text{loc}}(\Omega)$ satisfying (23) with $K_f(z) \le K < \infty$ is called a *K-quasiregular mappings*. If f is a homeomorphism, we call it *K-quasiconformal*. By Stoilow's factorization (Theorem A.9), any K-quasiregular mapping is a composition of holomorphic function and a K-quasiconformal mapping.

Remarks. 1. In this correspondence, ν is real valued if and only if the matrix A is symmetric.

2. A has determinant 1 if and only if $\nu = 0$ (this corresponds to the $\mathbb{C}$-linear Beltrami equation).

3. A is isotropic, that is, $A = \sigma(z)\mathbf{I}$ with $\sigma(z) \in \mathbb{R}_+$, if and only if $\mu(z) = 0$. For such A, the Beltrami equation (16) then takes the form
$$\frac{\partial f}{\partial \bar{z}} - \frac{1-\sigma}{1+\sigma} \overline{\frac{\partial f}{\partial z}} = 0. \tag{26}$$

2B. *Existence and uniqueness for nonlinear Beltrami equations.* Solutions to the Beltrami equation conformal near infinity are particularly useful in solving the equation.

When μ and ν as above have compact support and we have a $W^{1,2}_{\text{loc}}(\mathbb{C})$ solution to the Beltrami equation $f_{\bar{z}} = \mu f_z + \nu \overline{f_z}$ in $\mathbb{C}$, where $f_{\bar{z}} = \partial_{\bar{z}} f$ and $f_z = \partial_z f$, normalized by the condition
$$f(z) = z + \mathcal{O}(1/z)$$

near ∞, we call f a *principal solution*. Indeed, with the Cauchy and Beurling transform (see the Appendix) we have the identities
$$\frac{\partial f}{\partial z} = 1 + \mathcal{S}\frac{\partial f}{\partial \bar{z}} \tag{27}$$

and
$$f(z) = z + \mathcal{C}\left(\frac{\partial f}{\partial \bar{z}}\right)(z), \quad z \in \mathbb{C}. \tag{28}$$

Principal solutions are necessarily homeomorphisms. In fact we have the following fundamental *measurable Riemann mapping theorem*:

Theorem 2.4. Let $\mu(z)$ be compactly supported measurable function defined in $\mathbb{C}$ with $\|\mu\|_{L^\infty} \leq k < 1$. Then there is a unique principal solution to the Beltrami equation
$$\frac{\partial f}{\partial \bar{z}} = \mu(z) \frac{\partial f}{\partial z} \quad \text{for almost every } z \in \mathbb{C},$$
and the solution $f \in W^{1,2}_{\text{loc}}(\mathbb{C})$ is a K-quasiconformal homeomorphism of $\mathbb{C}$.

The result holds also for the general Beltrami equation with coefficients μ and ν; see Theorem 2.5 below.

In constructing the exponentially growing solutions to the divergence and Beltrami equations, the most powerful approach is by nonlinear Beltrami equations which we next discuss.

When one is looking for solutions to the general nonlinear elliptic systems
$$\frac{\partial f}{\partial \bar{z}} = H\!\left(z, f, \frac{\partial f}{\partial z}\right), \quad z \in \mathbb{C},$$
there are necessarily some constraints to be placed on the function H that we now discuss. We write
$$H : \mathbb{C} \times \mathbb{C} \times \mathbb{C} \to \mathbb{C}.$$
We will not strive for full generality, but settle for the following special case. For the most general existence results, with very weak assumptions on H, see [Astala et al. 2009]. Here we assume

(1) the homogeneity condition, that $f_{\bar{z}} = 0$ whenever $f_z = 0$, equivalently,
$$H(z, w, 0) \equiv 0, \quad \text{for almost every } (z, w) \in \mathbb{C} \times \mathbb{C};$$

(2) the uniform ellipticity condition, that for almost every $z, w \in \mathbb{C}$ and all $\zeta, \xi \in \mathbb{C}$,
$$|H(z, w, \zeta) - H(z, w, \xi)| \leq k |\zeta - \xi|, \quad 0 \leq k < 1; \tag{29}$$

(3) the Lipschitz continuity in the function variable,
$$|H(z, w_1, \zeta) - H(z, w_2, \zeta)| \leq C |\zeta| \, |w_1 - w_2|$$
for some absolute constant C independent of z and ζ.

Theorem 2.5. Suppose $H : \mathbb{C} \times \mathbb{C} \times \mathbb{C} \to \mathbb{C}$ satisfies the conditions (1)–(3) above and is compactly supported in the z-variable. Then the uniformly elliptic nonlinear differential equation
$$\frac{\partial f}{\partial \bar{z}} = H\!\left(z, f, \frac{\partial f}{\partial z}\right) \tag{30}$$
admits exactly one principal solution $f \in W^{1,2}_{\text{loc}}(\mathbb{C})$.

Sketch of the proof. (For complete proof, see [Astala et al. 2009, Chapter 8]) Uniqueness is easy. Suppose that both f and g are principal solutions to (30). Then

$$\frac{\partial f}{\partial \bar{z}} = H\left(z, f, \frac{\partial f}{\partial z}\right), \quad \frac{\partial g}{\partial \bar{z}} = H\left(z, g, \frac{\partial f}{\partial z}\right).$$

We set $F = f - g$ and estimate

$$\begin{aligned}|F_{\bar{z}}| &= |H(z, f, f_z) - H(z, g, g_z)| \\ &\leq |H(z, f, f_z) - H(z, f, g_z)| + |H(z, f, g_z) - H(z, g, g_z)| \\ &\leq k|f_z - g_z| + C\chi_R|g_z||f - g|,\end{aligned}$$

where χ_R denotes the characteristic function of the disk $\mathbb{D}(R)$ of radius R and center zero. It follows that F satisfies the differential inequality

$$|F_{\bar{z}}| \leq k|F_z| + C\chi_R|g_z||F|.$$

By assumption, the principal solutions f, g lie in $W^{1,2}_{\mathrm{loc}}(\mathbb{C})$ with

$$\lim_{z \to \infty} f(z) - g(z) = 0.$$

Once we observe that

$$\sigma = C\chi_R(z)|g_z| \in L^2(\mathbb{C})$$

has compact support, Liouville type results such as Theorem A.8 in the Appendix shows us that $F \equiv 0$, as desired.

The proof of existence we only sketch; for details, in the more general setup of Lusin measurable H, see [Astala et al. 2009, Chapter 8].

We look for a solution f in the form

$$f(z) = z + \mathcal{C}\phi, \quad \phi \in L^p(\mathbb{C}) \quad \text{of compact support}, \tag{31}$$

where the exponent $p > 2$. Note that

$$f_{\bar{z}} = \phi, \quad f_z = 1 + \mathcal{S}\phi.$$

Thus we need to solve only the integral equation

$$\phi = H(z, z + \mathcal{C}\phi, 1 + \mathcal{S}\phi). \tag{32}$$

To solve this equation we first associate with every given $\phi \in L^p(\mathbb{C})$ an operator $R : L^p(\mathbb{C}) \to L^p(\mathbb{C})$ defined by

$$R\Phi = H(z, z + \mathcal{C}\phi, 1 + \mathcal{S}\Phi)$$

Through the ellipticity hypothesis we observe that $\boldsymbol{R}$ is a contractive operator on $L^p(\mathbb{C})$. Indeed, from (29) we have the pointwise inequality

$$|\boldsymbol{R}\Phi_1 - \boldsymbol{R}\Phi_2| \le k\,|\mathscr{S}\Phi_1 - \mathscr{S}\Phi_2|.$$

Hence

$$\|\boldsymbol{R}\Phi_1 - \boldsymbol{R}\Phi_2\|_p \le k\,S_p\|\Phi_1 - \Phi_2\|_p, \qquad k\,S_p < 1,$$

for p sufficiently close to 2. By the Banach contraction principle, $\boldsymbol{R}$ has a unique fixed point $\Phi \in L^p(\mathbb{C})$. In other words, with each $\phi \in L^p(\mathbb{C})$ we can associate a unique function $\Phi \in L^p(\mathbb{C})$ such that

$$\Phi = H(z, z + \mathscr{C}\phi, 1 + \mathscr{S}\Phi) \tag{33}$$

In fact, the procedure (33), $\phi \mapsto \Phi$, gives a well-defined and nonlinear operator $T : L^p(\mathbb{C}) \to L^p(\mathbb{C})$ by simply requiring that $T\phi = \Phi$. Further, solving the original integral equation (32) means precisely that we have to find a fixed point for the operator T. This, however, is more involved than in the case of the contraction $\boldsymbol{R}$, and one needs to invoke the celebrated Schauder fixed-point theorem, see [Astala et al. 2009, Chapter 8] for details. □

2C. Complex geometric optics solutions. Below in this section, where we prove Theorem 2.1, *we will assume* that A is isotropic,

$$A(z) = \sigma(z)\mathbf{I}, \quad \sigma(z), \ \sigma(z) \in \mathbb{R}_+ \text{ and } c_1 \le \sigma(z) \le c_2 \text{ with } c_1, c_2 > 0.$$

Moreover, we will set

$$\nu = \frac{1 - \sigma(z)}{1 + \sigma(z)}.$$

We will use the convenient notation

$$e_\xi(z) = e^{i(z\xi + \bar{z}\bar{\xi})}, \qquad z, \xi \in \mathbb{C}. \tag{34}$$

The main emphasize in the analysis below is on isotropic conductivities, corresponding to the Beltrami equations of type (26). However, for later purposes it is useful to consider exponentially growing solutions to divergence equations with matrix coefficients, hence we are led to general Beltrami equations.

We will extend the coefficient matrix $A(z)$ to the entire plane $\mathbb{C}$ by requiring $A(z) \equiv \mathbf{I}$ when $|z| \ge 1$. Clearly, this keeps all ellipticity bounds. Moreover, then

$$\mu(z) \equiv \nu(z) \equiv 0, \qquad |z| \ge 1.$$

As a first step toward Theorem 2.1, we establish the existence of a family of special solutions to (16). These, called the complex geometric optics solutions, are specified by having the asymptotics

$$f_{\mu,\nu}(z,\xi) = e^{i\xi z} M_{\mu,\nu}(z,\xi), \tag{35}$$

where

$$M_{\mu,\nu}(z,\xi) - 1 = \mathcal{O}\left(\frac{1}{z}\right) \quad \text{as } |z| \to \infty. \tag{36}$$

Theorem 2.6. *Let μ and ν be functions supported in $\mathbb{D}$ that k in (24) satisfies $k < 1$. Then for each parameter $\xi \in \mathbb{C}$ and for each $2 \leq p < 1 + 1/k$, the equation*

$$\frac{\partial f}{\partial \bar{z}} = \mu(z)\frac{\partial f}{\partial z} + \nu(z)\overline{\frac{\partial f}{\partial z}} \tag{37}$$

admits a unique solution $f = f_{\mu,\nu} \in W^{1,p}_{\text{loc}}(\mathbb{C})$ that has the form (35) with (36) holding. In particular, $f(z,0) \equiv 1$.

Proof. Any solution to (37) is quasiregular. If $\xi = 0$, (35) and (36) imply that f is bounded, hence constant by the Liouville theorem.

If $\xi \neq 0$, we seek for a solution $f = f_{\mu,\nu}(z,\xi)$ of the form

$$f(z,\xi) = e^{i\xi \psi_\xi(z)}, \qquad \psi_\xi(z) = z + \mathcal{O}\left(\frac{1}{z}\right) \quad \text{as } |z| \to \infty \tag{38}$$

Substituting (38) into (37) indicates that ψ_ξ is the principal solution to the quasilinear equation

$$\frac{\partial}{\partial \bar{z}}\psi_\xi(z) = \mu(z)\frac{\partial}{\partial z}\psi_\xi(z) - \frac{\bar{\xi}}{\xi} e_{-\xi}(\psi_\xi(z))\, \nu(z)\, \overline{\frac{\partial}{\partial z}\psi_\xi(z)}. \tag{39}$$

The function $H(z,w,\zeta) = \mu(z)\zeta - (\bar{\xi}/\xi)\,\nu(z)\,e_\xi(w)\,\bar{\zeta}$ satisfies requirements (1)–(3) of Theorem 2.5. We thus obtain the existence and uniqueness of the principal solution ψ_ξ in $W^{1,2}_{\text{loc}}(\mathbb{C})$. Equation (39) and Theorem A.5 yield $\psi_\xi \in W^{1,p}_{\text{loc}}(\mathbb{C})$ for all $p < 1 + 1/k$ since $|\mu(z)| \leq k$ and e_ξ is unimodular.

Finally, to see the uniqueness of the complex geometric optics solution $f_{\mu,\nu}$, let $f \in W^{1,2}_{\text{loc}}(\mathbb{C})$ be a solution to (37) satisfying

$$f = \alpha e^{i\xi z}\left(1 + \mathcal{O}\left(\frac{1}{z}\right)\right) \quad \text{as } |z| \to \infty. \tag{40}$$

Denote then

$$\mu_1(z) = \mu(z)\frac{\overline{\partial_z f(z)}}{\partial_z f(z)}$$

where $\partial_z f(z) \neq 0$ and set $\mu_1 = 0$ elsewhere. Next, let φ be the unique principal solution to

$$\frac{\partial \varphi}{\partial \bar{z}} = \mu_1 \frac{\partial \varphi}{\partial z}. \tag{41}$$

Then the Stoilow factorization, Theorem A.9, gives $f = h \circ \varphi$, where $h : \mathbb{C} \to \mathbb{C}$ is an entire analytic function. But (40) shows that

$$\frac{h \circ \varphi(z)}{\exp(i\xi\varphi(z))} = \frac{f(z)}{\exp(i\xi\varphi(z))}$$

has the limit α when the variable $z \to \infty$. Thus

$$h(z) \equiv \alpha e^{i\xi z}.$$

Therefore $f(z) = \alpha \exp(i\xi\varphi(z))$. In particular, if we have two solutions f_1, f_2 satisfying (35), (36), then the argument gives

$$f_\varepsilon := f_1 - (1+\varepsilon)f_2 = \varepsilon e^{i\xi\varphi(z)},$$

The principal solutions are homeomorphisms with $\phi(z) = z + O(1/z)$ as $|z| \to \infty$, where the error term $O(1/z)$ is uniformly bounded by Koebe distortion (Lemma A.6). Letting now $\varepsilon \to 0$ gives $f_1 = f_2$. □

It is useful to note that if a function f satisfies (37), then if satisfies not the same equation but the equation where ν is replaced by $-\nu$. In terms of the real and imaginary parts of $f = u + iv$, we see that

$$\frac{\partial f}{\partial \bar{z}} = \mu(z) \frac{\partial f}{\partial z} + \nu(z) \overline{\frac{\partial f}{\partial z}} \quad \text{if and only if}$$

$$\nabla \cdot A(z) \nabla u = 0 \quad \text{and} \quad \nabla \cdot A^*(z) \nabla v = 0, \quad (42)$$

where

$$A^*(z) = *A(z)^{-1}* = \frac{1}{\det A} A.$$

In case $A(z) = \sigma(z)\mathbf{I}$ is isotropic (i.e., $\mu = 0$) and bounded by positive constants from above and below, we see that

$$\frac{\partial f}{\partial \bar{z}} = \frac{1-\sigma}{1+\sigma} \overline{\frac{\partial f}{\partial z}} \quad \Leftrightarrow \quad \nabla \cdot \sigma \nabla u = 0 \quad \text{and} \quad \nabla \cdot \frac{1}{\sigma} \nabla v = 0.$$

From these identities we obtain the complex geometric optics solutions also for the conductivity equation (1).

Corollary 2.7. *Suppose $A(z)$ is uniformly elliptic, so that (11) holds with $K \in L^\infty(\mathbb{D})$. Assume also that $A(z) = \mathbf{I}$ for $|z| \geq 1$.*
Then the equation $\nabla \cdot A(z) \nabla u(z) = 0$ admits a unique weak solution $u = u_\xi \in W^{1,2}_{\text{loc}}(\mathbb{C})$ such that

$$u(z, \xi) = e^{i\xi z}\left(1 + \mathbb{O}\left(\frac{1}{|z|}\right)\right) \quad \text{as } |z| \to \infty. \quad (43)$$

Proof. Let $f_{\mu,\nu}$ be the solution considered in Theorem 2.6. Using equivalence (42) we see that the function

$$u(z,\xi) = \operatorname{Re} f_{\mu,\nu} + i \operatorname{Im} f_{\mu,-\nu} \tag{44}$$

satisfies the requirements stated in the claim of the corollary.

To see the uniqueness, let $u \in W^{1,2}_{\text{loc}}$ be any function satisfying the divergence equation $\nabla \cdot A(z) \nabla u(z) = 0$ with (43). Then using Theorem 2.3 for the real and imaginary parts of u, we can write it as

$$u = \operatorname{Re} f_+ + i \operatorname{Im} f_- = \tfrac{1}{2}(f_+ + f_- + \overline{f_+} - \overline{f_-}),$$

where $f_\pm$ are quasiregular mappings with

$$\frac{\partial f_\pm}{\partial \bar{z}} = \mu(z) \frac{\partial f_\pm}{\partial z} \pm \nu(z) \overline{\frac{\partial f_\pm}{\partial z}}$$

and where μ, ν are given by (17). Given the asymptotics (43), it is not hard to see that both f_+ and f_- satisfy (35) with (36). Therefore $f_+ = f_{\mu,\nu}$ and $f_- = f_{\mu,-\nu}$. □

The exponentially growing solutions of Corollary 2.7 can be considered σ-harmonic counterparts of the usual exponential functions $e^{i\xi z}$. They are the building blocks of the *nonlinear Fourier transform* to be discussed in more detail in Section 2F.

2D. The Hilbert transform $\mathcal{H}_\sigma$. Assume that $u \in W^{1,2}(\mathbb{D})$ is a weak solution to $\nabla \cdot \sigma(z) \nabla u(z) = 0$. Then, by Theorem 2.3, u admits a conjugate function $v \in W^{1,2}(\mathbb{D})$ such that

$$\partial_x v = -\sigma \partial_y u, \quad \partial_y v = \sigma \partial_x u.$$

Let us now elaborate on the relationship between u and v. Since the function v is defined only up to a constant, we will normalize it by assuming

$$\int_{\partial \mathbb{D}} v \, ds = 0. \tag{45}$$

This way we obtain a unique map $\mathcal{H}_\mu : W^{1/2,2}(\partial \mathbb{D}) \to W^{1/2,2}(\partial \mathbb{D})$ by setting

$$\mathcal{H}_\mu : u\big|_{\partial \mathbb{D}} \mapsto v\big|_{\partial \mathbb{D}}. \tag{46}$$

In other words, $v = \mathcal{H}_\mu(u)$ if and only if $\int_{\partial \mathbb{D}} v \, ds = 0$, and $u + iv$ has a $W^{1,2}$-extension f to the disk $\mathbb{D}$ satisfying $f_{\bar{z}} = \mu \overline{f_z}$. We call $\mathcal{H}_\mu$ the *Hilbert transform* corresponding to (37).

Since the function $g = -if = v - iu$ satisfies $g_{\bar{z}} = -\mu \overline{g_z}$, we have

$$\mathcal{H}_\mu \circ \mathcal{H}_{-\mu} u = \mathcal{H}_{-\mu} \circ \mathcal{H}_\mu u = -u + \frac{1}{2\pi} \int_{\partial \mathbb{D}} u \, ds. \tag{47}$$

So far we have defined $\mathcal{H}_\mu(u)$ only for real-valued functions u. By setting

$$\mathcal{H}_\mu(iu) = i\mathcal{H}_{-\mu}(u),$$

we extend the definition of $\mathcal{H}_\mu(\cdot)$ to all $\mathbb{C}$-valued functions in $W^{1/2,2}(\partial \mathbb{D})$. Note, however, that $\mathcal{H}_\mu$ still remains only $\mathbb{R}$-linear.

As in the case of analytic functions, the Hilbert transform defines a projection, now on the "μ-analytic" functions. That is, we define

$$Q_\mu : W^{1/2,2}(\partial \mathbb{D}) \to W^{1/2,2}(\partial \mathbb{D})$$

by

$$Q_\mu(g) = \frac{1}{2}\left(g - i\mathcal{H}_\mu g\right) + \frac{1}{4\pi} \int_{\partial \mathbb{D}} g \, ds. \tag{48}$$

Then Q_μ is a projector in the sense that $Q_\mu^2 = Q_\mu$.

Lemma 2.8. *If $g \in W^{1/2,2}(\partial \mathbb{D})$, the following conditions are equivalent:*

(a) $g = f|_{\partial \mathbb{D}}$, *where $f \in W^{1,2}(\mathbb{D})$ satisfies $f_{\bar{z}} = \mu \overline{f_z}$.*

(b) $Q_\mu(g)$ *is a constant.*

Proof. Condition (a) holds if and only if $g = u + i\mathcal{H}_\mu u + ic$ for some real-valued $u \in W^{1/2,2}(\partial \mathbb{D})$ and real constant c. If g has this representation, then $Q_\mu(g) = \frac{1}{4\pi} \int_{\partial \mathbb{D}} u \, ds + ic$. On the other hand, if $Q_\mu(g)$ is a constant, then we put $g = u + iw$ into (48) and use (47) to show that $w = \mathcal{H}_\mu u + $ constant. This shows that (a) holds. $\square$

The Dirichlet-to-Neumann map (5) and the Hilbert transform (46) are closely related, as the next lemma shows.

Theorem 2.9. *Choose the counterclockwise orientation for $\partial \mathbb{D}$ and denote by ∂_T the tangential (distributional) derivative on $\partial \mathbb{D}$ corresponding to this orientation. We then have*

$$\partial_T \mathcal{H}_\mu(u) = \Lambda_\sigma(u). \tag{49}$$

In particular, the Dirichlet-to-Neumann map Λ_σ uniquely determines $\mathcal{H}_\mu$, $\mathcal{H}_{-\mu}$ and $\Lambda_{1/\sigma}$.

Proof. By the definition of Λ_σ we have

$$\int_{\partial \mathbb{D}} \varphi \Lambda_\sigma u \, ds = \int_{\mathbb{D}} \nabla \varphi \cdot \sigma \nabla u \, dm(x), \qquad \varphi \in C^\infty(\overline{\mathbb{D}}).$$

Thus, by (7) and integration by parts, we get

$$\int_{\partial \mathbb{D}} \varphi \Lambda_\sigma u \, ds = \int_{\mathbb{D}} \left(\partial_x \varphi \, \partial_y v - \partial_y \varphi \, \partial_x v \right) dm(x) = -\int_{\partial \mathbb{D}} v \, \partial_T \varphi \, ds,$$

and (49) follows. Next,

$$-\mu = (1 - 1/\sigma)/(1 + 1/\sigma),$$

and so $\Lambda_{1/\sigma}(u) = \partial_T \mathcal{H}_{-\mu}(u)$. Since by (47) $\mathcal{H}_\mu$ uniquely determines $\mathcal{H}_{-\mu}$, the proof is complete. $\square$

With these identities we can now show that, for the points z that lie outside $\mathbb{D}$, the values of the complex geometric optics solutions $f_\mu(z, \xi)$ and $f_{-\mu}(z, \xi)$ are determined by the Dirichlet-to-Neumann operator Λ_σ.

Theorem 2.10. *Let σ and $\widetilde{\sigma}$ be two conductivities satisfying the assumptions of Theorem 2.1 and assume $\Lambda_\sigma = \Lambda_{\widetilde{\sigma}}$. Then if μ and $\widetilde{\mu}$ are the corresponding Beltrami coefficients, we have*

$$f_\mu(z, \xi) = f_{\widetilde{\mu}}(z, \xi) \quad \text{and} \quad f_{-\mu}(z, \xi) = f_{-\widetilde{\mu}}(z, \xi) \qquad (50)$$

for all $z \in \mathbb{C} \setminus \overline{\mathbb{D}}$ and $\xi \in \mathbb{C}$.

Proof. By Theorem 2.9 the condition $\Lambda_\sigma = \Lambda_{\widetilde{\sigma}}$ implies that $\mathcal{H}_\mu = \mathcal{H}_{\widetilde{\mu}}$. In the same way Λ_σ determines $\Lambda_{\sigma^{-1}}$, and so it is enough to prove the first claim of (50).

Fix the value of the parameter $\xi \in \mathbb{C}$. From (48) we see that the projections $Q_\mu = Q_{\widetilde{\mu}}$, and thus by Lemma 2.8

$$Q_\mu(\widetilde{f}) = Q_{\widetilde{\mu}}(\widetilde{f}) \quad \text{is constant.}$$

Here we used the notation $\widetilde{f} = f_{\widetilde{\mu}}|_{\partial \mathbb{D}}$. Using Lemma 2.8 again, we see that there exists a function $G \in W^{1,2}(\mathbb{D})$ such that $G_{\bar{z}} = \mu \overline{G_z}$ in $\mathbb{D}$ and

$$G\big|_{\partial \mathbb{D}} = \widetilde{f}.$$

We then define $G(z) = f_{\widetilde{\mu}}(z, \xi)$ for z outside $\mathbb{D}$. Now $G \in W^{1,2}_{\text{loc}}(\mathbb{C})$, and it satisfies $G_{\bar{z}} = \mu \overline{G_z}$ in the whole plane. Thus it is quasiregular, and so $G \in W^{1,p}_{\text{loc}}(\mathbb{C})$ for all $2 \leq p < 2 + 1/k$, $k = \|\mu\|_\infty$. But now G is a solution to (35) and (36). By the uniqueness part of Theorem 2.6, we obtain $G(z) \equiv f_\mu(z, \xi)$. $\square$

Similarly, the Dirichlet-to-Neumann operator determines the complex geometric optics solutions to the conductivity equation at every point z outside the disk $\mathbb{D}$.

Corollary 2.11. *Let σ and $\tilde{\sigma}$ be two conductivities satisfying the assumptions of Theorem 2.1 and assume $\Lambda_\sigma = \Lambda_{\tilde{\sigma}}$. Then*
$$u_\sigma(z,\xi) = u_{\tilde{\sigma}}(z,\xi) \quad \text{for all } z \in \mathbb{C} \setminus \overline{\mathbb{D}} \text{ and } \xi \in \mathbb{C}.$$

Proof. The claim follows immediately from the previous theorem and the representation $u_\sigma(z,\xi) = \operatorname{Re} f_\mu(z,\xi) + i\operatorname{Im} f_{-\mu}(z,\xi)$. □

2E. *Dependence on parameters.* Our strategy will be to extend the identities $f_\mu(z,\xi) = f_{\tilde{\mu}}(z,\xi)$ and $u_\sigma(z,\xi) = u_{\tilde{\sigma}}(z,\xi)$ from outside the disk to points z inside $\mathbb{D}$. Once we do that, Theorem 2.1 follows via the equation $f_{\bar{z}} = \mu \overline{f_z}$.

For this purpose we need to understand the ξ-dependence in $f_\mu(z,\xi)$ and the quantities controlling it. In particular, we will derive equations relating the solutions and their derivatives with respect to the ξ-variable. For this purpose we prove the following theorem.

Theorem 2.12. *The complex geometric optics solutions $u_\sigma(z,\xi)$ and $f_\mu(z,\xi)$ are (Hölder-)continuous in z and C^∞-smooth in the parameter ξ.*

The continuity in the z-variable is clear since f_μ is a quasiregular function of z. However, for analyzing the ξ-dependence we need to realize the solutions in a different manner, by identities involving linear operators that depend smoothly on the variable ξ.

Let
$$f_\mu(z,\xi) = e^{i\xi z} M_\mu(z,\xi), \quad f_{-\mu}(z,\xi) = e^{i\xi z} M_{-\mu}(z,\xi)$$

be the solutions of Theorem 2.6 corresponding to conductivities σ and σ^{-1}, respectively. We can write (8), (35) and (36) in the form

$$\frac{\partial}{\partial \bar{z}} M_\mu = \mu(z) \overline{\frac{\partial}{\partial z}(e_\xi M_\mu)}, \quad M_\mu - 1 \in W^{1,p}(\mathbb{C}) \tag{51}$$

when $2 < p < 1 + 1/k$. By taking the Cauchy transform and introducing an $\mathbb{R}$-linear operator L_μ,

$$L_\mu g = \mathcal{C}\left(\mu \overline{\frac{\partial}{\partial \bar{z}}(e_{-\xi} \bar{g})}\right), \tag{52}$$

we see that (51) is equivalent to

$$(\mathbf{I} - L_\mu) M_\mu = 1. \tag{53}$$

Theorem 2.13. *Assume that $\xi \in \mathbb{C}$ and $\mu \in L^\infty(\mathbb{C})$ is compactly supported with $\|\mu\|_\infty \leq k < 1$. Then for $2 < p < 1 + 1/k$ the operator*

$$\mathbf{I} - L_\mu : W^{1,p}(\mathbb{C}) \oplus \mathbb{C} \to W^{1,p}(\mathbb{C}) \oplus \mathbb{C}$$

is bounded and invertible.

Here we denote by $W^{1,p}(\mathbb{C}) \oplus \mathbb{C}$ the Banach space consisting of functions of the form $f = \text{constant} + f_0$, where $f_0 \in W^{1,p}(\mathbb{C})$.

Proof. We write $L_\mu(g)$ as

$$L_\mu(g) = \mathscr{C}\big(\mu\, e_{-\xi}\,\overline{g_z} - i\overline{\xi}\,\mu\, e_{-\xi}\,\overline{g}\big). \tag{54}$$

Then Theorem A.2 shows that

$$L_\mu : W^{1,p}(\mathbb{C}) \oplus \mathbb{C} \to W^{1,p}(\mathbb{C}) \tag{55}$$

is bounded. Thus we need only establish invertibility.

To this end let us assume $h \in W^{1,p}(\mathbb{C})$. Consider the equation

$$(\mathbf{I} - L_\mu)(g + C_0) = h + C_1, \tag{56}$$

where $g \in W^{1,p}(\mathbb{C})$ and C_0, C_1 are constants. Then

$$C_0 - C_1 = g - h - L_\mu(g + C_0),$$

which by (55) gives $C_0 = C_1$. By differentiating and rearranging we see that (56) is equivalent to $g_{\bar{z}} - \mu(e_{-\xi}\overline{g})_{\bar{z}} = h_{\bar{z}} + \mu(\overline{C}_0 e_{-\xi})_{\bar{z}}$, or in other words, to

$$g_{\bar{z}} - (\mathbf{I} - \mu e_{-\xi}\overline{\mathscr{S}})^{-1}\big(\mu(e_{-\xi})_{\bar{z}}\overline{g}\big) = (\mathbf{I} - \mu e_{-\xi}\overline{\mathscr{S}})^{-1}\big(h_{\bar{z}} + \mu(\overline{C}_0 e_{-\xi})_{\bar{z}}\big). \tag{57}$$

We analyze this by using the real linear operator R defined by

$$R(g) = \mathscr{C}\left(\mathbf{I} - \nu\overline{\mathscr{S}}\right)^{-1}(\alpha\overline{g}),$$

where $\nu(z) = \mu e_{-\xi}$ satisfies $|\nu(z)| \le k\chi_{\mathbb{D}}(z)$ and α is defined by $\alpha = \mu(e_{-\xi})_{\bar{z}} = -i\overline{\xi}\,\mu\, e_{-\xi}$. According to Theorem A.4, $\mathbf{I} - \nu\overline{\mathscr{S}}$ is invertible in $L^p(\mathbb{C})$ when $1 + k < p < 1 + 1/k$, while the boundedness of the Cauchy transform requires $p > 2$. Therefore R is a well-defined and bounded operator on $L^p(\mathbb{C})$ for $2 < p < 1 + 1/k$.

Moreover, the right hand side of (57) belongs to $L^p(\mathbb{C})$ for each $h \in W^{1,p}(\mathbb{C})$. Hence this equation admits a unique solution $g \in W^{1,p}(\mathbb{C})$ if and only if the operator $\mathbf{I} - R$ is invertible in $L^p(\mathbb{C})$, $2 < p < 1 + 1/k$.

To get this we will use Fredholm theory. First, Theorem A.3 shows that R is a compact operator on $L^p(\mathbb{C})$ when $2 < p < 1 + 1/k$. Therefore it suffices to show that $\mathbf{I} - R$ is injective. Suppose now that $g \in L^p(\mathbb{C})$ satisfies

$$g = Rg = \mathscr{C}\left(\mathbf{I} - \nu\overline{\mathscr{S}}\right)^{-1}(\alpha\overline{g}).$$

Then $g \in W^{1,p}(\mathbb{C})$ by Theorem A.2 and $g_{\bar{z}} = \left(\mathbf{I} - \nu\overline{\mathscr{S}}\right)^{-1}(\alpha\overline{g})$. Equivalently,

$$g_{\bar{z}} - \nu\overline{g_z} = \alpha\overline{g} \tag{58}$$

Thus the assumptions of Theorem A.8 are fulfilled, and we must have $g \equiv 0$. Therefore $\mathbf{I} - R$ is indeed injective on $L^p(\mathbb{C})$. By the Fredholm alternative, it therefore is invertible in $L^p(\mathbb{C})$. Therefore the operator $\mathbf{I} - L_\mu$ is invertible in $W^{1,p}(\mathbb{C})$, $2 < p < 1 + 1/k$. □

A glance at (52) shows that $\xi \to L_\mu$ is an infinitely differentiable family of operators. Therefore, with Theorem 2.13, we see that $M_\mu = (\mathbf{I} - L_\mu)^{-1}1$ is C^∞-smooth in the parameter ξ. Thus we have obtained Theorem 2.12.

2F. Nonlinear Fourier transform. The idea of studying the $\bar{\xi}$-dependence of operators associated with complex geometric optics solutions was used by Beals and Coifman [1988] in connection with the inverse scattering approach to KdV-equations. Here we will apply this method to the solutions u_σ to the conductivity equation (1) and show that they satisfy a simple $\bar{\partial}$-equation with respect to the parameter ξ.

We start with the representation $u_\sigma(z, \xi) = \operatorname{Re} f_\mu(z, \xi) + i \operatorname{Im} f_{-\mu}(z, \xi)$, where $f_{\pm\mu}$ are the solutions to the corresponding Beltrami equations; in particular, they are analytic outside the unit disk. Hence with the asymptotics (36) they admit the following power series development,

$$f_{\pm\mu}(z, \xi) = e^{i\xi z}\left(1 + \sum_{n=1}^{\infty} b_n^{\pm}(\xi) z^{-n}\right), \qquad |z| > 1, \qquad (59)$$

where $b_n^+(\xi)$ and $b_n^-(\xi)$ are the coefficients of the series, depending on the parameter ξ. For the solutions to the conductivity equation, this gives

$$u_\sigma(z, \xi) = e^{i\xi z} + \frac{a(\xi)}{z} e^{i\xi z} + \frac{b(\xi)}{\bar{z}} e^{-i\bar{\xi}\bar{z}} + e^{i\xi z}\,\mathbb{O}\!\left(\frac{1}{|z|^2}\right)$$

as $z \to \infty$, where

$$a(\xi) = \frac{b_1^+(\xi) + b_1^-(\xi)}{2}, \qquad b(\xi) = \frac{\overline{b_1^+(\xi) - b_1^-(\xi)}}{2\bar{z}}. \qquad (60)$$

Fixing the z-variable, we take the $\partial_{\bar{\xi}}$-derivative of $u_\sigma(z, \xi)$ and get

$$\partial_{\bar{\xi}} u_\sigma(z, \xi) = -i\tau_\sigma(\xi)\, e^{-i\bar{\xi}\bar{z}}\left(1 + \mathbb{O}\!\left(\frac{1}{|z|}\right)\right), \qquad (61)$$

with the coefficient

$$\tau_\sigma(\xi) := \overline{b(\xi)}. \qquad (62)$$

However, the derivative $\partial_{\bar{\xi}} u_\sigma(z, \xi)$ is another solution to the conductivity equation! From the uniqueness of the complex geometric optics solutions under the given exponential asymptotics, Corollary 2.7, we therefore have the simple but important relation

$$\partial_{\bar{\xi}} u_\sigma(z,\xi) = -i\, \tau_\sigma(\xi)\, \overline{u_\sigma(z,\xi)} \quad \text{for all } \xi, z \in \mathbb{C}. \tag{63}$$

The remarkable feature of this relation is that the coefficient $\tau_\sigma(\xi)$ does not depend on the space variable z. Later, this phenomenon will become of crucial importance in solving the inverse problem.

In analogy with the one-dimensional scattering theory of integrable systems and associated inverse problems (see [Beals and Coifman 1988; Brown and Uhlmann 1997b; Nachman 1996b]), we call τ_σ the *nonlinear Fourier transform* of σ.

To understand the basic properties of the nonlinear Fourier transform, we need to return to the Beltrami equation. We will first show that the Dirichlet-to-Neumann data determines τ_σ. This is straightforward. Then the later sections are devoted to showing that the nonlinear Fourier transform τ_σ determines the coefficient σ almost everywhere. There does not seem to be any direct method for this, rather we will have to show that from τ_σ we can determine the exponentially growing solutions $f_{\pm\mu}$ defined in the entire plane. From this information the coefficient μ, and hence σ, can be found.

The nonlinear Fourier transform τ_σ has many properties which are valid for the linear Fourier transform. We have the usual transformation rules under scaling and translation:

$$\sigma_1(z) = \sigma(Rz) \;\Rightarrow\; \tau_{\sigma_1}(\xi) = \frac{1}{R}\tau_\sigma(\xi/R),$$

$$\sigma_2(z) = \sigma(z+p) \;\Rightarrow\; \tau_{\sigma_2}(\xi) = e^{i(p\xi + \bar{p}\bar{\xi})}\tau_\sigma(\xi).$$

However not much is known concerning questions such as the possibility of a Plancherel formula. However, some simple mapping properties of it can be proven. We will show that for σ as above, $\tau_\sigma \in L^\infty$. For this we need the following result, which is useful also elsewhere.

Here let $f_\mu(z,\xi) = e^{i\xi z} M_\mu(z,\xi)$ and $f_{-\mu}(z,\xi) = e^{i\xi z} M_{-\mu}(z,\xi)$ be the solutions of Theorem 2.6 corresponding to conductivities σ and σ^{-1}, respectively, which are holomorphic outside $\mathbb{D}$.

Theorem 2.14. *For every $\xi, z \in \mathbb{C}$ we have $M_{\pm\mu}(z,\xi) \neq 0$. Moreover,*

$$\operatorname{Re}\left(\frac{M_\mu(z,\xi)}{M_{-\mu}(z,\xi)}\right) > 0. \tag{64}$$

Proof. First, note that (8) implies, for $M_{\pm\mu}$,

$$\frac{\partial}{\partial \bar{z}} M_{\pm\mu} \mp \mu e_{-\xi} \overline{\frac{\partial}{\partial z} M_{\pm\mu}} = \mp i\bar{\xi}\mu e_{-\xi} \overline{M_{\pm\mu}}. \tag{65}$$

Thus we may apply Theorem A.8 to get
$$M_{\pm\mu}(z) = \exp(\eta_{\pm}(z)) \neq 0, \tag{66}$$
and consequently $M_\mu/M_{-\mu}$ is well defined. Second, if (64) is not true, the continuity of $M_{\pm\mu}$ and the fact $\lim_{z\to\infty} M_{\pm\mu}(z,\xi) = 1$ imply the existence of $z_0 \in \mathbb{C}$ such that
$$M_\mu(z_0,\xi) = it\, M_{-\mu}(z_0,\xi)$$
for some $t \in \mathbb{R}\setminus\{0\}$ and $\xi \in \mathbb{C}$. But then $g = M_\mu - itM_{-\mu}$ satisfies
$$\frac{\partial}{\partial \bar z} g = \mu(z) \overline{\frac{\partial}{\partial z}(e_\xi g)} \quad \text{and} \quad g(z) = 1 - it + \mathcal{O}\!\left(\frac{1}{z}\right) \quad \text{as } z \to \infty.$$
According to Theorem A.8, this implies
$$g(z) = (1 - it)\exp(\eta(z)) \neq 0,$$
contradicting the assumption $g(z_0) = 0$. □

The boundedness of the nonlinear Fourier transform is now a simple corollary of Schwarz's lemma.

Theorem 2.15. *The functions* $f_{\pm\mu}(z,\xi) = e^{i\xi z} M_{\pm\mu}(z,\xi)$ *satisfy, for* $|z| > 1$ *and for all* $\xi \in \mathbb{C}$,
$$|m(z)| \leq \frac{1}{|z|}, \quad \text{where } m(z) = \frac{M_\mu(z,\xi) - M_{-\mu}(z,\xi)}{M_\mu(z,\xi) + M_{-\mu}(z,\xi)}. \tag{67}$$
Moreover, for the nonlinear Fourier transform τ_σ, we have
$$|\tau_\sigma(\xi)| \leq 1 \quad \text{for all } \xi \in \mathbb{C}. \tag{68}$$

Proof. Fix the parameter $\xi \in \mathbb{C}$. Then, by Theorem 2.14, $|m(z)| < 1$ for all $z \in \mathbb{C}$. Moreover, m is holomorphic for $z \in \mathbb{C}\setminus\overline{\mathbb{D}}$, $m(\infty) = 0$, and thus by Schwarz's lemma we have $|m(z)| \leq 1/|z|$ for all $z \in \mathbb{C}\setminus\overline{\mathbb{D}}$.

On the other hand, from the development (59),
$$M_\mu(z,\xi) = 1 + \sum_{n=1}^\infty b_n(\xi) z^{-n} \quad \text{for } |z| > 1,$$
and similarly for $M_{-\mu}(z,\xi)$. We see that
$$\tau_\sigma(\xi) = \frac{1}{2}\overline{\bigl(b_1^+(\xi) - b_1^-(\xi)\bigr)} = \lim_{z\to\infty} \overline{z\, m(z)}.$$
Therefore the second claim also follows. □

With these results the Calderón problem reduces to the question whether we can invert the nonlinear Fourier transform.

Theorem 2.16. *The operator* Λ_σ *uniquely determines the nonlinear Fourier transform* τ_σ.

Proof. The claim follows immediately from Theorem 2.10, from the development (59) and from the definition (62) of τ_σ. □

From the relation $-\mu = \dfrac{1-1/\sigma}{1+1/\sigma}$ we see the symmetry

$$\tau_\sigma(\xi) = -\tau_{1/\sigma}(\xi).$$

It follows that the functions

$$u_1 = \operatorname{Re} f_\mu + i \operatorname{Im} f_{-\mu} = u_\sigma \quad \text{and} \quad u_2 = i \operatorname{Re} f_{-\mu} - \operatorname{Im} f_\mu = i u_{1/\sigma} \quad (69)$$

form a "primary pair" of complex geometric optics solutions:

Corollary 2.17. *The functions* $u_1 = u_\sigma$ *and* $u_2 = i u_{1/\sigma}$ *are complex-valued* $W^{1,2}_{\mathrm{loc}}(\mathbb{C})$-*solutions to the conductivity equations*

$$\nabla \cdot \sigma \nabla u_1 = 0 \quad \text{and} \quad \nabla \cdot \frac{1}{\sigma} \nabla u_2 = 0, \quad (70)$$

respectively. In the ξ-variable they are solutions to the same $\partial_{\bar\xi}$-equation,

$$\frac{\partial}{\partial \bar\xi} u_j(z,\xi) = -i\, \tau_\sigma(\xi) \overline{u_j}(z,\xi), \qquad j = 1, 2, \quad (71)$$

and their asymptotics, as $|z| \to \infty$, *are*

$$u_\sigma(z,\xi) = e^{i\xi z}\left(1 + \mathcal{O}\!\left(\frac{1}{|z|}\right)\right), \qquad u_{1/\sigma}(z,\xi) = e^{i\xi z}\left(i + \mathcal{O}\!\left(\frac{1}{|z|}\right)\right).$$

2G. *Subexponential growth.* A basic difficulty in the solution to Calderón's problem is to find methods to control the asymptotic behavior in the parameter ξ for complex geometric optics solutions. If we knew that the assumptions of the Liouville type Theorem A.8 were valid in (71), then the equation, hence the Dirichlet-to-Neumann map, would uniquely determine $u_\sigma(z,\xi)$ with $u_{1/\sigma}(z,\xi)$, and the inverse problem could easily be solved. However, we only know from Theorem 2.15 that $\tau_\sigma(\xi)$ is bounded in ξ. It takes considerably more effort to prove the counterpart of the Riemann–Lebesgue lemma, that is,

$$\tau_\sigma(\xi) \to 0, \quad \text{as } \xi \to \infty.$$

Indeed, this will be one of the consequences of the results in the present section.

It is clear that some control of the parameter ξ is needed for $u_\sigma(z,\xi)$. Within the category of conductivity equations with L^∞-coefficients σ, the complex analytic and quasiconformal methods provide by most powerful tools. Therefore we return to the Beltrami equation. The purpose of this section is to study the

ξ-behavior in the functions $f_\mu(z, \xi) = e^{i\xi z} M_\mu(z, \xi)$ and to show that for a fixed z, $M_\mu(z, \xi)$ grows at most subexponentially in ξ as $\xi \to \infty$. Subsequently, the result will be applied to $u_j(z, \xi)$.

For some later purposes we will also need to generalize the situation a bit by considering complex Beltrami coefficients μ_λ of the form $\mu_\lambda = \lambda \mu$, where the constant $\lambda \in \partial \mathbb{D}$ and μ is as before. Exactly as in Theorem 2.6, we can show the existence and uniqueness of $f_{\lambda\mu} \in W_{\text{loc}}^{1,p}(\mathbb{C})$ satisfying

$$\frac{\partial}{\partial \bar{z}} f_{\lambda\mu} = \lambda\mu \overline{\frac{\partial}{\partial z} f_{\lambda\mu}} \quad \text{in } \mathbb{C}, \tag{72}$$

$$f_{\lambda\mu}(z, \xi) = e^{i\xi z}\left(1 + \mathcal{O}\left(\frac{1}{z}\right)\right) \quad \text{as } |z| \to \infty. \tag{73}$$

In fact, we have that the function $f_{\lambda\mu}$ admits a representation of the form

$$f_{\lambda\mu}(z, \xi) = e^{i\xi \varphi_\lambda(z, \xi)}, \tag{74}$$

where for each fixed $\xi \in \mathbb{C} \setminus \{0\}$ and $\lambda \in \partial \mathbb{D}$, $\varphi_\lambda(z, \xi) = z + \mathcal{O}(1/z)$ for $z \to \infty$. The principal solution $\varphi = \varphi_\lambda(z, \xi)$ satisfies the nonlinear equation

$$\frac{\partial}{\partial \bar{z}} \varphi(z) = \kappa_{\lambda,\xi}\, e_{-\xi}(\varphi(z))\, \mu(z)\, \overline{\frac{\partial}{\partial z} \varphi(z)} \tag{75}$$

where $\kappa = \kappa_{\lambda,\xi} = -\lambda\, \bar{\xi}^2 |\xi|^{-2}$ is constant in z with $|\kappa_{\lambda,\xi}| = 1$.

The main goal of this section is to show the following theorem.

Theorem 2.18. *If $\varphi = \varphi_\lambda$ and $f_{\lambda\mu}$ are as in (72)–(75), then*

$$\lim_{\xi \to \infty} \varphi_\lambda(z, \xi) = z$$

uniformly in $z \in \mathbb{C}$ and $\lambda \in \partial \mathbb{D}$.

From the theorem we have this immediate consequence:

Corollary 2.19. *If $\sigma, \sigma^{-1} \in L^\infty(\mathbb{C})$ with $\sigma(z) = 1$ outside a compact set, then $\lim_{\xi \to \infty} \tau_\sigma(\xi) = 0$.*

Proof of Corollary 2.19. Let $\lambda = 1$. The principal solutions in (74) have the development

$$\varphi(z, \xi) = z + \sum_{n=1}^{\infty} \frac{c_n(\xi)}{z^n}, \quad |z| > 1.$$

By Cauchy integral formula and Theorem 2.18 we have $\lim_{\xi \to \infty} c_n(\xi) = 0$ for all $n \in \mathbb{N}$. Comparing now with (59)–(62) proves the claim. $\square$

It remains to prove Theorem 2.18, which will take up the rest of this section. We shall split the proof up into several lemmas.

Lemma 2.20. *Suppose $\varepsilon > 0$ is given. Suppose also that for $\mu_\lambda(z) = \lambda\mu(z)$, we have*

$$f_n = \mu_\lambda S_n \mu_\lambda S_{n-1} \mu_\lambda \cdots \mu_\lambda S_1 \mu_\lambda, \tag{76}$$

where $S_j : L^2(\mathbb{C}) \to L^2(\mathbb{C})$ are Fourier multiplier operators, each with a unimodular symbol. Then there is a number $R_n = R_n(k, \varepsilon)$ depending only on $k = \|\mu\|_\infty$, n and ε such that

$$|\widehat{f_n}(\eta)| < \varepsilon \quad \text{for } |\eta| > R_n. \tag{77}$$

Proof. It is enough to prove the claim for $\lambda = 1$. By assumption,

$$\widehat{S_j g}(\eta) = m_j(\eta)\widehat{g}(\eta),$$

where $|m_j(\eta)| = 1$ for $\eta \in \mathbb{C}$. We have by (76),

$$\|f_n\|_{L^2} \le \|\mu\|_{L^\infty}^n \|\mu\|_{L^2} \le \sqrt{\pi}k^{n+1} \tag{78}$$

since $\operatorname{supp}(\mu) \subset \mathbb{D}$. Choose ρ_n so that

$$\int_{|\eta|>\rho_n} |\widehat{\mu}(\eta)|^2 \, dm(\eta) < \varepsilon^2. \tag{79}$$

After this, choose $\rho_{n-1}, \rho_{n-2}, \ldots, \rho_1$ inductively so that for $l = n-1, \ldots, 1$,

$$\pi \int_{|\eta|>\rho_l} |\widehat{\mu}(\eta)|^2 \, dm(\eta) \le \varepsilon^2 \left(\prod_{j=l+1}^n \pi\rho_j\right)^{-2}. \tag{80}$$

Finally, choose ρ_0 so that

$$|\widehat{\mu}(\eta)| < \varepsilon \pi^{-n} \left(\prod_{j=1}^n \rho_j\right)^{-1} \quad \text{when } |\eta| > \rho_0. \tag{81}$$

All these choices are possible since $\mu \in L^1(\mathbb{C}) \cap L^2(\mathbb{C})$.

Now, we set $R_n = \sum_{j=0}^n \rho_j$ and claim that (77) holds for this choice of R_n. Hence assume that $|\eta| > \sum_{j=0}^n \rho_j$. We have

$$|\widehat{f_n}(\eta)| \le \left(\int_{|\eta-\zeta|\le\rho_n} + \int_{|\eta-\zeta|\ge\rho_n}\right) |\widehat{\mu}(\eta-\zeta)| |\widehat{f_{n-1}}(\zeta)| \, dm(\zeta). \tag{82}$$

But if $|\eta - \zeta| \le \rho_n$, then $|\zeta| > \sum_{j=0}^{n-1} \rho_j$. Thus, if we set

$$\Delta_n = \sup\left\{|\widehat{f_n}(\eta)| : |\eta| > \sum_{j=0}^n \rho_j\right\},$$

it follows from (82) and (78) that

$$\Delta_n \leq \Delta_{n-1}(\pi\rho_n^2)^{1/2}\|\mu\|_{L^2} + \left(\int_{|\zeta|\geq\rho_n}|\widehat{\mu}(\zeta)|^2\,dm(\zeta)\right)^{1/2}\|\widehat{f}_{n-1}\|_{L^2}$$

$$\leq \pi\rho_n k\,\Delta_{n-1} + k^n\left(\pi\int_{|\zeta|\geq\rho_n}|\widehat{\mu}(\zeta)|^2\,dm(\zeta)\right)^{1/2}$$

for $n \geq 2$. Moreover, the same argument shows that

$$\Delta_1 \leq \pi\rho_1 k\,\sup\{|\widehat{\mu}(\eta)| : |\eta| > \rho_0\} + k\left(\pi\int_{|\zeta|>\rho_1}|\widehat{\mu}(\zeta)|^2\,dm(\zeta)\right)^{1/2}.$$

In conclusion, after iteration we will have

$$\Delta_n \leq (k\pi)^n\left(\prod_{j=1}^n \rho_j\right)\sup\{|\widehat{\mu}(\eta)| : |\eta| > \rho_0\}$$

$$+ k^n \sum_{l=1}^n \left(\prod_{j=l+1}^n \pi\rho_j\right)\left(\pi\int_{|\zeta|>\rho_l}|\widehat{\mu}(\zeta)|^2\,dm(\zeta)\right)^{1/2}.$$

With the choices (79)–(81), this leads to

$$\Delta_n \leq (n+1)k^n\varepsilon \leq \frac{\varepsilon}{1-k},$$

which proves the claim. $\square$

Our next goal is to use Lemma 2.20 to prove the asymptotic result required in Theorem 2.18 for the solution of a closely related linear equation.

Theorem 2.21. *Suppose $\psi \in W^{1,2}_{\text{loc}}(\mathbb{C})$ satisfies*

$$\frac{\partial \psi}{\partial \bar{z}} = \kappa\,\mu(z)\,e_{-\xi}(z)\,\overline{\frac{\partial \psi}{\partial z}} \quad \text{in } \mathbb{C}, \tag{83}$$

$$\psi(z) = z + \mathcal{O}\left(\frac{1}{z}\right) \quad \text{as } z \to \infty, \tag{84}$$

where κ is a constant with $|\kappa| = 1$.

Then $\psi(z,\xi) \to z$, uniformly in $z \in \mathbb{C}$ and $\kappa \in \partial\mathbb{D}$, as $\xi \to \infty$.

To prove Theorem 2.21 we need some preparation. First, since the L^p-norm of the Beurling transform, denoted as S_p, tends to 1 when $p \to 2$, we can choose a $\delta_k > 0$ so that $k S_p < 1$ whenever $2 - \delta_k \leq p \leq 2 + \delta_k$.

Lemma 2.22. *Let $\psi = \psi(\cdot,\xi)$ be the solution of (83) and let $\varepsilon > 0$. Then $\psi_{\bar{z}}$ can be decomposed as $\psi_{\bar{z}} = g + h$, where*

(1) $\|h(\cdot,\xi)\|_{L^p} < \varepsilon$ *for* $2 - \delta_k \leq p \leq 2 + \delta_k$ *uniformly in ξ,*

(2) $\|g(\cdot,\xi)\|_{L^p} \leq C_0 = C_0(k)$ *uniformly in ξ,*

(3) $\widehat{g}(\eta,\xi) \to 0$ *as $\xi \to \infty$.*

In (3) *convergence is uniform on compact subsets of the η-plane and also uniform in $\kappa \in \partial \mathbb{D}$. Here, the Fourier transform is taken with respect to the first variable only.*

Proof. We may solve (83) using a Neumann series, which will converge in L^p:

$$\frac{\partial \psi}{\partial \bar{z}} = \sum_{n=0}^{\infty} (\kappa \mu e_{-\xi} \mathcal{S})^n (\kappa \mu e_{-\xi}).$$

Let

$$h = \sum_{n=n_0}^{\infty} (\kappa \mu e_{-\xi} \mathcal{S})^n (\kappa \mu e_{-\xi}).$$

Then

$$\|h\|_{L^p} \leq \pi^{1/p} \frac{k^{n_0+1} S_p^{n_0}}{1 - k S_p}$$

and we obtain the first statement by choosing n_0 large enough.

The remaining part clearly satisfies the second statement with a constant C_0 that is independent of ξ and λ. To prove (3) we first note that

$$\mathcal{S}(e_{-\xi} \phi) = e_{-\xi} S_\xi \phi,$$

where $\widehat{(S_\xi \phi)}(\eta) = m(\eta - \xi) \widehat{\phi}(\eta)$ and $m(\eta) = \eta/\bar{\eta}$. Consequently,

$$(\mu e_{-\xi} \mathcal{S})^n \mu e_{-\xi} = e_{-(n+1)\xi} \mu S_{n\xi} \mu S_{(n-1)\xi} \cdots \mu S_\xi \mu,$$

and so

$$g = \sum_{j=1}^{n_0} \kappa^j e_{-j\xi} \mu S_{(j-1)\xi} \mu \cdots \mu S_\xi \mu.$$

Therefore,

$$g = \sum_{j=1}^{n_0} e_{-j\xi} G_j,$$

where by Lemma 2.20, $|\widehat{G}_j(\eta)| < \tilde{\varepsilon}$ whenever $|\eta| > R = \max_{j \leq n_0} R_j$. As $\widehat{(e_{j\xi} G_j)}(\eta) = \widehat{G}_j(\eta + j\xi)$, for any fixed compact set K_0, we can take ξ so large that $j\xi + K_0 \subset \mathbb{C} \setminus \mathbb{D}(0, R)$ for each $1 \leq j \leq n_0$. Then

$$\sup_{\eta \in K_0} |\widehat{g}(\eta, \xi)| \leq n_0 \tilde{\varepsilon}.$$

This proves the lemma. □

Proof of Theorem 2.21. We show first that when $\xi \to \infty$, $\psi_{\bar{z}} \to 0$ weakly in $L^p(\mathbb{C})$, $2 - \delta_k \leq p \leq 2 + \delta_k$. For this suppose that $f_0 \in L^q(\mathbb{C})$, $q = p/(p-1)$, is fixed and choose $\varepsilon > 0$. Then there exists $f \in C_0^\infty(\mathbb{C})$ such that $\|f_0 - f\|_{L^q(\mathbb{C})} < \varepsilon$, and so by Lemma 2.22,

$$|\langle f_0, \psi_{\bar{z}}\rangle| \leq \varepsilon C_1 + \Big|\int_{\mathbb{C}} \widehat{f}(\eta)\widehat{g}(\eta, \xi)\, dm(\eta)\Big|.$$

First choose R so large that

$$\int_{\mathbb{C}\setminus\mathbb{D}(0,R)} |\widehat{f}(\eta)|^2\, dm(\eta) \leq \varepsilon^2$$

and then $|\xi|$ so large that $|\widehat{g}(\eta, \xi)| \leq \varepsilon/(\sqrt{\pi}R)$ for all $\eta \in \mathbb{D}(R)$. Now,

$$\Big|\int_{\mathbb{C}} \widehat{f}(\eta)\widehat{g}(\eta, \xi)\, d\eta\Big| \leq \int_{\mathbb{D}(R)} \widehat{f}(\eta)\widehat{g}(\eta, \xi)\, d\eta + \int_{\mathbb{C}\setminus\mathbb{D}(R)} \widehat{f}(\eta)\widehat{g}(\eta, \xi)\, d\eta$$
$$\leq \varepsilon(\|f\|_{L^2(\mathbb{C})} + \|g\|_{L^2(\mathbb{C})}) \leq C_2(f)\varepsilon. \tag{85}$$

The bound is the same for all κ, hence

$$\lim_{|\xi|\to\infty} \sup_{\kappa\in\partial\mathbb{D}} |\langle f_0, \psi_{\bar{z}}\rangle| = 0. \tag{86}$$

To prove the uniform convergence of ψ itself, we write

$$\psi(z,\xi) = z - \frac{1}{\pi}\int_{\mathbb{D}} \frac{1}{\zeta - z}\frac{\partial}{\partial\bar{\zeta}}\psi(\zeta,\xi)\, dm(\zeta). \tag{87}$$

Here note that $\mathrm{supp}(\psi_{\bar{z}}) \subset \mathbb{D}$ and $\chi_{\mathbb{D}}(\zeta)/(\zeta - z) \in L^q(\mathbb{C})$ for all $q < 2$. Thus by the weak convergence we have for each fixed $z \in \mathbb{C}$

$$\lim_{\xi\to\infty} \psi(z,\xi) = z, \quad \text{uniformly in } \kappa \in \partial\mathbb{D}. \tag{88}$$

On the other hand, as

$$\sup_\xi \Big\|\frac{\partial\psi}{\partial\bar{z}}\Big\|_{L^p(\mathbb{C})} \leq C_0 = C_0(p, \|\mu\|_\infty) < \infty$$

for all z sufficiently large, we have $|\psi(z,\xi) - z| < \varepsilon$ uniformly in $\xi \in \mathbb{C}$ and $\kappa \in \partial\mathbb{D}$. Moreover, (87) shows also that the family $\{\psi(\cdot,\xi) : \xi \in \mathbb{C}, \kappa \in \partial\mathbb{D}\}$ is equicontinuous. Combining all these observations shows that the convergence in (88) is uniform in $z \in \mathbb{C}$ and $\kappa \in \partial\mathbb{D}$. $\square$

Finally, we proceed to the nonlinear case: Assume that φ_λ satisfies (72) and (74). Since φ is a (quasiconformal) homeomorphism, we may consider its inverse $\psi_\lambda : \mathbb{C} \to \mathbb{C}$,

$$\psi_\lambda \circ \varphi_\lambda(z) = z, \tag{89}$$

which also is quasiconformal. By differentiating (89) with respect to z and $\bar{z}$ we find that ψ satisfies

$$\frac{\partial}{\partial \bar{z}} \psi_\lambda(z,\xi) = -\frac{\bar{\xi}}{\xi} \lambda \left(\mu(\psi_\lambda(z,\xi))\right) e_{-\xi}(z) \overline{\frac{\partial}{\partial z} \psi_\lambda(z,\xi)} \quad \text{and} \tag{90}$$

$$\psi_\lambda(z,\xi) = z + \mathcal{O}\left(\frac{1}{z}\right) \quad \text{as } z \to \infty. \tag{91}$$

Proof of Theorem 2.18. It is enough to show that

$$\lim_{\xi \to \infty} \psi_\lambda(z,\xi) = z. \tag{92}$$

uniformly in z and λ. For this we introduce the notation

$$\Sigma_k = \left\{ g \in W^{1,2}_{\text{loc}}(\mathbb{C}) : g_{\bar{z}} = \nu\, \overline{g_z},\ |\nu| \leq k \chi_\mathbb{D},\ g = z + \mathcal{O}\left(\frac{1}{z}\right) \text{ as } z \to \infty \right\}. \tag{93}$$

Note that all mappings $g \in \Sigma_k$ are principal solutions of Beltrami equations and hence homeomorphisms $g : \mathbb{C} \to \mathbb{C}$.

The support of the coefficient $\mu \circ \psi_\lambda$ in (90) need no longer be contained in $\mathbb{D}$. However, by the Koebe distortion theorem (see Lemma A.6 in the Appendix) $\varphi_\lambda(\mathbb{D}) \subset \mathbb{D}$ and thus $\operatorname{supp}(\mu \circ \psi_\lambda) \subset \mathbb{D}$. Accordingly, $\psi_\lambda \in \Sigma_k$.

Since normalized quasiconformal mappings form a normal family, we see that the family Σ_k is compact in the topology of uniform convergence. Given sequences $\xi_n \to \infty$ and $\lambda_n \in \partial\mathbb{D}$, we may pass to a subsequence and assume that

$$\kappa_{\lambda_n,\xi_n} = -\lambda_n \bar{\xi}_n^{\,2} |\xi_n|^{-2} \to \kappa \in \partial\mathbb{D}$$

as $n \to \infty$ and that the corresponding mappings satisfy $\lim_{n \to \infty} \psi_{\lambda_n}(\cdot,\xi_n) = \psi_\infty$ uniformly, where the limit satisfies $\psi_\infty \in \Sigma_k$. To prove Theorem 2.18 it is enough to show that for any such sequence $\psi_\infty(z) \equiv z$.

Let ψ_∞ be an arbitrary above obtained limit function. We consider the $W^{1,2}_{\text{loc}}$-solution $\Phi(z) = \Phi_\lambda(z,\xi)$ of

$$\frac{\partial \Phi}{\partial \bar{z}} = \kappa\, (\mu \circ \psi_\infty)\, e_{-\xi}\, \overline{\frac{\partial \Phi}{\partial \bar{z}}},$$

$$\Phi(z) = z + \mathcal{O}\left(\frac{1}{z}\right) \quad \text{as } z \to \infty.$$

Observe that this equation is a linear Beltrami equation which by Theorem 2.5 has a unique solution $\Phi \in \Sigma_k$ for each $\xi \in \mathbb{C}$ and $|\lambda| = 1$. According to Theorem 2.21,

$$\Phi_\lambda(z,\xi) \to z \quad \text{as } \xi \to \infty. \tag{94}$$

Further, when $2 < p < 1 + 1/k$, by Lemma A.7,

$$|\psi_{\lambda_n}(z, \xi_n) - \Phi_\lambda(z, \xi_n)|$$
$$= \frac{1}{\pi} \left| \int_\mathbb{D} \frac{1}{\zeta - z} \frac{\partial}{\partial \bar{z}} (\psi_{\lambda_n}(\zeta, \xi_n) - \Phi_\lambda(\zeta, \xi_n)) \, dm(\zeta) \right|$$
$$\leq C_1 \left\| \frac{\partial}{\partial \bar{z}} (\psi_{\lambda_n}(\zeta, \xi_n) - \Phi_\lambda(\zeta, \xi_n)) \right\|_{L^p(\mathbb{D}(2))}$$
$$\leq C_2 |\kappa_{\lambda_n, \xi_n} - \kappa|$$
$$+ C_2 \left(\int_{\mathbb{D}(2)} |\mu(\psi_{\lambda_n}(\zeta, \xi_n)) - \mu(\psi_\infty(\zeta))|^{\frac{p(1+\varepsilon)}{\varepsilon}} dm(\zeta) \right)^{\frac{\varepsilon}{p(1+\varepsilon)}}. \quad (95)$$

Finally, we apply the higher-integrability results for quasiconformal mappings, such as Theorem A.5: For all $2 < p < 1 + 1/k$ and $g = \psi^{-1}$, $\psi \in \Sigma_k$, we have the estimate for the Jacobian $J(z, g)$,

$$\int_\mathbb{D} J(z, g)^{p/2} \, dm(z) \leq \int_\mathbb{D} \left| \frac{\partial g}{\partial z} \right|^p dm(z) \leq C(k) < \infty, \quad (96)$$

where $C(k)$ depends only on k. We use this estimate in the cases $\psi(z)$ is equal to $\psi_{\lambda_n}(z, \xi_n)$ or ψ_∞. Then, we see for any $\gamma \in C_0^\infty(\mathbb{D})$ that

$$\int_{\mathbb{D}(2)} |\mu(\psi(y)) - \gamma(\psi(y))|^{\frac{p(1+\varepsilon)}{\varepsilon}} dy = \int_\mathbb{D} |\mu(z) - \gamma(z)|^{\frac{p(1+\varepsilon)}{\varepsilon}} J(z, g) \, dm(z)$$
$$\leq \left(\int_\mathbb{D} |\mu(z) - \gamma(z)|^{\frac{p^2(1+\varepsilon)}{\varepsilon(p-2)}} dm(z) \right)^{(p-2)/p} \left(\int_\mathbb{D} J(z, g)^{p/2} dm(z) \right)^{2/p}.$$

Since μ can be approximated in the mean by $\gamma \in C_0^\infty(\mathbb{D})$, the last term can be made arbitrarily small. By uniform convergence $\psi_{\lambda_n}(z, \xi_n) \to \psi_\infty(z)$ we see that $\gamma(\psi_{\lambda_n}(z, \xi_n)) \to \gamma(\psi_\infty(z))$ uniformly in z as $n \to \infty$. Also, $\kappa_{\lambda_n, \xi_n} \to \kappa$. Using these we see that right hand side of (95) converges to zero. In view of (94) and (95), we have established that

$$\lim_{n \to \infty} \psi_{\lambda_n}(z, \xi_n) = z$$

and thus $\psi_\infty(z) \equiv z$. The theorem is proved. $\square$

2H. *Completion of the proof of Theorem 2.1.* The Jacobian $J(z, f)$ of a quasi-regular map can vanish only on a set of Lebesgue measure zero. Since $J(z, f) = |f_z|^2 - |f_{\bar{z}}|^2 \leq |f_z|^2$, this implies that once we know the values $f_\mu(z, \xi)$ for every $z \in \mathbb{C}$, then we can recover from f_μ the values $\mu(z)$ and hence $\sigma(z)$ almost everywhere, by the formulas

$$\frac{\partial f_\mu}{\partial \bar{z}} = \mu(z) \overline{\frac{\partial f_\mu}{\partial z}} \quad \text{and} \quad \sigma = \frac{1 - \mu}{1 + \mu}. \quad (97)$$

On the other hand, considering the functions

$$u_1 := u_\sigma = \operatorname{Re} f_\mu + i \operatorname{Im} f_{-\mu} \quad \text{and} \quad u_2 := iu_{1/\sigma} = i \operatorname{Re} f_{-\mu} - \operatorname{Im} f_\mu$$

that were described in Corollary 2.17, it is clear that the pair $\{u_1(z,\xi), u_2(z,\xi)\}$ determines the pair $\{f_\mu(z,\xi), f_{-\mu}(z,\xi)\}$, and vice versa. Therefore to prove Theorem 2.1 it will suffice to establish the following result.

Theorem 2.23. *Assume that* $\Lambda_\sigma = \Lambda_{\tilde\sigma}$ *for two scalar conductivities* σ *and* $\tilde\sigma$ *for which* $\sigma, \tilde\sigma, 1/\sigma, 1/\tilde\sigma \in L^\infty(\mathbb{D})$. *Then for all* $z, \xi \in \mathbb{C}$,

$$u_\sigma(z,\xi) = u_{\tilde\sigma}(z,\xi) \quad \text{and} \quad u_{1/\sigma}(z,\xi) = u_{1/\tilde\sigma}(z,\xi).$$

For the proof of the theorem, our first task it to determine the asymptotic behavior of $u_\sigma(z,\xi)$. We state this as a separate result.

Lemma 2.24. *We have* $u_\sigma(z,\xi) \neq 0$ *for every* $(z,\xi) \in \mathbb{C} \times \mathbb{C}$. *Furthermore, for each fixed* $\xi \neq 0$, *we have with respect to* z

$$u_\sigma(z,\xi) = \exp(i\xi z + v(z)),$$

where $v = v_\xi \in L^\infty(\mathbb{C})$. *On the other hand, for each fixed* z *we have with respect to* ξ

$$u_\sigma(z,\xi) = \exp(i\xi z + \xi\varepsilon(\xi)), \tag{98}$$

where $\varepsilon(\xi) \to 0$ *as* $\xi \to \infty$.

Proof. For the first claim we write

$$u_\sigma = \tfrac{1}{2}(f_\mu + f_{-\mu} + \overline{f_\mu} - \overline{f_{-\mu}}) = f_\mu\left(1 + \frac{f_\mu - f_{-\mu}}{f_\mu + f_{-\mu}}\right)^{-1}\left(1 + \frac{\overline{f_\mu} - \overline{f_{-\mu}}}{f_\mu + f_{-\mu}}\right).$$

Each factor in the product is continuous and nonvanishing in z by Theorem 2.14. Taking the logarithm and using $f_{\pm\mu}(z,\xi) = e^{i\xi z}(1 + \mathcal{O}_\xi(1/z))$ we obtain

$$u_\sigma(z,\xi) = \exp\left(i\xi z + \mathcal{O}_\xi\!\left(\frac{1}{z}\right)\right).$$

Here, $\mathcal{O}_\xi(1/z))$ denotes a function $g(z,\xi)$ satisfying for each ξ an estimate $|g(z,\xi)| \leq C_\xi 1/|z|$ with some $C_\xi > 0$. For the ξ-asymptotics we apply Theorem 2.18, which governs the growth of the functions f_μ for $\xi \to \infty$. We see that for (98) it is enough to show that

$$\inf_t \left|\frac{f_\mu - f_{-\mu}}{f_\mu + f_{-\mu}} + e^{it}\right| \geq e^{-|\xi|\varepsilon(\xi)}. \tag{99}$$

For this, define

$$\Phi_t = e^{-it/2}(f_\mu \cos t/2 + if_{-\mu} \sin t/2).$$

Then for each fixed ξ,
$$\Phi_t(z,\xi) = e^{i\xi z}\left(1+\mathcal{O}_\xi\left(\frac{1}{z}\right)\right) \quad \text{as } z \to \infty,$$
and
$$\frac{\partial}{\partial \bar{z}} \Phi_t = \mu e^{-it} \overline{\frac{\partial}{\partial z} \Phi_t}.$$

Thus for $\lambda = e^{-it}$, the mapping $\Phi_t = f_{\lambda\mu}$ is precisely the exponentially growing solution satisfying the equations (72) and (73). A simple computation shows that
$$\frac{f_\mu - f_{-\mu}}{f_\mu + f_{-\mu}} + e^{it} = \frac{2e^{it}\Phi_t}{f_\mu + f_{-\mu}} = \frac{f_{\lambda\mu}}{f_\mu} \frac{2e^{it}}{1 + M_{-\mu}/M_\mu}. \tag{100}$$

By Theorem 2.18,
$$e^{-|\xi|\varepsilon_1(\xi)} \leq |M_{\pm\mu}(z,\xi)| \leq e^{|\xi|\varepsilon_1(\xi)} \tag{101}$$
and
$$e^{-|\xi|\varepsilon_2(\xi)} \leq \inf_{\lambda \in \partial \mathbb{D}} \left|\frac{f_{\lambda\mu}(z,\xi)}{f_\mu(z,\xi)}\right| \leq \sup_{\lambda \in \partial \mathbb{D}} \left|\frac{f_{\lambda\mu}(z,\xi)}{f_\mu(z,\xi)}\right| \leq e^{|\xi|\varepsilon_2(\xi)}, \tag{102}$$
where $\varepsilon_j(\xi) \to 0$ as $\xi \to \infty$. Since $\text{Re}\,(M_{-\mu}/M_\mu) > 0$, the inequality (99) follows. Thus the lemma is proven. $\square$

As discussed earlier, the functions $u_1 = u_\sigma$ and $u_2 = iu_{1/\sigma}$ satisfy a $\partial_{\bar\xi}$-equation as a function of the parameter ξ, but unfortunately, for a fixed z the asymptotics in (98) are not strong enough to determine the individual solution $u_j(z,\xi)$. However, if we consider the entire family $\{u_j(z,\xi) : z \in \mathbb{C}\}$, then, somewhat surprisingly, uniqueness properties do arise.

To consider the uniqueness properties, assume that the Dirichlet-to-Neumann operators are equal for the conductivities σ and $\tilde\sigma$. By Lemma 2.24, we have that $u_\sigma(z,\xi) \neq 0$ and $u_{\tilde\sigma}(z,\xi) \neq 0$ at every point (z,ξ). Therefore their logarithms, denoted by δ_σ and $\delta_{\tilde\sigma}$, respectively, are well defined. For each fixed $z \in \mathbb{C}$,
$$\delta_\sigma(z,\xi) = \log u_\sigma(z,\xi) = i\xi z + \xi\varepsilon_1(\xi), \tag{103}$$
$$\delta_{\tilde\sigma}(z,\xi) = \log u_{\tilde\sigma}(z,\xi) = i\xi z + \xi\varepsilon_2(\xi), \tag{104}$$
where $\varepsilon_j(\xi) \to 0$ as for $|\xi| \to \infty$. Moreover, by Theorem 2.6,
$$\delta_\sigma(z,0) \equiv \delta_{\tilde\sigma}(z,0) \equiv 0$$
for all $z \in \mathbb{C}$.

In addition, for each fixed $\xi \neq 0$ the function $z \mapsto \delta_\sigma(z, \xi)$ is continuous. By Lemma 2.24, we can write

$$\delta_\sigma(z, \xi) = i\xi z \left(1 + \frac{v_\xi(z)}{i\xi z}\right), \qquad (105)$$

where $v_\xi \in L^\infty(\mathbb{C})$ for each fixed $\xi \in \mathbb{C}$. This means that $\delta_\sigma(z, \xi)$ is close to a multiple of the identity for large $|z|$. Using an elementary homotopy argument, (105) yields that for any fixed $\xi \neq 0$ the map $z \mapsto \delta_\sigma(z, \xi)$ is surjective $\mathbb{C} \to \mathbb{C}$.

To prove the theorem it suffices to show that, if $\Lambda_\sigma = \Lambda_{\tilde{\sigma}}$, then

$$\delta_{\tilde{\sigma}}(z, \xi) \neq \delta_\sigma(w, \xi) \quad \text{for } z \neq w \text{ and } \xi \neq 0. \qquad (106)$$

Indeed, if the claim (106) is established, then (106) and the surjectivity of $z \mapsto \delta_\sigma(z, \xi)$ show that we necessarily have $\delta_\sigma(z, \xi) = \delta_{\tilde{\sigma}}(z, \xi)$ for all $\xi, z \in \mathbb{C}$. Hence $u_{\tilde{\sigma}}(z, \xi) = u_\sigma(z, \xi)$.

We are now at a point where the $\partial_{\bar{\xi}}$-method and (71) can be applied. Substituting $u_\sigma = \exp(\delta_\sigma)$ in this identity shows that $\xi \mapsto \delta_\sigma(z, \xi)$ and $\xi \mapsto \delta_{\tilde{\sigma}}(w, \xi)$ both satisfy the $\partial_{\bar{\xi}}$-equation

$$\frac{\partial \delta}{\partial \bar{\xi}} = -i\tau(\xi) e^{(\bar{\delta}-\delta)}, \qquad \xi \in \mathbb{C}, \qquad (107)$$

where by Theorem 2.10 and the assumption $\Lambda_\sigma = \Lambda_{\tilde{\sigma}}$, the coefficient $\tau(\xi)$ is the same for both functions δ_σ and $\delta_{\tilde{\sigma}}$. A simple computations shows then that the difference

$$g(\xi) := \delta_{\tilde{\sigma}}(w, \xi) - \delta_\sigma(z, \xi)$$

thus satisfies the identity

$$\frac{\partial g}{\partial \bar{\xi}} = -i\tau(\xi) e^{(\bar{\delta}-\delta)} \left[e^{(\bar{g}-g)} - 1\right].$$

In particular,

$$\left|\frac{\partial g}{\partial \bar{\xi}}\right| \leq |\bar{g} - g| \leq 2|g|. \qquad (108)$$

Using (103) we see that $g(\xi) = i(w-z)\xi + \xi\varepsilon(\xi)$ where $\varepsilon(\xi) \to 0$ as $\xi \to \infty$. Applying Theorem .9 (with respect to ξ) we see that for $w \neq z$ the function g vanishes only at $\xi = 0$. This proves (106).

According to Theorem 2.9 (or by the identity $\tau_\sigma = -\tau_{1/\sigma}$), if $\Lambda_\sigma = \Lambda_{\tilde{\sigma}}$, the same argument works to show that $u_{1/\tilde{\sigma}}(z, \xi) = u_{1/\sigma}(z, \xi)$ as well. Thus Theorem 2.23 is proved. As the pair $\{u_1(z, \xi), u_2(z, \xi)\}$ pointwise determines the pair $\{f_\mu(z, \xi), f_{-\mu}(z, \xi)\}$, we find via (97) that $\sigma \equiv \tilde{\sigma}$. Therefore the proof of Theorem 2.1 is complete. $\square$

3. Invisibility cloaking and the borderlines of visibility and invisibility

Next we consider the anisotropic conductivity equation in $\Omega \subset \mathbb{R}^2$,

$$\nabla \cdot \sigma \nabla u = \sum_{j,k=1}^{2} \frac{\partial}{\partial x^j}\left(\sigma^{jk}(x)\frac{\partial}{\partial x^k}u(x)\right) = 0 \text{ in } \Omega, \qquad (109)$$

where the conductivity $\sigma = [\sigma^{jk}(x)]_{j,k=1}^2$ is a measurable function whose values are symmetric, positive definite matrices. We say that a conductivity σ is *regular* if there are $c_1, c_2 > 0$ such that

$$c_1 \mathbf{I} \leq \sigma(x) \leq c_2 \mathbf{I}, \quad \text{for a.e. } x \in \Omega.$$

If conductivity is not regular, it is said to be *degenerate*. We will consider uniqueness results for the inverse problem in classes of degenerate conductivities both in the isotropic and the anisotropic case. We will also construct counterexamples for the uniqueness of the inverse problem having a close connection to the invisibility cloaking, a very topical subject in recent studies in mathematics, physics, and material science [Alu and Engheta 2005; Greenleaf et al. 2007; 2003c; Milton and Nicorovici 2006; Leonhardt 2006; Milton et al. 2009; Pendry et al. 2006; Weder 2008]. By invisibility cloaking we mean the possibility, both theoretical and practical, of shielding a region or object from detection via electromagnetic fields.

The counterexamples for inverse problems and the proposals for invisibility cloaking are closely related. In 2003, before the appearance of practical possibilities for cloaking, it was shown in [Greenleaf et al. 2003a; 2003c] that passive objects can be coated with a layer of material with a degenerate conductivity which makes the object undetectable by the electrostatic boundary measurements. These constructions were based on the blow up maps and gave counterexamples for the uniqueness of inverse conductivity problem in the three and higher-dimensional cases. In the two-dimensional case, the mathematical theory of the cloaking examples for conductivity equation have been studied in [Kohn et al. 2008; 2010; Lassas and Zhou 2011; Nguyen 2012].

Interest in cloaking was raised in particular in 2006 when it was realized that practical cloaking constructions are possible using so-called metamaterials which allow fairly arbitrary specification of electromagnetic material parameters. The construction of Leonhardt [2006] was based on conformal mapping on a nontrivial Riemannian surface. At the same time, Pendry et al. [2006] proposed a cloaking construction for Maxwell's equations using a blow up map and the idea was demonstrated in laboratory experiments [Schurig et al. 2006]. There are also other suggestions for cloaking based on active sources [Milton et al. 2009] or negative material parameters [Alu and Engheta 2005; Milton and Nicorovici 2006].

Let $\Sigma = \Sigma(\Omega)$ be the class of measurable matrix valued functions $\sigma : \Omega \to M$, where M is the set of symmetric nonnegative definite matrices. Instead of defining the Dirichlet-to-Neumann operator which may not be well defined for these conductivities, we consider the corresponding quadratic forms.

Definition 3.1. Let $h \in H^{1/2}(\partial\Omega)$. The Dirichlet-to-Neumann quadratic form corresponding to the conductivity $\sigma \in \Sigma(\Omega)$ is given by

$$Q_\sigma[h] = \inf A_\sigma[u], \quad \text{where } A_\sigma[u] = \int_\Omega \sigma(z)\nabla u(z) \cdot \nabla u(z)\, dm(z), \quad (110)$$

and the infimum is taken over real valued $u \in L^1(\Omega)$ such that $\nabla u \in L^1(\Omega)^3$ and $u|_{\partial\Omega} = h$. In the case where $Q_\sigma[h] < \infty$ and $A_\sigma[u]$ reaches its minimum at some u, we say that u is a $W^{1,1}(\Omega)$ solution of the conductivity problem.

When σ is smooth and bounded from below and above by positive constants, $Q_\sigma[h]$ is the quadratic form corresponding to the Dirichlet-to-Neumann map (5),

$$Q_\sigma[h] = \int_{\partial\Omega} h \Lambda_\sigma h\, ds, \quad (111)$$

where ds is the length measure on $\partial\Omega$. Physically, $Q_\sigma[h]$ corresponds to the power needed to keep voltage h at the boundary. As discussed above, for smooth conductivities bounded from below, for every $h \in H^{1/2}(\partial\Omega)$ the integral $A_\sigma[u]$ always has a unique minimizer $u \in H^1(\Omega)$ with $u|_{\partial\Omega} = h$. It is also a distributional solution to (4). Conversely, for functions $u \in H^1(\Omega)$ the traces lie in $H^{1/2}(\partial\Omega)$. As we mostly consider conductivities which are bounded from below and above near the boundary we chose to consider the $H^{1/2}$-boundary values also in the general case. We interpret that the Dirichlet-to-Neumann form corresponds to the idealization of the boundary measurements for $\sigma \in \Sigma(\Omega)$.

Next we present a few examples where the solutions u turn out to be nonsmooth or do not exist.

Example 1. Let us consider the one-dimensional conductivity equation on interval $I = [0, 1]$. Let $(q_j)_{j=1}^\infty$ be a sequence containing all rational numbers $\mathbb{Q} \cap (0, 1)$ so that each number appears only once in the sequence. Let $a_j = 2^{-1} j^{-4}$, $K_j = (q_j - 2^{-j-2}, q_j + 2^{-j-2}) \cap I$, and define the conductivity

$$\sigma(x) = 1 + \sum_{j=1}^\infty \sigma_j(x), \quad \sigma_j(x) = \frac{a_j}{|x - q_j|} \chi_{K_j}(x). \quad (112)$$

Note that the set K_j has the measure $|K_j| \leq 2^{-j-1}$. As $\left|\bigcup_{j \geq l} K_j\right| \leq 2^{-l}$, we see that the series (112) has only finitely many nonzero terms for $x \in \bigcap_{l \geq 1} \bigcup_{j \geq l} K_j$

and in particular, the sum $\sigma(x)$ is finite and positive function a.e. Now, assume that $u \in C^1(I)$ is a function for which

$$\int_I \sigma(x)|u'(x)|^2 dm(x) < \infty.$$

If $u'(q_j) \neq 0$, we see that there is an open nonempty interval $I_j \subset I$ containing q_j such that $|u'(x)| \geq t > 0$ for all $x \in I_j$, and

$$\int_I \sigma(x)|u'(x)|^2 dm(x) \geq \int_{I_j} \sigma_j(x)|u'(x)|^2 dm(x) = \infty.$$

This implies that $u'(q_j) = 0$ for all q_j, and as $\{q_j\}$ is dense in I, we see that u vanishes identically. Thus if the minimization (110) is taken only over $u \in C^1(I)$ with $u(0) = f_0$ and $u(1) = f_1$, the Dirichlet-to-Neumann form is infinite for all nonconstant boundary values $f_0 \neq f_1$. However, if the infimum is taken over all $u \in W^{1,1}(I)$, we see that the function

$$u_0(x) = \frac{|[0,x] \setminus K|}{|I \setminus K|}, \quad K = \bigcup_{j=1}^{\infty} K_j, \quad 0 < |K| < \tfrac{1}{2}$$

satisfies $u'(x) = 0$ for $x \in K$ and

$$\int_I \sigma(x)|u_0'(x)|^2 dm(x) = 1, \quad u_0(0) = 0, \ u_0(1) = 1.$$

Using functions $f_0 + (f_1 - f_0)u_0(x)$ we see that the Dirichlet-to-Neumann form for σ defined as a minimization over all $W^{1,1}$-functions is finite for all boundary values. Later we will show also examples of conductivities encountered in cloaking where the solution of the conductivity problem will be in $W^{1,p}$ for all $p < 2$ but not in $W^{1,2}$. This is another reason why $W^{1,1}$ is a convenient class to consider the minimization.

Example 2. Consider in the disc $\mathbb{D}(2)$ a strongly twisting map,

$$G(re^{i\theta}) = re^{i(\theta + t(r))}, \quad 0 < r \leq 2,$$

where $t(r) = \exp(r^{-1} - 2^{-1})$. When $\gamma = 1$ is the homogeneous conductivity, let σ be the conductivity in $\mathbb{D}(2)$ such that $\sigma = G_*\gamma$ in the set $\mathbb{D}(2) \setminus \{0\}$. We see that if the problem (110) has a minimizer $u \in W^{1,1}(D)$ with the boundary value f for which $A_\sigma(u) < \infty$, then it has to satisfy $\nabla \cdot \sigma \nabla u = 0$ in the set $\mathbb{D}(2) \setminus \{0\}$. Then $v = u \circ G$ is harmonic function in $\mathbb{D}(2) \setminus \{0\}$ having boundary value $f \in H^{1/2}(\partial \mathbb{D}(2))$ and finite norm in $H^1(\mathbb{D}(2) \setminus \{0\})$. This implies that v can be extended to a harmonic function in the whole disc $\mathbb{D}(2)$; see, e.g., [Kilpeläinen et al. 2000]. Thus, if the problem (110) has a minimizer $u \in W^{1,1}(\mathbb{D}(2))$ for $f(x_1, x_2) = x_1$ we see that $v(x_1, x_2) = x_1$ and $u = v \circ F$, where $F := G^{-1}$.

Then, by the chain rule we have $\nabla u(x) = DF(x)^t (Dv)(F(x)) \notin L^1(\mathbb{D}(2) \setminus \{0\})$. This shows that the minimizer u does not exists in the space $W^{1,1}(\mathbb{D}(2))$. Thus for a general degenerate conductivity it is reasonable to define the boundary measurements using the infimum of a quadratic form instead of a distributional solution of the differential equation $\nabla \cdot \sigma \nabla u = 0$.

Existence results for solutions with degenerate conductivities. As seen in the examples above, if σ is unbounded it is possible that $Q_\sigma[h] = \infty$. Moreover, even if $Q_\sigma[h]$ is finite, the minimization problem in (110) may generally have no minimizer and even if they exist the minimizers need not be distributional solutions to (4). However, if the singularities of σ are not too strong, minimizers satisfying (4) do always exist. Below we will consider singular conductivity of exponentially integrable ellipticity function $K_\sigma(z)$ and show that for such conductivities solutions exists. To study of these solutions, we consider the *regularity gauge*

$$Q(t) = \frac{t^2}{\log(e+t)}, \quad t \geq 0. \tag{113}$$

We say accordingly that f belongs to the Orlicz space $W^{1,Q}(\Omega)$ if f and its first distributional derivatives are in $L^1(\Omega)$ and

$$\int_\Omega \frac{|\nabla f(z)|^2}{\log(e+|\nabla f(z)|)} \, dm(z) < \infty.$$

In [Astala et al. 2011a] the following existence result for solutions corresponding to singular conductivity of exponentially integrable ellipticity is proven:

Theorem 3.2. *Let $\sigma(z)$ be a measurable symmetric matrix valued function. Suppose further that for some $p > 0$,*

$$\int_\Omega \exp(p\,[\mathrm{trace}(\sigma(z)) + \mathrm{trace}(\sigma(z)^{-1})])\, dm(z) = C_1 < \infty. \tag{114}$$

Then, if $h \in H^{1/2}(\partial\Omega)$ is such that $Q_\sigma[h] < \infty$ and $X = \{v \in W^{1,1}(\Omega); v|_{\partial\Omega} = h\}$, there is a unique $w \in X$ such that

$$A_\sigma[w] = \inf\{A_\sigma[v] \,;\, v \in X\}. \tag{115}$$

Moreover, w satisfies the conductivity equation

$$\nabla \cdot \sigma \nabla w = 0 \quad in\ \Omega \tag{116}$$

in sense of distributions, and it has the regularity $w \in W^{1,Q}(\Omega) \cap C(\Omega)$.

Let $F : \Omega_1 \to \Omega_2$, $y = F(x)$, be an orientation-preserving homeomorphism between domains $\Omega_1, \Omega_2 \subset \mathbb{C}$ for which F and its inverse F^{-1} are at least $W^{1,1}$-smooth and let $\sigma(x) = [\sigma^{jk}(x)]_{j,k=1}^2 \in \Sigma(\Omega_1)$ be a conductivity on Ω_1.

Then the map F pushes σ forward to a conductivity $(F_*\sigma)(y)$, defined on Ω_2 and given by

$$(F_*\sigma)(y) = \frac{1}{\det DF(x)} DF(x)\, \sigma(x)\, DF(x)^t, \quad x = F^{-1}(y). \quad (117)$$

The main methods for constructing counterexamples to Calderón's problem are based on the following principle.

Proposition 3.3. *Assume that $\sigma, \tilde{\sigma} \in \Sigma(\Omega)$ satisfy (114), and let $F : \Omega \to \Omega$ be a homeomorphism so that F and F^{-1} are $W^{1,Q}$-smooth and C^1-smooth near the boundary, and $F|_{\partial\Omega} = id$. Suppose that $\tilde{\sigma} = F_*\sigma$. Then $Q_\sigma = Q_{\tilde{\sigma}}$.*

This proposition generalizes the results from [Kohn and Vogelius 1984] to less smooth diffeomorphisms and conductivities.

Sketch of the proof. Two implications of the assumptions for F are essential in the proof. First one is that as F is a homeomorphism satisfying $F \in W^{1,Q}(\Omega)$, it satisfies the condition $\mathcal{N}$, that is, for any measurable set $E \subset \Omega$ we have $|E| = 0 \Rightarrow |F(E)| = 0$; see, e.g., [Astala et al. 2009, Theorem 19.3.2]. Also F^{-1} satisfies this condition. These imply that we have the area formula

$$\int_\Omega H(y)\, dm(y) = \int_\Omega H(F(x)) \det(DF(x))\, dm(x) \quad (118)$$

for $H \in L^1(\Omega)$.

The second implication is that, by the Gehring–Lehto theorem (see [Astala et al. 2009, Corollary 3.3.3]), a homeomorphism $F \in W^{1,1}_{\text{loc}}(\Omega)$ is differentiable almost everywhere in Ω, say in the set $\Omega \setminus A$, where A has Lebesgue measure zero. This pointwise differentiability at almost every point is essential in using the chain rule.

Let $h \in H^{1/2}(\partial\Omega)$ and assume that $Q_{\tilde{\sigma}}[h] < \infty$. By Theorem 3.2 there is $\tilde{u} \in W^{1,1}(\Omega)$ solving

$$\nabla \cdot \tilde{\sigma} \nabla \tilde{u} = 0, \quad \tilde{u}|_{\partial\Omega} = h. \quad (119)$$

We define $u = \tilde{u} \circ F : \Omega \to \mathbb{C}$. As F is C^1-smooth near the boundary we see that $u|_{\partial\Omega} = h$.

By the Stoilow factorization theorem (Theorem A.9), $\tilde{u}$ can be written in the form $\tilde{u} = \tilde{w} \circ \tilde{G}$ where $\tilde{w}$ is harmonic and $\tilde{G} \in W^{1,1}_{\text{loc}}(\mathbb{C})$ is a homeomorphism $\tilde{G} : \mathbb{C} \to \mathbb{C}$. By Gehring–Lehto theorem $\tilde{G}$ and the solution $\tilde{u}$ are differentiable almost everywhere, say in the set $\Omega \setminus A'$, where A' has Lebesgue measure zero.

Since F^{-1} has the property $\mathcal{N}$, we see that $A'' = A' \cup F^{-1}(A') \subset \Omega$ has measure zero, and for $x \in \Omega \setminus A''$ the chain rule gives

$$\nabla u(x) = DF(x)^t\, (\nabla \tilde{u})(F(x)). \quad (120)$$

Using this, the area formula and the definition (117) of $F_*\sigma$ one can show that

$$Q_\sigma[h]$$
$$= \int_\Omega \nabla u(x) \cdot \sigma(x) \nabla u(x) \, dm(x)$$
$$= \int_\Omega DF(x)^t \nabla \tilde{u}(F(x)) \cdot \frac{\sigma(x)}{\det(DF(x))} DF(x)^t \nabla \tilde{u}(F(x)) \det(DF(x)) \, dm(x)$$
$$= \int_\Omega \nabla \tilde{u}(y) \cdot \tilde{\sigma}(y) \nabla \tilde{u}(y) \, dy = Q_{\tilde{\sigma}}[h]. \qquad \square$$

Let us next consider various counterexamples for the solvability of inverse conductivity problem with degenerate conductivities.

Counterexample 1: invisibility cloaking. We consider here invisibility cloaking in general background σ, that is, we aim to coat an arbitrary body with a layer of exotic material so that the coated body appears in measurements the same as the background conductivity σ. Usually one is interested in the case when the background conductivity σ is equal to the constant $\gamma = 1$. However, we consider here a more general case and assume that σ is a L^∞-smooth conductivity in $\overline{\mathbb{D}}(2)$, $\sigma(z) \geq c_0 I$, $c_0 > 0$. Here, $\mathbb{D}(\rho)$ is an open two-dimensional disc of radius ρ and center zero and $\overline{\mathbb{D}}(\rho)$ is its closure. Consider a homeomorphism

$$F : \overline{\mathbb{D}}(2) \setminus \{0\} \to \overline{\mathbb{D}}(2) \setminus \mathcal{K}, \qquad (121)$$

where $\mathcal{K} \subset \mathbb{D}(2)$ is a compact set which is the closure of a smooth open set and suppose $F : \overline{\mathbb{D}}(2) \setminus \{0\} \to \overline{\mathbb{D}}(2) \setminus \mathcal{K}$ and its inverse F^{-1} are C^1-smooth in $\overline{\mathbb{D}}(2) \setminus \{0\}$ and $\overline{\mathbb{D}}(2) \setminus \mathcal{K}$, correspondingly. We also require that $F(z) = z$ for $z \in \partial \mathbb{D}(2)$. The standard example of invisibility cloaking is the case when $\mathcal{K} = \overline{\mathbb{D}}(1)$ and the map

$$F_0(z) = \left(\frac{|z|}{2} + 1\right) \frac{z}{|z|}. \qquad (122)$$

Using the map (121), we define a singular conductivity

$$\tilde{\sigma}(z) = \begin{cases} (F_*\sigma)(z) & \text{for } z \in \mathbb{D}(2) \setminus \mathcal{K}, \\ \eta(z) & \text{for } z \in \mathcal{K}, \end{cases} \qquad (123)$$

where $\eta(z) = [\eta^{jk}(x)]$ is any symmetric measurable matrix satisfying $c_1 I \leq \eta(z) \leq c_2 I$ with $c_1, c_2 > 0$. The conductivity $\tilde{\sigma}$ is called the cloaking conductivity obtained from the transformation map F and background conductivity σ and $\eta(z)$ is the conductivity of the cloaked (i.e. hidden) object.

In particular, choosing σ to be the constant conductivity $\sigma = 1$, $\mathcal{K} = \overline{\mathbb{D}}(1)$, and F to be the map F_0 given in (122), we obtain the standard example of the invisibility cloaking. In dimensions $n \geq 3$ it shown in 2003 in [Greenleaf et al. 2003a; Greenleaf et al. 2003c] that the Dirichlet-to-Neumann map corresponding to $H^1(\Omega)$ solutions for the conductivity (123) coincide with the Dirichlet-to-Neumann map for $\sigma = 1$. In 2008, the analogous result was proven in the two-dimensional case in [Kohn et al. 2008]. For cloaking results for the Helmholtz equation with frequency $k \neq 0$ and for Maxwell's system in dimensions $n \geq 3$, see results in [Greenleaf et al. 2007]. We note also that John Ball [1982] has used the push forward by the analogous radial blow-up maps to study the discontinuity of the solutions of partial differential equations, in particular the appearance of cavitation in the nonlinear elasticity.

In [Astala et al. 2011a] the following generalization of [Greenleaf et al. 2003a; 2003c; Kohn et al. 2008] is proven for cloaking in the context where measurements given in Definition 3.1.

Theorem 3.4. (i) *Let $\sigma \in L^\infty(\mathbb{D}(2))$ be a scalar conductivity, $\sigma(x) \geq c_0 > 0$, $\mathcal{K} \subset \mathbb{D}(2)$ be a relatively compact open set with smooth boundary and $F : \overline{\mathbb{D}}(2) \setminus \{0\} \to \overline{\mathbb{D}}(2) \setminus \mathcal{K}$ be a homeomorphism. Assume that F and F^{-1} are C^1-smooth in $\overline{\mathbb{D}}(2) \setminus \{0\}$ and $\overline{\mathbb{D}}(2) \setminus \mathcal{K}$, correspondingly and $F|_{\partial \mathbb{D}(2)} = id$. Moreover, assume there is $C_0 > 0$ such that $\|DF^{-1}(x)\| \leq C_0$ for all $x \in \overline{\mathbb{D}}(2) \setminus \mathcal{K}$. Let $\tilde{\sigma}$ be the conductivity defined in (123). Then the boundary measurements for $\tilde{\sigma}$ and σ coincide in the sense that $Q_{\tilde{\sigma}} = Q_\sigma$.*

(ii) *Let $\tilde{\sigma}$ be a cloaking conductivity of the form (123) obtained from the transformation map F and the background conductivity σ where F and σ satisfy the conditions in (i). Then*

$$\operatorname{trace}(\tilde{\sigma}) \notin L^1(\mathbb{D}(2) \setminus \mathcal{K}). \tag{124}$$

Sketch of the proof. We consider the case when $F = F_0$ is given by (122) and $\sigma = 1$ is constant function.

(i) For $0 \leq r \leq 2$ and a conductivity η we define the quadratic form $A_\eta^r : W^{1,1}(\mathbb{D}(2)) \to \mathbb{R}_+ \cup \{0, \infty\}$ by

$$A_\eta^r[u] = \int_{\mathbb{D}(2) \setminus \mathbb{D}(r)} \eta(x) \nabla u \cdot \nabla u \, dm(x).$$

Considering F_0 as a change of variables similarly to Proposition 3.3, we see that

$$A_{\tilde{\sigma}}^r[u] = A_\gamma^\rho[v], \quad u = v \circ F_0, \quad \rho = 2(r-1), \ r > 1.$$

Now for the conductivity $\gamma = 1$ the minimization problem (110) is solved by the unique minimizer u satisfying

$$\Delta u = 0 \quad \text{in } \mathbb{D}(2), \quad u|_{\partial \mathbb{D}(2)} = f.$$

The solution u is C^∞-smooth in $\mathbb{D}(2)$ and we see that $v = u \circ F_0$ is a $W^{1,1}$-function on $\mathbb{D}(2) \setminus \overline{\mathbb{D}}(1)$ which trace on $\partial\mathbb{D}(1)$ is equal to the constant function $h(x) = u(0)$ on $\partial\mathbb{D}(1)$. Let $\widetilde{v}$ be a function that is equal to v in $\mathbb{D}(2) \setminus \overline{\mathbb{D}}(1)$ and has the constant value $u(0)$ in $\overline{\mathbb{D}}(1)$. Then $\widetilde{v} \in W^{1,1}(\mathbb{D}(2))$ and

$$Q_{\widetilde{\sigma}}[f] \leq A_{\widetilde{\sigma}}^1[v] = \lim_{r \to 1} A_{\widetilde{\sigma}}^r[v] = \lim_{\rho \to 0} A_\gamma^\rho[u] = Q_\gamma[f]. \tag{125}$$

To construct an inequality opposite to (125), let η_ρ be a conductivity which coincides with $\widetilde{\sigma}$ in $\mathbb{D}(2) \setminus \mathbb{D}(\rho)$ and is 0 in $\mathbb{D}(\rho)$. For this conductivity the minimization problem (110) has a minimizer that in $\mathbb{D}(2) \setminus \overline{\mathbb{D}}(\rho)$ coincides with the solution of the boundary value problem

$$\Delta u = 0 \quad \text{in } \mathbb{D}(2) \setminus \overline{\mathbb{D}}(\rho), \quad u|_{\partial\mathbb{D}(2)} = f, \quad \partial_\nu u|_{\partial\mathbb{D}(\rho)} = 0$$

and is arbitrary $W^{1,1}$-smooth extension of u to $\mathbb{D}(\rho)$. Then $\eta_\rho(x) \leq \widetilde{\sigma}(x)$ for all $x \in \mathbb{D}(2)$ and thus $Q_{\eta_\rho}[f] \leq Q_{\widetilde{\sigma}}[f]$. It is not difficult to see that

$$\lim_{\rho \to 0} Q_{\eta_\rho}[f] = Q_\gamma[f],$$

that is, the effect of an insulating disc of radius ρ in the boundary measurements vanishes as $\rho \to 0$. These and (125) yield $Q_{\widetilde{\sigma}}[f] = Q_\gamma[f]$. This proves (i).

(ii) Assume that (124) is not valid, i.e., trace$(\widetilde{\sigma}) \in L^1(\mathbb{D}(2) \setminus \overline{\mathbb{D}}(1))$. As $\sigma = 1$ and det$(\widetilde{\sigma}) = 1$, simple linear algebra yields that $K_{\widetilde{\sigma}} \in L^1(\mathbb{D}(2) \setminus \overline{\mathbb{D}}(1))$ and

$$\|\widetilde{\sigma}(y)\| = \frac{\|DF(x) \cdot \sigma(x) \cdot DF(x)^t\|}{J(x,F)} \geq \frac{\|DF(x)\|^2}{J(x,F)} = K_F(x), \quad x = F^{-1}(y).$$

Then $G = F^{-1}$ satisfies $K_G = K_F \circ F^{-1} \in L^1(\mathbb{D}(2) \setminus \overline{\mathbb{D}}(1))$, which yields that $F \in W^{1,2}(\mathbb{D}(2) \setminus \{0\})$ and

$$\|DF\|_{L^2(\mathbb{D}(2) \setminus \{0\})} \leq 2 \|K_G\|_{L^1(\mathbb{D}(2) \setminus \overline{\mathbb{D}}(1))};$$

see, e.g., [Astala et al. 2009, Theorem 21.1.4]. By the removability of singularities in Sobolev spaces — see [Kilpeläinen et al. 2000] — this implies that the function $F : \mathbb{D}(2) \setminus \{0\} \to \mathbb{D}(2) \setminus \overline{\mathbb{D}}(1)$ can be extended to a function $F^{\text{ext}} : \mathbb{D}(2) \to \mathbb{C}$ in $W^{1,2}(\mathbb{D}(2))$. It follows from this and the continuity theorem of finite distortion maps [Astala et al. 2009, Theorem 20.1.1] that $F^{\text{ext}} : \mathbb{D}(2) \to \mathbb{C}$ is continuous, which is not possible. Thus (124) has to be valid. $\square$

The result (124) is optimal in the following sense. When F is the map F_0 in (122) and $\sigma = 1$, the eigenvalues of the cloaking conductivity $\widetilde{\sigma}$ in $\mathbb{D}(2) \setminus \overline{\mathbb{D}}(1)$ behaves asymptotically as $(|z|-1)$ and $(|z|-1)^{-1}$ as $|z| \to 1$. This cloaking conductivity has so strong degeneracy that (124) holds. On the other hand,

$$\text{trace}(\widetilde{\sigma}) \in L^1_{\text{weak}}(\mathbb{D}(2)). \tag{126}$$

where L^1_{weak} is the weak-L^1 space. We note that in the case when $\sigma = 1$, $\det(\tilde{\sigma})$ is identically 1 in $\mathbb{D}(2) \setminus \overline{\mathbb{D}}(1)$.

The formula (126) for the blow up map F_0 in (122) and Theorem 3.4 identify the *borderline of the invisibility* for the trace of the conductivity: Any cloaking conductivity $\tilde{\sigma}$ satisfies $\text{trace}(\tilde{\sigma}) \notin L^1(\mathbb{D}(2))$ and there is an example of a cloaking conductivity for which $\text{trace}(\tilde{\sigma}) \in L^1_{\text{weak}}(\mathbb{D}(2))$. Thus the borderline of invisibility is the same as the border between the space L^1 and the weak-L^1 space.

Counterexample 2: Illusion of a nonexistent obstacle. Next we consider new counterexamples for the inverse problem which could be considered as creating an illusion of a nonexisting obstacle. The example is based on a radial shrinking map, that is, a mapping $\mathbb{D}(2) \setminus \overline{\mathbb{D}}(1) \to \mathbb{D}(2) \setminus \{0\}$. The suitable maps are the inverse maps of the blow-up maps $F_1 : \mathbb{D}(2) \setminus \{0\} \to \mathbb{D}(2) \setminus \overline{\mathbb{D}}(1)$ which are constructed in [Iwaniec and Martin 2001] and have the optimal smoothness. Using the properties of these maps and defining a conductivity $\sigma_1 = (F_1^{-1})_*1$ on $\mathbb{D}(2) \setminus \{0\}$ we will later prove the following result.

Theorem 3.5. *Let γ_1 be a conductivity in $\mathbb{D}(2)$ which is identically 1 in $\mathbb{D}(2) \setminus \overline{\mathbb{D}}(1)$ and zero in $\mathbb{D}(1)$ and $\mathcal{A} : [1, \infty] \to [0, \infty]$ be any strictly increasing positive smooth function with $\mathcal{A}(1) = 0$ which is sublinear in the sense that*

$$\int_1^\infty \frac{\mathcal{A}(t)}{t^2} dt < \infty. \tag{127}$$

Then there is a conductivity $\sigma_1 \in \Sigma(B_2)$ satisfying $\det(\sigma_1) = 1$ and

$$\int_{\mathbb{D}(2)} \exp(\mathcal{A}(\text{trace}(\sigma_1) + \text{trace}(\sigma_1^{-1}))) \, dm(z) < \infty, \tag{128}$$

such that $Q_{\sigma_1} = Q_{\gamma_1}$, i.e., the boundary measurements corresponding to σ_1 and γ_1 coincide.

Sketch of the proof. Following [Iwaniec and Martin 2001, Sect. 11.2.1], there is $k(s)$ satisfies the relation

$$k(s) e^{\mathcal{A}(k(s))} = \frac{e}{s^2}, \quad 0 < s < 1$$

that is strictly decreasing function and satisfies $k(s) \leq s^{-1}$ and $k(1) = 1$. Then

$$\rho(t) = \exp\left(\int_0^t \frac{ds}{sk(s)} \right)$$

is a function for which $\rho(0) = 1$. Then, by defining the maps $h(t) = 2\rho(t/2)/\rho(1)$ and

$$F_h : \mathbb{D}(2) \setminus \{0\} \to \mathbb{D}(2) \setminus \overline{\mathbb{D}}(1), \quad F_h(x) = h(t)\frac{x}{|x|} \tag{129}$$

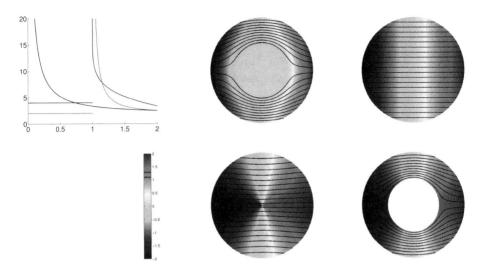

Figure 1. *Left:* trace(σ) of three radial and singular conductivities on the positive x axis. The curves correspond to the invisibility cloaking conductivity (red), with the singularity $\sigma^{22}(x,0) \sim (|x|-1)^{-1}$ for $|x| > 1$, a visible conductivity (blue) with a log log type singularity at $|x| = 1$, and an electric hologram (black) with the conductivity having the singularity $\sigma^{11}(x,0) \sim |x|^{-1}$. *Top two discs:* All measurements on the boundary of the invisibility cloak (left) coincide with the measurements for the homogeneous disc (right). The color shows the value of the solution u with the boundary value $u(x,y)|_{\partial \mathbb{D}(2)} = x$ and the black curves are the integral curves of the current $-\sigma \nabla u$. *Bottom disc and annulus:* All measurements on the boundary of the electric hologram (left) coincide with the measurements for an isolating disc covered with the homogeneous medium (right). The solutions and the current lines corresponding to the boundary value $u|_{\partial \mathbb{D}(2)} = x$ are shown.

and $\sigma_1 = (F_h)_* \gamma_1$, we obtain a conductivity that satisfies conditions of the statement.

Finally, the identity $Q_{\sigma_1} = Q_{\gamma_1}$ follows considering F_h as a change of variables similarly to the proof of Proposition 3.3. $\square$

We observe that for instance the function $\mathcal{A}_0(t) = t/(1+\log t)^{1+\varepsilon}$ satisfies (127) and for such weight function $\sigma_1 \in L^1(B_2)$.

Note that γ_1 corresponds to the case when $\mathbb{D}(2)$ is a perfect insulator which is surrounded with constant conductivity 1. Thus Theorem 3.5 can be interpreted by saying that there is a relatively weakly degenerated conductivity satisfying integrability condition (128) that creates in the boundary observations an illusion of an obstacle that does not exists. Thus the conductivity can be considered as "electric hologram". As the obstacle can be considered as a "hole" in the domain, we can say also that even the topology of the domain can not be detected. In other words, Calderón's program to image the conductivity inside a domain using the boundary measurements can not work within the class of degenerate conductivities satisfying (127) and (128).

3A. *Positive results for Calderón's inverse problem.* In this section we formulate positive results for uniqueness of the inverse problems. Proofs of the results can be found in [Astala et al. 2011a].

For inverse problems for anisotropic conductivities where both the trace and the determinant of the conductivity are degenerate the following result holds.

Theorem 3.6. *Let $\Omega \subset \mathbb{C}$ be a bounded simply connected domain with smooth boundary. Let $\sigma_1, \sigma_2 \in \Sigma(\Omega)$ be matrix valued conductivities in Ω which satisfy the integrability condition*

$$\int_\Omega \exp(p(\mathrm{trace}(\sigma(z)) + \mathrm{trace}(\sigma(z)^{-1}))) \, dm(z) < \infty$$

for some $p > 1$. Moreover, assume that

$$\int_\Omega \mathcal{E}(q \det \sigma_j(z)) \, dm(z) < \infty, \quad \text{for some } q > 0, \tag{130}$$

where $\mathcal{E}(t) = \exp(\exp(\exp(t^{1/2} + t^{-1/2})))$ and $Q_{\sigma_1} = Q_{\sigma_2}$. Then there is a $W^{1,1}_{\mathrm{loc}}$-homeomorphism $F : \Omega \to \Omega$ satisfying $F|_{\partial \Omega} = \mathrm{id}$ such that

$$\sigma_1 = F_* \sigma_2. \tag{131}$$

Equation (131) can be stated as saying that σ_1 and σ_2 are the same up to a change of coordinates, that is, the invariant manifold structures corresponding to these conductivities are the same. See [Lee and Uhlmann 1989; Lassas and Uhlmann 2001].

In the case when the conductivities are isotropic one can improve the result of Theorem 3.6 as follows.

Theorem 3.7. *Let $\Omega \subset \mathbb{C}$ be a bounded simply connected domain with smooth boundary. If $\sigma_1, \sigma_2 \in \Sigma(\Omega)$ are isotropic conductivities, i.e., $\sigma_j(z) = \gamma_j(z)I$, $\gamma_j(z) \in [0, \infty]$ satisfying for some $q > 0$*

$$\int_\Omega \exp\!\left(\exp\!\left[q(\gamma_j(z) + \frac{1}{\gamma_j(z)})\right]\right) dm(z) < \infty \tag{132}$$

and $Q_{\sigma_1} = Q_{\sigma_2}$, then $\sigma_1 = \sigma_2$.

Let us next consider anisotropic conductivities with bounded determinant but more degenerate ellipticity function $K_\sigma(z)$ and ask how far can we then generalize Theorem 3.6. Motivated by the counterexample given in Theorem 3.5 we consider the following class: We say that $\sigma \in \Sigma(\Omega)$ has an exponentially degenerated anisotropy with a weight $\mathcal{A}$ and denote $\sigma \in \Sigma_\mathcal{A} = \Sigma_\mathcal{A}(\Omega)$ if $\sigma(z) \in \mathbb{R}^{2\times 2}$ for a.e. $z \in \Omega$ and

$$\int_\Omega \exp(\mathcal{A}(\text{trace}(\sigma) + \text{trace}(\sigma^{-1}))) \, dm(z) < \infty. \tag{133}$$

In view of Theorem 3.5, for obtaining uniqueness for the inverse problem we need to consider weights that are strictly increasing positive smooth functions $\mathcal{A}: [1, \infty] \to [0, \infty]$, $\mathcal{A}(1) = 0$, with

$$\int_1^\infty \frac{\mathcal{A}(t)}{t^2} \, dt = \infty \quad \text{and} \quad t\mathcal{A}'(t) \to \infty, \text{ as } t \to \infty. \tag{134}$$

We say that $\mathcal{A}$ has almost linear growth if (134) holds.

Note in particular that affine weights $\mathcal{A}(t) = pt - p$, $p > 0$ satisfy the condition (134). To develop uniqueness results for inverse problems within the class $\Sigma_\mathcal{A}$, one needs to find the right Sobolev–Orlicz regularity for the solutions u of finite energy, i.e., for solutions satisfying $A_\sigma[u] < \infty$. For this, we use the counterpart of the gauge $Q(t)$ defined at (113). In the case of a general weight $\mathcal{A}$ we define

$$P(t) = \begin{cases} t^2, & \text{for } 0 \le t < 1, \\ \dfrac{t^2}{\mathcal{A}^{-1}(\log(t^2))}, & \text{for } t \ge 1 \end{cases} \tag{135}$$

where $\mathcal{A}^{-1}$ is the inverse function of $\mathcal{A}$. We note that the condition $\int_1^\infty \frac{\mathcal{A}(t)}{t^2} \, dt = \infty$ is equivalent to

$$\int_1^\infty \frac{P(t)}{t^3} \, dt = \frac{1}{2} \int_1^\infty \frac{\mathcal{A}'(t)}{t} \, dt = \frac{1}{2} \int_1^\infty \frac{\mathcal{A}(t)}{t^2} \, dt = \infty \tag{136}$$

where we have used the substitution $\mathcal{A}(s) = \log(t^2)$. A function $u \in W^{1,1}_{\text{loc}}(\Omega)$ is in the Orlicz space $W^{1,P}(\Omega)$ if

$$\int_\Omega P(|\nabla u(z)|) \, dm(z) < \infty.$$

When $\mathcal{A}$ satisfies the almost linear growth condition (134) and P is as above one can show for $\sigma \in \Sigma_\mathcal{A}(\Omega)$ and $u \in W^{1,1}_{\text{loc}}(\Omega)$ an inequality

$$\int_\Omega P(|\nabla u|) \, dm(z) \le 2 \int_\Omega e^{\mathcal{A}(\text{tr}\,\sigma + \text{tr}(\sigma^{-1}))} \, dm(z) + 2 \int_\Omega \nabla u \cdot \sigma \nabla u \, dm(z).$$

This implies that any solution u of the conductivity equation (4) with $\sigma \in \Sigma_\mathcal{A}(\Omega)$ satisfies $u \in W^{1,P}(\Omega)$.

The Sobolev–Orlicz gauge $P(t)$ is essential also in the study of the counterexamples for solvability of the inverse problem and the optimal smoothness of conductivities corresponding to electric holograms: Assume that $G : \mathbb{D}(2) \setminus \overline{\mathbb{D}}(1) \to \mathbb{D}(2) \setminus \{0\}$ is a homeomorphic map which produces a hologram conductivity $\widetilde{\sigma} = G_* 1$ in $\mathbb{D}(2) \setminus \{0\}$. Assume also that G and its inverse map, denoted $F = G^{-1}$, are $W^{1,1}_{\text{loc}}$-smooth. By the definition of the push forward of a conductivity (117), we see that

$$K_{\widetilde{\sigma}}(z) = K_F(z), \quad z \in \mathbb{D}(2) \setminus \{0\}.$$

This implies that F satisfies a Beltrami equation

$$\partial_{\bar{z}} F(z) = \widetilde{\mu}(z) \partial_z F(z), \quad z \in \mathbb{D}(2) \setminus \{0\}$$

where $K_{\widetilde{\mu}}(z) = K_{\widetilde{\sigma}}(z)$. By Theorem 2.3, the functions $w_1 = \operatorname{Re} F$ and $w_2 = \operatorname{Im} F$ satisfy a conductivity equation with a conductivity $A(z)$ with $K_A(z) = K_{\widetilde{\mu}}(z)$. Thus, if it happens that $\widetilde{\sigma} \in \Sigma_{\mathcal{A}}(\mathbb{D}(2))$ where $\mathcal{A}$ satisfies the almost linear growth condition (134), so that P satisfies condition (136), we see using (137) that $w_1, w_2 \in W^{1,P}(\mathbb{D}(2) \setminus \{0\})$. By using Stoilow factorization, Theorem A.9, we see that F can be written in the form $F(z) = \phi(f(z))$ where $f : \mathbb{C} \to \mathbb{C}$ is a homeomorphism and $\phi : f(\mathbb{D}(2) \setminus \{0\}) \to \mathbb{C}$ is analytic. As F and thus ϕ are bounded, we see that ϕ can be extended to an analytic function $\widetilde{\phi} : f(\mathbb{D}(2)) \to \mathbb{C}$ and thus also F can then be extended to a continuous function to $\widetilde{F} : \mathbb{D}(2) \to \mathbb{C}$. However, this is not possible as $F : \mathbb{D}(2) \setminus \{0\} \to \mathbb{D}(2) \setminus \overline{\mathbb{D}}(1)$ is a homeomorphism. This proves that no electric hologram conductivity $\widetilde{\sigma}$ can be in $\Sigma_{\mathcal{A}}(\mathbb{D}(2))$ where $\mathcal{A}$ satisfies the almost linear growth condition (134).

The above nonexistence of electric hologram conductivities in $\Sigma_{\mathcal{A}}(\mathbb{D}(2))$ motivates the following sharp result for the uniqueness of the inverse problem for singular anisotropic conductivities with a determinant bounded from above and below by positive constants.

Theorem 3.8. *Let $\Omega \subset \mathbb{C}$ be a bounded simply connected domain with smooth boundary and $\mathcal{A} : [1, \infty) \to [0, \infty)$ be a strictly increasing smooth function satisfying the almost linear growth condition (134). Let $\sigma_1, \sigma_2 \in \Sigma(\Omega)$ be matrix valued conductivities in Ω which satisfy the integrability condition*

$$\int_{\Omega} \exp(\mathcal{A}(\operatorname{trace}(\sigma(z))) + \operatorname{trace}(\sigma(z)^{-1}))) \, dm(z) < \infty. \tag{137}$$

Moreover, suppose that $c_1 \leq \det(\sigma_j(z)) \leq c_2$, $z \in \Omega$, $j = 1, 2$ for some $c_1, c_2 > 0$ and $Q_{\sigma_1} = Q_{\sigma_2}$. Then there is a $W^{1,1}_{\text{loc}}$-homeomorphism $F : \Omega \to \Omega$ satisfying $F|_{\partial \Omega} = \operatorname{id}$ such that

$$\sigma_1 = F_* \sigma_2.$$

We note that the determination of σ from Q_σ in Theorems 3.6, 3.7, and 3.8 is constructive in the sense that one can write an algorithm which constructs σ from Λ_σ. For example, for the nondegenerate scalar conductivities such a construction has been numerically implemented in [Astala et al. 2011b].

Let us next discuss the borderline of the visibility somewhat formally. Below we say that a conductivity is visible if there is an algorithm which reconstructs the conductivity σ from the boundary measurements Q_σ, possibly up to a change of coordinates. In other words, for visible conductivities one can use the boundary measurements to produce an image of the conductivity in the interior of Ω in some deformed coordinates. For simplicity, let us consider conductivities with $\det \sigma$ bounded from above and below. Then, Theorems 3.5 and 3.8 can be interpreted by saying that the almost linear growth condition (134) for the weight function $\mathcal{A}$ gives the *borderline of visibility* for the trace of the conductivity matrix: If $\mathcal{A}$ satisfies (134), the conductivities satisfying the integrability condition (137) are visible. However, if $\mathcal{A}$ does not satisfy (134) we can construct a conductivity in Ω satisfying the integrability condition (137) which appears as if an obstacle (which does not exist in reality) would have included in the domain.

Thus the borderline of the visibility is between any spaces $\Sigma_{\mathcal{A}_1}$ and $\Sigma_{\mathcal{A}_2}$ where $\mathcal{A}_1$ satisfies condition (134) and $\mathcal{A}_2$ does not. An example of such gauge functions is $\mathcal{A}_1(t) = t(1 + \log t)^{-1}$ and $\mathcal{A}_2(t) = t(1 + \log t)^{-1-\varepsilon}$ with $\varepsilon > 0$.

Summarizing, in terms of the trace of the conductivity, the above results identify the borderline of visible conductivities and the borderline of invisibility cloaking conductivities. Moreover, these borderlines are not the same and between the visible and the invisibility cloaking conductivities there are conductivities creating electric holograms.

Finally, let us comment the techniques needed to prove the above uniqueness results. The degeneracy of the conductivity causes that the exponentially growing solutions, the standard tools used to study Calderón's inverse problem, can not be constructed using purely microlocal or functional analytic methods. Instead, one needs to use the topological properties of the solutions: By Stoilow's theorem the solutions Beltrami equations are compositions of analytic functions and homeomorphisms. Using this, the continuity properties of the weakly monotone maps, and the Orlicz-estimates holding for homeomorphisms one can prove the existence of the exponentially growing solutions for Beltrami equations. Combining solutions of the appropriate Beltrami equations, see (44), one obtains exponentially growing solutions for conductivity equation in the Sobolev–Orlicz space $W^{1,Q}$ for isotropic conductivity and in $W^{1,P}$ for anisotropic conductivity.

Using these results one can obtain subexponential asymptotics for the families of exponentially growing solutions needed to apply similar $\bar\partial$ technique that were used to solve the inverse problem for the nondegenerate conductivity.

Appendix: The argument principle

The solution to the Calderón problem combines analysis with topological arguments that are specific to two dimensions. For instance, we need a version of the argument principle, which we here consider.

Theorem .9. *Let $F \in W^{1,p}_{\mathrm{loc}}(\mathbb{C})$ and $\gamma \in L^p_{\mathrm{loc}}(\mathbb{C})$ for some $p > 2$. Suppose that, for some constant $0 \le k < 1$, the differential inequality*

$$\left|\frac{\partial F}{\partial \bar{z}}\right| \le k \left|\frac{\partial F}{\partial z}\right| + \gamma(z)\,|F(z)| \tag{138}$$

holds for almost every $z \in \mathbb{C}$ and assume that, for large z, $F(z) = \lambda z + \varepsilon(z)z$, where the constant $\lambda \ne 0$ and $\varepsilon(z) \to 0$ as $|z| \to \infty$.

Then $F(z) = 0$ at exactly one point, $z = z_0 \in \mathbb{C}$.

Proof. The continuity of $F(z) = \lambda z + \varepsilon(z)z$ and an elementary topological argument show that F is surjective, and consequently there exists at least one point $z_0 \in \mathbb{C}$ such that $F(z_0) = 0$.

To show that F can not have more zeros, let $z_1 \in \mathbb{C}$ and choose a large disk $B = \mathbb{D}(R)$ containing both z_1 and z_0. If R is so large that $\varepsilon(z) < \lambda/2$ for $|z| = R$, then $F\big|_{\{|z|=R\}}$ is homotopic to the identity relative to $\mathbb{C} \setminus \{0\}$. Next, we express (138) in the form

$$\frac{\partial F}{\partial \bar{z}} = \nu(z)\frac{\partial F}{\partial z} + A(z)\,F, \tag{139}$$

where $|\nu(z)| \le k < 1$ and $|A(z)| \le \gamma(z)$ for almost every $z \in \mathbb{C}$. Now $A\chi_B \in L^r(\mathbb{C})$ for all $2 \le r \le p' = \min\{p, 1 + 1/k\}$, and we obtain from Theorem A.4 that $(\mathbf{I} - \nu\mathcal{S})^{-1}(A\chi_B) \in L^r(\mathbb{C})$ for all $p'/(p'-1) < r < p'$.

Next, we define $\eta = \mathcal{C}\big((\mathbf{I} - \nu\mathcal{S})^{-1}(A\chi_B)\big)$. Then by Theorem A.3 we have $\eta \in C_0(\mathbb{C})$, and we also have

$$\frac{\partial \eta}{\partial \bar{z}} - \nu\frac{\partial \eta}{\partial z} = A(z), \qquad z \in B. \tag{140}$$

Therefore simply by differentiation we see that the function

$$g = e^{-\eta} F \tag{141}$$

satisfies

$$\frac{\partial g}{\partial \bar{z}} - \nu\frac{\partial g}{\partial z} = 0, \qquad z \in B. \tag{142}$$

Since η has derivatives in $L^r(\mathbb{C})$, we have $g \in W^{1,r}_{\mathrm{loc}}(\mathbb{C})$. As $r \ge 2$, the mapping g is quasiregular in B. The Stoilow factorization theorem gives $g = h \circ \psi$, where $\psi : B \to B$ is a quasiconformal homeomorphism and h is holomorphic, both continuous up to the boundary.

Since η is continuous, (141) shows that $g\big|_{|z|=R}$ is homotopic to the identity relative to $\mathbb{C} \setminus \{0\}$, as is the holomorphic function h. Therefore the argument principle shows that h has precisely one zero in $B = \mathbb{D}(R)$. Already, $h(\psi(z_0)) = e^{-\eta(z_0)} F(z_0) = 0$, and there can be no further zeros for F either. This finishes the proof. $\square$

Appendix A. Some background in complex analysis and quasiconformal mappings.

Here we collect, without proof, some basic facts related to quasiconformal mappings. The proofs can be found in [Astala et al. 2009], for example.

We start with harmonic analysis, where we often need refine estimates of the Cauchy transform.

Definition A.1. The Cauchy transform is defined by the rule
$$(\mathscr{C}\phi)(z) := \frac{1}{\pi} \int_\mathbb{C} \frac{\phi(\tau)}{z - \tau} \, d\tau. \tag{143}$$

Theorem A.2. Let $1 < p < \infty$. If $\phi \in L^p(\mathbb{C})$ and $\phi(\tau) = 0$ for $|\tau| \geq R$, then

(1) $\|\mathscr{C}\phi\|_{L^p(D_{2R})} \leq 6R \|\phi\|_p$,

(2) $\|\mathscr{C}\phi(z) - \frac{1}{\pi z} \int \phi\|_{L^p(\mathbb{C} \setminus D_{2R})} \leq \frac{2R}{(p-1)^{1/p}} \|\phi\|_p$.

Thus, in particular, for $1 < p \leq 2$,
$$\|\mathscr{C}\phi\|_{L^p(\mathbb{C})} \leq \frac{8R}{(p-1)^{1/p}} \|\phi\|_p \quad \text{provided} \int_\mathbb{C} \phi(z) dm(z) = 0.$$

For $p > 2$ the vanishing condition for the integral over $\mathbb{C}$ is not needed, and we have
$$\|\mathscr{C}\phi\|_{L^p(\mathbb{C})} \leq (6 + 3(p-2)^{-1/p}) R \|\phi\|_p, \quad p > 2.$$

Concerning compactness, we have

Theorem A.3. Let Ω be a bounded measurable subset of $\mathbb{C}$. Then the following operators are compact

(1) $\chi_\Omega \circ \mathscr{C} : L^p(\mathbb{C}) \to C^\alpha(\Omega)$, or $2 < p \leq \infty$ and $0 \leq \alpha < 1 - \frac{2}{p}$

(2) $\chi_\Omega \circ \mathscr{C} : L^p(\mathbb{C}) \to L^s(\mathbb{C})$, for $1 \leq p \leq 2$, and $1 \leq s < \frac{2p}{2-p}$.

The fundamental operator in the theory of planar quasiconformal mappings is the Beurling transform,
$$(\mathscr{S}\phi)(z) := -\frac{1}{\pi} \iint_\mathbb{C} \frac{\phi(\tau)}{(z-\tau)^2} \, d\tau. \tag{144}$$

The importance of the Beurling transform in complex analysis is furnished by the identity

$$\mathcal{S} \circ \frac{\partial}{\partial \bar{z}} = \frac{\partial}{\partial z}, \qquad (145)$$

initially valid for functions contained in the space $C_0^\infty(\mathbb{C})$. Moreover, $\mathcal{S}$ extends to abounded operator on $L^p(\mathbb{C})$, $1 < p < \infty$; on $L^2(\mathbb{C})$ it is an isometry. We denote by

$$S_p := \|\mathcal{S}\|_{L^p(\mathbb{C}) \to L^p(\mathbb{C})}$$

the norm of this operator. By Riesz–Thorin interpolation, $S_p \to 1$ as $p \to 2$.

In other words, $\mathcal{S}$ intertwines the Cauchy–Riemann operators $\frac{\partial}{\partial \bar{z}}$ and $\frac{\partial}{\partial z}$, a fact that explains the importance of the operator in complex analysis. For instance we have [Astala et al. 2009, p.363] the following result.

Theorem A.4. *Let μ be measurable with $\|\mu\|_\infty \leq k < 1$. Then the operator $I - \mu \mathcal{S}$ is invertible on $L^p(\mathbb{C})$ whenever $\|\mu\|_\infty \leq k < 1$ and $1 + k < p < 1 + 1/k$.*

The result has important consequences on the regularity of elliptic systems. In fact, it is equivalent to the improved Sobolev regularity of quasiregular mappings.

Theorem A.5. *Let $\mu, \nu \in L^\infty(\mathbb{C})$ with $|\mu| + |\nu| \leq k < 1$ almost everywhere. Then the equation*

$$\frac{\partial f}{\partial \bar{z}} - \mu(z) \frac{\partial f}{\partial z} - \nu(z) \overline{\frac{\partial f}{\partial z}} = h(z)$$

has a solution f, locally integrable with gradient in $L^p(\mathbb{C})$, whenever $1 + k < p < 1 + 1/k$ and $h \in L^p(\mathbb{C})$. Further, f is unique up to an additive constant.

We will also need a simple version of the Koebe distortion theorem.

Lemma A.6 [Astala et al. 2009, p. 44]. *If $f \in W^{1,1}_{\mathrm{loc}}(\mathbb{C})$ is a homeomorphism analytic outside the disk $\mathbb{D}(r)$ with $|f(z) - z| = o(1)$ at ∞, then*

$$|f(z)| < |z| + 3r, \quad \text{for all } z \in \mathbb{C}. \qquad (146)$$

Next, we have the continuous dependence of the quasiconformal mappings on the complex dilatation.

Lemma A.7. *Suppose $|\mu|, |\nu| \leq k \chi_{\mathbb{D}(r)}$, where $0 \leq k < 1$. Let $f, g \in W^{1,2}_{\mathrm{loc}}(\mathbb{C})$ be the principal solutions to the equations*

$$\frac{\partial f}{\partial \bar{z}} = \mu(z) \frac{\partial f}{\partial z}, \qquad \frac{\partial g}{\partial \bar{z}} = \nu(z) \frac{\partial g}{\partial z}.$$

If for a number s we have $2 \leq p < ps < P(k)$, then

$$\|f_{\bar{z}} - g_{\bar{z}}\|_{L^p(\mathbb{C})} \leq C(p, s, k) \, r^{2/ps} \, \|\mu - \nu\|_{L^{ps/(s-1)}(\mathbb{C})}.$$

To prove uniqueness, Liouville type result are often valuable. Here we have collected a number of such results.

Theorem A.8. *Suppose that $F \in W^{1,q}_{\text{loc}}(\mathbb{C})$ satisfies the distortion inequality*

$$|F_{\bar{z}}| \leq k|F_z| + \sigma(z)|F|, \qquad 0 \leq k < 1, \tag{147}$$

where $\sigma \in L^2(\mathbb{C})$ and the Sobolev regularity exponent q lies in the critical interval $1 + k < q < 1 + 1/k$. Then $F = e^\theta g$, where g is quasiregular and $\theta \in VMO$. If $\sigma \in L^{2\pm}(\mathbb{C})$, then θ is continuous, and if furthermore F is bounded, then $F = C_1 e^\theta$.

In addition $F \equiv 0$ if one of the following additional hypotheses holds:

(1) *σ has compact support and $\lim_{z \to \infty} F(z) = 0$, or*

(2) *$F \in L^p(\mathbb{C})$ for some $p > 0$ and $\limsup_{z \to \infty} |F(z)| < \infty$.*

Here we used the notation

$$L^{2\pm}(\mathbb{C}) = \{f : f \in L^s(\mathbb{C}) \cap L^t(\mathbb{C}) \text{ for some } s < 2 < t\}.$$

Finally, we formulate a generalization of the Stoilow factorization theorem for the solutions of Beltrami equation in the space $W^{1,P}_{\text{loc}}(\Omega)$.

Theorem A.9. *Let $\mathcal{A}$ satisfy the almost linear growth condition (134). Suppose the Beltrami coefficient, with $|\mu(z)| < 1$ almost everywhere, is compactly supported and the associated distortion function $K_\mu(z) = \frac{1+|\mu(z)|}{1-|\mu(z)|}$ satisfies*

$$e^{\mathcal{A}(K_\mu(z))} \in L^1_{\text{loc}}(\mathbb{C}) \tag{148}$$

Then the Beltrami equation $f_{\bar{z}}(z) = \mu(z) f_z(z)$ admits a unique principal solution $f \in W^{1,P}_{\text{loc}}(\mathbb{C})$ with $P(t)$ as in (135). Moreover, any solution $h \in W^{1,P}_{\text{loc}}(\Omega)$ to this Beltrami equation in a domain $\Omega \subset \mathbb{C}$ admits a factorization

$$h = \phi \circ f,$$

where ϕ is holomorphic in $f(\Omega)$.

References

[Alu and Engheta 2005] A. Alu and N. Engheta, "Achieving transparency with plasmonic and metamaterial coatings", *Phys. Rev. E* **72** (2005), 016623.

[Astala and Päivärinta 2006] K. Astala and L. Päivärinta, "Calderón's inverse conductivity problem in the plane", *Ann. of Math.* (2) **163**:1 (2006), 265–299. MR 2007b:30019 Zbl 1111.35004

[Astala et al. 2005] K. Astala, L. Päivärinta, and M. Lassas, "Calderón's inverse problem for anisotropic conductivity in the plane", *Comm. Partial Differential Equations* **30**:1-3 (2005), 207–224. MR 2005k:35421

[Astala et al. 2009] K. Astala, T. Iwaniec, and G. Martin, *Elliptic partial differential equations and quasiconformal mappings in the plane*, Princeton Mathematical Series **48**, Princeton University Press, 2009. MR 2010j:30040 Zbl 1182.30001

[Astala et al. 2011a] K. Astala, M. Lassas, and L. Päivärinta, "The borderlines of the invisibility and visibility for Calderón's inverse problem", preprint, 2011. arXiv 1109.2749v1

[Astala et al. 2011b] K. Astala, J. L. Mueller, L. Päivärinta, A. Perämäki, and S. Siltanen, "Direct electrical impedance tomography for nonsmooth conductivities", *Inverse Probl. Imaging* **5**:3 (2011), 531–549. MR 2012g:65238 Zbl 1237.78014

[Ball 1982] J. M. Ball, "Discontinuous equilibrium solutions and cavitation in nonlinear elasticity", *Philos. Trans. Roy. Soc. London Ser. A* **306**:1496 (1982), 557–611. MR 84i:73041 Zbl 0513.73020

[Beals and Coifman 1988] R. Beals and R. Coifman, "The spectral problem for the Davey–Stewartson and Ishimori hierarchies", pp. 15–23 in *Nonlinear evolution equations: integrability and spectral methods*, Manchester University Press, 1988.

[Boiti et al. 1987] M. Boiti, J. J. Leon, M. Manna, and F. Pempinelli, "On a spectral transform of a KdV-like equation related to the Schrödinger operator in the plane", *Inverse Problems* **3**:1 (1987), 25–36. MR 88b:35167 Zbl 0624.35071

[Borcea 2002] L. Borcea, "Electrical impedance tomography", *Inverse Problems* **18**:6 (2002), R99–R136. MR 1955896 Zbl 1031.35147

[Brown 2001] R. M. Brown, "Estimates for the scattering map associated with a two-dimensional first-order system", *J. Nonlinear Sci.* **11**:6 (2001), 459–471. MR 2003b:34163 Zbl 0992.35024

[Brown and Torres 2003] R. M. Brown and R. H. Torres, "Uniqueness in the inverse conductivity problem for conductivities with $3/2$ derivatives in L^p, $p > 2n$", *J. Fourier Anal. Appl.* **9**:6 (2003), 563–574. MR 2004k:35392 Zbl 1051.35105

[Brown and Uhlmann 1997a] R. M. Brown and G. A. Uhlmann, "Uniqueness in the inverse conductivity problem for nonsmooth conductivities in two dimensions", *Comm. Partial Differential Equations* **22**:5-6 (1997), 1009–1027. MR 98f:35155 Zbl 0884.35167

[Brown and Uhlmann 1997b] R. M. Brown and G. A. Uhlmann, "Uniqueness in the inverse conductivity problem for nonsmooth conductivities in two dimensions", *Comm. Partial Differential Equations* **22**:5-6 (1997), 1009–1027. MR 98f:35155 Zbl 0884.35167

[Calderón 1980] A.-P. Calderón, "On an inverse boundary value problem", pp. 65–73 in *Seminar on numerical analysis and its applications to continuum physics* (Rio de Janeiro, 1980), Soc. Brasil. Mat., Rio de Janeiro, 1980. MR 81k:35160

[Cheney et al. 1999] M. Cheney, D. Isaacson, and J. C. Newell, "Electrical impedance tomography", *SIAM Rev.* **41**:1 (1999), 85–101. MR 99k:78017 Zbl 0927.35130

[Dijkstra et al. 1993] A. Dijkstra, B. Brown, N. Harris, D. Barber, and D. Endbrooke, "Review: Clinical applications of electrical impedance tomography", *J. Med. Eng. Technol.* **17** (1993), 89–98.

[Ferreira et al. 2009] D. Dos Santos Ferreira, C. E. Kenig, M. Salo, and G. Uhlmann, "Limiting Carleman weights and anisotropic inverse problems", *Invent. Math.* **178**:1 (2009), 119–171. MR 2010h:58033 Zbl 1181.35327

[Greenleaf et al. 2003a] A. Greenleaf, M. Lassas, and G. Uhlmann, "Anisotropic conductivities that cannot detected in EIT", *Physiolog. Meas.* **24** (2003), 413–420.

[Greenleaf et al. 2003b] A. Greenleaf, M. Lassas, and G. Uhlmann, "The Calderón problem for conormal potentials. I. Global uniqueness and reconstruction", *Comm. Pure Appl. Math.* **56**:3 (2003), 328–352. MR 2003j:35324

[Greenleaf et al. 2003c] A. Greenleaf, M. Lassas, and G. Uhlmann, "On nonuniqueness for Calderón's inverse problem", *Math. Res. Lett.* **10**:5-6 (2003), 685–693. MR 2005f:35316 Zbl 1054.35127

[Greenleaf et al. 2007] A. Greenleaf, Y. Kurylev, M. Lassas, and G. Uhlmann, "Full-wave invisibility of active devices at all frequencies", *Comm. Math. Phys.* **275**:3 (2007), 749–789. MR 2008g:78016 Zbl 1151.78006

[Haberman and Tataru 2011] B. Haberman and D. Tataru, "Uniqueness in Calderón's problem with Lipschitz conductivities", preprint, 2011. arXiv 1108.6068

[Imanuvilov et al. 2010] O. Y. Imanuvilov, G. Uhlmann, and M. Yamamoto, "The Calderón problem with partial data in two dimensions", *J. Amer. Math. Soc.* **23**:3 (2010), 655–691. MR 2012c:35472 Zbl 1201.35183

[Iwaniec and Martin 2001] T. Iwaniec and G. J. Martin, *Geometric function theory and nonlinear analysis*, Oxford University Press, 2001.

[Kenig et al. 2007] C. E. Kenig, J. Sjöstrand, and G. Uhlmann, "The Calderón problem with partial data", *Ann. of Math.* (2) **165**:2 (2007), 567–591. MR 2008k:35498 Zbl 1127.35079

[Kilpeläinen et al. 2000] T. Kilpeläinen, J. Kinnunen, and O. Martio, "Sobolev spaces with zero boundary values on metric spaces", *Potential Anal.* **12**:3 (2000), 233–247. MR 2000m:46071

[Kohn and Vogelius 1984] R. V. Kohn and M. Vogelius, "Identification of an unknown conductivity by means of measurements at the boundary", pp. 113–123 in *Inverse problems* (New York, 1983), SIAM-AMS Proc. **14**, Amer. Math. Soc., Providence, RI, 1984. MR 773707 Zbl 0573.35084

[Kohn et al. 2008] R. V. Kohn, H. Shen, M. S. Vogelius, and M. I. Weinstein, "Cloaking via change of variables in electric impedance tomography", *Inverse Problems* **24**:1 (2008), 015016. MR 2008m:78014 Zbl 1153.35406

[Kohn et al. 2010] R. V. Kohn, D. Onofrei, M. S. Vogelius, and M. I. Weinstein, "Cloaking via change of variables for the Helmholtz equation", *Comm. Pure Appl. Math.* **63**:8 (2010), 973–1016. MR 2011j:78004 Zbl 1194.35505

[Lassas and Uhlmann 2001] M. Lassas and G. Uhlmann, "On determining a Riemannian manifold from the Dirichlet-to-Neumann map", *Ann. Sci. École Norm. Sup.* (4) **34**:5 (2001), 771–787. MR 2003e:58037 Zbl 0992.35120

[Lassas and Zhou 2011] M. Lassas and T. Zhou, "Two dimensional invisibility cloaking for Helmholtz equation and non-local boundary conditions", *Math. Res. Lett.* **18**:3 (2011), 473–488. MR 2012d:35062 Zbl 1241.35045

[Lassas et al. 2007] M. Lassas, J. L. Mueller, and S. Siltanen, "Mapping properties of the nonlinear Fourier transform in dimension two", *Comm. Partial Differential Equations* **32**:4-6 (2007), 591–610. MR 2009b:81207 Zbl 1117.81133

[Lee and Uhlmann 1989] J. M. Lee and G. Uhlmann, "Determining anisotropic real-analytic conductivities by boundary measurements", *Comm. Pure Appl. Math.* **42**:8 (1989), 1097–1112. MR 91a:35166 Zbl 0702.35036

[Leonhardt 2006] U. Leonhardt, "Optical conformal mapping", *Science* **312**:5781 (2006), 1777–1780. MR 2237569 Zbl 1226.78001

[Milton and Nicorovici 2006] G. W. Milton and N.-A. P. Nicorovici, "On the cloaking effects associated with anomalous localized resonance", *Proc. R. Soc. Lond. Ser. A Math. Phys. Eng. Sci.* **462**:2074 (2006), 3027–3059. MR 2008e:78018 Zbl 1149.00310

[Milton et al. 2009] G. Milton, D. Onofrei, and F. Vasquez, "Active exterior cloaking for the 2D Laplace and Helmholtz equations", *Phys. Rev. Lett.* **103** (2009), 073901.

[Nachman 1988] A. I. Nachman, "Reconstructions from boundary measurements", *Ann. of Math.* (2) **128**:3 (1988), 531–576. MR 90i:35283 Zbl 0675.35084

[Nachman 1996a] A. I. Nachman, "Global uniqueness for a two-dimensional inverse boundary value problem", *Ann. of Math.* (2) **143**:1 (1996), 71–96. MR 96k:35189 Zbl 0857.35135

[Nachman 1996b] A. I. Nachman, "Global uniqueness for a two-dimensional inverse boundary value problem", *Ann. of Math.* (2) **143**:1 (1996), 71–96. MR 96k:35189 Zbl 0857.35135

[Nguyen 2012] H.-M. Nguyen, "Approximate cloaking for the Helmholtz equation via transformation optics and consequences for perfect cloaking", *Comm. Pure Appl. Math.* **65**:2 (2012), 155–186. MR 2855543 Zbl 1231.35310

[Päivärinta et al. 2003] L. Päivärinta, A. Panchenko, and G. Uhlmann, "Complex geometrical optics solutions for Lipschitz conductivities", *Rev. Mat. Iberoamericana* **19**:1 (2003), 57–72. MR 2004f:35187 Zbl 1055.35144

[Pendry et al. 2006] J. B. Pendry, D. Schurig, and D. R. Smith, "Controlling electromagnetic fields", *Science* **312**:5781 (2006), 1780–1782. MR 2237570 Zbl 1226.78003

[Schurig et al. 2006] D. Schurig, J. Mock, B. Justice, S. Cummer, J. Pendry, A. Starr, and D. Smith, "Metamaterial electromagnetic cloak at microwave frequencies", *Science* **314**:5801 (2006), 977–980.

[Sylvester 1990] J. Sylvester, "An anisotropic inverse boundary value problem", *Comm. Pure Appl. Math.* **43**:2 (1990), 201–232. MR 90m:35202 Zbl 0709.35102

[Sylvester and Uhlmann 1987] J. Sylvester and G. Uhlmann, "A global uniqueness theorem for an inverse boundary value problem", *Ann. of Math.* (2) **125**:1 (1987), 153–169. MR 88b:35205 Zbl 0625.35078

[Tao and Thiele 2003] T. Tao and C. Thiele, "Nonlinear Fourier analysis", lecture notes, LAS Park City Summer School, 2003. arXiv 1201.5129

[Tsai 1993] T.-Y. Tsai, "The Schrödinger operator in the plane", *Inverse Problems* **9**:6 (1993), 763–787. MR 94i:35154 Zbl 0797.35140

[Veselov and Novikov 1984] A. P. Veselov and S. P. Novikov, "Finite-gap two-dimensional potential Schrödinger operators. Explicit formulas and evolution equations", *Dokl. Akad. Nauk SSSR* **279**:1 (1984), 20–24. In Russian; translated in *Soviet Math. Dokl.* **30**:3 (1984), 588–591. MR 769198 (86d:58053)

[Weder 2008] R. Weder, "A rigorous analysis of high-order electromagnetic invisibility cloaks", *J. Phys. A* **41**:6 (2008), 065207, 21. MR 2009e:78009 Zbl 1132.35488

Department of Mathematics and Statistics, University of Helsinki,
P.O. Box 68 (Gustaf Hällströmin katu 2b), FI-00014 University of Helsinki, Finland
kari.astala@helsinki.fi   matti.lassas@helsinki.fi   lassi.paivarinta@rni.helsinki.fi

Resistor network approaches to electrical impedance tomography

LILIANA BORCEA, VLADIMIR DRUSKIN,
FERNANDO GUEVARA VASQUEZ
AND ALEXANDER V. MAMONOV

We review a resistor network approach to the numerical solution of the inverse problem of electrical impedance tomography (EIT). The networks arise in the context of finite volume discretizations of the elliptic equation for the electric potential, on sparse and adaptively refined grids that we call optimal. The name refers to the fact that the grids give spectrally accurate approximations of the Dirichlet to Neumann map, the data in EIT. The fundamental feature of the optimal grids in inversion is that they connect the discrete inverse problem for resistor networks to the continuum EIT problem.

1. Introduction

We consider the inverse problem of electrical impedance tomography (EIT) in two dimensions [Borcea 2002]. It seeks the scalar valued positive and bounded conductivity $\sigma(x)$, the coefficient in the elliptic partial differential equation for the potential $u \in H^1(\Omega)$,

$$\nabla \cdot [\sigma(x) \nabla u(x)] = 0, \qquad x \in \Omega. \tag{1-1}$$

The domain Ω is a bounded and simply connected set in $\mathbb{R}^2$ with smooth boundary $\mathcal{B}$. Because all such domains are conformally equivalent by the Riemann mapping theorem, we assume throughout that Ω is the unit disk,

$$\Omega = \{x = (r \cos \theta, r \sin \theta), \quad r \in [0, 1], \ \theta \in [0, 2\pi)\}. \tag{1-2}$$

The EIT problem is to determine $\sigma(x)$ from measurements of the Dirichlet to Neumann (DtN) map Λ_σ or equivalently, the Neumann to Dirichlet map $\Lambda_\sigma^\dagger$. We consider the *full boundary setup*, with access to the entire boundary, and the *partial measurement setup*, where the measurements are confined to an accessible subset $\mathcal{B}_A$ of $\mathcal{B}$, and the remainder $\mathcal{B}_I = \mathcal{B} \setminus \mathcal{B}_A$ of the boundary is grounded ($u|_{\mathcal{B}_I} = 0$).

The DtN map $\Lambda_\sigma : H^{1/2}(\mathcal{B}) \to H^{-1/2}(\mathcal{B})$ takes arbitrary boundary potentials $u_\mathcal{B}$ in the trace space $H^{1/2}(\mathcal{B})$ to normal boundary currents

$$\Lambda_\sigma u_\mathcal{B}(x) = \sigma(x)n(x) \cdot \nabla u(x), \qquad x \in \mathcal{B}, \tag{1-3}$$

where $n(x)$ is the outer normal at $x \in \mathcal{B}$ and $u(x)$ solves (1-1) with Dirichlet boundary conditions

$$u(x) = u_\mathcal{B}(x), \quad x \in \mathcal{B}. \tag{1-4}$$

Note that Λ_σ has a null space consisting of constant potentials and thus, it is invertible only on a subset $\mathcal{J}$ of $H^{-1/2}(\mathcal{B})$, defined by

$$\mathcal{J} = \left\{ J \in H^{-1/2}(\mathcal{B}) \text{ such that } \int_\mathcal{B} J(x) ds(x) = 0 \right\}. \tag{1-5}$$

Its generalized inverse is the NtD map $\Lambda_\sigma^\dagger : \mathcal{J} \to H^{1/2}(\mathcal{B})$, which takes boundary currents $J_\mathcal{B} \in \mathcal{J}$ to boundary potentials

$$\Lambda_\sigma^\dagger J_\mathcal{B}(x) = u(x), \qquad x \in \mathcal{B}. \tag{1-6}$$

Here u solves (1-1) with Neumann boundary conditions

$$\sigma(x) n(x) \cdot \nabla u(x) = J_\mathcal{B}(x), \qquad x \in \mathcal{B}, \tag{1-7}$$

and it is defined up to an additive constant, that can be fixed for example by setting the potential to zero at one boundary point, as if it were connected to the ground.

It is known that Λ_σ determines uniquely σ in the full boundary setup [Astala et al. 2005]. See also the earlier uniqueness results [Nachman 1996; Brown and Uhlmann 1997] under some smoothness assumptions on σ. Uniqueness holds for the partial boundary setup as well, at least for $\sigma \in C^{3+\epsilon}(\bar{\Omega})$ and $\epsilon > 0$, [Imanuvilov et al. 2008]. The case of real-analytic or piecewise real-analytic σ is resolved in [Druskin 1982; 1985; Kohn and Vogelius 1984; 1985].

However, the problem is exponentially unstable, as shown in [Alessandrini 1988; Barceló et al. 2001; Mandache 2001]. Given two sufficiently regular conductivities σ_1 and σ_2, the best possible stability estimate is of logarithmic type

$$\|\sigma_1 - \sigma_2\|_{L^\infty(\Omega)} \leq c \left| \log \|\Lambda_{\sigma_1} - \Lambda_{\sigma_2}\|_{H^{1/2}(\mathcal{B}), H^{-1/2}(\mathcal{B})} \right|^{-\alpha}, \tag{1-8}$$

with some positive constants c and α. This means that if we have noisy measurements, we cannot expect the conductivity to be close to the true one uniformly in Ω, unless the noise is exponentially small.

In practice the noise plays a role and the inversion can be carried out only by imposing some regularization constraints on σ. Moreover, we have finitely many measurements of the DtN map and we seek numerical approximations of σ with finitely many degrees of freedom (parameters). The stability of these approximations depends on the number of parameters and their distribution in the domain Ω.

It is shown in [Alessandrini and Vessella 2005] that if σ is piecewise constant, with a bounded number of unknown values, then the stability estimates on σ are no longer of the form (1-8), but they become of Lipschitz type. However, it is not really understood how the Lipschitz constant depends on the distribution of the unknowns in Ω. Surely, it must be easier to determine the features of the conductivity near the boundary than deep inside Ω.

Then, the question is how to parametrize the unknown conductivity in numerical inversion so that we can control its stability and we do not need excessive regularization with artificial penalties that introduce artifacts in the results. Adaptive parametrizations for EIT have been considered for example in [Isaacson 1986; MacMillan et al. 2004] and [Ben Ameur et al. 2002; Ben Ameur and Kaltenbacher 2002]. Here we review our inversion approach that is based on resistor networks that arise in finite volume discretizations of (1-1) on sparse and adaptively refined grids which we call *optimal*. The name refers to the fact that they give spectral accuracy of approximations of Λ_σ on finite volume grids. One of their important features is that they are refined near the boundary, where we make the measurements, and coarse away from it. Thus they capture the expected loss of resolution of the numerical approximations of σ.

Optimal grids were introduced in [Druskin and Knizhnerman 2000b; 2000a; Ingerman et al. 2000; Asvadurov et al. 2000; 2003] for accurate approximations of the DtN map in forward problems. Having such approximations is important for example in domain decomposition approaches to solving second order partial differential equations and systems, because the action of a subdomain can be replaced by the DtN map on its boundary [Quarteroni and Valli 1999]. In addition, accurate approximations of DtN maps allow truncations of the computational domain for solving hyperbolic problems. The studies in [Druskin and Knizhnerman 2000b; 2000a; Ingerman et al. 2000; Asvadurov et al. 2000; 2003] work with spectral decompositions of the DtN map, and show that by just placing grid points optimally in the domain, one can obtain exponential convergence rates of approximations of the DtN map with second order finite difference schemes. That is to say, although the solution of the forward problem is second order accurate inside the computational domain, the DtN map is approximated with spectral accuracy. Problems with piecewise constant and anisotropic coefficients are considered in [Druskin and Moskow 2002; Asvadurov et al. 2007].

Optimal grids are useful in the context of numerical inversion, because they resolve the inconsistency that arises from the exponential ill posedness of the problem and the second order convergence of typical discretization schemes applied to (1-1), on ad hoc grids that are usually uniform. The forward problem for the approximation of the DtN map is the inverse of the EIT problem, so it should converge exponentially. This can be achieved by discretizing on the optimal grids.

In this article we review the use of optimal grids in inversion, as it was developed over the last few years in [Borcea and Druskin 2002; Borcea et al. 2005; 2008; Guevara Vasquez 2006; Borcea et al. 2010a; 2010b; Mamonov 2010]. We present first, in Section 3, the case of layered conductivity $\sigma = \sigma(r)$ and full boundary measurements, where the DtN map has eigenfunctions $e^{ik\theta}$ and eigenvalues denoted by $f(k^2)$, with integer k. Then, the forward problem can be stated as one of rational approximation of $f(\lambda)$, for λ in the complex plane, away from the negative real axis. We explain in Section 3 how to compute the optimal grid from such rational approximants and also how to use it in inversion. The optimal grid depends on the type of discrete measurements that we make of Λ_σ (i.e., $f(\lambda)$) and so does the accuracy and stability of the resulting approximations of σ.

The two-dimensional problem $\sigma = \sigma(r, \theta)$ is reviewed in Sections 4 and 5. The easier case of full access to the boundary, and discrete measurements at n equally distributed points on $\mathcal{B}$ is in Section 4. There, the grids are essentially the same as in the layered case, and finite-volume discretization leads to circular networks with topology determined by the grids. We show how to use the discrete inverse problem theory for circular networks developed in [Curtis et al. 1994; 1998; Ingerman 2000; Colin de Verdière 1994; Colin de Verdière et al. 1996] for the numerical solution of the EIT problem. Section 5 considers the more difficult, partial boundary measurement setup, where the accessible boundary consists of either one connected subset of $\mathcal{B}$ or two disjoint subsets. There, the optimal grids are truly two-dimensional and cannot be computed directly from the layered case.

The theoretical review of our results is complemented by some numerical results. For brevity, all the results are in the noiseless case. We refer the reader to [Borcea et al. 2011] for an extensive study of noise effects on our inversion approach.

2. Resistor networks as discrete models for EIT

Resistor networks arise naturally in the context of finite volume discretizations of the elliptic equation (1-1) on staggered grids with interlacing primary and dual lines that may be curvilinear, as explained in Section 2.1. Standard finite volume discretizations use arbitrary, usually equidistant tensor product grids. We consider *optimal grids* that are designed to obtain very accurate approximations of the measurements of the DtN map, the data in the inverse problem. The geometry of these grids depends on the measurement setup. We describe in Section 2.2 the type of grids used for the full measurement case, where we have access to the entire boundary $\mathcal{B}$. The grids for the partial boundary measurement setup are discussed later, in Section 5.

2.1. Finite volume discretization and resistor networks. See Figure 1 for an illustration of a staggered grid. The potential $u(x)$ in (1-1) is discretized at the primary nodes $P_{i,j}$, the intersection of the primary grid lines, and the finite-volume method balances the fluxes across the boundary of the dual cells C_{ij},

$$\int_{C_{i,j}} \nabla \cdot [\sigma(x)\nabla u(x)] \, dx = \int_{\partial C_{i,j}} \sigma(x) n(x) \cdot \nabla u(x) \, ds(x) = 0. \quad (2\text{-}1)$$

A dual cell $C_{i,j}$ contains a primary point $P_{i,j}$, it has vertices (dual nodes) $P_{i \pm \frac{1}{2}, j \pm \frac{1}{2}}$, and boundary

$$\partial C_{i,j} = \Sigma_{i,j+\frac{1}{2}} \cup \Sigma_{i+\frac{1}{2},j} \cup \Sigma_{i,j-\frac{1}{2}} \cup \Sigma_{i-\frac{1}{2},j}, \quad (2\text{-}2)$$

the union of the dual line segments

$$\Sigma_{i,j \pm \frac{1}{2}} = (P_{i-\frac{1}{2}, j \pm \frac{1}{2}}, P_{i+\frac{1}{2}, j \pm \frac{1}{2}}) \quad \text{and} \quad \Sigma_{i \pm \frac{1}{2}, j} = (P_{i \pm \frac{1}{2}, j-\frac{1}{2}}, P_{i \pm \frac{1}{2}, j+\frac{1}{2}}).$$

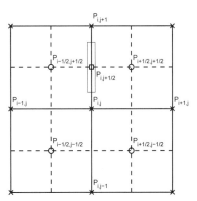

Figure 1. Finite-volume discretization on a staggered grid. The primary grid lines are solid and the dual ones are dashed. The primary grid nodes are indicated with × and the dual nodes with ○. The dual cell $C_{i,j}$, with vertices (dual nodes) $P_{i \pm \frac{1}{2}, j \pm \frac{1}{2}}$, surrounds the primary node $P_{i,j}$. A resistor is shown as a rectangle with axis along a primary line, that intersects a dual line at the point indicated with □.

Denote by $\mathcal{P} = \{P_{i,j}\}$ the set of primary nodes, and define the potential function $U : \mathcal{P} \to \mathbb{R}$ as the finite volume approximation of $u(x)$ at the points in $\mathcal{P}$:

$$U_{i,j} \approx u(P_{i,j}), \qquad P_{i,j} \in \mathcal{P}. \tag{2-3}$$

The set $\mathcal{P}$ is the union of two disjoint sets $\mathcal{P}_\mathcal{I}$ and $\mathcal{P}_\mathcal{B}$ of interior and boundary nodes, respectively. Adjacent nodes in $\mathcal{P}$ are connected by edges in the set $\mathcal{E} \subset \mathcal{P} \times \mathcal{P}$. We denote the edges by

$$E_{i,j\pm\frac{1}{2}} = (P_{i,j}, P_{i,j\pm1}) \quad \text{and} \quad E_{i\pm\frac{1}{2},j} = (P_{i\pm1,j}, P_{i,j}).$$

The finite volume discretization results in a system of linear equations for the potential:

$$\gamma_{i+\frac{1}{2},j}(U_{i+1,j} - U_{i,j}) + \gamma_{i-\frac{1}{2},j}(U_{i-1,j} - U_{i,j}) + \\ \gamma_{i,j+\frac{1}{2}}(U_{i,j+1} - U_{i,j}) + \gamma_{i,j-\frac{1}{2}}(U_{i,j-1} - U_{i,j}) = 0, \tag{2-4}$$

with terms given by approximations of the fluxes

$$\int_{\Sigma_{i,j\pm\frac{1}{2}}} \sigma(x)n(x) \cdot \nabla u(x)\, ds(x) \approx \gamma_{i,j\pm\frac{1}{2}}(U_{i,j\pm1} - U_{i,j}),$$
$$\int_{\Sigma_{i\pm\frac{1}{2},j}} \sigma(x)n(x) \cdot \nabla u(x)\, ds(x) \approx \gamma_{i\pm\frac{1}{2},j}(U_{i\pm1,j} - U_{i,j}). \tag{2-5}$$

Equations (2-4) are Kirchhoff's law for the interior nodes in a resistor network (Γ, γ) with graph $\Gamma = (\mathcal{P}, \mathcal{E})$ and conductance function $\gamma : \mathcal{E} \to \mathbb{R}^+$, that assigns to an edge like $E_{i\pm\frac{1}{2},j}$ in $\mathcal{E}$ a positive conductance $\gamma_{i\pm\frac{1}{2},j}$. At the boundary nodes we discretize either the Dirichlet conditions (1-4), or the Neumann conditions (1-7), depending on what we wish to approximate, the DtN or the NtD map.

To write the network equations in compact (matrix) form, let us number the primary nodes in some fashion, starting with the interior ones and ending with the boundary ones. Then we can write $\mathcal{P} = \{p_q\}$, where p_q are the numbered nodes. They correspond to points like $P_{i,j}$ in Figure 1. Let also $U_\mathcal{I}$ and $U_\mathcal{B}$ be the vectors with entries given by the potential at the interior nodes and boundary nodes, respectively. The vector of boundary fluxes is denoted by $J_\mathcal{B}$. We assume throughout that there are n boundary nodes, so $U_\mathcal{B}, J_\mathcal{B} \in \mathbb{R}^n$. The network equations are

$$KU = \begin{pmatrix} 0 \\ J_\mathcal{B} \end{pmatrix}, \quad U = \begin{pmatrix} U_\mathcal{I} \\ U_\mathcal{B} \end{pmatrix}, \quad K = \begin{pmatrix} K_{\mathcal{I}\mathcal{I}} & K_{\mathcal{I}\mathcal{B}} \\ K_{\mathcal{I}\mathcal{B}} & K_{\mathcal{B}\mathcal{B}} \end{pmatrix}, \tag{2-6}$$

where $\boldsymbol{K} = (K_{ij})$ is the Kirchhoff matrix with entries

$$K_{i,j} = \begin{cases} -\gamma(E) & \text{if } i \neq j \text{ and } E = (\boldsymbol{p}_i, \boldsymbol{p}_j) \in \mathcal{E}, \\ 0 & \text{if } i \neq j \text{ and } (\boldsymbol{p}_i, \boldsymbol{p}_j) \notin \mathcal{E}, \\ \sum_{k: E=(\boldsymbol{p}_i, \boldsymbol{p}_k) \in \mathcal{E}} \gamma(E) & \text{if } i = j. \end{cases} \quad (2\text{-}7)$$

In (2-6) we write it in block form, with $\boldsymbol{K}_{\mathcal{I}\mathcal{I}}$ the block with row and column indices restricted to the interior nodes, $\boldsymbol{K}_{\mathcal{I}\mathcal{B}}$ the block with row indices restricted to the interior nodes and column indices restricted to the boundary nodes, and so on. Note that $\boldsymbol{K}$ is symmetric, and its rows and columns sum to zero, which is just the condition of conservation of currents.

It is shown in [Curtis et al. 1994] that the potential U satisfies a discrete maximum principle. Its minimum and maximum entries are located on the boundary. This implies that the network equations with Dirichlet boundary conditions

$$\boldsymbol{K}_{\mathcal{I}\mathcal{I}} \boldsymbol{U}_{\mathcal{I}} = -\boldsymbol{K}_{\mathcal{I}\mathcal{B}} \boldsymbol{U}_{\mathcal{B}} \quad (2\text{-}8)$$

have a unique solution if $\boldsymbol{K}_{\mathcal{I}\mathcal{B}}$ has full rank. That is to say, $\boldsymbol{K}_{\mathcal{I}\mathcal{I}}$ is invertible and we can eliminate $\boldsymbol{U}_{\mathcal{I}}$ from (2-6) to obtain

$$\boldsymbol{J}_{\mathcal{B}} = (\boldsymbol{K}_{\mathcal{B}\mathcal{B}} - \boldsymbol{K}_{\mathcal{B}\mathcal{I}} \boldsymbol{K}_{\mathcal{I}\mathcal{I}}^{-1} \boldsymbol{K}_{\mathcal{I}\mathcal{B}}) \boldsymbol{U}_{\mathcal{B}} = \boldsymbol{\Lambda}_{\gamma} \boldsymbol{U}_{\mathcal{B}}. \quad (2\text{-}9)$$

The matrix $\boldsymbol{\Lambda}_{\gamma} \in \mathbb{R}^{n \times n}$ is the Dirichlet to Neumann map of the network. It takes the boundary potential $\boldsymbol{U}_{\mathcal{B}}$ to the vector $\boldsymbol{J}_{\mathcal{B}}$ of boundary fluxes, and is given by the Schur complement of the block $\boldsymbol{K}_{\mathcal{B}\mathcal{B}}$

$$\boldsymbol{\Lambda}_{\gamma} = \boldsymbol{K}_{\mathcal{B}\mathcal{B}} - \boldsymbol{K}_{\mathcal{B}\mathcal{I}} \boldsymbol{K}_{\mathcal{I}\mathcal{I}}^{-1} \boldsymbol{K}_{\mathcal{I}\mathcal{B}}. \quad (2\text{-}10)$$

The DtN map is symmetric, with nontrivial null space spanned by the vector $\mathbf{1}_{\mathcal{B}} \in \mathbb{R}^n$ of all ones. The symmetry follows directly from the symmetry of $\boldsymbol{K}$. Since the columns of $\boldsymbol{K}$ sum to zero, $\boldsymbol{K}\mathbf{1} = \mathbf{0}$, where $\mathbf{1}$ is the vector of all ones. Then, (2-9) gives $\boldsymbol{J}_{\mathcal{B}} = \mathbf{0} = \boldsymbol{\Lambda}_{\gamma} \mathbf{1}_{\mathcal{B}}$, which means that $\mathbf{1}_{\mathcal{B}}$ is in the null space of $\boldsymbol{\Lambda}_{\gamma}$.

The inverse problem for a network (Γ, γ) is to determine the conductance function γ from the DtN map $\boldsymbol{\Lambda}_{\gamma}$. The graph Γ is assumed known, and it plays a key role in the solvability of the inverse problem [Curtis et al. 1994; 1998; Ingerman 2000; Colin de Verdière 1994; Colin de Verdière et al. 1996]. More precisely, Γ must satisfy a certain criticality condition for the network to be uniquely recoverable from $\boldsymbol{\Lambda}_{\gamma}$, and its topology should be adapted to the type of measurements that we have. We review these facts in detail in Sections 3–5. We also show there how to relate the continuum DtN map $\boldsymbol{\Lambda}_{\sigma}$ to the discrete DtN map $\boldsymbol{\Lambda}_{\gamma}$. The inversion algorithms in this paper use the solution of the discrete

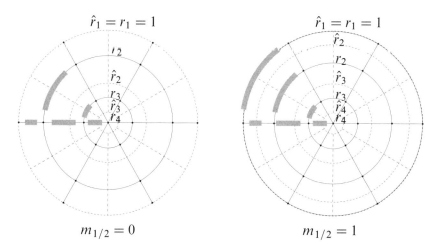

Figure 2. Examples of grids. The primary grid lines are solid and the dual ones are dotted. Both grids have $n = 6$ primary boundary points, and index of the layers $\ell = 3$. We have the type of grid indexed by $m_{1/2} = 0$ on the left and by $m_{1/2} = 1$ on the right.

inverse problem for networks to determine approximately the solution $\sigma(x)$ of the continuum EIT problem.

2.2. *Tensor product grids for the full boundary measurements setup.* In the full boundary measurement setup, we have access to the entire boundary $\mathcal{B}$, and it is natural to discretize the domain (1-2) with tensor product grids that are uniform in angle, as shown in Figure 2. Let

$$\theta_j = \frac{2\pi(j-1)}{n}, \qquad \hat{\theta}_j = \frac{2\pi(j-1/2)}{n}, \qquad j = 1, \ldots, n, \qquad (2\text{-}11)$$

be the angular locations of the primary and dual nodes. The radii of the primary and dual layers are denoted by r_i and $\hat{r}_i$, and we count them starting from the boundary. We can have two types of grids, so we introduce the parameter $m_{1/2} \in \{0, 1\}$ to distinguish between them. We have

$$1 = r_1 = \hat{r}_1 > r_2 > \hat{r}_2 > \cdots > r_\ell > \hat{r}_\ell > r_{\ell+1} \geq 0 \qquad (2\text{-}12)$$

when $m_{1/2} = 0$, and

$$1 = \hat{r}_1 = r_1 > \hat{r}_2 > r_2 > \cdots > r_\ell > \hat{r}_{\ell+1} > r_{\ell+1} \geq 0 \qquad (2\text{-}13)$$

for $m_{1/2} = 1$. In either case there are $\ell + 1$ primary layers and $\ell + m_{1/2}$ dual ones, as illustrated in Figure 2. We explain in Sections 3 and 4 how to place optimally in the interval [0, 1] the primary and dual radii, so that the finite volume discretization gives an accurate approximation of the DtN map Λ_σ.

The graph of the network is given by the primary grid. We follow [Curtis et al. 1994; 1998] and call it a circular network. It has n boundary nodes and $n(2\ell + m_{1/2} - 1)$ edges. Each edge is associated with an unknown conductance that is to be determined from the discrete DtN map Λ_γ, defined by measurements of Λ_σ, as explained in Sections 3 and 4. Since Λ_γ is symmetric, with columns summing to zero, it contains $n(n-1)/2$ measurements. Thus, we have the same number of unknowns as data points when

$$2\ell + m_{1/2} - 1 = \frac{n-1}{2}, \qquad n = \text{odd integer}. \qquad (2\text{-}14)$$

This condition turns out to be necessary and sufficient for the DtN map to determine uniquely a circular network, as shown in [Colin de Verdière et al. 1996; Curtis et al. 1998; Borcea et al. 2008]. We assume henceforth that it holds.

3. Layered media

In this section we assume a layered conductivity function $\sigma(r)$ in Ω, the unit disk, and access to the entire boundary $\mathcal{B}$. Then, the problem is rotation invariant and can be simplified by writing the potential as a Fourier series in the angle θ. We begin in Section 3.1 with the spectral decomposition of the continuum and discrete DtN maps and define their eigenvalues, which contain all the information about the layered conductivity. Then, we explain in Section 3.2 how to construct finite volume grids that give discrete DtN maps with eigenvalues that are accurate, rational approximations of the eigenvalues of the continuum DtN map. One such approximation brings an interesting connection between a classic Sturm–Liouville inverse spectral problem [Gel'fand and Levitan 1951; Chadan et al. 1997; Hochstadt 1973; Marchenko 1986; McLaughlin and Rundell 1987] and an inverse eigenvalue problem for Jacobi matrices [Chu and Golub 2002], as described in Sections 3.2.3 and 3.3. This connection allows us to solve the continuum inverse spectral problem with efficient, linear algebra tools. The resulting algorithm is the first example of resistor network inversion on optimal grids proposed and analyzed in [Borcea et al. 2005], and we review its convergence study in Section 3.3.

3.1. *Spectral decomposition of the continuum and discrete DtN maps.* Because Equation (1-1) is separable in layered media, we write the potential $u(r, \theta)$ as a Fourier series

$$u(r, \theta) = v_{\mathcal{B}}(0) + \sum_{\substack{k \in \mathbb{Z} \\ k \neq 0}} v(r, k) e^{ik\theta}, \qquad (3\text{-}1)$$

with coefficients $v(r,k)$ satisfying the differential equation

$$\frac{r}{\sigma(r)} \frac{d}{dr}\left[r\sigma(r) \frac{dv(r,k)}{dr}\right] - k^2 v(r,k) = 0, \quad r \in (0,1), \quad (3\text{-}2)$$

and the condition

$$v(0,k) = 0. \quad (3\text{-}3)$$

The first term $v_{\mathcal{B}}(0)$ in (3-1) is the average boundary potential

$$v_{\mathcal{B}}(0) = \frac{1}{2\pi} \int_0^{2\pi} u(1,\theta)\, d\theta. \quad (3\text{-}4)$$

The boundary conditions at $r = 1$ are Dirichlet or Neumann, depending on which map we consider, the DtN or the NtD map.

3.1.1. *The DtN map.* The DtN map is determined by the potential v satisfying (3-2)–(3-3), with Dirichlet boundary condition

$$v(1,k) = v_{\mathcal{B}}(k), \quad (3\text{-}5)$$

where $v_{\mathcal{B}}(k)$ are the Fourier coefficients of the boundary potential $u_{\mathcal{B}}(\theta)$. The normal boundary flux has the Fourier series expansion

$$\sigma(1) \frac{\partial u(1,\theta)}{\partial r} = \Lambda_\sigma u_{\mathcal{B}}(\theta) = \sigma(1) \sum_{k \in \mathbb{Z}, k \neq 0} \frac{dv(1,k)}{dr} e^{ik\theta}, \quad (3\text{-}6)$$

and we assume for simplicity that $\sigma(1) = 1$. Then, we deduce formally from (3-6) that $e^{ik\theta}$ are the eigenfunctions of the DtN map Λ_σ, with eigenvalues

$$f(k^2) = \frac{dv(1,k)}{dr} / v(1,k). \quad (3\text{-}7)$$

Note that $f(0) = 0$.

A similar diagonalization applies to the DtN map $\boldsymbol{\Lambda}_\gamma$ of networks arising in the finite volume discretization of (1-1) if the grids are equidistant in angle, as described in Section 2.2. Then, the resulting network is layered in the sense that the conductance function is rotation invariant. We can define various quadrature rules in (2-5), with minor changes in the results [Borcea et al. 2010a, Section 2.4]. In this section we use the definitions

$$\gamma_{j+\frac{1}{2},q} = \frac{h_\theta}{z(r_{j+1}) - z(r_j)} = \frac{h_\theta}{\alpha_j}, \quad \gamma_{j,q+\frac{1}{2}} = \frac{\hat{z}(\hat{r}_{j+1}) - \hat{z}(\hat{r}_j)}{h_\theta} = \frac{\hat{\alpha}_j}{h_\theta}, \quad (3\text{-}8)$$

derived in Appendix A, where $h_\theta = 2\pi/n$ and

$$z(r) = \int_r^1 \frac{dt}{t\sigma(t)}, \quad \hat{z}(r) = \int_r^1 \frac{\sigma(t)}{t} dt. \quad (3\text{-}9)$$

The network equations (2-4) become

$$\frac{1}{\hat{\alpha}_j}\left(\frac{U_{j+1,q}-U_{j,q}}{\alpha_j}-\frac{U_{j,q}-U_{j-1,q}}{\alpha_{j-1}}\right)-\frac{2U_{j,q}-U_{j,q+1}-U_{j,q-1}}{h_\theta^2}=0, \tag{3-10}$$

and we can write them in block form as

$$\frac{1}{\hat{\alpha}_j}\left(\frac{\boldsymbol{U}_{j+1}-\boldsymbol{U}_j}{\alpha_j}-\frac{\boldsymbol{U}_j-\boldsymbol{U}_{j-1}}{\alpha_{j-1}}\right)-[-\partial_\theta^2]\boldsymbol{U}_j=\boldsymbol{0}, \tag{3-11}$$

where

$$\boldsymbol{U}_j=(U_{j,1},\ldots,U_{j,n})^T, \tag{3-12}$$

and $[-\partial_\theta^2]$ is the circulant matrix

$$[-\partial_\theta^2]=\frac{1}{h_\theta^2}\begin{pmatrix} 2 & -1 & 0 & \ldots & \ldots & 0 & -1 \\ -1 & 2 & 1 & 0 & \ldots & 0 & 0 \\ \ddots & \ddots & \ddots & \ddots & \ddots & \ddots & \ddots \\ -1 & 0 & \ldots & \ldots & 0 & -1 & 2 \end{pmatrix}, \tag{3-13}$$

the discretization of the operator $-\partial_\theta^2$ with periodic boundary conditions. It has the eigenvectors

$$[e^{ik\theta}]=(e^{ik\theta_1},\ldots,e^{ik\theta_n})^T, \tag{3-14}$$

with entries given by the restriction of the continuum eigenfunctions $e^{ik\theta}$ at the primary grid angles. Here k is integer, satisfying $|k|\leq(n-1)/2$, and the eigenvalues are ω_k^2, where

$$\omega_k=|k|\left|\mathrm{sinc}\left(\frac{kh_\theta}{2}\right)\right|, \tag{3-15}$$

with $\mathrm{sinc}(x)=\sin(x)/x$. Note that $\omega_k^2\approx k^2$ only for $|k|\ll n$.

To determine the spectral decomposition of the discrete DtN map Λ_γ we proceed as in the continuum and write the potential $\boldsymbol{U}_j$ as a Fourier sum

$$\boldsymbol{U}_j=v_{\mathcal{B}}(0)\mathbf{1}_{\mathcal{B}}+\sum_{\substack{|k|\leq\frac{n-1}{2}\\k\neq 0}}V_j(k)[e^{ik\theta}], \tag{3-16}$$

where we recall that $\mathbf{1}_{\mathcal{B}}\in\mathbb{R}^n$ is a vector of all ones. We obtain the finite difference equation for the coefficients $V_j(k)$:

$$\frac{1}{\hat{\alpha}_j}\left(\frac{V_{j+1}(k)-V_j(k)}{\alpha_j}-\frac{V_j(k)-V_{j-1}(k)}{\alpha_{j-1}}\right)-\omega_k^2 V_j(k)=0, \tag{3-17}$$

where $j = 2, 3, \ldots, \ell$. It is the discretization of (3-2) that takes the form

$$\frac{d}{d\hat{z}}\left(\frac{dv(z,k)}{dz}\right) - k^2 v(z,k) = 0, \tag{3-18}$$

in the coordinates (3-9), where we let in an abuse of notation $v(r,k) \rightsquigarrow v(z,k)$. The boundary condition at $r = 0$ is mapped to

$$\lim_{z \to \infty} v(z,k) = 0, \tag{3-19}$$

and it is implemented in the discretization as $V_{\ell+1}(k) = 0$. At the boundary $r = 1$, where $z = 0$, we specify $V_1(k)$ as some approximation of $v_{\mathcal{B}}(k)$.

The discrete DtN map $\mathbf{\Lambda}_\gamma$ is diagonalized in the basis $\{[e^{ik\theta}]\}_{|k| \leq (n-1)/2}$, and we denote its eigenvalues by $F(\omega_k^2)$. Its definition depends on the type of grid that we use, indexed by $m_{1/2}$, as explained in Section 2.2. In the case $m_{1/2} = 0$, the first radius next to the boundary is r_2, and we define the boundary flux at $\hat{r}_1 = 1$ as $(V_1(k) - V_2(k))/\alpha_1$. When $m_{1/2} = 1$, the first radius next to the boundary is $\hat{r}_2$, so to compute the flux at $\hat{r}_1$ we introduce a ghost layer at $r_0 > 1$ and use (3-17) for $j = 1$ to define the boundary flux as

$$\frac{V_0(k) - V_1(k)}{\alpha_o} = \hat{\alpha}_1 \omega_k^2 V_1(k) + \frac{V_1(k) - V_2(k)}{\alpha_1}.$$

Therefore, the eigenvalues of the discrete DtN map are

$$F(\omega_k^2) = m_{1/2}\hat{\alpha}_1 \omega_k^2 + \frac{V_1(k) - V_2(k)}{\alpha_1 V_1(k)}. \tag{3-20}$$

3.1.2. *The NtD map.* The NtD map $\Lambda_\sigma^\dagger$ has eigenfunctions $e^{ik\theta}$ for $k \neq 0$ and eigenvalues $f^\dagger(k^2) = 1/f(k^2)$. Equivalently, in terms of the solution $v(z,k)$ of (3-18) with boundary conditions (3-19) and

$$-\frac{dv(0,k)}{dz} = \frac{1}{2\pi}\int_0^{2\pi} J_{\mathcal{B}}(\theta)e^{-ik\theta} d\theta = \varphi_{\mathcal{B}}(k), \tag{3-21}$$

we have

$$f^\dagger(k^2) = \frac{v(0,k)}{\varphi_{\mathcal{B}}(k)}. \tag{3-22}$$

In the discrete case, let us use the grids with $m_{1/2} = 1$. We obtain that the potential $V_j(k)$ satisfies (3-17) for $j = 1, 2, \ldots, \ell$, with boundary conditions

$$-\frac{V_1(k) - V_0(k)}{\alpha_0} = \Phi_{\mathcal{B}}(k), \qquad V_{\ell+1} = 0. \tag{3-23}$$

Here $\Phi_{\mathcal{B}}(k)$ is some approximation of $\varphi_{\mathcal{B}}(k)$. The eigenvalues of $\Lambda_\gamma^\dagger$ are

$$F^\dagger(\omega_k^2) = \frac{V_1(k)}{\Phi_{\mathcal{B}}(k)}. \quad (3\text{-}24)$$

3.2. Rational approximations, optimal grids and reconstruction mappings.
By analogy to (3-22) and (3-24) define the functions

$$f^\dagger(\lambda) = \frac{v(0)}{\varphi_{\mathcal{B}}}, \qquad F^\dagger(\lambda) = \frac{V_1}{\Phi_{\mathcal{B}}}, \quad (3\text{-}25)$$

where v solves Equation (3-18) with k^2 replaced by λ and V_j solves (3-17) with ω_k^2 replaced by λ. The spectral parameter λ may be complex, satisfying $\lambda \in \mathbb{C} \setminus (-\infty, 0]$. For simplicity, we suppress in the notation the dependence of v and V_j on λ. We consider in detail the discretizations on grids indexed by $m_{1/2} = 1$, but the results can be extended to the other type of grids, indexed by $m_{1/2} = 0$.

Lemma 1. *The function $f^\dagger(\lambda)$ is of form*

$$f^\dagger(\lambda) = \int_{-\infty}^0 \frac{d\mu(t)}{\lambda - t}, \quad (3\text{-}26)$$

where $\mu(t)$ is the positive spectral measure on $(-\infty, 0]$ of the differential operator $d_{\hat{z}} d_z$, with homogeneous Neumann condition at $z = 0$ and limit condition (3-19). The function $F^\dagger(\lambda)$ has a similar form

$$F^\dagger(\lambda) = \int_{-\infty}^0 \frac{d\mu^F(t)}{\lambda - t}, \quad (3\text{-}27)$$

where $\mu^F(t)$ is the spectral measure of the difference operator in (3-17) with boundary conditions (3-23).

Proof. The result (3-26) is shown in [Kac and Krein 1968] and it says that $f^\dagger(\lambda)$ is essentially a Stieltjes function. To derive the representation (3-27), we write our difference equations in matrix form for $V = (V_1, \ldots, V_\ell)^T$,

$$(A - \lambda I)V = -\frac{\Phi_{\mathcal{B}}(\lambda)}{\hat{\alpha}_1} e_1. \quad (3\text{-}28)$$

Here I is the $\ell \times \ell$ identity matrix, $e_1 = (1, \ldots, 0)^T \in \mathbb{R}^\ell$ and A is the tridiagonal matrix with entries

$$A_{ij} = \begin{cases} -\frac{1}{\hat{\alpha}_i}\left(\frac{1}{\alpha_i} + \frac{1}{\alpha_{i-1}}\right)\delta_{i,j} + \frac{1}{\hat{\alpha}_i \alpha_{i-1}}\delta_{i-1,j} + \frac{1}{\hat{\alpha}_i \alpha_i}\delta_{i+1,j} \\ \qquad\qquad\qquad\qquad\qquad\qquad \text{if } 1 < i \leq \ell,\ 1 \leq j \leq \ell, \\ -\frac{1}{\hat{\alpha}_1 \alpha_1}\delta_{1,j} + \frac{1}{\hat{\alpha}_1 \alpha_1}\delta_{2,j} \qquad\qquad \text{if } i = 1,\ 1 \leq j \leq \ell. \end{cases} \quad (3\text{-}29)$$

The Kronecker delta symbol $\delta_{i,j}$ is one when $i = j$ and zero otherwise. Note that A is a Jacobi matrix when it is defined on the vector space $\mathbb{R}^\ell$ with weighted inner product

$$\langle \boldsymbol{a}, \boldsymbol{b} \rangle = \sum_{j=1}^{\ell} \hat{\alpha}_j a_j b_j, \qquad \boldsymbol{a} = (a_1, \ldots, a_\ell)^T, \; \boldsymbol{b} = (b_1, \ldots, b_\ell)^T. \qquad (3\text{-}30)$$

That is to say,

$$\widetilde{A} = \operatorname{diag}(\hat{\alpha}_1^{1/2}, \ldots, \hat{\alpha}_\ell^{1/2}) A \operatorname{diag}(\hat{\alpha}_1^{-1/2}, \ldots, \hat{\alpha}_\ell^{-1/2}) \qquad (3\text{-}31)$$

is a symmetric, tridiagonal matrix, with negative entries on its diagonal and positive entries on its upper/lower diagonal. It follows from [Chu and Golub 2002] that A has simple, negative eigenvalues $-\delta_j^2$ and eigenvectors $Y_j = (Y_{1,j}, \ldots, Y_{\ell,j})^T$ that are orthogonal with respect to the inner product (3-30). We order the eigenvalues as

$$\delta_1 < \delta_2 < \cdots < \delta_\ell, \qquad (3\text{-}32)$$

and normalize the eigenvectors

$$\|Y_j\|^2 = \langle Y_j, Y_j \rangle = \sum_{p=1}^{\ell} \hat{\alpha}_p^2 Y_{p,j}^2 = 1. \qquad (3\text{-}33)$$

Then, we obtain from (3-25) and (3-28), after expanding V in the basis of the eigenvectors, that

$$F^\dagger(\lambda) = \sum_{j=1}^{\ell} \frac{Y_{1,j}^2}{\lambda + \delta_j^2}. \qquad (3\text{-}34)$$

This is precisely (3-27), for the discrete spectral measure

$$\mu^F(t) = -\sum_{j=1}^{\ell} \xi_j H(-t - \delta_j^2), \qquad \xi_j = Y_{1,j}^2, \qquad (3\text{-}35)$$

where H is the Heaviside step function. $\square$

Note that any function of the form (3-34) defines the eigenvalues $F^\dagger(\omega_k^2)$ of the NtD map $\Lambda_\gamma^\dagger$ of a finite-volume scheme with $\ell + 1$ primary radii and uniform discretization in angle. This follows from the decomposition in Section 3.1 and the results in [Kac and Krein 1968]. Note also that there is an explicit, continued fraction representation of $F^\dagger(\lambda)$, in terms of the network conductances, i.e., the parameters α_j and $\hat{\alpha}_j$,

$$F^\dagger(\lambda) = \cfrac{1}{\hat{\alpha}_1\lambda + \cfrac{1}{\alpha_1 + \cdots \cfrac{1}{\hat{\alpha}_\ell\lambda + \cfrac{1}{\alpha_\ell}}}}. \qquad (3\text{-}36)$$

This representation is known in the theory of rational function approximations [Nikishin and Sorokin 1991; Kac and Krein 1968] and its derivation is given in Appendix B.

Since both $f^\dagger(\lambda)$ and $F^\dagger(\lambda)$ are Stieltjes functions, we can design finite volume schemes (i.e., layered networks) with accurate, rational approximations $F^\dagger(\lambda)$ of $f^\dagger(\lambda)$. There are various approximants $F^\dagger(\lambda)$, with different rates of convergence to $f^\dagger(\lambda)$, as $\ell \to \infty$. We discuss two choices below, in Sections 3.2.2 and 3.2.3, but refer the reader to [Druskin and Knizhnerman 2000a; 2000b; Druskin and Moskow 2002] for details on various Padé approximants and the resulting discretization schemes. No matter which approximant we choose, we can compute the network conductances, i.e., the parameters α_j and $\hat{\alpha}_j$ for $j = 1, \ldots, \ell$, from 2ℓ measurements of $f^\dagger(\lambda)$. The type of measurements dictates the type of approximant, and only some of them are directly accessible in the EIT problem. For example, the spectral measure $\mu(\lambda)$ cannot be determined in a stable manner in EIT. However, we can measure the eigenvalues $f^\dagger(k^2)$ for integer k, and thus we can design a rational, multipoint Padé approximant.

Remark 1. We describe in detail in Appendix D how to determine the parameters $\{\alpha_j, \hat{\alpha}_j\}_{j=1,\ldots,\ell}$ from 2ℓ point measurements of $f^\dagger(\lambda)$, such as $f^\dagger(k^2)$, for $k = 1, \ldots, \frac{n-1}{2} = 2\ell$. The are two steps. The first is to write $F^\dagger(\lambda)$ as the ratio of two polynomials of λ, and determine the 2ℓ coefficients of these polynomials from the measurements $F^\dagger(\omega_k^2)$ of $f^\dagger(k^2)$, for $1 \leq k \leq \frac{n-1}{2}$. See Section 3.2.2 for examples of such measurements. The exponential instability of EIT comes into play in this step, because it involves the inversion of a Vandermonde matrix. It is known [Gautschi and Inglese 1988] that such matrices have condition numbers that grow exponentially with the dimension ℓ. The second step is to determine the parameters $\{\alpha_j, \hat{\alpha}_j\}_{j=1,\ldots,\ell}$ from the coefficients of the polynomials. This can be done in a stable manner with the Euclidean division algorithm.

The approximation problem can also be formulated in terms of the DtN map, with $F(\lambda) = 1/F^\dagger(\lambda)$. Moreover, the representation (3-36) generalizes to both types of grids, by replacing $\hat{\alpha}_1\lambda$ with $\hat{\alpha}_1 m_{1/2}\lambda$. Recall (3-20) and note the parameter $\hat{\alpha}_1$ does not play any role when $m_{1/2} = 0$.

3.2.1. Optimal grids and reconstruction mappings.
Once we have determined the network conductances, that is the coefficients

$$\alpha_j = \int_{r_{j+1}}^{r_j} \frac{dr}{r\sigma(r)}, \qquad \hat{\alpha}_j = \int_{\hat{r}_{j+1}}^{\hat{r}_j} \frac{\sigma(r)}{r} dr, \qquad j = 1, \ldots, \ell, \qquad (3\text{-}37)$$

we could determine the optimal placement of the radii r_j and $\hat{r}_j$, if we knew the conductivity $\sigma(r)$. But $\sigma(r)$ is the unknown in the inverse problem. The key idea behind the resistor network approach to inversion is that the grid depends only weakly on σ, and we can compute it approximately for the reference conductivity $\sigma^{(o)} \equiv 1$.

Let us denote by $f^{\dagger(o)}(\lambda)$ the analog of (3-25) for conductivity $\sigma^{(o)}$, and let $F^{\dagger(o)}(\lambda)$ be its rational approximant defined by (3-36), with coefficients $\alpha_j^{(o)}$ and $\hat{\alpha}_j^{(o)}$ given by

$$\alpha_j^{(o)} = \int_{r_{j+1}^{(o)}}^{r_j^{(o)}} \frac{dr}{r} = \log \frac{r_j^{(o)}}{r_{j+1}^{(o)}}, \qquad \hat{\alpha}_j^{(o)} = \int_{\hat{r}_{j+1}^{(o)}}^{\hat{r}_j^{(o)}} \frac{dr}{r} = \log \frac{\hat{r}_j^{(o)}}{\hat{r}_{j+1}^{(o)}}, \qquad j = 1, \ldots, \ell. \qquad (3\text{-}38)$$

Since $r_1^{(o)} = \hat{r}_1^{(o)} = 1$, we obtain

$$r_{j+1}^{(o)} = \exp\left(-\sum_{q=1}^{j} \alpha_q^{(o)}\right), \qquad \hat{r}_{j+1}^{(o)} = \exp\left(-\sum_{q=1}^{j} \hat{\alpha}_q^{(o)}\right), \qquad j = 1, \ldots, \ell. \qquad (3\text{-}39)$$

We call the radii (3-39) *optimal*. The name refers to the fact that finite volume discretizations on grids with such radii give an NtD map that matches the measurements of the continuum map $\Lambda^{\dagger}_{\sigma^{(o)}}$ for the reference conductivity $\sigma^{(o)}$.

Remark 2. It is essential that the parameters $\{\alpha_j, \hat{\alpha}_j\}$ and $\{\alpha_j^{(o)}, \hat{\alpha}_j^{(o)}\}$ are computed from the same type of measurements. For example, if we measure $f^{\dagger}(k^2)$, we compute $\{\alpha_j, \hat{\alpha}_j\}$ so that

$$F^{\dagger}(\omega_k^2) = f^{\dagger}(k^2),$$

and $\{\alpha_j^{(o)}, \hat{\alpha}_j^{(o)}\}$ so that

$$F^{\dagger(o)}(\omega_k^2) = f^{\dagger(o)}(k^2),$$

where $k = 1, \ldots, (n-1)/2$. This is because the distribution of the radii (3-39) in the interval $[0, 1]$ depends on what measurements we make, as illustrated with examples in Sections 3.2.2 and 3.2.3.

Now let us denote by $\mathcal{D}_n$ the set in $\mathbb{R}^{(n-1)/2}$ of measurements of $f^{\dagger}(\lambda)$, and introduce the reconstruction mapping $\mathcal{Q}_n$ defined on $\mathcal{D}_n$, with values in $\mathbb{R}_+^{(n-1)/2}$. It takes the measurements of $f^{\dagger}(\lambda)$ and returns the $(n-1)/2$ positive numbers

$$\sigma_{j+1-m_{1/2}} = \hat{\alpha}_j/\hat{\alpha}_j^{(o)}, \quad j = 2 - m_{1/2}, \ldots, \ell,$$
$$\hat{\sigma}_{j+m_{1/2}} = \alpha_j^{(o)}/\alpha_j, \quad j = 1, 2, \ldots, \ell, \quad (3\text{-}40)$$

where we recall the relation (2-14) between ℓ and n. We call $\mathcal{Q}_n$ a reconstruction mapping because if we take σ_j and $\hat{\sigma}_j$ as point values of a conductivity at nodes $r_j^{(o)}$ and $\hat{r}_j^{(o)}$, and interpolate them on the optimal grid, we expect to get a conductivity that is close to the interpolation of the true $\sigma(r)$. This is assuming that the grid does not depend strongly on $\sigma(r)$. The proof that the resulting sequence of conductivity functions indexed by ℓ converges to the true $\sigma(r)$ as $\ell \to \infty$ is carried out in [Borcea et al. 2005], given the spectral measure of $f^\dagger(\lambda)$. We review it in Section 3.3, and discuss the measurements in Section 3.2.3. The convergence proof for other measurements remains an open question, but the numerical results indicate that the result should hold. Moreover, the ideas extend to the two-dimensional case, as explained in detail in Sections 4 and 5.

3.2.2. *Examples of rational interpolation grids.* Let us begin with an example that arises in the discretization of the problem with lumped current measurements

$$J_q = \frac{1}{h_\theta} \int_{\hat{\theta}_q}^{\hat{\theta}_{q+1}} \Lambda_\sigma u_{\mathcal{B}}(\theta) d\theta,$$

for $h_\theta = \frac{2\pi}{n}$, and vector $U_{\mathcal{B}} = (u_{\mathcal{B}}(\theta_1), \ldots, u_{\mathcal{B}}(\theta_n))^T$ of boundary potentials. If we take harmonic boundary excitations $u_{\mathcal{B}}(\theta) = e^{ik\theta}$, the eigenfunction of Λ_σ for eigenvalue $f(k^2)$, we obtain

$$J_q = \frac{1}{h_\theta} \int_{\hat{\theta}_q}^{\hat{\theta}_{q+1}} \Lambda_\sigma e^{ik\theta} d\theta = f(k^2) \left| \mathrm{sinc}\left(\frac{kh_\theta}{2}\right) \right| e^{ik\theta_q} = \frac{f(k^2)}{|k|} \omega_k e^{ik\theta_q},$$
$$q = 1, \ldots, n. \quad (3\text{-}41)$$

These measurements, for all integers k satisfying $|k| \leq \frac{n-1}{2}$, define a discrete DtN map $M_n(\Lambda_\sigma)$. It is a symmetric matrix with eigenvectors $[e^{ik\theta}] = (e^{ik\theta_1}, \ldots, e^{ik\theta_n})^T$, and eigenvalues $(f(k^2)/|k|)\omega_k$.

The approximation problem is to find the finite volume discretization with DtN map $\Lambda_\gamma = M_n(\Lambda_\sigma)$. Since both Λ_γ and M_n have the same eigenvectors, this is equivalent to the rational approximation problem of finding the network conductances (3-8) (i.e., α_j and $\hat{\alpha}_j$), so that

$$F(\omega_k^2) = \frac{f(k^2)}{|k|} \omega_k, \quad k = 1, \ldots, \frac{n-1}{2}. \quad (3\text{-}42)$$

The eigenvalues depend only on $|k|$, and the case $k = 0$ gives no information, because it corresponds to constant boundary potentials that lie in the null space of the DtN map. This is why we take in (3-42) only the positive values of k, and

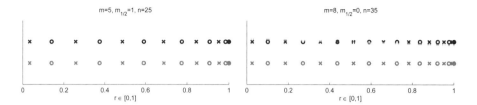

Figure 3. Examples of optimal grids with n equidistant boundary points and primary and dual radii shown with × and ∘. On the left we have $n = 25$ and a grid indexed by $m_{1/2} = 1$, with $\ell = m + 1 = 6$. On the right we have $n = 35$ and a grid indexed by $m_{1/2} = 0$, with $\ell = m + 1 = 8$. The bottom grid is computed with formulas (3-44). The top grid is obtained from the rational approximation (3-50).

obtain the same number $(n-1)/2$ of measurements as unknowns: $\{\alpha_j\}_{j=1,\ldots,\ell}$ and $\{\hat{\alpha}_j\}_{j=2-m_{1/2},\ldots,\ell}$.

When we compute the optimal grid, we take the reference $\sigma^{(o)} \equiv 1$, in which case $f^{(o)}(k^2) = |k|$. Thus, the optimal grid computation reduces to that of rational interpolation of $f(\lambda)$,

$$F^{(o)}(\omega_k^2) = \omega_k = f^{(o)}(\omega_k^2), \qquad k = 1, \ldots, \frac{n-1}{2}. \qquad (3\text{-}43)$$

This is solved explicitly in [Biesel et al. 2008]. For example, when $m_{1/2} = 1$, the coefficients $\alpha_j^{(o)}$ and $\hat{\alpha}_j^{(o)}$ are given by

$$\alpha_j^{(o)} = h_\theta \cot\left(\frac{h_\theta}{2}(2\ell - 2j + 1)\right), \quad \hat{\alpha}_j^{(o)} = h_\theta \cot\left(\frac{h_\theta}{2}(2\ell - 2j + 2)\right),$$
$$j = 1, 2, \ldots, \ell, \qquad (3\text{-}44)$$

and the radii follow from (3-39). They satisfy the interlacing relations

$$1 = \hat{r}_1^{(o)} = r_1^{(o)} > \hat{r}_2^{(o)} > r_2^{(o)} > \cdots > \hat{r}_{\ell+1}^{(o)} > r_{\ell+1}^{(o)} \geq 0, \qquad (3\text{-}45)$$

as can be shown easily using the monotonicity of the cotangent and exponential functions. We show an illustration of the resulting grids in red, in Figure 3. Note the refinement toward the boundary $r = 1$ and the coarsening toward the center $r = 0$ of the disk. Note also that the dual points shown with ∘ are almost half way between the primary points shown with ×. The last primary radii $r_{\ell+1}^{(o)}$ are small, but the points do not reach the center of the domain at $r = 0$.

In Sections 4 and 5 we work with slightly different measurements of the DtN map $\Lambda_\gamma = M_n(\Lambda_\sigma)$, with entries defined by

$$(\Lambda_\gamma)_{p,q} = \int_0^{2\pi} \chi_p(\theta) \Lambda_\sigma \chi_q(\theta) \, d\theta, \qquad p \neq q,$$
$$(\Lambda_\gamma)_{p,p} = -\sum_{q \neq p} (\Lambda_\gamma)_{p,q}, \tag{3-46}$$

using nonnegative measurement (electrode) functions $\chi_q(\theta)$, that are compactly supported in $(\hat{\theta}_q, \hat{\theta}_{q+1})$ and are normalized by

$$\int_0^{2\pi} \chi_q(\theta) \, d\theta = 1.$$

For example, we can take

$$\chi_q(\theta) = \begin{cases} 1/h_\theta & \text{if } \hat{\theta}_q < \theta < \hat{\theta}_{q+1}, \\ 0 & \text{otherwise,} \end{cases}$$

and obtain after a calculation given in Appendix C that the entries of Λ_γ are given by

$$(\Lambda_\gamma)_{p,q} = \frac{1}{2\pi} \sum_{k \in \mathbb{Z}} e^{ik(\theta_p - \theta_q)} f(k^2) \operatorname{sinc}^2\left(\frac{kh_\theta}{2}\right), \qquad p, q, = 1, \ldots, n. \tag{3-47}$$

We also show in Appendix C that

$$\Lambda_\gamma [e^{ik\theta}] = \frac{1}{h_\theta} \tilde{F}(\omega_k^2)[e^{ik\theta}], \qquad |k| \leq \frac{n-1}{2}, \tag{3-48}$$

with eigenvectors $[e^{ik\theta}]$ defined in (3-14) and scaled eigenvalues

$$\tilde{F}(\omega_k^2) = f(k^2) \operatorname{sinc}^2\left(\frac{kh_\theta}{2}\right) = F(\omega_k^2) \left|\operatorname{sinc}\left(\frac{kh_\theta}{2}\right)\right|. \tag{3-49}$$

Here we recalled (3-42) and (3-15).

There is no explicit formula for the optimal grid satisfying

$$\tilde{F}^{(o)}(\omega_k^2) = F^{(o)}(\omega_k^2) \left|\operatorname{sinc}\left(\frac{kh_\theta}{2}\right)\right| = \omega_k \left|\operatorname{sinc}\left(\frac{kh_\theta}{2}\right)\right|, \tag{3-50}$$

but we can compute it as explained in Remark 1 and Appendix D. We show in Figure 3 two examples of the grids, and note that they are very close to those obtained from the rational interpolation (3-43). This is not surprising because the sinc factor in (3-50) is not significantly different from 1 over the range $|k| \leq \frac{n-1}{2}$,

$$\frac{2}{\pi} < \frac{\sin\left[\frac{\pi}{2}\left(1-\frac{1}{n}\right)\right]}{\frac{\pi}{2}\left(1-\frac{1}{n}\right)} \leq \left|\text{sinc}\left(\frac{kh_\theta}{2}\right)\right| \leq 1.$$

Thus, many eigenvalues $\widetilde{F}^{(o)}(\omega_k^2)$ are approximately equal to ω_k, and this is why the grids are similar.

3.2.3. *Truncated measure and optimal grids.* Another example of rational approximation arises in a modified problem, where the positive spectral measure μ in Lemma 1 is discrete:

$$\mu(t) = -\sum_{j=1}^{\infty} \xi_j H(-t - \delta_j^2). \tag{3-51}$$

This does not hold for (3-2) or equivalently (3-18), where the origin of the disc $r = 0$ is mapped to ∞ in the logarithmic coordinates $z(r)$, and the measure $\mu(t)$ is continuous. To obtain a measure like (3-51), we change the problem here and in the next section to

$$\frac{r}{\sigma(r)} \frac{d}{dr}\left[r\sigma(r)\frac{dv(r)}{dr}\right] - \lambda v(r) = 0, \qquad r \in (\epsilon, 1), \tag{3-52}$$

with $\epsilon \in (0, 1)$ and boundary conditions

$$\frac{\partial v(o)}{\partial r} = \varphi_{\mathcal{B}}, \qquad v(\epsilon) = 0. \tag{3-53}$$

The Dirichlet boundary condition at $r = \epsilon$ may be realized if we have a perfectly conducting medium in the disk concentric with Ω and of radius ϵ. Otherwise, $v(\epsilon) = 0$ gives an approximation of our problem, for small but finite ϵ.

Coordinate change and scaling. It is convenient here and in the next section to introduce the scaled logarithmic coordinate

$$\zeta(r) = \frac{z^{(o)}(r)}{Z} = \frac{1}{Z}\int_r^1 \frac{dt}{t}, \qquad Z = -\log(\epsilon) = z^{(o)}(\epsilon), \tag{3-54}$$

and write (3-9) in the scaled form

$$\frac{z(r)}{Z} = \int_0^\zeta \frac{dt}{\sigma(r(t))} = z'(\zeta), \qquad \frac{\hat{z}(r)}{Z} = \int_0^\zeta \sigma(r(t))\, dt = \hat{z}'(\zeta). \tag{3-55}$$

The conductivity function in the transformed coordinates is

$$\sigma'(\zeta) = \sigma(r(\zeta)), \qquad r(\zeta) = e^{-Z\zeta}, \tag{3-56}$$

and the potential
$$v'(z') = \frac{v(r(z'))}{\varphi_{\mathcal{B}}} \tag{3-57}$$

satisfies the scaled equations
$$\frac{d}{d\hat{z}'}\left(\frac{dv'}{dz'}\right) - \lambda' v' = 0, \quad z' \in (0, L'),$$
$$\frac{dv(0)}{dz'} = -1, \quad v(L') = 0, \tag{3-58}$$

where we let $\lambda' = \lambda/Z^2$ and
$$L' = z'(1) = \int_0^1 \frac{dt}{\sigma'(t)}. \tag{3-59}$$

Remark 3. We assume in the remainder of this section and in Section 3.3 that we work with the scaled equations (3-58) and drop the primes for simplicity of notation.

The inverse spectral problem. The differential operator $\frac{d}{d\hat{z}}\frac{d}{dz}$ acting on the vector space of functions with homogeneous Neumann conditions at $z = 0$ and Dirichlet conditions at $z = L$ is symmetric with respect to the weighted inner product
$$(a, b) = \int_0^{\hat{L}} a(z) b(z) \, d\hat{z} = \int_0^1 a(z(\zeta)) b(z(\zeta)) \sigma(\zeta) \, d\zeta, \quad \hat{L} = \hat{z}(1). \tag{3-60}$$

It has negative eigenvalues $\{-\delta_j^2\}_{j=1,2,\ldots}$, the points of increase of the measure (3-51), and eigenfunctions $y_j(z)$. They are orthogonal with respect to the inner product (3-60), and we normalize them by
$$\|y_j\|^2 = (y_j, y_j) = \int_0^{\hat{L}} y_j^2(z) \, d\hat{z} = 1. \tag{3-61}$$

The weights ξ_j in (3-51) are defined by
$$\xi_j = y_j^2(0). \tag{3-62}$$

For the discrete problem we assume in the remainder of the section that $m_{1/2} = 1$, and work with the NtD map, that is with $F^\dagger(\lambda)$ represented in Lemma 1 in terms of the discrete measure $\mu^F(t)$. Comparing (3-51) and (3-35), we note that we ask that $\mu^F(t)$ be the truncated version of $\mu(t)$, given the first ℓ weights ξ_j and eigenvalues $-\delta_j^2$, for $j = 1, \ldots, \ell$. We arrived at the classic *inverse spectral problem* [Gel'fand and Levitan 1951; Chadan et al. 1997; Hochstadt 1973; Marchenko 1986; McLaughlin and Rundell 1987], that seeks an approximation of the conductivity σ from the truncated measure. We can solve it using the theory of resistor networks, via an *inverse eigenvalue problem* [Chu and Golub 2002] for the Jacobi like matrix A defined in (3-29). The key ingredient in the connection between the continuous and discrete eigenvalue problems is the optimal grid, as was first noted in [Borcea and Druskin 2002] and proved in [Borcea et al. 2005]. We review this result in Section 3.3.

The truncated measure optimal grid. The optimal grid is obtained by solving the discrete inverse problem with spectral data for the reference conductivity $\sigma^{(o)}(\zeta)$:

$$\mathcal{D}_n^{(o)} = \left\{ \xi_j^{(o)} = 2, \ \delta_j^{(o)} = \pi\left(j - \frac{1}{2}\right), \ j = 1, \ldots, \ell \right\}. \qquad (3\text{-}63)$$

The parameters $\{\alpha_j^{(o)}, \hat{\alpha}_j^{(o)}\}_{j=1,\ldots,\ell}$ can be determined from $\mathcal{D}_n^{(o)}$ with the Lanczos algorithm [Trefethen and Bau 1997; Chu and Golub 2002], which is reviewed briefly in Appendix E. The grid points are given by

$$\zeta_{j+1}^{(o)} = \alpha_j^{(o)} + \zeta_j^{(o)} = \sum_{q=1}^{j} \alpha_q^{(o)}, \quad \hat{\zeta}_{j+1}^{(o)} = \hat{\alpha}_j^{(o)} + \hat{\zeta}_j^{(o)} = \sum_{q=1}^{j} \hat{\alpha}_q^{(o)}, \quad j = 1, \ldots, \ell, \qquad (3\text{-}64)$$

where $\zeta_1^{(o)} = \hat{\zeta}_1^{(o)} = 0$. This is in the logarithmic coordinates that are related to the optimal radii as in (3-56). The grid is calculated explicitly in [Borcea et al. 2005, Appendix A]. We summarize its properties in the next lemma, for large ℓ.

Lemma 2. *The steps* $\{\alpha_j^{(o)}, \hat{\alpha}_j^{(o)}\}_{j=1,\ldots,\ell}$ *of the truncated measure optimal grid satisfy the monotone relation*

$$\hat{\alpha}_1^{(o)} < \alpha_1^{(o)} < \hat{\alpha}_2^{(o)} < \alpha_2^{(o)} < \cdots < \hat{\alpha}_k^{(o)} < \alpha_k^{(o)}. \qquad (3\text{-}65)$$

Moreover, for large ℓ, the primary grid steps are

$$\alpha_j^{(o)} = \begin{cases} \dfrac{2 + O\left[(\ell - j)^{-1} + j^{-2}\right]}{\pi\sqrt{\ell^2 - j^2}} & \text{if } 1 \leq j \leq \ell - 1, \\[2mm] \dfrac{\sqrt{2} + O(\ell^{-1})}{\sqrt{\pi\ell}} & \text{if } j = \ell, \end{cases} \qquad (3\text{-}66)$$

and the dual grid steps are

$$\hat{\alpha}_j^{(o)} = \frac{2 + O\left[(\ell + 1 - j)^{-1} + j^{-2}\right]}{\pi\sqrt{\ell^2 - (j - 1/2)^2}}, \quad 1 \leq j \leq \ell. \qquad (3\text{-}67)$$

We show in Figure 4 an example for the case $\ell = 6$. To compare it with the grid in Figure 3, we plot in Figure 5 the radii given by the coordinate transformation (3-56), for three different parameters ϵ. Note that the primary and dual points are interlaced, but the dual points are not half way between the primary points,

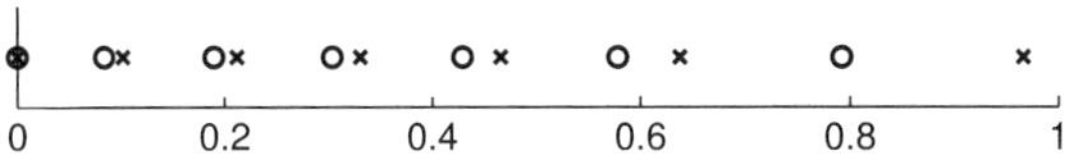

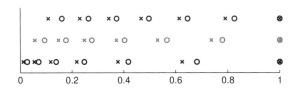

Figure 4. Example of a truncated measure optimal grid with $\ell = 6$. This is in the logarithmic scaled coordinates $\zeta \in [0, 1]$. The primary points are denoted with × and the dual ones with ○.

Figure 5. The radial grid obtained with the coordinate change $r = e^{-Z\zeta}$. The scale $Z = -\log \epsilon$ affects the distribution of the radii. The choice $\epsilon = 0.1$ is on top, $\epsilon = 0.05$ is in the middle and $\epsilon = 0.01$ is at the bottom. The primary radii are indicated with × and the dual ones with ○.

as was the case in Figure 3. Moreover, the grid is not refined near the boundary at $r = 1$. In fact, there is accumulation of the grid points near the center of the disk, where we truncate the domain. The smaller the truncation radius ϵ, the larger the scale $Z = -\log \epsilon$, and the more accumulation near the center.

Intuitively, we can say that the grids in Figure 3 are much superior to the ones computed from the truncated measure, for both the forward and inverse EIT problem. Indeed, for the forward problem, the rate of convergence of $F^\dagger(\lambda)$ to $f^\dagger(\lambda)$ on the truncated measure grids is algebraic [Borcea et al. 2005]

$$|f^\dagger(\lambda) - F^\dagger(\lambda)| = \left| \sum_{j=\ell+1}^{\infty} \frac{\xi_j}{\lambda + \delta_j^2} \right| = O\left(\sum_{j=\ell+1}^{\infty} \frac{1}{j^2} \right) = O\left(\frac{1}{\ell} \right).$$

The rational interpolation grids described in Section 3.2.2 give exponential convergence of $F^\dagger(\lambda)$ to $f^\dagger(\lambda)$ [Mamonov 2009]. For the inverse problem, we expect that the resolution of reconstructions of σ decreases rapidly away from the boundary where we make the measurements, so it makes sense to invert on grids like those in Figure 3, that are refined near $r = 1$.

The examples in Figures 3 and 5 show the strong dependence of the grids on the measurement setup. Although the grids in Figure 5 are not good for the EIT problem, they are optimal for the inverse spectral problem. The optimality is in the sense that the grids give an exact match of the spectral measurements (3-63) of the NtD map for conductivity $\sigma^{(o)}$. Furthermore, they give a very good match of the spectral measurements (3-68) for the unknown σ, and the reconstructed conductivity on them converges to the true σ, as we show next.

3.3. *Continuum limit of the discrete inverse spectral problem on optimal grids.*
Let $\mathfrak{A}_n : \mathfrak{D}_n \to \mathbb{R}_+^{2\ell}$ be the reconstruction mapping that takes the data

$$\mathfrak{D}_n = \{\xi_j, \delta_j, \quad j = 1, \ldots, \ell\} \tag{3-68}$$

to the $2\ell = \dfrac{n-1}{2}$ positive values $\{\sigma_j, \hat{\sigma}_j\}_{j=1,\ldots,\ell}$ given by

$$\sigma_j = \frac{\hat{\alpha}_j}{\hat{\alpha}_j^{(o)}}, \qquad \hat{\sigma}_{j+1} = \frac{\alpha_j^{(o)}}{\alpha_j}, \quad j = 1, 2, \ldots, \ell. \tag{3-69}$$

The computation of $\{\alpha_j, \hat{\alpha}_j\}_{j=1,\ldots,\ell}$ requires solving the discrete inverse spectral problem with data $\mathfrak{D}_n$, using for example the Lanczos algorithm reviewed in Appendix E. We define the *reconstruction* $\sigma^\ell(\zeta)$ of the conductivity as the piecewise constant interpolation of the point values (3-69) on the optimal grid (3-64). We have

$$\sigma^\ell(\zeta) = \begin{cases} \sigma_j & \text{if } \zeta \in [\zeta_j^{(o)}, \hat{\zeta}_{j+1}^{(o)}), \ j = 1, \ldots, \ell, \\ \hat{\sigma}_j & \text{if } \zeta \in [\hat{\zeta}_j^{(o)}, \zeta_j^{(o)}), \ j = 2, \ldots, \ell+1, \\ \hat{\sigma}_{\ell+1} & \text{if } \zeta \in [\zeta_{l+1}^{(o)}, 1], \end{cases} \tag{3-70}$$

and we discuss here its convergence to the true conductivity function $\sigma(\zeta)$, as $\ell \to \infty$.

To state the convergence result, we need some assumptions on the decay with j of the perturbations of the spectral data

$$\Delta \delta_j = \delta_j - \delta_j^{(o)}, \qquad \Delta \xi_j = \xi_j - \xi_j^{(o)}. \tag{3-71}$$

The asymptotic behavior of δ_j and ξ_j is well known, under various smoothness requirements on $\sigma(z)$ [McLaughlin and Rundell 1987; Pöschel and Trubowitz 1987; Coleman and McLaughlin 1993]. For example, if $\sigma(\zeta) \in H^3[0, 1]$, we have

$$\Delta \delta_j = \delta_j - \delta_j^{(o)} = \frac{\int_0^1 q(\zeta)\, d\zeta}{(2j-1)\pi} + O(j^{-2}) \text{ and } \Delta \xi_j = \xi_j - \xi_j^{(o)} = O(j^{-2}), \tag{3-72}$$

where $q(\zeta)$ is the Schrödinger potential

$$q(\zeta) = \sigma(\zeta)^{-\frac{1}{2}} \frac{d^2 \sigma(\zeta)^{\frac{1}{2}}}{d\zeta^2}. \tag{3-73}$$

We have the following convergence result proved in [Borcea et al. 2005].

Theorem 1. *Suppose that $\sigma(\zeta)$ is a positive and bounded scalar conductivity function, with spectral data satisfying the asymptotic behavior*

$$\Delta \delta_j = O\left(\frac{1}{j^s \log(j)}\right), \qquad \Delta \xi_j = O\left(\frac{1}{j^s}\right), \qquad \text{for some } s > 1, \text{ as } j \to \infty. \tag{3-74}$$

Then $\sigma^\ell(\zeta)$ converges to $\sigma(\zeta)$ as $\ell \to \infty$, pointwise and in $L^1[0,1]$.

Before we outline the proof, let us note that it appears from (3-72) and (3-74) that the convergence result applies only to the class of conductivities with zero mean potential. However, if

$$\bar{q} = \int_0^1 q(\zeta)\, d\zeta \neq 0, \tag{3-75}$$

we can modify the point values (3-69) of the reconstruction $\sigma^\ell(\zeta)$ by replacing $\alpha_j^{(o)}$ and $\hat{\alpha}_j^{(o)}$ with $\alpha_j^{(\bar{q})}$ and $\hat{\alpha}_j^{(\bar{q})}$, for $j = 1, \ldots, \ell$. These are computed by solving the discrete inverse spectral problem with data

$$\mathcal{D}_n^{(\bar{q})} = \{\xi_j^{(\bar{q})}, \delta_j^{(\bar{q})}, \ j = 1, \ldots, \ell\},$$

for the conductivity function

$$\sigma^{(\bar{q})}(\zeta) = \tfrac{1}{4}\left(e^{\sqrt{\bar{q}}\,\zeta} + e^{-\sqrt{\bar{q}}\,\zeta}\right)^2. \tag{3-76}$$

This conductivity satisfies the initial value problem

$$\frac{d^2 \sqrt{\sigma^{(\bar{q})}(\zeta)}}{d\zeta^2} = \bar{q}\sqrt{\sigma^{(\bar{q})}(\zeta)} \quad \text{for} \quad 0 < \zeta \leq 1,$$

$$\frac{d\sigma^{(\bar{q})}(0)}{d\zeta} = 0 \quad \text{and} \quad \sigma^{(\bar{q})}(0) = 1, \tag{3-77}$$

and we assume that

$$\bar{q} > -\frac{\pi^2}{4}, \tag{3-78}$$

so that (3-76) stays positive for $\zeta \in [0,1]$.

As seen from (3-72), the perturbations $\delta_j - \delta_j^{(\bar{q})}$ and $\xi_j - \xi_j^{(\bar{q})}$ satisfy the assumptions (3-74), so Theorem 1 applies to reconstructions on the grid given by $\sigma^{(\bar{q})}$. We show below in Corollary 1 that this grid is asymptotically the same as the *optimal grid*, calculated for $\sigma^{(o)}$. Thus, the convergence result in Theorem 1 applies after all, without changing the definition of the reconstruction (3-70).

3.3.1. *The case of constant Schrödinger potential.* The equation (3-58) for $\sigma \rightsquigarrow \sigma^{(\bar{q})}$ can be transformed to Schrödinger form with constant potential $\bar{q}$

$$\frac{d^2 w(\zeta)}{d\zeta^2} - (\lambda + \bar{q})w(\zeta) = 0, \qquad \zeta \in (0,1),$$
$$\frac{dw(0)}{d\zeta} = -1, \qquad w(1) = 0,$$
(3-79)

by letting $w(\zeta) = v(\zeta)\sqrt{\sigma^{(\bar{q})}(\zeta)}$. Thus, the eigenfunctions $y_j^{(\bar{q})}(\zeta)$ of the differential operator associated with $\sigma^{(\bar{q})}(\zeta)$ are related to $y_j^{(o)}(\zeta)$, the eigenfunctions for $\sigma^{(o)} \equiv 1$, by

$$y_j^{(\bar{q})}(\zeta) = \frac{y_j^{(o)}(\zeta)}{\sqrt{\sigma^{(\bar{q})}(\zeta)}}.$$
(3-80)

They satisfy the orthonormality condition

$$\int_0^1 y_j^{(\bar{q})}(\zeta) y_p^{(\bar{q})}(\zeta) \sigma^{(\bar{q})}(\zeta)\, d\zeta = \int_0^1 y_j^{(o)}(\zeta) y_p^{(o)}(\zeta)\, d\zeta = \delta_{jp},$$
(3-81)

and since $\sigma^{(\bar{q})}(0) = 1$,

$$\xi_j^{(\bar{q})} = \left[y_j^{(\bar{q})}(0)\right]^2 = \left[y_j^{(o)}(0)\right]^2 = \xi_j^{(o)}, \qquad j = 1, 2, \ldots$$
(3-82)

The eigenvalues are shifted by $\bar{q}$:

$$-\left(\delta_j^{(\bar{q})}\right)^2 = -\left(\delta_j^{(o)}\right)^2 - \bar{q}, \qquad j = 1, 2, \ldots$$
(3-83)

Let $\{\alpha_j^{(\bar{q})}, \hat{\alpha}_j^{(\bar{q})}\}_{j=1,\ldots,\ell}$ be the parameters obtained by solving the discrete inverse spectral problem with data $\mathcal{D}_n^{(\bar{q})}$. The reconstruction mapping

$$\mathcal{Q}_n : \mathcal{D}_n^{(\bar{q})} \to \mathbb{R}^{2\ell}$$

gives the sequence of $2\ell = \frac{n-1}{2}$ pointwise values

$$\sigma_j^{(\bar{q})} = \frac{\hat{\alpha}_j^{(\bar{q})}}{\hat{\alpha}_j^{(o)}}, \qquad \hat{\sigma}_{j+1}^{(\bar{q})} = \frac{\alpha_j^{(o)}}{\alpha_j^{(\bar{q})}}, \qquad j = 1, \ldots, \ell.$$
(3-84)

We have the following result stated and proved in [Borcea et al. 2005]. See the review of the proof in Appendix F.

Lemma 3. *The point values $\sigma_j^{(\bar{q})}$ satisfy the finite difference discretization of initial value problem* (3-77), *on the optimal grid, namely*

$$\frac{1}{\hat{\alpha}_j^{(o)}} \left(\frac{\sqrt{\sigma_{j+1}^{(\bar{q})}} - \sqrt{\sigma_j^{(\bar{q})}}}{\alpha_j^{(o)}} - \frac{\sqrt{\sigma_j^{(\bar{q})}} - \sqrt{\sigma_{j-1}^{(\bar{q})}}}{\alpha_{j-1}^{(o)}} \right) - \bar{q}\sqrt{\sigma_j^{(\bar{q})}} = 0 \qquad (3\text{-}85)$$

for $j = 2, 3, \ldots, \ell$, and

$$\frac{1}{\hat{\alpha}_1^{(o)}} \frac{\sqrt{\sigma_2^{(\bar{q})}} - \sqrt{\sigma_1^{(\bar{q})}}}{\alpha_1^{(o)}} - \bar{q}\sqrt{\sigma_1^{(\bar{q})}} = 0, \qquad \sigma_1^{(\bar{q})} = 1. \qquad (3\text{-}86)$$

Moreover, $\hat{\sigma}_{j+1}^{(\bar{q})} = \sqrt{\sigma_j^{(\bar{q})} \sigma_{j+1}^{(\bar{q})}}$, for $j = 1, \ldots, \ell$.

The convergence of the reconstruction $\sigma^{(\bar{q}),\ell}(\zeta)$ follows from this lemma and a standard finite-difference error analysis [Godunov and Ryabenki 1964] on the optimal grid satisfying Lemma 2. The reconstruction is defined as in (3-70), by the piecewise constant interpolation of the point values (3-84) on the optimal grid.

Theorem 2. *As $\ell \to \infty$ we have*

$$\max_{1 \le j \le \ell} \left| \sigma_j^{(\bar{q})} - \sigma^{(\bar{q})}(\zeta_j^{(o)}) \right| \to 0 \quad \text{and} \quad \max_{1 \le j \le \ell} \left| \hat{\sigma}_{j+1}^{(\bar{q})} - \sigma^{(\bar{q})}(\hat{\zeta}_{j+1}^{(o)}) \right| \to 0,$$

and the reconstruction $\sigma^{(\bar{q}),\ell}(\zeta)$ converges to $\sigma^{(\bar{q})}(\zeta)$ in $L^\infty[0, 1]$.

As a corollary to this theorem, we can now obtain that the grid induced by $\sigma^{(\bar{q})}(\zeta)$, with primary nodes $\zeta_j^{(\bar{q})}$ and dual nodes $\hat{\zeta}_j^{(\bar{q})}$, is asymptotically close to the optimal grid. The proof is in Appendix F.

Corollary 1. *The grid induced by $\sigma^{(\bar{q})}(\zeta)$ is defined by the equations*

$$\int_0^{\zeta_{j+1}^{(\bar{q})}} \frac{d\zeta}{\sigma^{(\bar{q})}(\zeta)} = \sum_{p=1}^{j} \alpha_p^{(\bar{q})}, \qquad \int_0^{\hat{\zeta}_{j+1}^{(\bar{q})}} \sigma^{(\bar{q})}(\zeta) \, d\zeta = \sum_{p=1}^{j} \hat{\alpha}_p^{(\bar{q})}, \qquad j = 1, \ldots, \ell,$$

$$\zeta_1^{(\bar{q})} = \hat{\zeta}_1^{(\bar{q})} = 0, \qquad (3\text{-}87)$$

and satisfies

$$\max_{1 \le j \le \ell+1} \left| \zeta_j^{(\bar{q})} - \zeta_j^{(o)} \right| \to 0, \qquad \max_{1 \le j \le \ell+1} \left| \hat{\zeta}_j^{(\bar{q})} - \hat{\zeta}_j^{(o)} \right| \to 0, \quad \text{as } \ell \to \infty. \quad (3\text{-}88)$$

3.3.2. *Outline of the proof of Theorem 1.* The proof given in detail in [Borcea et al. 2005] has two main steps. The first step is to establish the compactness of the set of reconstructed conductivities. The second step uses the established compactness and the uniqueness of solution of the continuum inverse spectral problem to get the convergence result.

Step 1: Compactness. We show here that the sequence $\{\sigma^\ell(\zeta)\}_{\ell \geq 1}$ of reconstructions (3-70) has bounded variation.

Lemma 4. *The sequence* $\{\sigma_j, \hat{\sigma}_{j+1}\}_{j=1,\ldots,\ell}$ *of* (3-69) *returned by the reconstruction mapping* $\mathcal{Q}_n$ *satisfies*

$$\sum_{j=1}^{\ell} \left|\log \hat{\sigma}_{j+1} - \log \sigma_j\right| + \sum_{j=1}^{\ell} \left|\log \hat{\sigma}_{j+1} - \log \sigma_{j+1}\right| \leq C, \quad (3\text{-}89)$$

where the constant C is independent of ℓ. Therefore the sequence of reconstructions $\{\sigma^\ell(\zeta)\}_{\ell \geq 1}$ *has uniformly bounded variation.*

Our original formulation is not convenient for proving (3-89), because when written in Schrödinger form, it involves the second derivative of the conductivity, as seen from (3-73). Thus, we rewrite the problem in first order system form, which involves only the first derivative of $\sigma(\zeta)$, which is all we need to show (3-89). At the discrete level, the linear system of ℓ equations

$$AV - \lambda V = -\frac{e_1}{\hat{\alpha}_1} \quad (3\text{-}90)$$

for the potential $V = (V_1, \ldots, V_\ell)^T$ is transformed into the system of 2ℓ equations

$$BH^{\frac{1}{2}}W - \sqrt{\lambda}H^{\frac{1}{2}}W = -\frac{e_1}{\sqrt{\lambda\hat{\alpha}_1}} \quad (3\text{-}91)$$

for the vector $W = (W_1, \hat{W}_2, \ldots, W_\ell, \hat{W}_{\ell+1})^T$ with components

$$W_j = \sqrt{\sigma_j}V_j, \quad \hat{W}_{j+1} = \frac{\hat{\sigma}_{j+1}}{\sqrt{\lambda\sigma_j}} \frac{V_{j+1} - V_j}{\alpha_j^{(o)}}, \quad j = 1, \ldots, \ell. \quad (3\text{-}92)$$

Here $H = \text{diag}(\hat{\alpha}_1^{(o)}, \alpha_1^{(o)}, \ldots, \hat{\alpha}_\ell^{(o)}, \alpha_\ell^{(o)})$ and B is the tridiagonal, skew-symmetric matrix

$$B = \begin{pmatrix} 0 & \beta_1 & 0 & 0 & \cdots \\ -\beta_1 & 0 & \beta_2 & 0 & \cdots \\ 0 & -\beta_2 & 0 & \ddots & \vdots \\ \vdots & & & & \\ 0 & \cdots & & -\beta_{2\ell-1} & 0 \end{pmatrix} \quad (3\text{-}93)$$

with entries

$$\beta_{2p} = \frac{1}{\sqrt{\alpha_p \hat{\alpha}_{p+1}}} = \frac{1}{\sqrt{\alpha_p^{(o)} \hat{\alpha}_p^{(o)} + 1}} \sqrt{\frac{\hat{\sigma}_{p+1}}{\sigma_p}} = \beta_{2p}^{(o)} \sqrt{\frac{\hat{\sigma}_{p+1}}{\sigma_{p+1}}} \quad (3\text{-}94)$$

and

$$\beta_{2p-1} = \frac{1}{\sqrt{\alpha_p \hat{\alpha}_p}} = \frac{1}{\sqrt{\alpha_p^{(o)} \hat{\alpha}_p^{(o)}}} \sqrt{\frac{\hat{\sigma}_{p+1}}{\sigma_p}} = \beta_{2p-1}^{(o)} \sqrt{\frac{\hat{\sigma}_{p+1}}{\sigma_p}}. \qquad (3\text{-}95)$$

Note that we have

$$\sum_{p=1}^{2\ell-1} \left| \log \frac{\beta_p}{\beta_p^{(o)}} \right|$$

$$= \frac{1}{2} \sum_{p=1}^{\ell} \left| \log \hat{\sigma}_{p+1} - \log \sigma_p \right| + \frac{1}{2} \sum_{p=1}^{\ell} \left| \log \hat{\sigma}_{p+1} - \log \sigma_{p+1} \right|, \qquad (3\text{-}96)$$

and we can prove (3-89) using the method of small perturbations. Recall definitions (3-71) and let

$$\Delta \delta_j^r = r \Delta \delta_j, \qquad \Delta \xi_j^r = r \Delta \xi_j, \qquad j = 1, \ldots, \ell, \qquad (3\text{-}97)$$

where $r \in [0, 1]$ is an arbitrary continuation parameter. Let also β_j^r be the entries of the tridiagonal, skew-symmetric matrix $\boldsymbol{B}^r$ determined by the spectral data $\delta_j^r = \delta_j^{(o)} + \Delta \delta_j^r$ and $\xi_j^r = \xi_j^{(o)} + \Delta \xi_j^r$, for $j = 1, \ldots, \ell$. We explain in Appendix G how to obtain explicit formulae for the perturbations $d \log \beta_j^r$ in terms of the eigenvalues and eigenvectors of matrix $\boldsymbol{B}^r$ and perturbations $d\delta_j^r = \Delta \delta_j dr$ and $d\xi_j^r = \Delta \xi_j dr$. These perturbations satisfy the uniform bound

$$\sum_{j=1}^{2\ell-1} \left| d \log \beta_j^r \right| \le C_1 |dr|, \qquad (3\text{-}98)$$

with constant C_1 independent of ℓ and r. Then,

$$\log \frac{\beta_j}{\beta_j^{(o)}} = \int_0^1 d \log \beta_j^r \qquad (3\text{-}99)$$

satisfies the uniform bound $\sum_{j=1}^{2\ell-1} \left| \log \frac{\beta_j}{\beta_j^{(o)}} \right| \le C_1$ and (3-89) follows from (3-96).

Step 2: Convergence. Recall from Section 3.2 that the eigenvectors Y_j of A are orthonormal with respect to the weighted inner product (3-30). Then, the matrix $\widetilde{Y}$ with columns $\operatorname{diag}(\hat{\alpha}_1^{1/2}, \ldots, \hat{\alpha}_\ell^{1/2}) Y_j$ is orthogonal and we have

$$(\widetilde{Y} \widetilde{Y}^T)_{11} = \hat{\alpha}_1 \sum_{j=1}^{\ell} \xi_j = 1. \qquad (3\text{-}100)$$

Similarly

$$\hat{\alpha}_1^{(o)} \sum_{j=1}^{\ell} \xi_j^{(o)} = 2\ell \hat{\alpha}_1^{(o)} = 1, \quad (3\text{-}101)$$

where we used (3-63), and since $\Delta \xi_j$ are summable by assumption (3-74),

$$\sigma_1 = \frac{\hat{\alpha}_1}{\hat{\alpha}_1^{(o)}} = \left(1 + \hat{\alpha}_1^{(o)} \sum_{j=1}^{\ell} \Delta \xi_j\right)^{-1} = 1 + O(\hat{\alpha}_1^{(o)}) = 1 + O\left(\frac{1}{\ell}\right). \quad (3\text{-}102)$$

But $\sigma^\ell(0) = \sigma_1$, and since $\sigma^\ell(\zeta)$ has bounded variation by Lemma 4, we conclude that $\sigma^\ell(\zeta)$ is uniformly bounded in $\zeta \in [0, 1]$.

Now, to show that $\sigma^\ell(\zeta) \to \sigma(\zeta)$ in $L^1[0, 1]$, suppose for contradiction that it does not. Then, there exists $\varepsilon > 0$ and a subsequence σ^{ℓ_k} such that

$$\|\sigma^{\ell_k} - \sigma\|_{L^1[0,1]} \geq \varepsilon.$$

But since this subsequence is bounded and has bounded variation, we conclude from Helly's selection principle and the compactness of the embedding of the space of functions of bounded variation in $L^1[0, 1]$ (see [Natanson 1955]) that it has a subsequence that converges pointwise and in $L^1[0, 1]$. Call again this subsequence σ^{ℓ_k} and denote its limit by $\sigma^\star \neq \sigma$. Since the limit is in $L^1[0, 1]$, we have by definitions (3-55) and Remark 3,

$$\begin{aligned} z(\zeta; \sigma^{\ell_k}) &= \int_0^\zeta \frac{dt}{\sigma^{\ell_k}(t)} \to z(\zeta; \sigma) = \int_0^\zeta \frac{dt}{\sigma^\star(t)}, \\ \hat{z}(\zeta; \sigma^{\ell_k}) &= \int_0^\zeta \sigma^{\ell_k}(t)\, dt \to \hat{z}(\zeta; \sigma^\star) = \int_0^\zeta \sigma(t)\, dt. \end{aligned} \quad (3\text{-}103)$$

The continuity of $f^\dagger$ with respect to the conductivity gives $f^\dagger(\lambda; \sigma^{\ell_k}) \to f^\dagger(\lambda; \sigma^\star)$. However, Lemma 1 and (3-51) show that $f^\dagger(\lambda; \sigma^\ell) \to f^\dagger(\lambda; \sigma)$ by construction, and since the inverse spectral problem has a unique solution [Gel'fand and Levitan 1951; Levitan 1987; Coleman and McLaughlin 1993; Pöschel and Trubowitz 1987], we must have $\sigma^\star = \sigma$. We have reached a contradiction, so $\sigma^\ell(\zeta) \to \sigma(\zeta)$ in $L^1[0, 1]$. The pointwise convergence can be proved analogously.

Remark 4. All the elements of the proof presented here, except for establishing the bound (3-98), apply to any measurement setup. The challenge in proving convergence of inversion on optimal grids for general measurements lies entirely in obtaining sharp stability estimates of the reconstructed sequence with respect to perturbations in the data. The inverse spectral problem is stable, and this is why we could establish the bound (3-98). The EIT problem is exponentially unstable, and it remains an open problem to show the compactness of the function space of reconstruction sequences σ^ℓ from measurements such as (3-49).

4. Two-dimensional media and full boundary measurements

We now consider the two-dimensional EIT problem, where $\sigma = \sigma(r, \theta)$ and we cannot use separation of variables as in Section 3. More explicitly, we cannot reduce the inverse problem for resistor networks to one of rational approximation of the eigenvalues of the DtN map. We start by reviewing in Section 4.1 the conditions of unique recovery of a network (Γ, γ) from its DtN map Λ_γ, defined by measurements of the continuum Λ_σ. The approximation of the conductivity σ from the network conductance function γ is described in Section 4.2.

4.1. *The inverse problem for planar resistor networks.* The unique recoverability from Λ_γ of a network (Γ, γ) with known circular planar graph Γ is established in [Colin de Verdière 1994; Colin de Verdière et al. 1996; Curtis et al. 1994; Curtis et al. 1998]. A graph $\Gamma = (\mathcal{P}, \mathcal{E})$ is called circular and planar if it can be embedded in the plane with no edges crossing and with the boundary nodes lying on a circle. We call by association the networks with such graphs circular planar. The recoverability result states that if the data is *consistent* and the graph Γ is *critical* then the DtN map Λ_γ determines uniquely the conductance function γ. By consistent data we mean that the measured matrix Λ_γ belongs to the set of DtN maps of circular planar resistor networks.

A graph is critical if and only if it is *well-connected* and the removal of any edge breaks the well-connectedness. A graph is well-connected if all its *circular pairs* (P, Q) are *connected*. Let P and Q be two sets of boundary nodes with the same cardinality $|P| = |Q|$. We say that (P, Q) is a circular pair when the nodes in P and Q lie on disjoint segments of the boundary $\mathcal{B}$. The pair is *connected* if there are $|P|$ disjoint paths joining the nodes of P to the nodes of Q.

A symmetric $n \times n$ real matrix Λ_γ is the DtN map of a circular planar resistor network with n boundary nodes if its rows sum to zero $\Lambda_\gamma \mathbf{1} = \mathbf{0}$ (conservation of currents) and all its *circular minors* $(\Lambda_\gamma)_{P,Q}$ have nonpositive determinant. A circular minor $(\Lambda_\gamma)_{P,Q}$ is a square submatrix of Λ_γ defined for a circular pair (P, Q), with row and column indices corresponding to the nodes in P and Q, ordered according to a predetermined orientation of the circle $\mathcal{B}$. Since subsets of P and Q with the same cardinality also form circular pairs, the determinantal inequalities are equivalent to requiring that all circular minors be totally nonpositive. A matrix is totally nonpositive if all its minors have nonpositive determinant.

Examples of critical networks were given in Section 2.2, with graphs Γ determined by tensor product grids. Criticality of such networks is proved in

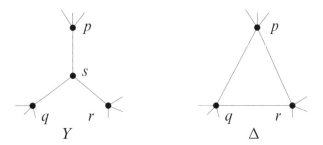

Figure 6. Given some conductances in the Y network, there is a choice of conductances in the Δ network for which the two networks are indistinguishable from electrical measurements at the nodes p, q and r.

[Curtis et al. 1994] for an odd number n of boundary points. As explained in Section 2.2 (see in particular (2-14)), criticality holds when the number of edges in $\mathscr{E}$ is equal to the number $n(n-1)/2$ of independent entries of the DtN map $\mathbf{\Lambda}_\gamma$.

The discussion in this section is limited to the tensor product topology, which is natural for the full boundary measurement setup. Two other topologies admitting critical networks (pyramidal and two-sided), are discussed in more detail in Sections 5.2.1 and 5.2.2. They are better suited for partial boundary measurements setups [Borcea et al. 2010b; 2011].

Remark 5. It is impossible to recover both the topology and the conductances from the DtN map of a network. An example of this indetermination is the so-called $Y - \Delta$ transformation given in Figure 6. A critical network can be transformed into another by a sequence of $Y-\Delta$ transformations without affecting the DtN map [Curtis et al. 1998].

4.1.1. *From the continuum to the discrete DtN map.* Ingerman and Morrow [1998] showed that pointwise values of the kernel of Λ_σ at any n distinct nodes on $\mathscr{B}$ define a matrix that is consistent with the DtN map of a circular planar resistor network, as defined above. We consider a generalization of these measurements, taken with electrode functions $\chi_q(\theta)$, as given in (3-46). It is shown in [Borcea et al. 2008] that the measurement operator $\boldsymbol{M}_n$ in (3-46) gives a matrix $\boldsymbol{M}_n(\Lambda_\sigma)$ that belongs to the set of DtN maps of circular planar resistor networks. We can equate therefore

$$\boldsymbol{M}_n(\Lambda_\sigma) = \mathbf{\Lambda}_\gamma, \tag{4-1}$$

and solve the inverse problem for the network (Γ, γ) to determine the conductance γ from the data $\mathbf{\Lambda}_\gamma$.

4.2. *Solving the 2D problem with optimal grids.* To approximate $\sigma(x)$ from the network conductance γ we modify the reconstruction mapping introduced in Section 3.2 for layered media. The approximation is obtained by interpolating the output of the reconstruction mapping on the optimal grid computed for the reference $\sigma^{(o)} \equiv 1$. This grid is described in Sections 2.2 and 3.2.2. But which interpolation should we take? If we could have grids with as many points as we wish, the choice of the interpolation would not matter. This was the case in Section 3.3, where we studied the continuum limit $n \to \infty$ for the inverse spectral problem. The EIT problem is exponentially unstable and the whole idea of our approach is to have a sparse parametrization of the unknown σ. Thus, n is typically small, and the approximation of σ should go beyond ad-hoc interpolations of the parameters returned by the reconstruction mapping. We show in Section 4.2.3 how to approximate σ with a Gauss–Newton iteration preconditioned with the reconstruction mapping. We also explain briefly how one can introduce prior information about σ in the inversion method.

4.2.1. *The reconstruction mapping.* The idea behind the reconstruction mapping is to interpret the resistor network (Γ, γ) determined from the measured $\Lambda_\gamma = M_n(\Lambda_\sigma)$ as a finite-volume discretization of the equation (1-1) on the optimal grid computed for $\sigma^{(o)} \equiv 1$. This is what we did in Section 3.2 for the layered case, and the approach extends to the two-dimensional problem.

The conductivity is related to the conductances $\gamma(E)$, for $E \in \mathcal{E}$, via quadrature rules that approximate the current fluxes (2-5) through the dual edges. We could use for example the quadrature in [Borcea et al. 2010a; 2010b; Mamonov 2010], where the conductances are

$$\gamma_{a,b} = \sigma(P_{a,b}) \frac{L(\Sigma_{a,b})}{L(E_{a,b})}, \quad (a,b) \in \{(i, j \pm \tfrac{1}{2}), (i \pm \tfrac{1}{2}, j)\}, \qquad (4\text{-}2)$$

where L denotes the arc length of the primary and dual edges E and Σ (see Section 2.1 for the indexing and edge notation). Another example of quadrature is given in [Borcea et al. 2008]. It is specialized to tensor product grids in a disk, and it coincides with the quadrature (3-8) in the case of layered media. For inversion purposes, the difference introduced by different quadrature rules is small (see [Borcea et al. 2010a, Section 2.4]).

To define the reconstruction mapping $\mathcal{Q}_n$, we solve two inverse problems for resistor networks. One with the measured data $\Lambda_\gamma = M_n(\Lambda_\sigma)$, to determine the conductance γ, and one with the computed data $\Lambda_{\gamma^{(o)}} = M_n(\Lambda_{\sigma^{(o)}})$, for the reference $\sigma^{(o)} \equiv 1$. The latter gives the reference conductance $\gamma^{(o)}$ which we associate with the geometrical factor in (4-2)

$$\gamma_{a,b}^{(o)} \approx \frac{L(\Sigma_{a,b})}{L(E_{a,b})}, \qquad (4\text{-}3)$$

so that we can write

$$\tilde{\sigma}(P_{a,b}) \approx \sigma_{a,b} = \frac{\gamma_{a,b}}{\gamma_{a,b}^{(o)}}. \qquad (4\text{-}4)$$

Note that (4-4) becomes (3-40) in the layered case, where (3-8) gives $\alpha_j = h_\theta/\gamma_{j+\frac{1}{2},q}$ and $\hat{\alpha}_j = h_\theta \gamma_{j,q+\frac{1}{2}}$. The factors h_θ cancel out.

Let us call $\mathcal{D}_n$ the set in $\mathbb{R}^e$ of $e = n(n-1)/2$ independent measurements in $M_n(\Lambda_\sigma)$, obtained by removing the redundant entries. Note that there are e edges in the network, as many as the number of the data points in $\mathcal{D}_n$, given for example by the entries in the upper triangular part of $M_n(\Lambda_\sigma)$, stacked column by column in a vector in $\mathbb{R}^e$. By the consistency of the measurements (Section 4.1.1), $\mathcal{D}_n$ coincides with the set of the strictly upper triangular parts of the DtN maps of circular planar resistor networks with n boundary nodes. The mapping $\mathcal{Q}_n : \mathcal{D}_n \to \mathbb{R}^e_+$ associates to the measurements in $\mathcal{D}_n$ the e positive values $\sigma_{a,b}$ in (4-4).

We illustrate in Figure 7(b) the output of the mapping $\mathcal{Q}_n$, linearly interpolated on the optimal grid. It gives a good approximation of the conductivity that is improved further in Figure 7(c) with the Gauss–Newton iteration described below. The results in Figure 7 are obtained by solving the inverse problem for the networks with a fast layer peeling algorithm [Curtis et al. 1994]. Optimization can also be used for this purpose, at some additional computational cost. In any case, because we have relatively few $n(n-1)/2$ parameters, the cost is negligible compared to that of solving the forward problem on a fine grid.

4.2.2. *The optimal grids and sensitivity functions.* The definition of the tensor product optimal grids considered in Sections 2.2 and 3 does not extend to partial boundary measurement setups or to nonlayered reference conductivity functions. We present here an alternative approach to determining the location of the points $P_{a,b}$ at which we approximate the conductivity in the output (4-4) of the reconstruction mapping. This approach extends to arbitrary setups, and it is based on the sensitivity analysis of the conductance function γ to changes in the conductivity [Borcea et al. 2010b].

The *sensitivity grid* points are defined as the maxima of the sensitivity functions $D_\sigma \gamma_{a,b}(x)$. They are the points at which the conductances $\gamma_{a,b}$ are most sensitive to changes in the conductivity. The sensitivity functions $D_\sigma \gamma(x)$ are obtained by differentiating the identity $\Lambda_{\gamma(\sigma)} = M_n(\Lambda_\sigma)$ with respect to σ:

$$(D_\sigma \gamma)(x) = \left(D_\gamma \Lambda_\gamma \big|_{\Lambda_\gamma = M_n(\Lambda_\sigma)} \right)^{-1} \mathrm{vec}\left(M_n(D\mathcal{H}_\sigma)(x) \right), \quad x \in \Omega. \quad (4\text{-}5)$$

The left-hand side is a vector in $\mathbb{R}^e$. Its k-th entry is the Fréchet derivative of conductance γ_k with respect to changes in the conductivity σ. The entries of the

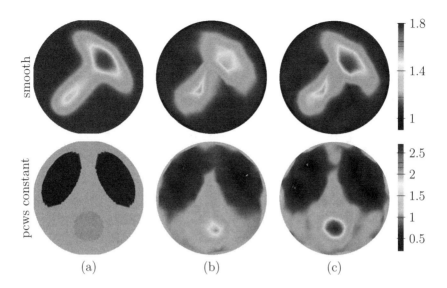

Figure 7. (a) True conductivity phantoms. (b) The output of the reconstruction mapping $\mathfrak{Q}_n$, linearly interpolated on a grid obtained for layered media as in Section 3.2.2. (c) One step of Gauss–Newton improves the reconstructions.

Jacobian $D_\gamma \Lambda_\gamma \in \mathbb{R}^{e \times e}$ are

$$\left(D_\gamma \Lambda_\gamma\right)_{jk} = \left(\text{vec}\left(\frac{\partial \Lambda_\gamma}{\partial \gamma_k}\right)\right)_j, \tag{4-6}$$

where vec(A) denotes the operation of stacking in a vector in $\mathbb{R}^e$ the entries in the strict upper triangular part of a matrix $A \in \mathbb{R}^{n \times n}$. The last factor in (4-5) is the sensitivity of the measurements to changes of the conductivity, given by

$$(M_n(D\mathcal{K}_\sigma))_{ij}(x) = \begin{cases} \displaystyle\int_{\mathcal{B} \times \mathcal{B}} \chi_i(x) D\mathcal{K}_\sigma(x; x, y) \chi_j(y)\, dx\, dy, & i \neq j, \\[1em] -\displaystyle\sum_{k \neq i} \int_{\mathcal{B} \times \mathcal{B}} \chi_i(x) D\mathcal{K}_\sigma(x; x, y) \chi_k(y)\, dx\, dy, & i = j. \end{cases} \tag{4-7}$$

Here $\mathcal{K}_\sigma(x, y)$ is the kernel of the DtN map evaluated at points x and $y \in \mathcal{B}$. Its Jacobian to changes in the conductivity is

$$D\mathcal{K}_\sigma(x; x, y) = \sigma(x)\sigma(y)\{\nabla_x(n(x) \cdot \nabla_x G(x, x))\} \cdot \{\nabla_x(n(y) \cdot \nabla_y G(x, y))\}, \tag{4-8}$$

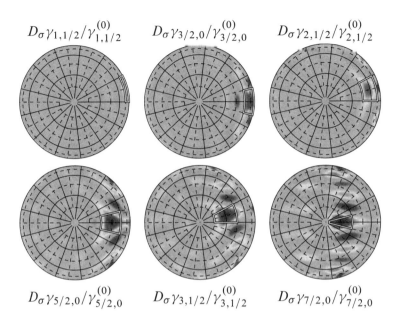

Figure 8. Sensitivity functions $\text{diag}(1/\gamma^{(0)})D_\sigma \gamma$ computed around the conductivity $\sigma = 1$ for $n = 13$. The images have a linear scale from dark blue to dark red spanning $\pm$ their maximum in absolute value. Light green corresponds to zero. We only display 6 sensitivity functions, the other ones can be obtained by integer multiple of $2\pi/13$ rotations. The primary grid is displayed in solid lines and the dual grid in dotted lines. The maxima of the sensitivity functions are very close to those of the optimal grid (intersection of solid and dotted lines).

where G is the Green's function of the differential operator $u \to \nabla \cdot (\sigma \nabla u)$ with Dirichlet boundary conditions, and $\boldsymbol{n}(x)$ is the outer unit normal at $x \in \mathscr{B}$. For more details on the calculation of the sensitivity functions see [Borcea et al. 2010b, Section 4].

The definition of the sensitivity grid points is

$$P_{a,b} = \arg\max_{x \in \Omega} (D_\sigma \gamma_{a,b})(x), \quad \text{evaluated at } \sigma = \sigma^{(o)} \equiv 1. \quad (4\text{-}9)$$

We display in Figure 8 the sensitivity functions with the superposed optimal grid obtained as in Section 3.2.2. Note that the maxima of the sensitivity functions are very close to the optimal grid points in the full measurements case.

4.2.3. *The preconditioned Gauss–Newton iteration.* Since the reconstruction mapping $\mathcal{Q}_n$ gives good reconstructions when properly interpolated, we can think of it as an approximate inverse of the forward map $M_n(\Lambda_\sigma)$ and use it as a *nonlinear* preconditioner. Instead of minimizing the misfit in the data, we solve the optimization problem

$$\min_{\sigma>0} \|\mathfrak{Q}_n(\text{vec}(M_n(\Lambda_\sigma))) - \mathfrak{Q}_n(\text{vec}(M_n(\Lambda_{\sigma_*})))\|_2^2. \qquad (4\text{-}10)$$

Here σ_* is the conductivity that we would like to recover. For simplicity the minimization (4-10) is formulated with noiseless data and no regularization. We refer to [Borcea et al. 2011] for a study of the effect of noise and regularization on the minimization (4-10).

The positivity constraints in (4-10) can be dealt with by solving for the log-conductivity $\kappa = \ln(\sigma)$ instead of the conductivity σ. With this change of variable, the residual in (4-10) can be minimized with the standard Gauss–Newton iteration, which we write in terms of the sensitivity functions (4-5) evaluated at $\sigma^{(j)} = \exp \kappa^{(j)}$:

$$\kappa^{(j+1)} = \kappa^{(j)} - \left(\text{diag}(1/\gamma^{(0)}) \, D_\sigma \gamma \, \text{diag}(\exp \kappa^{(j)})\right)^\dagger$$
$$\times \left[\mathfrak{Q}_n(\text{vec}(M_n(\Lambda_{\exp \kappa^{(j)}}))) - \mathfrak{Q}_n(\text{vec}(M_n(\Lambda_{\sigma_*})))\right]. \qquad (4\text{-}11)$$

The superscript $\dagger$ denotes the Moore–Penrose pseudoinverse and the division is understood componentwise. We take as initial guess the log-conductivity $\kappa^{(0)} = \ln \sigma^{(0)}$, where $\sigma^{(0)}$ is given by the linear interpolation of $\mathfrak{Q}_n(\text{vec}(M_n(\Lambda_{\sigma_*})))$ on the optimal grid (i.e., the reconstruction from Section 4.2.1). Having such a good initial guess helps with the convergence of the Gauss–Newton iteration. Our numerical experiments indicate that the residual in (4-10) is mostly reduced in the first iteration [Borcea et al. 2008]. Subsequent iterations do not change significantly the reconstructions and result in negligible reductions of the residual in (4-10). Thus, for all practical purposes, the preconditioned problem is linear. We have also observed in [Borcea et al. 2008; 2011] that the conditioning of the linearized problem is significantly reduced by the preconditioner $\mathfrak{Q}_n$.

Remark 6. The conductivity obtained after one step of the Gauss–Newton iteration is in the span of the sensitivity functions (4-5). The use of the sensitivity functions as an optimal parametrization of the unknown conductivity was studied in [Borcea et al. 2011]. Moreover, the same preconditioned Gauss–Newton idea was used in [Guevara Vasquez 2006] for the inverse spectral problem of Section 3.2.

We illustrate the improvement of the reconstructions with one Gauss–Newton step in Figure 7 (c). If prior information about the conductivity is available, it can be added in the form of a regularization term in (4-10). An example using total variation regularization is given in [Borcea et al. 2008].

5. Two-dimensional media and partial boundary measurements

In this section we consider the two-dimensional EIT problem with partial boundary measurements. As mentioned in Section 1, the boundary $\mathcal{B}$ is the union of the accessible subset $\mathcal{B}_A$ and the inaccessible subset $\mathcal{B}_I$. The accessible boundary $\mathcal{B}_A$ may consist of one or multiple connected components. We assume that the inaccessible boundary is grounded, so the partial boundary measurements are a set of Cauchy data $\{u|_{\mathcal{B}_A}, (\sigma \mathbf{n} \cdot \nabla u)|_{\mathcal{B}_A}\}$, where u satisfies (1-1) and $u|_{\mathcal{B}_I} = 0$. The inverse problem is to determine σ from these Cauchy data.

Our inversion method described in the previous sections extends to the partial boundary measurement setup. But there is a significant difference concerning the definition of the optimal grids. The tensor product grids considered so far are essentially one-dimensional, and they rely on the rotational invariance of the problem for $\sigma^{(o)} \equiv 1$. This invariance does not hold for the partial boundary measurements, so new ideas are needed to define the optimal grids. We present two approaches in Sections 5.1 and 5.2. The first one uses circular planar networks with the same topology as before, and mappings that take uniformly distributed points on $\mathcal{B}$ to points on the accessible boundary $\mathcal{B}_A$. The second one uses networks with topologies designed specifically for the partial boundary measurement setups. The underlying two-dimensional optimal grids are defined with sensitivity functions.

5.1. Coordinate transformations for the partial data problem.

The idea of the approach described in this section is to map the partial data problem to one with full measurements at equidistant points, where we know from Section 4 how to define the optimal grids. Since Ω is a unit disk, we can do this with diffeomorphisms of the unit disk to itself.

Let us denote such a diffeomorphism by F and its inverse F^{-1} by G. If the potential u satisfies (1-1), then the transformed potential $\tilde{u}(x) = u(F(x))$ solves the same equation with conductivity $\tilde{\sigma}$ defined by

$$\tilde{\sigma}(x) = \left. \frac{G'(y)\sigma(y)(G'(y))^T}{|\det G'(y)|} \right|_{y=F(x)}, \quad (5\text{-}1)$$

where G' denotes the Jacobian of G. The conductivity $\tilde{\sigma}$ is the *push forward* of σ by G, and it is denoted by $G_*\sigma$. Note that if $G'(y)(G'(y))^T \neq I$ and $\det G'(y) \neq 0$, then $\tilde{\sigma}$ is a symmetric positive definite tensor. If its eigenvalues are distinct, then the push forward of an isotropic conductivity is anisotropic.

The push forward $g_* \Lambda_\sigma$ of the DtN map is written in terms of the restrictions of diffeomorphisms G and F to the boundary. We call these restrictions $g = G|_{\mathcal{B}}$

and $f = F|_\mathcal{B}$ and write

$$((g_*\Lambda_\sigma)u_\mathcal{B})(\theta) = (\Lambda_\sigma(u_\mathcal{B} \circ g))(\tau)|_{\tau=f(\theta)}, \qquad \theta \in [0, 2\pi), \qquad (5\text{-}2)$$

for $u_\mathcal{B} \in H^{1/2}(\mathcal{B})$. It is shown in [Sylvester 1990] that the DtN map is *invariant* under the push forward in the following sense

$$g_*\Lambda_\sigma = \Lambda_{G_*\sigma}. \qquad (5\text{-}3)$$

Therefore, given (5-3) we can compute the push forward of the DtN map, solve the inverse problem with data $g_*\Lambda_\sigma$ to obtain $G_*\sigma$, and then map it back using the inverse of (5-2). This requires the full knowledge of the DtN map. However, if we use the discrete analogue of the above procedure, we can transform the discrete measurements of Λ_σ on $\mathcal{B}_A$ to discrete measurements at equidistant points on $\mathcal{B}$, from which we can estimate $\tilde{\sigma}$ as described in Section 4.

There is a major obstacle to this procedure: The EIT problem is uniquely solvable only for isotropic conductivities. Anisotropic conductivities are determined by the DtN map only up to a boundary-preserving diffeomorphism [Sylvester 1990]. Two distinct approaches to overcome this obstacle are described in Sections 5.1.1 and 5.1.2. The first uses conformal mappings F and G that preserve the isotropy of the conductivity at the expense of rigid placement of the measurement points. The second approach uses extremal quasiconformal mappings that minimize the artificial anisotropy of $\tilde{\sigma}$ introduced by the placement at our will of the measurement points in $\mathcal{B}_A$.

5.1.1. *Conformal mappings.* The push forward $G_*\sigma$ of an isotropic σ is isotropic if G and F satisfy $G'((G')^T) = I$ and $F'((F')^T) = I$. This means that the diffeomorphism is *conformal* and the push forward is simply

$$G_*\sigma = \sigma \circ F. \qquad (5\text{-}4)$$

Since all conformal mappings of the unit disk to itself belong to the family of Möbius transforms [Lavrentiev and Shabat 1987], F must be of the form

$$F(z) = e^{i\omega}\frac{z-a}{1-\overline{a}z}, \qquad z \in \mathbb{C}, \ |z| \leq 1, \ \omega \in [0, 2\pi), \ a \in \mathbb{C}, \ |a| < 1, \qquad (5\text{-}5)$$

where we associate $\mathbb{R}^2$ with the complex plane $\mathbb{C}$. Note that the group of transformations (5-5) is extremely rigid, its only degrees of freedom being the numerical parameters a and ω.

To use the full data discrete inversion procedure from Section 4 we require that G maps the accessible boundary segment $\mathcal{B}_A = \{e^{i\tau} \mid \tau \in [-\beta, \beta]\}$ to the whole boundary with the exception of one segment between the equidistant measurement points θ_k, $k = (n+1)/2, (n+3)/2$ as shown in Figure 9. This determines completely the values of the parameters a and ω in (5-5) which in

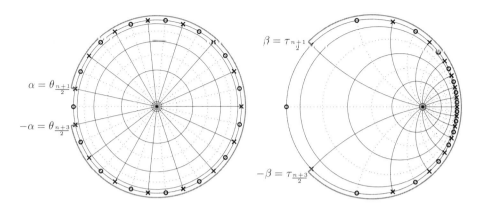

Figure 9. The optimal grid in the unit disk (left) and its image under the conformal mapping F (right). Primary grid lines are solid black, dual grid lines are dotted black. Boundary grid nodes: primary ×, dual ○. The accessible boundary segment $\mathcal{B}_A$ is indicated by the outermost arc (thick solid line).

turn determine the mapping f on the boundary. Thus, we have no further control over the positioning of the measurement points $\tau_k = f(\theta_k)$, $k = 1, \ldots, n$.

As shown in Figure 9 the lack of control over τ_k leads to a grid that is highly nonuniform in angle. In fact it is demonstrated in [Borcea et al. 2010a] that as n increases there is no asymptotic refinement of the grid away from the center of $\mathcal{B}_A$, where the points accumulate. However, since the limit $n \to \infty$ is unattainable in practice due to the severe ill-conditioning of the problem, the grids obtained by conformal mapping can still be useful in practical inversion. We show reconstructions with these grids in Section 5.3.

5.1.2. *Extremal quasiconformal mappings.* To overcome the issues with conformal mappings that arise due to the inherent rigidity of the group of conformal automorphisms of the unit disk, we use here quasiconformal mappings. A quasiconformal mapping F obeys a Beltrami equation in Ω

$$\frac{\partial F}{\partial \bar{z}} = \mu(z) \frac{\partial F}{\partial z}, \quad \|\mu\|_\infty < 1, \tag{5-6}$$

with a Beltrami coefficient $\mu(z)$ that measures how much F differs from a conformal mapping. If $\mu \equiv 0$, then (5-6) reduces to the Cauchy–Riemann equation and F is conformal. The magnitude of μ also provides a measure of the anisotropy κ of the push forward of σ by F. The anisotropy is defined by

$$\kappa(F_*\sigma, z) = \frac{\sqrt{\lambda_1(z)/\lambda_2(z)} - 1}{\sqrt{\lambda_1(z)/\lambda_2(z)} + 1}, \tag{5-7}$$

where $\lambda_1(z)$ and $\lambda_2(z)$ are the largest and smallest eigenvalues of $F_*\sigma$ respectively. The connection between μ and κ is given by

$$\kappa(F_*\sigma, z) = |\mu(z)|, \tag{5-8}$$

and the maximum anisotropy is

$$\kappa(F_*\sigma) = \sup_z \kappa(F_*\sigma, z) = \|\mu\|_\infty. \tag{5-9}$$

Since the unknown conductivity is isotropic, we would like to minimize the amount of artificial anisotropy that we introduce into the reconstruction by using F. This can be done with extremal quasiconformal mappings, which minimize $\|\mu\|_\infty$ under constraints that fix $f = F|_{\mathcal{B}}$, thus allowing us to control the positioning of the measurement points $\tau_k = f(\theta_k)$, for $k = 1, \ldots, n$.

For sufficiently regular boundary values f there exists a unique extremal quasiconformal mapping that is known to be of a Teichmüller type [Strebel 1976]. Its Beltrami coefficient satisfies

$$\mu(z) = \|\mu\|_\infty \frac{\overline{\phi(z)}}{|\phi(z)|}, \tag{5-10}$$

for some holomorphic function $\phi(z)$ in Ω. Similarly, we can define the Beltrami coefficient for G, using a holomorphic function ψ. It is established in [Reich 1976] that F admits a decomposition

$$F = \Psi^{-1} \circ A_K \circ \Phi, \tag{5-11}$$

where

$$\Phi(z) = \int \sqrt{\phi(z)} dz, \qquad \Psi(\zeta) = \int \sqrt{\psi(\zeta)} d\zeta, \tag{5-12}$$

are conformal away from the zeros of ϕ and ψ, and

$$A_K(x + iy) = Kx + iy \tag{5-13}$$

is an affine stretch, the only source of anisotropy in (5-11):

$$\kappa(F_*\sigma) = \|\mu\|_\infty = \left|\frac{K-1}{K+1}\right|. \tag{5-14}$$

Since only the behavior of f at the measurement points θ_k is of interest to us, it is possible to construct explicitly the mappings Φ and Ψ [Borcea et al. 2010a]. They are Schwartz–Christoffel conformal mappings of the unit disk to polygons of special form, as shown in Figure 10. See [Borcea et al. 2010a, Section 3.4] for more details.

We demonstrate the behavior of the optimal grids under the extremal quasiconformal mappings in Figure 11. We present the results for two different values

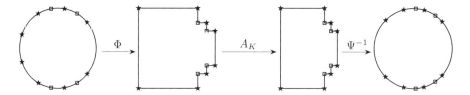

Figure 10. Teichmüller mapping decomposed into conformal mappings Φ and Ψ, and an affine transform A_K. The poles of ϕ and ψ and their images under Φ and Ψ are ★, the zeros of ϕ and ψ and their images under Φ and Ψ are □.

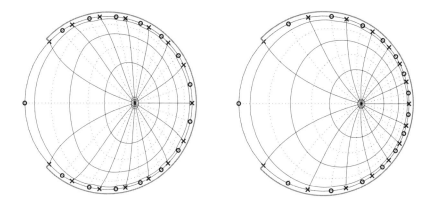

Figure 11. The optimal grid under the quasiconformal Teichmüller mappings F with different K. Left: $K = 0.8$ (smaller anisotropy); right: $K = 0.66$ (higher anisotropy). Primary grid lines are solid black, dual grid lines are dotted black. Boundary grid nodes: primary ×, dual ○. The accessible boundary segment $\mathcal{B}_A$ is indicated by the outermost arc (thick solid line).

of the affine stretching constant K. As we increase the amount of anisotropy from $K = 0.8$ to $K = 0.66$, the distribution of the grid nodes becomes more uniform. The price to pay for this more uniform grid is an increased amount of artificial anisotropy, which may detriment the quality of the reconstruction, as shown in the numerical examples in Section 5.3.

5.2. *Special network topologies for the partial data problem.* The limitations of the construction of the optimal grids with coordinate transformations can be attributed to the fact that there is no nonsingular mapping between the full boundary $\mathcal{B}$ and its proper subset $\mathcal{B}_A$. Here we describe an alternative approach, that avoids these limitations by considering networks with different topologies, constructed specifically for the partial measurement setups. The one-sided case, with the accessible boundary $\mathcal{B}_A$ consisting of one connected segment, is in Section 5.2.1. The two sided case, with $\mathcal{B}_A$ the union of two disjoint segments, is in Section 5.2.2. The optimal grids are constructed using the sensitivity analysis of the discrete and continuum problems, as explained in Sections 4.2.2 and 5.2.3.

5.2.1. *Pyramidal networks for the one-sided problem.* We consider here the case of $\mathcal{B}_A$ consisting of one connected segment of the boundary. The goal is to choose a topology of the resistor network based on the flow properties of the continuum partial data problem. Explicitly, we observe that since the potential excitation is supported on $\mathcal{B}_A$, the resulting currents should not penetrate deep into Ω, away from $\mathcal{B}_A$. The currents are so small sufficiently far away from $\mathcal{B}_A$ that in the discrete (network) setting we can ask that there is no flow escaping the associated nodes. Therefore, these nodes are interior ones. A suitable choice of networks that satisfy such conditions was proposed in [Borcea et al. 2010b]. We call them *pyramidal* and denote their graphs by Γ_n, with n the number of boundary nodes.

We illustrate two pyramidal graphs in Figure 12, for $n = 6$ and 7. Note that it is not necessary that n be odd for the pyramidal graphs Γ_n to be critical, as was the case in the previous sections. In what follows we refer to the edges of Γ_n as vertical or horizontal according to their orientation in Figure 12. Unlike the circular networks in which all the boundary nodes are in a sense adjacent, there is a gap between the boundary nodes v_1 and v_n of a pyramidal network. This gap is formed by the bottommost $n - 2$ interior nodes that enforce the condition of zero normal flux, the approximation of the lack of penetration of currents away from $\mathcal{B}_A$.

It is known from [Curtis et al. 1998; Borcea et al. 2010b] that the pyramidal networks are critical and thus uniquely recoverable from the DtN map. Similar to the circular network case, pyramidal networks can be recovered using a layer

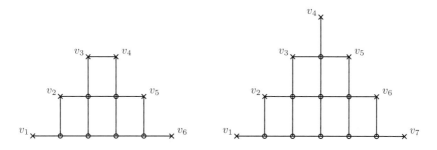

Figure 12. Pyramidal networks Γ_n for $n = 6, 7$. The boundary nodes v_j, $j = 1, \ldots, n$ are indicated with $\times$ and the interior nodes with $\circ$.

peeling algorithm in a finite number of algebraic operations. We recall such an algorithm below, from [Borcea et al. 2010b], in the case of even $n = 2m$. A similar procedure can also be used for odd n.

Algorithm 1. To determine the conductance γ of the pyramidal network (Γ_n, γ) from the DtN map $\Lambda^{(n)}$, perform the following steps:

(1) To compute the conductances of horizontal and vertical edges emanating from the boundary node v_p, for each $p = 1, \ldots, 2m$, define the following sets:
$Z = \{v_1, \ldots, v_{p-1}, v_{p+1}, \ldots, v_m\}$, $C = \{v_{m+2}, \ldots, v_{2m}\}$,
$H = \{v_1, \ldots, v_p\}$ and $V = \{v_p, \ldots, v_{m+1}\}$, in the case $p \leq m$.
$Z = \{v_{m+1}, \ldots, v_{p-1}, v_{p+1}, \ldots, v_{2m}\}$, $C = \{v_1, \ldots, v_{m-1}\}$,
$H = \{v_p, \ldots, v_{2m}\}$ and $V = \{v_m, \ldots, v_p\}$, for $m+1 \leq p \leq 2m$.

(2) Compute the conductance $\gamma(E_{p,h})$ of the horizontal edge emanating from v_p using
$$\gamma(E_{p,h}) = \left(\Lambda^{(n)}_{p,H} - \Lambda^{(n)}_{p,C}\left(\Lambda^{(n)}_{Z,C}\right)^{-1}\Lambda^{(n)}_{Z,H}\right)\mathbf{1}_H; \quad (5\text{-}15)$$
compute the conductance $\gamma(E_{p,v})$ of the vertical edge emanating from v_p using
$$\gamma(E_{p,v}) = \left(\Lambda^{(n)}_{p,V} - \Lambda^{(n)}_{p,C}\left(\Lambda^{(n)}_{Z,C}\right)^{-1}\Lambda^{(n)}_{Z,V}\right)\mathbf{1}_V, \quad (5\text{-}16)$$
where $\mathbf{1}_V$ and $\mathbf{1}_H$ are column vectors of all ones.

(3) Once $\gamma(E_{p,h})$, $\gamma(E_{p,v})$ are known, peel the outer layer from Γ_n to obtain the subgraph Γ_{n-2} with the set $\mathcal{S} = \{w_1, \ldots, w_{2m-2}\}$ of boundary nodes. Assemble the blocks $K_{\mathcal{SS}}$, $K_{\mathcal{SB}}$, $K_{\mathcal{BS}}$, $K_{\mathcal{BB}}$ of the Kirchhoff matrix of (Γ_n, γ), and compute the updated DtN map $\Lambda^{(n-2)}$ of the smaller network (Γ_{n-2}, γ), as follows
$$\Lambda^{(n-2)} = -K'_{\mathcal{SS}} - K_{\mathcal{SB}}P^T\left(P\left(\Lambda^{(n)} - K_{\mathcal{BB}}\right)P^T\right)^{-1}PK_{\mathcal{BS}}. \quad (5\text{-}17)$$
Here $P \in \mathbb{R}^{(n-2) \times n}$ is a projection operator: $PP^T = I_{n-2}$, and $K'_{\mathcal{SS}}$ is a part of $K_{\mathcal{SS}}$ that only includes the contributions from the edges connecting $\mathcal{S}$ to $\mathcal{B}$.

(4) If $m = 1$ terminate. Otherwise, decrease m by 1, update $n = 2m$ and go back to step 1.

Similar to the layer peeling method in [Curtis et al. 1994], Algorithm 1 is based on the construction of special solutions. In steps 1 and 2 the special solutions are constructed implicitly, to enforce a unit potential drop on edges $E_{p,h}$ and $E_{p,v}$ emanating from the boundary node v_p. Since the DtN map is known, so is the current at v_p, which equals to the conductance of an edge due to a unit potential drop on that edge. Once the conductances are determined for all the edges adjacent to the boundary, the layer of edges is peeled off and the DtN map of a smaller network Γ_{n-2} is computed in step 3. After m layers have been peeled off, the network is completely recovered. The algorithm is studied in detail in [Borcea et al. 2010b], where it is also shown that all matrices that are inverted in (5-15), (5-16) and (5-17) are nonsingular.

Remark 7. The DtN update formula (5-17) provides an interesting connection to the layered case. It can be viewed as a matrix generalization of the continued fraction representation (3-36). The difference between the two formulas is that (3-36) expresses the eigenvalues of the DtN map, while (5-17) gives an expression for the DtN map itself.

5.2.2. *The two-sided problem.* We call the problem two-sided when the accessible boundary $\mathcal{B}_A$ consists of two disjoint segments of $\mathcal{B}$. A suitable network topology for this setting is that of a *two-sided* graph T_n shown in Figure 13. The number of boundary nodes n is assumed even $n = 2m$. Half of the nodes are on one segment of the boundary and the other half on the other, as illustrated in the figure. Similar to the one-sided case, the two groups of m boundary nodes are separated by the outermost interior nodes, which model the lack of penetration of currents away from the accessible boundary segments. One can verify that the two-sided network is critical, and thus it can be uniquely recovered from the DtN map by the Algorithm 2 given below.

When referring to either the horizontal or vertical edges of a two sided network, we use their orientation in Figure 13.

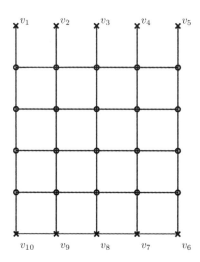

Figure 13. Two-sided network T_n for $n = 10$. Boundary nodes v_j, $j = 1, \ldots, n$ are ×, interior nodes are ∘.

Algorithm 2. To determine the conductance γ of the two-sided network (T_n, γ) from the DtN map Λ_γ, perform the following steps:

(1) Peel the lower layer of horizontal resistors:
For $p = m+2, m+3, \ldots, 2m$ define the sets
$$Z = \{p+1, p+2, \ldots, p+m-1\} \text{ and } C = \{p-2, p-3, \ldots, p-m\}.$$
The conductance of the edge $E_{p,q,h}$ between v_p and v_q, where $q = p-1$, is given by
$$\gamma(E_{p,q,h}) = -\Lambda_{p,q} + \Lambda_{p,C}(\Lambda_{Z,C})^{-1}\Lambda_{Z,q}. \tag{5-18}$$
Assemble a symmetric tridiagonal matrix A with off-diagonal entries
$$-\gamma(E_{p,p-1,h})$$
and rows summing to zero. Update the lower right m-by-m block of the DtN map by subtracting A from it.

(2) Let $s = m-1$.

(3) Peel the top and bottom layers of vertical resistors:
For $p = 1, 2, \ldots, 2m$ define the sets $L = \{p-1, p-2, \ldots, p-s\}$ and $R = \{p+1, p+2, \ldots, p+s\}$. If $p < m/2$ for the top layer, or $p > 3m/2$ for the bottom layer, set $Z = L, C = R$. Otherwise let $Z = R, C = L$. The conductance of the vertical edge emanating from v_p is given by
$$\gamma(E_{p,v}) = \Lambda_{p,p} - \Lambda_{p,C}(\Lambda_{Z,C})^{-1}\Lambda_{Z,p}. \tag{5-19}$$
Let $D = \text{diag}(\gamma(E_{p,v}))$ and update the DtN map
$$\Lambda_\gamma = -D - D(\Lambda_\gamma + D)^{-1} D. \tag{5-20}$$

(4) If $s = 1$ go to step (7). Otherwise decrease s by 2.

(5) Peel the top and bottom layers of horizontal resistors:
For $p = 1, 2, \ldots, 2m$ define the sets $L = \{p-1, p-2, \ldots, p-s\}$ and $R = \{p+2, p+3, \ldots, p+s+1\}$. If $p < m/2$ for the top layer, or $p < 3m/2$ for the bottom layer, set $Z = L, C = R, q = p+1$. Otherwise let $Z = R, C = L, q = p-1$. The conductance of the edge connecting v_p and v_q is given by (5-18). Update the upper left and lower right blocks of the DtN map as in step (1).

(6) If $s = 0$ go to step (7), otherwise go to (3).

(7) Determine the last layer of resistors. If m is odd the remaining vertical resistors are the diagonal entries of the DtN map. If m is even, the remaining resistors are horizontal. The leftmost of the remaining horizontal resistors

$\gamma(E_{1,2,h})$ is determined from (5-18) with $p = 1$, $q = m+1$, $C = \{1, 2\}$, $Z = \{m+1, m+2\}$ and a change of sign. The rest are determined by

$$\gamma(E_{p,p+1,h}) = \left(\Lambda_{p,H} - \Lambda_{p,C}(\Lambda_{Z,C})^{-1}\Lambda_{Z,H}\right)\mathbf{1}, \qquad (5\text{-}21)$$

where $p = 2, 3, \ldots, m-1$,

$C = \{p-1, p, p+1\}, \qquad Z = \{p+m-1, p+m, p+m+1\},$
$H = \{p+m-1, p+m\}, \quad \mathbf{1} = (1, 1)^T.$

Similar to Algorithm 1, Algorithm 2 is based on the construction of special solutions examined in [Curtis et al. 1994; Curtis and Morrow 2000]. These solutions are designed to localize the flow on the outermost edges, whose conductance we determine first. In particular, formulas (5-18) and (5-19) are known as the "boundary edge" and "boundary spike" formulas [Curtis and Morrow 2000, Corollaries 3.15 and 3.16].

5.2.3. *Sensitivity grids for pyramidal and two-sided networks.* The underlying grids of the pyramidal and two-sided networks are truly two-dimensional, and they cannot be constructed explicitly as in Section 3 by reducing the problem to a one-dimensional one. We define the grids with the sensitivity function approach described in Section 4.2.2. The computed sensitivity grid points are presented in Figure 14, and we observe a few important properties. First, the neighboring points corresponding to the same type of resistors (vertical or horizontal) form

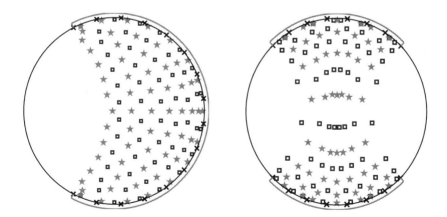

Figure 14. Sensitivity optimal grids in the unit disk for the pyramidal network Γ_n (left) and the two-sided network T_n (right) with $n = 16$. The accessible boundary segments $\mathcal{B}_A$ are solid red. The symbol $\square$ corresponds to vertical edges, $\star$ corresponds to horizontal edges, and measurement points are marked with $\times$.

rather regular virtual quadrilaterals. Second, the points corresponding to different types of resistors interlace in the sense of lying inside the virtual quadrilaterals formed by the neighboring points of the other type. Finally, while there is some refinement near the accessible boundary (more pronounced in the two-sided case), the grids remain quite uniform throughout the covered portion of the domain.

Note from Figure 13 that the graph T_n lacks upside-down symmetry. Thus, it is possible to come up with two sets of optimal grid nodes, by fitting the measured DtN map $M_n(\Lambda_\sigma)$ once with a two-sided network and the second time with the network turned upside-down. This way the number of nodes in the grid is essentially doubled, thus doubling the resolution of the reconstruction. However, this approach can only improve resolution in the direction transversal to the depth.

5.3. *Numerical results.* We present in this section numerical reconstructions with partial boundary measurements. The reconstructions with the four methods from Sections 5.1.1, 5.1.2, 5.2.1 and 5.2.2 are compared row by row in Figure 15. We use the same two test conductivities as in Figure 7(a). Each row in Figure 15 corresponds to one method. For each test conductivity, we show first the piecewise linear interpolation of the entries returned by the reconstruction mapping $\mathfrak{Q}_n$, on the optimal grids (first and third column in Figure 15). Since these grids do not cover the entire Ω, we display the results only in the subset of Ω populated by the grid points. We also show the reconstructions after one-step of the Gauss–Newton iteration (4-11) (second and fourth columns in Figure 15).

As expected, reconstructions with the conformal mapping grids are the worst. The highly nonuniform conformal mapping grids cannot capture the details of the conductivities away from the middle of the accessible boundary. The reconstructions with quasiconformal grids perform much better, capturing the details of the conductivities much more uniformly throughout the domain. Although the piecewise linear reconstructions $\mathfrak{Q}_n$ have slight distortions in the geometry, these distortions are later removed by the first step of the Gauss–Newton iteration. The piecewise linear reconstructions with pyramidal and two-sided networks avoid the geometrical distortions of the quasiconformal case, but they are also improved after one step of the Gauss–Newton iteration.

Note that while the Gauss–Newton step improves the geometry of the reconstructions, it also introduces some spurious oscillations. This is more pronounced for the piecewise constant conductivity phantom (fourth column in Figure 15). To overcome this problem one may consider regularizing the Gauss–Newton iteration (4-11) by adding a penalty term of some sort. For example, for the piecewise constant phantom, we could penalize the total variation of the reconstruction, as was done in [Borcea et al. 2008].

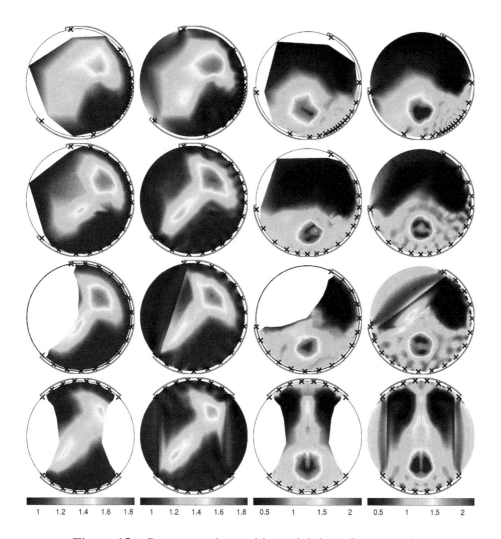

Figure 15. Reconstructions with partial data. Same conductivities are used as in Figure 7. Two leftmost columns: smooth conductivity. Two rightmost columns: piecewise constant chest phantom. Columns 1 and 3: piecewise linear reconstructions. Columns 2 and 4: reconstructions after one step of Gauss–Newton iteration (4-11). Rows from top to bottom: conformal mapping, quasiconformal mapping, pyramidal network, two-sided network. The accessible boundary $\mathcal{B}_A$ is indicated by solid arcs exterior to each disk. The centers of supports of measurement (electrode) functions χ_q are marked with ×.

6. Summary

We presented a discrete approach to the numerical solution of the inverse problem of electrical impedance tomography (EIT) in two dimensions. Due to the severe ill-posedness of the problem, it is desirable to parametrize the unknown conductivity $\sigma(x)$ with as few parameters as possible, while still capturing the best attainable resolution of the reconstruction. To obtain such a parametrization, we used a discrete, model reduction formulation of the problem. The discrete models are resistor networks with special graphs.

We described in detail the solvability of the model reduction problem. First, we showed that boundary measurements of the continuum Dirichlet to Neumann (DtN) map Λ_σ for the unknown $\sigma(x)$ define matrices that belong to the set of discrete DtN maps for resistor networks. Second, we described the types of network graphs appropriate for different measurement setups. By appropriate we mean those graphs that ensure unique recoverability of the network from its DtN map. Third, we showed how to determine the networks.

We established that the key ingredient in the connection between the discrete model reduction problem (inverse problem for the network) and the continuum EIT problem is the optimal grid. "Optimal" refers to the fact that finite-volumes discretization on these grids give spectrally accurate approximations of the DtN map, the data in EIT. We defined reconstructions of the conductivity using the optimal grids, and studied them in detail in three cases: (1) layered media and full boundary measurements, where the problem can be reduced to one dimension via Fourier transforms; (2) two-dimensional media with measurement access to the full boundary; and (3) two-dimensional media with access to a subset of the boundary.

Finally, we illustrated the approach's performance with numerical simulations.

Appendix A. The quadrature formulas

To understand definitions (3-8), recall Figure 1. Take for example the dual edge

$$\Sigma_{i-\frac{1}{2},j} = (P_{i-\frac{1}{2},j-\frac{1}{2}}, P_{i-\frac{1}{2},j-\frac{1}{2}}),$$

where $P_{i-\frac{1}{2},j-\frac{1}{2}} = \hat{r}_i(\cos\hat{\theta}_j, \sin\hat{\theta}_j)$. We have from (2-5) and the change of variables to $z(r)$ that

$$\int_{\Sigma_{i-\frac{1}{2},j}} \sigma(x)n(x)\cdot\nabla u(x)\,ds(x) = \int_{\hat{\theta}_j}^{\hat{\theta}_{j+1}} \hat{r}_{i-1}\sigma(\hat{r}_{i-1})\frac{\partial u(\hat{r}_{i-1},\theta)}{\partial r}\,d\theta$$

$$\approx -h_\theta \frac{\partial u(\hat{r}_{i-1},\hat{\theta}_j)}{\partial z} \approx \frac{h_\theta\left(U_{i-1,j} - U_{i,j}\right)}{z(r_i) - z(r_{i-1})},$$

which gives the first equation in (3-8). Similarly, the flux across
$$\Sigma_{i,j+\frac{1}{2}} = (P_{i-\frac{1}{2},j+\frac{1}{2}}, P_{i+\frac{1}{2},j+\frac{1}{2}})$$
is given by
$$\int_{\Sigma_{i,j+\frac{1}{2}}} \sigma(\mathbf{x})\mathbf{n}(\mathbf{x}) \cdot \nabla u(\mathbf{x})\, ds(\mathbf{x}) = \int_{\hat{r}_i}^{\hat{r}_{i-1}} \frac{\sigma(r)}{r} \frac{\partial u(r, \hat{\theta}_{j+1})}{\partial \theta}\, dr$$
$$\approx \frac{\partial u(r_i, \hat{\theta}_{j+1})}{\partial \theta} \int_{\hat{r}_i}^{\hat{r}_{i-1}} \frac{\sigma(r)}{r}\, dr$$
$$\approx (\hat{z}(\hat{r}_i) - \hat{z}(\hat{r}_{i-1})) \frac{U_{i,j+1} - U(i,j)}{h_\theta},$$
which gives the second equation in (3-8).

Appendix B. Continued fraction representation

Let us begin with the system of equations satisfied by the potential V_j, which we rewrite as
$$b_j = b_{j+1} + \hat{\alpha}_{j+1} \lambda V_{j+1} \quad (0, 1, \ldots, \ell), \quad b_0 = \Phi_{\mathcal{B}}, \quad V_{\ell+1} = 0, \quad \text{(B-1)}$$
where we let
$$V_j = V_{j+1} + \alpha_j b_j. \quad \text{(B-2)}$$
Combining the first equation in (B-1) with (B-2), we obtain the recursive relation
$$\frac{b_j}{V_j} = \frac{1}{\alpha_j + \dfrac{1}{\hat{\alpha}_{j+1}\lambda + \dfrac{b_{j+1}}{V_{j+1}}}}, \quad j = 1, 2, \ldots, \ell, \quad \text{(B-3)}$$
which we iterate for j decreasing from $j = \ell - 1$ to 1, and starting with
$$\frac{b_\ell}{V_\ell} = \frac{1}{\alpha_\ell}. \quad \text{(B-4)}$$
The latter follows from the first equation in (B-1) evaluated at $j = \ell$, and boundary condition $V_{\ell+1} = 0$. We obtain that
$$F^\dagger(\lambda) = V_1/\Phi_{\mathcal{B}} = \frac{V_1}{b_0} = \frac{V_1}{b_1 + \hat{\alpha}_1 \lambda V_1} = \frac{1}{\hat{\alpha}_1 \lambda + \dfrac{b_1}{V_1}} \quad \text{(B-5)}$$
has the continued fraction representation (3-36).

Appendix C. Derivation of (3-47) and (3-48)

To derive (3-47) let us begin with the Fourier series of the electrode functions

$$\chi_q(\theta) = \sum_{k \in \mathbb{Z}} C_q(k) e^{ik\theta} = \sum_{k \in \mathbb{Z}} \overline{C_q(k)} e^{-ik\theta}, \qquad \text{(C-1)}$$

where the bar denotes complex conjugate and the coefficients are

$$C_q(\theta) = \frac{1}{2\pi} \int_0^{2\pi} \chi_q(\theta) e^{ik\theta} d\theta = \frac{e^{ik\theta_q}}{2\pi} \operatorname{sinc}\left(\frac{kh_\theta}{2}\right). \qquad \text{(C-2)}$$

Then, we have

$$(\Lambda_\gamma)_{p,q} = \int_0^{2\pi} \chi_p(\theta) \Lambda_\sigma \chi_q(\theta) \, d\theta = \sum_{k,k' \in \mathbb{Z}} C_p(k) \overline{C_q(k')} \int_0^{2\pi} e^{ik\theta} \Lambda_\sigma e^{-ik'\theta} d\theta$$

$$= \frac{1}{2\pi} \sum_{k \in \mathbb{Z}} e^{ik(\theta_p - \theta_q)} f(k^2) \operatorname{sinc}^2\left(\frac{kh_\theta}{2}\right), \qquad p \neq q. \qquad \text{(C-3)}$$

The diagonal entries are

$$(\Lambda_\gamma)_{p,p} = -\sum_{q \neq p} (\Lambda_\gamma)_{p,q}$$

$$= -\frac{1}{2\pi} \sum_{k \in \mathbb{Z}} e^{ik\theta_p} f(k^2) \operatorname{sinc}^2\left(\frac{kh_\theta}{2}\right) \sum_{q \neq p} e^{-ik\theta_q}. \qquad \text{(C-4)}$$

But

$$\sum_{q \neq p} e^{-ik\theta_q} = \sum_{q=1}^n e^{-i\frac{2\pi k}{n}(q-1)} - e^{-ik\theta_p}$$

$$= e^{i\pi k(1 - 1/n)} \frac{\sin(\pi k)}{\sin(\pi k/n)} - e^{-ik\theta_p} = n\delta_{k,0} - e^{ik\theta_p}. \qquad \text{(C-5)}$$

Since $f(0) = 0$, we obtain from (C-4) and (C-5) that (C-3) holds for $p = q$ as well. This is the result (3-47). Moreover, (3-48) follows from

$$(\Lambda_\gamma[e^{ik\theta}])_p = \sum_{q=1}^n (\Lambda_\gamma)_{p,q} e^{ik\theta_q}$$

$$= \frac{1}{2\pi} \sum_{k_1 \in \mathbb{Z}} e^{ik_1\theta_p} f(k_1^2) \operatorname{sinc}^2\left(\frac{k_1 h_\theta}{2}\right) \sum_{q=1}^n e^{i(k-k_1)\theta_q}, \qquad \text{(C-6)}$$

and the identity

$$\sum_{q=1}^{n} e^{i(k-k_1)\theta_q} = \sum_{q=1}^{n} e^{i\frac{2\pi(k-k_1)}{n}(q-1)} = n\delta_{k,k_1}. \quad \text{(C-7)}$$

Appendix D. Rational interpolation and Euclidean division

Consider the case $m_{1/2} = 1$, where $F(\lambda) = 1/F^{\dagger}(\lambda)$ follows from (3-36). We rename the coefficients as

$$\kappa_{2j-1} = \hat{\alpha}_j, \quad \kappa_{2j} = \alpha_j, \quad j = 1, \ldots, \ell, \quad \text{(D-1)}$$

and let $\lambda = x^2$ to obtain

$$\frac{F(x^2)}{x} = \kappa_1 x + \cfrac{1}{\kappa_2 x + \cfrac{\ddots}{\kappa_{2\ell-1} x + \cfrac{1}{\kappa_{2\ell} x}}}. \quad \text{(D-2)}$$

To determine κ_j, for $j = 1, \ldots, 2\ell$, we write first (D-2) as the ratio of two polynomials of x, $P_{2\ell}(x)$ and $Q_{2\ell-1}(x)$ of degrees 2ℓ and $2\ell - 1$ respectively, and seek their coefficients c_j,

$$\frac{F(x^2)}{x} = \frac{P_{2\ell}(x)}{Q_{2\ell-1}(x)} = \frac{c_{2\ell} x^{2\ell} + c_{2(\ell-1)} x^{2(\ell-1)} + \cdots + c_2 x^2 + c_0}{c_{2\ell-1} x^{2\ell-1} + c_{2\ell-3} x^{2\ell-3} + \ldots + c_1 x}. \quad \text{(D-3)}$$

We normalize the ratio by setting $c_0 = -1$.

Now suppose that we have measurements of F at $\lambda_k = x_k^2$, for $k = 1, \ldots, 2\ell$, and introduce the notation

$$\frac{F(x_k^2)}{x_k} = D_k. \quad \text{(D-4)}$$

We obtain from (D-3) the following linear system of equations for the coefficients

$$P_{2\ell}(x_k) - D_k Q_{2\ell-1}(x_k) = 0, \quad k = 1, \ldots, 2\ell, \quad \text{(D-5)}$$

or in matrix form

$$\begin{pmatrix} -D_1 x_1 & x_1^2 & -D_1 x_1^3 & \cdots & -D_1 x_1^{2\ell-1} & x_1^{2\ell} \\ -D_2 x_2 & x_2^2 & -D_2 x_2^3 & \cdots & -D_2 x_2^{2\ell-1} & x_2^{2\ell} \\ & & & \vdots & & \\ -D_{2\ell} x_{2\ell} & x_{2\ell}^2 & -D_{2\ell} x_{2\ell}^3 & \cdots & -D_{2\ell} x_{2\ell}^{2\ell-1} & x_{2\ell}^{2\ell} \end{pmatrix} \begin{pmatrix} c_1 \\ c_2 \\ \vdots \\ c_{2\ell} \end{pmatrix} = \mathbf{1}, \quad \text{(D-6)}$$

with right-hand side a vector of all ones. The coefficients are obtained by inverting the Vandermonde-like matrix in (D-6). In the special case of the rational interpolation (3-43), it is precisely a Vandermonde matrix. Since the condition number of such matrices grows exponentially with their size [Gautschi and Inglese 1988], the determination of $\{c_j\}_{j=1,\ldots,2\ell}$ is an ill-posed problem, as stated in Remark 1.

Once we have determined the polynomials $P_{2\ell}(x)$ and $Q_{2\ell-1}(x)$, we can obtain $\{\kappa_j\}_{j=1,\ldots 2\ell}$ by Euclidean polynomial division. Explicitly, let us introduce a new polynomial $P_{2\ell-2}(x) = \tilde{c}_{2\ell-2} x^{2\ell-2} + \cdots + \tilde{c}_0$, such that

$$\kappa_2 x + \cfrac{1}{\kappa_3 x + \cfrac{1}{\ddots\, \kappa_{2\ell-1} x + \cfrac{1}{\kappa_{2\ell} x}}} = \frac{Q_{2\ell-1}(x)}{P_{2\ell-2}(x)}, \qquad \text{(D-7)}$$

$$\kappa_1 x + \frac{P_{2\ell-2}(x)}{Q_{2\ell-1}(x)} = \frac{P_{2\ell}(x)}{Q_{2\ell-1}(x)}.$$

Equating powers of x we get

$$\kappa_1 = \frac{c_{2\ell}}{c_{2\ell-1}}, \qquad \text{(D-8)}$$

and the coefficients of the polynomial $P_{2\ell-2}(x)$ are determined by

$$\tilde{c}_{2j} = c_{2j} - \kappa_1 c_{2j-1}, \qquad j = 1, \ldots, \ell-1, \qquad \text{(D-9)}$$
$$\tilde{c}_0 = c_0. \qquad \text{(D-10)}$$

Then, we proceed similarly to get κ_2, and introduce a new polynomial $Q_{2\ell-3}(x)$ such that

$$\kappa_3 x + \cfrac{1}{\kappa_4 x + \cfrac{1}{\ddots\, \kappa_{2\ell-1} x + \cfrac{1}{\kappa_{2\ell} x}}} = \frac{P_{2\ell-2}(x)}{Q_{2\ell-3}(x)}, \quad \kappa_2 x + \frac{Q_{2\ell-3}(x)}{P_{2\ell-2}(x)} = \frac{Q_{2\ell-1}(x)}{P_{2\ell-2}(x)}.$$

Equating powers of x we get $\kappa_2 = c_{2\ell-1}/\tilde{c}_{2\ell-2}$ and the polynomial $Q_{2\ell-3}(x)$ and so on.

Appendix E. The Lanczos iteration

Let us write the Jacobi matrix (3-31) as

$$\tilde{A} = \begin{pmatrix} -a_1 & b_1 & 0 & \ldots & \ldots & 0 & 0 \\ b_1 & -a_2 & b_2 & 0 & \ldots & 0 & 0 \\ \ddots & \ddots & \ddots & \ddots & \ddots & \ddots & \ddots \\ 0 & 0 & \ldots & \ldots & 0 & b_{\ell-1} & -a_\ell \end{pmatrix}, \qquad \text{(E-1)}$$

where $-a_j$ are the negative diagonal entries and b_j the positive off-diagonal ones. Let also
$$-\Delta = \operatorname{diag}(-\delta_1^2, \ldots, -\delta_\ell^2) \tag{E-2}$$
be the diagonal matrix of the eigenvalues and
$$\tilde{Y}_j = \operatorname{diag}(\sqrt{\hat{\alpha}_1}, \ldots, \sqrt{\hat{\alpha}_\ell}) Y_j \tag{E-3}$$
the eigenvectors. They are orthonormal and the matrix $\tilde{Y} = (\tilde{Y}_1, \ldots, \tilde{Y}_\ell)$ is orthogonal
$$\tilde{Y}\tilde{Y}^T = \tilde{Y}^T\tilde{Y} = I. \tag{E-4}$$
The spectral theorem gives that $\tilde{A} = -\tilde{Y}\Delta\tilde{Y}^T$ or, equivalently,
$$\tilde{A}\tilde{Y} = -\tilde{Y}\Delta. \tag{E-5}$$

The Lanczos iteration [Trefethen and Bau 1997; Chu and Golub 2002] determines the entries a_j and b_j in $\tilde{A}$ by taking equations (E-5) row by row.

Let us denote the rows of $\tilde{Y}$ by
$$W_j = e_j^T \tilde{Y}, \quad j = 1, \ldots, \ell, \tag{E-6}$$
and observe from (E-5) that they are orthonormal
$$W_j W_q = \delta_{j,q} \tag{E-7}$$
We get for $j = 1$ that
$$\|W_1\|^2 = \sum_{j=1}^{\ell} \hat{\alpha}_1 Y_{1,j}^2 = \hat{\alpha}_1 \sum_{j=1}^{\ell} \xi_j = 1, \tag{E-8}$$
which determines $\hat{\alpha}_1$, and we can set
$$W_1 = \sqrt{\hat{\alpha}_1}(\sqrt{\xi_1}, \ldots, \sqrt{\xi_\ell}). \tag{E-9}$$
The first row in (E-5) gives
$$-a_1 W_1 + b_1 W_2 = -W_1 \Delta, \tag{E-10}$$
and using the orthogonality in (E-7), we obtain
$$a_1 = W_1 \Delta W_1^T = \sum_{j=1}^{\ell} \delta_j^2 \xi_j, \quad b_1 = \|a_1 W_1 - W_1 \Delta\|, \tag{E-11}$$
and
$$W_2 = b_1^{-1}(a_1 W_1 - W_1 \Delta). \tag{E-12}$$

The second row in (E-5) gives

$$b_1 W_1 - a_2 W_2 + b_2 W_3 = -W_2 \Delta, \tag{E-13}$$

and we can compute a_2 and b_2 as follows,

$$a_2 = W_2 \Delta W_2^T, \qquad b_2 = \|a_2 W_2 - W_2 \Delta - b_1 W_1\|. \tag{E-14}$$

Moreover,

$$W_3 = b_2^{-1} (a_2 W_2 - W_2 \Delta - b_1 W_1), \tag{E-15}$$

and the equation continues to the next row.

Once we have determined $\{a_j\}_{j=1,\ldots,\ell}$ and $\{b_j\}_{1,\ldots,\ell-1}$ with the Lanczos iteration described above, we can compute $\{\alpha_j, \hat{\alpha}_j\}_{j=1,\ldots,\ell}$. We already have from (E-8) that

$$\hat{\alpha}_1 = 1 / \sum_{j=1}^{\ell} \xi_j. \tag{E-16}$$

The remaining parameters are determined from the identities

$$a_j = \frac{1}{\hat{\alpha}_1 \alpha_1} \delta_{j,1} + (1 - \delta_{j,1}) \frac{1}{\hat{\alpha}_j} \left(\frac{1}{\alpha_j} + \frac{1}{\alpha_{j-1}} \right), \qquad b_j = \frac{1}{\alpha_j \sqrt{\hat{\alpha}_j \hat{\alpha}_{j+1}}}. \tag{E-17}$$

Appendix F. Proofs of Lemma 3 and Corollary 1

To prove Lemma 3, let $A^{(\bar{q})}$ be the tridiagonal matrix with entries defined by $\{\alpha_j^{(\bar{q})}, \hat{\alpha}_j^{(\bar{q})}\}_{j=1,\ldots,\ell}$, like in (3-29). It is the discretization of the operator in (3-58) with $\sigma \rightsquigarrow \sigma^{(\bar{q})}$. Similarly, let $A^{(o)}$ be the matrix defined by $\{\alpha_j^{(o)}, \hat{\alpha}_j^{(o)}\}_{j=1,\ldots,\ell}$, the discretization of the second derivative operator for conductivity $\sigma^{(o)}$. By the uniqueness of solution of the inverse spectral problem and (3-82)–(3-83), the matrices $A^{(\bar{q})}$ and $A^{(o)}$ are related by

$$\mathrm{diag}\left(\sqrt{\frac{\hat{\alpha}_1^{(\bar{q})}}{\hat{\alpha}_1^{(o)}}}, \ldots, \sqrt{\frac{\hat{\alpha}_\ell^{(\bar{q})}}{\hat{\alpha}_\ell^{(o)}}} \right) A^{(\bar{q})} \mathrm{diag}\left(\sqrt{\frac{\hat{\alpha}_1^{(o)}}{\hat{\alpha}_1^{(\bar{q})}}}, \ldots, \sqrt{\frac{\hat{\alpha}_\ell^{(o)}}{\hat{\alpha}_\ell^{(\bar{q})}}} \right) = A^{(o)} - \bar{q} I. \tag{F-1}$$

They have eigenvectors $Y_j^{(\bar{q})}$ and $Y_j^{(o)}$ respectively, related by

$$\mathrm{diag}\left(\sqrt{\hat{\alpha}_1^{(\bar{q})}}, \ldots, \sqrt{\hat{\alpha}_\ell^{(\bar{q})}} \right) Y_j^{(\bar{q})} = \mathrm{diag}\left(\sqrt{\hat{\alpha}_1^{(o)}}, \ldots, \sqrt{\hat{\alpha}_\ell^{(o)}} \right) Y_j^{(o)},$$

$$j = 1, \ldots, \ell, \tag{F-2}$$

and the matrix $\widetilde{Y}$ with columns (F-2) is orthogonal. Thus, we have the identity

$$(\widetilde{Y}\widetilde{Y}^T)_{11} = \hat{\alpha}_1^{(\bar{q})} \sum_{j=1}^{\ell} \xi^{(\bar{q})} = \hat{\alpha}_1^{(o)} \sum_{j=1}^{\ell} \xi^{(o)} = 1, \tag{F-3}$$

which gives $\hat{\alpha}_1^{(\bar{q})} = \hat{\alpha}_1^{(o)}$ by (3-82) or, equivalently

$$\sigma_1^{(\bar{q})} = \frac{\hat{\alpha}_1^{(\bar{q})}}{\hat{\alpha}_1^{(o)}} = 1 = \sigma^{(\bar{q})}(0). \tag{F-4}$$

Moreover, straightforward algebraic manipulations of the equations in (F-1) and definitions (3-84) give the finite difference equations (3-85) and (3-86). □

To prove Corollary 1, recall the definitions (3-84) and (3-87) to write

$$\sum_{p=1}^{j} \hat{\alpha}_j^{(\bar{q})} = \int_0^{\hat{\xi}_{j+1}^{(\bar{q})}} \sigma^{(\bar{q})}(\zeta)\, d\zeta = \sum_{p=1}^{j} \hat{\alpha}_p^{(o)} \sigma_p^{(\bar{q})} = \sum_{p=1}^{j} \hat{\alpha}_p^{(o)} \sigma^{(\bar{q})}(\zeta_p^{(o)}) + o(1). \tag{F-5}$$

Here we used the convergence result in Theorem 2 and denote by $o(1)$ a negligible residual in the limit $\ell \to \infty$. We have

$$\int_{\hat{\xi}_{j+1}^{(o)}}^{\hat{\xi}_{j+1}^{(\bar{q})}} \sigma^{(\bar{q})}(\zeta)\, d\zeta = \sum_{p=1}^{j} \hat{\alpha}_p^{(o)} \sigma^{(\bar{q})}(\zeta_p^{(o)}) - \int_0^{\hat{\xi}_{j+1}^{(o)}} \sigma^{(\bar{q})}(\zeta)\, d\zeta + o(1), \tag{F-6}$$

and therefore

$$|\hat{\xi}_{j+1}^{(\bar{q})} - \hat{\xi}_{j+1}^{(o)}| \leq C \left| \int_0^{\hat{\xi}_{j+1}^{(o)}} \sigma^{(\bar{q})}(\zeta)\, d\zeta - \sum_{p=1}^{j} \hat{\alpha}_p^{(o)} \sigma^{(\bar{q})}(\zeta_p^{(o)}) \right| + o(1),$$
$$C = 1/\min_\zeta \sigma^{(\bar{q})}(\zeta). \tag{F-7}$$

But the first term in the bound is just the error of the quadrature on the optimal grid, with nodes at $\zeta_j^{(o)}$ and weights $\hat{\alpha}_j^{(o)} = \hat{\xi}_{j+1}^{(o)} - \hat{\xi}_j^{(o)}$, and it converges to zero by the properties of the optimal grid stated in Lemma 2 and the smoothness of $\sigma^{(\bar{q})}(\zeta)$. Thus, we have shown that

$$|\hat{\xi}_{j+1}^{(\bar{q})} - \hat{\xi}_{j+1}^{(o)}| \to 0 \quad \text{as } \ell \to \infty, \tag{F-8}$$

uniformly in j. The proof for the primary nodes $\zeta_j^{(\bar{q})}$ is similar.

Appendix G. Perturbation analysis

It is shown in [Borcea et al. 2005, Appendix B] that the skew-symmetric matrix B given in (3-93) has eigenvalues $\pm i\delta_j$ and eigenvectors

$$\mathcal{Y}(\pm\delta_j) = \frac{1}{\sqrt{2}}(\mathcal{Y}_1(\delta_j), \pm i\widehat{\mathcal{Y}}_1(\delta_j), \ldots, \mathcal{Y}_\ell(\delta_j), \pm i\widehat{\mathcal{Y}}_\ell(\delta_j))^T, \qquad \text{(G-1)}$$

with

$$\begin{aligned}(\mathcal{Y}_1(\delta_j), \ldots, \mathcal{Y}_\ell(\delta_j))^T &= \mathrm{diag}(\hat{\alpha}_1^{\frac{1}{2}}, \ldots, \hat{\alpha}_\ell^{\frac{1}{2}}) Y_j, \\ (\widehat{\mathcal{Y}}_1(\delta_j), \ldots, \widehat{\mathcal{Y}}_\ell(\delta_j))^T &= \mathrm{diag}(\alpha_1^{\frac{1}{2}}, \ldots, \alpha_\ell^{\frac{1}{2}}) \widehat{Y}_j. \end{aligned} \qquad \text{(G-2)}$$

Here $Y_j = (Y_{1,j}, \ldots, Y_{\ell,j})^T$ are the eigenvectors of matrix A for the eigenvalues $-\delta_j^2$, and $\widehat{Y}_j = (\widehat{Y}_{1,j}, \ldots, \widehat{Y}_{\ell,j})^T$ is the vector with entries

$$\widehat{Y}_{p,j} = \frac{Y_{p+1,j} - Y_{p,j}}{\delta_j \alpha_j}. \qquad \text{(G-3)}$$

G.1. *Discrete Gel'fand–Levitan formulation.* It is difficult to carry a precise perturbation analysis of the recursive Lanczos iteration that gives B from the spectral data. We use instead the following discrete Gel'fand–Levitan formulation due to Natterer [1994].

Consider the "reference" matrix B^r, for an arbitrary, but fixed $r \in [0, 1]$, and define the lower triangular, transmutation matrix G, satisfying

$$EGB = EB^r G, \quad e_1^T G = e_1^T, \qquad \text{(G-4)}$$

where $E = I - e_{2\ell} e_{2\ell}^T$. Clearly, if $B = B^r$, then $G = G^r = I$, the identity. In general G is lower triangular and it is uniquely defined as shown with a Lanczos iteration argument in [Borcea et al. 2005, Section 6.2].

Next, consider the initial value problem

$$EB\boldsymbol{\phi}(\lambda) = i\lambda E\boldsymbol{\phi}(\lambda), \quad e_1^T \boldsymbol{\phi}(\lambda) = 1, \qquad \text{(G-5)}$$

which has a unique solution $\boldsymbol{\phi}(\lambda) \in \mathbb{C}^{2\ell}$, as shown in [Borcea et al. 2005, Section 6.2]. When $\lambda = \pm\delta_j$, one of the eigenvalues of B, we have

$$\boldsymbol{\phi}(\pm\delta_j) = \frac{\sqrt{2}}{\mathcal{Y}_1(\delta_j)} \mathcal{Y}(\pm\delta_j) = \sqrt{\frac{2}{\hat{\alpha}_1 \xi_j}} \mathcal{Y}(\pm\delta_j), \qquad \text{(G-6)}$$

and (G-5) holds even for E replaced by the identity matrix. The analogue of (G-5) for B^r is

$$EB^r \boldsymbol{\phi}^r(\lambda) = i\lambda E\boldsymbol{\phi}^r(\lambda), \quad e_1^T \boldsymbol{\phi}^r(\lambda) = 1, \qquad \text{(G-7)}$$

and, using (G-4) and the lower triangular structure of G, we obtain

$$\phi^r(\pm\delta_j) = G\phi(\pm\delta_j), \quad 1 \le j \le \ell. \tag{G-8}$$

Equivalently, in matrix form (G-8) and (G-6) give

$$\Phi^r = G\Phi = G\mathcal{Y}S, \tag{G-9}$$

where Φ is the matrix with columns (G-6), $\mathcal{Y}$ is the orthogonal matrix of eigenvectors of B with columns (G-1), and S is the diagonal scaling matrix

$$S = \sqrt{\frac{2}{\hat{\alpha}_1}} \mathrm{diag}\left(\xi_1^{-1/2}, \xi_1^{-1/2}, \ldots, \xi_\ell^{-1/2}, \xi_\ell^{-1/2}\right). \tag{G-10}$$

Then, letting

$$F = \Phi^r S^{-1} = G\mathcal{Y} \tag{G-11}$$

and using the orthogonality of $\mathcal{Y}$ we get

$$F\overline{F}^T = GG^T, \tag{G-12}$$

where the bar denotes complex conjugate. Moreover, Equation (G-4) gives

$$EB^r F = EB^r G\mathcal{Y} = EGB\mathcal{Y} = iEG\mathcal{Y}D = iEFD, \tag{G-13}$$

where $iD = i\,\mathrm{diag}\,(\delta_1, -\delta_1, \ldots, \delta_\ell, -\delta_\ell)$ is the matrix of the eigenvalues of B.

The discrete Gel'fand–Levitan inversion method proceeds as follows: Start with a known reference matrix B^r, for some $r \in [0, 1]$. The usual choice is $B^0 = B^{(o)}$, the matrix corresponding to the constant coefficient $\sigma^{(o)} \equiv 1$. Determine Φ^r from (G-7), with a Lanczos iteration as explained in [Borcea et al. 2005, Section 6.2]. Then, $F = \Phi^r S^{-1}$ is determined by the spectral data δ_j^r and ξ_j^r, for $1 \le j \le \ell$. The matrix G is obtained from (G-12) by a Cholesky factorization, and B follows by solving (G-4), using a Lanczos iteration.

G.2. Perturbation estimate.
Consider perturbations $d\delta_j = \Delta\delta_j dr$ and $d\xi_j = \Delta\xi_j dr$ of the spectral data of reference matrix B^r. We denote the perturbed quantities with a tilde as in

$$\tilde{D} = D^r + dD, \quad \tilde{S} = S^r + dS, \quad \tilde{\mathcal{Y}} = \mathcal{Y}^r + d\mathcal{Y}, \quad \tilde{F} = \mathcal{Y}^r + dF, \tag{G-14}$$

with D, S, $\mathcal{Y}$ and F defined above. Note that $F^r = \mathcal{Y}^r$, because $G^r = I$. Substituting (G-14) in (G-13) and using (G-7), we get

$$EB^r dF = iE\mathcal{Y}^r dD + iE\,dF\,D^r. \tag{G-15}$$

Now multiply by $\overline{\mathcal{Y}^r}^T$ on the right and use that $D^r \overline{\mathcal{Y}^r}^T = -i \overline{\mathcal{Y}^r}^T B^r$ to obtain that $dW - dF \overline{\mathcal{Y}^r}^T$ satisfies

$$EB^r dW - E \, dW \, B^r = i E \, \mathcal{Y}^r dD \, \overline{\mathcal{Y}^r}^T, \tag{G-16}$$

with initial condition

$$e_1^T dW = e_1^T dF \, \overline{\mathcal{Y}^r}^T$$

$$= \left(d\sqrt{\frac{\hat{\alpha}_1 \xi_1}{2}}, d\sqrt{\frac{\hat{\alpha}_1 \xi_1}{2}}, \ldots, d\sqrt{\frac{\hat{\alpha}_\ell \xi_\ell}{2}}, d\sqrt{\frac{\hat{\alpha}_\ell \xi_\ell}{2}} \right) \overline{\mathcal{Y}^r}^T. \tag{G-17}$$

Similarly, we get from (G-4) and $G^r = I$ that

$$E \, dB + E \, dG \, B^r = EB^r dG, \quad e_1^T dG = 0. \tag{G-18}$$

Furthermore, Equation (G-12) and $F^r = \mathcal{Y}^r$ give

$$dF \, \overline{\mathcal{Y}^r}^T + \mathcal{Y}^r \, dF = dW + \overline{dW}^T = dG + dG^T. \tag{G-19}$$

Equations (G-16)–(G-19) allow us to estimate $d\beta_j / \beta_j^r$. Indeed, consider the $j, j+1$ component in (G-18) and use (G-19) and the structure of G, dG and B^r to get

$$\frac{d\beta_j}{\beta_j^r} = dG_{j+1,j+1} - dG_{j,j} = dW_{j+1,j+1} - dW_{j,j},$$

$$j = 1, \ldots, 2\ell - 1. \tag{G-20}$$

The right-hand side is given by the components of dW satisfying (G-16)–(G-17) and calculated explicitly in [Borcea et al. 2005, Appendix C] in terms of the eigenvalues and eigenvectors of B^r. Then, the estimate

$$\sum_{j=1}^{2\ell-1} \left| \frac{d\beta_j}{\beta_j^r} \right| \leq C_1 dr \tag{G-21}$$

which is equivalent to (3-98) follows after some calculation given in [Borcea et al. 2005, Section 6.3], using the assumptions (3-74) on the asymptotic behavior of $\Delta \delta_j$ and $\Delta \xi_j$, i.e., of $\delta_j^r - \delta_j^{(o)} = r \Delta \delta_j$ and $\xi_j^r - \xi_j^{(o)} = r \Delta \xi_j$.

Acknowledgements

The work of L. Borcea was partially supported by the National Science Foundation grants DMS-0934594, DMS-0907746 and by the Office of Naval Research grant N000140910290. The work of F. Guevara Vasquez was partially supported by the National Science Foundation grant DMS-0934664. The work of A.V. Mamonov was partially supported by the National Science Foundation grants DMS-0914465 and DMS-0914840. LB, FGV and AVM were also partially supported by the National Science Foundation and the National Security Agency, during the Fall 2010 special semester on Inverse Problems at MSRI.

References

[Alessandrini 1988] G. Alessandrini, "Stable determination of conductivity by boundary measurements", *Appl. Anal.* **27**:1-3 (1988), 153–172. MR 89f:35195

[Alessandrini and Vessella 2005] G. Alessandrini and S. Vessella, "Lipschitz stability for the inverse conductivity problem", *Adv. in Appl. Math.* **35**:2 (2005), 207–241. MR 2006d:35289 Zbl 1095.35058

[Astala et al. 2005] K. Astala, L. Päivärinta, and M. Lassas, "Calderón's inverse problem for anisotropic conductivity in the plane", *Comm. Partial Differential Equations* **30**:1-3 (2005), 207–224. MR 2005k:35421

[Asvadurov et al. 2000] S. Asvadurov, V. Druskin, and L. Knizhnerman, "Application of the difference Gaussian rules to solution of hyperbolic problems", *J. Comput. Phys.* **158**:1 (2000), 116–135. MR 2000j:76109 Zbl 0955.65063

[Asvadurov et al. 2003] S. Asvadurov, V. Druskin, M. N. Guddati, and L. Knizhnerman, "On optimal finite-difference approximation of PML", *SIAM J. Numer. Anal.* **41**:1 (2003), 287–305. MR 2004e:65138 Zbl 1047.65063

[Asvadurov et al. 2007] S. Asvadurov, V. Druskin, and S. Moskow, "Optimal grids for anisotropic problems", *Electron. Trans. Numer. Anal.* **26** (2007), 55–81. MR 2009a:65286 Zbl 1124.65102

[Barceló et al. 2001] J. A. Barceló, T. Barceló, and A. Ruiz, "Stability of the inverse conductivity problem in the plane for less regular conductivities", *J. Differential Equations* **173**:2 (2001), 231–270. MR 2002h:35325 Zbl 0986.35126

[Ben Ameur and Kaltenbacher 2002] H. Ben Ameur and B. Kaltenbacher, "Regularization of parameter estimation by adaptive discretization using refinement and coarsening indicators", *J. Inverse Ill-Posed Probl.* **10**:6 (2002), 561–583. MR 2004b:65140 Zbl 1038.35156

[Ben Ameur et al. 2002] H. Ben Ameur, G. Chavent, and J. Jaffré, "Refinement and coarsening indicators for adaptive parametrization: application to the estimation of hydraulic transmissivities", *Inverse Problems* **18**:3 (2002), 775–794. MR 2003c:76118 Zbl 0999.35105

[Biesel et al. 2008] O. D. Biesel, D. V. Ingerman, J. A. Morrow, and W. T. Shore, "Layered networks, the discrete laplacian and a continued fraction identity", 2008, Available at http://www.math.washington.edu/~reu/papers/2008/william/layered.pdf.

[Borcea 2002] L. Borcea, "Electrical impedance tomography", *Inverse Problems* **18**:6 (2002), R99–R136. MR 1955896 Zbl 1031.35147

[Borcea and Druskin 2002] L. Borcea and V. Druskin, "Optimal finite difference grids for direct and inverse Sturm–Liouville problems", *Inverse Problems* **18**:4 (2002), 979–1001. MR 2003h:65097 Zbl 1034.34028

[Borcea et al. 2005] L. Borcea, V. Druskin, and L. Knizhnerman, "On the continuum limit of a discrete inverse spectral problem on optimal finite difference grids", *Comm. Pure Appl. Math.* **58**:9 (2005), 1231–1279. MR 2006d:65078 Zbl 1079.65084

[Borcea et al. 2008] L. Borcea, V. Druskin, and F. Guevara Vasquez, "Electrical impedance tomography with resistor networks", *Inverse Problems* **24**:3 (2008), 035013. MR 2009e:78027 Zbl 1147.78002

[Borcea et al. 2010a] L. Borcea, V. Druskin, and A. V. Mamonov, "Circular resistor networks for electrical impedance tomography with partial boundary measurements", *Inverse Problems* **26**:4 (2010), 045010. MR 2011f:65239 Zbl 1191.78050

[Borcea et al. 2010b] L. Borcea, V. Druskin, A. V. Mamonov, and F. Guevara Vasquez, "Pyramidal resistor networks for electrical impedance tomography with partial boundary measurements", *Inverse Problems* **26**:10 (2010), 105009. MR 2011f:65238 Zbl 1201.65016

[Borcea et al. 2011] L. Borcea, F. Guevara Vasquez, and A. V. Mamonov, "Study of noise effects in electrical impedance tomography with resistor networks", preprint, 2011. arXiv 1105.1183

[Brown and Uhlmann 1997] R. M. Brown and G. A. Uhlmann, "Uniqueness in the inverse conductivity problem for nonsmooth conductivities in two dimensions", *Comm. Partial Differential Equations* **22**:5-6 (1997), 1009–1027. MR 98f:35155 Zbl 0884.35167

[Chadan et al. 1997] K. Chadan, D. Colton, L. Päivärinta, and W. Rundell, *An introduction to inverse scattering and inverse spectral problems*, SIAM Monographs on Mathematical Modeling and Computation, SIAM, Philadelphia, 1997. MR 1445771 (98a:34017)

[Chu and Golub 2002] M. T. Chu and G. H. Golub, "Structured inverse eigenvalue problems", *Acta Numer.* **11** (2002), 1–71. MR 2005b:65040 Zbl 1105.65326

[Coleman and McLaughlin 1993] C. F. Coleman and J. R. McLaughlin, "Solution of the inverse spectral problem for an impedance with integrable derivative, I and II", *Comm. Pure Appl. Math.* **46**:2 (1993), 145–212. MR 93m:34017

[Colin de Verdière 1994] Y. Colin de Verdière, "Réseaux électriques planaires, I", *Comment. Math. Helv.* **69**:3 (1994), 351–374. MR 96k:05131 Zbl 0816.05052

[Colin de Verdière et al. 1996] Y. Colin de Verdière, I. Gitler, and D. Vertigan, "Réseaux électriques planaires, II", *Comment. Math. Helv.* **71**:1 (1996), 144–167. MR 98a:05054 Zbl 0853.05074

[Curtis and Morrow 2000] E. B. Curtis and J. A. Morrow, *Inverse problems for electrical networks*, World Scientific, 2000.

[Curtis et al. 1994] E. Curtis, E. Mooers, and J. Morrow, "Finding the conductors in circular networks from boundary measurements", *RAIRO Modél. Math. Anal. Numér.* **28**:7 (1994), 781–814. MR 96i:65110 Zbl 0820.94028

[Curtis et al. 1998] E. B. Curtis, D. Ingerman, and J. A. Morrow, "Circular planar graphs and resistor networks", *Linear Algebra Appl.* **283**:1-3 (1998), 115–150. MR 99k:05096 Zbl 0931.05051

[Druskin 1982] V. Druskin, "The unique solution of the inverse problem of electrical surveying and electrical well-logging for piecewise-continuous conductivity", *Izv. Earth Physics* **18** (1982), 51–3.

[Druskin 1985] V. Druskin, "On uniqueness of the determination of the three-dimensional underground structures from surface measurements with variously positioned steady-state or monochromatic field sources", *Sov. Phys.–Solid Earth* **21** (1985), 210–4.

[Druskin and Knizhnerman 2000a] V. Druskin and L. Knizhnerman, "Gaussian spectral rules for the three-point second differences. I. A two-point positive definite problem in a semi-infinite domain", *SIAM J. Numer. Anal.* **37**:2 (2000), 403–422. MR 2000i:65163

[Druskin and Knizhnerman 2000b] V. Druskin and L. Knizhnerman, "Gaussian spectral rules for second order finite-difference schemes", *Numer. Algorithms* **25**:1-4 (2000), 139–159. MR 2002c:65177 Zbl 0978.65092

[Druskin and Moskow 2002] V. Druskin and S. Moskow, "Three-point finite-difference schemes, Padé and the spectral Galerkin method, I: One-sided impedance approximation", *Math. Comp.* **71**:239 (2002), 995–1019. MR 2003f:65142

[Gautschi and Inglese 1988] W. Gautschi and G. Inglese, "Lower bounds for the condition number of Vandermonde matrices", *Numer. Math.* **52**:3 (1988), 241–250. MR 89b:65108 Zbl 0646.15003

[Gel'fand and Levitan 1951] I. M. Gel'fand and B. M. Levitan, "On the determination of a differential equation from its spectral function", *Izvestiya Akad. Nauk SSSR. Ser. Mat.* **15** (1951), 309–360. In Russian. MR 13,558f

[Godunov and Ryabenki 1964] S. K. Godunov and V. S. Ryabenki, *Theory of difference schemes: an introduction*, North-Holland, Amsterdam, 1964. MR 31 #5346 Zbl 0116.33102

[Guevara Vasquez 2006] F. Guevara Vasquez, *On the parametrization of ill-posed inverse problems arising from elliptic partial differential equations*, Ph.D. thesis, Rice University, Houston, TX, 2006.

[Hochstadt 1973] H. Hochstadt, "The inverse Sturm–Liouville problem", *Comm. Pure Appl. Math.* **26** (1973), 715–729. MR 48 #8944 Zbl 0281.34015

[Imanuvilov et al. 2008] O. Imanuvilov, G. Uhlmann, and M. Yamamoto, "Global uniqueness from partial Cauchy data in two dimensions", preprint, 2008. arXiv 0810.2286

[Ingerman 2000] D. V. Ingerman, "Discrete and continuous Dirichlet-to-Neumann maps in the layered case", *SIAM J. Math. Anal.* **31**:6 (2000), 1214–1234. MR 2001g:39043 Zbl 0972.35183

[Ingerman and Morrow 1998] D. Ingerman and J. A. Morrow, "On a characterization of the kernel of the Dirichlet-to-Neumann map for a planar region", *SIAM J. Math. Anal.* **29**:1 (1998), 106–115. MR 99c:35258 Zbl 0924.35193

[Ingerman et al. 2000] D. Ingerman, V. Druskin, and L. Knizhnerman, "Optimal finite difference grids and rational approximations of the square root. I. Elliptic problems", *Comm. Pure Appl. Math.* **53**:8 (2000), 1039–1066. MR 2001d:65140

[Isaacson 1986] D. Isaacson, "Distinguishability of conductivities by electric current computed tomography", *IEEE transactions on medical imaging* **5**:2 (1986), 91–95.

[Kac and Krein 1968] I. S. Kac and M. G. Krein, pp. 648–737 (Appendix II) in Дискретные и непрерывные граничные задачи, by F. V. Atkinson, Mir, Moscow, 1968. In Russian; translated in *Amer. Math. Soc. Transl* **103**:2 (1974), 19–102. Zbl 0164.29702

[Kohn and Vogelius 1984] R. Kohn and M. Vogelius, "Determining conductivity by boundary measurements", *Comm. Pure Appl. Math.* **37**:3 (1984), 289–298. MR 85f:80008 Zbl 0586.35089

[Kohn and Vogelius 1985] R. V. Kohn and M. Vogelius, "Determining conductivity by boundary measurements, II: Interior results", *Comm. Pure Appl. Math.* **38**:5 (1985), 643–667. MR 86k:35155

[Lavrentiev and Shabat 1987] M. A. Lavrentiev and B. V. Shabat, Методы теории функций комплексного переменного, Nauka, Moscow, 1987. Spanish translation: *Métodos de la teoría de las funciones de una variable compleja*, Mir, Moscow, 1991.

[Levitan 1987] B. M. Levitan, *Inverse Sturm–Liouville problems*, VSP, Zeist, 1987. MR 89b:34001 Zbl 0749.34001

[MacMillan et al. 2004] H. R. MacMillan, T. A. Manteuffel, and S. F. McCormick, "First-order system least squares and electrical impedance tomography", *SIAM J. Numer. Anal.* **42**:2 (2004), 461–483. MR 2005h:65195 Zbl 1073.78020

[Mamonov 2009] A. V. Mamonov, "Resistor network approaches to the numerical solution of electrical impedance tomography with partial boundary measurements", Master's thesis, Rice University, Houston, TX, 2009.

[Mamonov 2010] A. V. Mamonov, *Resistor networks and optimal grids for the numerical solution of electrical impedance tomography with partial boundary measurements*, Ph.D. thesis, Rice University, Houston, TX, 2010, Available at http://search.proquest.com/docview/756838036.

[Mandache 2001] N. Mandache, "Exponential instability in an inverse problem for the Schrödinger equation", *Inverse Problems* **17**:5 (2001), 1435–1444. MR 2002h:35339 Zbl 0985.35110

[Marchenko 1986] V. A. Marchenko, *Sturm–Liouville operators and applications*, Birkhäuser, Boston, 1986. Revised ed., AMS Chelsea, 2011. MR 2012b:34002 Zbl 0592.34011

[McLaughlin and Rundell 1987] J. R. McLaughlin and W. Rundell, "A uniqueness theorem for an inverse Sturm–Liouville problem", *J. Math. Phys.* **28**:7 (1987), 1471–1472. MR 88k:34026 Zbl 0633.34016

[Nachman 1996] A. I. Nachman, "Global uniqueness for a two-dimensional inverse boundary value problem", *Ann. of Math.* (2) **143**:1 (1996), 71–96. MR 96k:35189 Zbl 0857.35135

[Natanson 1955] I. P. Natanson, *Theory of functions of a real variable*, vol. 1, Ungar, New York, 1955. MR 16,804c Zbl 0064.29102

[Natterer 1994] F. Natterer, "A discrete Gelfand–Levitan theory", Institut für Numerische und instrumentelle Mathematik, Münster, 1994, Available at http://wwwmath.uni-muenster.de/num/Preprints/1998/natterer_3/index.html.

[Nikishin and Sorokin 1991] E. M. Nikishin and V. N. Sorokin, *Rational approximations and orthogonality*, Translations of Mathematical Monographs **92**, American Mathematical Society, Providence, RI, 1991. MR 92i:30037 Zbl 0733.41001

[Pöschel and Trubowitz 1987] J. Pöschel and E. Trubowitz, *Inverse spectral theory*, Pure and Applied Mathematics **130**, Academic Press, Boston, 1987. MR 89b:34061 Zbl 0623.34001

[Quarteroni and Valli 1999] A. Quarteroni and A. Valli, *Domain decomposition methods for partial differential equations*, Oxford Univ. Press, New York, 1999. MR 2002i:65002 Zbl 0931.65118

[Reich 1976] E. Reich, "Quasiconformal mappings of the disk with given boundary values", pp. 101–137 in *Advances in complex function theory* (College Park, MD, 1973–1974), Lecture Notes in Math. **505**, Springer, Berlin, 1976. MR 54 #7776 Zbl 0328.30018

[Strebel 1976] K. Strebel, "On the existence of extremal Teichmueller mappings", *J. Analyse Math.* **30** (1976), 464–480. MR 55 #12912 Zbl 0334.30013

[Sylvester 1990] J. Sylvester, "An anisotropic inverse boundary value problem", *Comm. Pure Appl. Math.* **43**:2 (1990), 201–232. MR 90m:35202 Zbl 0709.35102

[Trefethen and Bau 1997] L. N. Trefethen and D. Bau, III, *Numerical linear algebra*, Society for Industrial and Applied Mathematics (SIAM), Philadelphia, PA, 1997. MR 98k:65002 Zbl 0874.65013

borcea@caam.rice.edu	*Computational and Applied Mathematics, Rice University, MS 134, Houston, TX 77005-1892, United States*
druskin1@slb.com	*Schlumberger Doll Research Center, One Hampshire St., Cambridge, MA 02139-1578, United States*
fguevara@math.utah.edu	*Department of Mathematics, University of Utah, 155 S 1400 E RM 233, Salt Lake City, UT 84112-0090, United States*
mamonov@ices.utexas.edu	*Institute for Computational Engineering and Sciences, University of Texas at Austin, 201 East 24th St., Stop C0200, Austin, TX 78712, United States*

The Calderón inverse problem in two dimensions

COLIN GUILLARMOU AND LEO TZOU

We review recent progress on the two-dimensional Calderón inverse problem, that is, the uniqueness of coefficients of an elliptic equation on a domain of $\mathbb{C}$ (or a surface with boundary) from Cauchy data at the boundary.

1. The Calderón problem

The global uniqueness for inverse boundary value problems of elliptic equations at fixed frequency in dimension $n=2$ is quite particular and remained open for many years. Now these problems are well understood, with a variety of results appearing in the last 10 or 15 years, essentially all using the complex structure $\mathbb{R}^2 \simeq \mathbb{C}$ and $\bar{\partial}$-techniques. This is therefore a good time to write a short survey on the subject. Although we tried to cover as much as we can, we do not pretend to be exhaustive and we apologize in advance for any forgotten reference, which is not a decision made on purpose but rather a sign of our ignorance. We have decided to give more details about the proofs of recent results based on Bukhgeim's idea [2008], for there is already a survey by Uhlmann [2003] on the subject about older results. The results of Astala, Lassas, and Päivärinta using quasiconformal methods are the subject of a separate survey in this volume [Astala et al. 2013]. Finally, we do not discuss questions about stability and reconstruction, nor inverse scattering results.

1A. *The inverse problem for the conductivity.* Let $\Omega \subset \mathbb{C}$ be a bounded domain with boundary (say smooth boundary) and let $\gamma \in L^\infty(\Omega, S^2_+(\Omega))$ be a field of positive definite symmetric matrices on Ω. The Dirichlet-to-Neumann map is the operator

$$\mathcal{N}_\gamma : H^{\frac{1}{2}}(\partial\Omega) \to H^{-\frac{1}{2}}(\partial\Omega)$$

defined by

$$\langle f_1, \mathcal{N}_\gamma f_2 \rangle := \int_\Omega \gamma \nabla u_1 . \nabla u_2,$$

where $f_1, f_2 \in H^{\frac{1}{2}}(\partial\Omega)$, $\langle \cdot, \cdot \rangle$ is the pairing between $H^{\frac{1}{2}}(\partial\Omega)$ and $H^{-\frac{1}{2}}(\partial\Omega)$, u_2 is the $H^1(\Omega)$ solution of the elliptic equation

$$\text{div}(\gamma \nabla u) = 0, \quad u|_{\partial\Omega} = f_2 \tag{1}$$

and u_1 is any H^1 function with trace f_1 on $\partial\Omega$. Equivalently, $\mathcal{N}_\gamma f_2 = \gamma \nabla u_2 . \nu$, where u is the solution of (1) and ν is the normal outward pointing vector field to the boundary. The operator $\mathcal{N}_\gamma$ is a nonlocal operator, in fact it is an elliptic pseudodifferential operator of order 1 on $\partial\Omega$, at least when γ is smooth. Its dependence on γ is nonlinear. The problem asked by Calderón [1980] is the following:

$$\text{Is the map } \gamma \to \mathcal{N}_\gamma \text{ injective?} \tag{Q1}$$

The conductivity is called *isotropic* when $\gamma = \gamma(x)\text{Id}$ for some function $\gamma(x)$. If Ω is an inhomogeneous body with conductivity γ, then $\mathcal{N}_\gamma f$ is the current flux at the boundary corresponding to a voltage potential f on $\partial\Omega$. The Dirichlet-to-Neumann operator represents the information which can be obtained from static voltage and current measurements at the boundary, and (Q1) is a question about uniqueness of a media giving rise to a given (infinite) set of measurements. The graph of $\mathcal{N}_\gamma$ is called the *Cauchy data space*.

1B. *The inverse problem for metrics.* An alternative and quite similar problem is as follows. Let M be a surface with boundary and g is a Riemannian metric on M, one can define the Dirichlet-to-Neumann operator associated to (M, g) by

$$\mathcal{N}_{(M,g)} : H^{\frac{1}{2}}(\partial\Omega) \to H^{-\frac{1}{2}}(\partial\Omega), \quad f \mapsto \partial_\nu u|_{\partial M}$$

where u is the unique solution of the elliptic equation

$$\Delta_g u = 0, \quad u|_{\partial M} = f,$$

here $\Delta_g = d^*d$ where d is the exterior derivative and d^* its adjoint for the Riemannian L^2 product $\langle u, v \rangle = \int_M u\bar{v} \, dv_g$. Then we ask

$$\text{Is the map } (M, g) \to \mathcal{N}_{(M,g)} \text{ injective ?} \tag{Q2}$$

Here M runs over the set of Riemannian surfaces with a given fixed boundary $\partial M = N$.

1C. *Gauge invariance.* The obvious answer one can give for both (Q1) and (Q2) is "No". Indeed, if $\psi : \Omega \to \Omega$ and $\varphi : M \to \varphi(M)$ are two diffeomorphisms which satisfy $\psi|_{\partial\Omega} = \text{Id}$ and $\varphi|_{\partial M} = \text{Id}$, then

$$\mathcal{N}_{\psi_*\gamma} = \mathcal{N}_\gamma, \quad \mathcal{N}_{(M,g)} = \mathcal{N}_{(\varphi(M), \varphi_*g)}$$

where φ_*g is the pushforward of the metric g by φ and
$$\psi_*\gamma(x) := \left(\frac{d\psi^t \gamma d\psi}{|\det d\psi|}\right)(\psi^{-1}(x)).$$
In fact, for the metric case, there is another invariance, which comes from the conformal covariance of the Laplacian in 2 dimensions: since $\Delta_g = e^{2\omega}\Delta_{e^{2\omega}g}$ for all smooth function ω, one easily deduces that for all function ω which satisfies $\omega|_{\partial M} = 0$, then
$$\mathcal{N}_{(M,g)} = \mathcal{N}_{(\varphi(M),e^{2\omega}\varphi_*g)}.$$
The good questions to ask are then

Does $\mathcal{N}_{\gamma_1} = \mathcal{N}_{\gamma_2}$ imply $\exists \psi : \Omega \to \Omega$ (diffeo) s.t. $\psi|_{\partial\Omega} = \mathrm{Id}$, $\psi_*\gamma_1 = \gamma_2$? (Q1')

and

Do $\mathcal{N}_{(M_1,g_1)} = \mathcal{N}_{(M_2,g_2)}$ and $\partial M_1 = \partial M_2$ imply $\exists \psi M_1 \to M_2$ (diffeo) and $\omega : M_2 \to \mathbb{R}$ s.t. $\psi|_{\partial M_1} = \mathrm{Id}$ and $\psi_*g_1 = e^{2\omega}g_2$? (Q2')

1D. *The inverse problem for potentials.* We conclude by another similar problem for Schrödinger operators. If (M, g) is a fixed compact Riemannian surface with boundary, and $V \in L^\infty(M)$ is a potential such that $\Delta_g + V$ has no element in its kernel vanishing at ∂M, then the Dirichlet-to-Neumann operator associated to V is defined as before by
$$\mathcal{N}_V : H^{\frac{1}{2}}(\partial M) \to H^{-\frac{1}{2}}(\partial M), \quad f \mapsto \partial_\nu u|_{\partial M}$$
where u is the unique solution of the elliptic equation
$$(\Delta_g + V)u = 0, \quad u|_{\partial M} = f.$$
The uniqueness question in this case is

$$\text{Does } \mathcal{N}_{V_1} = \mathcal{N}_{V_2} \text{ imply } V_2 = V_1? \qquad (Q3)$$

1E. *Relation between isotropic conductivity and potentials.* There is an easy remark that one can do about the relation between the isotropic conductivity problem and the potential problem: indeed, setting $u = \gamma^{-1/2}v$ shows that u is a solution of $\mathrm{div}(\gamma \nabla u) = 0$ if and only if v is a solution of $(\Delta + V_\gamma)v = 0$ with
$$V_\gamma = -\frac{\Delta \gamma^{1/2}}{\gamma^{1/2}}.$$
Therefore, if $\gamma \in W^{2,\infty}(\Omega)$ and if γ is supposed to be known at the boundary, a resolution of the problem (Q3) implies the resolution of (Q1) if the conductivity is isotropic.

1F. *Other related problems.* Another natural problem is to identify up to gauge a magnetic field or a Hermitian connection $\nabla^X = d + iX$ on a complex line bundle E over a surface with boundary (or more simply a domain) from the Cauchy data space of the connection Laplacian $L_X := (\nabla^X)^* \nabla^X$. More generally one can add a potential V to L_X and try to identify X up to gauge and V from the Cauchy data space of $L_{X,V} = L_X + V$. The question also makes sense for connections on complex vector bundles, where one has to deal with elliptic systems, and for Dirac type operators.

1G. *Partial data problems.* Practically, there are many situations where we have measurements of the currents on only a small piece $\Gamma \subset \partial M$ of the boundary, it is therefore important to see what can be obtained from the Dirichlet-to-Neumann operator acting on functions supported in Γ. For instance, a natural question is to take two open sets Γ_+, Γ_- of the boundary, and see if the partial Cauchy data set

$$\{(u|_{\Gamma_+}, \partial_\nu u|_{\Gamma_-}); (\Delta + V)u = 0, \ u \in H^1(M), \ u|_{\partial M \setminus \Gamma_+} = 0\}$$

determines the potential.

1H. *Why these problems are not simple.* Let H_h be a family of elliptic operators of order 2, depending on a small parameter $h \in (0, h_0)$, and of the form $H_h = h^2 H + V_h$ where V_h is a family of real potentials depending smoothly in $h \in [0, h_0)$ and H an elliptic self-adjoint operator of order 2 with principal symbol p. The semiclassical theory tells us that, when there is a characteristic set $\{(m, \xi) \in T^*\mathbb{R}^2 \setminus \{\xi = 0\}; \ p(m, \xi) + V_0(m) = 0\} \neq \varnothing$, the solutions of $H_h u = 0$ are microlocalized near this set and oscillating with frequency of order $1/h$ as $h \to 0$, moreover the microlocal concentration is characterized by the flow of the Hamiltonian vector field associated to the Hamiltonian $p + V_0$. In particular, if one know something about this concentration on the boundary of the domain, one can expect to propagate it through this flow to say something in the interior of the domain. A typical example would be if we know the Dirichlet-to-Neumann operators $\mathcal{N}_{(M,g)}(\lambda)$ for the equation $(\Delta_g - \lambda^2)u = 0$ for all $\lambda > 0$, since one could set $\lambda = 1/h$. In the Calderón problem, we only know an information at 0 (or fixed) frequency, which makes the problem much more complicated. In a way, the solutions of this problem are often based on complexifying the frequencies to see high frequencies phenomena.

Notation. We shall use the complex variable $z = x + iy$ for $\mathbb{C}$ and the variable $w = (x, y)$ for $\mathbb{R}^2$ in what follows.

2. Local uniqueness

2A. Kohn–Vogelius local uniqueness. For all the problems above, the Dirichlet-to-Neumann operator is a nonlocal operator with singular integral kernel and the singularities of its kernel at the diagonal contain local information about the coefficients of the elliptic equation in the interior. Actually, it is shown in all the cases above that it determines the Taylor expansion of the coefficients at the boundary, say when those coefficients are smooth. This was apparently first observed by Kohn and Vogelius [1984]

Theorem 2.1 [Kohn and Vogelius 1984]. *Let γ_1, γ_2 be two smooth isotropic conductivities on a smooth domain Ω, and assume that $\mathcal{N}_{\gamma_1} = \mathcal{N}_{\gamma_2}$. Then for all $k \geq 0$, we have $\partial_\nu^k \gamma_1 = \partial_\nu^k \gamma_2$ everywhere on $\partial\Omega$.*

In fact, the proof is a local determination and the assumption that $\mathcal{N}_{\gamma_1} = \mathcal{N}_{\gamma_2}$ can be replaced by

$$\langle \mathcal{N}_{\gamma_1} f, f \rangle = \langle \mathcal{N}_{\gamma_2} f, f \rangle \quad \text{for all } f \in C_0^\infty(\Gamma)$$

where $\Gamma \subset \partial\Omega$ is an open set, and this would show that $\partial_\nu^k \gamma_1 = \partial_\nu^k \gamma_2$ on Γ. Notice that this allows to say that the Dirichlet-to-Neumann map determines real analytic isotropic conductivities by analytic continuation from the boundary.

Idea of proof. The idea is to construct solutions u_h depending on a small parameter $h > 0$ such that their boundary values f_h are supported in an h-neighborhood of a point $x_0 \in \partial\Omega$, and that

$$\|f_h\|_{H^{1/2+\ell}(\partial\Omega)} = O(h^{-\ell}) \quad \text{for any } \ell > -M$$

for some $M > 0$ chosen arbitrarily large. Then one can show that if $U \subset \Omega$ is an open set with $d(\partial\Omega, \partial U) > 0$ and W is an open neighborhood of x_0,

$$\|\nabla u_h\|_{L^2(U)} = O(h^M) \quad \text{and} \quad \|\rho^m \nabla u_h\|_{L^2(W)} = O(h^{-(1+\epsilon)m}) \quad (2)$$

for some small $\epsilon > 0$ if $\rho(x) = \operatorname{dist}(x, \partial\Omega)$. Assuming that $\partial_\nu^m \gamma_1 \neq \partial_\nu^m \gamma_2$ near x_0 for some $m \in \mathbb{N}$, then by writing the Taylor expansion in normal coordinates to the boundary, this means that either $\gamma_1 - \gamma_2 \geq C\rho^m$ or $\gamma_2 - \gamma_1 \geq C\rho^m$ in a neighborhood of x_0, for some $C > 0$. Let us assume the first case. From the estimates (2) above and taking $M \gg m$ and h very small, this gives

$$\int_\Omega \gamma_1 |\nabla u_h|^2 \geq \int_W \gamma_1 |\nabla u_h|^2 \geq \int_W \gamma_2 |\nabla u_h|^2 + O(h^{-(1+\epsilon)m}) \geq \int_\Omega \gamma_2 |\nabla u_h|^2$$

which contradicts $\langle \mathcal{N}_{\gamma_1} f_h, f_h \rangle = \langle \mathcal{N}_{\gamma_2} f, f \rangle$. $\square$

2B. *Further results*. The argument of Kohn–Vogelius can be extended easily to recover k derivatives of $\gamma_1 - \gamma_2$ when $\gamma_1 - \gamma_2$ has $1 + k$ derivatives (by some L^2-Sobolev embeddings). The identification of the boundary value of an isotropic conductivity has been improved (in terms of regularity) for smooth domains by Sylvester and Uhlmann [1988] to continuous conductivities with an estimate

$$\|\gamma_1 - \gamma_2\|_{L^\infty(\partial\Omega)} \leq C \|\mathcal{N}_{\gamma_1} - \mathcal{N}_{\gamma_2}\|_{H^{\frac{1}{2}}(\partial\Omega) \to H^{-\frac{1}{2}}(\partial\Omega)}.$$

The uniqueness is also local in the sense that one only needs to know the Dirichlet-to-Neumann map in an open set to determine the conductivity at a point of this set. Alessandrini [1990] proved that for Lipschitz domains, if γ_j are Lipschitz and $\gamma_1 - \gamma_2$ is C^k in a neighborhood of $\partial\Omega$ then $\mathcal{N}_{\gamma_1} = \mathcal{N}_{\gamma_2}$ implies $\partial^\alpha(\gamma_1 - \gamma_2) = 0$ on $\partial\Omega$ for all $|\alpha| \leq k$. Brown [2001] proved a result for continuous conductivities on Lipschitz domains. For smooth metrics, Lee and Uhlmann [1989] proved that the full symbol of the Dirichlet-to-Neumann operator (as a classical pseudodifferential operator of order 1) determines the Taylor expansion to all order of the metric at the boundary.

3. The method of complex geometric optic solutions

The first approach to recover a conductivity from boundary data was to reduce the problem to the potential problem, as explained above and to use particular solutions of the Schrödinger equation $(\Delta + V)u = 0$ where V is a real potential of the form $-\Delta \gamma^{\frac{1}{2}}/\gamma^{\frac{1}{2}}$. The advantage of reducing the problem to $\Delta + V$ is that one has to identify a term of order 0 in the equation while for the conductivity problem, γ is contained in the principal symbol of the operator. The first observation one can make using this fact is the following: if u_1 and u_2 are solutions of

$$(\Delta + V_j)u_j = 0, \quad u_j|_{\partial\Omega} = f_j, \quad j = 1, 2$$

in a domain $\Omega \in \mathbb{R}^2$, then Green's formula yields the integral identity

$$\int_\Omega (V_2 - V_1)u_1 u_2 = \int_\Omega \Delta u_1 . u_2 - u_1 . \Delta u_2$$
$$= \int_{\partial\Omega} \partial_\nu u_1 . u_2 - u_1 . \partial_\nu u_2 = \int_{\partial\Omega} \mathcal{N}_{V_1} f_1 . f_2 - f_1 \mathcal{N}_{V_2} f_2;$$

that is,

$$\int_\Omega (V_2 - V_1)u_1 u_2 = \int_{\partial\Omega} (\mathcal{N}_{V_1} - \mathcal{N}_{V_2}) f_1 . f_2, \tag{3}$$

where we have used the symmetry of the Dirichlet-to-Neumann map when the potential is real, which is a consequence of Green's formula again: for any solutions w_1, w_2 of $(\Delta + V)w = 0$ with boundary values f_1, f_2, we have

$$0 = \int_\Omega (\Delta + V)w_1.u_2 - w_1.(\Delta + V)w_2 = \int_{\partial\Omega} \mathcal{N}_V f_1.f_2 - f_1.\mathcal{N}_V f_2 .$$

The integral identity (3) shows that if $\mathcal{N}_{V_1} = \mathcal{N}_{V_2}$, then $V_1 - V_2$ is orthogonal in $L^2(\Omega)$ to the product of solutions of $(\Delta+V_1)u=0$ with solutions of $(\Delta+V_2)u = 0$. The idea initiated by Calderón was then to construct certain families of solutions with contain high oscillations to give enough information on $V_1 - V_2$ when one integrates against those.

3A. *The linearized Calderón problem.* Calderón [1980] considered the linearized problem at the potential $V = 0$ as follows: if V_t is a one parameter family of potentials ($t \in (-\epsilon, \epsilon)$) such that $V_0 = 0$ and $\partial_t \mathcal{N}_{V_t}|_{t=0} = 0$, then one has, for all u_t, v_t satisfying $(\Delta + V_t)u_t = (\Delta + V_t)v_t = 0$ with respective fixed boundary value f, g,

$$\int_\Omega \nabla u_t.\nabla v_t + V_t u_t.v_t = \int_{\partial\Omega} \mathcal{N}_t f.g$$

therefore differentiating at $t = 0$ (which we denote by a dot)

$$0 = \int_\Omega \nabla u_0.\nabla \dot{v}_0 + \dot{V}_0 u_0.v_0 = \int_{\partial\Omega} \partial_\nu f.\dot{v}_0|_{\partial\Omega} + \int_\Omega \dot{V}_0 u_0.v_0 = \int_\Omega \dot{V}_0 u_0.v_0.$$

The element $\dot{V}_0$ in the kernel of the linearization of $V \to \mathcal{N}_V$ is orthogonal to the product of harmonic functions. Calderón's idea was to use particular solutions, in fact exponentials of linear holomorphic functions (recall that $z = x + iy \in \mathbb{C}$ denotes the complex variable)

$$u_0(z) = e^{z\zeta}, \quad v_0(z) = e^{\overline{z\zeta}} \quad \text{with } \zeta \in \mathbb{C}.$$

These are clearly harmonic, since holomorphic and antiholomorphic, and therefore one obtains (recall the notation $w = (x, y) \in \mathbb{R}^2$)

$$\int_\Omega e^{2i \text{Im}(z\zeta)} \dot{V}_0(z) = 0 = \int_\Omega e^{2iw.\zeta^T} \dot{V}_0(w)$$

where $\zeta^T := (\text{Im}\,\zeta, \text{Re}\,\zeta) \in \mathbb{R}^2$ which implies that the Fourier transform of $\mathbb{1}_\Omega \dot{V}_0$ at ζ^T is 0 for all $\zeta \in \mathbb{C}$, and thus that $\dot{V}_0 = 0$. The linearized Dirichlet-to-Neumann operator at 0 is injective, but Calderón [1980] observed that its range is not closed and therefore we cannot use the local inverse theorem to consider the nonlinear problem.

3B. The nonlinear case.

The construction of solutions u_h of the Schrödinger equation $(\Delta + V)u_h = 0$ which grow exponentially as $h \to 0$ appeared first in the work of Faddeev [1965; 1974] and were later used to solve the Calderón problem in dimension 2 for isotropic conductivities by Sylvester and Uhlmann [1986], under the assumption that the conductivity γ is close to 1, and then by Nachman [1996] to solve the problem unconditionally (except for regularity conditions on γ). For the inverse scattering problem in dimension 2, this was used by Novikov [1992] to solve the problem under smallness assumptions on the potential. These solutions with complex phases depending on a parameter are called *complex geometric optic solutions* (CGO in short) or *Faddeev type solutions*, the first terminology obviously arising from the analogy with geometric optic solutions with real phases used in the WKB approximation of solutions of hyperbolic partial differential equations.

Definition 3.1. More precisely, we will say that a family of solutions u_h (with $h \in (0, h_0)$ small) of $(\Delta + V)u_h = 0$ are complex geometric optic solutions with phase Φ if there exists a complex-valued function Φ and some functions $a \in L^2$ independent of h and $r_h \in L^2$ such that

$$u_h = e^{\Phi/h}(a + r_h), \quad \|r_h\|_{L^2} \to 0 \text{ as } h \to 0.$$

Practically, these solutions will have their maximum (of the modulus) localized on the boundary $\partial \Omega$ and pairing with a function V will concentrate for small h the value of V at the maximum, roughly speaking. However they will have an oscillating phase (given by $e^{i \text{Im}(\Phi)/h}$) and this term can provide us with information on V in Ω, as in the linearized case. The construction of CGO as defined above is in fact not a very complicated thing to do if one thinks in terms of Carleman estimates (this will be developed below), but there is a complication: indeed an observation of the integral identity (3) shows that if V_1, V_2 are bounded potentials on Ω, say, and if $\mathcal{N}_{V_1} = \mathcal{N}_{V_2}$, then

$$\int_\Omega (V_1 - V_2) u_1 u_2 = 0$$

for all $u_1, u_2 \in H^2(\Omega)$ s.t. $(\Delta + V_1)u_1 = 0 = (\Delta + V_2)u_2$. (4)

In particular, we see that if we expect to obtain information on $V_1 - V_2$ in the interior Ω from plugging CGO $u_1 = u_h$ with phase Φ_1 and $u_2 = v_h$ with phase Φ_2 we should ask that $\text{Re}(\Phi_1) = -\text{Re}(\Phi_2)$, which turns out to be a much more restrictive condition.

The phases which appeared in Sylvester and Uhlmann [1986] are linear and the existence of CGO for the isotropic conductivity equation was proved under the assumption that $\|1 - \gamma\|_{W^{3,\infty}} \le \epsilon$ where $\epsilon > 0$ is small depending only on Ω.

Let us instead give the main technical result[1] of [Sylvester and Uhlmann 1987], which consists in showing the existence of CGO with linear phases without smallness assumptions on γ or V.

Theorem 3.2 [Sylvester and Uhlmann 1987]. *Let $V \in C^\infty(\Omega)$ where $\Omega \subset \mathbb{C}$ is a domain with smooth boundary. For any $s > 1$, there exist constants C_1, C_2 such that if $h > 0$ and $\zeta \in \mathbb{C}^2$, $\frac{1}{2} \leq |\zeta| \leq 2$, possibly depending on h, are such that*

$$\zeta.\zeta = 0, \quad h^{-1} \geq C_1 \|V\|_{H^s(\Omega)}$$

then there exists $u_h \in H^s(\Omega)$ satisfying $(\Delta + V)u_h = 0$ of the form

$$u_h(w) = e^{\frac{w.\zeta}{h}}(1 + r_h(\zeta, w)), \quad \text{with } \|r_h\|_{H^s(\Omega)} \leq C_2 h \|V\|_{H^s(\Omega)}.$$

We will give an idea of how to construct these CGO a bit later. First, let us check what we can deduce from the existence of such solutions, in comparison to the linearized case where we obtained the Fourier transform of the difference of potential. If u_h, v_h are solutions of $(\Delta + V_1)u_h = 0 = (\Delta + V_2)v_h$ of the form

$$\left.\begin{array}{l} u_h(w) = e^{\frac{w.\zeta}{h}}(1 + r_h(\zeta, w)) \\ v_h(w) = e^{\frac{w.\eta}{h}}(1 + s_h(\eta, w)) \end{array}\right\} \quad \text{with } \|r_h\|_{L^2} + \|s_h\|_{L^2} = O(h),$$

$$\zeta = \alpha + i\mu, \ \eta = -\alpha + i\nu, \ \alpha, \mu, \nu, \in \mathbb{R}^2, \ \zeta.\zeta = \eta.\eta = 0, \ |\zeta| = |\eta| = 1,$$

given by Theorem 3.2, this implies that $\mu = \pm J\alpha$ and $\alpha = \pm J\nu$, where J is the rotation of angle $\pi/2$ in $\mathbb{R}^2$. By (4) with $u_1 = u_h$ and $u_2 = v_h$, we deduce

$$\int_\Omega (V_1 - V_2) e^{2i \frac{w.\mu}{h}} + O(h) = 0, \quad \text{if } 0 < h^{-1} \leq C \max_{i=1,2} \|V_i\|_{H^1}.$$

We see that from this identity, we *cannot* show that $V_1 = V_2$ since as $h \to 0$ this equality does not say anything if $V_1 - V_2$ has a bit of regularity. It could however say something about the singularity of $V_1 - V_2$, for instance if the potentials have conormal singularities somewhere in the domain.

3C. *Comparison with higher dimensions.* In higher dimensions, $n > 2$, it turns out that CGO with linear phases give enough information to identify a potential and thus an isotropic conductivity. Indeed, applying Theorem 3.2 (recall that ζ there can also depend on h) we obtain

$$\left.\begin{array}{l} u_h(w) = e^{\frac{w.\zeta}{h}}(1 + r_h(\zeta, w)) \\ v_h(w) = e^{\frac{w.\eta}{h}}(1 + s_h(\eta, w)) \end{array}\right\} \quad \text{with } \|r_h\|_{L^2} + \|s_h\|_{L^2} = O(h),$$

$$\zeta = (\alpha + kh) + i(\mu + kh), \quad \eta = -(\alpha + kh) + i(-\mu + kh),$$

---
[1] The construction of the CGO in [Sylvester and Uhlmann 1987] holds in any dimension.

where $\alpha, \mu, k \in \mathbb{R}^n$ are chosen such that $\alpha.k = \alpha.\mu = k.\mu = 0$, and $\frac{1}{2} < |\alpha| = |\mu| < 2$, in order that $\zeta.\zeta - \eta.\eta = 0$; here $h > 0$ is taken very small. Of course, here we use that there are at least 3 orthogonal directions to define α, k, μ. Plugging those in the integral identity (4), this yields

$$0 = \int_\Omega e^{iw.k}(V_1 - V_2) + O(h)$$

and by letting $h \to 0$, we see that $V_1 = V_2$ since its Fourier transform is 0. The element which tells us information is somehow the leading term $e^{w.(\pm k + ik)}$ in the amplitudes of the CGO. This is summarized as follows:

Theorem 3.3 [Sylvester and Uhlmann 1986]. *Let $V_1, V_2 \in C^\infty(\Omega)$, where Ω is a domain in $\mathbb{R}^n$ with smooth boundary, with $n \geq 3$. If the Dirichlet-to-Neumann for the Schrödinger equations $(\Delta + V_i)u = 0$ agree, i.e., $\mathcal{N}_{V_1} = \mathcal{N}_{V_2}$, then $V_1 = V_2$.*

It can be noticed from the proof of their paper that the smoothness assumption on V_1, V_2 can be relaxed to $W^{2,\infty}(\Omega)$ regularity.

On the other hand, in dimension $n = 2$, Sylvester and Uhlmann [1986] were able to prove by using the CGO with linear phases that $\mathcal{N}_\gamma$ determine locally a conductivity close to 1. See Theorem 4.1.

3D. *Constructing CGO in dimension 2.*

3D.1. *For linear phases, a direct approach using Fourier transform.* Let us first explain the method used in [Sylvester and Uhlmann 1987] to construct CGO with linear phases for the Schrödinger equation $(\Delta + V)u = 0$ on a domain $\Omega \subset \mathbb{C}$. Of course, here the characteristic variety for the conjugated Laplacian is much simpler than in higher dimension, which makes the proof easier, as we shall now see. We search for solutions of the form

$$u(w) = e^{\frac{w.\zeta}{h}}(1 + r_h(\zeta, w)), \quad \zeta \in \mathbb{C}^2, \zeta.\zeta = 0$$

where ζ may depend on h but $\frac{1}{2} \leq |\zeta| \leq 2$. Then r_h needs to solve

$$(h^2\Delta + h^2 V - 2h\zeta\nabla)r_h = -h^2 V. \tag{5}$$

If we think in terms of semiclassical calculus, one has an operator $P_h = h^2\Delta + h^2 V - 2h\zeta\nabla$ to invert on the right, and its semiclassical principal symbol is $p_h(w, \xi) = \xi^2 - 2i\zeta.\xi$. Writing $\zeta = \mu + i\nu$, we have by an elementary calculation (splitting self-adjoint and anti self-adjoint components)

$$\|P_h u\|_{L^2} \geq \|(h^2\Delta - 2h\zeta.\nabla)u\|_{L^2} - h^2\|Vu\|_{L^2}$$

and

$$\|(h^2\Delta - 2h\zeta\nabla)u\|_{L^2}^2 = \|(h^2\Delta - 2hi\nu.\nabla)u\|_{L^2}^2 + \|h\mu.\nabla u\|_{L^2}^2.$$

Observe that $|v|^{-1}(v, \mu)$ is an orthonormal basis of $\mathbb{C}$ if $\zeta = \mu + iv$ solves $\zeta.\zeta = 0$. Fourier transforming, we obtain

$$\|(h^2\Delta - 2h\zeta\nabla)u\|_{L^2}^2 = \int_{\mathbb{R}^2} |h\xi|^2 |h\xi - 2v|^2 |\hat{u}(\xi)|^2 d\xi$$
$$= h^{-2} \int_{\mathbb{R}^2} |\xi|^2 |\xi - 2v|^2 |\hat{u}(\xi/h)|^2 d\xi.$$

Let $\chi_1^2 + \chi_2^2 + \chi_3^2 = 1$ be a partition of unity on $(-1, \infty)$, with $\chi_1 \in C_0^\infty(-1, \frac{1}{2})$, $\chi_2 \in C_0^\infty(\frac{1}{4}, 3)$ and χ_3 with support in $(2, \infty]$. We write

$$\hat{u} = \sum_{i=1}^{3} \hat{u}_i,$$

where $\hat{u}_i(\xi) := \chi_i(|\xi|)\hat{u}(\xi/h)$. Since $\chi_3(|v|) = 0 = \chi_3(0)$, we clearly have

$$h^{-2} \int_{\mathbb{R}^2} |\xi|^2 |\xi - 2v|^2 |\hat{u}_3(\xi)|^2 d\xi \geq Ch^{-2} \|\hat{u}_3\|^2. \tag{6}$$

Now, observe by integrating by parts, we have for any $v \in C_0^\infty(\mathbb{R}^n)$

$$2n\int_{\mathbb{R}^n} |v|^2 = -\int_{\mathbb{R}^n} \nabla|v|^2.\nabla|\xi|^2 d\xi = -\int_{\mathbb{R}^n} \nabla|v|^2.\nabla|\xi|^2 d\xi$$
$$= -4\operatorname{Re} \int_{\mathbb{R}^n} \bar{v}\nabla v.\xi d\xi$$
$$\leq \frac{4}{\epsilon} \int_{\mathbb{R}^n} |\xi|^2|v|^2 d\xi + 4\epsilon \int_{\mathbb{R}^n} |\nabla v|^2 \tag{7}$$

for any $\epsilon > 0$. We apply this with $v = \hat{u}_1$ after observing that $u(w)$ is supported in $|w| \leq R/h$ for some $R > 0$ and thus the H^1 norm of $\hat{u}_1$ is controlled by $\|u\|$ as follows:

$$\int_{\mathbb{R}^2} |\nabla \hat{u}_1|^2 = h^4 \int_{\mathbb{R}^2} |w|^2 |\hat{\chi}_1 \star u(hw)|^2 dw$$
$$\leq h^2 \|u\|_{L^2}^2 \left(\int |\hat{\chi}_1(w)|(|w| + R/h)\, dw \right)^2 \leq C\|u\|_{L^2}^2.$$

This implies, by taking $\epsilon = \delta h^2$ with δ small,

$$h^{-2}\int_{\mathbb{R}^2} |\xi|^2|\xi - 2v|^2|\hat{u}_1(\xi)|^2 d\xi \geq Ch^{-2}\int_{\mathbb{R}^n}|\xi|^2|\hat{u}_1|^2 d\xi$$
$$\geq C\delta(\|\hat{u}_1\|_{L^2}^2 - C\delta h^2\|u\|_{L^2}^2). \tag{8}$$

For dealing with $\hat{u}_2$, we use the same argument after the change of variables $\xi \to \xi + 2\nu$:

$$h^{-2}\int_{\mathbb{R}^2}|\xi|^2|\xi-2\nu|^2|\hat{u}_2(\xi)|^2 d\xi$$

$$= h^{-2}\int_{\mathbb{R}^2}|\xi+2\nu|^2|\xi|^2|\hat{u}(\xi+2\nu)|^2\chi_2(|\xi+2\nu|)^2 d\xi$$

$$\geq Ch^{-2}\int_{\mathbb{R}^2}|\xi|^2|\hat{u}(\xi+2\nu)|^2\chi_2(|\xi+2\nu|)^2 d\xi$$

$$\geq C\delta(\|\hat{u}_2\|_{L^2}^2 - C\delta h^2\|u\|_{L^2}^2) \tag{9}$$

for some small $\delta > 0$. We conclude by taking δ small enough and combining (8), (9) and (6) that

$$\|(h^2\Delta - 2h\zeta\nabla)u\|_{L^2}^2 \geq C\delta\left(\int_{\mathbb{R}^2}|u(\xi/h)|^2 d\xi - C\delta h^2\|u\|_{L^2}^2\right) \geq C\delta h^2\|u\|_{L^2}^2$$

and thus (fixing δ)

$$\|P_h u\|_{L^2} \geq Ch\|u\|_{L^2}.$$

By the Riesz representation theorem (or Lax–Milgram), it is clear that P_h^* has a right bounded inverse mapping $L^2(\Omega')$ to $H_0^1(\Omega') \cap H^2(\Omega')$ with norm $\mathcal{O}_{L^2 \to L^2}(h^{-1})$, but P_h has exactly the same form as P_h with $\overline{V}$ instead of V and $-\bar{\zeta}$ instead of ζ, we can then apply the same argument to say that P_h has a right inverse $G_h = L^2(\Omega') \to H_0^1(\Omega') \cap H^2(\Omega')$ with norm $\mathcal{O}_{L^2 \to L^2}(h^{-1})$. Equation (5) is solved by setting

$$r_h = -h^2 G_h V$$

which has norm $\|r_h\|_{L^2} = \mathcal{O}(h)$. The Sobolev norm $\|\nabla r_h\|_{L^2} = \mathcal{O}(1)$ can also be obtained easily from this proof above.

3D.2. *For holomorphic phases.* We will now give a more direct argument based on Carleman estimates for general holomorphic phases without critical points. This follows the method of Imanuvilov, Uhlmann, and Yamamoto [Imanuvilov et al. 2010b]; see also [Guillarmou and Tzou 2009].

Lemma 3.4. *Let φ be a harmonic function on Ω and $V \in L^\infty$. The following estimate holds for all $u \in C_0^\infty(\Omega)$ and $h > 0$:*

$$\|e^{-\varphi/h}\Delta e^{\varphi/h}u\|_{L^2} \geq C\left(\frac{1}{h}\||\nabla\varphi|u\|_{L^2} + \|\nabla u\|_{L^2}\right).$$

In particular, if φ has no critical points, then for small $h > 0$

$$\|e^{-\varphi/h}(\Delta+V)e^{\varphi/h}u\|_{L^2} \geq C\left(\frac{1}{h}\|u\|_{L^2} + \|\nabla u\|_{L^2}\right).$$

Proof. It suffices to prove this for real-valued u. We use that $\Delta = -\partial_{\bar{z}}\partial_z$ where $z = x + iy$ is the complex coordinate, and assume Ω is simply connected so that there exists ψ harmonic with $\Phi = \varphi + i\psi$ holomorphic,

$$e^{-\varphi/h}\Delta e^{\varphi/h} = -e^{i\psi/h}\partial_{\bar{z}}e^{-i\psi/h}e^{i\psi/h}\partial_z e^{i\psi/h}$$

and computing explicitly and integrating by parts ($u \in C_0^\infty(\Omega)$)

$$\|e^{-i\psi/h}\partial_z e^{i\psi/h}u\|_{L^2}^2$$
$$= \left\|\partial_z u + iu\frac{\partial_z\psi}{h}\right\|_{L^2}^2 = \int_\Omega \left(\partial_x u + u\frac{\partial_y\psi}{h}\right)^2 + \left(\partial_y u - u\frac{\partial_x\psi}{h}\right)^2$$
$$= \|\nabla u\|^2 + \frac{1}{h^2}\|u\nabla\psi\|^2 + \frac{1}{h}\int_\Omega \partial_x(u^2)\partial_y\psi - \partial_y(u^2)\partial_x\psi$$
$$= \|\nabla u\|^2 + \frac{1}{h^2}\|u\nabla\psi\|^2. \tag{10}$$

We can now use the Poincaré inequality: if $v \in C_0^\infty(\Omega)$,

$$\|e^{-i\psi/h}\partial_{\bar{z}}(e^{i\psi/h}v)\|_{L^2} = \|\partial_{\bar{z}}(e^{i\psi/h}v)\|_{L^2} = C\|\nabla(e^{i\psi/h}v)\| \geq C\|v\|_{L^2}^2, \tag{11}$$

where the second equality uses integration by parts and the fact that v is compactly supported. Combining (11) and (10), this proves the Lemma. If the domain is not simply connected, the proof works the same for harmonic functions with a harmonic conjugate, but in fact by using local Carleman estimates and convexification arguments (see [Guillarmou and Tzou 2009]) this even works for all harmonic functions without critical points. $\square$

Again, using Riesz representation theorem, this construct a right inverse $G_h^\pm$ on $L^2(\Omega)$ for $e^{\mp\Phi/h}(\Delta + V)e^{\pm\Phi/h}$ with $L^2 \to L^2$ norm $\mathcal{O}(h)$ and allows to construct complex geometric optic solutions of $(\Delta + V)u = 0$ by setting

$$u_h = e^{\Phi/h}(a + r_h), \quad \partial_{\bar{z}}a = 0, \quad r_h = -G_h^+(Va) = \mathcal{O}_{L^2}(h)$$

since $ae^{\Phi/h}$ is a solution of $\Delta(ae^{\Phi/h}) = 0$. The same obviously holds if we take antiholomorphic phases $\overline{\Phi}$ instead of Φ. The proof we just described is simpler than the Fourier transform approach above, but it is very particular to dimension 2 while the other one can be adapted to higher dimensions (for linear phases).

As for linear phases in 2 dimensions, it seems difficult to get enough information from these CGO. Indeed, if $u_h = e^{\Phi/h}(a+r_h)$ is a solution of $(\Delta+V_1)u = 0$ and $v_h = e^{-\Phi/h}(b + s_h)$ a solutions of $(\Delta + V_2)v = 0$ with a, b holomorphic and $\|r_h\|_{L^2} + \|s_h\|_{L^2} = \mathcal{O}(h)$, we deduce from the integral identity (3) by letting $h \to 0$ that

$$\int_\Omega (V_1 - V_2)ab = 0,$$

which means that $V_1 - V_2$ is orthogonal to antiholomorphic functions. Similarly, we can show it is orthogonal to holomorphic functions, but that does not show that $V_1 = V_2$. If instead we take $v_h = e^{-\bar{\Phi}/h}(\bar{b} + s_h)$, we obtain (with $\psi = \text{Im}(\Phi)$)

$$\int_\Omega (V_1 - V_2) e^{2i\psi/h} a\bar{b} = \mathcal{O}(h),$$

but the oscillating term is decaying in h by nonstationary phase (we know $V_1 = V_2$ on the boundary, by boundary uniqueness); thus we do not obtain anything interesting.

Remark. Carleman estimates have been used extensively to prove unique continuation for solutions of PDE (or differential inequalities), they turn out to be very powerful. In control theory, they are also a strong tool. We refer the interested reader to [Lebeau and Le Rousseau 2011]. In inverse problems, they have been apparently first used for the Calderón problem in [Bukhgeim and Uhlmann 2002], and then developed in [Kenig et al. 2007; Dos Santos Ferreira et al. 2009]. The property of holomorphic phases (without critical points) to be good weights for Carleman estimates was observed in [Dos Santos Ferreira et al. 2009] and [Bukhgeim 2008]. The first of these two papers studied in general what the authors call *limiting Carleman weights*: weights φ such that a Carleman estimate holds for both φ and $-\varphi$ with $d\varphi$ never vanishing. In dimension 2, they showed that harmonic functions with no critical points verify this. In [Bukhgeim 2008], the author used holomorphic phases with critical points to solve the inverse problem for a potential (this will be explained further down.)

4. The inverse problem for conductivities

4A. *Local uniqueness near constant conductivities.* We start with the first result obtained for conductivities in dimension 2:

Theorem 4.1 [Sylvester and Uhlmann 1986]. *Let $\Omega \subset \mathbb{R}^2$ be a domain with smooth boundary. There exists $\epsilon > 0$ depending on Ω such that if $\mathcal{N}_{\gamma_1} = \mathcal{N}_{\gamma_2}$ and $\|1 - \gamma_j\|_{W^{3,\infty}} \leq \epsilon$ for $j = 1, 2$, then $\gamma_1 = \gamma_2$.*

The method is based on the construction of CGO with linear phases; we refer the reader to the original paper for details.

4B. *Global uniqueness in a particular case.*

Theorem 4.2 [Sun 1990]. *If $\gamma_1, \gamma_2 \in C^4(\Omega)$ with Ω simply connected and $\Delta \log(\gamma_1) = 0$ or $\Delta \gamma_1^m = 0$ for some $m \neq 0$, then $\mathcal{N}_{\gamma_1} = \mathcal{N}_{\gamma_2}$ implies $\gamma_1 = \gamma_2$.*

4C. *The theorem of Nachman: global uniqueness.* Here is the first definitive result for conductivity in 2 dimensions:

Theorem 4.3 [Nachman 1996]. *Let $\Omega \subset \subset \mathbb{C}$ be a Lipschitz bounded domain and $\gamma_1, \gamma_2 \in W^{2,p}(\Omega)$ for $p > 1$ be two positive functions $\inf_\Omega \gamma_i > 0$. If $\mathcal{N}_{\gamma_1} = \mathcal{N}_{\gamma_2}$, then $\gamma_1 = \gamma_2$.*

Idea of proof. Nachman's approach is based on some sort of scattering theory for complex frequencies. This uses Faddeev Green's functions [Faddeev 1974] and some $\bar\partial$ methods, which appeared first for one-dimensional inverse scattering in the work of Beals and Coifman [1981; 1988] and later in 2 dimensions in [Ablowitz et al. 1983; Grinevich and Manakov 1986; Grinevich and Novikov 1988b; Novikov 1986; 1992]. We have seen in Theorem 3.2 that linear CGO for $\Delta + V$ can be constructed for large complex frequencies ζ (i.e., $h = |\zeta|^{-1}$ small); this is one of the difficulties to recover γ and somehow this is what Nachman achieves.

By boundary uniqueness, we can always extend the conductivities outside Ω, so that they agree outside Ω and are equal to 1 outside a large ball of $\mathbb{C}$. The problem of solving the equation $(\Delta + V)u = 0$ can then be considered in the whole complex plane $\mathbb{C}$. Nachman actually proves that if V comes from a conductivity, that is $V_\gamma = -\Delta \gamma^{\frac{1}{2}}/\gamma^{\frac{1}{2}}$, then it possible to construct CGO for all complex frequencies, not only for large ones, and with uniqueness if one assumes some decay at infinity. More precisely he shows that if $V_\gamma \in L^p$ for $1 < p < 2$ for any $\zeta \in \{\zeta \in \mathbb{C}^2 \setminus \{0\}; \zeta.\zeta = 0\}$, there is a unique solution u_ζ of $(\Delta + V_\gamma)u = 0$ which satisfies as a function of $w \in \mathbb{R}^2$

$$r_\zeta(w) := e^{-i\zeta.w} u_\zeta(w) - 1 \in L^q(\mathbb{R}^2) \cap L^\infty(\mathbb{R}^2), \quad \tfrac{1}{p} - \tfrac{1}{q} = \tfrac{1}{2}.$$

Essentially, to solve this problem for large ζ we have seen that it amounts to invert on the right the operator $\Delta - 2\zeta.\nabla$ acting on functions compactly supported (or decaying at infinity), with a decay estimate $\mathcal{O}(|\zeta|^{-1})$ in L^2 for the operator norm when $|\zeta| \to \infty$. This can be done when $|\zeta|$ is large enough. Nachman manages to show that the solution is unique under the decay condition at infinity, and using Fredholm theory, he manages to deal with small $|\zeta|$. The estimate he obtains on the CGO is

$$\|r_\zeta\|_{L^q} \leq C|\zeta|^{-1}\|V\|_{L^p} \text{ for large } |\zeta|.$$

Nachman shows that $\mathcal{N}_\gamma$ determines $r_\zeta|_{\partial\Omega}$ by using a sort of scattering operator S_k: he shows that $u_\zeta|_{\partial\Omega}$ is the solution of the integral equation (we use $z = x+iy$ and $w = (x, y)$ to identify $\mathbb{C}$ with $\mathbb{R}^2$)

$$u_\zeta(z) = e^{ikz} - (S_k(\mathcal{N}_\gamma - \mathcal{N}_1)u_\zeta)(z) \text{ on } \partial\Omega, \text{ if } \zeta = (k, ik) \text{ with } k \in \mathbb{C}$$

where S_k is the operator with integral kernel given by the trace at the boundary of the Faddeev Green's function:

$$S_k f(w) := \int_{\partial\Omega} G_k(w - w') f(w') \, dw',$$

$$G_k(w) := \frac{e^{izk}}{4\pi^2} \int_{\mathbb{R}^2} \frac{e^{iw\cdot\xi}}{|\xi|^2 + 2k(\xi_1 + i\xi_2)} \, d\xi,$$

and $\mathcal{N}_1$ is the Dirichlet-to-Neumann operator for the conductivity $\gamma_0 = 1$. Defining the scattering transform $t(k) \in \mathbb{C}$ to be

$$t(k) := \int_{\partial\Omega} e^{i\overline{z}\overline{k}} (\mathcal{N}_\gamma - \mathcal{N}_1) u_\zeta |_{\partial\Omega}, \quad \zeta = (k, ik)$$

one sees that $t(k)$ is determined by $\mathcal{N}_\gamma$. The crucial observation of Nachman is that $\mu_k(z) := e^{-ikz} u_\zeta(z)$ solves a $\bar{\partial}$ equation in the frequency k parameter involving $t(k)$

$$\partial_{\bar{k}} \mu_k(z) = \frac{1}{4\pi \bar{k}} t(k) e^{-2i\mathrm{Re}(kz)} \overline{\mu_k(z)}.$$

in a certain weighted Sobolev space in z. Notice that this $\bar{\partial}$-type equation in the frequency was also previously in the works [Grinevich and Manakov 1986; Grinevich and Novikov 1988b; Novikov 1992; Beals and Coifman 1988]. Nachman then shows that such equation have unique solutions by using a sort of Liouville theorem for pseudo-analytic functions, at least if we know that μ_k is bounded for k near 0 and $t(k)/\bar{k}$ is not too singular at $k = 0$. The boundedness in $k \to 0$ is shown in [Nachman 1996], using in particular the fact that Faddeev Green's function does not degenerate too much as $k \to 0$ (essentially by a $\log k$). The function μ_k turns out to be the solution of the integral equation

$$\mu_k(z) = 1 + \frac{1}{8\pi^2 i} \int_{\mathbb{C}} \frac{t(k')}{(k' - k)\bar{k}'} e^{-2i\mathrm{Re}(kz)} \overline{\mu_{k'}(z)} \, dk' \wedge d\bar{k}'$$

in the weighted Sobolev space and it also satisfies $\mu_k(z) \to \gamma^{1/2}$ as $k \to 0$. In particular μ_k is determined by the scattering transform $t(k)$ and thus by $\mathcal{N}_\gamma$. Letting $k \to 0$ in $\mu_k(z)$ determines γ. $\square$

4D. *The Brown–Uhlmann and Beals–Coifman results.* The regularity assumption was weakened to $\gamma \in W^{1,p}(\Omega)$ by Brown and Uhlmann [1997], who modified and simplified Nachman's proof using the $\bar{\partial}$-method of [Beals and Coifman 1988]. It turns out that the result of Beals and Coifman for the Davey–Stewartson equation, when interpreted in the right way, proves Nachman's result if one assumes smooth conductivities. (Amusingly, it seems almost a decade elapsed before someone made this observation.)

Theorem 4.4 [Brown and Uhlmann 1997]. *Let $\Omega \subset \mathbb{C}$ be a Lipschitz bounded domain and $\gamma_1, \gamma_2 \in W^{1,p}(\Omega)$ for $p > 2$ be two positive functions $\inf_\Omega \gamma_i > 0$. If $\mathcal{N}_{\gamma_1} = \mathcal{N}_{\gamma_2}$, then $\gamma_1 = \gamma_2$.*

Idea of proof. As mentioned, the idea is to use the $\bar{\partial}$-method of [Beals and Coifman 1988]: u is a solution of $\text{div}(\gamma \nabla u) = 0$ if and only if

$$\left[\begin{pmatrix} \bar{\partial} & 0 \\ 0 & \partial \end{pmatrix} - \begin{pmatrix} 0 & V \\ \bar{V} & 0 \end{pmatrix}\right] \cdot \begin{pmatrix} v \\ v' \end{pmatrix} = 0, \quad \text{with } v := \gamma^{\frac{1}{2}} \partial u, \ v' := \gamma^{\frac{1}{2}} \bar{\partial} u$$

and the potential is $V(z) := -\frac{1}{2} \partial_z \log \gamma$ involves only one derivative of the conductivity. The operator above also acts on 2×2 matrices and the CGO in this setting is given by a

$$u_k(z) = m_k(z) \begin{pmatrix} e^{izk} & 0 \\ 0 & e^{-i\bar{z}k} \end{pmatrix}, \quad k \in \mathbb{C} \setminus \{0\}$$

where $m_k(z)$ is a matrix-valued function such that $m_k(z) \to \text{Id}$ as $|z| \to \infty$, and $m_k - \to \text{Id} \in L^q$ for some $q > 2$. It can be also be shown that $\|m_k(z) - \text{Id}\|_{L^q(k \in \mathbb{C})} \leq C$ uniformly in z for $q = p/(p-2)$. The argument is then similar to what we explained above in Nachman's result: one uses the fact (proved in [Beals and Coifman 1988]) that $m_k(z)$ satisfies a $\bar{\partial}$ equation in k:

$$\partial_{\bar{k}} m_k(z) = m_{\bar{k}}(z) E_k S_k, \quad \text{with } E_k := \begin{pmatrix} e^{2i\text{Re}(z\bar{k})} & 0 \\ 0 & e^{-2i\text{Re}(zk)} \end{pmatrix}$$

where S_k is the scattering data (which we do not define but is analogous to the Nachman scattering transform), shown to be determined by $\mathcal{N}_\gamma$. Then certain linear combinations $\omega_k(z)$ of the coefficients of $m_k(z)$ satisfy a pseudo-analytic equation of the form $\partial_{\bar{k}} \omega_k(z) = r(k) \overline{\omega_k(z)}$ with $r \in L^2$ and Brown and Uhlmann showed the following Liouville-type result: if $\omega \in L^p \cap L^2_{\text{loc}}(\mathbb{C})$ is a solution of $\partial_{\bar{k}} \omega = a\omega + b\bar{\omega}$ with $a, b \in L^2$, then $u = 0$. This implies that $m_k(z)$ is determined by $\mathcal{N}_\gamma$ as in Nachman's paper. Now to recover the potential V, it suffices to notice that $D_k m_k = V m_k$ where

$$D_k := D + \frac{k}{2} \begin{pmatrix} -i & 0 \\ 0 & i \end{pmatrix}$$

and the potential V can be recovered by the expression

$$V(z) \text{Vol } B_R(0) = -\lim_{k_0 \to \infty} \int_{|k-k_0| \leq R} D_k m_k(z) \frac{dk \wedge d\bar{k}}{2i}, \quad R > 0 \text{ fixed,}$$

since $m_k(z) \to 1$ as $k \to \infty$ in a sufficiently uniform way. $\square$

4E. *The theorem of Astala–Päivärinta.* Astala and Päivärinta found a new approach related to quasiconformal techniques to show the uniqueness of an L^∞ conductivity from the Dirichlet-to-Neumann operator.

Theorem 4.5 [Astala and Päivärinta 2006]. *Let $\Omega \subset \mathbb{C}$ be a simply connected bounded domain and $\gamma_1, \gamma_2 \in L^\infty(\Omega)$ be two positive functions $\inf_\Omega \gamma_i > 0$. If $\mathcal{N}_{\gamma_1} = \mathcal{N}_{\gamma_2}$, then $\gamma_1 = \gamma_2$.*

Idea of proof. To avoid using regularity on the conductivity, the idea is to transform the equation $\operatorname{div}(\gamma \nabla u) = 0$ into a Beltrami equation. They show that if $u \in H^1(\Omega)$ is a solution of $\operatorname{div}(\gamma \nabla u) = 0$, then there exists a unique function $v \in H^1(\Omega)$ such that $f = u + iv$ solves the Beltrami equation

$$\partial_{\bar{z}} f = \mu \overline{\partial_z f}, \quad \text{where } \mu := \frac{1-\gamma}{1+\gamma}. \tag{12}$$

Conversely, if $f = u + iv$ is a solution of the Beltrami equation with $\|\mu\|_{L^\infty} < 1 - \epsilon$ for some $\epsilon > 0$, then

$$\operatorname{div}(\gamma \nabla u) = 0 \quad \text{and} \quad \operatorname{div}(\gamma^{-1} \nabla v) = 0,$$

where $\gamma = (1-\mu)/(1+\mu)$. The map $\mathcal{H}_\mu : u|_{\partial\Omega} \to v|_{\partial\Omega}$ is the μ-Hilbert transform, and Astala and Päivärinta show that the $\mathcal{N}_\gamma$ determine the $\mathcal{H}_\mu$ and conversely (they also determine the $\mathcal{H}_{-\mu}$ and the $\mathcal{N}_{\gamma^{-1}}$). Similarly to the results discussed in Sections 4C and 4D, the authors show the existence of $m_k(z)$ such that $f_k(z) = e^{izk} m_k(z)$ solves (12) and $m_k(z) - 1 = \mathcal{O}(1/|z|)$ as $|z| \to \infty$. If $g_k = e^{izk} m'_k(z)$ denotes a solution for the Beltrami coefficient $-\mu$ instead of μ, then $h_+ := (f_k + f'_k)/2$ and $h_- := i(\bar{f}_k - \bar{f}'_k)/2$ solves pseudo-analytic equations in the frequency parameter

$$\partial_{\bar{k}} h_+(k) = \tau(k) \overline{h_-(k)}, \quad \partial_{\bar{k}} h_-(k) = \tau(k) \overline{h_+(k)}$$

where $\tau(k)$ is the scattering coefficient

$$\overline{\tau(k)} := \frac{i}{4\pi} \int_\mathbb{C} \partial_{\bar{z}}(m_k(z) - m'_k(z)) \, dz \wedge d\bar{z}.$$

Like their predecessors, Astala and Päivärinta show that $\mathcal{N}_\gamma$ determines the coefficient $\tau(k)$. The main difficulty in their proof is to study the behavior of $m_k(z) - 1$ as $k \to \infty$, and the decay of this function was a fundamental tool in [Nachman 1996; Brown and Uhlmann 1997] to prove that once we know the scattering coefficient, $m_k(z)$ is determined uniquely. The decay was $m_k(z) - 1$ was of order $\mathcal{O}(1/|k|)$ when the conductivity was regular enough, but in the L^∞ case Astala–Päivärinta show that $f_k(z) = e^{ik\varphi_k(z)}$ with $\varphi_k(z) - z = o(1)$ as $|k| \to \infty$ uniformly in $z \in \mathbb{C}$, which implies that $m_k(z) - 1 \to 0$ as $|k| \to \infty$ uniformly in z. □

Remark. Astala, Lassas and Päivärinta [2011] posted recently a paper where they describe in a way the classes of conductivities which can be determined by the Dirichlet-to-Neumann map and find examples which are invisible (related to "cloaking"). The sharp class for isotropic conductivities that can be identified are those γ with values in $[0, \infty]$ such that

$$\int_\Omega \exp(\exp(q\gamma(w) + q/\gamma(w)))\, dw < \infty \quad \text{for some } q > 0.$$

They also have sharp criteria for anisotropic cases in terms of the regularity of $\mathrm{Tr}(\gamma), \mathrm{Tr}(\gamma^{-1})$ and $\det(\gamma), \det(\gamma^{-1})$.

4F. *From isotropic to anisotropic conductivity.* When the conductivity is anisotropic, there is a way in dimension 2 (for domains of $\mathbb{C}$) to reduce the problem to the isotropic case by using isothermal coordinates. For a metric

$$\gamma = E\, dx^2 + G\, dy^2 + 2F\, dx\, dy = \lambda |dz + \mu\, d\bar{z}|^2$$

on a domain $\Omega \subset \mathbb{C}$, where

$$\lambda := \tfrac{1}{4}\left(E + G + 2\sqrt{EG - F^2}\right) \quad \text{and} \quad \mu := \frac{E - G + 2iF}{\lambda},$$

there is a diffeomorphism $\Phi: \Omega \to \Omega'$ such that $\Phi_*\gamma = e^\omega |dz|^2$ is conformal to the Euclidean metric (ω is some function), and Φ is a complex-valued function solving the Beltrami equation

$$\partial_{\bar{z}}\Phi = \mu \partial_z \Phi.$$

An anisotropic conductivity γ is a positive definite symmetric (with respect to the Euclidean metric) endomorphism acting on 1-forms and the anisotropic conductivity equation is $d\gamma du = 0$. The push forward by a diffeomorphism Φ is defined in that setting by $(\Phi_*\gamma)\alpha := \Phi_*(\gamma(\Phi^*\alpha))$ if α is any 1-form. Using those isothermal coordinates, we have:

Theorem 4.6 [Sylvester 1990]. *Let γ_1, γ_2 be two anisotropic C^3 conductivities, viewed as endomorphisms acting on 1-forms, where $\Omega \subset \mathbb{C}$ is a domain with C^3 boundary. Then if there exists a C^3 diffeomorphism $\phi: \partial\Omega \to \partial\Omega$ such that $\phi_*\mathcal{N}_{\gamma_1} = \mathcal{N}_{\gamma_2}$, then there exists a conformal diffeomorphism $\Phi: \Omega \to \Omega$ which satisfies*

$$\Phi_*\gamma_1 = \left(\frac{\det(\gamma_1 \circ \Phi^{-1})}{\det \gamma_2}\right)^{\frac{1}{2}} \gamma_2.$$

Idea of proof. The method is to extend the conductivities γ_j outside Ω so that it is the identity outside a compact set of $\mathbb{C}$ and they agree outside Ω (which is possible by Kohn–Vogelius boundary uniqueness) and then use the isothermal coordinates on $\mathbb{C}$, with the condition that the diffeomorphism Φ_j pushing the conductivity γ_j to an isotropic one is asymptotically equal to Id near infinity and is the unique solution of a Beltrami equation with this asymptotic condition. Then the final step is to show that $\Phi_1 = \Phi_2$ in $\mathbb{C} \setminus \Omega$ by using equality of Dirichlet-to-Neumann operators. To that end, Sylvester [1990] uses CGO of the form

$$u_j(z;k) = e^{ik\Phi_j(z)} \det(\gamma_j)^{-1/4}(1+r^j(z;k)), \quad r_j(z;k) \to 0 \text{ as } k \to \infty,$$

which are uniquely determined by their asymptotics

$$u_j(z;k) \sim e^{ik\Phi_j(z)} \det(\gamma_j)^{-1/4}$$

as $|z| \to \infty$. Then he uses that $\lim_{|k|\to\infty} \log(u(z;k))/k = \Phi_j(z)$ and the fact that $\mathcal{N}_{\gamma_1} = \mathcal{N}_{\gamma_2}$ implies that $u_1(z;k) = u_2(z;k)$ in $\mathbb{C} \setminus \Omega$ (by unique continuation in that set) to conclude that $\Phi_1(z) = \Phi_2(z)$ in $\mathbb{C} \setminus \Omega$. □

Remarks. Now, since γ_2 in the theorem can be pushed forward into an isotropic conductivity $e^{\omega}\text{Id}$ for some function ω using isothermal coordinates, we can use Nachman's theorem to deduce directly that $\mathcal{N}_{\gamma_1} = \mathcal{N}_{\gamma_2}$ implies $\gamma_1 = \Phi_*\gamma_2$ for some diffeomorphism Φ which is the identity on the boundary. The regularity was improved to $\gamma_j \in W^{1,p}$ with $p > 2$ in [Sun and Uhlmann 2003] and then to $\gamma_j \in L^\infty$ (with the condition that $C\text{Id} \geq \gamma_j \geq C^{-1}\text{Id}$ on Ω for some $C > 0$) in [Astala et al. 2005].

5. The inverse problem for potentials and magnetic field in domains of $\mathbb{C}$

5A. *The case with a potential.* As we have seen before, the CGO with linear phase do not provide enough information to be able to recover a general potential in 2 dimensions. But we have the following result for generic potentials:

Theorem 5.1 [Sun and Uhlmann 1991]. *Let $\Omega \subset \mathbb{R}^2$ be domain with smooth boundary. There exists an open dense set $\mathcal{O} \subset W^{1,\infty}(\Omega) \times W^{1,\infty}(\Omega)$ such that if $\mathcal{N}_{V_1} = \mathcal{N}_{V_2}$ and $(V_1, V_2) \in \mathcal{O}$, then $V_1 = V_2$.*

The proof is based on construction of CGO with linear phases, combined with analytic Fredholm theory. We do not discuss it further and refer the interested reader to the original paper.

Grinevich and Novikov also showed local uniqueness for potentials close to positive constants, and later Novikov extended this to potentials close to nonzero constants:[2]

---

[2]The proofs dealt with the scattering problem, but the result for the bounded domain setting follows directly by using boundary uniqueness and extending the potentials to $\mathbb{R}^2$.

Theorem 5.2 [Grinevich and Novikov 1988a; Novikov 1992]. *Let $\Omega \subset \mathbb{R}^2$ be domain with smooth boundary and $E \in \mathbb{R} \setminus \{0\}$. Then there exists $C_E > 0$ depending on $|E|$ such that for V_1, V_2 satisfying $\|V_j - E\|_{L^\infty} \leq C_E$, then $\mathcal{N}_{V_1} = \mathcal{N}_{V_2}$ implies $V_1 = V_2$. The constant C_E tend to 0 as $|E| \to 0$ and to $+\infty$ as $|E| \to \infty$.*

The general case has been recently tackled by Bukhgeim. His new idea is to use Morse holomorphic phases with a critical point where one wants to identify the potential:[3]

Theorem 5.3 [Bukhgeim 2008]. *Let Ω be a domain in $\mathbb{C}$ and $V_1, V_2 \in W^{1,p}(\Omega)$ for $p > 2$. If $\mathcal{N}_{V_1} = \mathcal{N}_{V_2}$, then $V_1 = V_2$.*

Proof. As we have seen in Lemma 3.4, holomorphic and antiholomorphic functions are Carleman weights in dimension 2. Before we start the proof, let us recall that if for a holomorphic function Φ, one can construct $u_1 = e^{\Phi/h}(1 + r_h)$ and $u_2 = e^{-\bar{\Phi}/h}(1 + s_h)$ some CGO which solve $(\Delta + V_j)u_j = 0$ with $\|s_h\|_{L^2} + \|r_h\|_{L^2} = o(1)$ as $h \to 0$, the integral identity (4) tells us as $h \to 0$ that

$$\int_\Omega (V_1 - V_2) e^{2i \operatorname{Im}(\Phi)} + \mathcal{O}(\|r_h\|_{L^2} + \|s_h\|_{L^2}) = 0. \tag{13}$$

Of course, the oscillating term will be decreasing very fast as $h \to 0$ if the phase is nonstationary, and therefore we won't get any good information; we are instead tempted to take Φ with a nondegenerate critical point $z_0 \in \Omega$ and apply stationary phase to deduce the value of $V_1 - V_2$ at z_0. This is essentially the main idea of the proof. However, by inspecting Lemma 3.4, the remainder terms r_h, s_h cannot reasonably be smaller than $\mathcal{O}_{L^2}(h)$ and the terms obtained by stationary phase in (13) is also of order h, which makes the recovery of $(V_1 - V_2)(z_0)$ quite tricky.

For constructing CGO, Bukhgeim makes a reduction of the problem to a $(\partial, \bar{\partial})$-system. Let $\Phi = \phi + i\psi$ be a holomorphic function on Ω with a unique critical point at z_0, which is nondegenerate in the sense $\partial_z^2 \Phi(z_0) \neq 0$. Although here 1-forms and functions are easily identified on a domain Ω, we prefer to keep in mind that the operator $\bar{\partial} = \partial_{\bar{z}} = \frac{1}{2}(\partial_x + i\partial_y)$ maps functions $f(z)$ to $(0, 1)$-forms $(\partial_{\bar{z}} f) d\bar{z}$, since we will later discuss the same problems for Riemann surfaces. We denote by $\Lambda^k(\Omega)$ the bundle of k-forms and by $\Lambda^{0,1}(\Omega)$ (resp. $\Lambda^{1,0}(\Omega)$) the bundle whose sections are of the form $f(z) d\bar{z}$, or equivalently in $(T^*\Omega)^{0,1}$ (resp. of the form $f(z) dz$, or equivalently in $(T^*\Omega)^{1,0}$). The operator $\bar{\partial}^*$ is the adjoint of $\bar{\partial}$ and is given by $\bar{\partial}^* = -2i * \partial$ where $*$ is the Hodge star operator mapping $\Lambda^{1,1}(\Omega)$ to $\Lambda^0(\Omega)$ by $*(dz \wedge d\bar{z}) = -2i$, ∂ maps $\Lambda^{(0,1)}(\Omega)$

---
[3] In [Bukhgeim 2008], it is claimed that a potential in $L^p(\Omega)$ with $p > 2$ can be identified with this method, but the argument of the paper does not seem to imply directly that such regularity can be dealt with.

to $\Lambda^{(1,1)}(\Omega)$ by $\partial(f(z)\,d\bar{z}) = (\partial_z f(z))\,dz \wedge d\bar{z}$ if $\partial_z = \frac{1}{2}(\partial_x - i\partial_y)$. Recall that the Laplace operator is $\Delta = \bar{\partial}^*\bar{\partial}$, therefore setting $\bar{\partial}u = w$, we can reduce the equation $(\Delta + V)u = 0$ to the equivalent first order system

$$\begin{pmatrix} V & \bar{\partial}^* \\ \bar{\partial} & -1 \end{pmatrix} \begin{pmatrix} u \\ w \end{pmatrix} = 0.$$

and $\mathcal{N}_V$ determine the boundary values of this system. In fact, as we shall see, the Cauchy data at the boundary

$$\mathcal{C}_Q := \{F|_{\partial\Omega}; (D+Q)F = 0\}$$

for first order systems of the form $(D+Q)F = 0$ (with the notation (14)) determine any matrix potential $Q \in W^{1,p}(\Omega)$ with $p > 2$, where

$$D = \begin{pmatrix} 0 & \bar{\partial}^* \\ \bar{\partial} & 0 \end{pmatrix} \quad \text{and} \quad Q = \begin{pmatrix} q & 0 \\ 0 & q' \end{pmatrix} \tag{14}$$

are the Dirac type operator and the matrix potential, $q, q' \in W^{1,p}(M_0)$ (with $p > 2$) being complex-valued (both acting on sections of $\Sigma := \Lambda^0(\Omega) \oplus \Lambda^{0,1}(\Omega)$ over Ω).

(i) *Construction of CGO.* The goal is to construct complex geometrical optics $F \in W^{1,p}(\Omega)$ that solve the equation

$$(D+Q)F = 0$$

on Ω. It is clear that

$$D = \begin{pmatrix} e^{-\bar{\Phi}/h} & 0 \\ 0 & e^{-\Phi/h} \end{pmatrix} D \begin{pmatrix} e^{\Phi/h} & 0 \\ 0 & e^{\bar{\Phi}/h} \end{pmatrix}$$

and thus

$$\begin{pmatrix} e^{-\bar{\Phi}/h} & 0 \\ 0 & e^{-\Phi/h} \end{pmatrix} (D+Q) \begin{pmatrix} e^{\Phi/h} & 0 \\ 0 & e^{\bar{\Phi}/h} \end{pmatrix} = D + Q_h,$$

where

$$Q_h = \begin{pmatrix} e^{2i\psi/h}q & 0 \\ 0 & e^{-2i\psi/h}q' \end{pmatrix}.$$

We want to construct solutions F_h of $(D+Q_h)F_h = 0$ having the form

$$F_h = \begin{pmatrix} a + r_h \\ b + s_h \end{pmatrix} =: A + R_h, \tag{15}$$

where a is some holomorphic function on Ω, b some antiholomorphic 1-form, and $R_h = (r_h, s_h)$ an element of $W^{1,p}(\Omega)$ that decays appropriately as $h \to 0$.

In particular, we need to solve the system

$$(D + V_h) R_h = -Q_h A = -\begin{pmatrix} e^{2i\psi/h} qa \\ e^{-2i\psi/h} q'b \end{pmatrix}.$$

One can now use a right inverse for the operator D, by taking the Cauchy integral kernel:

$$\bar{\partial}^{-1} : f(z) \, dz \mapsto \left(z \to \frac{1}{\pi} \int_\Omega \frac{f(z')}{z - z'} \, dx' \, dy' \right), \qquad z' = x' + iy',$$

$$\partial^{-1} : f(z) \, dz \wedge d\bar{z} \mapsto \left(z \to \frac{d\bar{z}}{\pi} \int_\Omega \frac{f(z')}{\bar{z} - \bar{z}'} \, dx' \, dy' \right), \qquad z' = x' + iy'.$$

We next define the operators D^{-1}

$$D^{-1} := \begin{pmatrix} 0 & \bar{\partial}^{-1} \\ \bar{\partial}^{*-1} & 0 \end{pmatrix}, \quad \text{with } \bar{\partial}^{*-1} = -(2i)^{-1} \partial^{-1} *$$

which satisfy $DD^{-1} = \mathrm{Id}$ on $L^q(\Omega)$ for all $q \in (1, \infty)$. Similarly,

$$D_h^{-1} := \begin{pmatrix} 0 & \bar{\partial}_h^{-1} \\ \bar{\partial}_h^{*-1} & 0 \end{pmatrix}, \quad \text{with } \bar{\partial}_h^{-1} = \bar{\partial}^{-1} e^{-2i\psi/h}, \ \bar{\partial}_h^{*-1} = \bar{\partial}^{*-1} e^{2i\psi/h}.$$

To construct R_h solving $(D + V_h) R_h = -Q_h A$ in Ω, it then suffices to solve

$$(\mathrm{Id} + D_h^{-1} Q) R_h = -D_h^{-1} Q A.$$

Writing the components of this system explicitly we get

$$\begin{aligned} r_h + \bar{\partial}_h^{-1}(q' s_h) &= -\bar{\partial}_h^{-1}(q' b), \\ s_h + \bar{\partial}_h^{*-1}(q r_h) &= -\bar{\partial}_h^{*-1}(q a). \end{aligned} \qquad (16)$$

Since we are allowed to choose any holomorphic function a and antiholomorphic 1-form b, we may set $a = 0$ in (16) and solve for r_h to get

$$(I - S_h) r_h = -\bar{\partial}_h^{-1}(q' b) \quad \text{with } S_h := \bar{\partial}_h^{-1} q' \bar{\partial}_h^{*-1} q. \qquad (17)$$

where q, q' are viewed as multiplication operators. Now we want to estimate the norm of S_h, and in that aim we can use the following crucial operator bound whose proof we give in detail since it is the main technical point of Bukhgeim's paper.[4]

Lemma 5.4 (The key estimate). *There exist $\epsilon > 0$, $h_0 > 0$ and $C > 0$ such that for all $h \in (0, h_0)$ and all $u \in W^{1,p}(\Omega, \Lambda^1(\Omega))$ and $v \in W^{1,p}(\Omega)$*

$$\|\bar{\partial}_h^{-1} u\|_{L^2(\Omega)} \le C h^{\frac{1}{2} + \epsilon} \|u\|_{W^{1,p}(\Omega)}, \quad \|\bar{\partial}_h^{*-1} v\|_{L^2(\Omega)} \le C h^{\frac{1}{2} + \epsilon} \|v\|_{W^{1,p}(\Omega)}.$$

---
[4] The proof presented here is not exactly Bukhgeim's proof but the idea essentially the same; we took it from [Guillarmou and Tzou 2011b].

Proof of the lemma. We only give a proof for $\bar\partial_h^{-1}$, the other case is exactly the same. The $L^?$ estimate will be obtained by interpolation between $L^q, q < 2$ estimates and $L^q, q > 2$ estimates. By standard Calderón–Zygmund theory for singular integral kernels, the operators $\bar\partial^{-1}$ and $\bar\partial^{*-1}$ map $L^q(\Omega) \to W^{1,q}(\Omega)$ for all $q \in (1, \infty)$.

(i) Case $q < 2$. Let $\varphi \in C^\infty(\bar\Omega)$ be a function which is equal to 1 for $|z - z_0| > 2\delta$ and to 0 in $|z - z_0| \leq \delta$, where $\delta > 0$ is a parameter that will be chosen later (it will depend on h). Using the Minkowski inequality, one can write (with $z' = x' + iy'$)

$$\|\bar\partial^{-1}((1-\varphi)e^{-2i\psi/h}f)\|_{L^q(\Omega)}$$
$$\leq \int_\Omega \left\|\frac{1}{|\cdot - z'|}\right\|_{L^q(\Omega)} |(1-\varphi(z'))f(z')|\,dx'dy'$$
$$\leq C\|f\|_{L^\infty} \int_\Omega |(1-\varphi(z'))|\,dx'\,dy' \leq C\delta^2 \|f\|_{L^\infty} \quad (18)$$

and we know by Sobolev embedding that $\|f\|_{L^\infty} \leq C\|f\|_{W^{1,p}}$. On the support of φ, we observe that since $\varphi = 0$ near z_0, we can use

$$\bar\partial^{-1}(e^{-2i\psi/h}\varphi f) = \frac{1}{2}ih\left[e^{-2i\psi/h}\frac{\varphi f}{\bar\partial\psi} - \bar\partial^{-1}\left(e^{-2i\psi/h}\bar\partial\left(\frac{\varphi f}{\bar\partial\psi}\right)\right)\right]$$

and the boundedness of $\bar\partial^{-1}$ on L^q to deduce that for any $q < 2$

$$\|\bar\partial^{-1}(\varphi e^{-2i\psi/h}f)\|_{L^q(\Omega)}$$
$$\leq Ch\left(\left\|\frac{\varphi f}{\bar\partial\psi}\right\|_{L^q} + \left\|\frac{f\bar\partial\varphi}{\bar\partial\psi}\right\|_{L^q} + \left\|\frac{\varphi\bar\partial f}{\bar\partial\psi}\right\|_{L^q} + \left\|\frac{f\varphi}{(\bar\partial\psi)^2}\right\|_{L^q}\right). \quad (19)$$

The first term is clearly bounded by $\delta^{-1}\|f\|_{L^\infty}$ due to the fact that ψ is Morse. For the last term, observe that since ψ is Morse, we have $1/|\partial\psi| \leq c/|z - z_0|$ near z_0; therefore

$$\left\|\frac{f\varphi}{(\bar\partial\psi)^2}\right\|_{L^q} \leq C\|f\|_{L^\infty}\left(\int_\delta^1 r^{1-2q}\,dr\right)^{1/q} \leq C\delta^{\frac{2}{q}-2}\|f\|_{L^\infty}.$$

The second term can be bounded by

$$\left\|\frac{f\bar\partial\varphi}{\bar\partial\psi}\right\|_{L^q} \leq \|f\|_{L^\infty}\left\|\frac{\bar\partial\varphi}{\bar\partial\psi}\right\|_{L^q}.$$

Observe that while $\|\bar{\partial}\varphi/\bar{\partial}\psi\|_{L^\infty}$ grows as δ^{-2}, $\bar{\partial}\varphi$ is only supported in a neighborhood of radius 2δ. Therefore we obtain

$$\left\|\frac{f\bar{\partial}\varphi}{\bar{\partial}\psi}\right\|_{L^q} \leq \delta^{2/q-2}\|f\|_{L^\infty}.$$

The third term can be estimated by

$$\left\|\frac{\varphi\bar{\partial}f}{\bar{\partial}\psi}\right\|_{L^q} \leq C\|\bar{\partial}f\|_{L^p}\left\|\frac{\varphi}{\bar{\partial}\psi}\right\|_{L^\infty} \leq C\delta^{-1}\|\bar{\partial}f\|_{L^p}.$$

Combining these four estimates with (19) we obtain

$$\|\bar{\partial}^{-1}(\varphi e^{-2i\psi/h}f)\|_{L^q(\Omega)} \leq h\|f\|_{W^{1,p}}(\delta^{-1} + \delta^{2/q-2}).$$

Combining this and (18) and optimizing by taking $\delta = h^{1/3}$, we deduce that

$$\|\bar{\partial}^{-1}(e^{-2i\psi/h}f)\|_{L^q(\Omega)} \leq h^{2/3}\|f\|_{W^{1,p}} \qquad (20)$$

(ii) <u>Case $q \geq 2$.</u> One can use the boundedness of $\bar{\partial}^{-1}$ on L^q to obtain

$$\|\bar{\partial}^{-1}((1-\varphi)e^{-2i\psi/h}f)\|_{L^q(\Omega)} \leq \|(1-\varphi)e^{-2i\psi/h}f\|_{L^q(\Omega)}$$
$$\leq C\delta^{\frac{2}{q}}\|f\|_{L^\infty}. \qquad (21)$$

Now since $\varphi = 0$ near z_0, we can use the identity

$$\bar{\partial}^{-1}(e^{-2i\psi/h}\varphi f) = \tfrac{1}{2}ih\left[e^{-2i\psi/h}\frac{\varphi f}{\partial_{\bar{z}}\psi} - \bar{\partial}^{-1}\left(e^{-2i\psi/h}\partial_{\bar{z}}\left(\frac{\varphi f}{\partial_{\bar{z}}\psi}\right)\right)\right]$$

and the boundedness of $\bar{\partial}^{-1}$ on L^q to deduce that for any $q \leq p$, (19) holds again with all the terms satisfying the same estimates as before, so that

$$\|\bar{\partial}^{-1}(e^{-2i\psi/h}\varphi f)\|_{L^q} \leq Ch\|f\|_{W^{1,p}}(\delta^{2/q-2} + \delta^{-1}) \leq Ch\delta^{2/q-2}\|f\|_{W^{1,p}}$$

since now $q \geq 2$. Now combine the above estimate with (21) and take $\delta = h^{\frac{1}{2}}$ we get

$$\|\bar{\partial}^{-1}(e^{-2i\psi/h}f)\|_{L^q} \leq h^{1/q}\|f\|_{W^{1,p}}$$

for $2 \leq q \leq p$. The estimate claimed in the lemma is obtained by interpolating the case $q < 2$ with $q > 2$. $\square$

From this lemma, we see directly that

$$\|S_h\|_{L^2 \to L^2} \leq Ch^{1/2+\epsilon}$$

for some $\epsilon > 0$ when the potential Q is $W^{1,p}(\Omega)$ with $p > 2$. Therefore (17) can be solved by using Neumann series by setting (for small $h > 0$)

$$r_h := -\sum_{j=0}^{\infty} S_h^j \bar{\partial}_h^{-1}(q'b) \tag{22}$$

as an element of any $L^2(\Omega)$. Substituting this expression for r into (16) when $a = 0$, we get

$$s_h = -\bar{\partial}_h^{*-1}(q r_h). \tag{23}$$

(ii) *Identification of the potential.* Using boundary identification as in Section 2, one can assume that $Q_1 - Q_2 = 0$ at $\partial \Omega$ if the Cauchy data at $\partial \Omega$ agree. Let F_h^1, F_h^2 be some CGO solutions of the form (15) constructed as above for respectively the operators $(D + Q_1)$ and $(D + Q_2^*)$, where Q_j are diagonal matrices defined as in(14) for some $q_j, q'_j \in W^{1,p}(\Omega)$. Assume that the boundary values of solutions of the equations $(D+Q_1)F = 0$ and $(D+Q_2)F = 0$ coincide. Then there exists a solution F_h of $(D + V_2)F = 0$ such that $F_h|_{\partial\Omega} = F_h^1|_{\partial\Omega}$; therefore $(D + V_2)(F_h^1 - F_h) = (Q_2 - Q_1)F_h^1$, and using Green's formula and the vanishing of $F_h^1 - F_h$ on the boundary,

$$0 = \int_\Omega \langle (D + Q_2)(F_h^1 - F_h), F_h^2 \rangle = \int_\Omega \langle (Q_2 - Q_1)F_h^1, F_h^2 \rangle. \tag{24}$$

This gives

$$0 = \int_\Omega (q'_2 - q'_1)e^{-2i\psi/h}\big(|b|^2 + \langle b, s_h^2 \rangle + \langle s_h^1, b \rangle\big) + (q_2 - q_1)e^{2i\psi/h} r_h^1 \overline{r_h^2}. \tag{25}$$

The last term is $\mathcal{O}(h^{1+2\epsilon})$ by Lemma 5.4. Using boundary identification as in Section 2, one has $Q_1 - Q_2 = 0$ at $\partial\Omega$ if the Cauchy data at $\partial\Omega$ agree. The stationary phase gives an asymptotic expansion as $h \to 0$

$$\int_\Omega (q'_2 - q'_1)e^{-2i\psi/h}|b|^2 = C_{z_0}|b(z_0)|^2(q'_2(z_0) - q'_1(z_0))h + o(h)$$

for some $C_{z_0} \neq 0$. Now for the remainder, we use (23) so that

$$\int_\Omega (q'_2 - q'_1)e^{-2i\psi/h}\langle b, s_h^2 \rangle = -\int_\Omega e^{2i\psi/h}\bar{\partial}_h^{-1}(b(q'_2 - q'_1))q_2 \overline{r_h^2}$$

and by Lemma 5.4, this is a $\mathcal{O}(h^{1+2\epsilon})$. The same argument applies for the last term and this shows that $q'_1 = q'_2$ since z_0 can be chosen arbitrarily in Ω. One can prove the same thing for $q_1 = q_2$ with the same argument but taking $b = 0$ and $a(z_0) \neq 0$ in the CGO. □

5B. *The case with a magnetic field.* A natural problem to consider is the inverse problem for the magnetic Schrödinger Laplacian on a smooth domain $\Omega \subset \mathbb{C}$:

$$L_{X,V} = (d+iX)^*(d+iX) + V,$$

where $V \in L^\infty(\Omega)$ is a potential and $X \in W^{1,\infty}(\Omega; \Lambda^1(\Omega))$ is a real-valued 1-form, called the magnetic potential, and $*$ denotes the adjoint with respect to the Euclidean L^2 product. The natural Cauchy data space associated to (X, V) in this case is

$$\mathcal{C}_{X,V} := \{(u, \partial_\nu u + iX(\nu)u)|_{\partial\Omega}; L_{X,V}u = 0, u \in H^1(\Omega)\}.$$

The magnetic field is the exact 2-form dX and it is easily seen that there is a gauge invariance since

$$\mathcal{C}_{X,V} = \mathcal{C}_{X+d\varphi,V} \tag{26}$$

for any function $\varphi \in W^{2,\infty}(\Omega)$ such that $\varphi = 0$ on $\partial\Omega$, simply by observing that $d + i(X + d\varphi) = e^{-i\varphi}(d+iX)e^{i\varphi}$. In fact, if φ is function in $\mathbb{R}/2\pi\mathbb{Z}$ with $e^{i\varphi}|_{\partial\Omega} = 1$, then (26) holds true as well and one can therefore identify at best $X + F^{-1}dF$ where F is any S^1-valued functions which are equal to 1 on the boundary. When Ω is simply connected, this is the same as identifying the magnetic field dX (or equivalently the curvature of the connection $d + iX$).

Theorem 5.5 [Sun 1993]. *Let $\Omega \subset \mathbb{C}$ be a smooth domain, then there exists an open dense set $\mathcal{O} \subset W^{1,\infty}(\Omega)$ such that for any $V \in \mathcal{O}$, there is a neighborhood $\mathcal{O}_V$ of V in $\mathcal{O}$ and a constant $C > 0$ such that if $X_1, X_2 \in W^{3,\infty}(\Omega, \Lambda^1(\Omega))$ are two magnetic potentials satisfying $\|dX_j\|_{W^{2,\infty}} \leq C$ and if $V_1, V_2 \in \mathcal{O}_V$, then*

$$\mathcal{C}_{X_1,V_1} = \mathcal{C}_{X_2,V_2} \implies d(X_1 - X_2) = 0 \text{ and } V_1 = V_2.$$

Another result in this direction, given for smooth data, is the following:

Theorem 5.6 [Kang and Uhlmann 2004]. *Let $\Omega \subset \mathbb{C}$ be a smooth domain, and let X_1, X_2, and V_1, V_2 be smooth 1-forms and potentials. Let $p > 2$, then there exists ϵ such that if $\|V_1\|_{W^{1,p}} \leq \epsilon$ and $\mathcal{C}_{X_1,V_1} = \mathcal{C}_{X_2,V_2}$, then $d(X_1 - X_2) = 0$ and $V_1 = V_2$.*

In particular, when $V_1 = V_2 = 0$, this allows one to identify a smooth magnetic potential up to gauge in a simply connected domain.

Further results. For partial data measurement, Imanuvilov, Uhlmann and Yamamoto [Imanuvilov et al. 2010a] obtain identification of magnetic field (up to gauge) and potential in a domain. Guillarmou and Tzou [2011b] recover a connection (i.e., a magnetic potential) up to gauge and the potential on a Riemann surface from full Cauchy data. See below for those results.

6. Inverse problems on Riemann surfaces

6A. *The metric problem.* As we mentioned in Section 1B, the Dirichlet-to-Neumann operator $\mathcal{N}_{M,g}$ associated to a Riemannian metric g on a surface M does not determine the isometry class, but can at best determine the conformal class of g (and g on the boundary). Recall that a conformal class on an oriented surface M is equivalent to a complex structure, that is a atlas with biholomorphic changes of charts in $\mathbb{C}$.

This problem has been solved by Lassas and Uhlmann [2001], with a simpler proof by Lassas, Taylor and Uhlmann, who were also able to determine the topology of M from $\mathcal{N}_g$:

Theorem 6.1 [Lassas et al. 2003]. *Let $(M_1, g_1), (M_2, g_1)$ be two Riemannian surfaces with the same boundaries Z. If $\mathcal{N}_{(M_1,g_1)} = \mathcal{N}_{(M_2,g_2)}$. There exists a diffeomorphism $\Phi : M_1 \to M_2$ such that $\Phi^* g_2$ is conformal to g_1 and $\Phi|_Z = \mathrm{Id}$.*

Sketch of proof. The idea is to use the Green's function and show that for analytic manifolds, the Green's function determines the manifold and the conformal class in dimension 2. Near a point $p \in Z$ of the boundary, there exists a neighborhood U_i in M_i and some diffeomorphisms $\psi_i : B \to U_i$ where B are half balls in the upper half plane $B = B(0, 1) \cap \{\mathrm{Im}(z) > 0\}$, such that $\psi_i(0) = p$ and $\psi_i^* g_i = e^{2f_i}|dz|^2$ for some functions f_i (isothermal coordinates). The associated Laplacians in these coordinates are given by $e^{-2f_i}\Delta$ where $\Delta = -\partial_z\partial_{\bar{z}}$ is the flat Laplacian in $\mathbb{R}^2$. The Green's functions $G_i(m, m')$ with Dirichlet conditions on M_i are the L^1 functions such that in the distribution sense $\Delta_{g_i} G_i(m, m') = \delta(m - m')$ is the distribution kernel of the Identity with $G_i(m, m') = 0$ if m or m' (but $m \neq m'$) is on the boundary Z. In general, we have following property as an application of Green's formula (this is standard, see for instance [Guillarmou and Sá Barreto 2009, Lemma 3.1]):

Lemma 6.2. *For a metric g on a manifold M, the Schwartz kernel of $\mathcal{N}_{(M,g)}$ is a singular integral kernel given for $y \neq y' \in \partial M$ by*

$$\mathcal{N}_{(M,g)}(y, y') = \partial_\nu \partial_{\nu'} G(m, m')|_{(m,m')=(y,y')}$$

where G is the Green kernel with Dirichlet conditions and $\partial_\nu, \partial_{\nu'}$ denote the normal derivative to the boundary in the left and right variable.

Therefore we can deduce that if $\mathcal{N}_{(M_1,g_1)} = \mathcal{N}_{(M_2,g_2)}$ then $G_1(m, m') = G_2(m, m')$ when viewed in the chart B, since they solve the same elliptic problem with the same local Cauchy data (Dirichlet and Neumann), this is by unique continuation for a scalar elliptic PDE. The idea of [Lassas et al. 2003] is to define an analytic embedding of M to a Hilbert space through the Green's function:

they define
$$\mathcal{G}_i : M_i \to L^2(B), \quad \mathcal{G}_i(m) := G_i(m, \psi_i(\cdot)).$$

Notice by the remark above that $\mathcal{G}_1 \circ \psi_1 = \mathcal{G}_2 \circ \psi_2$. Now, the crucial fact is that each M_i has a real analytic atlas (induced by the complex local coordinates z), the metric in these charts is conformal to the Euclidean metric $|dz|^2$, and since the Green's function solves $\partial_z \partial_{\bar{z}} G_i(z, z') = 0$ when $z \neq z'$, the functions G_i are analytic outside the diagonal (with respect to the analytic structure on M_i). This implies that $\mathcal{G}_i$ are analytic maps and in fact it can be shown that they are embeddings. Indeed, assume that $\nabla \mathcal{G}_i(m) = 0$ at some $m \in M_i$, then $\nabla_m G_i(m; m') = 0$ for all $m' \in U_i$, which by analyticity gives $\nabla_m G_i(m; m') = 0$ for all $m' \neq m$, but that contradicts the asymptotic behavior of G_i at the diagonal (in local complex coordinates z)
$$G_i(z, z') = \frac{1}{2\pi} \log(|z - z'|) + C(z) + o(1) \quad \text{as } |z - z'| \to 0$$

for some function $C(z)$, and thus $\mathcal{G}_i$ is a local analytic diffeomorphism; it is also injective since $G_i(m_1, m') = G_i(m_2, m')$ for all $m' \in U_i$ implies the same for all $m' \in M_i$ and the asymptotic at the diagonal implies $m_1 = m_2$. It remains to show that $\mathcal{G}_1(M_1) = \mathcal{G}_2(M_2)$ and $\mathcal{G}_2^{-1} \circ \mathcal{G}_1$ is a conformal map. We already know that $\mathcal{G}_1(U_1) = \mathcal{G}_2(U_2)$ and it can be proved by analytic continuation and the implicit function theorem that the set $\{m \in M_1; \mathcal{G}_1(m) \in \mathcal{G}_2(M_2)\}$ is the whole of M_1. Essentially, this amounts to say that two submanifolds with boundary of $L^2(B)$ obtained by analytic embeddings are the same if they are equal on an open set. The map $J := \mathcal{G}_2^{-1} \circ \mathcal{G}_1$ is analytic and is such that $G_1(m, m') = G_2(J(m), J(m'))$ for $m, m' \in U_1$ and $m \neq m'$ since $\mathcal{G}_1 \circ \psi_1 = \mathcal{G}_2 \circ \psi_2$. But this identity extends to $M_1 \times M_1 \setminus \{m = m'\}$ by analytic continuation. Then by looking at the asymptotic of the Green's kernels near a diagonal point $(m, m) \in M_1 \times M_1$, this implies that there exist functions $C_1(m), C_2(m)$ such that

$$\log(d_{g_1}(m, m_t)) + C_1(m) = \log(d_{g_2}(J(m), J(m_t))) + C_2(J(m)) + o(1)$$

as $t \to 0$, if $t \mapsto m_t$ is a smooth curve such that $m_0 = m$ and d_{g_i} denotes the Riemannian distance. Writing this equation in terms of the metric (here we use the notation $\dot{m}_s = \partial_s m(s)$) we get

$$\log(|\dot{m}_0|_{g_1(m)}) + C_1(m) = \log(|\dot{m}_0|_{J^* g_2(m)}) + C_2(J(m)) + o(1) \quad \text{as } t \to 0,$$

and since $\dot{m}_0$ can be chosen arbitrarily, one deduces $g_1 = e^\omega J^* g_2$ for some function $\omega \in C^\infty(M_1)$. □

Remarks and further results. In fact, the agreement of the Dirichlet-to-Neumann operators on only a open set $\Gamma \subset Z = \partial M_i$ is sufficient to run this argument, as was shown in [Lassas and Uhlmann 2001]. The identification for Riemann surfaces was later shown using a different approach by Belishev [2003]. Then Henkin and Michel [2007] gave another proof of this result by using embedding of the surface into $\mathbb{C}^n$, and we should point out moreover that the method in [Henkin and Michel 2007] gives a reconstruction procedure for the Riemann surface, which is not quite the case with the previous results. In fact, Henkin and Michel [2007] show that the action of the Dirichlet-to-Neumann map $\mathcal{N}_g$ on 3 generic functions (u_0, u_1, u_2) on ∂M is sufficient to determine the Riemann surface (M, g) (as a surface with complex structure). More precisely:

Theorem 6.3 [Henkin and Michel 2007]. *Let (M_1, g_1), (M_2, g_2) be two Riemann surfaces with the same boundary Z, such that there exist some real-valued smooth function $u = (u_0, u_1, u_2) : Z^3 \to \mathbb{R}$ with $\mathcal{N}_{g_1} u = \mathcal{N}_{g_2} u$ and such that*

$$\Phi : Z \to \mathbb{C}^2, \quad \Phi : m \mapsto \left(\frac{(\mathcal{N}_{g_1} - i\partial_\tau) u_1}{(\mathcal{N}_{g_1} - i\partial_\tau) u_0}, \frac{(\mathcal{N}_{g_1} - i\partial_\tau) u_2}{(\mathcal{N}_{g_1} - i\partial_\tau) u_0} \right)$$

is an embedding, where ∂_τ is an positively oriented length one tangent vector field to the boundary Z. Then (M_1, g_1) is isomorphic to (M_2, g_2) as a Riemann surface.

The condition about the embedding is claimed to be generic, in a way, by the authors. The proof of this theorem is based on complex geometric arguments.

Finally, we refer to [Salo 2013] in this volume for a survey on the inverse problem on manifolds in dimension $n > 2$.

6B. *Identification of a conductivity or a potential on a fixed Riemann surface.* The same type of problem can be considered when the background setting is not a domain of $\mathbb{C}$ but a Riemann surface with boundary. Thus, we fix a Riemann surface with boundary (M, g), a conductivity is a positive symmetric endomorphism γ of TM and we consider the elliptic equation

$$\operatorname{div}_g (\gamma \nabla^g u) = 0$$

where $\nabla^g u$ is the gradient of u defined by $g(\nabla^g u, X) = X(u)$ for all vector field X, and $\mathcal{L}_X(\operatorname{dvol}_g) = \operatorname{div}_g(X) \operatorname{dvol}_g$ if $\mathcal{L}_X$ is the Lie derivative with respect to X. The Dirichlet-to-Neumann operator associated to γ is still denoted $\mathcal{N}_\gamma$ and defined by $\mathcal{N}_\gamma f := g(\gamma \nabla^g u, \nu)$ where ν is the normal outward pointing vector field at the boundary, and we want to see if $\gamma \to \mathcal{N}_\gamma$ is injective up to gauge. An equivalent question is to consider the elliptic equation

$$d(\gamma du) = 0, \quad u|_{\partial M} = f$$

where $\gamma : M \to \mathrm{End}(T^*M)$ is a section of positive symmetric (with respect to a given metric g) endomorphisms on 1-forms and the Dirichlet-to-Neumann map is $\mathcal{N}_\gamma f := (\gamma(du)(v))|_{\partial M} \in \Lambda^1(M)|_{\partial M}$.

Viewing M_0 as a subset of a closed Riemann surface M of genus g, Henkin and Michel [2008] consider $g + 2$ points $A = \{A_1, \ldots, A_{g+2}\}$ in $M \setminus M_0$ and use an immersion $j = (f_1, f_2) \colon M \subset A \to \mathbb{C}^2$ of M into $\mathbb{C}^2$ using 2 independent meromorphic functions f_1, f_2 on M with poles at A and they assume the complex curve $j(M)$ can be written under the form $j(M) = \{(z_1, z_2) \in \mathbb{C}^2; P(z_1, z_2) = 0\}$ with P a homogeneous holomorphic polynomial such that $\nabla P \neq 0$ on $j(M)$, i.e., M can be viewed as a regular complex algebraic curve. They proved:

Theorem 6.4 [Henkin and Michel 2008]. *If $\gamma \in C^3(M_0)$ is an isotropic conductivity on a fixed Riemann surface M_0 with boundary, which can be embedded in $\mathbb{C}^2$ as a subset of a regular complex algebraic curve as described above, then the Dirichlet-to-Neumann map $\mathcal{N}_\gamma$ determines γ and γ can be reconstructed by an explicit procedure.*

This result has been extended to anisotropic conductivities:

Theorem 6.5 [Henkin and Santacesaria 2010]. *Let M be a C^3 surface with boundary, and let γ_1, γ_2 be two C^3 positive definite symmetric endomorphisms of TM (i.e., two anisotropic conductivities on M). If the Dirichlet-to-Neumann operators agree, $\mathcal{N}_{\gamma_1} = \mathcal{N}_{\gamma_2}$, then there exists a C^3 diffeomorphism $F \colon M \to M$ such that $F|_{\partial M} = \mathrm{Id}$ and $F_* \gamma_1 = \gamma_2$.*

As for the flat case, the case where γ is isotropic (i.e., when it is of the form $\gamma \mathrm{Id}$ for some function γ) can be reduced to the case of the equation $\Delta_g + V$ with $V = -\Delta_g \gamma^{1/2}/\gamma^{1/2}$. We have:

Theorem 6.6 [Guillarmou and Tzou 2009]. *If V_1, V_2 are two potentials[5] in $W^{1,p}(M)$ with $p > 2$ on a Riemann surface (M, g) with boundary, and the map $\mathcal{N}_{V_1} = \mathcal{N}_{V_2}$ agree, then $V_1 = V_2$.*

In particular this allows to identify isotropic conductivities in $W^{3,p}(M_0)$ on the Riemann surface and to identify a metric in its conformal class.

Arguments in the proof. The proof is based on the Bukhgeim method, as described above, but in this geometric setting one needs to find holomorphic Morse functions with prescribed critical points (the function $(z - z_0)^2$ does not quite make sense anymore). Let us discuss the construction of the phase in this setting.

---

[5]The regularity of the potential is stated to be C^∞ in [Guillarmou and Tzou 2009] for convenience of exposition, but the $W^{1,p}(M)$ regularity result follows from [Guillarmou and Tzou 2011b].

Riemann surfaces and complex structure. A conformal class $[g]$ on an oriented closed surface M makes M into a closed Riemann surface, i.e., a closed surface equipped with a complex structure via holomorphic charts $z_\alpha : U_\alpha \to \mathbb{C}$. The Hodge star operator $*$ acts on the cotangent bundle T^*M, its eigenvalues are $\pm i$ and the respective eigenspace $T^*_{1,0}M := \ker(*+i\operatorname{Id})$ and $T^*_{0,1}M := \ker(*-i\operatorname{Id})$ are subbundles of the complexified cotangent bundle $\mathbb{C}T^*M$, and we have a splitting $\mathbb{C}T^*M = T^*_{1,0}M \oplus T^*_{0,1}M$. The Hodge $*$ operator is conformally invariant on 1-forms on M, the complex structure depends only on the conformal class of g (and orientation). In holomorphic coordinates $z = x + iy$ in a chart U_α, one has $\star(u\,dx + v\,dy) = -v\,dx + u\,dy$ and

$$T^*_{1,0}M|_{U_\alpha} \simeq \mathbb{C}\,dz, \quad T^*_{0,1}M|_{U_\alpha} \simeq \mathbb{C}\,d\bar{z}$$

where $dz = dx + i\,dy$ and $d\bar{z} = dx - i\,dy$. We define the natural projections induced by the splitting of $\mathbb{C}T^*M$

$$\pi_{1,0} : \mathbb{C}T^*M \to T^*_{1,0}M, \quad \pi_{0,1} : \mathbb{C}T^*M \to T^*_{0,1}M.$$

The exterior derivative d defines the De Rham complex $0 \to \Lambda^0 \to \Lambda^1 \to \Lambda^2 \to 0$ where $\Lambda^k := \Lambda^k T^*M$ denotes the real bundle of k-forms on M. Let us denote $\mathbb{C}\Lambda^k$ the complexification of Λ^k, then the ∂ and $\bar{\partial}$ operators can be defined as differential operators $\partial : \mathbb{C}\Lambda^0 \to T^*_{1,0}M$ and $\bar{\partial} : \mathbb{C}\Lambda_0 \to T^*_{0,1}M$ by

$$\partial f := \pi_{1,0} df, \quad \bar{\partial} := \pi_{0,1} df,$$

they satisfy $d = \partial + \bar{\partial}$ and are expressed in holomorphic coordinates by

$$\partial f = \partial_z f\,dz, \quad \bar{\partial} f = \partial_{\bar{z}} f\,d\bar{z}.$$

with $\partial_z := \frac{1}{2}(\partial_x - i\partial_y)$ and $\partial_{\bar{z}} := \frac{1}{2}(\partial_x + i\partial_y)$. Similarly, one can define the ∂ and $\bar{\partial}$ operators from $\mathbb{C}\Lambda^1$ to $\mathbb{C}\Lambda^2$ by setting

$$\partial(\omega_{1,0} + \omega_{0,1}) := d\omega_{0,1}, \quad \bar{\partial}(\omega_{1,0} + \omega_{0,1}) := d\omega_{1,0}$$

if $\omega_{0,1} \in T^*_{0,1}M$ and $\omega_{1,0} \in T^*_{1,0}M$. In coordinates this is simply

$$\partial(u\,dz + v\,d\bar{z}) = \partial v \wedge d\bar{z}, \quad \bar{\partial}(u\,dz + v\,d\bar{z}) = \bar{\partial} u \wedge dz.$$

The Laplacian acting on functions is defined by

$$\Delta f := -2i * \bar{\partial}\partial f = d^*d$$

where d^* is the adjoint of d through the metric g and $*$ is the Hodge star operator mapping Λ^2 to Λ^0 and induced by g.

The holomorphic phase. To construct holomorphic functions with prescribed critical points, we use the Riemann–Roch theorem: a divisor D on M is an element

$$D = ((p_1, n_1), \ldots, (p_k, n_k)) \in (M \times \mathbb{Z})^k, \quad \text{where } k \in \mathbb{N}$$

which will also be denoted $D = \prod_{i=1}^k p_i^{n_i}$ or $D = \prod_{p \in M} p^{\alpha(p)}$ where $\alpha(p) = 0$ for all p except $\alpha(p_i) = n_i$. The inverse divisor of D is defined to be $D^{-1} := \prod_{p \in M} p^{-\alpha(p)}$ and the degree of the divisor D is defined by $\deg(D) := \sum_{i=1}^k n_i = \sum_{p \in M} \alpha(p)$. A meromorphic function on M is said to have divisor D if $(f) := \prod_{p \in M} p^{\operatorname{ord}(p)}$ is equal to D, where $\operatorname{ord}(p)$ denotes the order of p as a pole or zero of f (with positive sign convention for zeros). Notice that in this case we have $\deg(f) = 0$. For divisors $D' = \prod_{p \in M} p^{\alpha'(p)}$ and $D = \prod_{p \in M} p^{\alpha(p)}$, we say that $D' \geq D$ if $\alpha'(p) \geq \alpha(p)$ for all $p \in M$. The same exact notions apply for meromorphic 1-forms on M. Then we define for a divisor D

$$r(D) := \dim(\{f \text{ meromorphic function on } M; (f) \geq D\} \cup \{0\}),$$

$$i(D) := \dim(\{u \text{ meromorphic 1 form on } M; (u) \geq D\} \cup \{0\}).$$

The Riemann–Roch theorem states the following identity: for any divisor D on the closed Riemann surface M of genus g,

$$r(D^{-1}) = i(D) + \deg(D) - g + 1. \tag{27}$$

Notice also that for any divisor D with $\deg(D) > 0$, one has $r(D) = 0$ since $\deg(f) = 0$ for all f meromorphic. By [Farkas and Kra 1992, p. 70, Theorem], let D be a divisor, then for any nonzero meromorphic 1-form ω on M, one has

$$i(D) = r(D(\omega)^{-1}) \tag{28}$$

which is thus independent of ω. For instance, if $D = 1$, we know that the only holomorphic function on M is 1 and one has $1 = r(1) = r((\omega)^{-1}) - g + 1$ and thus $r((\omega)^{-1}) = g$ if ω is a nonzero meromorphic 1 form. Now if $D = (\omega)$, we obtain again from (27)

$$g = r((\omega)^{-1}) = 2 - g + \deg((\omega))$$

which gives $\deg((\omega)) = 2(g-1)$ for any nonzero meromorphic 1-form ω. In particular, if D is a divisor such that $\deg(D) > 2(g-1)$, then we get $\deg(D(\omega)^{-1}) = \deg(D) - 2(g-1) > 0$ and thus $i(D) = r(D(\omega)^{-1}) = 0$, which implies by (27)

$$\deg(D) > 2(g-1) \implies r(D^{-1}) = \deg(D) - g + 1 \geq g. \tag{29}$$

In particular, taking M_0 as a subset of a closed Riemann surface M with genus g (for instance by doubling the surface along the boundary and extending the conformal class smoothly), we know that by assigning a pole at $p \in M \setminus M_0$ of large order $N > 2(g-1)$, there exists a vector space V of dimension $N - 2(g-1)$ of functions holomorphic on M_0 meromorphic in M with a unique pole of order at most N at p (take the divisor $D = p^N$). This implies by dimension count that for all $z_0 \in M_0$

(i) there exists a vector subspace $V(z_0) \subset$ of dimension $\geq N - 2$ of holomorphic functions with a zero of order 2 at z_0.

This creates critical points everywhere we want, but unfortunately the functions will not a priori be Morse. Similarly, taking the divisors $D_1 = z_0^{-1} p^N$ and $D_2 = z_0^{-2} p^N$ and since their degree is larger than $2(g-1)$, one has $r(D_1^{-1}) - r(D_2^{-1}) = 1$ and thus

(ii) $\exists f \in V$, holomorphic on M_0 with a zero of order exactly 1 at z_0.

Using (i) and (ii), we show:

Lemma 6.7. *There exists a dense set of points z_0 in M_0 for which there exists a Morse holomorphic function with a critical point at z_0.*

By Cauchy–Riemann equations, a holomorphic function is Morse if and only if its real part is Morse. To prove this, one can use some transversality arguments of Uhlenbeck [1976]. Take the real vector space $H = \text{Re}(V)$ and define the map

$$F : H \times M_0 \to T^* M_0, \quad F : (u, m) \mapsto (m, du(m)).$$

A real function u is Morse if $F_u := F(u, \cdot)$ is transverse to the zero section $T_0^* M_0 = \{(m, 0) \in T^* M_0\}$ in $T^* M_0$ in the sense that

$$\text{Im}(dF_u(m)) + T_{F_u(m)} T_0^* M_0 = T^*_{F_u(p)} M_0$$

for all m such that $F(u, m) = 0$. Now an application of Sard's theorem gives that if F is transverse to $T_0^* M_0$ (in the sense above but with $\text{Im}(dF)(u, m)$ instead of $\text{Im}(dF_u(m))$) then

$$\{u \in H; F_u \text{ is transverse to } T_0^* M_0\}$$

has Lebesgue measure 0 in H. A little inspection shows that the transversality of F with respect to $T_0^* M_0$ can be proved if we can show that at any z_0 then there exist a function $f \in V$ such that $\partial_{z_0} f \neq 0$. But this is insured by (ii), and we conclude that the set of Morse functions in V has a complement in V of 0 measure (thus is dense in V). Taking $z_0 \in M_0$ and a function f in $V(z_0)$ of (i), then for any $\epsilon > 0$ there exists a function Φ in V (thus Morse) with $|\Phi - f| \leq \epsilon$ with respect to any norm on the finite dimensional space V, and thus in particular by Rouché theorem its critical point is going to be in a neighborhood of size $o(1)$ of z_0 as $\epsilon \to 0$. This concludes the proof of the lemma and gives us a phase for constructing CGO.

The rest of the proof. This goes similarly to what we explained about the Bukhgeim theorem. The main technical differences here are that we need to construct a global right inverse for $\bar{\partial}$ and its adjoint and since Φ may have several critical points, we choose the amplitudes (which are holomorphic or antiholomorphic functions and 1-forms) to vanish at all critical points except the point z_0 where we want to identify the potential, so that those other points do not contribute to the stationary phase. We refer to [Guillarmou and Tzou 2009; 2011b] for more details. □

Remark. The regularity $W^{1,p}$ with $p > 2$ on surfaces was proved in [Guillarmou and Tzou 2011b] for the potential using Bukhgeim method and was improved later to C^α for all $\alpha > 0$ for domains of $\mathbb{C}$ by Imanuvilov and Yamamoto [2011]. See also [Blasten 2011] for the $W^{1,p}$ regularity for domains.

6C. *Inverse problems for systems and magnetic Schrödinger operators.* Let $\pi : E \to M$ be a Hermitian complex vector bundle on a Riemann surface with boundary and ∇ be a Hermitian connection on E. Such a bundle is trivializable and the connection in a trivialization is of the form $\nabla = d + iX$ for self-adjoint matrix-valued 1-form X. We can define the magnetic Schrödinger operator $L_{\nabla,V}$ (or connection Laplacian) by

$$L_{\nabla,V} := \nabla^*\nabla + V$$

where V is a section of the endomorphism on E (i.e., a potential). This elliptic operator has Cauchy data defined by

$$\mathscr{C}_{\nabla,V} := \{(u|_{\partial M}, \nabla_\nu u|_{\partial M}); L_{\nabla,V}u = 0, u \in H^1(M, E)\}.$$

It is then natural to ask whether $\mathscr{C}_{\nabla,V}$ uniquely determines the connection ∇ and the endomorphism V. The answer is no since there is a gauge invariance. Indeed, consider a section F of $\mathrm{End}(E)$ satisfying $F^* = F^{-1}$ and $F|_{\partial M} = \mathrm{Id}$. Then it is easy to see that $\mathscr{C}_{\nabla,V} = \mathscr{C}_{F^*\nabla F, F^*VF}$. Therefore we can at best hope to identify ∇ and V up to gauge. The following result is proved:

Theorem 6.8 [Albin et al. 2011]. *Let ∇^1 and ∇^2 be two Hermitian connections on a smooth Hermitian bundle E, of complex dimension n and let V_1, V_2 be two sections of the bundle $\mathrm{End}(E)$. We assume that the connection forms of ∇^j have the regularity $C^r \cap W^{s,p}(M)$ with $0 < r < s$, $p \in (1, \infty)$ satisfy $r + s > 1$, $r \in \mathbb{N}$, $sp > 2n + 2$ and that $V_j \in W^{1,q}(M)$ with $q > 2$. Let $L_j := (\nabla^j)^*\nabla^j + V_j$ and assume that the Cauchy data spaces agree $\mathscr{C}_{\nabla^1,V_1} = \mathscr{C}_{\nabla^2,V_2}$, then there exists a* unitary endomorphism $F \in C^1(M; \text{End}(E))$, satisfying $F|_{\partial M} = Id$, such that $\nabla^1 = F^*\nabla^2 F$ and $V_1 = F^*V_2 F$.

Sketch of proof. One uses reduction to a $\bar{\partial}$-system and the Bukhgeim method [2008]. Indeed, if we denote by $A_j = \pi_{0,1} X_j$ and $A_j^* = \pi_{1,0} X_j$ the $(T^{0,1}M)^*$ and $(T^{1,0}M)^*$ component of the connection 1-form X_j of ∇^j, (in a fixed trivialization), we have

$$\nabla^j = d + iX_j = (\bar{\partial} + iA_j) + (\partial + iA_j^*).$$

Thus, if u_j is a solution to $L_{\nabla^j, V_j} u_j = 0$ we have

$$\begin{pmatrix} 0 & (\bar{\partial}+iA_j)^* \\ (\bar{\partial}+iA_j) & 0 \end{pmatrix} \begin{pmatrix} u_j \\ \omega_j \end{pmatrix} + \begin{pmatrix} *(dX_j + X_j \wedge X_j) + V_j & 0 \\ 0 & -1 \end{pmatrix} \begin{pmatrix} u_j \\ \omega_j \end{pmatrix}$$
$$= 0. \quad (30)$$

if we set $\omega_j := (\bar{\partial} + iA_j)u_j$ It is clear that knowledge of the Cauchy data for the second order equation is equivalent to knowledge of the Cauchy data for (30). We would like to transform this problem, via conjugation, to the type of system considered by Bukhgeim (and explained above for Riemann surfaces) where only the $\bar{\partial}$ and $\bar{\partial}^*$ operator appear on the off diagonal.

As shown in [Kobayashi 1987, Chapter 1, Proposition 3.7], the operator $\partial + iA_j^*$ induces a holomorphic structure on E and this structure is holomorphically trivializable since M has nonempty boundary; see [Forster 1991, Theorems 30.1 and 30.4]. This means that there exists an invertible section F_j of $\text{End}(E)$ that is annihilated by the operator $\partial + iA_j^*$. More precisely, $\partial F_j = -iA_j^* F_j$. Taking adjoint of both sides we get that $(F_j^*)^{-1}\bar{\partial} F_j^* = iA_j$. Therefore, $(F_j^*)^{-1}\bar{\partial} F_j^* u = (\bar{\partial} + iA_j)u$ for all smooth sections u of E. We would like to remark that such endomorphisms are by no means unique.

We are now in a position to transform system (30) into a simplified $\bar{\partial}$ system. Set $(\tilde{u}_j, \tilde{\omega}_j) := (F_j^* u_j, F_j^{-1}\omega_j)$ then system (30) is equivalent to

$$\begin{pmatrix} 0 & \bar{\partial}^* \\ \bar{\partial} & 0 \end{pmatrix} \begin{pmatrix} \tilde{u}_j \\ \tilde{\omega}_j \end{pmatrix} + \begin{pmatrix} F_j^{-1}(*(dX_j + X_j \wedge X_j) + V_j)(F_j^*)^{-1} & 0 \\ 0 & -F_j^* F_j \end{pmatrix} \begin{pmatrix} \tilde{u}_j \\ \tilde{\omega}_j \end{pmatrix}$$
$$= 0. \quad (31)$$

However, the fact that the systems (30) have identical Cauchy data does not a priori ensure that the systems (31) have the same Cauchy data for $j = 1, 2$ since the conjugation factors F_j may not necessarily agree on the boundary. An important part of the resolution of the problem is to show that the F_j can be chosen to agree on the boundary if the Cauchy data agree.

Lemma 6.9. *If the Cauchy data for the systems* (30) *agree for* $j = 1, 2$, *then there exist invertible sections* F_j *of* $\mathrm{End}(E)$ *satisfying* $(F_j^*)^{-1} \bar{\partial} F_j^* = i A_j$ *and* $F_1|_{\partial M} = F_2|_{\partial M}$

Idea of proof in the case of a line bundle. (Things are similar for higher-rank bundles.) The idea is to use CGO for the system (30) of the form

$$U_h^1 = \begin{pmatrix} e^{\Phi/h}(F_1^*)^{-1}(a + r_h^1) \\ e^{\overline{\Phi}/h} F_1 s_h^1 \end{pmatrix}, \quad U_h^2 = \begin{pmatrix} e^{-\Phi/h}(F_2^*)^{-1} r_h^2 \\ e^{-\overline{\Phi}/h} F_2 (b + s_h^2) \end{pmatrix}$$

with $\bar{\partial} a = 0, \bar{\partial}^* b = 0$, and writing this system as

$$(D + Q_j) U = 0, \quad \text{with } D = \begin{pmatrix} 0 & \bar{\partial}^* \\ \bar{\partial} & 0 \end{pmatrix},$$

the following integral identity follows from the equality of Cauchy data spaces:

$$0 = \int_M \langle (Q_2 - Q_1) U_h^1, U_h^2 \rangle$$
$$= i \int_M \langle (A_2 - A_1)(F_1^*)^{-1}(a + r_h^1), F_2(b + s_h^2) \rangle + o(1)$$
$$= i \int_M \langle F_2^*(A_2 - A_1)(F_1^*)^{-1} a, b \rangle + o(1) = \int_M \langle \bar{\partial}(F_2^*(F_1^*)^{-1} a), b \rangle + o(1)$$

as $h \to 0$. Letting $h \to 0$, and applying Stokes one gets $0 = \int_{\partial M} F_2^*(F_1^*)^{-1} a i_{\partial M}^* \bar{b}$, for all b antiholomorphic 1-form, but by Hodge theory [Guillarmou and Tzou 2011b, Lemma 4.1], this means exactly that $F_2^*(F_1^*)^{-1} a|_{\partial M}$ is the boundary value of a holomorphic function. Taking $a = 1$, we set F the holomorphic function with restriction $F_2^*(F_1^*)^{-1}$ at ∂M, this is invertible since one can apply the same argument by switching the role of $j = 1, 2$ and this gives a holomorphic function which multiplied by F is equal to 1 on ∂M, thus is equal to 1 on M. Now modify F_1^* by multiplying it by F so that $(F_1 F^*)^*$ and F_2^* agree at ∂M and $F_1 F^*$ and F_2 play the role of F_1 and F_2 in the statement of the lemma. □

This lemma allows us to choose the conjugation factors for $j = 1, 2$ such that the conjugated systems (31) for $j = 1$ and $j = 2$ has the same Cauchy data. We have now reduced the problem to one or the type considered in [Bukhgeim 2008], but with higher rank. The same techniques used in that paper and adapted to Riemann surfaces in [Guillarmou and Tzou 2011b] can then be applied to deduce that

$$F_1^{-1}(*(dX_1 + X_1 \wedge X_1) + V_1)(F_1^*)^{-1} = F_2^{-1}(*(dX_2 + X_2 \wedge X_2) + V_2)(F_2^*)^{-1} \tag{32}$$

and
$$F_1^* F_1 = F_2^* F_2$$
with the boundary condition $F_1 = F_2$ on ∂M. We then set $F = (F_1^*)^{-1} F_2^* \in \operatorname{End}(E)$ which by using $F = F_1 F_2^{-1}$ satisfies $F^* = F^{-1}$ and moreover $F = \operatorname{Id}$ on ∂M. Then it is easy to see that F solves the homogeneous elliptic equation
$$\bar{\partial} F + i A_1 F - i F A_2 = 0,$$
and by taking adjoint and using $F^* = F^{-1}$ this implies the equality of the connections $F^{-1}(d + i X_1) F = d + i X_2$, and thus $F^{-1} V_1 F = V_2$ by (32). □

Remark. The proof just described also allows one to identify zeroth-order terms up to gauge in a first order system
$$\begin{pmatrix} 0 & \bar{\partial}^* \\ \bar{\partial} & 0 \end{pmatrix} + \begin{pmatrix} Q_+ & A' \\ A & Q_- \end{pmatrix}.$$
See [Albin et al. 2011, Proposition 3] for details and precise statements.

7. Inverse problems with partial data measurements

7A. Identification of a potential from partial measurements. An important question related to Calderón's problem is to identify a conductivity or a potential from measurements on an open subset of the boundary instead of the whole boundary. The first partial data result seems to be in dimension $n > 2$ by [Bukhgeim and Uhlmann 2002], where Carleman estimates were fundamental to approach this problem. In dimension $n = 2$, Imanuvilov, Uhlmann and Yamamoto gave a proof by combining the method of [Bukhgeim 2008] with Carleman estimates:

Theorem 7.1 [Imanuvilov et al. 2010b]. *Let $\Omega \subset \mathbb{C}$ be a domain and let $V_1, V_2 \in C^{2,\alpha}(\Omega)$ be two potentials. Let $\Gamma \subset \partial\Omega$, and consider the partial Cauchy data on Γ,*
$$\mathcal{C}_{V_i}^\Gamma := \{(u, \partial_\nu u)|_\Gamma ; (\Delta + V_i) u = 0, u \in H^1(\Omega), u = 0 \text{ in } \partial\Omega \setminus \Gamma\}. \quad (33)$$
If $\mathcal{C}_{V_1}^\Gamma = \mathcal{C}_{V_2}^\Gamma$ then $V_1 = V_2$.

We later extended this result to Riemann surfaces:

Theorem 7.2 [Guillarmou and Tzou 2011a]. *Let (M, g) be a Riemann surface with smooth boundary and let $\Gamma \subset \partial M$ be any open subset of the boundary. Suppose V_1 and V_2 are $C^{1,\alpha}(M)$ potentials, for some $\alpha > 0$, such that $\mathcal{C}_{V_1}^\Gamma = \mathcal{C}_{V_2}^\Gamma$ with the notation similar to (33), then $V_1 = V_2$.*

Sketch of proof. We follow the method of [Imanuvilov et al. 2010b], but use geometric arguments to construct the phases of the CGO and we make adaptations for constructing appropriate CGO on Riemann surfaces.

To motivate the proof, we observe that since $\mathscr{C}_{V_1,\Gamma} = \mathscr{C}_{V_2,\Gamma}$, for any two solutions of $(\Delta_g + V_j)u_j = 0$ such that $u_1|_{\partial M}, u_2|_{\partial M} \in C_0^\infty(\Gamma)$, we have the boundary integral identity

$$\int_M \overline{u_1}(V_1 - V_2)u_2 = 0. \tag{34}$$

We wish to extend Bukhgeim's idea of using stationary phase expansion discussed in Section 5. To this end, we look to construct solutions of the type discussed in Section 3 having the additional property that when restricted to the boundary they are compactly supported in Γ.

The idea of [Imanuvilov et al. 2010b] is to first construct harmonic functions of exponential type which vanish on $\partial M \setminus \Gamma$, and correct them using a Carleman estimate. As such, suppose we have a holomorphic and Morse function $\Phi = \phi + i\psi$ with prescribed critical points which is purely real on $\partial M \setminus \Gamma$, and let a be a holomorphic function purely real on Γ. Then the harmonic function defined by $u_0 := e^{\Phi/h}a - e^{\overline{\Phi}/h}\overline{a}$ is a harmonic function which vanishes outside of Γ and we will search for solutions of $(\Delta + V)u = 0$ of the form $u = u_0 + e^{\phi/h}r_h$ where r_h will be small in L^2 as $h \to 0$. Therefore, the first key step in constructing CGO which vanish on $\partial M \setminus \Gamma$ is to construct such a holomorphic function Φ on a general Riemann surface. Observe that since the potentials V_j are assumed to be continuous, it suffices to prove that $V_1(p) = V_2(p)$ for a dense subset of $p \in M$. In [Guillarmou and Tzou 2011a], we proved:

Lemma 7.3. *Let (M, g) be a Riemann surface with boundaries with $\Gamma \subset \partial M$ be an open subset. Then there exists a dense subset $\mathcal{A} \subset M$ such that for all $p \in \mathcal{A}$ there exists a holomorphic Morse function Φ which is purely real on $\partial M \setminus \Gamma$ such that $\partial \Phi(p) = 0$.*

The existence of Φ is also proved in [Imanuvilov et al. 2010b] for all points when $M = \Omega$ is a domain of $\mathbb{C}$, using variational methods.

Outline of proof of Lemma 7.3. We discuss briefly the procedure for constructing such holomorphic phase functions. For a detailed outline, we refer to [Guillarmou and Tzou 2011a] for the geometric approach which works for general surfaces and [Imanuvilov et al. 2010b] for the variational approach which works for planar domains. In this survey we will describe the geometric approach. To this end, we view holomorphic functions as sections of the trivial line bundle $E := \mathbb{C} \times M$ which is annihilated by the linear elliptic operator $\bar{\partial}$. And the fact that Φ is purely real on $\partial M \setminus \Gamma$ will be interpreted as $\Phi|_{\partial M}$ being a section of the totally real rank 1 subbundle[6] $F \subset E|_{\partial M}$ over ∂M such that

$$F|_{\partial M \setminus \Gamma} = \mathbb{R} \times \partial M \setminus \Gamma. \tag{35}$$

Note that while F needs to be $\mathbb{R} \times \partial M \setminus \Gamma$ on $\partial M \setminus \Gamma$, we have the freedom to choose it as we wish on Γ as long as F remains a smooth bundle of real rank one. The question is, what choices of F satisfying condition (35) makes the $\bar\partial$ operator acting on

$$H^k(M, F) = \{u \in H^k(M); u|_{\partial M} \in F\}$$

a Fredholm operator? And how does the winding number (half of Maslov index) of F affect the Fredholm index of $\bar\partial$? This type of problem is well known in Floer homology and J-holomorphic curves. The following Riemann–Roch theorem is shown in [McDuff and Salamon 2004]:

Theorem 7.4 (Riemann–Roch with boundary). *The operator $\bar\partial$ acting on $H^k(F)$, denoted $\bar\partial_F$, is Fredholm and its index is the sum of the Euler characteristic and twice the winding number of F. Furthermore, if the sum of the winding number of F and the Euler characteristic of M is larger than zero, then the operator $\bar\partial_F$ is surjective and consequently the dimension of $\ker \bar\partial_F$ is the sum of the Euler characteristic of M plus twice the winding number of F.*

With this theorem we are now ready to construct F so that it satisfies condition (35) and at the same time having $\bar\partial_F$ possess the desirable Fredholm properties. Assume for simplicity that M has only a single boundary component, so $\partial M \simeq S^1$, which we parametrize by $\theta \in [0, 2\pi)$ so that $\Gamma = (0, \theta_0)$. Let ϕ_N be a smooth function in $[0, 2\pi)$ such that $\phi_N(0) = 0$, $\phi_N(\theta) = 2\pi N$ for $\theta > \theta_0$. We then define the boundary real rank 1 subbundle F_N fiberwise by

$$F_N(\theta) := e^{i\phi_N(\theta)} \mathbb{R}.$$

By construction, the winding number of F_N is N and taking this parameter large enough we can apply Theorem 7.4 to conclude that

$$\dim \ker \bar\partial_{F_N} = 2N + \chi(M) \tag{36}$$

where $\chi(M)$ is the Euler characteristic of the surface.
It remains now to prescribe critical points to these holomorphic functions. Indeed, if $p \in M$ is an interior point we consider the map

$$\Phi \in \ker \bar\partial_{F_N} \mapsto \partial \Phi(p).$$

---
[6]We mean a subbundle of real rank 1 of the real rank 2 bundle $\mathbb{C} \times \partial M$ where $\mathbb{C} \simeq \mathbb{R}^2$ is viewed as a real vector space.

By (36) this is a linear map from an $2N + \chi(M)$ dimensional subspace to a real 2 dimensional subspace and therefore has a nontrivial kernel. Consequently, there exists an $N - 2$ dimensional subspace of holomorphic functions in $H^k(M, F_N)$ with a critical point at p. The holomorphic functions with prescribed critical point and the desirable boundary conditions is however, not a priori Morse. To remedy this fact we use the same type of arguments as those of the proof of Theorem 6.6, based on a transversality property: this shows that Morse holomorphic functions are dense within the space of holomorphic functions. More precisely:

Lemma 7.5 [Guillarmou and Tzou 2011a]. *Let Φ be a holomorphic function on M which is purely real on $\partial M \setminus \Gamma$. Then there exists a sequence of Morse holomorphic functions Φ_j which are purely real on $\partial M \setminus \Gamma$ such that $\Phi_j \to \Phi$ in $C^k(\overline{M})$ for all $k \in \mathbb{N}$.*

Starting with a holomorphic function Φ, purely real on Γ, with a critical point at p, one finds a Morse sequence Φ_j approaching Φ, and by Cauchy's argument principle we deduce easily that the critical points of Φ_j approach p. This proves Lemma 7.3. □

Carleman estimate and construction of remainder terms. We have so far constructed harmonic functions $u_0 := e^{\Phi/h}a - \overline{e^{\Phi/h}a}$ of exponential type that vanish on $\partial M \setminus \Gamma$. It remains to show that we can construct the suitable remainder term r_h so that $u = u_0 + e^{\phi/h}r_h$ is a family of solutions to $(\Delta_g + V)u = 0$ such that u vanishes on $\partial M \setminus \Gamma$ with r_h satisfying suitable decaying properties as $h \to 0$.

The method here was developed [Imanuvilov et al. 2010b]; it is a combination of ideas from earlier works on partial data [Bukhgeim and Uhlmann 2002; Kenig et al. 2007] (e.g., Carleman estimates) with the idea of Bukhgeim [2008] of using phases with critical points. We present a sketch of proof which adapts to Riemann surface (details are in [Guillarmou and Tzou 2011a]), but the original proof for domains can be found in [Imanuvilov et al. 2010b].

We will split r_h into two parts, $r_h = r_1 + r_2$. The first term r_1 will be constructed using a Cauchy–Riemann operator $\overline{\partial}^{-1}$ (a right inverse for $\overline{\partial}$) to solve the approximate equation

$$e^{-\Phi/h}(\Delta_g + V)e^{\Phi/h}(a + r_1) = \mathcal{O}_{L^2}(h|\log h|)$$

with the boundary condition $e^{\Phi/h}r_1 \in \mathbb{R}$ on $\partial M \setminus \Gamma$. We then subtract its complex conjugate to obtain

$$e^{-\phi/h}(\Delta_g + V)(e^{\Phi/h}(a + r_1) - \overline{e^{\Phi/h}(a + r_1)}) = \mathcal{O}_{L^2}(h|\log h|) \quad (37)$$

with $(e^{\Phi/h}(a + r_1) - \overline{e^{\Phi/h}(a + r_1)}) = 0$ on $\partial M \setminus \Gamma$.

To go from an approximate solution to a full solution, we can use a Carleman estimate with boundary terms. This first appeared in [Imanuvilov et al. 2010b] on domains.

Proposition 7.6. *Let (M, g) be a smooth Riemann surface with boundary, and let $\phi : M_0 \to \mathbb{R}$ be a $C^k(M)$ harmonic Morse function for k large. Then for all $V \in L^\infty(M)$ there exists an $h_0 > 0$ such that for all $h \in (0, h_0)$ and $u \in C^\infty(M)$ with $u|_{\partial M} = 0$, we have*

$$\frac{1}{h}\|u\|^2_{L^2(M)} + \|u\|^2_{H^1(M)} + \|\partial_\nu u\|^2_{L^2(\Gamma)}$$
$$\leq C\Big(\|e^{-\phi/h}(\Delta_g + V)e^{\phi/h}u\|^2_{L^2(M)} + \frac{1}{h}\|\partial_\nu u\|^2_{L^2(\partial M \setminus \Gamma)}\Big), \quad (38)$$

where ∂_ν is the inward pointing normal along the boundary.

The Carleman estimate of Lemma 3.4 is a simpler version of this estimate. Let us simply indicate why the exponent $h^{-1}\|u\|^2_{L^2}$ is coming in (38), instead of the term $h^{-2}\||\nabla\phi|u\|^2_{L^2}$ of Lemma 3.4: for $u \in C_0^\infty$ supported in a small neighborhood of a critical point (given by $z = 0$ in local coordinates), we have $|\nabla\phi(z)|^2 \geq C|z|^2$ for some C and using (7) with $\epsilon = \alpha h$ for some small α independent of h, we obtain

$$\frac{1}{h^2}\||\nabla\phi|u\|^2_{L^2} + \|\nabla u\|^2_{L^2} \geq \frac{C}{h}\|u\|^2_{L^2} + C\|\nabla u\|^2_{L^2}$$

for some $C > 0$. Gluing local estimates with convexified weight methods, one obtain something of the form (38) for $u \in C_0^\infty(M)$, and for functions supported near the boundary, an estimate like this is also available but with boundary terms coming from integrating by parts.

A direct application of the Riesz theorem and Proposition 7.6 gives the following solvability result:

Proposition 7.7. *Let $f \in L^2(M)$, then for all $h < h_0$ there exists a solution r to*

$$e^{-\phi/h}(\Delta_g + V)e^{\phi/h}r = f$$

with $r|_{\partial M \setminus \Gamma} = 0$ satisfying the estimate $\|r\|_{L^2} \leq \sqrt{h}\|f\|_{L^2}$.

With this proposition we can solve for the $O_{L^2}(h \log h)$ remainder in (37) and obtain a solution to $(\Delta_g + V)u = 0$ of the form

$$u = (e^{\Phi/h}(a + r_1) - \overline{e^{\Phi/h}(a + r_1)}) + e^{\phi/h}r_2$$

satisfying $u|_{\partial M \setminus \Gamma} = 0$

Identification of the potential. Once we have constructed CGO u_j for the equation $(\Delta + V_j)u_j = 0$, having the form

$$u_1 = e^{\Phi/h}(a + r_1) + \overline{e^{\Phi/h}(a + r_1)} + e^{\varphi/h}r_2,$$
$$u_2 = e^{-\Phi/h}(a + s_1) + \overline{e^{-\Phi/h}(a + s_1)} + e^{-\varphi/h}s_2,$$

with $u_1|_{\partial M \setminus \Gamma} = u_2|_{\partial M \setminus \Gamma} = 0$, we plug this into the integral identity (34) and obtain (with some work to deal with the crossed terms involving r_1 and s_1)

$$0 = \int_M (V_1 - V_2)(a^2 + \bar{a}^2) + 2\mathrm{Re} \int_M e^{2i\mathrm{Im}(\Phi)/h}(V_1 - V_2)|a|^2 + o(h).$$

and applying stationary phase (knowing that $(V_1 - V_2)|_{\partial M} = 0$ by boundary local uniqueness) we deduce that $V_1(p) = V_2(p)$ at the critical point p of Φ as $h \to 0$. Notice that Φ can have several critical points, in which case we choose the amplitude a to vanish at all critical points except the point p where one wants to recover the potential. This ends the proof of Theorem 7.1. □

7B. *Identification of first-order terms.* Imanuvilov, Uhlmann and Yamamoto [2010a] studied the partial data problem for general second order elliptic operators on a domain $\Omega \subset \mathbb{C}$, i.e., operators of the form

$$L = \Delta_g + A\partial_z + B\partial_{\bar{z}} + V \tag{39}$$

where A, B, V are complex-valued functions and g is a metric. They have a general result we shall only state the case of $g = \mathrm{Id}$ since the general statement is quite complicated (see their paper for the general case).

Theorem 7.8 [Imanuvilov et al. 2010a]. *Let*

$$(A_j, B_j, V_j) \in C^{5,\alpha}(\Omega) \times C^{5,\alpha}(\Omega) \times C^{4,\alpha}(\Omega)$$

for $j = 1, 2$ and $\alpha > 0$. Assume there exists an open set $\Gamma \subset \partial\Omega$ such that

$$\{(u, \partial_\nu u)|_\Gamma; L_1 u = 0, u \in H^1, u|_{\partial\Omega \setminus \Gamma} = 0\}$$
$$= \{(u, \partial_\nu u)|_\Gamma; L_1 u = 0, u \in H^1, u|_{\partial\Omega \setminus \Gamma} = 0\},$$

where L_j is defined as in (39) with $(A, B, V) = (A_j, B_j, V_j)$ and $g = \mathrm{Id}$. Then there exists a function $\eta \in C^{6,\alpha}(\Omega)$ such that $\eta|_\gamma = \partial_\nu \eta|_\Gamma = 0$ and

$$L_1 = e^{-\eta} L_2 e^\eta.$$

The multiplication by e^η is the gauge invariance in this setting (here there is no "topology" so the gauge is exactly conjugation by exponential of a function η with value in $\mathbb{C}$, while in theorem [Albin et al. 2011] it is not always the case). The proof is long and technical, and builds on previous work [Imanuvilov et al. 2010b]. The results are summarized and announced in [Imanuvilov et al. 2011a].

7C. Disjoint measurements. We conclude with a recent result.

Theorem 7.9 [Imanuvilov et al. 2011b]. *Let $\Omega \subset \mathbb{C}$ be a connected and simply connected domain with smooth boundary. Assume that $\partial\Omega = \overline{\Gamma_+} \cup \overline{\Gamma_-} \cup \overline{\Gamma_0}$ where $\Gamma_\pm, \Gamma_0$ are open disjoint subsets of $\partial\Omega$ such that $\overline{\Gamma_+} \cap \overline{\Gamma_-} = \varnothing$, $\Gamma_\pm$ have two connected components $\Gamma_\pm^1, \Gamma_\pm^2$ and each of the 4 connected components of Γ_0 has its closure intersecting both Γ_+ and Γ_-. If $V_1, V_2 \in C^{2+\alpha}(\Omega)$ for some $\alpha > 0$ and if the partial Cauchy data*

$$\mathcal{C}_{V_j}^\Gamma := \{(u|_{\Gamma_+}, \partial_\nu u|_{\Gamma_-}); (\Delta + V_j)u = 0, u \in H^1(\Omega), u|_{\partial\Omega \setminus \Gamma_+} = 0\}$$

agree, i.e., $\mathcal{C}_{V_1}^\Gamma = \mathcal{C}_{V_2}^\Gamma$, then $V_1 = V_2$.

8. Open problems in 2 dimensions

(1) Prove the L^∞ conductivity result of Astala–Päivärinta in the context of Riemann surfaces.

(2) Find a better regularity assumption for the potential in Bukhgeim's result.

(3) Recover a metric g on a Riemann surface M up to isometry equal to Id on the boundary from the Cauchy data space for the equation $\Delta_g - \lambda$, where $\lambda \neq 0$.

(4) More generally, see what can be recovered from the Cauchy data space for the equation $\Delta_g + V$ where V is a potential and g a metric. On the region where $V = 0$, only the conformal class can be obtained.

(5) Find a "good" partial data measurement for d-bar type elliptic systems $Pu = 0$ which allows one to recover the coefficients inside the surface. That is, we search for natural subspaces of the full Cauchy data space which determine the coefficients inside the surface, for instance in terms of measurements on pieces of the boundary of certain components of the vector-valued u (as in [Salo and Tzou 2010], for instance).

(6) Solve the general disjoint partial data measurements for $\Delta_g + V$, that is, when the Dirichlet data is measured on Γ_1 and the Neumann data is measured on Γ_2, with $\Gamma_1 \cap \Gamma_2 = \varnothing$.

(7) Find a reconstruction method for partial data measurement, as in [Nachman and Street 2010], for higher dimensions.

(8) Solve the inverse problem for the elasticity system of [Nakamura and Uhlmann 1993].

(9) Solve the Calderón problem for L^∞ complex-valued conductivities.

References

[Ablowitz et al. 1983] M. J. Ablowitz, D. Bar Yaacov, and A. S. Fokas, "On the inverse scattering transform for the Kadomtsev–Petviashvili equation", *Stud. Appl. Math.* **69**:2 (1983), 135–143. MR 85h:35179 Zbl 0527.35080

[Albin et al. 2011] P. Albin, C. Guillarmou, L. Tzou, and G. Uhlmann, "Inverse boundary problems for systems in two dimensions", preprint, 2011. arXiv 1105.4565

[Alessandrini 1990] G. Alessandrini, "Singular solutions of elliptic equations and the determination of conductivity by boundary measurements", *J. Differential Equations* **84**:2 (1990), 252–272. MR 91e:35210 Zbl 0778.35109

[Astala and Päivärinta 2006] K. Astala and L. Päivärinta, "Calderón's inverse conductivity problem in the plane", *Ann. of Math.* (2) **163**:1 (2006), 265–299. MR 2007b:30019 Zbl 1111.35004

[Astala et al. 2005] K. Astala, L. Päivärinta, and M. Lassas, "Calderón's inverse problem for anisotropic conductivity in the plane", *Comm. Partial Differential Equations* **30**:1-3 (2005), 207–224. MR 2005k:35421

[Astala et al. 2011] K. Astala, M. Lassas, and L. Päivärinta, "The borderlines of the invisibility and visibility for Calderón's inverse problem", preprint, 2011. arXiv 1109.2749

[Astala et al. 2013] K. Astala, M. Lassas, and L. Päivärinta, "Calderón's inverse problem: imaging and invisibility", pp. 1–54 in *Inverse problems and applications: Inside out II*, edited by G. Uhlmann, Publ. Math. Sci. Res. Inst. **60**, Cambridge University Press, New York, 2013.

[Beals and Coifman 1981] R. Beals and R. Coifman, "Scattering, transformations spectrales et équations d'évolution non linéaires", Exp. No. XXII in *Goulaouic–Meyer–Schwartz Seminar, 1980–1981*, École Polytech., Palaiseau, 1981. MR 84d:35129

[Beals and Coifman 1988] R. Beals and R. Coifman, "The spectral problem for the Davey–Stewartson and Ishimori hierarchies", pp. pages 15–23 in *Nonlinear evolution equations: integrability and spectral methods*, edited by A. Degasperis et al., Manchester University Press, 1988.

[Belishev 2003] M. I. Belishev, "The Calderon problem for two-dimensional manifolds by the BC-method", *SIAM J. Math. Anal.* **35**:1 (2003), 172–182. MR 2004f:58029 Zbl 1048.58019

[Blasten 2011] E. Blasten, "The inverse problem of the Shrödinger equation in the plane: A dissection of Bukhgeim's result", preprint, 2011. arXiv 1103.6200

[Brown 2001] R. M. Brown, "Recovering the conductivity at the boundary from the Dirichlet to Neumann map: a pointwise result", *J. Inverse Ill-Posed Probl.* **9**:6 (2001), 567–574. MR 2003a:35196 Zbl 0991.35104

[Brown and Uhlmann 1997] R. M. Brown and G. A. Uhlmann, "Uniqueness in the inverse conductivity problem for nonsmooth conductivities in two dimensions", *Comm. Partial Differential Equations* **22**:5-6 (1997), 1009–1027. MR 98f:35155 Zbl 0884.35167

[Bukhgeim 2008] A. L. Bukhgeim, "Recovering a potential from Cauchy data in the two-dimensional case", *J. Inverse Ill-Posed Probl.* **16**:1 (2008), 19–33. MR 2008m:30049

[Bukhgeim and Uhlmann 2002] A. L. Bukhgeim and G. Uhlmann, "Recovering a potential from partial Cauchy data", *Comm. Partial Differential Equations* **27**:3-4 (2002), 653–668. MR 2003d:35262 Zbl 0998.35063

[Calderón 1980] A.-P. Calderón, "On an inverse boundary value problem", pp. 65–73 in *Seminar on Numerical Analysis and its Applications to Continuum Physics* (Rio de Janeiro, 1980), Soc. Brasil. Mat., Rio de Janeiro, 1980. MR 81k:35160

[Dos Santos Ferreira et al. 2009] D. Dos Santos Ferreira, C. E. Kenig, M. Salo, and G. Uhlmann, "Limiting Carleman weights and anisotropic inverse problems", *Invent. Math.* **178**:1 (2009), 119–171. MR 2010h:58033 Zbl 1181.35327

[Faddeev 1965] L. Faddeev, "Growing solutions of the Schrödinger equation", *Dokl. Akad. Nauk SSSR* **165** (1965), 514–517. In Russian. Translated in *Soviet Phys. Dokl.*, **10** (1965), 1033–1035.

[Faddeev 1974] L. D. Faddeev, "The inverse problem in the quantum theory of scattering, II", pp. 93–180 in *Current problems in mathematics*, vol. 3, Akad. Nauk SSSR Vsesojuz. Inst. Naučn. i Tehn. Informacii, Moscow, 1974. In Russian. MR 58 #25585 Zbl 0299.35027

[Farkas and Kra 1992] H. M. Farkas and I. Kra, *Riemann surfaces*, 2nd ed., Graduate Texts in Mathematics **71**, Springer, New York, 1992. MR 93a:30047 Zbl 0764.30001

[Forster 1991] O. Forster, *Lectures on Riemann surfaces*, Graduate Texts in Mathematics **81**, Springer, New York, 1991. MR 93h:30061 Zbl 0475.30002

[Grinevich and Manakov 1986] P. G. Grinevich and S. V. Manakov, "Inverse problem of scattering theory for the two-dimensional Schrödinger operator, the $\bar{\partial}$-method and nonlinear equations", *Funktsional. Anal. i Prilozhen.* **20**:2 (1986), 14–24, 96. MR 88g:35197

[Grinevich and Novikov 1988a] P. G. Grinevich and S. P. Novikov, "Inverse scattering problem for the two-dimensional Schrödinger operator at a fixed negative energy and generalized analytic functions", pp. 58–85 in *Plasma theory and nonlinear and turbulent processes in physics* (Kiev, 1987), edited by V. G. Baryakhtar et al., World Sci. Publishing, Singapore, 1988. MR 90c:35199 Zbl 0704.35137

[Grinevich and Novikov 1988b] P. G. Grinevich and S. P. Novikov, "A two-dimensional "inverse scattering problem" for negative energies, and generalized-analytic functions, I: Energies lower than the ground state", *Funktsional. Anal. i Prilozhen.* **22**:1 (1988), 23–33, 96. MR 90a:35181

[Guillarmou and Sá Barreto 2009] C. Guillarmou and A. Sá Barreto, "Inverse problems for Einstein manifolds", *Inverse Probl. Imaging* **3**:1 (2009), 1–15. MR 2010m:58040 Zbl 1229.58025

[Guillarmou and Tzou 2009] C. Guillarmou and L. Tzou, "Calderón inverse problem for the Schrödinger operator on Riemann surfaces", pp. 129–142 in *The AMSI-ANU Workshop on Spectral Theory and Harmonic Analysis*, Proc. Centre Math. Appl. **44**, Australian National Univ., Canberra, 2009.

[Guillarmou and Tzou 2011a] C. Guillarmou and L. Tzou, "Calderón inverse problem with partial data on Riemann surfaces", *Duke Math. J.* **158**:1 (2011), 83–120. MR 2012f:35574 Zbl 1222.35212

[Guillarmou and Tzou 2011b] C. Guillarmou and L. Tzou, "Identification of a connection from Cauchy data on a Riemann surface with boundary", *Geom. Funct. Anal.* **21**:2 (2011), 393–418. MR 2795512 Zbl 05902965

[Henkin and Michel 2007] G. Henkin and V. Michel, "On the explicit reconstruction of a Riemann surface from its Dirichlet–Neumann operator", *Geom. Funct. Anal.* **17**:1 (2007), 116–155. MR 2009b:58051 Zbl 1118.32009

[Henkin and Michel 2008] G. Henkin and V. Michel, "Inverse conductivity problem on Riemann surfaces", *J. Geom. Anal.* **18**:4 (2008), 1033–1052. MR 2010b:58031 Zbl 1151.35101

[Henkin and Santacesaria 2010] G. Henkin and M. Santacesaria, "Gel'fand–Calderón's inverse problem for anisotropic conductivities on bordered surfaces in $\mathbb{R}^3$", preprint. arXiv 1006.0647

[Imanuvilov and Yamamoto 2011] O. Imanuvilov and M. Yamamoto, "Inverse boundary value problem for Schrödinger equation in two dimensions", preprint, 2011. arXiv 1105.2850

[Imanuvilov et al. 2010a] O. Imanuvilov, G. Uhlmann, and M. Yamamoto, "Partial Cauchy data for general second-order elliptic operators in two dimensions", preprint, 2010. arXiv 1010.5791

[Imanuvilov et al. 2010b] O. Y. Imanuvilov, G. Uhlmann, and M. Yamamoto, "The Calderón problem with partial data in two dimensions", *J. Amer. Math. Soc.* **23**:3 (2010), 655–691. MR 2012c:35472 Zbl 1201.35183

[Imanuvilov et al. 2011a] O. Y. Imanuvilov, G. Uhlmann, and M. Yamamoto, "Determination of second-order elliptic operators in two dimensions from partial Cauchy data", *Proc. Natl. Acad. Sci. USA* **108**:2 (2011), 467–472. MR 2012a:35364

[Imanuvilov et al. 2011b] O. Y. Imanuvilov, G. Uhlmann, and M. Yamamoto, "Inverse boundary value problem by measuring Dirichlet data and Neumann data on disjoint sets", *Inverse Problems* **27**:8 (2011), 085007. MR 2012c:78002 Zbl 1222.35213

[Kang and Uhlmann 2004] H. Kang and G. Uhlmann, "Inverse problems for the Pauli Hamiltonian in two dimensions", *J. Fourier Anal. Appl.* **10**:2 (2004), 201–215. MR 2005d:81135 Zbl 1081.35141

[Kenig et al. 2007] C. E. Kenig, J. Sjöstrand, and G. Uhlmann, "The Calderón problem with partial data", *Ann. of Math.* (2) **165**:2 (2007), 567–591. MR 2008k:35498 Zbl 1127.35079

[Kobayashi 1987] S. Kobayashi, *Differential geometry of complex vector bundles*, Publications of the Mathematical Society of Japan **15**, Princeton University Press, 1987. MR 89e:53100 Zbl 0708.53002

[Kohn and Vogelius 1984] R. Kohn and M. Vogelius, "Determining conductivity by boundary measurements", *Comm. Pure Appl. Math.* **37**:3 (1984), 289–298. MR 85f:80008 Zbl 0586.35089

[Lassas and Uhlmann 2001] M. Lassas and G. Uhlmann, "On determining a Riemannian manifold from the Dirichlet-to-Neumann map", *Ann. Sci. École Norm. Sup.* (4) **34**:5 (2001), 771–787. MR 2003e:58037 Zbl 0992.35120

[Lassas et al. 2003] M. Lassas, M. Taylor, and G. Uhlmann, "The Dirichlet-to-Neumann map for complete Riemannian manifolds with boundary", *Comm. Anal. Geom.* **11**:2 (2003), 207–221. MR 2004h:58033 Zbl 1077.58012

[Lebeau and Le Rousseau 2011] G. Lebeau and J. Le Rousseau, "Introduction aux inégalités de Carleman pour les opérateurs elliptiques et paraboliques. Applications au prolongement unique et au contrôle des équations paraboliques", *ESAIM Control Optim. Calc. Var.* (2011).

[Lee and Uhlmann 1989] J. M. Lee and G. Uhlmann, "Determining anisotropic real-analytic conductivities by boundary measurements", *Comm. Pure Appl. Math.* **42**:8 (1989), 1097–1112. MR 91a:35166 Zbl 0702.35036

[McDuff and Salamon 2004] D. McDuff and D. Salamon, *J-holomorphic curves and symplectic topology*, American Mathematical Society Colloquium Publications **52**, American Mathematical Society, Providence, RI, 2004. MR 2004m:53154 Zbl 1064.53051

[Nachman 1996] A. I. Nachman, "Global uniqueness for a two-dimensional inverse boundary value problem", *Ann. of Math.* (2) **143**:1 (1996), 71–96. MR 96k:35189 Zbl 0857.35135

[Nachman and Street 2010] A. Nachman and B. Street, "Reconstruction in the Calderón problem with partial data", *Comm. Partial Differential Equations* **35**:2 (2010), 375–390. MR 2012b:35368 Zbl 1186.35242

[Nakamura and Uhlmann 1993] G. Nakamura and G. Uhlmann, "Identification of Lamé parameters by boundary measurements", *Amer. J. Math.* **115**:5 (1993), 1161–1187. MR 94k:35328 Zbl 0803.35164

[Novikov 1986] R. G. Novikov, "Reconstruction of a two-dimensional Schrödinger operator from the scattering amplitude in the presence of fixed energy", *Funktsional. Anal. i Prilozhen.* **20**:3 (1986), 90–91. MR 88g:35198

[Novikov 1992] R. G. Novikov, "The inverse scattering problem on a fixed energy level for the two-dimensional Schrödinger operator", *J. Funct. Anal.* **103**:2 (1992), 409–463. MR 93e:35080 Zbl 0762.35077

[Salo 2013] M. Salo, "The Calderón problem on Riemannian manifolds", pp. 167–247 in *Inverse problems and applications: Inside out II*, edited by G. Uhlmann, Publ. Math. Sci. Res. Inst. **60**, Cambridge University Press, New York, 2013.

[Salo and Tzou 2010] M. Salo and L. Tzou, "Inverse problems with partial data for a Dirac system: a Carleman estimate approach", *Adv. Math.* **225**:1 (2010), 487–513. MR 2011g:35432 Zbl 1197.35329

[Sun 1990] Z. Q. Sun, "The inverse conductivity problem in two dimensions", *J. Differential Equations* **87**:2 (1990), 227–255. MR 92a:35163 Zbl 0716.35080

[Sun 1993] Z. Q. Sun, "An inverse boundary value problem for the Schrödinger operator with vector potentials in two dimensions", *Comm. Partial Differential Equations* **18**:1-2 (1993), 83–124. MR 94e:35145 Zbl 0781.35073

[Sun and Uhlmann 1991] Z. Q. Sun and G. Uhlmann, "Generic uniqueness for an inverse boundary value problem", *Duke Math. J.* **62**:1 (1991), 131–155. MR 92b:35172 Zbl 0728.35132

[Sun and Uhlmann 2003] Z. Sun and G. Uhlmann, "Anisotropic inverse problems in two dimensions", *Inverse Problems* **19**:5 (2003), 1001–1010. MR 2004k:35415 Zbl 1054.35139

[Sylvester 1990] J. Sylvester, "An anisotropic inverse boundary value problem", *Comm. Pure Appl. Math.* **43**:2 (1990), 201–232. MR 90m:35202 Zbl 0709.35102

[Sylvester and Uhlmann 1986] J. Sylvester and G. Uhlmann, "A uniqueness theorem for an inverse boundary value problem in electrical prospection", *Comm. Pure Appl. Math.* **39**:1 (1986), 91–112. MR 87j:35377 Zbl 0611.35088

[Sylvester and Uhlmann 1987] J. Sylvester and G. Uhlmann, "A global uniqueness theorem for an inverse boundary value problem", *Ann. of Math.* (2) **125**:1 (1987), 153–169. MR 88b:35205 Zbl 0625.35078

[Sylvester and Uhlmann 1988] J. Sylvester and G. Uhlmann, "Inverse boundary value problems at the boundary—continuous dependence", *Comm. Pure Appl. Math.* **41**:2 (1988), 197–219. MR 89f:35213

[Uhlenbeck 1976] K. Uhlenbeck, "Generic properties of eigenfunctions", *Amer. J. Math.* **98**:4 (1976), 1059–1078. MR 57 #4264 Zbl 0355.58017

[Uhlmann 2003] G. Uhlmann, "Inverse boundary problems in two dimensions", pp. 183–203 in *Function spaces, differential operators and nonlinear analysis* (Teistungen, 2001), Birkhäuser, Basel, 2003. MR 2004h:35225 Zbl 1054.35140

DMA, U.M.R. 8553 CNRS, École Normale Supérieure, 45 rue d'Ulm, F-75230 Paris cedex 05, France
cguillar@dma.ens.fr

Department of Mathematics, University of Helsinki, PO Box 68, 00014 Helsinki, Finland
ltzou@gmail.com

The Calderón problem on Riemannian manifolds

MIKKO SALO

We discuss recent developments in Calderón's inverse problem on Riemannian manifolds (the anisotropic Calderón problem) in three and higher dimensions. The topics considered include the relevant Riemannian geometry background, limiting Carleman weights on manifolds, a Fourier analysis proof of Carleman estimates on product type manifolds, and uniqueness results for inverse problems based on complex geometrical optics solutions and the geodesic ray transform.

Preface	167
1. Introduction	169
2. Riemannian geometry	173
2A. Smooth manifolds	173
2B. Riemannian manifolds	181
2C. Laplace–Beltrami operator	188
2D. DN map	193
2E. Geodesics and covariant derivative	196
3. Limiting Carleman weights	202
3A. Motivation and definition	203
3B. Characterization	208
3C. Geometric interpretation	213
4. Carleman estimates	216
4A. Motivation and main theorem	217
4B. Proof of the estimates	219
5. Uniqueness result	230
5A. Complex geometrical optics solutions	231
5B. Geodesic ray transform	237
Acknowledgements	246
References	246

Preface

This text is an introduction to Calderón's inverse conductivity problem on Riemannian manifolds. This problem arises as a model for electrical imaging in anisotropic media, and it is one of the most basic inverse problems in a geometric setting. The problem is still largely open, but we will discuss recent developments based on complex geometrical optics and the geodesic X-ray transform in the case where one restricts to a fixed conformal class of conductivities.

This work is based on lectures for courses given at the University of Helsinki in 2010 and at Universidad Autónoma de Madrid in 2011. It has therefore the feeling of a set of lecture notes for a graduate course on the topic, together with exercises and also some problems which are open at the time of writing this. The main focus is on manifolds of dimension three and higher, where one has to rely on real variable methods instead of using complex analysis. The text can be considered as an introduction to geometric inverse problems, but also as an introduction to the use of real analysis methods in the setting of Riemannian manifolds.

Section 1 is an introduction to the Calderón problem on manifolds, stating the main questions studied in this text. Section 2 reviews basic facts on smooth and Riemannian manifolds, also discussing the Laplace–Beltrami operator and geodesics. Limiting Carleman weights, which turn out to exist on manifolds with a certain product structure, are treated in Section 3. Section 4 then proves Carleman estimates on manifolds with product structure. The proof uses a combination of the Fourier transform and eigenfunction expansions. Finally, in Section 5 we prove a uniqueness result for the inverse problem in certain geometries, based on inverting the geodesic X-ray transform.

As prerequisites for reading these notes, basic knowledge of real analysis, Riemannian geometry, and elliptic partial differential equations would be helpful. Familiarity with [Rudin 1986], [Lee 1997, Chapters 1-5], and [Evans 2010, Chapters 5-6] should be sufficient.

References. For a more thorough discussion on Calderón's inverse problem on manifolds and for references to known results, we refer to the introduction in [Dos Santos Ferreira et al. 2009]. In particular, that paper includes precise references to results in two dimensions and on real-analytic manifolds that we have omitted in this presentation. See also the survey by Guillarmou and Tzou in this same volume for recent work on the Schrödinger equation in two dimensions.

General references for Section 2 include [Lee 2003] for smooth manifolds, [Lee 1997] for Riemannian manifolds, and [Taylor 1996] for the Laplace–Beltrami operator. Section 3 on limiting Carleman weights mostly follows [Dos Santos Ferreira et al. 2009, Section 2].

To motivate the definition of limiting Carleman weights, we use a little bit of semiclassical symbol calculus (for differential operators, not pseudodifferential ones). This is not covered in these notes, but on the other hand it is only used in

Section 3A for motivation. See the lecture notes [Zworski 2012] for details on this topic.

The Fourier analysis proof of the Carleman estimates given in Section 4 is taken from [Kenig et al. 2011]. Section 5, with the proof of the uniqueness result, follows [Dos Santos Ferreira et al. 2009, Sections 5 and 6]. For more details on the geodesic X-ray transform we refer the reader to [Sharafutdinov 1994] and [Dos Santos Ferreira et al. 2009, Section 7].

1. Introduction

To motivate the problems studied in this text, we start with the classical inverse conductivity problem of Calderón. This problem asks to determine the interior properties of a medium by making electrical measurements on its boundary.

In mathematical terms, one considers a bounded open set $\Omega \subseteq \mathbb{R}^n$ with smooth (C^∞) boundary, with electrical conductivity given by the matrix

$$\gamma(x) = (\gamma^{jk}(x))_{j,k=1}^n.$$

We assume that the functions γ^{jk} are smooth in $\overline{\Omega}$, and for each x the matrix $\gamma(x)$ is positive definite and symmetric. If $\gamma(x) = \sigma(x)I$ for some scalar function σ we say that the medium is *isotropic*, otherwise it is *anisotropic*. The electrical properties of anisotropic materials depend on direction. This is common in many applications such as in medical imaging (for instance cardiac muscle has a fiber structure and is an anisotropic conductor).

We seek to find the conductivity γ by prescribing different voltages on $\partial\Omega$ and by measuring the resulting current fluxes. If there are no sources or sinks of current in Ω, a boundary voltage f induces an electrical potential u which satisfies the conductivity equation

$$\begin{cases} \operatorname{div}(\gamma \nabla u) = 0 & \text{in } \Omega, \\ u = f & \text{on } \partial\Omega. \end{cases} \quad (1\text{-}1)$$

Since γ is positive definite this equation is elliptic and has a unique weak solution for any reasonable f (say in the L^2-based Sobolev space $H^{1/2}(\partial\Omega)$). The current flux on the boundary is given by the conormal derivative (where ν is the outer unit normal vector on $\partial\Omega$)

$$\Lambda_\gamma f := \gamma \nabla u \cdot \nu|_{\partial\Omega}.$$

The last expression is well defined also when γ is a matrix, and a suitable weak formulation shows that Λ_γ becomes a bounded map $H^{1/2}(\partial\Omega) \to H^{-1/2}(\partial\Omega)$.

The map Λ_γ is called the *Dirichlet-to-Neumann map*, DN map for short, since it maps the Dirichlet boundary value of a solution to what is essentially the Neumann boundary value. The DN map encodes the electrical boundary measurements (in the idealized case where we have infinite precision measurements for all possible data). The inverse problem is to find information about the conductivity matrix γ from the knowledge of the map Λ_γ.

The first important observation is that if γ is anisotropic, the full conductivity matrix can not be determined from Λ_γ. This is due to a transformation law for the conductivity equation under diffeomorphisms (that is, bijective maps F such that both F and F^{-1} are smooth up to the boundary).

Lemma. *If $F : \overline{\Omega} \to \overline{\Omega}$ is a diffeomorphism and if $F|_{\partial\Omega} = \mathrm{Id}$, then*

$$\Lambda_{F_*\gamma} = \Lambda_\gamma.$$

Here $F_*\gamma$ is the pushforward of γ, defined by

$$F_*\gamma(\tilde{x}) = \frac{(DF)\gamma(DF)^t}{|\det(DF)|}\bigg|_{F^{-1}(\tilde{x})}$$

where $DF = (\partial_k F_j)_{j,k=1}^n$ is the Jacobian matrix.

Exercise 1.1. Prove the lemma. (Hint: if u solves $\mathrm{div}(\gamma \nabla u) = 0$, show that $u \circ F^{-1}$ solves the analogous equation with conductivity $F_*\gamma$.)

The following conjecture for $n \geq 3$ is one of the most important open questions related to the inverse problem of Calderón. It has only been proved when $n = 2$.

Question 1.1 (anisotropic Calderón problem). Let γ_1, γ_2 be two smooth positive definite symmetric matrices in $\overline{\Omega}$. If $\Lambda_{\gamma_1} = \Lambda_{\gamma_2}$, show that $\gamma_2 = F_*\gamma_1$ for some diffeomorphism $F : \overline{\Omega} \to \overline{\Omega}$ with $F|_{\partial\Omega} = \mathrm{Id}$.

In fact, the anisotropic Calderón problem is a question of geometric nature and can be formulated more generally on any Riemannian manifold. To do this, we replace the set $\overline{\Omega} \subseteq \mathbb{R}^n$ by a compact n-dimensional manifold M with smooth boundary ∂M, and the conductivity matrix γ by a smooth Riemannian metric g on M. On such a Riemannian manifold (M, g) there is a canonical second-order elliptic operator Δ_g called the *Laplace–Beltrami operator*. In local coordinates

$$\Delta_g u = |g|^{-1/2} \frac{\partial}{\partial x_j}\left(|g|^{1/2} g^{jk} \frac{\partial u}{\partial x_k}\right).$$

We have written $g = (g_{jk})$ for the metric in local coordinates, $g^{-1} = (g^{jk})$ for its inverse matrix, and $|g|$ for $\det(g_{jk})$.

The Dirichlet problem for Δ_g analogous to (1-1) is

$$\begin{cases} \Delta_g u = 0 & \text{in } M, \\ u = f & \text{on } \partial M. \end{cases} \tag{1-2}$$

The boundary measurements are given by the DN map

$$\Lambda_g f := \partial_\nu u|_{\partial M}$$

where $\partial_\nu u$ is the Riemannian normal derivative, given in local coordinates by $g^{jk}(\partial_{x_j} u)\nu_k$ where ν is the outer unit normal vector on ∂M. The inverse problem is to determine information on g from the DN map Λ_g.

There is a similar obstruction to uniqueness as for the conductivity equation, which is given by diffeomorphisms.

Lemma. *If $F : M \to M$ is a diffeomorphism and if $F|_{\partial M} = \mathrm{Id}$, then*

$$\Lambda_{F^*g} = \Lambda_g.$$

Here F^*g is the pullback of g, defined in local coordinates by

$$F^*g(x) = DF(x)^t g(F(x)) DF(x).$$

Exercise 1.2. Prove the lemma.

The geometric formulation of the anisotropic Calderón problem is as follows. We only state the question for $n \geq 3$, since again the two dimensional case is known (also the formulation for $n = 2$ would look slightly different since Δ_g has an additional conformal invariance then).

Question 1.2 (anisotropic Calderón problem). Let (M, g_1) and (M, g_2) be two compact Riemannian manifolds of dimension $n \geq 3$ with smooth boundary, and assume that $\Lambda_{g_1} = \Lambda_{g_2}$. Show that $g_2 = F^*g_1$ for some diffeomorphism $F : M \to M$ with $F|_{\partial M} = \mathrm{Id}$.

A function u satisfying $\Delta_g u = 0$ is called a harmonic function in (M, g). Note that if M is a subset of $\mathbb{R}^n$ with Euclidean metric, then this just gives the usual harmonic functions. Since $(u|_{\partial M}, \partial_\nu u|_{\partial M})$ is the Cauchy data of a function u, and since metrics satisfying $g_2 = F^*g_1$ are isometric in the sense of Riemannian geometry, the anisotropic Calderón problem reduces to the question: *Do the Cauchy data of all harmonic functions in (M, g) determine the manifold up to isometry?*

Exercise 1.3. Show that a positive answer to Question 1.2 would imply a positive answer to Question 1.1 when $n \geq 3$. (Hint: assume the boundary determination result that $\Lambda_{\gamma_1} = \Lambda_{\gamma_2}$ implies $\det(\gamma_1^{jk}) = \det(\gamma_2^{jk})$ on $\partial \Omega$ [Lee and Uhlmann 1989].)

Instead of the full anisotropic Calderón problem, we will consider the simpler problem where the manifolds are assumed to be in the same conformal class. This means that the metrics g_1 and g_2 in M satisfy $g_2 = cg_1$ for some smooth positive function c on M. In this problem there is only one underlying metric

g_1, and one is looking to determine a scalar function c. This covers the case of isotropic conductivities in Euclidean space, but if the metric is not Euclidean the problem still requires substantial geometric arguments.

The relevant question is as follows. It is known that any diffeomorphism $F : M \to M$ which satisfies $F|_{\partial M} = \text{Id}$ and $F^* g_1 = c g_1$ must be the identity [Lionheart 1997], so in this case there is no ambiguity arising from diffeomorphisms.

Question 1.3 (anisotropic Calderón problem in a conformal class). Let (M, g_1) and (M, g_2) be two compact Riemannian manifolds of dimension $n \geq 3$ with smooth boundary which are in the same conformal class. If $\Lambda_{g_1} = \Lambda_{g_2}$, show that $g_1 = g_2$.

Exercise 1.4. Using the fact on diffeomorphisms given above, show that a positive answer to Question 1.2 implies a positive answer to Question 1.3.

Finally, let us formulate one more question which will imply Question 1.3 but which is somewhat easier to study. This last question will be the one that the rest of these notes is devoted to.

The main point is the observation that the Laplace–Beltrami operator transforms under conformal scalings of the metric by

$$\Delta_{cg} u = c^{-\frac{n+2}{4}} (\Delta_g + q)(c^{\frac{n-2}{4}} u)$$

where $q = c^{\frac{n-2}{4}} \Delta_{cg}(c^{-\frac{n-2}{4}})$. It can be shown that for any smooth positive function c with $c|_{\partial M} = 1$ and $\partial_\nu c|_{\partial M} = 0$, one has

$$\Lambda_{cg} = \Lambda_{g,-q}$$

where $\Lambda_{g,V} : f \mapsto \partial_\nu u|_{\partial M}$ is the DN map for the Schrödinger equation

$$\begin{cases} (-\Delta_g + V)u = 0 & \text{in } M, \\ u = f & \text{on } \partial M. \end{cases} \quad (1\text{-}3)$$

For general V this last Dirichlet problem may not be uniquely solvable, but for $V = -q$ it is and the DN map is well defined since the Dirichlet problem for Δ_{cg} is uniquely solvable. We will make the standing assumption that all potentials V are such that (1-3) is uniquely solvable (this assumption could easily be removed by using Cauchy data sets). Then the last question is as follows. It is also of independent interest and a solution would have important consequences for the anisotropic Calderón problem, inverse problems for Maxwell equations, and inverse scattering theory.

Question 1.4. Let (M, g) be a compact Riemannian manifold with smooth boundary, and let V_1 and V_2 be two smooth functions on M. If $\Lambda_{g,V_1} = \Lambda_{g,V_2}$, show that $V_1 = V_2$.

Exercise 1.5. Prove the above identities for Δ_{cg} and Λ_{cg}. Show that a positive answer to Question 1.4 implies a positive answer to Question 1.3. (You may assume the boundary determination result that $\Lambda_{cg} = \Lambda_g$ implies $c|_{\partial M} = 1$ and $\partial_\nu c|_{\partial M} = 0$ [Lee and Uhlmann 1989].)

2. Riemannian geometry

2A. Smooth manifolds.

Manifolds. We recall some basic definitions from the theory of smooth manifolds. We will consistently also consider manifolds with boundary.

Definition. A *smooth n-dimensional manifold* is a second countable Hausdorff topological space together with an open cover $\{U_\alpha\}$ and homeomorphisms $\varphi_\alpha : U_\alpha \to \tilde{U}_\alpha$ such that each $\tilde{U}_\alpha$ is an open set in $\mathbb{R}^n$, and $\varphi_\beta \circ \varphi_\alpha^{-1} : \varphi_\alpha(U_\alpha \cap U_\beta) \to \varphi_\beta(U_\alpha \cap U_\beta)$ is a smooth map whenever $U_\alpha \cap U_\beta$ is nonempty.

Any family $\{(U_\alpha, \varphi_\alpha)\}$ as above is called an *atlas*. Any atlas gives rise to a maximal atlas, called a *smooth structure*, which is not strictly contained in any other atlas. We assume that we are always dealing with the maximal atlas. The pairs $(U_\alpha, \varphi_\alpha)$ are called *charts*, and the maps φ_α are called *local coordinate systems* (one usually writes $x = \varphi_\alpha$ and thus identifies points $p \in U_\alpha$ with points $x(p) \in \tilde{U}_\alpha$ in $\mathbb{R}^n$).

Definition. A *smooth n-dimensional manifold with boundary* is a second countable Hausdorff topological space together with an open cover $\{U_\alpha\}$ and homeomorphisms
$$\varphi_\alpha : U_\alpha \to \tilde{U}_\alpha$$
such that each $\tilde{U}_\alpha$ is an open set in $\mathbb{R}^n_+ := \{x \in \mathbb{R}^n \,;\, x_n \geq 0\}$, and
$$\varphi_\beta \circ \varphi_\alpha^{-1} : \varphi_\alpha(U_\alpha \cap U_\beta) \to \varphi_\beta(U_\alpha \cap U_\beta)$$
is a smooth map whenever $U_\alpha \cap U_\beta$ is nonempty.

Here, if $A \subseteq \mathbb{R}^n$ we say that a map $F : A \to \mathbb{R}^n$ is smooth if it extends to a smooth map $\tilde{A} \to \mathbb{R}^n$ where $\tilde{A}$ is an open set in $\mathbb{R}^n$ containing A.

If M is a manifold with boundary we say that p is a boundary point if $\varphi(p) \in \partial \mathbb{R}^n_+$ for some chart φ, and an interior point if $\varphi(p) \in \text{int}(\mathbb{R}^n_+)$ for some φ. We write ∂M for the set of boundary points and M^{int} for the set of interior points. Since M is not assumed to be embedded in any larger space, these definitions may differ from the usual ones in point set topology.

Exercise 2.1. If M is a manifold with boundary, show that the sets M^{int} and ∂M are always disjoint.

To clarify the relations between the definitions, note that a manifold is always a manifold with boundary (the boundary being empty), but a manifold with boundary is a manifold if and only if the boundary is empty (by the above exercise). However, we will loosely refer to manifolds both with and without boundary as manifolds.

We have the following classes of manifolds:

- A *closed manifold* is compact, connected, and has no boundary.
    - Examples: the sphere S^n, the torus $T^n = \mathbb{R}^n/\mathbb{Z}^n$
- An *open manifold* has no boundary and no component is compact.
    - Examples: open subsets of $\mathbb{R}^n$, strict open subsets of a closed manifold
- A *compact manifold with boundary* is a manifold with boundary which is compact as a topological space.
    - Examples: the closures of bounded open sets in $\mathbb{R}^n$ with smooth boundary, the closures of open sets with smooth boundary in closed manifolds

Smooth maps.

Definition. Let $f : M \to N$ be a map between two manifolds. We say that f is *smooth* near a point p if $\psi \circ f \circ \varphi^{-1} : \varphi(U) \to \psi(V)$ is smooth for some charts (U, φ) of M and (V, ψ) of N such that $p \in U$ and $f(U) \subseteq V$. We say that f is *smooth* in a set $A \subseteq M$ if it is smooth near any point of A. The set of all maps $f : M \to N$ which are smooth in A is denoted by $C^\infty(A, N)$. If $N = \mathbb{R}$ we write $C^\infty(A, N) = C^\infty(A)$.

Summation convention. Below and throughout these notes we will apply the *Einstein summation convention*: repeated indices in lower and upper position are summed. For instance, the expression

$$a_{jkl} b^j c^k$$

is shorthand for

$$\sum_{j,k} a_{jkl} b^j c^k.$$

The summation indices run typically from 1 to n, where n is the dimension of the manifold.

Tangent bundle.

Definition. Let $p \in M$. A *derivation* at p is a linear map $v : C^\infty(M) \to \mathbb{R}$ which satisfies $v(fg) = (vf)g(p) + f(p)(vg)$. The *tangent space* $T_p M$ is the vector space consisting of all derivations at p. Its elements are called *tangent vectors*.

The tangent space T_pM is an n-dimensional vector space when $\dim(M) = n$. If x is a local coordinate system in a neighborhood U of p, the *coordinate vector fields* ∂_j are defined for any $q \in U$ to be the derivations

$$\partial_j|_q f := \frac{\partial}{\partial x_j}(f \circ x^{-1})(x(q)), \quad j = 1, \ldots, n.$$

Then $\{\partial_j|_q\}$ is a basis of T_qM, and any $v \in T_qM$ may be written as $v = v^j \partial_j$. The *tangent bundle* is the disjoint union

$$TM := \bigvee_{p \in M} T_p M.$$

The tangent bundle has the structure of a $2n$-dimensional manifold defined as follows. For any chart (U, x) of M we represent elements of T_qM for $q \in U$ as $v = v^j(q)\partial_j|_q$, and define a map $\tilde{\varphi} : TU \to \mathbb{R}^{2n}$, $\tilde{\varphi}(q, v) = (x(q), v^1(q), \ldots, v^n(q))$. The charts $(TU, \tilde{\varphi})$ are called the *standard charts* of TM and they define a smooth structure on TM.

Exercise 2.2. Prove that T_pM is an n-dimensional vector space spanned by $\{\partial_j\}$ also when M is a manifold with boundary.

Cotangent bundle. The dual space of a vector space V is

$$V^* := \{u : V \to \mathbb{R} \, ; \, u \text{ linear}\}.$$

The dual space of T_pM is denoted by T_p^*M and is called the *cotangent space* of M at p. Let x be local coordinates in U, and let ∂_j be the coordinate vector fields that span T_qM for $q \in U$. We denote by dx^j the elements of the dual basis of T_q^*M, so that any $\xi \in T_q^*M$ can be written as $\xi = \xi_j \, dx^j$. The dual basis is characterized by

$$dx^j(\partial_k) = \delta_{jk}.$$

The *cotangent bundle* is the disjoint union

$$T^*M = \bigvee_{p \in M} T_p^*M.$$

This becomes a $2n$-dimensional manifold by defining for any chart (U, φ) of M a chart $(T^*U, \tilde{\varphi})$ of T^*M by $\tilde{\varphi}(q, \xi_j \, dx^j) = (\varphi(q), \xi_1, \ldots, \xi_n)$.

Tensor bundles. If V is a finite dimensional vector space, the space of (covariant) k-tensors on V is

$$T^k(V) := \{u : \underbrace{V \times \ldots \times V}_{k \text{ copies}} \to \mathbb{R} \, ; \, u \text{ linear in each variable}\}.$$

The k-tensor bundle on M is the disjoint union

$$T^k M = \bigvee_{p \in M} T^k(T_pM).$$

If x are local coordinates in U and dx^j is the basis for T_q^*M, then each $u \in T^k(T_qM)$ for $q \in U$ can be written as

$$u = u_{j_1\cdots j_k} dx^{j_1} \otimes \ldots \otimes dx^{j_k}$$

Here $\otimes$ is the *tensor product*

$$T^k(V) \times T^{k'}(V) \to T^{k+k'}(V), \quad (u, u') \mapsto u \otimes u',$$

where for $v \in V^k$, $v' \in V^{k'}$ we have

$$(u \otimes u')(v, v') := u(v)u'(v').$$

It follows that the elements $dx^{j_1} \otimes \ldots \otimes dx^{j_k}$ span $T^k(T_qM)$. Similarly as above, $T^k M$ has the structure of a smooth manifold (of dimension $n + n^k$).

Exterior powers. The space of alternating k-tensors is

$$A^k(V) := \{u \in T^k(V) \,;\, u(v_1, \ldots, v_k) = 0 \text{ if } v_i = v_j \text{ for some } i \neq j\}.$$

This gives rise to the bundle

$$\Lambda^k(M) := \bigvee_{p \in M} A^k(T_pM).$$

To describe a basis for $A^k(T_pM)$, we introduce the *wedge product*

$$A^k(V) \times A^{k'}(V) \to A^{k+k'}(V), \quad (\omega, \omega') \mapsto \omega \wedge \omega' := \frac{(k+k')!}{k!\,(k')!} \mathrm{Alt}(\omega \otimes \omega'),$$

where $\mathrm{Alt} : T^k(V) \to A^k(V)$ is the projection to alternating tensors:

$$\mathrm{Alt}(T)(v_1, \ldots, v_k) = \frac{1}{k!} \sum_{\sigma \in S_k} \mathrm{sgn}(\sigma) T(v_{\sigma(1)}, \ldots, v_{\sigma(k)}).$$

We have written S_k for the group of permutations of $\{1, \ldots, k\}$, and $\mathrm{sgn}(\sigma)$ for the signature of $\sigma \in S_k$.

If x is a local coordinate system in U, then a basis of $A^k(T_pM)$ is given by

$$\{dx^{j_1} \wedge \ldots dx^{j_k}\}_{1 \leq j_1 < j_2 < \cdots < j_k \leq n}.$$

Again, $\Lambda^k(M)$ is a smooth manifold (of dimension $n + \binom{n}{k}$).

Exercise 2.3. Show that Alt maps $T^k(V)$ into $A^k(V)$ and that $(\mathrm{Alt})^2 = \mathrm{Alt}$.

Smooth sections. The above constructions of the tangent bundle, cotangent bundle, tensor bundles, and exterior powers are all examples of *vector bundles* with base manifold M. We will not need a precise definition here, but just note that in each case there is a natural vector space over any point $p \in M$ (called the *fiber over p*). A *smooth section* of a vector bundle E over M is a smooth map $s: M \to E$ such that for each $p \in M$, $s(p)$ belongs to the fiber over p. The space of smooth sections of E is denoted by $C^\infty(M, E)$.

We have the following terminology:

- $C^\infty(M, TM)$ is the set of *vector fields* on M,
- $C^\infty(M, T^*M)$ is the set of *1-forms* on M,
- $C^\infty(M, T^k M)$ is the set of *k-tensor fields* on M,
- $C^\infty(M, \Lambda^k M)$ is the set of *(differential) k-forms* on M.

Let x be local coordinates in a set U, and let ∂_j and dx^j be the coordinate vector fields and 1-forms in U which span $T_q M$ and $T_q^* M$, respectively, for $q \in U$. In these local coordinates,

- a vector field X has the expression $X = X^j \partial_j$,
- a 1-form α has expression $\alpha = \alpha_j \, dx^j$,
- a k-tensor field u can be written as

$$u = u_{j_1 \cdots j_k} dx^{j_1} \otimes \ldots \otimes dx^{j_k},$$

- a k-form ω has the form

$$\omega = \omega_I \, dx^I$$

where $I = (i_1, \ldots, i_k)$ and $dx^I = dx^{i_1} \wedge \ldots \wedge dx^{i_k}$, with the sum being over all I such that $1 \leq i_1 < i_2 < \ldots < i_k \leq n$.

Here, the component functions $X^j, \alpha_j, u_{j_1 \cdots j_k}, \omega_I$ are all smooth real-valued functions in U.

Note that a vector field $X \in C^\infty(M, TM)$ gives rise to a linear map $X: C^\infty(M) \to C^\infty(M)$ via $Xf(p) = X(p)f$.

Example. Some examples of the smooth sections that will be encountered in this text are:

- Vector fields: the gradient vector field $\mathrm{grad}(f)$ for $f \in C^\infty(M)$, coordinate vector fields ∂_j in a chart U
- One-forms: the exterior derivative df for $f \in C^\infty(M)$
- 2-tensor fields: Riemannian metrics g, Hessians $\mathrm{Hess}(f)$ for $f \in C^\infty(M)$

- k-forms: the volume form dV in Riemannian manifold (M, g), the volume form dS of the boundary ∂M

Changes of coordinates. We consider the transformation law for k-tensor fields under changes of coordinates, or more generally under pullbacks by smooth maps. If $F : M \to N$ is a smooth map, the *pullback* by F is the map $F^* : C^\infty(N, T^k N) \to C^\infty(M, T^k M)$,

$$(F^*u)_p(v_1, \ldots, v_k) = u_{F(p)}(F_*v_1, \ldots, F_*v_k)$$

where $v_1, \ldots, v_k \in T_p \tilde{M}$. Here $F_* : T_p M \to T_{F(p)} N$ is the pushforward, defined by $(F_*v)f = v(f \circ F)$ for $v \in T_p M$ and $f \in C^\infty(N)$. Clearly F^* pulls back k-forms on N to k-forms on M.

The pullback satisfies

- $F^*(fu) = (f \circ F) F^* u$
- $F^*(u \otimes u') = F^* u \otimes F^* u'$
- $F^*(\omega \wedge \omega') = F^* \omega \wedge F^* \omega'$

In terms of local coordinates, the pullback acts by

- $F^* f = f \circ F$ if f is a smooth function (0-form)
- $F^*(\alpha_j \, dx^j) = (\alpha_j \circ F) \, d(x^j \circ F)$ if α is a 1-form

and it has similar expressions for higher order tensors.

Exterior derivative. The exterior derivative d is a first-order differential operator mapping differential k-forms to $k+1$-forms. It can be defined first on 0-forms (that is, smooth functions f) by the local coordinate expression

$$df := \frac{\partial f}{\partial x_j} dx^j.$$

In general, if $\omega = \omega_I \, dx^I$ is a k-form we define

$$d\omega := d\omega_I \wedge dx^I.$$

It turns out that this definition is independent of the choice of coordinates, and one obtains a linear map $d : C^\infty(M, \Lambda^k) \to C^\infty(M, \Lambda^{k+1})$. It has the properties

- $d^2 = 0$
- $d = 0$ on n-forms
- $d(\omega \wedge \omega') = d\omega \wedge \omega' + (-1)^k \omega \wedge d\omega'$ for a k-form ω, k'-form ω'
- $F^* d\omega = d F^* \omega$

Exercise 2.4. If f is a smooth function and $V = (V_1, V_2, V_3)$ is a smooth vector field on $\mathbb{R}^3$, show that the exterior derivative is related to the gradient, curl, and divergence by

$$df = (\nabla f)_j \, dx^j,$$
$$d(V_j \, dx^j) = (\nabla \times V)_j \, dx^{\widehat{j}},$$
$$d(V_j \, dx^{\widehat{j}}) = (\nabla \cdot V) \, dx^1 \wedge dx^2 \wedge dx^3,$$
$$d(f \, dx^1 \wedge dx^2 \wedge dx^3) = 0.$$

Here $dx^{\widehat{1}} := dx^2 \wedge dx^3$, $dx^{\widehat{2}} := dx^3 \wedge dx^1$, $dx^{\widehat{3}} := dx^1 \wedge dx^2$.

Partition of unity. A major reason for including the condition of second countability in the definition of manifolds is to ensure the existence of partitions of unity. These make it possible to make constructions in local coordinates and then glue them together to obtain a global construction.

Theorem 2.1. *Let M be a manifold and let $\{U_\alpha\}$ be an open cover. There exists a family of C^∞ functions $\{\chi_\alpha\}$ on M such that $0 \leq \chi_\alpha \leq 1$, $\mathrm{supp}(\chi_\alpha) \subseteq U_\alpha$, any point of M has a neighborhood which intersects only finitely many of the sets $\mathrm{supp}(\chi_\alpha)$, and further*

$$\sum_\alpha \chi_\alpha = 1 \quad \text{in } M.$$

Integration on manifolds. The natural objects that can be integrated on an n-dimensional manifold are the differential n-forms. This is due to the transformation law for n-forms in $\mathbb{R}^n$ under smooth diffeomorphisms F in $\mathbb{R}^n$,

$$F^*(dx^1 \wedge \cdots \wedge dx^n) = (\det DF) dx^1 \wedge \cdots \wedge dx^n.$$

This is almost the same as the transformation law for integrals in $\mathbb{R}^n$ under changes of variables, the only difference being that in the latter the factor $|\det DF|$ instead $\det DF$ appears. To define an invariant integral, we therefore need to make sure that all changes of coordinates have positive Jacobian.

Definition. If M admits a smooth nonvanishing n-form we say that M is *orientable*. An *oriented manifold* is a manifold together with a given nonvanishing n-form.

If M is oriented with a given n-form Ω, a basis $\{v_1, \ldots, v_n\}$ of $T_p M$ is called *positive* if $\Omega(v_1, \ldots, v_n) > 0$. There are many n-forms on an oriented manifold which give the same positive bases; we call any such n-form an *orientation form*. If (U, φ) is a connected coordinate chart, we say that this chart is *positive* if the coordinate vector fields $\{\partial_1, \ldots, \partial_n\}$ form a positive basis of $T_q M$ for all $q \in M$.

A map $F: M \to N$ between two oriented manifolds is said to be *orientation preserving* if it pulls back an orientation form on N to an orientation form of M. In terms of local coordinates given by positive charts, one can see that a map is orientation preserving if and only if its Jacobian determinant is positive.

Example. The standard orientation of $\mathbb{R}^n$ is given by the n-form $dx^1 \wedge \cdots \wedge dx^n$, where x are the Cartesian coordinates.

If ω is a compactly supported n-form in $\mathbb{R}^n$, we may write
$$\omega = f\, dx^1 \wedge \cdots \wedge dx^n$$
for some smooth compactly supported function f. Then the integral of ω is defined by
$$\int_{\mathbb{R}^n} \omega := \int_{\mathbb{R}^n} f(x)\, dx^1 \cdots dx^n.$$

If ω is a smooth 1-form in a manifold M whose support is compactly contained in U for some positive chart (U, φ), then the integral of ω over M is defined by
$$\int_M \omega := \int_{\varphi(U)} ((\varphi)^{-1})^* \omega.$$

Finally, if ω is a compactly supported n-form in a manifold M, the integral of ω over M is defined by
$$\int_M \omega := \sum_j \int_{U_j} \chi_j \omega.$$
where $\{U_j\}$ is some open cover of $\text{supp}(\omega)$ by positive charts, and $\{\chi_j\}$ is a partition of unity subordinate to the cover $\{U_j\}$.

Exercise 2.5. Prove that the definition of the integral is independent of the choice of positive charts and the partition of unity.

The following result is a basic integration by parts formula which implies the usual theorems of Gauss and Green.

Theorem 2.2 (Stokes' theorem). *If M is an oriented manifold with boundary and if ω is a compactly supported $(n-1)$-form on M, then*
$$\int_M d\omega = \int_{\partial M} i^* \omega$$
where $i: \partial M \to M$ is the natural inclusion.

Here, if M is an oriented manifold with boundary, then ∂M has a natural orientation defined as follows: for any point $p \in \partial M$, a basis $\{E_1, \ldots, E_{n-1}\}$ of $T_p(\partial M)$ is defined to be positive if $\{N_p, E_1, \ldots, E_{n-1}\}$ is a positive basis of

T_pM where N is some outward pointing vector field near ∂M (that is, there is a smooth curve $\gamma : [0, \varepsilon) \to M$ with $\gamma(0) = p$ and $\dot{\gamma}(0) = -N_p$).

Exercise 2.6. Prove that any manifold with boundary has an outward pointing vector field, and show that the above definition gives a valid orientation on ∂M.

2B. Riemannian manifolds.

Riemannian metrics. If u is a 2-tensor field on M, we say that u is *symmetric* if $u(v, w) = u(w, v)$ for any tangent vectors v, w, and that u is *positive definite* if $u(v, v) > 0$ unless $v = 0$.

Definition. Let M be a manifold. A *Riemannian metric* is a symmetric positive definite 2-tensor field g on M. The pair (M, g) is called a *Riemannian manifold*.

If g is a Riemannian metric on M, then $g_p : T_pM \times T_pM$ is an inner product on T_pM for any $p \in M$. We will write

$$\langle v, w \rangle := g(v, w), \qquad |v| := \langle v, v \rangle^{1/2}.$$

In local coordinates, a Riemannian metric is just a positive definite symmetric matrix. To see this, let (U, x) be a chart of M, and write $v, w \in T_qM$ for $q \in U$ in terms of the coordinate vector fields ∂_j as $v = v^j \partial_j$, $w = w^j \partial_j$. Then

$$g(v, w) = g(\partial_j, \partial_k) v^j w^k.$$

This shows that g has the local coordinate expression

$$g = g_{jk} dx^j \otimes dx^k$$

where $g_{jk} := g(\partial_j, \partial_k)$ and the matrix $(g_{jk})_{j,k=1}^n$ is symmetric and positive definite. We will also write $(g^{jk})_{j,k=1}^n$ for the inverse matrix of (g_{jk}), and $|g| := \det(g_{jk})$ for the determinant.

Example. Some examples of Riemannian manifolds:

1. (Euclidean space) If Ω is a bounded open set in $\mathbb{R}^n$, then (Ω, e) is a Riemannian manifold if e is the *Euclidean metric* for which $e(v, w) = v \cdot w$ is the Euclidean inner product of $v, w \in T_p\Omega \approx \mathbb{R}^n$. In Cartesian coordinates, e is just the identity matrix.

2. If Ω is as above, then more generally (Ω, g) is a Riemannian manifold if $g(x) = (g_{jk}(x))_{j,k=1}^n$ is any family of positive definite symmetric matrices whose elements depend smoothly on $x \in \Omega$.

3. If Ω is a bounded open set in $\mathbb{R}^n$ with smooth boundary, then $(\overline{\Omega}, g)$ is a compact Riemannian manifold with boundary if $g(x)$ is a family of positive definite symmetric matrices depending smoothly on $x \in \overline{\Omega}$.

4. (hypersurfaces) Let S be a smooth hypersurface in $\mathbb{R}^n$ such that $S = f^{-1}(0)$ for some smooth function $f : \mathbb{R}^n \to \mathbb{R}$ which satisfies $\nabla f \neq 0$ when $f = 0$. Then S is a smooth manifold of dimension $n - 1$, and the tangent space $T_p S$ for any $p \in S$ can be identified with $\{v \in \mathbb{R}^n ; v \cdot \nabla f(p) = 0\}$. Using this identification, we define an inner product $g_p(v, w)$ on $T_p S$ by taking the Euclidean inner product of v and w interpreted as vectors in $\mathbb{R}^n$. Then (S, g) is a Riemannian manifold, and g is called the *induced Riemannian metric* on S (this metric being induced by the Euclidean metric in $\mathbb{R}^n$).

5. (model spaces) The model spaces of Riemannian geometry are the Euclidean space $(\mathbb{R}^n, e)$, the sphere (S^n, g) where S^n is the unit sphere in $\mathbb{R}^{n+1}$ and g is the induced Riemannian metric, and the hyperbolic space (H^n, g) which may be realized by taking H^n to be the unit ball in $\mathbb{R}^n$ with metric

$$g_{jk}(x) = \frac{4}{(1 - |x|^2)^2} \delta_{jk}.$$

The Riemannian metric allows to measure lengths and angles of tangent vectors on a manifold, the *length* of a vector $v \in T_p M$ being $|v|$ and the *angle* between two vectors $v, w \in T_p M$ being the number $\theta(v, w) \in [0, \pi]$ which satisfies

$$\cos \theta(v, w) := \frac{\langle v, w \rangle}{|v||w|}. \tag{2-1}$$

Physically, one may think of a Riemannian metric g as the resistivity of a conducting medium (in the introduction, the conductivity matrix (γ^{jk}) corresponded formally to $(|g|^{1/2} g^{jk})$), or as the inverse of sound speed squared in a medium where acoustic waves propagate (if a medium $\Omega \subseteq \mathbb{R}^n$ has scalar sound speed $c(x)$ then a natural Riemannian metric is $g_{jk}(x) = c(x)^{-2} \delta_{jk}$). In the latter case, regions where g is large (resp. small) correspond to low velocity regions (resp. high velocity regions). We will later define geodesics, which are length-minimizing curves on a Riemannian manifold, and these tend to avoid low velocity regions as one would expect.

Exercise 2.7. Use a partition of unity to prove that any smooth manifold M admits a Riemannian metric.

Raising and lowering of indices. On a Riemannian manifold (M, g) there is a canonical way of converting tangent vectors into cotangent vectors and vice versa. We define a map

$$T_p M \to T_p^* M, \quad v \mapsto v^\flat$$

by requiring that $v^\flat(w) = \langle v, w \rangle$. This map (called the *flat operator*) is an isomorphism, which is given in local coordinates by

$$(v^j \partial_j)^\flat = v_j\, dx^j, \quad \text{where } v_j := g_{jk} v^k.$$

We say that $v^\flat$ is the cotangent vector obtained from v by *lowering indices*. The inverse of this map is the *sharp operator*

$$T_p^* M \to T_p M, \quad \xi \mapsto \xi^\sharp$$

given in local coordinates by

$$(\xi_j\, dx^j)^\sharp = \xi^j \partial_j, \quad \text{where } \xi^j := g^{jk} \xi_k.$$

We say that $\xi^\sharp$ is obtained from ξ by *raising indices* with respect to the metric g.

A standard example of this construction is the *metric gradient*. If $f \in C^\infty(M)$, the metric gradient of f is the vector field

$$\operatorname{grad}(f) := (df)^\sharp.$$

In local coordinates, $\operatorname{grad}(f) = g^{jk}(\partial_j f)\partial_k$.

Inner products of tensors. If (M, g) is a Riemannian manifold, we can use the Riemannian metric g to define inner products of differential forms and other tensors in a canonical way. We will mostly use the inner product of 1-forms, defined via the sharp operator by

$$\langle \alpha, \beta \rangle := \langle \alpha^\sharp, \beta^\sharp \rangle.$$

In local coordinates one has $\langle \alpha, \beta \rangle = g^{jk} \alpha_j \beta_k$ and $g^{jk} = \langle dx^j, dx^k \rangle$.

More generally, if u and v are k-tensor fields with local coordinate representations $u = u_{i_1 \cdots i_k}\, dx^{i_1} \otimes \cdots \otimes dx^{i_k}$, $v = v_{i_1 \cdots i_k}\, dx^{i_1} \otimes \cdots \otimes dx^{i_k}$, we define

$$\langle u, v \rangle := g^{i_1 j_1} \cdots g^{i_k j_k} u_{i_1 \cdots i_k} v_{j_1 \cdots j_k}. \tag{2-2}$$

This definition turns out to be independent of the choice of coordinates, and it gives a valid inner product on k-tensor fields.

Orthonormal frames. If U is an open subset of M, we say that a set $\{E_1, \ldots, E_n\}$ of vector fields in U is a *local orthonormal frame* if $\{E_1(q), \ldots, E_n(q)\}$ forms an orthonormal basis of $T_q M$ for any $q \in U$.

Lemma 2.3 (local orthonormal frame). *If (M, g) is a Riemannian manifold, then for any point $p \in M$ there is a local orthonormal frame in some neighborhood of p.*

If $\{E_j\}$ is a local orthonormal frame, the dual frame $\{\varepsilon^j\}$ which is characterized by $\varepsilon^j(E_k) = \delta_{jk}$ gives an orthonormal basis of $T_q^* M$ for any q near p. The inner product in (2-2) is the unique inner product on k-tensor fields such that $\{\varepsilon^{i_1} \otimes \cdots \otimes \varepsilon^{i_k}\}$ gives an orthonormal basis of $T^k(T_q M)$ for q near p whenever $\{\varepsilon^j\}$ is a local orthonormal frame of 1-forms near p.

Exercise 2.8. Prove the lemma by applying the Gram-Schmidt orthonormalization procedure to a basis $\{\partial_j\}$ of coordinate vector fields, and prove the statements after the lemma.

Volume form, integration, and L^2 Sobolev spaces. From this point on, *all Riemannian manifolds will be assumed to be oriented*. Clearly near any point p in (M, g) there is a positive local orthonormal frame (that is, a local orthonormal frame $\{E_j\}$ which gives a positive orthonormal basis of $T_q M$ for q near p).

Lemma 2.4 (volume form). *Let (M, g) be a Riemannian manifold. There is a unique n-form on M, denoted by dV and called the* volume form, *such that $dV(E_1, \ldots, E_n) = 1$ for any positive local orthonormal frame $\{E_j\}$. In local coordinates*
$$dV = |g|^{1/2} \, dx^1 \wedge \ldots \wedge dx^n.$$

Exercise 2.9. Prove this lemma.

If f is a function on (M, g), we can use the volume form to obtain an n-form $f \, dV$. The integral of f over M is then defined to be the integral of the n-form $f \, dV$. Thus, on a Riemannian manifold there is a canonical way to integrate functions (instead of just n-forms).

If $u, v \in C^\infty(M)$ are complex-valued functions, we define the L^2 inner product by
$$(u, v) = (u, v)_{L^2(M)} := \int_M u\bar{v} \, dV.$$

The completion of $C^\infty(M)$ with respect to this inner product is a Hilbert space denoted by $L^2(M)$ or $L^2(M, dV)$. It consists of square integrable functions defined almost everywhere on M with respect to the measure dV. The L^2 norm is defined by
$$\|u\| = \|u\|_{L^2(M)} := (u, u)_{L^2(M)}^{1/2}.$$

Similarly, we may define the spaces of square integrable k-forms or k-tensor fields, denoted by $L^2(M, \Lambda^k M)$ or $L^2(M, T^k M)$, by using the inner product
$$(u, v) := \int_M \langle u, \bar{v}\rangle \, dV, \quad u, v \in C^\infty(M, T^k M) \text{ complex-valued.}$$

We may use the above inner products to give a definition of low order Sobolev spaces on Riemannian manifolds which does not involve local coordinates. We

define the $H^1(M)$ inner product
$$(u, v)_{H^1(M)} := (u, v) + (du, dv), \quad u, v \in C^\infty(M) \text{ complex-valued.}$$
The space $H^1(M)$ (resp. $H_0^1(M)$) is defined to be the completion of $C^\infty(M)$ (resp. $C_c^\infty(M^{\text{int}})$) with respect to this inner product. These are subspaces of $L^2(M)$ which have first-order weak derivatives in $L^2(M)$, and they coincide with the spaces defined in the usual way by using local coordinates. Also, we define $H^{-1}(M)$ to be the dual space of $H_0^1(M)$.

Codifferential. Using the inner product on k-forms, we can define the codifferential operator δ as the adjoint of the exterior derivative via the relation
$$(\delta u, v) = (u, dv)$$
where $u \in C^\infty(M, \Lambda^k)$ and $v \in C_c^\infty(M^{\text{int}}, \Lambda^{k-1})$. It can be shown that δ gives a well-defined map
$$\delta : C^\infty(M, \Lambda^k) \to C^\infty(M, \Lambda^{k-1}).$$
We will only use δ for 1-forms, and in this case the operator can be easily defined by a local coordinate expression. Let α be a 1-form in M, let (U, x) be a chart and let $\varphi \in C_c^\infty(U)$. One computes in local coordinates
$$(\alpha, dv) = \int_U \langle \alpha, d\bar{v} \rangle \, dV = \int_U g^{jk} \alpha_j \overline{\partial_k v} \, |g|^{1/2} \, dx$$
$$= -\int_U |g|^{-1/2} \partial_k(|g|^{1/2} g^{jk} \alpha_j) \bar{v} \, dV.$$
This computation shows that the function $\delta \alpha$, defined in local coordinates by
$$\delta \alpha := -|g|^{-1/2} \partial_j(|g|^{1/2} g^{jk} \alpha_k),$$
is a smooth function in M and satisfies $(\delta \alpha, v) = (\alpha, dv)$.

It follows that $\delta \alpha$ is related to the divergence of vector fields by $\delta \alpha = -\text{div}(\alpha^\sharp)$, where the divergence is defined by
$$\text{div}(X) := |g|^{-1/2} \partial_j(|g|^{1/2} X^j).$$

Exercise 2.10 (Hodge star operator). Let (M, g) be a Riemannian manifold of dimension n. If ω and η are k-forms on M, show that the identity
$$\omega \wedge *\eta = \langle \omega, \eta \rangle \, dV$$
determines uniquely a linear operator (called the *Hodge star operator*)
$$* : C^\infty(M, \Lambda^k) \to C^\infty(M, \Lambda^{n-k}).$$
Prove the following properties:

- $** = (-1)^{k(n-k)}$ on k-forms
- $*1 = dV$
- $*(\varepsilon^1 \wedge \ldots \wedge \varepsilon^k) = \varepsilon^{k+1} \wedge \ldots \wedge \varepsilon^n$ whenever $(\varepsilon^1, \ldots, \varepsilon^n)$ is a positive local orthonormal frame on T^*M
- $\langle *\omega, \eta \rangle = -\langle \omega, *\eta \rangle$ when ω, η are 1-forms and $\dim(M) = 2$ (that is, on 2D manifolds the Hodge star on 1-forms corresponds to rotation by $90°$)

Prove that the operator
$$\delta := (-1)^{(k-1)(n-k)-1} *d* \quad \text{on } k\text{-forms}$$
gives a map $\delta : C^\infty(M, \Lambda^k) \to C^\infty(M, \Lambda^{k-1})$ satisfying $(\delta u, v) = (u, dv)$ for compactly supported v, and thus gives a valid definition of the codifferential on forms of any order.

Conformality. As the last topic in this section, we discuss the notion of conformality of manifolds.

Definition. Two metrics g_1 and g_2 on a manifold M are called *conformal* if $g_2 = cg_1$ for a smooth positive function c on M. A diffeomorphism $f : (M, g) \to (M', g')$ is called a *conformal transformation* if f^*g' is conformal to g, that is,
$$f^*g' = cg.$$
Two Riemannian manifolds are called conformal if there is a conformal transformation between them.

We relate this definition of conformality to the standard one in complex analysis via the concept of angle $\theta(v, w) = \theta_g(v, w) \in [0, \pi]$ defined in (2-1).

Lemma 2.5 (conformal = angle-preserving). *Let $f : (M, g) \to (M', g')$ be a diffeomorphism. The following are equivalent.*

(1) *f is a conformal transformation.*

(2) *f preserves angles in the sense that $\theta_g(v, w) = \theta_{g'}(f_*v, f_*w)$.*

Exercise 2.11. Prove the lemma.

It follows that f is a conformal transformation if and only if for any point p and tangent vectors v and w, and for any curves γ_v and γ_w with $\dot{\gamma}_v(0) = v$, $\dot{\gamma}_w(0) = w$, the curves $f \circ \gamma_v$ and $f \circ \gamma_w$ intersect in the same angle as γ_v and γ_w. This corresponds to the standard interpretation of conformality.

The two dimensional case is special because of the classical fact that orientation preserving conformal maps are holomorphic. The proof is given for completeness.

Lemma 2.6 (conformal = holomorphic). *Let Ω and $\tilde{\Omega}$ be open sets in $\mathbb{R}^2$. An orientation preserving map $f : (\Omega, e) \to (\tilde{\Omega}, e)$ is conformal if and only if it is holomorphic and bijective.*

Proof. We use complex notation and write $z = x + iy$, $f = u + iv$. If f is conformal then it is bijective and $f^*e = ce$. The last condition means that for all $z \in \Omega$ and for $v, w \in \mathbb{R}^2$,
$$c(z) v \cdot w = (f_* v) \cdot (f_* w) = Df(z) v \cdot Df(z) w = Df(z)^t Df(z) v \cdot w.$$
Since $Df(z) = \begin{pmatrix} u_x & u_y \\ v_x & v_y \end{pmatrix}$, this implies
$$\begin{pmatrix} u_x^2 + v_x^2 & u_x u_y + v_x v_y \\ u_x u_y + v_x v_y & u_y^2 + v_y^2 \end{pmatrix} = \begin{pmatrix} c & 0 \\ 0 & c \end{pmatrix}.$$
Thus the vectors $(u_x\ v_x)^t$ and $(u_y\ v_y)^t$ are orthogonal and have the same length. Since f is orientation preserving so $\det Df > 0$, we must have
$$u_x = v_y, \quad u_y = -v_x.$$
This shows that f is holomorphic. The converse follows by running the argument backwards. $\square$

It follows from the existence of *isothermal coordinates* that any 2D Riemannian manifold is locally conformal to a set in Euclidean space. The conformal structure of manifolds with dimension $n \geq 3$ is much more complicated. However, the model spaces are locally conformally Euclidean.

Lemma 2.7. (1) *Let (S^n, g) be the unit sphere in $\mathbb{R}^{n+1}$ with its induced metric, and let $N = e_{n+1}$ be the north pole. Then the stereographic projection*
$$f : (S^n \smallsetminus \{N\}, g) \to (\mathbb{R}^n, e), \quad f(y, y_{n+1}) := \frac{y}{1 - y_{n+1}}$$
is a conformal transformation.

(2) *Hyperbolic space (H^n, g), where H^n is the unit ball B in $\mathbb{R}^n$ and $g_{jk}(x) = \frac{4}{(1-|x|^2)^2} \delta_{jk}$, is conformal to (B, e).*

Exercise 2.12. Prove the lemma.

Finally, we mention Liouville's theorem which characterizes all conformal transformations in $\mathbb{R}^n$ for $n \geq 3$. This result shows that up to translation, scaling, and rotation, the only conformal transformations are the identity map and Kelvin transform (this is in contrast to the 2D case where there is a rich family of conformal transformations, the holomorphic bijective maps). See [Iwaniec and Martin 2001] for a proof.

Theorem (Liouville). *If $\Omega, \tilde{\Omega} \subseteq \mathbb{R}^n$ with $n \geq 3$, then an orientation preserving diffeomorphism $f : (\Omega, e) \to (\tilde{\Omega}, e)$ is conformal if and only if*

$$f(x) = \alpha A h(x - x_0) + b$$

where $\alpha \in \mathbb{R}$, A is an $n \times n$ orthogonal matrix, $h(x) = x$ or $h(x) = \dfrac{x}{|x|^2}$, $x_0 \in \mathbb{R}^n \smallsetminus \Omega$, and $b \in \mathbb{R}^n$.

2C. Laplace–Beltrami operator.

Definition. In this section we will see that on any Riemannian manifold there is a canonical second-order elliptic operator, called the Laplace–Beltrami operator, which is an analog of the usual Laplacian in $\mathbb{R}^n$.

Motivation. Let first Ω be a bounded domain in $\mathbb{R}^n$ with smooth boundary, and consider the Laplace operator

$$\Delta = \sum_{j=1}^n \frac{\partial^2}{\partial x_j^2}.$$

Solutions of the equation $\Delta u = 0$ are called harmonic functions, and by standard results for elliptic PDE [Evans 2010, Section 6], for any $f \in H^1(\Omega)$ there is a unique solution $u \in H^1(\Omega)$ of the Dirichlet problem

$$\begin{cases} -\Delta u = 0 & \text{in } \Omega, \\ u = f & \text{on } \partial\Omega. \end{cases} \tag{2-3}$$

The last line means that $u - f \in H_0^1(\Omega)$.

One way to produce the solution of (2-3) is based on variational methods and Dirichlet's principle [Evans 2010, Section 2]. We define the Dirichlet energy

$$E(v) := \frac{1}{2} \int_\Omega |\nabla v|^2 \, dx, \qquad v \in H^1(\Omega).$$

If we define the admissible class

$$\mathcal{A}_f := \{v \in H^1(\Omega) \,;\, v = f \text{ on } \partial\Omega\},$$

then the solution of (2-3) is the unique function $u \in \mathcal{A}_f$ which minimizes the Dirichlet energy:

$$E(u) \leq E(v) \quad \text{for all } v \in \mathcal{A}_f.$$

The heuristic idea is that the solution of (2-3) represents a physical system in equilibrium, and therefore should minimize a suitable energy functional. The point is that one can start from the energy functional $E(\cdot)$ and conclude that any minimizer u must satisfy $\Delta u = 0$, which gives another way to define the Laplace operator.

From this point on, let (M, g) be a compact Riemannian manifold with smooth boundary. Although there is no obvious analog of the coordinate definition of Δ in $\mathbb{R}^n$, there is a natural analog of the Dirichlet energy. It is given by

$$E(v) := \frac{1}{2} \int_M |dv|^2 \, dV, \qquad v \in H^1(M).$$

Here $|dv|$ is the Riemannian length of the 1-form dv, and dV is the volume form.

We wish to find a differential equation which is satisfied by minimizers of $E(\cdot)$. Suppose $u \in H^1(M)$ is a minimizer which satisfies $E(u) \leq E(u + t\varphi)$ for all $t \in \mathbb{R}$ and all $\varphi \in C_c^\infty(M^{\text{int}})$. We have

$$E(u + t\varphi) = \frac{1}{2} \int_M \langle d(u + t\varphi), d(u + t\varphi) \rangle \, dV$$
$$= E(u) + t \int_M \langle du, d\varphi \rangle \, dV + t^2 E(\varphi).$$

Since $I_\varphi(t) := E(u + t\varphi)$ is a smooth function of t for fixed φ, and since $I_\varphi(0) \leq I_\varphi(t)$ for $|t|$ small, we must have $I_\varphi'(0) = 0$. This shows that if u is a minimizer, then

$$\int_M \langle du, d\varphi \rangle \, dV = 0$$

for any choice of $\varphi \in C_c^\infty(M^{\text{int}})$. By the properties of the codifferential δ, this implies that

$$\int_M (\delta du)\varphi \, dV = 0$$

for all $\varphi \in C_c^\infty(M^{\text{int}})$. Thus any minimizer u has to satisfy the equation

$$\delta du = 0 \quad \text{in } M.$$

We have arrived at the definition of the Laplace–Beltrami operator.

Definition. If (M, g) is a compact Riemannian manifold (with or without boundary), the Laplace–Beltrami operator is defined by

$$\Delta_g u := -\delta du.$$

The next result implies, in particular, that in Euclidean space Δ_g is just the usual Laplacian.

Lemma 2.8. *In local coordinates*

$$\Delta_g u = |g|^{-1/2} \partial_j (|g|^{1/2} g^{jk} \partial_k u)$$

where, as before, $|g| = \det(g_{jk})$ is the determinant of g.

Proof. Follows from the coordinate expression for δ. □

Weak solutions. We move on to the question of finding weak solutions to the Dirichlet problem
$$\begin{cases} -\Delta_g u = F & \text{in } M, \\ u = 0 & \text{on } \partial M. \end{cases} \tag{2-4}$$

Here $F \in H^{-1}(M)$ (thus F is a bounded linear functional on $H_0^1(M)$). By definition, a weak solution is a function $u \in H_0^1(M)$ which satisfies
$$\int_M \langle du, d\varphi \rangle \, dV = F(\varphi) \quad \text{for all } \varphi \in H_0^1(M).$$

We will have use of the following compactness result also later.

Theorem (Rellich–Kondrachov compact embedding theorem). *Let (M, g) be a compact Riemannian manifold with smooth boundary. Then the natural inclusion $i : H^1(M) \to L^2(M)$ is a compact operator.*

Proof. See [Evans 2010, Chapter 5] for the Euclidean case and [Taylor 1996] for the Riemannian case. □

The solvability of (2-4) will be a consequence of the following inequality.

Theorem (Poincaré inequality). *There is $C > 0$ such that*
$$\|u\|_{L^2(M)} \leq C \|du\|_{L^2(M)}, \quad u \in H_0^1(M).$$

Proof. Suppose the claim is false. Then there is a sequence $(u_k)_{k=1}^\infty$ with $u_k \in H_0^1(M)$ and
$$\|u_k\|_{L^2(M)} > k \|du_k\|_{L^2(M)}.$$
Letting $v_k = u_k / \|u_k\|_{L^2(M)}$, we have $\|v_k\|_{L^2(M)} = 1$ and
$$\|dv_k\|_{L^2(M)} < \frac{1}{k}.$$
Thus (v_k) is a bounded sequence in $H_0^1(M)$, and therefore it has a subsequence (also denoted by (v_k)) which converges weakly to some $v \in H_0^1(M)$. The compact embedding $H^1(M) \hookrightarrow L^2(M)$ implies that
$$v_k \to v \quad \text{in } L^2(M).$$
It follows that $dv_k \to dv$ in $H^{-1}(M)$. But also $dv_k \to 0$ in $L^2(M)$, and uniqueness of limits shows that $dv = 0$. Now any function $v \in H^1(M)$ with $dv = 0$ must be constant on each connected component of M (this follows from the corresponding result in $\mathbb{R}^n$), and since $v \in H_0^1(M)$ we get that $v = 0$. This contradicts the fact that $\|v_k\|_{L^2(M)} = 1$. □

It follows from the Poincaré inequality that for $u \in H_0^1(M)$,
$$\|du\|_{L^2(M)}^2 \leq \|u\|_{H^1(M)}^2 = \|u\|_{L^2(M)}^2 + \|du\|_{L^2(M)}^2 \leq C\|du\|_{L^2(M)}^2.$$
Consequently the norms $\|\cdot\|_{H^1(M)}$ and $\|d\cdot\|_{L^2(M)}$ are equivalent norms on $H_0^1(M)$ (they induce the same topology). We can now prove the solvability of the Dirichlet problem.

Proposition 2.9 (existence of weak solutions). *The problem (2-4) has a unique weak solution $u \in H_0^1(M)$ for any $F \in H^{-1}(M)$. The solution operator*
$$G : H^{-1}(M) \to H_0^1(M), \quad F \mapsto u,$$
is a bounded linear operator.

Proof. Consider the bilinear form
$$B[u, v] := \int_M \langle du, dv \rangle \, dV, \quad u, v \in H_0^1(M).$$
This satisfies $B[u, v] = B[v, u]$, $|B[u, u]| \leq \|u\|_{H_0^1(M)} \|v\|_{H_0^1(M)}$, and
$$B[u, u] = \int_M |du|^2 \, dV = \|du\|_{L^2(M)}^2 \geq C\|u\|_{H^1(M)}^2$$
by using the equivalent norms on $H_0^1(M)$. Thus $H_0^1(M)$ equipped with the inner product $B[\cdot, \cdot]$ is the same Hilbert space as $H_0^1(M)$ equipped with the usual inner product $(\cdot, \cdot)_{H^1(M)}$. Since F is an element of the dual of $H_0^1(M)$, the Riesz representation theorem shows that there is a unique $u \in H_0^1(M)$ with
$$B[u, \varphi] = F(\varphi), \quad \varphi \in H_0^1(M).$$
This is the required unique weak solution. Writing $u = GF$, the boundedness of G follows from the estimate $\|u\|_{H^1(M)} \leq \|F\|_{H^{-1}(M)}$ also given by the Riesz representation theorem. $\square$

Corollary 2.10 (existence of weak solutions). *The problem*
$$\begin{cases} -\Delta_g u = 0 & \text{in } M, \\ u = f & \text{on } \partial M. \end{cases} \quad (2\text{-}5)$$
has a unique weak solution $u \in H^1(M)$ for any $f \in H^1(M)$, and the solution satisfies $\|u\|_{H^1(M)} \leq C\|f\|_{H^1(M)}$.

Proof. Let $F = \Delta_g f \in H^{-1}(M)$ (one defines $F(\varphi) := -(df, d\varphi)$). Then (2-5) is equivalent with
$$\begin{cases} -\Delta_g(u - f) = F & \text{in } M, \\ u - f = 0 & \text{on } \partial M. \end{cases}$$

This has a unique solution $u_0 = GF$ with $\|u_0\|_{H^1(M)} \leq C\|f\|_{H^1(M)}$, and one can take $u = u_0 + f$. □

Spectral theory. Combined with the spectral theorem for compact operators, the previous results show that the spectrum of $-\Delta_g$ consists of a discrete set of eigenvalues and there is an orthonormal basis of $L^2(M)$ consisting of eigenfunctions of $-\Delta_g$.

Proposition 2.11 (spectral theory for $-\Delta_g$). *Let (M, g) be a compact Riemannian manifold with smooth boundary. There exist numbers $0 < \lambda_1 \leq \lambda_2 \leq \cdots$ and an orthonormal basis $\{\phi_l\}_{l=1}^{\infty}$ of $L^2(M)$ such that*

$$\begin{cases} -\Delta_g \phi_l = \lambda_l \phi_l & \text{in } M, \\ \phi_l \in H_0^1(M). \end{cases}$$

We write $\mathrm{Spec}(-\Delta_g) = \{\lambda_1, \lambda_2, \ldots\}$. *If $\lambda \notin \mathrm{Spec}(-\Delta_g)$, then $-\Delta_g - \lambda$ is an isomorphism from $H_0^1(M)$ onto $H^{-1}(M)$.*

Before giving the proof, we note that by standard Hilbert space theory any function $f \in L^2(M)$ can be written as an L^2-convergent *Fourier series*

$$f = \sum_{l=1}^{\infty} (f, \phi_l)_{L^2(M)} \phi_l$$

where (f, ϕ_l) is the lth *Fourier coefficient*. These eigenfunction (or Fourier) expansions can sometimes be used as a substitute in M for the Fourier transform in Euclidean space, as we will see in Section 4.

Proof of Proposition 2.11. Let $G : H^{-1}(M) \to H_0^1(M)$ be the solution operator from Proposition 2.9. By compact embedding, we have that $G : L^2(M) \to L^2(M)$ is compact. It is also self-adjoint and positive semidefinite, since for $f, h \in L^2(M)$ (with $u = Gf$)

$$(Gf, h) = (u, -\Delta_g Gh) = (du, dGh) = (-\Delta_g u, Gh) = (f, Gh),$$
$$(Gf, f) = (Gf, -\Delta_g Gf) = (dGf, dGf) \geq 0.$$

By the spectral theorem for compact operators, there exist $\mu_1 \geq \mu_2 \geq \cdots$ with $\mu_j \to 0$ and $\phi_l \in L^2(M)$ with $G\phi_l = \mu_l \phi_l$ such that $\{\phi_l\}_{l=1}^{\infty}$ is an orthonormal basis of $L^2(M)$. Note that 0 is not in the spectrum of G, since $Gf = 0$ implies $f = 0$. Taking $\lambda_l = 1/\mu_l$ gives $-\Delta_g \phi_l = \lambda_l \phi_l$. If $\lambda \neq \lambda_l$ for all l then for $F \in H^{-1}(M)$,

$$(-\Delta_g - \lambda)u = F \iff u = G(F + \lambda u) \iff \left(\tfrac{1}{\lambda}\mathrm{Id} - G\right)u = \tfrac{1}{\lambda}GF.$$

Since $\tfrac{1}{\lambda} \neq \mu_l$ for all l, $\tfrac{1}{\lambda}\mathrm{Id} - G$ is invertible and we see that $-\Delta_g - \lambda$ is bijective and bounded, therefore an isomorphism. □

We conclude the section with an analog of Proposition 2.11 where the Laplace–Beltrami operator is replaced by the Schrödinger operator $-\Delta_g + V$. The proof is the same except for minor modifications and is left as an exercise. The main point is that for λ outside the discrete set $\text{Spec}(-\Delta_g + V)$, this result implies unique solvability for the Dirichlet problem

$$\begin{cases} (-\Delta_g + V - \lambda)u = 0 & \text{in } M, \\ u = f & \text{on } \partial M \end{cases}$$

with the norm estimate $\|u\|_{H^1(M)} \leq C\|f\|_{H^1(M)}$.

Proposition 2.12 (spectral theory for $-\Delta_g + V$). *Let (M, g) be a compact Riemannian manifold with smooth boundary, and assume that $V \in L^\infty(M)$ is real-valued. There exist numbers $\lambda_1 \leq \lambda_2 \leq \cdots$ and an orthonormal basis $\{\psi_l\}_{l=1}^\infty$ of $L^2(M)$ such that*

$$\begin{cases} (-\Delta_g + V)\psi_l = \lambda_l \psi_l & \text{in } M, \\ \psi_l \in H_0^1(M). \end{cases}$$

We write

$$\text{Spec}(-\Delta_g + V) = \{\lambda_1, \lambda_2, \ldots\}.$$

If $\lambda \notin \text{Spec}(-\Delta_g + V)$, then $-\Delta_g + V - \lambda$ is an isomorphism from $H_0^1(M)$ onto $H^{-1}(M)$.

Exercise 2.13. Prove this result by first showing an analog of Proposition 2.9 where $-\Delta_g$ is replaced by $-\Delta_g + V + k_0$ for k_0 sufficiently large, and then by following the proof of Proposition 2.11 where G is replaced by the inverse operator for $-\Delta_g + V + k_0$.

2D. *DN map.*

Definition. We now rigorously define the Dirichlet-to-Neumann map, or DN map for short, discussed in the introduction. Let (M, g) be a compact manifold with smooth boundary, and let $V \in L^\infty(M)$. Proposition 2.12 shows that the Dirichlet problem

$$\begin{cases} (-\Delta_g + V)u = 0 & \text{in } M, \\ u = f & \text{on } \partial M \end{cases} \tag{2-6}$$

has a unique solution $u \in H^1(M)$ for any $f \in H^1(M)$, provided that 0 is not a Dirichlet eigenvalue (meaning that $0 \notin \text{Spec}(-\Delta_g + V)$). We make the standing assumption that all Schrödinger operators are such that

$$0 \text{ is not a Dirichlet eigenvalue of } -\Delta_g + V.$$

As mentioned in the introduction, it would be easy to remove this assumption by using so called Cauchy data sets instead of the DN map.

If 0 is not a Dirichlet eigenvalue, then (2-6) is uniquely solvable for any $f \in H^1(M)$. If $f \in H_0^1(M)$ then $u = 0$ is a solution (since then $u - f \in H_0^1(M)$), which means that the solution with boundary value f coincides with the solution with boundary value $f + \varphi$ where $\varphi \in H_0^1(M)$. Motivated by this, we define the quotient space

$$H^{1/2}(\partial M) := H^1(M)/H_0^1(M).$$

This is a Hilbert space which can be identified with a space of functions on ∂M which have $1/2$ derivatives in $L^2(\partial M)$, but the abstract setup will be enough for us. We also define $H^{-1/2}(\partial M)$ as the dual space of $H^{1/2}(\partial M)$.

By the above discussion, the Dirichlet problem (2-6) is well posed for boundary values $f \in H^{1/2}(M)$. Denoting the solution by u_f, the DN map is formally defined as the map

$$\Lambda_{g,V} : f \mapsto \partial_\nu u_f|_{\partial M}.$$

Here, for sufficiently smooth u, the normal derivative is defined by

$$\partial_\nu u|_{\partial M} := \langle \nabla u, \nu \rangle|_{\partial M}.$$

To find a rigorous definition of Λ_g we will use an integration by parts formula.

Theorem (Green's formula). *If $u, v \in C^2(M)$ then*

$$\int_{\partial M} (\partial_\nu u) v \, dS = \int_M (\Delta_g u) v \, dV + \int_M \langle du, dv \rangle \, dV.$$

Exercise 2.14. Prove this formula by using Stokes' theorem.

Let now $f, h \in H^{1/2}(\partial M)$, let u_f be the solution of (2-6), and let e_h be any function in $H^1(M)$ with $e_h|_{\partial M} = h$ (with natural interpretations). Then, again purely formally,

$$\langle \Lambda_{g,V} f, h \rangle = \int_{\partial M} (\partial_\nu u_f) e_h \, dS = \int_M (\Delta_g u_f) e_h \, dV + \int_M \langle du_f, de_h \rangle \, dV$$
$$= \int_M \left[\langle du_f, de_h \rangle + V u_f e_h \right] dV.$$

We have finally arrived at the precise definition of $\Lambda_{g,V}$.

Definition. $\Lambda_{g,V}$ is the linear map from $H^{1/2}(\partial \Omega)$ to $H^{-1/2}(\partial \Omega)$ defined via the bilinear form

$$\langle \Lambda_{g,V} f, h \rangle = \int_M \left[\langle du_f, de_h \rangle + V u_f e_h \right] dV, \quad f, h \in H^{1/2}(\partial M),$$

where u_f and e_h are as above.

Exercise 2.15. Prove that the bilinear form indeed gives a well defined map $H^{1/2}(\partial\Omega) \to H^{-1/2}(\partial\Omega)$.

The DN map is also self-adjoint:

Lemma 2.13. *If V is real-valued, then*
$$\langle \Lambda_{g,V} f, h \rangle = \langle f, \Lambda_{g,V} h \rangle, \quad f, h \in H^{1/2}(\partial M).$$

Exercise 2.16. Prove the lemma.

Integral identity. The main point in this section is an integral identity which relates the difference of two DN maps to an integral over M involving the difference of two potentials. This identity is the starting point for recovering interior information (the potentials in M) from boundary measurements (the DN maps on ∂M).

Proposition 2.14 (integral identity). *Let (M, g) be a compact Riemannian manifold with smooth boundary, and let $V_1, V_2 \in L^\infty(M)$ be real-valued. Then*
$$\langle (\Lambda_{g,V_1} - \Lambda_{g,V_2}) f_1, f_2 \rangle = \int_M (V_1 - V_2) u_1 u_2 \, dV, \quad f_1, f_2 \in H^{1/2}(\partial M),$$
where $u_j \in H^1(M)$ are the solutions of $(-\Delta_g + V_j) u_j = 0$ in M with $u_j|_{\partial M} = f_j$.

Proof. By definition and by self-adjointness of Λ_{g,V_2},
$$\langle \Lambda_{g,V_1} f_1, f_2 \rangle = \int_M \left(\langle du_1, du_2 \rangle + V_1 u_1 u_2 \right) dV,$$
$$\langle \Lambda_{g,V_2} f_1, f_2 \rangle = \langle f_1, \Lambda_{g,V_2} f_2 \rangle = \int_M \left(\langle du_1, du_2 \rangle + V_2 u_1 u_2 \right) dV.$$

The result follows by subtracting the two identities. □

In this text we are interested in uniqueness results, where one would like to show that $\Lambda_{g,V_1} = \Lambda_{g,V_2}$ implies $V_1 = V_2$. For this purpose, the following corollary is appropriate. It shows that if two DN maps coincide, then the integral of the difference of potentials against the product of *any* two solutions (with no requirements for their boundary values) vanishes.

Corollary 2.15 (integral identity). *Let (M, g) be a compact Riemannian manifold with smooth boundary, and let $V_1, V_2 \in L^\infty(M)$ be real-valued. If $\Lambda_{g,V_1} = \Lambda_{g,V_2}$, then*
$$\int_M (V_1 - V_2) u_1 u_2 \, dV = 0$$
for any $u_j \in H^1(M)$ which satisfy $(-\Delta_g + V_j) u_j = 0$ in M.

2E. Geodesics and covariant derivative.

In this section we let (M, g) be a connected Riemannian manifold *without* boundary (for our purposes, geodesics and the Levi-Civita connection on manifolds with boundary can be defined by embedding into a compact manifold without boundary).

Lengths of curves. For the analysis of the Calderón problem on manifolds we will need to introduce some basic properties of geodesics. These are locally length-minimizing curves on (M, g), so we begin by discussing lengths of curves.

Definition. A smooth map $\gamma : [a, b] \to M$ whose tangent vector $\dot{\gamma}(t)$ is always nonzero is called a *regular curve*. The *length* of γ is defined by

$$L(\gamma) := \int_a^b |\dot{\gamma}(t)| \, dt.$$

The length of a piecewise regular curve is defined as the sum of lengths of the regular parts. The *Riemannian distance* between two points $p, q \in M$ is defined by

$$d(p, q) := \inf\{L(\gamma) \,;\, \gamma : [a, b] \to M \text{ is a piecewise regular curve with}$$
$$\gamma(a) = p \text{ and } \gamma(b) = q\}.$$

Exercise 2.17. Show that $L(\gamma)$ is independent of the way the curve γ is parametrized, and that we may always parametrize γ by *arc length* so that $|\dot{\gamma}(t)| = 1$ for all t.

Exercise 2.18. Show that d is a metric distance function on M, and that (M, d) is a metric space whose topology is the same as the original topology on M.

Geodesic equation. We now wish to show that any length-minimizing curve satisfies a certain ordinary differential equation. Suppose that $\gamma : [a, b] \to M$ is a length-minimizing curve between two points p and q parametrized by arc length, and let $\gamma_s : [a, b] \to M$ be a family of curves from p to q such that $\gamma_0(t) = \gamma(t)$ and $\Gamma(s, t) := \gamma_s(t)$ depends smoothly on $s \in (-\varepsilon, \varepsilon)$ and on $t \in [a, b]$. We assume for simplicity that each γ_s is regular and contained in a coordinate neighborhood of M, and write $x_s(t) = (x_s^1(t), \ldots, x_s^n(t))$ and $x(t) = x_0(t)$ instead of $\gamma_s(t)$ and $\gamma(t)$ in local coordinates.

Lemma 2.16. *The length-minimizing curve $x(t)$ satisfies the so-called* geodesic equation

$$\ddot{x}^l(t) + \Gamma_{jk}^l(x(t))\dot{x}^j(t)\dot{x}^k(t) = 0, \quad 1 \leq l \leq n,$$

where Γ_{jk}^l is the Christoffel symbol

$$\Gamma_{jk}^l = \tfrac{1}{2} g^{lm}(\partial_j g_{km} + \partial_k g_{jm} - \partial_m g_{jk}).$$

Proof. Since γ minimizes length from p to q, we have
$$L(\gamma_0) \leq L(\gamma_s), \qquad s \in (-\varepsilon, \varepsilon).$$
Define
$$I(s) := L(\gamma_s) = \int_a^b (g_{pq}(x_s(t))\dot{x}_s^p(t)\dot{x}_s^q(t))^{1/2}\, dt.$$
Since I is smooth and $I(0) \leq I(s)$ for $|s| < \varepsilon$, we must have $I'(0) = 0$. To prepare for computing the derivative, define two vector fields
$$T(t) := \partial_t x_s(t)|_{s=0}, \qquad V(t) := \partial_s x_s(t)|_{s=0}.$$
Using that $|\dot{\gamma}_0(t)| = 1$ and (g_{jk}) is symmetric, we have
$$I'(0) = \frac{1}{2}\int_a^b \left(\partial_r g_{pq}(x(t))V^r(t)T^p(t)T^q(t) + 2g_{pq}(x(t))\dot{V}^p(t)T^q(t)\right) dt.$$
Integrating by parts in the last term, this shows that
$$I'(0) = \int_a^b \left(\tfrac{1}{2}\partial_r g_{pq}(x)T^p T^q - \partial_m g_{rq}(x)T^m T^q - g_{rq}(x)\dot{T}^q\right)V^r\, dt.$$
The last expression vanishes for all possible vector fields $V(t)$ obtained as $\partial_s x_s(t)|_{s=0}$. It can be seen that any vector field with $V(a) = V(b) = 0$ arises as $V(t)$ for some family of curves $\gamma_s(t)$. This implies that
$$\tfrac{1}{2}\partial_r g_{pq}(x)T^p T^q - \partial_m g_{rq}(x)T^m T^q - g_{rq}(x)\dot{T}^q = 0, \ t \in [a,b], 1 \leq r \leq n.$$
Multiplying this by g^{lr} and summing over r, and using that
$$\partial_m g_{rq}(x)T^m T^q = \tfrac{1}{2}(\partial_m g_{rq}(x) + \partial_q g_{rm}(x))T^m T^q,$$
gives the geodesic equation upon relabeling indices. $\square$

Covariant derivative. It would be possible to develop the theory of geodesics based on the ODE derived in Lemma 2.16. However, it will be very useful to be able to do computations such as those in Lemma 2.16 in an invariant way, without resorting to local coordinates. For this purpose we want to be able to take derivatives of vector fields in a way which is compatible with the Riemannian inner product $\langle \cdot, \cdot \rangle$.

We first recall the commutator of vector fields. Any smooth vector field on M gives rise to a first-order differential operator $X: C^\infty(M) \to C^\infty(M)$ by
$$Xf(p) = X(p)f.$$

If X and Y are vector fields, their commutator $[X,Y]$ is the differential operator acting on smooth functions by
$$[X,Y]f := X(Yf) - Y(Xf).$$
The commutator of two vector fields is itself a vector field.

The next result is sometimes called the fundamental lemma of Riemannian geometry.

Theorem (Levi-Civita connection). *On any Riemannian manifold (M,g) there is a unique $\mathbb{R}$-bilinear map*
$$D: C^\infty(M,TM) \times C^\infty(M,TM) \to C^\infty(M,TM), \quad (X,Y) \mapsto D_X Y,$$
which satisfies

(1) $D_{fX} Y = f D_X Y$ (*linearity*),
(2) $D_X(fY) = f D_X Y + (Xf)Y$ (*Leibniz rule*),
(3) $D_X Y - D_Y X = [X,Y]$ (*symmetry*),
(4) $X\langle Y, Z\rangle = \langle D_X Y, Z\rangle + \langle Y, D_X Z\rangle$ (*metric connection*).

Here X, Y, Z are vector fields and f is a smooth function on M.

Proof. See [Lee 1997].   □

The map D is called the *Levi-Civita connection* of (M,g). The expression $D_X Y$ is called the *covariant derivative* of the vector field Y in direction X.

Example. In $(\mathbb{R}^n, e)$ the Levi-Civita connection is given by
$$D_X Y = X^j (\partial_j Y^k) \partial_k.$$
This is just the natural derivative of Y in direction X.

Example. On a general manifold (M, g), one has
$$D_X Y = X^j (\partial_j Y^k) \partial_k + X^j Y^k \Gamma_{jk}^l \partial_l$$
where Γ_{jk}^l are the Christoffel symbols from Lemma 2.16, and they also satisfy
$$D_{\partial_j} \partial_k = \Gamma_{jk}^l \partial_l.$$

Covariant derivative of tensors. At this point we will define the connection and covariant derivatives also for other tensor fields. Let X be a vector field on M. The covariant derivative of 0-tensor fields is given by
$$D_X f := Xf.$$

For k-tensor fields u, the covariant derivative is defined by

$$D_X u(Y_1, \ldots, Y_k) := X(u(Y_1, \ldots, Y_k)) - \sum_{j=1}^{k} u(Y_1, \ldots, D_X Y_j, \ldots, Y_k).$$

Exercise 2.19. Show that these formulas give a well defined covariant derivative

$$D_X : C^\infty(M, T^k M) \to C^\infty(M, T^k M).$$

Example. The main example of the above construction is the covariant derivative of 1-forms, which is uniquely specified by the identity

$$D_{\partial_j} dx^k = -\Gamma^k_{jl} dx^l.$$

By using D_X on tensors, it is possible to define the *total covariant derivative* as the map

$$D : C^\infty(M, T^k M) \to C^\infty(M, T^{k+1} M),$$
$$Du(X, Y_1, \ldots, Y_k) := D_X u(Y_1, \ldots, Y_k).$$

Example. On 0-forms $Df = df$.

Example. The most important use for the total covariant derivative in these notes is the *covariant Hessian*. If f is a smooth function, then the covariant Hessian of f is

$$\text{Hess}(f) := D^2 f.$$

In local coordinates it is given by

$$D^2 f = (\partial_j \partial_k f - \Gamma^l_{jk} \partial_l f) \, dx^j \otimes dx^k.$$

Finally, we mention that the total covariant derivative can be used to define higher order Sobolev spaces invariantly on a Riemannian manifold.

Definition. If $k \geq 0$, consider the inner product on $C^\infty(M)$ given by

$$(u, v)_{H^k(M)} := \sum_{j=0}^{k} (D^j u, D^j v)_{L^2(M)}.$$

Here the L^2 norm is the natural one using the inner product on tensors. The Sobolev space $H^k(M)$ is defined to be the completion of $C^\infty(M)$ with respect to this inner product.

Geodesics. Let us return to length-minimizing curves. If $\gamma : [a,b] \to M$ is a curve and $X : [a,b] \to TM$ is a smooth vector field along γ (meaning that $X(t) \in T_{\gamma(t)} M$), we define the derivative of X along γ by

$$D_{\dot\gamma} X := D_{\dot\gamma} \tilde X$$

where $\tilde X$ is any vector field defined in a neighborhood of $\gamma([a,b])$ such that $\tilde X_{\gamma(t)} = X_{\gamma(t)}$. It is easy to see that this does not depend on the choice of $\tilde X$. The relation to geodesics now comes from the fact that in local coordinates, if $\gamma(t)$ corresponds to $x(t)$,

$$\begin{aligned} D_{\dot\gamma} \dot\gamma &= D_{\dot x^j \partial_j} (\dot x^k \partial_k) \\ &= (\ddot x^l + \Gamma^l_{jk}(x) \dot x^j \dot x^k) \partial_l. \end{aligned}$$

Thus the geodesic equation is satisfied if and only if $D_{\dot\gamma} \dot\gamma = 0$. We now give the precise definition of a geodesic.

Definition. A regular curve γ is called a geodesic if $D_{\dot\gamma} \dot\gamma = 0$.

The arguments above give evidence to the following result, which is proved for instance in [Lee 1997].

Theorem (geodesics minimize length). *If γ is a piecewise regular length-minimizing curve from p to q, then γ is regular and $D_{\dot\gamma} \dot\gamma = 0$. Conversely, if γ is a regular curve and $D_{\dot\gamma} \dot\gamma = 0$, then γ minimizes length at least locally.*

We next list some basic properties of geodesics.

Theorem (properties of geodesics). *Let (M,g) be a Riemannian manifold without boundary. Then*

(1) *for any $p \in M$ and $v \in T_p M$, there is an open interval I containing 0 and a geodesic $\gamma_v : I \to M$ with $\gamma_v(0) = p$ and $\dot\gamma_v(0) = v$,*

(2) *any two geodesics with $\gamma_1(0) = \gamma_2(0)$ and $\dot\gamma_1(0) = \dot\gamma_2(0)$ agree in their common domain,*

(3) *any geodesic satisfies $|\dot\gamma(t)| = const$,*

(4) *if M is compact then any geodesic γ can be uniquely extended as a geodesic defined on all of $\mathbb{R}$.*

Exercise 2.20. Prove this theorem by using the existence and uniqueness of solutions to ordinary differential equations.

By (3) in the theorem, we may (and will) always assume that geodesics are parametrized by arc length and satisfy $|\dot\gamma| = 1$. Part (4) says that the maximal domain of any geodesic on a closed manifold is $\mathbb{R}$, where the maximal domain is the largest interval to which the geodesic can be extended. We will always assume that the geodesics are defined on their maximal domain.

Normal coordinates. The following important concept enables us to parametrize a manifold locally by its tangent space.

Definition. If $p \in M$ let $\mathscr{E}_p := \{v \in T_pM \, ; \, \gamma_v \text{ is defined on } [0,1]\}$, and define the *exponential map*
$$\exp_p : \mathscr{E}_p \to M, \quad \exp_p(v) = \gamma_v(1).$$

This is a smooth map and satisfies $\exp_p(tv) = \gamma_v(t)$. Thus, the exponential map is obtained by following radial geodesics starting from the point p. This parametrization also gives rise to a very important system of coordinates on Riemannian manifolds.

Theorem (normal coordinates). *For any $p \in M$, $\exp_p$ is a diffeomorphism from some neighborhood V of 0 in T_pM onto a neighborhood of p in M. If $\{e_1, \ldots, e_n\}$ is an orthonormal basis of T_pM and we identify T_pM with $\mathbb{R}^n$ via $v^j e_j \leftrightarrow (v^1, \ldots, v^n)$, then there is a coordinate chart (U, φ) such that $\varphi = \exp_p^{-1} : U \to \mathbb{R}^n$ and*

(1) $\varphi(p) = 0$,

(2) *if $v \in T_pM$ then $\varphi(\gamma_v(t)) = (tv^1, \ldots, tv^n)$,*

(3) *one has*
$$g_{jk}(0) = \delta_{jk}, \quad \partial_l g_{jk}(0) = 0, \quad \Gamma^l_{jk}(0) = 0.$$

Proof. See [Lee 1997]. □

The local coordinates in the theorem are called *normal coordinates* at p. In these coordinates geodesics through p correspond to rays through the origin. Further, by (3) the metric and its first derivatives have a simple form at 0. This fact is often exploited when proving an identity where both sides are invariantly defined, and thus it is enough to verify the identity in some suitable coordinate system. The properties given in (3) sometimes simplify these local coordinate computations dramatically.

Finally, we will need the fact that when switching to polar coordinates in a normal coordinate system, the metric has special form in a full neighborhood of 0 instead of just at the origin.

Theorem (polar normal coordinates). *Let (U, φ) be normal coordinates at p. If (r, θ) are the corresponding polar coordinates (thus $r(q) = |\varphi(q)| > 0$ and $\theta(q)$ is the corresponding direction in S^{n-1}), then the metric has the form*
$$(g_{jk}(r, \theta)) = \begin{pmatrix} 1 & 0 \\ 0 & g_{\alpha\beta}(r, \theta) \end{pmatrix}.$$

This means that $|\partial/\partial r| = 1$, $\langle \partial/\partial r, \partial/\partial \theta \rangle = 0$, and $r(q) = d(p, q)$.

3. Limiting Carleman weights

In this section we will establish a starting point for solving some of the problems mentioned in the introduction. The approach taken here is to construct special solutions to the Schrödinger equation (or special harmonic functions if there is no potential) in (M, g), in such a way that the products of these special solutions are dense in $L^1(M)$.

The exact form of the special solutions is motivated by developments in $\mathbb{R}^n$, where harmonic exponential functions $e^{\rho \cdot x}$ with $\rho \in \mathbb{C}^n$ and $\rho \cdot \rho = 0$ have been successful in the solution of inverse problems. On a Riemannian manifold there is no immediate analog for the linear phase function $\rho \cdot x$ (one can always find such a function in local coordinates, but not globally in general). We will instead look for general phase functions φ which are expected to have desirable properties for the purposes of constructing special solutions. Such phase functions will be called *limiting Carleman weights* (LCWs).

The main result is a geometric characterization of those manifolds which admit LCWs. It makes use of the crucial fact that the existence of LCWs only depends on the conformal class of the manifold. The result is stated in terms of the existence of a parallel vector field in some conformal manifold.

Theorem 3.1 (manifolds that admit LCWs). *Let (M, g) be a simply connected open Riemannian manifold. Then (M, g) admits an LCW if and only if some conformal multiple of g admits a parallel unit vector field.*

Intuitively, the geometric condition means that up to a conformal factor there has to be a Euclidean direction on the manifold.

At this point we also mention a few open questions related to the theorem. The notation will be explained below. The first question asks to show that in dimensions $n \geq 3$ most metrics do not admit LCWs even locally (in fact, it would be interesting to prove the existence of even one metric which does not admit LCWs).

Question 3.1 (counterexamples). *If M is a smooth manifold of dimension $n \geq 3$ and if $p \in M$, show that a generic metric near p does not admit an LCW.* [1]

We will show later that if φ is an LCW, then one has a suitable Carleman estimate for the conjugated Laplace–Beltrami operators $P_{\pm\varphi}$. The next question is asking for a converse.

Question 3.2 (Carleman estimates imply LCW). *If (M, g) is an open manifold and φ is such that for any $M_1 \Subset M$ there are $C_0, h_0 > 0$ for which*

$$h\|u\|_{L^2(M_1)} \leq C \|P_{\pm\varphi} u\|_{L^2(M_1)}, \quad u \in C_c^\infty(M_1^{\mathrm{int}}), \ 0 < h < h_0,$$

---

[1] A positive answer to this question was recently given in [Liimatainen and Salo 2012].

then φ is an LCW. [2]

The last question asks to find an analog in dimensions $n \geq 3$ of the Carleman weights with critical points which have recently been very successful in 2D inverse problems.

Question 3.3. Find an analog in dimensions $n \geq 3$ of Bukhgeim-type weights φ in 2D manifolds which satisfy a Carleman estimate of the type

$$h^{3/2}\|u\| \leq C\|P_{\pm\varphi}u\|$$

for $u \in C_c^\infty(M^{\text{int}})$ and $0 < h < h_0$.

In this section we mostly follow [Dos Santos Ferreira et al. 2009, Section 2].

3A. *Motivation and definition.* Let (M, g) be a compact Riemannian manifold with boundary, and let $V_1, V_2 \in C^\infty(M)$. As always, we assume that the Dirichlet problems for $-\Delta_g + V_j$ in M are uniquely solvable, so that the DN maps Λ_{g,V_j} are well defined. Assume that $\Lambda_{g,V_1} = \Lambda_{g,V_2}$, that is, the two potentials V_1 and V_2 result in identical boundary measurements. Then we know that

$$\int_M (V_1 - V_2) u_1 u_2 \, dV = 0$$

for any solutions $u_j \in H^1(M)$ which satisfy $(-\Delta_g + V_j)u_j = 0$ in M. To solve the inverse problem of proving that $V_1 = V_2$, it is therefore enough to show that the set of products of solutions

$$\{u_1 u_2 \; ; \; u_j \in H^1(M) \text{ and } (-\Delta_g + V_j)u_j = 0 \text{ in } M\}$$

is dense in $L^1(M)$.

In Euclidean space in dimensions $n \geq 3$, the density of solutions can be proved based on harmonic complex exponentials. The following argument is from [Sylvester and Uhlmann 1987] and is explained in detail in [Salo 2008, Chapter 3].

Motivation. Let $(M, g) = (\bar{\Omega}, e)$ where Ω is a bounded open subset of $\mathbb{R}^n$ with C^∞ boundary. In this setting we have special harmonic functions

$$u_0(x) = e^{\rho \cdot x}, \quad \rho \in \mathbb{C}^n, \; \rho \cdot \rho = 0. \tag{3-1}$$

Clearly $\Delta u_0 = (\rho \cdot \rho) u_0 = 0$. By [Sylvester and Uhlmann 1987], if $|\rho|$ is large there exist solutions to Schrödinger equations which look like these harmonic

---

[2] A positive answer was outlined in lectures of Dos Santos Ferreira [Ferreira 2011].

exponentials and have the form

$$u_1 = e^{\rho \cdot x}(a_1 + r_1),$$
$$u_2 = e^{-\rho \cdot x}(a_2 + r_2),$$

where a_j are certain explicit functions and r_j are correction terms which are small when $|\rho|$ is large, in the sense that $\|r_j\|_{L^2(\Omega)} \leq C/|\rho|$. We have chosen one solution with $e^{\rho \cdot x}$ and the other solution with $e^{-\rho \cdot x}$ so that the exponential factors will cancel in the product $u_1 u_2$, thus making it possible to take the limit as $|\rho| \to \infty$ which will get rid of the correction terms r_j.

The density of products of solutions in this case can be proved as follows. We fix $\xi \in \mathbb{R}^n$ and choose $a_1 = e^{ix \cdot \xi}$, $a_2 = 1$. If $n \geq 3$ then there exists a family of complex vectors ρ with $\rho \cdot \rho = 0$ and $|\rho| \to \infty$ such that solutions with the above properties can be constructed. To show density of the set $\{u_1 u_2\}$ for solutions of this type, we take $V \in L^\infty(\Omega)$ and assume that

$$\int_\Omega V u_1 u_2 \, dx = 0$$

for all u_1 and u_2 as above. Then

$$\int_\Omega V(e^{ix \cdot \xi} + r_1 + e^{ix \cdot \xi} r_2 + r_1 r_2) \, dx = 0.$$

By the L^2 estimates for r_j we may take the limit as $|\rho| \to \infty$, which will imply that $\int_\Omega V e^{ix \cdot \xi} \, dx = 0$. Since this is true for any fixed $\xi \in \mathbb{R}^n$, it follows from the uniqueness of the Fourier transform that $V = 0$ as required.

After having discussed the proof in the Euclidean case, we move on to the setting on Riemannian manifolds and try to see if a similar argument could be achieved. If (M, g) is a compact Riemannian manifold with boundary, we first seek approximate solutions satisfying $\Delta_g u_0 \approx 0$ (in some sense) having the form

$$u_0 = e^{-\varphi/h} m.$$

Here φ is assumed to be a smooth real-valued function on M, $h > 0$ will be a small parameter, and $m \in C^\infty(M)$ is some complex function. In the Euclidean case this corresponds to (3-1) by taking

$$h = 1/|\rho|, \quad \varphi(x) = -\text{Re}(\rho/|\rho|) \cdot x, \quad m(x) = e^{\text{Im}(\rho) \cdot x}.$$

Loosely speaking, φ will be a limiting Carleman weight if such approximate solutions with weight $\pm \varphi$ can always be converted into exact solutions of $\Delta_g u = 0$ (we can forget the potential V at this point). More precisely, we would like the following condition to be satisfied:

(∗) For any function $u_0 = e^{\mp \varphi/h} m \in C^\infty(M)$ there is a solution
$$u = e^{\mp \varphi/h}(m+r)$$
of $\Delta_g u = 0$ in M such that $\|r\|_{L^2(M)} \leq Ch\|e^{\pm \varphi/h}\Delta_g u_0\|_{L^2(M)}$ for h small.

To find conditions on φ which would guarantee that this is possible, we introduce the conjugated Laplace–Beltrami operator
$$P_\varphi := e^{\varphi/h}(-h^2 \Delta_g) e^{-\varphi/h}. \tag{3-2}$$

Note that if $u = e^{\mp \varphi/h}(m+r)$, then
$$\Delta_g u = 0 \iff e^{\pm \varphi/h}(-h^2 \Delta_g)e^{\mp \varphi/h}(m+r) = 0$$
$$\iff P_{\pm\varphi} r = -P_{\pm\varphi} m.$$

Thus (∗) would follow if for any $f \in L^2(M)$ there is a function v satisfying for h small
$$P_{\pm\varphi} v = f \quad \text{in } M,$$
$$h\|v\|_{L^2(M)} \leq C\|f\|_{L^2(M)}.$$

One approach for proving existence of solutions to the last equation, or more generally an inhomogeneous equation $Tv = f$, is to prove uniqueness of solutions to the homogeneous adjoint equation $T^*v = 0$. This follows the general principle
$$\begin{cases} T^* \text{ injective} \\ \text{range of } T^* \text{ closed} \end{cases} \implies T \text{ surjective.}$$

Exercise 3.1. Find out why this principle holds for $m \times n$ matrices, for operators $T = \text{Id} + K$ where K is a compact operator on a Hilbert space, or for bounded operators T between two Hilbert spaces.

Since $P^*_{\pm\varphi} = P_{\mp\varphi}$, injectivity and closed range for the adjoint operator would be a consequence of the *a priori* estimate
$$h\|u\|_{L^2(M)} \leq C\|P_{\pm\varphi} u\|_{L^2(M)}, \quad u \in C_c^\infty(M^{\text{int}}), \ h \text{ small}. \tag{3-3}$$

This is called a *Carleman estimate* (that is, a norm estimate with exponential weights depending on a parameter). Estimates of this type have turned out to be very useful in unique continuation for solutions of partial differential equations, control theory, and inverse problems.

We will look for conditions on φ which would imply the Carleman estimate (3-3). The following decomposition of P_φ into its self-adjoint part A and skew-adjoint part iB will be useful.

Lemma 3.2. $P_\varphi = A + iB$ where A and B are the formally self-adjoint operators (in the $L^2(M)$ inner product)
$$A := -h^2 \Delta_g - |d\varphi|^2,$$
$$B := \frac{h}{i}\left(2\langle d\varphi, d\,\cdot\,\rangle + \Delta_g \varphi\right).$$

Proof. The quickest way to see this is a computation in local coordinates. We write $D_j = -i\partial_{x_j}$, and note that
$$e^{\varphi/h} h D_j e^{-\varphi/h} = h D_j + i\varphi_{x_j}.$$

Then
$$\begin{aligned}P_\varphi u &= e^{\varphi/h}(-h^2\Delta_g)e^{-\varphi/h}u \\ &= |g|^{-1/2} e^{\varphi/h} h D_j (e^{-\varphi/h}|g|^{1/2} g^{jk} e^{\varphi/h} h D_k (e^{-\varphi/h} u)) \\ &= |g|^{-1/2}(hD_j + i\varphi_{x_j})|g|^{1/2} g^{jk}(hD_k + i\varphi_{x_k})u \\ &= -h^2 \Delta_g u + h g^{jk} \varphi_{x_j} u_{x_k} + h|g|^{-1/2}\partial_j(|g|^{1/2} g^{jk}\varphi_{x_k} u) - g^{jk}\varphi_{x_j}\varphi_{x_k} u \\ &= -h^2 \Delta_g u + h\left(2\langle d\varphi, du\rangle + (\Delta_g \varphi)u\right) - |d\varphi|^2 u.\end{aligned}$$

The result follows immediately upon checking that A and B are formally self-adjoint. $\square$

Exercise 3.2. Check that A and B are formally self-adjoint.

Next we give a basic computation in the proof of a Carleman estimate such as (3-3), evaluating the square of the right hand side.

Lemma 3.3. *If* $u \in C_c^\infty(M^{\text{int}})$ *then*
$$\|P_\varphi u\|^2 = \|Au\|^2 + \|Bu\|^2 + (i[A, B]u, u).$$

Proof. Since $P_\varphi = A + iB$,
$$\begin{aligned}\|P_\varphi u\|^2 &= (P_\varphi u, P_\varphi u) = ((A + iB)u, (A + iB)u) \\ &= (Au, Au) + i(Bu, Au) - i(Au, Bu) + (Bu, Bu) \\ &= \|Au\|^2 + \|Bu\|^2 + (i[A, B]u, u).\end{aligned}$$

We used that A and B are formally self-adjoint. $\square$

Thus $\|P_\varphi u\|^2$ can be written as the sum of two nonnegative terms $\|Au\|^2$ and $\|Bu\|^2$ and a third term which involves the commutator $[A, B] := AB - BA$. The only negative contribution may come from the commutator term. Therefore, a positivity condition for $i[A, B]$ would be helpful for proving the Carleman estimate (3-3) for P_φ. We will state such a positivity condition on the level of principal symbols.

Lemma 3.4. *The principal symbols of A and B are*
$$a(x,\xi) := |\xi|^2 - |d\varphi|^2,$$
$$b(x,\xi) := 2\langle d\varphi, \xi\rangle.$$
The principal symbol of $i[A, B]$ is the Poisson bracket $h\{a, b\}$.

Proof. The principal symbol of A is obtained by writing A in some local coordinates and by looking at the symbol of the corresponding operator in $\mathbb{R}^n$. But in local coordinates
$$A = g^{jk} h D_j h D_k - g^{jk} \varphi_{x_j} \varphi_{x_k} + h(|g|^{-1/2} D_j(|g|^{1/2} g^{jk}) D_k).$$
The last term is lower order, hence does not affect the principal symbol. The symbol of $g^{jk} h D_j h D_k - g^{jk} \varphi_{x_j} \varphi_{x_k}$ is $g^{jk} \xi_j \xi_k - g^{jk} \varphi_{x_j} \varphi_{x_k}$, so we may take the invariantly defined function $a(x,\xi) := |\xi|^2 - |d\varphi|^2$ on T^*M as the principal symbol. A similar argument works for B, and the claim for $i[A, B]$ is a general fact. □

Given this information, the positivity condition that we will require of $i[A, B]$ is the following condition for the principal symbol:
$$\{a, b\} \geq 0 \text{ when } a = b = 0.$$
More precisely, we ask that $\{a, b\}(x, \xi) \geq 0$ for any $(x, \xi) \in T^*M$ for which $a(x, \xi) = b(x, \xi) = 0$. The idea is that in Lemma 3.3 one has the nonnegative terms $\|Au\|^2$ and $\|Bu\|^2$, and if either of these is large then it may cancel a negative contribution from the commutator term. On the level of symbols, one therefore only needs positivity of $\{a, b\}$ when the principal symbols of A and B vanish.

Recall that one wants the estimate (3-3) also for $P_{-\varphi}$. Changing φ to $-\varphi$ in Lemma 3.2, we see that $P_{-\varphi} = A - iB$. As in Lemma 3.3 one then asks a positivity condition for $i[A, -B]$, which has principal symbol $-\{a, b\}$. Thus, we also require that
$$\{a, b\} \leq 0 \text{ when } a = b = 0.$$

Combining the above conditions for $\{a, b\}$, we have finally arrived at the definition of limiting Carleman weights. The definition is most naturally stated on open manifolds, and it includes the useful additional condition that φ should have nonvanishing gradient.

Definition. Let (M, g) be an open Riemannian manifold. We say that a smooth real-valued function φ in M is a *limiting Carleman weight* (LCW) if $d\varphi \neq 0$ in M and
$$\{a, b\} = 0 \quad \text{when} \quad a = b = 0.$$

Example. Let $(M, g) = (\Omega, e)$ where Ω is an open set in $\mathbb{R}^n$. We will verify that the linear function $\varphi(x) = \alpha \cdot x$, where $\alpha \in \mathbb{R}^n$ is a nonzero vector, is an LCW. Indeed, one has $\nabla \varphi = \alpha$ and the principal symbols are
$$a(x, \xi) = |\xi|^2 - |\alpha|^2,$$
$$b(x, \xi) = 2\alpha \cdot \xi.$$

Since a and b are independent of x, the Poisson bracket is
$$\{a, b\} = \nabla_\xi a \cdot \nabla_x b - \nabla_x a \cdot \nabla_\xi b \equiv 0.$$

Thus φ is an LCW.

Exercise 3.3. If $(M, g) = (\Omega, e)$ and $0 \notin \overline{\Omega}$, verify that $\varphi(x) = \log |x|$ and $\varphi(x) = \alpha \cdot x / |x|^2$ are LCWs. Here $\alpha \in \mathbb{R}^n$ is a fixed vector.

3B. *Characterization.* In the previous section, after a long motivation we ended up with a definition of LCWs involving a rather abstract vanishing condition for a certain Poisson bracket. Here we give a geometric meaning to this condition, and also prove Theorem 3.1 which characterizes all Riemannian manifolds which admit LCWs. We recall the statement.

Theorem 3.1 (manifolds which admit LCWs). *Let (M, g) be a simply connected open Riemannian manifold. Then (M, g) admits an LCW if and only if some conformal multiple of g admits a parallel unit vector field.*

Recall that a vector field X is parallel if $D_V X = 0$ for any vector field V. Also recall that a manifold is simply connected if it is connected and if every closed curve is homotopic to a point. An explanation of the geometric condition, including examples of manifolds which satisfy it, is given in the next section.

We now begin the proof of Theorem 3.1. Let (M, g) be an open manifold. Recall that $\varphi \in C^\infty(M; \mathbb{R})$ is an LCW if $d\varphi \neq 0$ in M and
$$\{a, b\} = 0 \quad \text{when} \quad a = b = 0.$$

Here $a(x, \xi) = |\xi|^2 - |\nabla \varphi|^2$ and $b(x, \xi) = 2\langle d\varphi, \xi \rangle$ are smooth functions in T^*M. The first step is to find an expression for the Poisson bracket $\{a, b\}$, defined in local coordinates by $\{a, b\} := \nabla_\xi a \cdot \nabla_x b - \nabla_x a \cdot \nabla_\xi b$.

Motivation. We compute the Poisson bracket in $\mathbb{R}^n$. Then $a(x,\xi) = |\xi|^2 - |\nabla\varphi|^2$ and $b(x,\xi) = 2\nabla\varphi \cdot \xi$, and writing φ'' for the Hessian matrix $(\varphi_{x_j x_k})_{j,k=1}^n$ we have

$$\{a,b\} = \nabla_\xi a \cdot \nabla_x b - \nabla_x a \cdot \nabla_\xi b$$
$$= 2\xi \cdot 2\varphi''\xi - (-2\varphi''\nabla\varphi) \cdot 2\nabla\varphi$$
$$= 4\varphi''\xi \cdot \xi + 4\varphi''\nabla\varphi \cdot \nabla\varphi.$$

A computation in normal coordinates will show that a similar expression, now involving the covariant Hessian, holds on a Riemannian manifold.

Lemma 3.5 (expression for Poisson bracket). *The Poisson bracket is given by*

$$\{a,b\}(x,\xi) = 4D^2\varphi(\xi^\sharp, \xi^\sharp) + 4D^2\varphi(\nabla\varphi, \nabla\varphi).$$

Proof. Both sides are invariantly defined functions on T^*M, so it is enough to check the identity in some local coordinates at a given point. Fix $p \in M$, let x be normal coordinates centered at p, and let (x,ξ) be the associated local coordinates in T^*M near p. Then

$$a(x,\xi) = g^{jk}\xi_j\xi_k - g^{jk}\varphi_{x_j}\varphi_{x_k},$$
$$b(x,\xi) = 2g^{jk}\varphi_{x_j}\xi_k.$$

Using that $g^{jk}|_p = \delta^{jk}$ and $\partial_l g^{jk}|_p = \Gamma^l_{jk}|_p = 0$, we have

$$\{a,b\}(x,\xi)|_p = \sum_{l=1}^n \left(\partial_{\xi_l} a \partial_{x_l} b - \partial_{x_l} a \partial_{\xi_l} b\right)\Big|_p$$
$$= \sum_{l=1}^n \left((2g^{jl}\xi_l)(2g^{jk}\varphi_{x_j x_l}\xi_k) - (-2g^{jk}\varphi_{x_j x_l}\varphi_{x_k})(2g^{jl}\varphi_{x_j})\right)\Big|_p$$
$$= \sum_{j,l=1}^n \left(4\varphi_{x_j x_l}\xi_j\xi_l + 4\varphi_{x_j x_l}\varphi_{x_j}\varphi_{x_l}\right)\Big|_p$$
$$= (4D^2\varphi(\xi^\sharp, \xi^\sharp) + 4D^2\varphi(\nabla\varphi, \nabla\varphi))|_p,$$

since $D^2\varphi|_p = \varphi_{x_j x_l} dx^j \otimes dx^l|_p$. $\square$

This immediately implies a condition for LCWs which is easier to work with than the original one.

Corollary 3.6. *φ is an LCW if and only if $d\varphi \neq 0$ in M and*

$$D^2\varphi(X,X) + D^2\varphi(\nabla\varphi, \nabla\varphi) = 0 \text{ when } |X| = |\nabla\varphi| \text{ and } \langle X, \nabla\varphi\rangle = 0.$$

We can now give a full characterization of LCWs in two dimensions. To do this, recall that the trace of a 2-tensor S on an n-dimensional manifold (N, g) is (analogously to the trace of an $n \times n$ matrix) defined by

$$\mathrm{Tr}(S)|_p := \sum_{j=1}^n S(e_j, e_j)$$

where $\{e_1, \ldots, e_n\}$ is any orthonormal basis of $T_p N$. The trace of the Hessian is just the Laplace–Beltrami operator, as may be seen by a computation in normal coordinates at p:

$$\mathrm{Tr}(D^2\varphi)|_p = \sum_{j=1}^n D^2\varphi(\partial_j, \partial_j)|_p = \sum_{j=1}^n \varphi_{x_j x_j}|_p = \Delta_g \varphi|_p.$$

Proposition 3.7 (LCWs in 2D). *The LCWs in a 2D manifold (M, g) are exactly the harmonic functions with nonvanishing differential.*

Proof. If $|X| = |\nabla \varphi|$ and $\langle X, \nabla \varphi \rangle = 0$, then $\{X/|\nabla \varphi|, \nabla \varphi / |\nabla \varphi|\}$ is an orthonormal basis of the tangent space. Then

$$D^2\varphi(X, X) + D^2\varphi(\nabla \varphi, \nabla \varphi) = |\nabla \varphi|^2 \mathrm{Tr}(D^2 \varphi) = |\nabla \varphi|^2 \Delta_g \varphi.$$

By Corollary 3.6, φ is an LCW if and only if $\Delta_g \varphi = 0$ and $d\varphi \neq 0$. □

After having characterized the situation in two dimensions, we move on to the case $n \geq 3$. The crucial fact here is that the existence of LCWs is a conformally invariant condition.

Proposition 3.8 (existence of LCWs only depends on conformal class). *If φ is an LCW in (M, g), then φ is an LCW in (M, cg) for any smooth positive function c.*

Proof. Suppose φ is an LCW in (M, g), and let $\tilde{g} = cg$. Then the symbols $\tilde{a}$ and $\tilde{b}$ for the metric $\tilde{g}$ are

$$\tilde{a} = \tilde{g}^{jk} \xi_j \xi_k - \tilde{g}^{jk} \varphi_{x_j} \varphi_{x_k} = c^{-1}(g^{jk} \xi_j \xi_k - g^{jk} \varphi_{x_j} \varphi_{x_k}) = c^{-1} a,$$
$$\tilde{b} = 2\tilde{g}^{jk} \varphi_{x_j} \xi_k = 2c^{-1} g^{jk} \varphi_{x_j} \xi_k = c^{-1} b.$$

Since c^{-1} does not depend on ξ, it follows that

$$\{\tilde{a}, \tilde{b}\} = \{c^{-1}a, c^{-1}b\} = c^{-1} \nabla_\xi a \cdot \nabla_x (c^{-1} b) - c^{-1} \nabla_x (c^{-1} a) \cdot \nabla_\xi b$$
$$= c^{-2}\{a, b\} + c^{-1} b \{a, c^{-1}\} + c^{-1} a \{c^{-1}, b\}.$$

Suppose that $\tilde{a} = \tilde{b} = 0$. Then $a = b = 0$, and using that φ is an LCW it follows that $\{a, b\} = 0$. Consequently $\{\tilde{a}, \tilde{b}\} = 0$ when $\tilde{a} = \tilde{b} = 0$, showing that φ is an LCW in $(M, \tilde{g})$. □

At this point we record a lemma which expresses relations between the Hessian and the covariant derivative.

Lemma 3.9. *If $\varphi \in C^\infty(M)$ then*
$$D^2\varphi(X, Y) = \langle D_X \nabla\varphi, Y \rangle,$$
$$D^2\varphi(X, \nabla\varphi) = \langle D_X \nabla\varphi, \nabla\varphi \rangle = \tfrac{1}{2} X(|\nabla\varphi|^2),$$
$$D^2\varphi(\dot\gamma(t), \dot\gamma(t)) = \frac{d^2}{dt^2}\varphi(\gamma(t))$$
for any X, Y and for any geodesic γ.

Proof. The first identity follows from a computation in normal coordinates. The second identity follows from the first one and the metric property of D. The third identity holds since
$$\frac{d^2}{dt^2}\varphi(\gamma(t)) = \frac{d}{dt}\langle \nabla\varphi(\gamma(t)), \dot\gamma(t) \rangle = \langle D_{\dot\gamma(t)} \nabla\varphi(\gamma(t)), \dot\gamma(t) \rangle$$
$$= D^2\varphi(\dot\gamma(t), \dot\gamma(t))$$
by the first identity. Here we used that $D_{\dot\gamma(t)} \dot\gamma(t) = 0$ since γ is a geodesic. □

Using the second identity in the previous lemma, we now observe that if φ is an LCW and additionally $|\nabla\varphi| = 1$, then the second term in Corollary 3.6 vanishes:
$$D^2\varphi(\nabla\varphi, \nabla\varphi) = \tfrac{1}{2} \nabla\varphi(|\nabla\varphi|^2) = 0.$$
A smooth function which satisfies $|\nabla\varphi| = 1$ is called a *distance function* (since any such function is locally given by the Riemannian distance to a point or submanifold, but we will not need this fact). If one is given an LCW φ in (M, g), one can always reduce to the case where the LCW is a distance function by using the following conformal transformation.

Lemma 3.10 (conformal normalization). *If φ is a smooth function in (M, g) and if $\tilde g = |\nabla\varphi|^2 g$, then $|\nabla_{\tilde g}\varphi|_{\tilde g} = 1$.*

Proof. $|\nabla_{\tilde g}\varphi|^2_{\tilde g} = \tilde g^{jk}\varphi_{x_j}\varphi_{x_k} = |\nabla\varphi|^{-2} g^{jk}\varphi_{x_j}\varphi_{x_k} = 1$. □

We have an important characterization of LCWs which are also distance functions.

Lemma 3.11 (LCWs which are distance functions). *Let $\varphi \in C^\infty(M)$ and $|\nabla\varphi| = 1$. The following conditions are equivalent*:

(1) φ *is an LCW.*

(2) $D^2\varphi \equiv 0$.

(3) $\nabla\varphi$ *is parallel.*

(4) If $p \in M$ and if v is in the domain of $\exp_p$, then
$$\varphi(\exp_p(v)) = \varphi(p) + \langle \nabla\varphi(p), v \rangle$$

Proof. Since $|\nabla\varphi| = 1$ we have $D^2\varphi(\nabla\varphi, \nabla\varphi) = 0$. Thus by Corollary 3.6, φ is an LCW if and only if
$$D^2\varphi(X, X) = 0 \text{ when } |X| = 1 \text{ and } \langle X, \nabla\varphi \rangle = 0.$$

Since $D^2\varphi$ is bilinear we may drop the condition $|X| = 1$, and the condition for LCW becomes
$$D^2\varphi(X, X) = 0 \text{ when } \langle X, \nabla\varphi \rangle = 0.$$

(1) $\Longrightarrow$ (2): Suppose φ is an LCW. Fix $p \in M$ and choose an orthonormal basis $\{e_1, \ldots, e_n\}$ of $T_p M$ such that $e_1 = \nabla\varphi$. Then, by the above discussion,
$$D^2\varphi(e_1, e_1) = 0,$$
$$D^2\varphi(e_j, e_k) = 0 \quad \text{for } 2 \le j, k \le n.$$

By Lemma 3.9 we also have $D^2\varphi(X, \nabla\varphi) = \frac{1}{2}X(|\nabla\varphi|^2) = 0$ for any X, therefore
$$D^2\varphi(e_j, e_1) = 0 \quad \text{for } 2 \le j \le n.$$

Since $D^2\varphi$ is bilinear and symmetric, we obtain $D^2\varphi \equiv 0$.

(2) $\Longrightarrow$ (1): This is immediate.

(2) $\Longleftrightarrow$ (3): Follows from $D^2\varphi(X, Y) = \langle D_X \nabla\varphi, Y \rangle$.

(2) $\Longleftrightarrow$ (4): Let $\gamma_v(t) = \exp_p(tv)$. Then
$$\frac{d}{dt}\varphi(\gamma_v(t)) = \langle \nabla\varphi(\gamma_v(t)), \dot\gamma_v(t) \rangle,$$
$$\frac{d^2}{dt^2}\varphi(\gamma_v(t)) = D^2\varphi(\dot\gamma_v(t), \dot\gamma_v(t)).$$

If $D^2\varphi \equiv 0$ then the second derivative of $\varphi(\gamma_v(t))$ vanishes, therefore $\varphi(\gamma_v(t)) = a_0 + b_0 t$ for some real constants a_0, b_0. Evaluating $\varphi(\gamma_v(t))$ and its derivative at $t = 0$ gives
$$\varphi(\exp_p(tv)) = \varphi(p) + \langle \nabla\varphi(p), v \rangle t.$$

Conversely, if the last identity is valid then the second derivative of $\varphi(\gamma_v(t))$ vanishes, which implies $D^2\varphi \equiv 0$. $\square$

Remarks. 1. Condition (4) indicates that LCWs which are also distance functions (normalized so that $\varphi(p) = 0$) are the analog on Riemannian manifolds of the linear Carleman weights in Euclidean space.

2. If φ is an LCW and a distance function, the above lemma shows that the Poisson bracket $\{a, b\}$ vanishes on all of T^*M instead of just on the submanifold where $a = b = 0$.

We have now established all the statements needed for the proof of Theorem 3.1, except for the fact that any parallel vector field in a simply connected manifold is a gradient field. Leaving this fact to the next section, we give the proof of the main theorem.

Proof of Theorem 3.1. Let (M, g) be simply connected and open.

For the forward implication, suppose φ is an LCW in (M, g). By conformal invariance (Proposition 3.8) we know that φ is an LCW in $(M, \tilde{g})$ where $\tilde{g} = |\nabla \varphi|^2 g$. Lemma 3.10 shows that φ is also a distance function in $(M, \tilde{g})$. Then Lemma 3.11 applies, and we see that $\nabla_{\tilde{g}} \varphi$ is a unit parallel vector field in $(M, \tilde{g})$.

For the converse, assume that X is a unit parallel vector field in (M, cg) where $c > 0$. Since M is simply connected, the fact mentioned just before this proof shows that $X = \nabla_{cg} \varphi$ for some smooth function φ. Since $\nabla_{cg} \varphi$ is parallel and $|\nabla_{cg} \varphi|_{cg} = 1$, Lemma 3.11 implies that φ is an LCW in (M, cg). By conformal invariance φ is then an LCW also in (M, g). $\square$

3C. Geometric interpretation. The geometric meaning of having a parallel unit vector field is given in the following result.

Lemma 3.12 (parallel field $\Leftrightarrow$ product structure). *Let X be a unit parallel vector field in (M, g). Near any point of M there exist local coordinates $x = (x_1, x')$ such that $X = \partial_1$ and*

$$g(x_1, x') = \begin{pmatrix} 1 & 0 \\ 0 & g_0(x') \end{pmatrix}, \text{ for some metric } g_0 \text{ in the } x' \text{ variables.}$$

Conversely, if g is of this form then ∂_1 is a unit parallel vector field.

This says that the existence of a unit parallel vector field X implies that M is locally isometric to a subset of $(\mathbb{R}, e) \times (M_0, g_0)$ for some $(n-1)$-dimensional manifold (M_0, g_0). One can think of the direction of X as being a Euclidean direction on the manifold.

Note that any parallel vector field X has constant length on each component of M, since $V(|X|^2) = 2\langle D_V X, X \rangle = 0$ for any vector field V. Thus the existence of any nontrivial parallel vector field implies a product structure.

Theorem 3.1 now says that (M, g) admits an LCW if and only if up to a conformal factor there is a Euclidean direction on the manifold. More precisely:

Lemma 3.13 (LCWs in local coordinates). *Let φ be an LCW in (M, g). Near any point of M there are local coordinates $x = (x_1, x')$ such that in these* *coordinates $\varphi(x) = x_1$ and*

$$g(x_1, x') = c(x) \begin{pmatrix} 1 & 0 \\ 0 & g_0(x') \end{pmatrix}$$

where c is a positive function and g_0 is some metric in the x' variables. Conversely, any metric of this form has the LCW $\varphi(x) = x_1$.

Exercise 3.4. Prove this lemma.

Example. Manifolds which admit LCWs include the following:

1. Euclidean space $\mathbb{R}^n$ since any constant vector field is parallel,
2. all open subsets of the model spaces $\mathbb{R}^n$, $S^n \smallsetminus \{p_0\}$, and H^n since these are conformal to Euclidean space,
3. more general manifolds locally conformal to $\mathbb{R}^n$, such as symmetric spaces in 3D, admit LCWs locally,
4. all 2D manifolds admit LCWs at least locally by Proposition 3.7,
5. (Ω, g) admits an LCW if $\Omega \subseteq \mathbb{R}^n$ and if in some coordinates $x = (x_1, x')$ the metric g has the form

$$g(x_1, x') = c(x) \begin{pmatrix} 1 & 0 \\ 0 & g_0(x') \end{pmatrix}$$

for some positive function c and some $(n-1)$-dimensional metric g_0.

The rest of this section is devoted to the proofs of Lemma 3.12 and the fact which was used in the proof of Theorem 3.1. We start with the latter.

Lemma 3.14. *If M is a manifold with $H^1_{\mathrm{dR}}(M) = \{0\}$, then any parallel unit vector field on M is a gradient field.*

Proof. Let X be a parallel unit vector field on M. We choose $\omega = X^\flat$ to be the 1-form corresponding to X. It is enough to prove that $d\omega = 0$, since then the condition on the first de Rham cohomology group implies that $\omega = d\varphi$ for some smooth function φ and consequently $X = (d\varphi)^\sharp = \nabla \varphi$.

The fact that $d\omega = 0$ follows from the general identity

$$d(X^\flat)(Y, Z) = \langle D_Y X, Z \rangle - \langle D_Z X, Y \rangle$$

since $D_V X = 0$ for any V. $\square$

Exercise 3.5. Show the identity used in the proof.

To prove Lemma 3.12 we need a version of the Frobenius theorem. For this purpose we introduce some terminology; see [Lee 2003, Section 14] for more details. A *k-plane field* on a manifold M is a rule Γ which associates to each point p in M a k-dimensional subspace Γ_p of $T_p M$, such that Γ_p varies smoothly with p. A vector field X on M is called a *section of* Γ if $X(p) \in \Gamma_p$ for any p. A k-plane field Γ is called *involutive* if for any V, W which are sections of Γ, also the Lie bracket $[V, W]$ is a section of Γ.

Theorem (Frobenius). *If Γ is an involutive k-plane field, then through any point p in M there is an integral manifold S of Γ (that is, S is a k-dimensional submanifold of M with $\Gamma|_S = TS$).*

The other tool that is needed is a special local coordinate system called *semigeodesic coordinates*. The usual geodesic normal coordinates are obtained by following geodesic rays starting at a given point. Semigeodesic coordinates are instead obtained by following geodesics which are normal to a given hypersurface S. On manifolds with boundary, semigeodesic coordinates where S is part of the boundary are called *boundary normal coordinates*.

Lemma 3.15 (semigeodesic coordinates). *Let $p \in M$ and let S be a hypersurface through p. There is a chart (U, x) at p such that $S \cap U = \{x_1 = 0\}$, the curves $x_1 \mapsto (x_1, x')$ correspond to normal geodesics starting from S, and the metric has the form*

$$g(x_1, x') = \begin{pmatrix} 1 & 0 \\ 0 & g_0(x_1, x') \end{pmatrix}.$$

The inverse of the map $(x_1, x') \mapsto \exp_{q(x')}(x_1 N(q(x')))$ gives such a chart, where $x' \mapsto q(x')$ is a parametrization of S near p and N is a unit normal vector field of S.

Exercise 3.6. Prove this lemma.

Proof of Lemma 3.12. For the forward implication, let Let X be unit parallel, and let Γ be the $(n-1)$-plane field consisting of vectors orthogonal to X. If V, W are vector fields orthogonal to X then

$$\langle [V, W], X \rangle = \langle D_V W - D_W V, X \rangle = V \langle W, X \rangle - W \langle V, X \rangle = 0$$

using the symmetry and metric property of the Levi-Civita connection and the fact that X is parallel. This shows that Γ is an involutive $(n-1)$-plane field.

Fix $p \in M$, and use the Frobenius theorem to find a hypersurface S through p such that X is normal to S. If $x' \mapsto q(x')$ parametrizes S near p, then $(x_1, x') \mapsto \exp_{q(x')}(x_1 X(q(x')))$ gives semigeodesic coordinates near p such that ∂_1 is the tangent vector of a normal geodesic to S and

$$g(x_1, x') = \begin{pmatrix} 1 & 0 \\ 0 & g_0(x_1, x') \end{pmatrix}.$$

Now the integral curves of X are geodesics, because $\dot{\gamma}(t) = X(\gamma(t))$ implies $D_{\dot{\gamma}(t)} \dot{\gamma}(t) = D_{\dot{\gamma}(t)} X(\gamma(t)) = 0)$. This shows that $X = \partial_1$. It remains to prove

that $g_0(x_1, x')$ is independent of x_1. But for $j, k \geq 2$ we have
$$\partial_1 g_{jk} = \partial_1 \langle \partial_j, \partial_k \rangle = \langle D_{\partial_1} \partial_j, \partial_k \rangle + \langle \partial_j, D_{\partial_1} \partial_k \rangle$$
$$= \langle D_{\partial_j} \partial_1, \partial_k \rangle + \langle \partial_j, D_{\partial_k} \partial_1 \rangle = 0$$
since $D_{\partial_1} \partial_l - D_{\partial_l} \partial_1 = [\partial_1, \partial_l] = 0$ and since $\partial_1 = X$ is parallel.
The converse if left as an exercise. □

Exercise 3.7. Prove the converse direction in Lemma 3.12.

4. Carleman estimates

In the previous chapter we introduced limiting Carleman weights (LCWs), motivated by the possibility of constructing special solutions to the Schrödinger equation $(-\Delta_g + V)u = 0$ in M having the form
$$u = e^{\pm \varphi / h}(a + r)$$
where φ is an LCW, $h > 0$ is a small parameter, and the correction term r converges to zero as $h \to 0$. The arguments involved solving inhomogeneous equations of the type
$$e^{\pm \varphi / h}(-\Delta_g + V)e^{\mp \varphi / h} r = f \quad \text{in } M \tag{4-1}$$
with the norm estimate
$$\|r\|_{L^2(M)} \leq Ch \|f\|_{L^2(M)}, \quad 0 < h < h_0.$$
We then gave a definition of LCWs based on an abstract condition on the vanishing of a Poisson bracket and proved that on a simply connected open manifold (M, g), by Theorem 3.1 and Lemma 3.13,

φ is an LCW in (M, g)

$\iff \nabla_{\tilde{c} g} \varphi$ is unit parallel in $(M, \tilde{c} g)$ for some $\tilde{c} > 0$

$\implies$ locally in some coordinates $\varphi(x) = x_1$ and $g = c(e \oplus g_0)$.

On the last line, the notation means that $c^{-1} g$ is the product of the Euclidean metric e on $\mathbb{R}$ and some $(n-1)$-dimensional metric g_0.

In this section we will show that the existence of an LCW indeed implies the solvability of the inhomogeneous equation (4-1) with the right norm estimates. We will prove this under the extra assumption that the metric has the product structure $g = c(e \oplus g_0)$ *globally* instead of just locally. Following [Kenig et al. 2011], this assumption makes it possible to use Fourier analysis to write down the solutions in a rather explicit way. See [Dos Santos Ferreira et al. 2009, Section 4] for a different (though less explicit) proof based on integration by parts arguments as in Section 3A, which does not require the extra assumption on global structure of g.

4A. *Motivation and main theorem.* As usual, we will first consider solvability of the inhomogeneous equation in the Euclidean case. Here and below we will consider a large parameter $\tau = 1/h$ instead of a small parameter. This is just a matter of notation, and this choice will be slightly more transparent (also the Fourier analysis proof will allow us to avoid semiclassical symbol calculus for which a small parameter would be more natural).

Motivation. Consider the analog of the Equation (4-1) in $\mathbb{R}^n$ with the LCW $\varphi(x) = x_1$ and with $V = 0$,

$$e^{\tau x_1}(-\Delta)e^{-\tau x_1}u = f \quad \text{in } \mathbb{R}^n.$$

Noting that $e^{\tau x_1} D e^{-\tau x_1} = D + i\tau e_1$ where $D = -i\nabla$, we compute

$$e^{\tau x_1}(-\Delta)e^{-\tau x_1} = (D + i\tau e_1)^2 = -\Delta + 2\tau\partial_1 - \tau^2.$$

The equation becomes

$$(-\Delta + 2\tau\partial_1 - \tau^2)u = f \quad \text{in } \mathbb{R}^n.$$

The operator on the left has constant coefficients, and one can try to find a solution by taking the Fourier transform of both sides. Since $(D_j u)\hat{}(\xi) = \xi_j \hat{u}(\xi)$, this gives the equation

$$(|\xi|^2 + 2i\tau\xi_1 - \tau^2)\hat{u}(\xi) = \hat{f}(\xi) \quad \text{in } \mathbb{R}^n.$$

Thus, one formally obtains the solution

$$u = \mathcal{F}^{-1}\left\{\frac{1}{p(\xi)}\hat{f}(\xi)\right\}$$

where $p(\xi) := |\xi|^2 - \tau^2 + 2i\tau\xi_1$. The problem is that the symbol $p(\xi)$ has zeros, and it is not immediately obvious if one can divide by $p(\xi)$. In fact the zero set of the symbol is a codimension 2 manifold,

$$p^{-1}(0) = \{\xi \in \mathbb{R}^n ; |\xi| = |\tau|, \xi_1 = 0\}.$$

It was shown in [Sylvester and Uhlmann 1987] after a careful analysis that one can indeed justify the division by $p(\xi)$ if the functions are in certain weighted L^2 spaces. Define for $\delta \in \mathbb{R}$ the space

$$L^2_\delta(\mathbb{R}^n) := \{f \in L^2_{\text{loc}}(\mathbb{R}^n) ; (1 + |x|^2)^{\delta/2} f \in L^2(\mathbb{R}^n)\}.$$

The result of [Sylvester and Uhlmann 1987] states that if $-1 < \delta < 0$, then for any $f \in L^2_{\delta+1}(\mathbb{R}^n)$ this argument gives a unique solution $u \in L^2_\delta(\mathbb{R}^n)$ with the right norm estimates.

It turns out that a similar Fourier analysis argument will also work in the Riemannian case if the metric is related to the product metric on $\mathbb{R} \times M_0$. One can then use the ordinary Fourier transform on $\mathbb{R}$, but on the transversal manifold M_0 the Fourier transform is replaced by eigenfunction expansions. Also, since the spectrum in the transversal directions is discrete, it turns out we can easily avoid the problem of dividing by zero just by imposing a harmless extra condition on the large parameter τ.

In this section we will be working in a cylinder $T := \mathbb{R} \times M_0$ with metric $g := c(e \oplus g_0)$, where (M_0, g_0) is a compact $(n-1)$-dimensional manifold with boundary and $c > 0$ is a smooth positive function. We will write points of T as (x_1, x') where x_1 is the Euclidean coordinate on $\mathbb{R}$ and x' are local coordinates on M_0. Thus g has the form

$$g(x_1, x') = c(x) \begin{pmatrix} 1 & 0 \\ 0 & g_0(x') \end{pmatrix}.$$

Note that these coordinates and the representation of the metric are valid globally in x_1 and locally in M_0.

We denote by $L^2(T) = L^2(T, dV_g)$ the natural L^2 space on (T, g). The local L^2 space is

$$L^2_{\text{loc}}(T) := \{f \,;\, f \in L^2([-R, R] \times M_0) \text{ for all } R > 0\}.$$

Writing $\langle x \rangle = (1 + |x|^2)^{1/2}$, we define for any $\delta \in \mathbb{R}$ the polynomially weighted (in the x_1 variable) spaces

$$L^2_\delta(T) := \{f \in L^2_{\text{loc}}(T) \,;\, \langle x_1 \rangle^\delta f \in L^2(T)\},$$
$$H^1_\delta(T) := \{f \in L^2_\delta(T) \,;\, df \in L^2_\delta(T)\},$$
$$H^1_{\delta,0}(T) := \{f \in H^1_\delta(T) \,;\, f|_{\mathbb{R} \times M_0} = 0\}.$$

These have natural norms

$$\|f\|_{L^2_\delta(T)} := \|\langle x_1 \rangle^\delta f\|_{L^2(T)},$$
$$\|f\|_{H^1_\delta(T)} := \|\langle x_1 \rangle^\delta f\|_{L^2(T)} + \|\langle x_1 \rangle^\delta df\|_{L^2(T)}.$$

More precisely, $L^2_\delta(T)$ and $H^1_\delta(T)$ are the completions in the respective norms of the space $\{f \in C^\infty(T) \,;\, f(x_1, x') = 0 \text{ for } |x_1| \text{ large}\}$, and $H^1_{\delta,0}(T)$ is the completion of $C^\infty_c(T^{\text{int}})$ in the $H^1_\delta(T)$ norm.

If g has the special form given above, $\varphi(x) = x_1$ is a natural LCW. We denote by Δ_g and Δ_{g_0} the Laplace–Beltrami operators in (T, g) and (M_0, g_0), respectively. The main result is as follows.

Theorem 4.1 (solvability and norm estimates). *Assume that $c(x_1, x') = 1$ for $|x_1|$ large. Let $\delta > 1/2$, and let V be a complex function in T with $\langle x_1 \rangle^{2\delta} V$ in $L^\infty(T)$. There exist $C_0, \tau_0 > 0$ such that whenever*

$$|\tau| \geq \tau_0 \quad \text{and} \quad \tau^2 \notin \mathrm{Spec}(-\Delta_{g_0}),$$

then for any $f \in L^2_\delta(T)$ there is a unique solution $u \in H^1_{-\delta,0}(T)$ of the equation

$$e^{\tau x_1}(-\Delta_g + V)e^{-\tau x_1} u = f \quad \text{in } T.$$

This solution satisfies

$$\|u\|_{L^2_{-\delta}(T)} \leq \frac{C_0}{|\tau|} \|f\|_{L^2_\delta(T)},$$
$$\|u\|_{H^1_{-\delta}(T)} \leq C_0 \|f\|_{L^2_\delta(T)}.$$

Here $\mathrm{Spec}(-\Delta_{g_0})$ is the discrete set of Dirichlet eigenvalues of $-\Delta_{g_0}$ in (M_0, g_0). The extra restriction $\tau^2 \notin \mathrm{Spec}(-\Delta_{g_0})$ allows us to avoid the problem of dividing by zero. One can always find a sequence of τ's converging to plus or minus infinity which satisfies this restriction, which is all that we will need for the applications to inverse problems. Typically, if we consider an inverse problem in a compact manifold (M, g) with boundary, Theorem 4.1 will be used by embedding (M, g) in a cylinder (T, g) of the above type and then solving the inhomogeneous equations in the larger manifold (T, g).

Let us formulate some open questions related to the above theorem (some of these questions should be quite doable).

Question 4.1. By using slightly different function spaces, prove an analog of Theorem 4.1 without the restriction $\tau^2 \notin \mathrm{Spec}(-\Delta_{g_0})$.

Question 4.2 (existence of LCW implies global product structure). Find conditions on a manifold (M, g) such that the existence of an LCW on (M, g) would imply that $(M, g) \Subset (T, g)$ for a cylinder as above.

Question 4.3 (operators with first-order terms). Prove an analog of Theorem 4.1 when the operator $-\Delta_g + V$ is replaced by $-\Delta_g + 2X + V$ where X is a vector field on T with suitable regularity and decay.

4B. *Proof of the estimates.* We begin the proof of Theorem 4.1. The first step is to observe that it is enough to prove the result for $c \equiv 1$. Note that the metric in T is of the form $c\tilde{g}$ where $\tilde{g} = e \oplus g_0$ is a product metric.

Lemma 4.2 (Schrödinger equation under conformal scaling). *If c is a positive function in $(M, \tilde{g})$ and V is a function in M then*

$$c^{\frac{n+2}{4}}(-\Delta_{c\tilde{g}} + V)(c^{-\frac{n-2}{4}}v) = \left(-\Delta_{\tilde{g}} + cV - c^{\frac{n+2}{4}}\Delta_g(c^{-\frac{n-2}{4}})\right)v.$$

Exercise 4.1. Prove the lemma.

Suppose now that Theorem 4.1 has been proved for the metric $\tilde{g} = e \oplus g_0$. For the general case $g = c\tilde{g}$, we need to produce a solution of

$$e^{\tau x_1}(-\Delta_{c\tilde{g}} + V)e^{-\tau x_1}u = f \quad \text{in } T.$$

We try $u = c^{-\frac{n-2}{4}}v$ for some v. By Lemma 4.2, it is enough to solve

$$e^{\tau x_1}\left(-\Delta_{\tilde{g}} + cV - c^{\frac{n+2}{4}}\Delta_g(c^{-\frac{n-2}{4}})\right)e^{-\tau x_1}v = c^{\frac{n+2}{4}}f \quad \text{in } T.$$

But since $c = 1$ for $|x_1|$ large, the potential $\tilde{V} := cV - c^{\frac{n+2}{4}}\Delta_g(c^{-\frac{n-2}{4}})$ has the same decay properties as V (that is, $\tilde{V} \in \langle x_1 \rangle^{2\delta} L^\infty(T)$). The right hand side $\tilde{f} := c^{\frac{n+2}{4}}f$ is also in $L^2_\delta(T)$ like f, so Theorem 4.1 for $\tilde{g}$ implies the existence of a unique solution v. Since $u = c^{-\frac{n-2}{4}}v$ the solution u belongs to the same function spaces and satisfies similar estimates as v, and Theorem 4.1 follows in full generality.

From now on we will assume that $c \equiv 1$ and that we are working in (T, g) where $g = e \oplus g_0$, or in local coordinates

$$g(x_1, x') = \begin{pmatrix} 1 & 0 \\ 0 & g_0(x') \end{pmatrix}.$$

Since $|g|$ only depends on x', the Laplace–Beltrami operator splits as

$$\Delta_g = \partial_1^2 + \Delta_{g_0}.$$

Similarly, using that $e^{\tau x_1} D_1 e^{-\tau x_1} = D_1 + i\tau$, the conjugated Laplace–Beltrami operator has the expression

$$e^{\tau x_1}(-\Delta_g)e^{-\tau x_1} = (D_1 + i\tau)^2 - \Delta_{g_0}$$
$$= -\partial_1^2 + 2\tau\partial_1 - \tau^2 - \Delta_{g_0}.$$

Assuming that $V = 0$ for the moment, the equation that we need to solve has now the form

$$(-\partial_1^2 + 2\tau\partial_1 - \tau^2 - \Delta_{g_0})u = f \quad \text{in } T. \tag{4-2}$$

As mentioned above, we will employ eigenfunction expansions in the manifold M_0 to solve the equation. Let $0 < \lambda_1 \leq \lambda_2 \leq \cdots$ be the Dirichlet eigenvalues of the

Laplace–Beltrami operator $-\Delta_{g_0}$ in (M_0, g_0), and let ϕ_l be the corresponding Dirichlet eigenfunctions so that

$$-\Delta_{g_0}\phi_l = \lambda_l \phi_l \text{ in } M, \quad \phi_l \in H_0^1(M_0).$$

We assume that $\{\phi_l\}_{l=1}^{\infty}$ is an orthonormal basis of $L^2(M_0)$. Then, if f is a function on T such $f(x_1, \cdot) \in L^2(M_0)$ for almost every x_1, we define the partial Fourier coefficients

$$\tilde{f}(x_1, l) := \int_{M_0} f(x_1, x')\phi_l(x')\, dV_{g_0}(x'). \tag{4-3}$$

One has the eigenfunction expansion

$$f(x_1, x') = \sum_{l=1}^{\infty} \tilde{f}(x_1, l)\phi_l(x')$$

with convergence in $L^2(M_0)$ for almost every x_1.

Motivation. Formally, the proof of Theorem 4.1 now proceeds as follows. We consider eigenfunction expansions

$$u(x_1, x') = \sum_{l=1}^{\infty} \tilde{u}(x_1, l)\phi_l(x'), \quad f(x_1, x') = \sum_{l=1}^{\infty} \tilde{f}(x_1, l)\phi_l(x').$$

Inserting these expansions in (4-2) and using that $-\Delta_{g_0}\phi_l = \lambda_l \phi_l$ results in the following ODEs for the partial Fourier coefficients:

$$(-\partial_1^2 + 2\tau\partial_1 - \tau^2 + \lambda_l)\tilde{u}(\cdot, l) = \tilde{f}(\cdot, l) \quad \text{for all } l. \tag{4-4}$$

The easiest way to prove uniqueness of solutions is to take Fourier transforms in the x_1 variable. If the ODEs (4-4) are satisfied with zero right hand side, then with obvious notations

$$(\xi_1^2 + 2i\tau\xi_1 - \tau^2 + \lambda_l)\hat{u}(\xi_1, l) = 0 \quad \text{for all } l.$$

Now if the symbol $p(\xi_1, l) := \xi_1^2 + 2i\tau\xi_1 - \tau^2 + \lambda_l$ would be zero, looking at real and imaginary parts would imply $\xi_1 = 0$ and $\tau^2 = \lambda_l$. But the condition $\tau^2 \notin \text{Spec}(-\Delta_{g_0})$ shows that this is not possible. Thus $p(\xi_1, l)$ is nonvanishing, and we obtain $\hat{u}(\xi_1, l) \equiv 0$ and consequently $u \equiv 0$. This proves uniqueness.

To show existence with the right norm estimates we observe that

$$-\partial_1^2 + 2\tau\partial_1 - \tau^2 = -(\partial_1 - \tau)^2,$$

and we factor (4-4) as

$$(\partial_1 - \tau - \sqrt{\lambda_l})(\partial_1 - \tau + \sqrt{\lambda_l})\tilde{u}(\cdot, l) = -\tilde{f}(\cdot, l) \quad \text{for all } l.$$

The Fourier coefficients of the solution u are then obtained from the Fourier coefficients of f by solving two ODEs of first-order.

After this formal discussion, we will give the rigorous arguments which lie behind these ideas. Let us begin with uniqueness.

Proposition 4.3 (uniqueness for $V = 0$). *Let $u \in H^1_{\delta,0}(T)$ for some $\delta \in \mathbb{R}$, let $\tau^2 \notin \mathrm{Spec}(-\Delta_{g_0})$, and assume that u satisfies*
$$(-\partial_1^2 + 2\tau\partial_1 - \tau^2 - \Delta_{g_0})u = 0 \quad \text{in } T.$$
Then $u = 0$.

Proof. The condition that u is a solution implies that
$$\int_T u(-\partial_1^2 - 2\tau\partial_1 - \tau^2 - \Delta_{g_0})\psi \, dV_g = 0$$
for any $\psi \in C_c^\infty(T^{\mathrm{int}})$. We make the choice $\psi(x_1, x') = \chi(x_1)\phi_{lj}(x')$ where $\chi \in C_c^\infty(\mathbb{R})$ and $\phi_{lj} \in C_c^\infty(M_0^{\mathrm{int}})$ with $\phi_{lj} \to \phi_l$ in $H^1(M_0)$ as $j \to \infty$. The last fact is possible since $\phi_l \in H_0^1(M_0)$. Now $g = e \oplus g_0$, so we have for any w
$$\int_T w \, dV_g = \int_{-\infty}^\infty \int_{M_0} w(x_1, x') \, dV_{g_0}(x') \, dx_1.$$

Thus, with this choice of ψ we obtain that
$$\int_{-\infty}^\infty \left(\int_{M_0} u(x_1, \cdot)\phi_{lj} \, dV_{g_0}\right)(-\partial_1^2 - 2\tau\partial_1 - \tau^2)\chi(x_1) \, dx_1$$
$$+ \int_{-\infty}^\infty \left(\int_{M_0} u(x_1, \cdot)(-\Delta_{g_0}\phi_{lj}) \, dV_{g_0}\right)\chi(x_1) \, dx_1 = 0. \quad (4\text{-}5)$$

Note that $u(x_1, \cdot) \in H_0^1(M_0)$ for almost every x_1, because of the assumption $u \in H^1_{\delta,0}(T)$ and the facts
$$\int_{-\infty}^\infty \langle x_1 \rangle^{2\delta} \|u(x_1, \cdot)\|^2_{L^2(M_0)} \, dx_1 = \|u\|_{L^2_\delta(T)} < \infty,$$
$$\int_{-\infty}^\infty \langle x_1 \rangle^{2\delta} \|\nabla_{g_0} u(x_1, \cdot)\|^2_{L^2(M_0)} \, dx_1 = \|\nabla_{g_0} u\|_{L^2_\delta(T)} < \infty.$$

Since $-\Delta_{g_0}$ is an isomorphism $H_0^1(M_0) \to H^{-1}(M_0)$, we have
$$\int_{M_0} u(x_1, \cdot)\phi_{lj} \, dV_{g_0} \to \tilde{u}(x_1, l),$$
$$\int_{M_0} u(x_1, \cdot)(-\Delta_{g_0}\phi_{lj}) \, dV_{g_0} \to \lambda_l \tilde{u}(x_1, l)$$

as $j \to \infty$ for any x_1 such that $u(x_1, \cdot) \in H_0^1(M_0)$. Dominated convergence shows that we may take the limit in (4-5) and obtain

$$\int_{-\infty}^{\infty} \tilde{u}(x_1, l)(-\partial_1^2 - 2\tau\partial_1 - \tau^2 + \lambda_l)\chi(x_1)\, dx_1 = 0 \quad \text{for all } l.$$

The condition $u \in L_\delta^2(T)$ ensures that $\tilde{u}(\cdot, l) \in \langle \cdot \rangle^{-\delta} L^2(\mathbb{R})$, and the last identity implies

$$(-\partial_1^2 + 2\tau\partial_1 - \tau^2 + \lambda_l)\tilde{u}(\cdot, l) = 0 \quad \text{for all } l.$$

It only remains to take the Fourier transform in x_1 (which can be done in the sense of tempered distributions on $\mathbb{R}$), which gives

$$(\xi_1^2 + 2i\tau\xi_1 - \tau^2 + \lambda_l)\hat{u}(\cdot, l) = 0 \quad \text{for all } l.$$

The symbol $\xi_1^2 + 2i\tau\xi_1 - \tau^2 + \lambda_l$ is never zero because $\tau^2 \notin \text{Spec}(-\Delta_{g_0})$. Thus $\tilde{u}(\cdot, l) = 0$ for all l, showing that $u(x_1, \cdot) = 0$ for almost every x_1 and consequently $u = 0$. $\square$

As discussed above, the existence of solutions will be established via certain first-order ODEs. The next result gives the required solvability results and norm estimates. Here $L_\delta^2(\mathbb{R})$ is the space defined via the norm $\|f\|_{L_\delta^2(\mathbb{R})} := \|\langle x \rangle^\delta f\|_{L^2(\mathbb{R})}$, and $\mathcal{S}'(\mathbb{R})$ is the space of tempered distributions in $\mathbb{R}$.

Proposition 4.4 (solvability and norm estimates for an ODE). *Let a be a nonzero real number, and consider the equation*

$$u' - au = f \quad \text{in } \mathbb{R}.$$

For any $f \in \mathcal{S}'(\mathbb{R})$ there is a unique solution $u \in \mathcal{S}'(\mathbb{R})$. Writing $S_a f := u$, we have the mapping properties

$$S_a : L_\delta^2(\mathbb{R}) \to L_\delta^2(\mathbb{R}) \quad \text{for all } \delta \in \mathbb{R},$$
$$S_a : L^1(\mathbb{R}) \to L^\infty(\mathbb{R}),$$

and the norm estimates

$$\|S_a f\|_{L_\delta^2} \leq \frac{C_\delta}{|a|} \|f\|_{L_\delta^2} \quad \text{if } |a| \geq 1 \text{ and } \delta \in \mathbb{R},$$
$$\|S_a f\|_{L_{-\delta}^2} \leq C_\delta \|f\|_{L_\delta^2} \quad \text{if } a \neq 0 \text{ and } \delta > 1/2,$$
$$\|S_a f\|_{L^\infty} \leq \|f\|_{L^1}.$$

Proof. **Step 1.** Let us first consider solvability in $\mathscr{S}'(\mathbb{R})$. Taking Fourier transforms, we have

$$u' - au = f \iff (i\xi - a)\hat{u} = \hat{f}$$
$$\iff u = \mathscr{F}^{-1}\{m(\xi)\hat{f}(\xi)\}$$

with $m(\xi) := (i\xi - a)^{-1}$. Since $a \neq 0$ the function m is smooth and its derivatives are given by $m^{(k)}(\xi) = (-i)^k k!(i\xi - a)^{-k-1}$. Therefore

$$\|m^{(k)}\|_{L^\infty} \leq k! |a|^{-k-1}, \quad k = 0, 1, 2, \ldots. \tag{4-6}$$

Thus m has bounded derivatives and $v \mapsto mv$ is continuous on $\mathscr{S}'(\mathbb{R})$. It follows that $S_a f := \mathscr{F}^{-1}\{m(\xi)\hat{f}(\xi)\}$ produces for any $f \in \mathscr{S}'(\mathbb{R})$ a unique solution in $\mathscr{S}'(\mathbb{R})$ to the given ODE.

Step 2. Let $f \in L^2_\delta(\mathbb{R})$ where $\delta \in \mathbb{R}$. We will use the following Sobolev space facts on $\mathbb{R}$: if $\|m\|_{W^{k,\infty}} := \sum_{j=0}^k \|m^{(j)}\|_{L^\infty}$ then

$$\|v\|_{H^\delta} = \|\langle \cdot \rangle^\delta \hat{v}\|_{L^2} = \|\hat{v}\|_{L^2_\delta}, \tag{4-7}$$
$$\|mv\|_{H^\delta} \leq C_\delta \|m\|_{W^{k,\infty}} \|v\|_{H^\delta} \quad \text{when } k \geq |\delta|. \tag{4-8}$$

Then for $k \geq |\delta|$

$$\|S_a f\|_{L^2_\delta} = \|\mathscr{F}^{-1}\{S_a f\}\|_{H^\delta} = (2\pi)^{-1} \|(m\hat{f})(-\cdot)\|_{H^\delta}$$
$$\leq C_\delta (2\pi)^{-1} \|m\|_{W^{k,\infty}} \|\hat{f}(-\cdot)\|_{H^\delta}$$
$$= C_\delta \|m\|_{W^{k,\infty}} \|f\|_{L^2_\delta}.$$

This proves that S_a maps L^2_δ to itself. If additionally $|a| \geq 1$, the estimates (4-6) imply

$$\|S_a f\|_{L^2_\delta} \leq \frac{C_\delta}{|a|} \|f\|_{L^2_\delta}.$$

Step 3. Let $f \in L^1(\mathbb{R})$, and let $a > 0$ (the case $a < 0$ is analogous). To prove the $L^1 \to L^\infty$ bounds we will work on the spatial side and solve the ODE by using the standard method of integrating factors. In the sense of distributions

$$u' - au = f \iff u'e^{-at} - aue^{-at} = fe^{-at}$$
$$\iff (ue^{-at})' = fe^{-at}.$$

Integrating both sides from x to ∞ (here we use that $a > 0$ so e^{-at} is decreasing as $t \to \infty$), we define

$$u(x) := -\int_x^\infty f(t) e^{-a(t-x)} \, dt.$$

Since $|e^{-a(t-x)}| \leq 1$ for $t \geq x$, uniformly over $a > 0$, we see that $\|u\|_{L^\infty} \leq \|f\|_{L^1}$. Since u clearly solves the ODE we have $u = S_a f$ by uniqueness of solutions. This shows the mapping property and norm estimates of S_a on L^1.

Step 4. Finally, let $f \in L^2_\delta(\mathbb{R})$ with $\delta > 1/2$. It remains to convert the $L^1 \to L^\infty$ estimate to a weighted L^2 estimate. Using that

$$c_\delta := \left(\int_{-\infty}^\infty \langle t \rangle^{-2\delta} \right)^{1/2} < \infty$$

for $\delta > 1/2$, we have

$$\|S_a f\|_{L^2_{-\delta}} = \left(\int_{-\infty}^\infty \langle t \rangle^{-2\delta} |S_a f(t)|^2 \, dt \right)^{1/2} \leq c_\delta \|S_a f\|_{L^\infty} \leq c_\delta \|f\|_{L^1}$$

$$= c_\delta \int_{-\infty}^\infty \langle t \rangle^{-\delta} \langle t \rangle^\delta |f(t)| \, dt \leq c_\delta^2 \|f\|_{L^2_\delta}.$$

The last inequality follows by Cauchy-Schwarz. $\square$

Exercise 4.2. Verify the Sobolev space facts (4-7), (4-8).

Remark 4.5. We will employ the $L^2_\delta \to L^2_\delta$ estimate when $|a| \geq 1$. The proof shows that when a is small then the constant in this estimate blows up. This is why we need the $L^2_\delta \to L^2_{-\delta}$ estimate for $\delta > \frac{1}{2}$, with constant independent of a. The method for converting an $L^1 \to L^\infty$ estimate to a weighted L^2 estimate arises in Agmon's scattering theory for short range potentials. The weighted L^2 estimate is more convenient for our purposes than the stronger $L^1 \to L^\infty$ estimate since the weighted L^2 spaces will make it possible to use orthogonality.

We can now show the existence of solutions to the inhomogeneous equation with no potential.

Proposition 4.6 (existence for $V = 0$). *Let $f \in L^2_\delta(T)$ where $\delta > 1/2$. There is $C_0 > 0$ such that whenever $|\tau| \geq 1$ and $\tau^2 \notin \mathrm{Spec}(-\Delta_{g_0})$, then the equation*

$$(-\partial_1^2 + 2\tau \partial_1 - \tau^2 - \Delta_{g_0}) u = f \quad \text{in } T \tag{4-9}$$

has a solution $u \in H^1_{-\delta,0}(T)$ satisfying

$$\|u\|_{L^2_{-\delta}(T)} \leq \frac{C_0}{|\tau|} \|f\|_{L^2_\delta(T)},$$

$$\|u\|_{H^1_{-\delta}(T)} \leq C_0 \|f\|_{L^2_\delta(T)}.$$

Proof. Step 1. We begin with a remark on orthogonality. Since $f \in L^2_\delta(T)$, we know that $f(x_1, \cdot) \in L^2(M_0)$ for almost every x_1. Then for such x_1 the

Parseval identity implies

$$\int_{L^2(M_0)} |f(x_1, x')|^2 \, dV_{g_0}(x') = \sum_{l=1}^{\infty} |\tilde{f}(x_1, l)|^2.$$

Here $\tilde{f}(x_1, l)$ are the Fourier coefficients (4-3). It follows that

$$\|f\|_{L^2_\delta(T)}^2 = \int_{-\infty}^{\infty} \langle x_1 \rangle^{2\delta} \left(\int_{M_0} |f(x_1, x')|^2 \, dV_{g_0}(x') \right) dx_1$$

$$= \int_{-\infty}^{\infty} \langle x_1 \rangle^{2\delta} \left(\sum_{l=1}^{\infty} |\tilde{f}(x_1, l)|^2 \right) dx_1$$

$$= \sum_{l=1}^{\infty} \|\tilde{f}(\cdot, l)\|_{L^2_\delta(\mathbb{R})}^2.$$

In the last equality, we used Fubini's theorem which is valid since the integrand is nonnegative. In particular, this argument shows that $\tilde{f}(\cdot, l) \in L^2_\delta(\mathbb{R})$ for all l, and that the last sum converges.

Step 2. From now on we assume that $\tau > 0$ (the case $\tau < 0$ is analogous). Motivated by the discussion before (4-4), we will show that for any l there is a solution $\tilde{u}(\cdot, l) \in L^2_{-\delta}(\mathbb{R})$ of the ODE

$$(-\partial_1^2 + 2\tau \partial_1 - \tau^2 + \lambda_l) \tilde{u}(\cdot, l) = \tilde{f}(\cdot, l) \tag{4-10}$$

satisfying the norm estimate

$$\|\tilde{u}(\cdot, l)\|_{L^2_{-\delta}(\mathbb{R})} \leq \frac{C_0}{\tau + \sqrt{\lambda_l}} \|\tilde{f}(\cdot, l)\|_{L^2_\delta(\mathbb{R})}. \tag{4-11}$$

In fact, using the factorization to first-order equations given after (4-4), the ODE for $\tilde{u}(\cdot, l)$ becomes

$$(\partial_1 - \tau - \sqrt{\lambda_l})(\partial_1 - \tau + \sqrt{\lambda_l}) \tilde{u}(\cdot, l) = -\tilde{f}(\cdot, l).$$

Since $\tilde{f}(\cdot, l) \in L^2_\delta(\mathbb{R})$, Proposition 4.4 shows there is a unique solution given by

$$\tilde{u}(\cdot, l) := -S_{\tau - \sqrt{\lambda_l}} S_{\tau + \sqrt{\lambda_l}} \tilde{f}(\cdot, l). \tag{4-12}$$

Since $\tau - \sqrt{\lambda_l} \neq 0$ and $\tau + \sqrt{\lambda_l} \geq 1$ by the assumptions on τ, the estimates in Proposition 4.4 yield (4-11).

THE CALDERÓN PROBLEM ON RIEMANNIAN MANIFOLDS

Step 3. With $\tilde{u}(\,\cdot\,,l)$ as above, define

$$u_N(x_1, x') := \sum_{l=1}^{N} \tilde{u}(x_1, l)\phi_l(x').$$

Our objective is to show that as $N \to \infty$, u_N converges in $L^2_{-\delta}(T)$ to a function u which is a weak solution of (4-9) and satisfies

$$\|u\|_{L^2_{-\delta}(T)} \leq \frac{C_0}{\tau} \|f\|_{L^2_{\delta}(T)}.$$

If $N' > N$, the orthogonality argument in Step 1 and the estimate (4-11) show that

$$\|u_{N'} - u_N\|^2_{L^2_{-\delta}(T)} = \sum_{l=N}^{N'-1} \|\tilde{u}(\,\cdot\,,l)\|^2_{L^2_{-\delta}(\mathbb{R})} \leq \left(\frac{C_0}{\tau}\right)^2 \sum_{l=N}^{N'-1} \|\tilde{f}(\,\cdot\,,l)\|^2_{L^2_{\delta}(\mathbb{R})}.$$

Since $f \in L^2_{\delta}(T)$ the last expression converges to zero as $N, N' \to \infty$. This shows that (u_N) is a Cauchy sequence in $L^2_{-\delta}(T)$, hence converges to a function $u \in L^2_{-\delta}(T)$.

Using that $-\Delta_{g_0}\phi_l = \lambda_l \phi_l$, we have by (4-10)

$$(-\partial_1^2 + 2\tau\partial_1 - \tau^2 - \Delta_{g_0})u_N = \sum_{l=1}^{N}(-\partial_1^2 + 2\tau\partial_1 - \tau^2 + \lambda_l)\tilde{u}(x_1,l)\phi_l(x')$$

$$= \sum_{l=1}^{N} \tilde{f}(x_1,l)\phi_l(x').$$

The right hand side converges to f in $L^2_{\delta}(T)$ as $N \to \infty$. Integrating against a test function in $C_c^{\infty}(T^{\text{int}})$, we see that u is indeed a weak solution of (4-9). The norm estimate follows from orthogonality and (4-11):

$$\|u\|^2_{L^2_{-\delta}(T)} = \sum_{l=1}^{\infty} \|\tilde{u}(\,\cdot\,,l)\|^2_{L^2_{-\delta}(\mathbb{R})} \leq \left(\frac{C_0}{\tau}\right)^2 \sum_{l=1}^{\infty} \|\tilde{f}(\,\cdot\,,l)\|^2_{L^2_{\delta}(\mathbb{R})}$$

$$\leq \left(\frac{C_0}{\tau}\right)^2 \|f\|^2_{L^2_{\delta}(T)}.$$

Step 4. It remains to show that $u \in H^1_{-\delta,0}(T)$ and

$$\|u\|_{H^1_{-\delta}(T)} \leq C_0 \|f\|_{L^2_{\delta}(T)}.$$

This can be done by looking at the first-order derivatives in x_1 and x' separately. By the definition (4-12) of $\tilde{u}(\,\cdot\,,l)$ (where of course $S_{\tau-\sqrt{\lambda_l}}$ and $S_{\tau+\sqrt{\lambda_l}}$ can be interchanged) and the definition of S_a, we have

$$\partial_1 \tilde{u}(\cdot, l) = (\tau + \sqrt{\lambda_l})\tilde{u}(\cdot, l) - S_{\tau - \sqrt{\lambda_l}} \tilde{f}(\cdot, l).$$

Then (4-11) and Proposition 4.4 imply

$$\|\partial_1 \tilde{u}(\cdot, l)\|_{L^2_{-\delta}(\mathbb{R})} \leq C_0 \|\tilde{f}(\cdot, l)\|_{L^2_\delta(\mathbb{R})}.$$

Orthogonality shows that $\|\partial_1 u\|_{L^2_{-\delta}(T)} \leq C_0 \|f\|_{L^2_\delta(T)}$.

For the x' derivatives we use the exterior derivative $d_{x'}$ in (M_0, g_0). Since u_N vanishes on $\mathbb{R} \times \partial M_0$, we have

$$\|d_{x'} u_N\|^2_{L^2_{-\delta}(T)} = \int_{-\infty}^\infty \langle x_1 \rangle^{-2\delta} \langle d_{x'} u_N, d_{x'} \bar{u}_N \rangle_{M_0} \, dx_1$$

$$= \int_{-\infty}^\infty \langle x_1 \rangle^{-2\delta} \langle (-\Delta_{g_0} u_N), \bar{u}_N \rangle_{M_0} \, dx_1$$

$$= \int_{-\infty}^\infty \sum_{l=1}^N \langle x_1 \rangle^{-2\delta} \lambda_l |\tilde{u}(\cdot, l)|^2 \, dx_1$$

$$= \sum_{l=1}^N \lambda_l \|\tilde{u}(\cdot, l)\|^2_{L^2_{-\delta}(\mathbb{R})}.$$

Orthogonality and (4-11) give the estimate

$$\|d_{x'} u_N\|^2_{L^2_{-\delta}(T)} \leq C_0 \|f\|^2_{L^2_\delta(T)}.$$

An argument using Cauchy sequences shows that $d_{x'} u_N$ converges in $L^2_{-\delta}(T)$, hence also $d_{x'} u \in L^2_{-\delta}(T)$ and $\|d_{x'} u\|_{L^2_{-\delta}(T)} \leq C_0 \|f\|_{L^2_\delta(T)}$.

We have proved that $u \in H^1_{-\delta}(T)$ with the right norm estimate. It is now enough to note that $u_N \in H^1_{-\delta,0}(T)$, and the same is true for the limit u since this space is closed in $H^1_{-\delta}(T)$. □

We have now completed the proof of Theorem 4.1 in the case where $c = 1$ and $V = 0$. In fact, the combination of Propositions 4.3 and 4.6 immediately shows the existence of a solution operator G_τ for the conjugated Laplace–Beltrami equation with metric $g = e \oplus g_0$.

Proposition 4.7 (solution operator for $V = 0$). *Let $\delta > 1/2$. If $|\tau| \geq 1$ and $\tau^2 \notin \mathrm{Spec}(-\Delta_{g_0})$, there is a bounded operator*

$$G_\tau : L^2_\delta(T) \to H^1_{-\delta,0}(T)$$

such that $u = G_\tau f$ is the unique solution in $H^1_{-\delta,0}(T)$ of the equation

$$e^{\tau x_1}(-\Delta_g) e^{-\tau x_1} u = f \quad \text{in } T.$$

This operator satisfies

$$\|G_\tau f\|_{L^2_{-\delta}(T)} \leq \frac{C_0}{|\tau|}\|f\|_{L^2_\delta(T)},$$

$$\|G_\tau f\|_{H^1_{-\delta}(T)} \leq C_0\|f\|_{L^2_\delta(T)}.$$

It is now an easy matter to prove Theorem 4.1 also with a nonzero potential V by using a perturbation argument.

Proof of Theorem 4.1. We assume, as we may, that $c \equiv 1$. Let us first consider uniqueness. Assume that $u \in H^1_{-\delta,0}(T)$ satisfies

$$e^{\tau x_1}(-\Delta_g + V)e^{-\tau x_1}u = 0 \quad \text{in } T.$$

This can be written as

$$e^{\tau x_1}(-\Delta_g)e^{-\tau x_1}u = -Vu \quad \text{in } T.$$

By the assumption $\langle x_1 \rangle^{2\delta} V \in L^\infty(T)$, the right hand side is in $L^2_\delta(T)$. The uniqueness part of Proposition 4.7 implies

$$u = -G_\tau(Vu).$$

The norm estimates for G_τ give

$$\|u\|_{L^2_{-\delta}(T)} \leq \frac{C_0\|\langle x_1 \rangle^{2\delta} V\|_{L^\infty(T)}}{|\tau|}\|u\|_{L^2_{-\delta}(T)}.$$

Thus, if we choose

$$\tau_0 := \max(2C_0\|\langle x_1 \rangle^{2\delta} V\|_{L^\infty(T)}, 1), \tag{4-13}$$

then the condition $|\tau| \geq \tau_0$ will imply $\|u\|_{L^2_{-\delta}(T)} \leq \frac{1}{2}\|u\|_{L^2_{-\delta}(T)}$, showing that $u \equiv 0$.

As for existence, we seek a solution of the equation

$$e^{\tau x_1}(-\Delta_g + V)e^{-\tau x_1}u = f \quad \text{in } T$$

in the form $u = G_\tau \tilde{f}$ for some $\tilde{f} \in L^2_\delta(T)$. Inserting this expression in the equation and using that G_τ is the inverse of the conjugated Laplace–Beltrami operator, we see that $\tilde{f}$ should satisfy

$$(\text{Id} + VG_\tau)\tilde{f} = f \quad \text{in } T.$$

Now if $|\tau| \geq \tau_0$ with τ_0 as in (4-13), we have

$$\|VG_\tau\|_{L^2_\delta(T) \to L^2_\delta(T)} \leq \frac{C_0\|\langle x_1 \rangle^{2\delta} V\|_{L^\infty(T)}}{|\tau|} \leq \tfrac{1}{2}.$$

Thus $\mathrm{Id} + VG_\tau$ is invertible on $L^2_\delta(T)$, with norm of the inverse ≤ 2. It follows that $u := G_\tau(\mathrm{Id} + VG_\tau)^{-1}f$ is a solution with the required properties. □

Exercise 4.3. Prove that the solution construction in Theorem 4.1 is in fact in $H^2_{-\delta}(T)$ and satisfies $\|u\|_{H^2_{-\delta}(T)} \leq C_0|\tau|\|f\|_{L^2_\delta(T)}$.

Exercise 4.4. Prove Theorem 4.1 in the more general case where the Schrödinger operator $-\Delta_g + V$ with $\langle x_1 \rangle^{2\delta} V \in L^\infty(T)$ is replaced by a Helmholtz operator $-\Delta_g + V - k^2$ where $k > 0$ is fixed.

5. Uniqueness result

In this section we will prove a uniqueness result for the inverse problem considered in the introduction. The result will be proved for the case of the Schrödinger equation in a compact manifold (M, g). The method, as discussed in Section 3, is to show that the set of products $\{u_1 u_2\}$ of solutions to two Schrödinger equations is dense in $L^1(M)$. The special solutions which will be used to prove the density statement have the form
$$u = e^{\pm \tau \varphi}(m + r_0).$$

The starting point for constructing such solutions is an LCW φ. For this reason we will need to work in manifolds which admit LCWs. Thus we will assume that (M, g) is contained in a cylinder (T, g) where $T = \mathbb{R} \times M_0$ and $g = c(e \oplus g_0)$, which is roughly equivalent to M having an LCW by the results in Section 3.

However, the existence of an LCW is only a starting point for the solution of the inverse problem. One also needs construct the term m so that $e^{\pm \tau \varphi} m$ is an approximate solution, which can be corrected into an exact solution by the term r_0 obtained from solving an inhomogeneous equation as in Section 4. Finally, one needs to do this construction so that the density of the products $\{u_1 u_2\}$ can be proved by using the special solutions. In Euclidean space one typically employs the Fourier transform, which is not immediately available in (M, g).

We will use a hybrid method which involves the Fourier transform in the x_1 variable where it is available, and integrals over geodesics in the x' variables. In fact, we will choose the functions m to concentrate near fixed geodesics in (M_0, g_0). The uniqueness theorem will then rely on the result that a function in M_0 can be determined from its integrals over geodesics. At present, such a result is only known under strong restrictions on the geodesic flow of (M_0, g_0). One such restriction is that (M_0, g_0) is *simple*, meaning roughly that any two points can be connected by a unique length-minimizing geodesic.

Leaving the precise definition of simple manifolds to Section 5B, we now define the class of admissible manifolds for which we can prove uniqueness results for inverse problems. There are three conditions: the first one requiring the dimension to be at least three (the case of 2D manifolds requires quite different methods), the second stating that the manifold should admit an LCW, and the third stating that the transversal manifold (M_0, g_0) satisfies a restriction ensuring that functions are determined by their integrals over geodesics.

Definition. A compact manifold (M, g) with smooth boundary is called *admissible* if

(a) $\dim(M) \geq 3$,

(b) $(M, g) \Subset (T, g)$ where $T = \mathbb{R} \times M_0$ and $g = c(e \oplus g_0)$ with $c > 0$ a smooth positive function and e the Euclidean metric on $\mathbb{R}$, and

(c) (M_0, g_0) is a simple $(n-1)$-dimensional manifold.

The main uniqueness result is as follows. Recall that we implicitly assume that all DN maps are well defined.

Theorem 5.1 (global uniqueness). *Let (M, g) be an admissible manifold, and assume that V_1 and V_2 are continuous functions on M. If $\Lambda_{g,V_1} = \Lambda_{g,V_2}$, then $V_1 = V_2$.*

In fact, it is enough to prove the theorem for admissible manifolds where the conformal factor is constant and V_1 and V_2 are in $C_c(M^{\text{int}})$. In the proofs below, we will work under these assumptions. We now give a sketch how to make this reduction.

Suppose (M, g) is admissible and $g = c\tilde{g}$ with $\tilde{g} = e \oplus g_0$, and assume that $\Lambda_{g,V_1} = \Lambda_{g,V_2}$. Note that we are free to assume that $c = 1$ outside a small neighborhood of M in T. A boundary determination result [Dos Santos Ferreira et al. 2009, Theorem 8.4] shows that $V_1|_{\partial M} = V_2|_{\partial M}$. Extending V_1, V_2 to a slightly larger admissible manifold $(\tilde{M}, g)$ so that $c = 1$ and $V_1 = V_2 = 0$ near $\partial \tilde{M}$, it is not hard to see that $\Lambda_{g,V_1} = \Lambda_{g,V_2}$ for the DN maps in $(\tilde{M}, g)$. Now by the conformal scaling law for Δ_g, it holds that

$$\Lambda_{c\tilde{g},V_j} = \Lambda_{\tilde{g},c(V_j-q_c)}$$

where $q_c = c^{\frac{n-2}{4}} \Delta_{c\tilde{g}}(c^{-\frac{n-2}{4}})$. Thus $\Lambda_{\tilde{g},V_1} = \Lambda_{\tilde{g},V_2}$ for the DN maps in $(\tilde{M}, \tilde{g})$, which completes the reduction.

5A. Complex geometrical optics solutions. Here we will construct the special solutions, also called *complex geometrical optics solutions*, to the Schrödinger equation. The first step is to construct approximate solutions

$$u_0 = e^{-\tau \Phi} a$$

where $\tau > 0$ is a large parameter, $\Phi \in C^\infty(M)$ is a complex function (the complex phase), and a is smooth complex function on M (the complex amplitude). Note that we have replaced the real function φ with a complex function Φ. In fact the real part of Φ is later taken to be an LCW φ.

We extend the inner product $\langle \cdot, \cdot \rangle$ as a $\mathbb{C}$-bilinear form to complex tangent and cotangent vectors. This means that for $\xi, \eta, \xi', \eta' \in T_p^* M$,

$$\langle \xi + i\eta, \xi' + i\eta' \rangle := \langle \xi, \xi' \rangle - \langle \eta, \eta' \rangle + i(\langle \xi, \eta' \rangle + \langle \eta, \xi' \rangle).$$

Note that $\langle \cdot, \cdot \rangle$ is not a Hermitian inner product, since there are nonzero complex vectors whose inner product with itself is zero.

With this notation, we have the following analog of the computation in Lemma 3.2 (just replace φ by Φ).

Lemma 5.2 (expression for conjugated Schrödinger operator).

$$e^{\tau \Phi}(-\Delta_g + V)e^{-\tau \Phi}v$$
$$= -\tau^2 \langle d\Phi, d\Phi \rangle v + \tau(2\langle d\Phi, dv \rangle + (\Delta_g \Phi)v) + (-\Delta_g + V)v.$$

This result gives an expansion of the conjugated operator $e^{\tau \Phi}(-\Delta_g + V)e^{-\tau \Phi}$ in terms of powers of τ. We will look for approximate solutions $u_0 = e^{-\tau \Phi} a$ such that the terms with highest powers of τ go away. This leads to equations for Φ and a, and also an equation for the correction term r_0 when one looks for the exact solution u corresponding to u_0. The next result follows from Lemma 5.2.

Proposition 5.3 (equations). *Let (M, g) be a compact manifold with boundary and let $V \in L^\infty(M)$. The function $u = e^{-\tau \Phi}(a + r_0)$ is a solution of*

$$(-\Delta_g + V)u = 0$$

in M, provided that in M

$$\langle d\Phi, d\Phi \rangle = 0, \tag{5-1}$$

$$2\langle d\Phi, da \rangle + (\Delta_g \Phi)a = 0, \tag{5-2}$$

$$e^{\tau \Phi}(-\Delta_g + V)e^{-\tau \Phi}r_0 = (\Delta_g - V)a. \tag{5-3}$$

The last result is analogous the (real) geometrical optics method, or the WKB method, for constructing solutions to various equations. The main difference to the standard setting is that we need to consider complex quantities. Here (5-1) is called a *complex eikonal equation*, that is, a certain nonlinear first-order equation for the complex phase Φ. Equation (5-2) is a *complex transport equation*, which is a linear first-order equation for the amplitude a. The last equation, (5-3), is an inhomogeneous equation for the correction term r_0.

Writing $\Phi = \varphi + i\psi$ where φ and ψ are real, the Equation (5-3) becomes

$$e^{\tau\varphi}(-\Delta_g + V)e^{-\tau\varphi}(e^{-i\tau\psi}r_0) = e^{-i\tau\psi}(\Delta_g - V)a.$$

This equation can be solved by Theorem 4.1 if φ is an LCW and the manifold has an underlying product structure. Using that ψ is real we have $\|e^{-i\tau\psi}v\|_{L^2(M)} = \|v\|_{L^2(M)}$, so the terms $e^{-i\tau\psi}$ will not change the resulting L^2 estimates.

We now assume that (M, g) is admissible, and further that $c \equiv 1$ which is possible by the reduction above. Thus (M, g) is embedded in the cylinder (T, g) where $T = \mathbb{R} \times M_0$ and $g = e \oplus g_0$, and further $(M_0, g_0) \Subset (U, g_0)$ with $(\overline{U}, g_0)$ simple. In the coordinates $x = (x_1, x')$,

$$g(x_1, x') = \begin{pmatrix} 1 & 0 \\ 0 & g_0(x') \end{pmatrix}.$$

We also assume that $\mathrm{Re}(\Phi) = \varphi$ where $\varphi(x_1, x') := x_1$ is the natural LCW in the cylinder.

Eikonal equation. Writing $\Phi = \varphi + i\psi$ where φ and ψ are real-valued, the complex eikonal equation (5-1) becomes the pair of equations

$$|d\psi|^2 = |d\varphi|^2, \quad \langle d\varphi, d\psi \rangle = 0. \tag{5-4}$$

Using that $\varphi(x) = x_1$ and the special form of the metric, these equations become

$$|d\psi|^2 = 1, \quad \partial_1\psi = 0.$$

The second equation just means that ψ should be independent of x_1, that is, $\psi = \psi(x')$. Thus we have reduced matters to solving a (real) eikonal equation in M_0:

$$|d\psi|^2_{g_0} = 1 \quad \text{in } M_0.$$

Such an equation does not have global smooth solutions on a general manifold (M_0, g_0). However, in our case where (M_0, g_0) is assumed to be simple (see Section 5B), there are many global smooth solutions. It is enough to choose some point $\omega \in U \smallsetminus M_0$ and to take

$$\psi(x_1, r, \theta) = \psi_\omega(x_1, r, \theta) := r$$

where (r, θ) are polar normal coordinates in (U, g_0) with center ω. Since $|dr|_{g_0} = 1$ on the maximal domain where polar normal coordinates are defined (excluding the center), this gives a smooth solution in M.

In fact, if $x = (x_1, r, \theta)$ are coordinates in T where (r, θ) are polar normal coordinates in (U, g_0) with center ω, then the form of the metric g_0 in polar normal coordinates shows that

$$g(x_1, r, \theta) = \begin{pmatrix} 1 & 0 & 0 \\ 0 & 1 & 0 \\ 0 & 0 & g_1(r, \theta) \end{pmatrix} \tag{5-5}$$

for some $(n-2) \times (n-2)$ positive definite matrix g_1. This gives the coordinate representation

$$\Phi(x_1, r, \theta) = \Phi_\omega(x_1, r, \theta) := x_1 + ir.$$

Remark 5.4. For $n = 2$ the complex eikonal equation, which is equivalent to the pair (5-4), just says that φ and ψ should be (anti)conjugate harmonic functions, so that Φ should be (anti)holomorphic. In dimensions $n \geq 3$ solutions of the complex eikonal equation can be considered as analogs in a certain sense of (anti)holomorphic functions. In our setting, using the given coordinates, Φ is just $x_1 + ir$ which can be considered as a complex variable z and hence also as a holomorphic function.

Transport equation. Having obtained the complex phase $\Phi = \varphi + i\psi = x_1 + ir$, it is not difficult to solve the complex transport equation. Using the coordinates (x_1, r, θ) and the special form (5-5) for the metric, we see that

$$\langle d\Phi, da \rangle = g^{jk} \partial_j \Phi \partial_k a = (\partial_1 + i\partial_r)a$$

and

$$\Delta_g \Phi = |g|^{-1/2} \partial_j (|g|^{1/2} g^{jk} \partial_k (x_1 + ir))$$
$$= |g|^{-1/2} \partial_r (|g|^{1/2} i)$$
$$= \tfrac{1}{2}(\partial_1 + i\partial_r)(\log |g|).$$

The transport equation (5-2) now has the form

$$(\partial_1 + i\partial_r)a + (\partial_1 + i\partial_r)(\log |g|^{1/4})a = 0.$$

Multiplying by the integrating factor $|g|^{1/4}$, we obtain the equivalent equation

$$(\partial + i\partial_r)(|g|^{1/4} a) = 0.$$

Thus the complex amplitudes satisfying (5-2) have the form

$$a(x_1, r, \theta) = |g|^{-1/4} a_0(x_1, r, \theta)$$

where a_0 is a smooth function in M satisfying $(\partial_1 + i\partial_r)a_0 = 0$.

Inhomogeneous equation. Given Φ and a, the final equation (5-3) in the present setting becomes

$$e^{\tau x_1}(-\Delta_g + V)e^{-\tau x_1}(e^{-i\tau r}r_0) = f \quad \text{in } M$$

where $f := e^{-i\tau r}(\Delta_g - V)a$. We extend V and f by zero to T, and consider the equation

$$e^{\tau x_1}(-\Delta_g + V)e^{-\tau x_1}v = f \quad \text{in } T.$$

If $|\tau|$ is large and $\tau^2 \notin \text{Spec}(-\Delta_{g_0})$, this equation has a unique solution $v \in H^1_{-\delta,0}(T)$ by Theorem 4.1. It satisfies for any $\delta > 1/2$

$$\|v\|_{L^2_{-\delta}(T)} \leq \frac{C_0}{|\tau|}\|f\|_{L^2_\delta(T)}.$$

Define $r_0 := e^{i\tau r}v|_M$. Then $r_0 \in H^1(M)$ and

$$\|r_0\|_{L^2(M)} \leq \frac{C_0}{|\tau|}\|a\|_{H^2(M)}.$$

Also, r_0 satisfies (5-3) by construction.

We collect the results of the preceding arguments in the next proposition.

Proposition 5.5 (complex geometrical optics solutions). *Assume (M,g) is an admissible manifold embedded in (T,g), where $T = \mathbb{R} \times M_0$ and $g = e \oplus g_0$ and where $(M_0, g_0) \Subset (\overline{U}, g_0)$ are simple manifolds. Let also $V \in L^\infty(M)$. There are $C_0, \tau_0 > 0$ such that whenever*

$$|\tau| \geq \tau_0 \quad \text{and} \quad \tau^2 \notin \text{Spec}(-\Delta_{g_0}),$$

then for any $\omega \in U \smallsetminus M_0$ and for any smooth function a_0 in M with $(\partial_1 + i\partial_r)a_0 = 0$, where (x_1, r, θ) are coordinates in M such that (r, θ) are polar normal coordinates in (U, g_0) with center ω, there is a solution

$$u = e^{-\tau(x_1 + ir)}(|g|^{-1/4}a_0 + r_0)$$

of the equation $(-\Delta_g + V)u = 0$ in M, such that

$$\|r_0\|_{L^2(M)} \leq \frac{C_0}{|\tau|}\|a_0\|_{H^2(M)}.$$

We can now complete the proof of Theorem 5.1, modulo the following statement on the attenuated geodesic ray transform which will be discussed in the next section.

Theorem (injectivity for the attenuated geodesic ray transform). *Let (M_0, g_0) be a simple manifold. There exists $\varepsilon > 0$ such that for any $\lambda \in (-\varepsilon, \varepsilon)$, if a*

function $f \in C(M_0)$ satisfies
$$\int_\gamma e^{-\lambda t} f(\gamma(t))\, dt$$
for any maximal geodesic γ going from ∂M_0 into M_0, then $f \equiv 0$.

Proof of Theorem 5.1. We make the reduction described after Theorem 5.1 to the case where $c \equiv 1$ and $V_1, V_2 \in C_c(M^{\text{int}})$. The assumption that $\Lambda_{g,V_1} = \Lambda_{g,V_2}$ implies that
$$\int_M (V_1 - V_2) u_1 u_2 \, dV = 0$$
for any $u_j \in H^1(M)$ with $(-\Delta_g + V_j) u_j = 0$ in M.

We use Proposition 5.5 and choose u_j to be solutions of the following form. Let ω be a fixed point in $U \smallsetminus M_0$, let (x_1, r, θ) be coordinates near M such (r, θ) are polar normal coordinates in (U, g_0) with center ω, and let λ be a fixed real number and $b = b(\theta) \in C^\infty(S^{n-2})$ a fixed function. Then, for $\tau > 0$ large enough and outside a discrete set, we can choose u_j of the form
$$u_1 = e^{-\tau(x_1 + ir)} (|g|^{-1/4} e^{i\lambda(x_1 + ir)} b(\theta) + r_1),$$
$$u_2 = e^{\tau(x_1 + ir)} (|g|^{-1/4} + r_2).$$

Note that the functions $e^{i\lambda(x_1 + ir)} b(\theta)$ and 1 are holomorphic in the (x_1, r) variables, so we indeed have solutions of this form. Further, $\|r_j\|_{L^2(M)} \leq C/\tau$.

Inserting the solutions in the integral identity and letting $\tau \to \infty$ outside a discrete set, we obtain
$$\int_M (V_1 - V_2) |g|^{-1/2} e^{i\lambda(x_1 + ir)} b(\theta) \, dV_g = 0.$$

Since V_1 and V_2 are compactly supported, the integral can be taken over the cylinder T. Using the (x_1, r, θ) coordinates in T and the fact that $dV_g = |g(x_1, r, \theta)|^{1/2} dx_1 \, dr \, d\theta$, this implies that
$$\int_{S^{n-2}} \left(\int_{-\infty}^\infty \int_0^\infty (V_1 - V_2)(x_1, r, \theta) e^{i\lambda(x_1 + ir)} \, dx_1 \, dr \right) b(\theta) \, d\theta = 0.$$

The last statement is valid for any fixed $b \in C^\infty(S^{n-2})$. We can choose b to resemble a delta function at a fixed direction θ_0 in S^{n-2}, and varying b will then imply that the quantity in brackets vanishes for all θ_0. This is the point where we have chosen the solution u_1 to approximately concentrate near a fixed geodesic, corresponding to a fixed direction in S^{n-2}, in the transversal manifold (M_0, g_0).

We have proved that

$$\int_0^\infty e^{-\lambda r}\left(\int_{-\infty}^\infty (V_1-V_2)(x_1,r,\theta)e^{i\lambda x_1}\,dx_1\right)dr = 0, \quad \text{for all }\theta.$$

Denote the quantity in brackets by $f_\lambda(r,\theta)$. Then f_λ is a smooth function in (M_0,g_0) compactly supported in M_0^{int}, and the curve $\gamma_{\omega,\theta}: r \mapsto (r,\theta)$ is a geodesic in (U,g_0) issued from the point ω in direction θ. This shows that

$$\int_0^\infty e^{-\lambda r} f_\lambda(\gamma_{\omega,\theta}(r))\,dr = 0$$

for all $\omega \in U \smallsetminus M_0$ and for all directions θ. Letting ω approach the boundary of M_0 and varying θ, the last result implies that

$$\int_\gamma e^{-\lambda t} f_\lambda(\gamma(t))\,dt = 0$$

for all geodesics γ starting from points of ∂M_0 which are maximal in the sense that γ is defined for the maximal time until it exits M_0.

The injectivity result for the attenuated geodesic ray transform, stated just before this proof, shows that there is $\varepsilon > 0$ such that for any $\lambda \in (-\varepsilon,\varepsilon)$, the function f_λ is identically zero on M_0. Thus for $|\lambda| < \varepsilon$,

$$\int_{-\infty}^\infty (V_1-V_2)(x_1,r,\theta)e^{i\lambda x_1}\,dx_1 = 0, \quad \text{for any fixed }r,\theta.$$

If (r,θ) is fixed then the function $x_1 \mapsto (V_1-V_2)(x_1,r,\theta)$ is compactly supported on the real line, and the last result says that its Fourier transform vanishes for $|\lambda| < \varepsilon$. But by the Paley-Wiener theorem the Fourier transform is analytic, which is only possible if $(V_1-V_2)(\,\cdot\,,r,\theta) = 0$ on the real line. This is true for any fixed (r,θ), showing that $V_1 = V_2$ as required. □

5B. Geodesic ray transform. In this section we will give some arguments related to the injectivity result for the attenuated geodesic ray transform, which was used in the proof of the global uniqueness theorem. The treatment will be very sketchy and not self-contained, but hopefully it will give an idea about why such a result would be true.

Explicit inversion methods. To set the stage and to obtain some intuition to the problem, we first consider the classical question of inverting the Radon transform in $\mathbb{R}^2$. This is the transform which integrates a function $f \in C_c^\infty(\mathbb{R}^2)$ over all lines, and can be expressed as follows:

$$Rf(s,\omega) := \int_{-\infty}^\infty f(s\omega^\perp + t\omega)\,dt, \quad s \in \mathbb{R}, \omega \in S^1.$$

Here $\omega^\perp$ is the vector in S^1 obtained by rotating ω counterclockwise by $90°$.

There is a well-known relation between Rf and the Fourier transform $\hat{f}$. We denote by $\widehat{Rf}(\,\cdot\,, \omega)$ the Fourier transform of Rf with respect to s.

Proposition 5.6 (Fourier slice theorem).
$$\widehat{Rf}(\sigma, \omega) = \hat{f}(\sigma\omega^\perp).$$

Proof. Parametrizing $\mathbb{R}^2$ by $y = s\omega^\perp + t\omega$, we have
$$\widehat{Rf}(\sigma, \omega) = \int_{-\infty}^{\infty} e^{-i\sigma s} \int_{-\infty}^{\infty} f(s\omega^\perp + t\omega)\, dt\, ds = \int_{\mathbb{R}^2} e^{-i\sigma y \cdot \omega^\perp} f(y)\, dy$$
$$= \hat{f}(\sigma\omega^\perp). \qquad \square$$

This result gives the first proof of injectivity of Radon transform: if $f \in C_c^\infty(\mathbb{R}^2)$ is such that $Rf \equiv 0$, then $\hat{f} \equiv 0$ and consequently $f \equiv 0$. To obtain a different inversion formula, and for later purposes, we will consider the adjoint of R. This is obtained by computing for $f \in C_c^\infty(\mathbb{R}^2)$ and $h \in C^\infty(\mathbb{R} \times S^1)$ that

$$(Rf, h)_{\mathbb{R} \times S^1} = \int_{-\infty}^{\infty} \int_{S^1} Rf(s, \omega) h(s, \omega)\, d\omega\, ds$$
$$= \int_{-\infty}^{\infty} \int_{S^1} \int_{-\infty}^{\infty} f(s\omega^\perp + t\omega) h(s, \omega)\, dt\, d\omega\, ds$$
$$= \int_{\mathbb{R}^2} f(y) \left(\int_{S^1} h(y \cdot \omega^\perp, \omega)\, d\omega \right) dy.$$

Thus the adjoint of R is the operator
$$R^* : C^\infty(\mathbb{R} \times S^1) \to C^\infty(\mathbb{R}^2), \quad R^*h(y) = \int_{S^1} h(y \cdot \omega^\perp, \omega)\, d\omega.$$

Proposition 5.7 (Fourier transform of R^*). *Letting $\hat{\xi} = \xi/|\xi|$,*
$$(R^*h)\hat{\,}(\xi) = \frac{2\pi}{|\xi|}\left(\hat{h}(|\xi|, -\hat{\xi}^\perp) + \hat{h}(-|\xi|, \hat{\xi}^\perp) \right).$$

Proof. We will make a formal computation (which is not difficult to justify). Using again the parametrization $y = s\omega^\perp + t\omega$,

$$(R^*h)\hat{\,}(\xi) = \int_{\mathbb{R}^2} \int_{S^1} e^{-iy\cdot\xi} h(y \cdot \omega^\perp, \omega)\, d\omega\, dy$$
$$= \int_{-\infty}^{\infty} \int_{-\infty}^{\infty} \int_{S^1} e^{-is\omega^\perp \cdot \xi} e^{-it\omega \cdot \xi} h(s, \omega)\, d\omega\, ds\, dt$$
$$= \int_{S^1} \hat{h}(\omega^\perp \cdot \xi, \omega) \left(\int_{-\infty}^{\infty} e^{-it\omega \cdot \xi}\, dt \right) d\omega.$$

The quantity in the parentheses is just $(2\pi/|\xi|)\delta_0(\omega\cdot\hat{\xi})$ where δ_0 is the Dirac delta function at the origin. Since $\omega\cdot\hat{\xi}$ is zero exactly when $\omega = \pm\hat{\xi}^\perp$, the result follows. $\square$

The Radon transform in $\mathbb{R}^2$ satisfies the symmetry $Rf(-s,-\omega) = Rf(s,\omega)$, and the Fourier slice theorem implies

$$(R^*Rf)\hat{}(\xi) = \frac{4\pi}{|\xi|}\widehat{Rf}(|\xi|,-\hat{\xi}^\perp) = \frac{4\pi}{|\xi|}\hat{f}(\xi).$$

This shows that the normal operator R^*R is a classical pseudodifferential operator of order -1 in $\mathbb{R}^2$, and also gives an inversion formula.

Proposition 5.8 (normal operator). *One has*

$$R^*R = 4\pi(-\Delta)^{-1/2},$$

and f can be recovered from Rf by the formula

$$f = \frac{1}{4\pi}(-\Delta)^{1/2}R^*Rf.$$

The last result is an example of an explicit inversion method for the Radon transform in the Euclidean plane, based on the Fourier transform. Similar methods are available for the Radon transform on manifolds with many symmetries where variants of the Fourier transform exist (see [Helgason 1999] and other books of Helgason for results of this type). However, for manifolds which do not have symmetries, such as small perturbations of the Euclidean metric, explicit transforms are usually not available and other inversion methods are required.

Pseudodifferential methods. Let (M,g) be a compact manifold with smooth boundary, assumed to be embedded in a compact manifold (N,g) without boundary. We parametrize geodesics by points in the *unit sphere bundle*, defined by

$$SM := \bigvee_{x\in M} S_xM, \quad S_xM := \{\xi \in T_xM \,;\, |\xi| = 1\}.$$

If $(x,\xi) \in SM$ we denote by $\gamma(t,x,\xi)$ the geodesic in N which starts at the point x in direction ξ, that is,

$$D_{\dot\gamma}\dot\gamma = 0, \quad \gamma(0,x,\xi) = x, \quad \dot\gamma(0,x,\xi) = \xi.$$

Let $\tau(x,\xi)$ be the first time when $\gamma(t,x,\xi)$ exits M,

$$\tau(x,\xi) := \inf\{t > 0 \,;\, \gamma(t,x,\xi) \in N \smallsetminus M\}.$$

We assume that (M,g) is *nontrapping*, meaning that $\tau(x,\xi)$ is finite for any $(x,\xi) \in SM$.

The *geodesic ray transform* of a function $f \in C^\infty(M)$ is defined by
$$If(x,\xi) := \int_0^{\tau(x,\xi)} f(\gamma(t,x,\xi))\,dt, \quad (x,\xi) \in \partial(SM).$$

Thus, If gives the integral of f over any maximal geodesic in M starting from ∂M, such geodesics being parametrized by points of
$$\partial(SM) = \{(x,\xi) \in SM \,;\, x \in \partial M\}.$$

So far, we have not imposed any restrictions on the behavior of geodesics in (M, g) other than the nontrapping condition. However, injectivity and inversion results for If are only known under strong geometric restrictions. One class of manifolds where such results have been proved is the following. From now on the treatment will be sketchy, and we refer to [Dos Santos Ferreira et al. 2009; Dairbekov et al. 2007; Sharafutdinov 1994] for more details.

Definition. A compact manifold (M, g) with boundary is called *simple* if

(a) for any point $p \in M$, the exponential map $\exp_p$ is a diffeomorphism from its maximal domain in $T_p M$ onto M, and

(b) the boundary ∂M is strictly convex.

Several remarks are in order. A diffeomorphism is, as earlier, a homeomorphism which together with its inverse is smooth up to the boundary. The maximal domain of $\exp_p$ is starshaped, and the fact that $\exp_p$ is a diffeomorphism onto M thus implies that M is diffeomorphic to a closed ball. The last fact uses that τ is smooth in $S(M^{\text{int}})$. This is a consequence of strict convexity, which is precisely defined as follows:

Definition. Let (M, g) be a compact manifold with boundary. We say that ∂M is *stricly convex* if the second fundamental form $l_{\partial M}$ is positive definite. Here $l_{\partial M}$ is the 2-tensor on ∂M defined by
$$l_{\partial M}(X, Y) = -\langle D_X \nu, Y \rangle, \quad X, Y \in C^\infty(\partial M, T(\partial M)),$$
where ν is the outer unit normal to ∂M.

Alternatively, the boundary is strictly convex if and only if any geodesic in N starting from a point $x \in \partial M$ in a direction tangent to ∂M stays outside M for small positive and negative times. This implies that any maximal geodesic going from ∂M into M stays inside M except for its endpoints, which corresponds to the usual notion of strict convexity.

If (M, g) is simple, one can always find an open manifold (U, g) such that $(M, g) \Subset (U, g)$ where $(\overline{U}, g)$ is simple. We will always understand that (M, g) and (U, g) are related in this way.

Intuitively, a manifold is simple if the boundary is strictly convex and if the whole manifold can be parametrized by geodesic rays starting from any fixed point. The last property can be thought of as an analog for the parametrization $y = s\omega^\perp + t\omega$ of $\mathbb{R}^2$ used in the discussion of the Radon transform in the plane. These parametrizations can be used to prove the analog of the first part of Proposition 5.8 on a simple manifold.

Proposition 5.9 (normal operator). *If (M, g) is a simple manifold, then $\tilde{I}^* \tilde{I}$ is an elliptic pseudodifferential operator of order -1 in U where $\tilde{I}$ is the geodesic ray transform in $(\overline{U}, g)$.*

It is well known that elliptic pseudodifferential operators can be inverted up to smoothing (and thus compact) operators. This implies an inversion formula as in Proposition 5.8 which however contains a compact error term (resulting in a Fredholm problem). If g is *real-analytic* in addition to being simple then this error term can be removed by the methods of analytic microlocal analysis, thus proving injectivity of I in this case.

For general simple metrics one does not obtain injectivity in this way, but invertibility up to a compact operator implies considerable stability properties for this problem. In particular, if I is known to be injective in (M, g), then suitable small perturbations of I are also injective: it follows from the results of [Frigyik et al. 2008] that injectivity of I implies the injectivity of the attenuated transform in Section 5A for sufficiently small λ. Thus, it remains to prove in some way the injectivity of the unattenuated transform I on simple manifolds.

Energy estimates. The most general known method for proving injectivity of the geodesic ray transform, in the absence of symmetries or real-analyticity, is based on energy estimates. Typically these estimates allow to bound some norm of a function u by some norm of Pu where P is a differential operator, or to prove the uniqueness result that $u = 0$ whenever $Pu = 0$. Such estimates are often proved by integration by parts.

Motivation. Let us consider a very simple energy estimate for the Laplace operator in a bounded open set $\Omega \subseteq \mathbb{R}^2$ with smooth boundary. Suppose that $u \in C^2(\overline{\Omega})$ and $-\Delta u = 0$ in Ω, $u|_{\partial\Omega} = 0$. We wish to show that $u = 0$. To do this, we integrate the equation $-\Delta u = 0$ against the test function u and use the Gauss-Green formula:

$$0 = \int_\Omega (-\Delta u) u \, dx = -\int_{\partial\Omega} \frac{\partial u}{\partial \nu} u \, dS + \int_\Omega |\nabla u|^2 \, dx.$$

Since $u|_{\partial\Omega} = 0$ it follows that $\int_\Omega |\nabla u|^2 \, dx = 0$, showing that u is constant on each component and consequently $u = 0$.

We will now proceed to prove an energy estimate for the geodesic ray transform in the case $(M, g) = (\overline{\Omega}, e)$ where $\Omega \subset \mathbb{R}^2$ is a bounded open set with strictly convex boundary and e is the Euclidean metric. This will give an alternative proof of the injectivity result for the Radon transform in $\mathbb{R}^2$, the point being that this proof only uses integration by parts and can be generalized to other geometries.

Suppose $f \in C_c^{\infty}(M^{\text{int}})$ and $If \equiv 0$. The first step is to relate the integral operator I to a differential operator. This is the standard reduction of the integral geometry problem to a transport equation. We identify SM with $M \times S^1$ and vectors $\omega_\theta = (\cos\theta, \sin\theta) \in S^1$ with the angle $\theta \in [0, 2\pi)$. Consider the function u defined as the integral of f over lines,

$$u(x, \theta) := \int_0^{\tau(x,\theta)} f(x + t\omega_\theta)\,dt, \quad x \in M,\ \theta \in [0, 2\pi).$$

The *geodesic vector field* is the differential operator on SM defined for $v \in C^{\infty}(SM)$ by

$$\mathcal{H}v(x, \theta) := \frac{\partial}{\partial s} v(x + s\omega_\theta, \theta)\bigg|_{s=0} = \omega_\theta \cdot \nabla_x v(x, \theta).$$

Since u is the integral of f over lines and $\mathcal{H}$ differentiates along lines, it is not surprising that

$$\mathcal{H}u(x, \theta) = \frac{\partial}{\partial s} \int_0^{\tau(x,\theta)-s} f(x + (s+t)\omega_\theta)\,dt\bigg|_{s=0}$$

$$= \int_0^{\tau(x,\theta)} \frac{\partial}{\partial t} f(x + t\omega_\theta)\,dt = -f(x).$$

Here we used the rule for differentiating under the integral sign.

Thus, if $f \in C_c^{\infty}(M^{\text{int}})$ and $If \equiv 0$, then u as defined above is a smooth function in SM and satisfies the following boundary value problem for the transport equation involving $\mathcal{H}$:

$$\begin{cases} \mathcal{H}u = -f & \text{in } SM, \\ u = 0 & \text{on } \partial(SM). \end{cases} \tag{5-6}$$

Further, since f does not depend on θ, we can take the derivative in θ and obtain

$$\begin{cases} \partial_\theta \mathcal{H}u = 0 & \text{in } SM, \\ u = 0 & \text{on } \partial(SM). \end{cases} \tag{5-7}$$

We will prove an energy estimate which shows that any smooth solution u of this problem must be identically zero. By (5-6) this will imply that $f \equiv 0$, proving that I is injective (at least on smooth compactly supported functions, which we assume for simplicity).

To establish the energy estimate, we use $\partial_\theta \mathcal{H} u$ as a test function and integrate (5-7) against this function, and then apply integration by parts to identify some positive terms and to show that some terms are zero. This will make use of the following special identity.

Proposition 5.10 (Pestov identity in $\mathbb{R}^2$). *For smooth $u = u(x, \theta)$, one has the identity*
$$|\partial_\theta \mathcal{H} u|^2 = |\mathcal{H} \partial_\theta u|^2 + \mathrm{div}_h(V) + \mathrm{div}_v(W)$$
where for smooth $X = (X^1(x, \theta), X^2(x, \theta))$, the horizontal and vertical divergences are defined by
$$\mathrm{div}_h(X) := \nabla_x \cdot X(x, \theta),$$
$$\mathrm{div}_v(X) := \nabla_\xi \cdot \left(X\left(x, \frac{\xi}{|\xi|}\right) \right)\bigg|_{\xi = \omega_\theta} = \omega_\theta^\perp \cdot \partial_\theta X(x, \theta)$$
and the vector fields V and W are given by
$$V := \left((\omega_\theta^\perp \cdot \nabla_x u)\omega_\theta - (\omega_\theta \cdot \nabla_x u)\omega_\theta^\perp \right) \partial_\theta u,$$
$$W := (\omega_\theta \cdot \nabla_x u) \nabla_x u.$$

Once the identity is known, the proof is in fact a direct computation and is left as an exercise. Let us now show how the Pestov identity can be used to prove that the only solution to (5-7) is the zero function. Note how the divergence terms are converted to boundary terms by integration by parts, and how one term vanishes because of the boundary condition and the other term is nonnegative.

Proposition 5.11. *If $u \in C^\infty(SM)$ solves (5-7), then $u \equiv 0$.*

Proof. As promised, we integrate (5-7) against the test function $\partial_\theta \mathcal{H} u$ and use the Pestov identity:
$$0 = \int_M \int_{S^1} |\partial_\theta \mathcal{H} u|^2 \, d\theta \, dx$$
$$= \int_M \int_{S^1} \left(|\mathcal{H} \partial_\theta u|^2 + \mathrm{div}_h(V) + \mathrm{div}_v(W) \right) d\theta \, dx.$$

Here
$$\int_M \mathrm{div}_h(V) \, dx = \int_M \nabla_x \cdot V(x, \theta) \, dx = \int_{\partial M} \nu \cdot V(x, \theta) \, dS(x) = 0$$

since $V(x,\theta) = [\,\cdot\,]\partial_\theta u(x,\theta) = 0$ for $x \in \partial M$ by the boundary condition for u. Also, integrating by parts on S^1,

$$\int_{S^1} \mathrm{div}_v(W)\,d\theta = \int_{S^1} \omega_\theta^\perp \cdot \partial_\theta W\,d\theta = -\int_{S^1} \partial_\theta(-\sin\theta, \cos\theta) \cdot W\,d\theta$$
$$= \int_{S^1} \omega_\theta \cdot W\,d\theta = \int_{S^1} |\omega_\theta \cdot \nabla_x u|^2\,d\theta.$$

This shows that

$$\int_M \int_{S^1} \left(|\mathcal{H}\partial_\theta u|^2 + |\omega_\theta \cdot \nabla_x u|^2 \right) d\theta\,dx = 0.$$

Since the integrand is nonnegative, we see that $\omega_\theta \cdot \nabla_x u = 0$ on SM. Thus $u(\,\cdot\,,\theta)$ is constant along lines with direction ω_θ, and the boundary condition implies that $u = 0$ as required. $\square$

This concludes the energy estimate proof of the injectivity of the ray transform in bounded domains in $\mathbb{R}^2$. A similar elementary argument can be used to show that the geodesic ray transform is injective on simple domains in $\mathbb{R}^2$, see [Bal 2012] or [Sharafutdinov 1994].

Let us finish by sketching the proof of the injectivity result for the geodesic ray transform on simple manifolds of any dimension $n \geq 2$. For details see [Sharafutdinov 1994] and [Dos Santos Ferreira et al. 2009, Section 7] in particular.

Proposition 5.12 (injectivity of the geodesic ray transform). *Let (M, g) be a simple n-manifold, let $f \in C_c^\infty(M^{\mathrm{int}})$, and suppose that $If \equiv 0$. Then $f \equiv 0$.*

Sketch of proof. If (M, g) and f are as in the statement, then as in the $\mathbb{R}^2$ case we define a function $u \in C^\infty(SM)$ by

$$u(x, \xi) := \int_0^{\tau(x,\xi)} f(\gamma(t, x, \xi))\,dt, \quad (x, \xi) \in SM.$$

The geodesic vector field acting on smooth functions $v \in C^\infty(SM)$ is given by

$$\mathcal{H}v(x, \xi) := \frac{\partial}{\partial t} v(\gamma(t, x, \xi), \dot\gamma(t, x, \xi))\Big|_{t=0}.$$

Since $If \equiv 0$, we obtain as above that u solves the transport equation

$$\begin{cases} \mathcal{H}u = -f & \text{in } SM, \\ u = 0 & \text{on } \partial(SM). \end{cases} \tag{5-8}$$

At this point we would like to differentiate the equation in the angular variable ξ to remove the f term. To do this, we need to introduce the horizontal and vertical gradients ∇ and ∂, which are invariantly defined differential operators on so-called *semibasic tensors* on SM. For smooth functions $v \in C^\infty(SM)$, they are defined by

$$\nabla_j u(x, \xi) := \frac{\partial}{\partial x_j}(u(x, \xi/|\xi|)) - \Gamma_{jk}^l \xi^k \partial_l u(x, \xi),$$

$$\partial_j u(x, \xi) := \frac{\partial}{\partial \xi_j}(u(x, \xi/|\xi|)).$$

The geodesic vector field can be defined on semibasic tensor fields via $\mathcal{H} := \xi^j \nabla_j$. We also define $|\partial v|^2 := g^{jk} \partial_j v \partial_k v$, etc. One then has the following general Pestov identity whose proof is again a direct computation (which uses basic properties of ∇ and ∂). A major difference to the Euclidean case is the appearance of a curvature term.

Proposition 5.13 (Pestov identity). *For (M, g) an n-manifold and $u \in C^\infty(SM)$, one has the identity*

$$|\partial \mathcal{H} u|^2 = |\mathcal{H} \partial u|^2 + \mathrm{div}_h(V) + \mathrm{div}_v(W) - R(\partial u, \xi, \xi, \partial u)$$

where the horizontal and vertical divergence are defined by

$$\mathrm{div}_h(X) := \nabla_j X^j, \quad \mathrm{div}_v(X) := \partial_j X^j,$$

and V and W are given by

$$V^j := \langle \partial u, \nabla u \rangle \xi^j - (\mathcal{H} u) \partial^j u, \quad W^j := (\mathcal{H} u) \nabla^j u.$$

Also, R is the Riemann curvature tensor.

We now take the vertical gradient in (5-8) and obtain

$$\begin{cases} \partial \mathcal{H} u = 0 & \text{in } SM, \\ u = 0 & \text{on } \partial(SM). \end{cases} \quad (5\text{-}9)$$

Similarly as in the $\mathbb{R}^2$ case, we pair this equation against $\partial \mathcal{H} u$, integrate over SM and use the Pestov identity to obtain that

$$\int_{SM} \left(|\mathcal{H} \partial u|^2 + \mathrm{div}_h(V) + \mathrm{div}_v(W) - R(\partial u, \xi, \xi, \partial u) \right) d(SM) = 0.$$

Integrating by parts, the $\mathrm{div}_h(V)$ term vanishes and the $\mathrm{div}_v(W)$ term gives a positive contribution as in the Euclidean case. One eventually gets that

$$\int_{SM} \left(|\mathcal{H} \partial u|^2 - R(\partial u, \xi, \xi, \partial u) \right) d(SM) + (n-1) \int_{SM} |\mathcal{H} u|^2 d(SM) = 0.$$

The first term is related to the *index form* for a geodesic $\gamma = \gamma(\,\cdot\,, x, \xi)$ in (M, g), which is given by

$$I(X, X) := \int_0^{\tau(x,\xi)} (|D_{\dot\gamma} X|^2 - R(X, \dot\gamma, \dot\gamma, X))\, dt$$

for vector fields X on γ with $X(0) = X(\tau(x, \xi)) = 0$. If (M, g) is simple, or more generally if no geodesic in (M, g) has conjugate points, then the index form is known to be always nonnegative. This implies that the first term above is nonnegative, showing that $\mathcal{H}u = 0$ and $u = 0$ as required. From (5-8) one obtains that $f \equiv 0$. □

Acknowledgements

I would like to thank the audience in the courses given in Helsinki and Madrid for useful questions and comments which have improved the presentation considerably. I would also like to thank MSRI for a wonderful semester program in inverse problems.

References

[Bal 2012] G. Bal, "Introduction to inverse problems", lecture notes, 2012, available at http://www.columbia.edu/~gb2030/PAPERS/IntroductionInverseProblems.pdf.

[Dairbekov et al. 2007] N. S. Dairbekov, G. P. Paternain, P. Stefanov, and G. Uhlmann, "The boundary rigidity problem in the presence of a magnetic field", *Adv. Math.* **216**:2 (2007), 535–609. MR 2008m:37107 Zbl 1131.53047

[Dos Santos Ferreira et al. 2009] D. Dos Santos Ferreira, C. E. Kenig, M. Salo, and G. Uhlmann, "Limiting Carleman weights and anisotropic inverse problems", *Invent. Math.* **178**:1 (2009), 119–171. MR 2010h:58033 Zbl 1181.35327

[Evans 2010] L. C. Evans, *Partial differential equations*, 2nd ed., Graduate Studies in Mathematics **19**, American Mathematical Society, Providence, RI, 2010. MR 2597943 (2011c:35002)

[Ferreira 2011] D. D. S. Ferreira, "Microlocal analysis and inverse problems", lecture notes, 2011, available at http://www.uam.es/gruposinv/inversos/WEBpage/dossantos.html.

[Frigyik et al. 2008] B. Frigyik, P. Stefanov, and G. Uhlmann, "The X-ray transform for a generic family of curves and weights", *J. Geom. Anal.* **18**:1 (2008), 89–108. MR 2008j:53128 Zbl 1148.53055

[Helgason 1999] S. Helgason, *The Radon transform*, 2nd ed., Progress in Mathematics **5**, Birkhäuser, Boston, MA, 1999. MR 2000m:44003 Zbl 0932.43011

[Iwaniec and Martin 2001] T. Iwaniec and G. Martin, *Geometric function theory and non-linear analysis*, New York, 2001. MR 2003c:30001

[Kenig et al. 2011] C. E. Kenig, M. Salo, and G. Uhlmann, "Inverse problems for the anisotropic Maxwell equations", *Duke Math. J.* **157**:2 (2011), 369–419. MR 2012d:35408 Zbl 1226.35086

[Lee 1997] J. M. Lee, *Riemannian manifolds: An introduction to curvature*, Graduate Texts in Mathematics **176**, Springer, New York, 1997. MR 98d:53001 Zbl 0905.53001

[Lee 2003] J. M. Lee, *Introduction to smooth manifolds*, Graduate Texts in Mathematics **218**, Springer, New York, 2003. MR 2003k:58001 Zbl 06034615

[Lee and Uhlmann 1989] J. M. Lee and G. Uhlmann, "Determining anisotropic real-analytic conductivities by boundary measurements", *Comm. Pure Appl. Math.* **42**:8 (1989), 1097–1112. MR 91a:35166 Zbl 0702.35036

[Liimatainen and Salo 2012] T. Liimatainen and M. Salo, "Nowhere conformally homogeneous manifolds and limiting Carleman weights", *Inverse Probl. Imaging* **6** (2012), 523–530.

[Lionheart 1997] W. R. B. Lionheart, "Conformal uniqueness results in anisotropic electrical impedance imaging", *Inverse Problems* **13**:1 (1997), 125–134. MR 98c:78025 Zbl 0868.35140

[Rudin 1986] W. Rudin, *Real and complex analysis*, 3rd ed., McGraw-Hill, New York, 1986. MR 0210528 (35 #1420)

[Salo 2008] M. Salo, "Calderón problem", lecture notes, 2008, http://www.rni.helsinki.fi/~msa/teaching/calderon/calderon_lectures.pdf.

[Sharafutdinov 1994] V. A. Sharafutdinov, *Integral geometry of tensor fields*, VSP, Utrecht, 1994. MR 97h:53077 Zbl 0883.53004

[Sylvester and Uhlmann 1987] J. Sylvester and G. Uhlmann, "A global uniqueness theorem for an inverse boundary value problem", *Ann. of Math.* (2) **125**:1 (1987), 153–169. MR 88b:35205 Zbl 0625.35078

[Taylor 1996] M. E. Taylor, *Partial differential equations, I: Basic theory*, Applied Mathematical Sciences **115**, Springer, New York, 1996. MR 98b:35002b

[Zworski 2012] M. Zworski, *Semiclassical analysis*, Graduate Studies in Mathematics **138**, American Mathematical Society, Providence, RI, 2012.

mikko.j.salo@jyu.fi

Department of Mathematics and Statistics, University of Helsinki, P.O. Box 68 (Gustaf Hällströmin katu 2b), FI-00014 University of Helsinki, Finland

Enclosure methods for Helmholtz-type equations

JENN-NAN WANG AND TING ZHOU

The inverse problem under consideration is to reconstruct the shape information of obstacles or inclusions embedded in the (inhomogeneous) background medium from boundary measurements of propagating waves. This article is a survey of enclosure-type methods implementing exponential complex geometrical optics waves as boundary illumination. The equations for acoustic waves, electromagnetic waves and elastic waves are considered for a medium with impenetrable obstacles and penetrable inclusions (characterized by a jump discontinuity in the parameters). We also outlined some open problems along this direction of research.

1. Introduction

This paper serves as a survey of enclosure-type methods used to determine the obstacles or inclusions embedded in the background medium from the near-field measurements of propagating waves. A type of complex geometric optics waves that exhibits exponential decay with distance from some critical level surfaces (hyperplanes, spheres or other types of level sets of phase functions) are sent to probe the medium. One can easily manipulate the speed of decay such that the waves can only detect the material feature that is close enough to the level surfaces. As a result of sending such waves with level surfaces moving along each direction, one should be able to pick out those that enclose the inclusion.

The problem that Calderón proposed [1980] was whether one can determine the electrical conductivity by making voltage and current measurements at the boundary of the medium. Such electrical methods are also known as electrical impedance tomography (EIT) and have broad applications in medical imaging, geophysics and so on. A breakthrough in solving the problem was due to Sylvester and Uhlmann [1987], who constructed complex geometric optics (CGO) solutions to the conductivity equation and proved the unique determination of C^∞ isotropic conductivity from the boundary measurements in three- and higher-dimensional spaces. The result has been extended to Lipschitz conductivities [Haberman

---

The first author was supported in part by the National Science Council of Taiwan.

and Tataru 2011] in three dimensions and L^∞ conductivities in two dimensions [Astala and Päivärinta 2006].

The inverse problem in this paper concerns reconstructing an obstacle or a jump-type inclusion (in three dimensions) embedded in a known background medium, which is not included in the previous results when considering electrostatics. Several methods are proposed to solve the problem based on utilizing, generally speaking, two special types of solutions. The Green's type solutions were considered first in [Isakov 1990], and several sampling methods [Cakoni and Colton 2006; Kirsch and Grinberg 2008; Arens 2004; Arens and Lechleiter 2009] and probing methods [Ikehata 1998; Potthast 2001] were developed. On the other hand, with the CGO solutions at disposal, the enclosure method was introduced by Ikehata [1999a; 2000] with the idea as described in the first paragraph. Another method worth mentioning uses the oscillating-decaying type of solutions and was proved valid for elasticity systems [Nakamura et al. 2005]. It is the enclosure type of methods that is of the presenting paper's interest.

Here we aim to discuss the enclosure method for Helmholtz-type equations. For the enclosure method in the static equations, we refer to [Ikehata 1999a; 2000; Ide et al. 2007; Uhlmann and Wang 2008; Takuwa et al. 2008] for the conductivity equation, to [Uhlmann and Wang 2007; Uhlmann et al. 2009] for the isotropic elasticity. The major difference between the static equations and Helmholtz-type equations is the loss of positivity in the latter equations. It turns out we have to analyze the effect of the reflected solution due to the existence of lower order term in Helmholtz-type equations. For the acoustic equation outside of a cavity having a C^2 boundary (representing and impenetrable obstacle), one can overcome the difficulty by the Sobolev embedding theorem, see [Nakamura and Yoshida 2007] (also see [Ikehata 1999b] for a similar idea). Such a result can be generalized for Maxwell's equations to determine impenetrable electromagnetic obstacles [Zhou 2010]. However, in the inclusion case, i.e., penetrable obstacles, the coefficient is merely *piecewise* smooth. The Sobolev embedding theorem does not work because the solution is not smooth enough. To tackle the problem, a Hölder type estimate for the second order elliptic equation with coefficients having jump discontinuity based on the result of Li and Vogelius [2000] was developed by Nagayasu, Uhlmann, and the first author in [Nagayasu et al. 2011]. Later, this result was improved by Sini and Yoshida [2010] using L^p estimate for the second order elliptic equation in divergence form developed by Meyers [1963]. Recently, Kuan [2012] extended Sini and Yoshida's method to the elastic wave equations.

The paper is organized as follows. In Section 2, we discuss the enclosure method for the acoustic and electromagnetic equations with impenetrable obstacles. In Section 3, we to survey results in the inclusion case (penetrable obstacle) for the acoustic and elastic waves. Some open problems are listed in Section 4.

2. Enclosing obstacles using acoustic and electromagnetic waves

In this section, we give more precise descriptions of the enclosure methods to identify impenetrable obstacles of acoustic or electromagnetic equations. In particular, we are interested in the results in [Ikehata 1999a] and [Nakamura and Yoshida 2007] for both convex and nonconvex sound hard obstacles using complex geometrical optics (CGO) solutions for Helmholtz equations and the result in [Zhou 2010] for perfect magnetic conducting (PMC) obstacles using CGO solutions for Maxwell's equations.

2A. *Nonconvex sound hard obstacles.* In [Ikehata 1999a] and [Nakamura and Yoshida 2007], the authors consider the inverse scattering problem of identifying a sound hard obstacle $D \subset \mathbb{R}^n$, $n \geq 2$ in a homogeneous medium from the far field pattern. It can be reformulated as an equivalent inverse boundary value problem with near-field measurements described as follows. Given a bounded domain $\Omega \subset \mathbb{R}^n$ with smooth boundary and such that $\overline{D} \subset \Omega$ and $\Omega \setminus \overline{D}$ is connected, the underlying boundary value problem for acoustic wave propagation in the known homogeneous medium in $\Omega \setminus \overline{D}$ with no source is given by

$$\begin{cases} (\Delta + k^2)u = 0 & \text{in } \Omega \setminus \overline{D}, \\ u\big|_{\partial \Omega} = f, \\ \partial_\nu u\big|_{\partial D} = 0 \end{cases} \tag{2-1}$$

where $k > 0$ is the wave number and ν denotes the unit outer normal of ∂D. At this point, we assume that ∂D is C^2. Suppose k is not a Dirichlet eigenvalue of Laplacian. Given each prescribed boundary sound pressure $f \in H^{1/2}(\partial \Omega)$, there exists a unique solution $u(x) \in H^1(\Omega \setminus \overline{D})$ to (2-1). The inverse boundary value problem consists of reconstructing the obstacle D from the full boundary data that can be encoded in the Dirichlet-to-Neumann (DN) map on $\partial \Omega$:

$$\begin{aligned} \Lambda_D : H^{1/2}(\partial \Omega) &\to H^{-1/2}(\partial \Omega), \\ f &\mapsto \partial_\nu u\big|_{\partial \Omega}. \end{aligned} \tag{2-2}$$

In particular, the enclosure method utilizes the measurements (DN map) for those f taking the traces of CGO solutions to $(\Delta + k^2)u = 0$ in the background domain Ω

$$u_0 = e^{\tau(\varphi(x)-t)+i\psi(\tau;x)}(a(x) + r(x;\tau)), \tag{2-3}$$

where $r(x; \tau)$ and its first derivatives are uniformly bounded in τ. As $\tau \to \infty$, u_0 evolves vertical slope at the level set $\{x \mid \varphi(x) = t\}$ for $t \in \mathbb{R}$. Physically speaking, such evanescent waves couldn't detect the change of the material, namely the presence of D in Ω, happening relatively far from the level set. Hence, there is little gap between the associated energies of domains with and without D. On the other hand, if $\bar{D}$ ever intersects the level set, the energy gap is going to be significant for large τ. This implies that the geometric relation between D and the level set $\{x \mid \varphi(x) = t\}$ can be read from the following indicator function describing the energy gap associated to the input $f = u_0|_{\partial\Omega}$:

$$I(\tau, t) := \int_{\partial\Omega} (\Lambda_D - \Lambda_\varnothing)(u_0|_{\partial\Omega}), \overline{u_0}|_{\partial\Omega}\, dS, \qquad (2\text{-}4)$$

where $\Lambda_\varnothing$ represents the DN map associated to the background domain Ω without D, hence $\Lambda_\varnothing(u_0|_{\partial\Omega}) = \partial_\nu u_0|_{\partial\Omega}$. When the linear phase $\varphi(x) = x \cdot \omega$ is used, where $\omega \in \mathbb{S}^{n-1}$, the CGO solution (2-3) is the exponential function

$$u_0(x) = e^{\tau(x\cdot\omega - t) + i\sqrt{\tau^2 + k^2}\, x \cdot \omega^\perp}$$

where $\omega^\perp \in \mathbb{S}^{n-1}$ satisfies $\omega \cdot \omega^\perp = 0$. The discussion above is verified in the following result by Ikehata to enclose the convex hull of D by reconstructing the support function

$$h_D(\omega) := \sup_{x \in D} x \cdot \omega.$$

Theorem 2.1 [Ikehata 1999a]. *Assume that the set $\{x \in \mathbb{R}^n \mid x \cdot \omega = h_D(\omega)\} \cap \partial D$ consists of one point and the Gaussian curvature of ∂D is not vanishing at that point. Then the support function $h_D(\omega)$ can be reconstructed by the formula*

$$h_D(\omega) = \inf\{t \in \mathbb{R} \mid \lim_{\tau \to \infty} I(\tau, t) = 0\}. \qquad (2\text{-}5)$$

This result shows that a strictly convex obstacle can be identified by an envelope surface of planes. Geometrically, this appears as the planes are enclosing the obstacle from every direction, justifying the name "enclosure method".

It is natural to expect that the method can be generalized to recover some nonconvex part of the shape of D by using CGO solutions with nonlinear phase. Based on a Carleman estimate approach, such solutions were constructed in [Kenig et al. 2007] (or see [Dos Santos Ferreira et al. 2007]) for the Schrödinger operator (or the conductivity operator) in $\mathbb{R}^3$, with φ being one of a few limiting Carleman weights (LCW)

$$\varphi(x) = \ln|x - x_0|, \quad x_0 \in \mathbb{R}^3 \setminus \overline{\Omega},$$

which bears spherical level sets, and therefore were called complex spherical waves (CSW). Then such CSW were used into the enclosure method in [Ide et al. 2007] to identify nonconvex inclusions in a conductive medium. In $\mathbb{R}^2$, there are more candidates for limiting Carleman weights than in $\mathbb{R}^3$: all harmonic functions with nonvanishing gradient are LCW (see [Dos Santos Ferreira et al. 2009] for more descriptions of LCW). Then the similar reconstruction scheme is available in [Uhlmann and Wang 2008] for more generalized two-dimensional systems by using level curves of harmonic polynomials.

Below we present the result of Nakamura and Yoshida that adopts the CSW described in the following proposition to enclose a nonconvex sound hard obstacle.

Proposition 2.2 [Dos Santos Ferreira et al. 2007]. *Choose* $x_0 \in \mathbb{R}^n \setminus \overline{\Omega}$ *and let* $\omega_0 \in \mathbb{S}^{n-1}$ *be a vector such that*

$$\{x \in \mathbb{R}^n \mid x - x_0 = m\omega_0,\ m \in \mathbb{R}\} \cap \partial\Omega = \varnothing.$$

Then there exists a solution to the Helmholtz equation in Ω *of the form*

$$u_0(x; \tau, t, x_0, \omega_0) = e^{\tau(t - \ln|x - x_0|) - i\tau\psi(x)}\bigl(a(x) + r(x; \tau, t, x_0, \omega_0)\bigr), \quad (2\text{-}6)$$

where $\tau > 0$ *and* $t \in \mathbb{R}$ *are parameters,* $a(x)$ *is a smooth function on* $\overline{\Omega}$ *and* $\psi(x)$ *is a function defined by*

$$\psi(x) := d_{\mathbb{S}^{n-1}}\left(\frac{x - x_0}{|x - x_0|}, \omega_0\right),$$

with the metric function $d_{\mathbb{S}^{n-1}}(\cdot, \cdot)$ *on* $\mathbb{S}^{n-1}$. *Moreover, the remainder function r is in* $H^1(\Omega)$ *and satisfies*

$$\|r\|_{H^1(\Omega)} = O(\tau^{-1}) \quad \text{as } \tau \to \infty.$$

The corresponding support function is given by

$$h_D(x_0) = \inf_{x \in D} \ln|x - x_0|, \quad x_0 \in \mathbb{R}^n \setminus \overline{\Omega},$$

and can be reconstructed based on the following result.

Theorem 2.3 [Nakamura and Yoshida 2007]. *Let* $x_0 \in \mathbb{R}^n \setminus \overline{\Omega}$. *Assume that the set* $\{x \in \mathbb{R}^n \mid |x - x_0| = e^{h_D(x_0)}\} \cap \partial D$ *consists of finitely many points and the relative curvatures of* ∂D *at these points are positive. Then there are two characterizations of* $h_D(x_0)$:

$$h_D(x_0) = \sup\{t \in \mathbb{R} \mid \liminf_{\tau \to \infty} |I(\tau, t)| = 0\} \qquad (2\text{-}7)$$

and

$$t - h_D(x_0) = \lim_{\tau \to \infty} \frac{\ln|I(\tau, t)|}{2\tau}, \qquad (2\text{-}8)$$

where $I(\tau, t)$ *is defined by* (2-4) *with* u_0 *given by* (2-6).

Remark. The relative curvature in the theorem refers to the Gaussian curvature after a change of coordinates that stretches the sphere onto flat space. For a more rigorous definition, we refer to [Nakamura and Yoshida 2007].

For completeness, we provide briefly the steps of the proof. The proof of (2-7) involves showing that

$$\lim_{\tau \to \infty} |I(\tau,t)| = 0 \quad \text{when } t < h_D(x_0), \tag{2-9}$$

that is, when the level sphere $S_{t,x_0} := \{x \in \mathbb{R}^n \mid |x-x_0| = e^t\}$ has no intersection with $\bar{D}$, and showing that

$$\liminf_{\tau \to \infty} |I(\tau,t)| > C > 0 \quad \text{when } t \geq h_D(x_0), \tag{2-10}$$

namely, when S_{t,x_0} intersects $\bar{D}$. These two statements can be shown by establishing proper upper and lower bounds of $I(\tau,t)$ from the key equality

$$-I(\tau,t)$$
$$= \int_{\Omega \setminus \bar{D}} |\nabla w|^2 \, dx + \int_D |\nabla u_0|^2 \, dx - k^2 \int_{\Omega \setminus \bar{D}} |w|^2 \, dx - k^2 \int_D |u_0|^2 \, dx, \tag{2-11}$$

where $w := u - u_0$ is the reflected solution and u is the solution to (2-1) with $f = u_0|_{\partial\Omega}$. Since w is a solution to

$$\begin{cases} (\Delta + k^2)w = 0 & \text{in } \Omega \setminus \bar{D}, \\ w|_{\partial\Omega} = 0, \\ \partial_\nu w|_{\partial D} = -\partial_\nu u_0|_{\partial D}, \end{cases} \tag{2-12}$$

and by (2-11), one has the upper bound

$$|I(\tau,t)| \leq C \|u_0\|_{H^1(D)}^2$$

for some constant $C > 0$ (throughout the article we use the same letter C to denote various constants). As a consequence of plugging in the CGO solution (2-6), the first statement (2-9) is obtained since

$$|I(\tau,t)| \leq C\tau^2 \int_D e^{2\tau(t - \ln|x-x_0|)} \, dx \quad (\tau \gg 1).$$

However, difficulties arise in dealing with the second statement, (2-10). Due to the loss of positivity for the associated bilinear form, two negative terms are present in (2-11), which implies that to find a nonvanishing (as $\tau \to \infty$) lower bound for $I(\tau,t)$ is not as easy as the case of the conductivity equations, in which we have $I(\tau,t) \geq C \int_D |\nabla u_0|^2 \, dx$. As a remedy, one needs to show that the two negative terms can be absorbed by the positive terms for τ large. To be more specific, first it is not hard to see

$$I(\tau,t) = e^{2\tau(t - h_D(x_0))} I(\tau, h_D(x_0)). \tag{2-13}$$

This implies that it is sufficient to show (2-10) for $t = h_D(x_0)$, which in turn can be derived from (2-11) and the following two inequalities when $t = h_D(x_0)$:

$$\liminf_{\tau \to \infty} \int_D |\nabla u_0|^2 \, dx > C > 0 \tag{2-14}$$

and

$$\frac{k^2 \int_{\Omega \setminus \bar{D}} |w|^2 \, dx + k^2 \int_D |u_0|^2 \, dx}{\int_D |\nabla u_0|^2 \, dx} < \delta < 1 \quad (\tau \gg 1). \tag{2-15}$$

Equation (2-14) is true since

$$\int_D |\nabla u_0|^2 \, dx \geq C\tau^2 \int_D e^{-2\tau(\ln|x-x_0| - h_D(x_0))} \, dx$$

$$\geq \begin{cases} O(\tau^{1/2}) & n = 2 \\ O(1) & n = 3 \end{cases} \quad (\tau \gg 1), \tag{2-16}$$

given the geometric assumption of the positive relative curvature of ∂D.

As for (2-15), the difficult part is to show that

$$\liminf_{\tau \to \infty} \frac{k^2 \int_{\Omega \setminus \bar{D}} |w|^2 \, dx}{\int_D |\nabla u_0|^2 \, dx} = 0 \tag{2-17}$$

since the property of CGO solutions gives

$$\frac{k^2 \int_D |u_0|^2 \, dx}{\int_D |\nabla u_0|^2 \, dx} = O(\tau^{-2}) \quad (\tau \gg 1).$$

In both [Ikehata 1999a] and [Nakamura and Yoshida 2007], (2-17) is proved by establishing the following estimate.

Lemma 2.4. *Let* $S_{h_D(x_0), x_0} \cap \partial D = \{x_1, \ldots, x_N\}$ *and define for* $\alpha \in (0, 1)$

$$I_{x_j, \alpha} := \int_{\partial D} |\partial_\nu u_0| \, |x - x_j|^\alpha \, dS, \quad j = 1, \ldots, N.$$

Then

$$\|w\|^2_{L^2(\Omega \setminus \bar{D})} \leq C \left(\sum_{j=1}^N I^2_{x_j, \alpha} + \|u_0\|^2_{L^2(D)} \right), \quad \alpha \in (0, 1) \tag{2-18}$$

Remark. The proof of Lemma 2.4 is based on H^2-regularity theory and the Sobolev embedding theorem for an auxiliary boundary value problem

$$\begin{cases} (\Delta + k^2) p = \bar{w} & \text{in } \Omega \setminus \bar{D}, \\ p|_{\partial \Omega} = 0, \\ \partial_\nu p|_{\partial D} = 0. \end{cases}$$

Such estimates of the reflected solution $\|w\|_{L^2(\Omega\setminus\bar{D})}$ for the impenetrable obstacle case and $\|w\|_{L^2(\Omega)}$ for the penetrable inclusion case, which will be reviewed in the next section, are usually crucial for the justification of the enclosure methods. Several improvements of the result and removal of geometric assumptions are basically due to the development of different estimates, which we will see shortly.

In particular, choosing $\alpha = \frac{1}{2}$ for $n = 3$ and $\alpha = \frac{3}{4}$ when $n = 2$, one can show that

$$I^2_{x_j,\alpha} \leq \begin{cases} \sqrt{\varepsilon}\, O(\tau^{1/2}) & \text{if } n = 2, \\ O(\tau^{-1/2}) & \text{if } n = 3, \end{cases}$$

for arbitrary small ε, again by the assumption that the relative curvature is positive. Combined with (2-16), this immediately yields (2-15).

Lastly, the formula (2-8) is directly derived from (2-13) and the fact that

$$|I(\tau, h_D(x_0))| \leq C\tau^2, \quad (\tau \gg 1).$$

Remark. The result can be easily extended to the case with inhomogeneous background medium in $\Omega \setminus \bar{D}$, where the CSW in Proposition 2.2 is available.

2B. *Electromagnetic PMC obstacles.* This section is devoted to reviewing the enclosure method for Maxwell's equations [Zhou 2010] to identify perfect magnetic conducting (PMC) obstacles. The same reconstruction scheme works for identifying perfect electric conducting (PEC) obstacles and more generalized impenetrable obstacles.

In a bounded domain $\Omega \subset \mathbb{R}^3$ with an obstacle D such that $\bar{D} \subset \Omega$ with ∂D being C^2 and $\Omega \setminus \bar{D}$ connected, the electric-magnetic field (E, H) satisfies the Maxwell equations

$$\begin{cases} \nabla \times E = ik\mu H, \quad \nabla \times H = -ik\varepsilon E, & \text{in } \Omega \setminus \bar{D}, \\ \nu \times E|_{\partial\Omega} = f, \\ \nu \times H|_{\partial D} = 0 & \text{(PMC condition)}, \end{cases} \quad (2\text{-}19)$$

where k is the frequency and $\mu(x)$ and $\varepsilon(x)$ describe the isotropic (inhomogeneous) background electromagnetic medium and satisfy the following assumptions: there are positive constants $\varepsilon_m, \varepsilon_M, \mu_m, \mu_M, \varepsilon_c$ and μ_c such that for all $x \in \Omega$

$$\varepsilon_m \leq \varepsilon(x) \leq \varepsilon_M, \quad \mu_m \leq \mu(x) \leq \varepsilon_M, \quad \sigma(x) = 0$$

and $\varepsilon - \varepsilon_c, \mu - \mu_c \in C_0^3(\Omega)$. Given that k is not a resonant frequency, we have a well-defined boundary impedance map

$$\Lambda_D : TH^{1/2}(\partial\Omega) \to TH^{1/2}(\partial\Omega),$$
$$f = \nu \times E|_{\partial\Omega} \mapsto \nu \times H|_{\partial\Omega}.$$

To show that D can be determined by the impedance map Λ_D using the enclosure method, we first notice an analogue of the identity (2-11) for Maxwell's equations:

$$i\omega \int_{\partial\Omega} (\nu \times E_0) \cdot \left(\overline{(\Lambda_D - \Lambda_\varnothing)(\nu \times E_0)} \times \nu\right) dS$$
$$= \int_{\Omega \setminus \bar{D}} \mu |\tilde{H}|^2 - \omega^2 \varepsilon |\tilde{E}|^2 \, dx + \int_D \mu |H_0|^2 - \omega^2 \varepsilon |E_0|^2 \, dx, \quad (2\text{-}20)$$

where $(\tilde{E}, \tilde{H}) := (E - E_0, H - H_0)$ denotes the reflected solutions, (E, H) is the solution to (2-19), (E_0, H_0) is the solution to the Maxwell's equations

$$\nabla \times E_0 = ik\mu H_0, \quad \nabla \times H_0 = -ik\varepsilon E_0 \quad \text{in } \Omega, \quad (2\text{-}21)$$

and $\nu \times E|_{\partial\Omega} = \nu \times E_0|_{\partial\Omega}$.

One would encounter the same difficulty as that for Helmholtz equations due to the loss of positivity of the system. We recall that this was actually overcome by the property that the CGO solution u_0 shares different asymptotic speed (τ^2 slower) from ∇u_0. More specifically, this is because of the H^1 boundedness of the remainder $r(x; \tau)$ with respect to τ in (2-3). The natural question to ask is then whether this key ingredient: such CGO type of solutions, can be constructed for the background Maxwell's system.

The construction of CGO solutions for the Maxwell's equations has been extensively studied in [Ola et al. 1993; Ola and Somersalo 1996; Colton and Päivärinta 1992]. The work in [Zhou 2010] adopts the construction approach in [Ola and Somersalo 1996] by reducing the Maxwell's equations into a matrix Schrödinger equation. Finally, to guarantee that the CGO solution for the reduced matrix Schrödinger operator derives the CGO solution (E_0, H_0) for the Maxwell's equations and at the same time that the electric field E_0 and H_0 have different asymptotic speeds as $\tau \to \infty$, the incoming constant field corresponding to $a(x)$ in (2-3) has to be chosen very carefully. To summarize, one has

Proposition 2.5. *Let $\omega, \omega^\perp \in \mathbb{S}^2$ with $\omega \cdot \omega^\perp = 0$. Set*

$$\zeta = -i\tau\omega + \sqrt{\tau^2 + k^2}\omega^\perp$$

where $k_1 = k(\varepsilon_0\mu_0)^{1/2}$. Choose $a \in \mathbb{R}^3$ such that

$$a \perp \omega, \quad a \perp \omega^\perp \quad \text{and} \quad b = \frac{1}{\sqrt{2}}\overline{(-i\omega + \omega^\perp)}.$$

Then, given

$$\theta := \frac{1}{|\zeta|}(-(\zeta\cdot a)\zeta - k_1\zeta \times b + k_1^2 a), \quad \eta := \frac{1}{|\zeta|}(k_1\zeta \times a - (\zeta\cdot b)\zeta + k_1^2 b),$$

for $t \in \mathbb{R}$ and $\tau > 0$ large enough, there exists a unique complex geometric optics solution $(E_0, H_0) \in H^1(\Omega)^3 \times H^1(\Omega)^3$ of Maxwell's equations (2-21) having the form

$$E_0 = \varepsilon(x)^{-1/2} e^{\tau(x\cdot\omega - t) + i\sqrt{\tau^2 + k^2}\, x\cdot\omega^\perp}(\eta + R(x)),$$

$$H_0 = \mu(x)^{-1/2} e^{\tau(x\cdot\omega - t) + i\sqrt{\tau^2 + k^2}\, x\cdot\omega^\perp}(\theta + Q(x)).$$

Moreover, we have

$$\eta = \mathcal{O}(1), \quad \theta = \mathcal{O}(\tau) \quad \text{for } \tau \gg 1,$$

and $R(x)$ and $Q(x)$ are bounded in $(L^2(\Omega))^3$ for $\tau \gg 1$.

Plugging (E_0, H_0) into the indicator function defined by

$$I(\tau, t) := i\omega \int_{\partial\Omega} (\nu \times E_0) \cdot \left(\overline{(\Lambda_D - \Lambda_\varnothing)(\nu \times E_0)} \times \nu\right) dS,$$

a similar argument as for Helmholtz equations follows using identity (2-20) and we have:

Theorem 2.6 [Zhou 2010]. *There is a subset $\Sigma \subset \mathbb{S}^2$ of measure zero such that, when $\omega \in \mathbb{S}^2 \setminus \Sigma$, the support function*

$$h_D(\omega) := \sup_{x \in D} x \cdot \omega$$

can be recovered by

$$h_D(\omega) = \inf\{t \in \mathbb{R} \mid \lim_{\tau \to \infty} I(\tau, t) = 0\}.$$

Moreover, if D is strictly convex, one can reconstruct D.

On the other hand, the construction of a proper CGO solution with nonlinear weight for the Maxwell's equations has not been successful based on the Carleman estimate argument. An alternative approach to reconstruct nonconvex part of the shape of D would be to introduce some transformation. For example, one can utilize the Kelvin transformation

$$T_{x_0,R} : x \mapsto R^2 \frac{x - x_0}{|x - x_0|^2} + x_0 := y,$$

which is the inversion transformation with respect to the sphere $S(x_0, R)$ for $R > 0$ and $x_0 \in \mathbb{R}^3 \setminus \overline{\Omega}$. $T_{x_0,R}$ maps generalized spheres (spheres and planes) into generalized spheres. Geometrically, fixing a reference circle $S(x_0, R)$, enclosing D with spheres passing through x_0 corresponds to enclosing $\hat{D}_{x_0,R} = T_{x_0,R}(D)$ with planes, where the reconstruction scheme in Theorem 2.6 applies. A rigorous proof consists of showing that the Maxwell's equations are invariant under the transformation and computing the impedance map $\hat{I}(\tau, t)$ associated to the image domain. It is worth mentioning the byproduct of this method is the complex spherical wave

$$\hat{E}(y) = \hat{E}_j dy^j = \left((DT_{x_0,R}^{-1})_j^k(y) E_k(T_{x_0,R}^{-1}(y))\right) dy^j, \quad y = T_{x_0,R}(x),$$

with nonlinear limiting Carleman weight

$$\varphi(x) = \left(R^2 \frac{x - x_0}{|x - x_0|^2} + x_0\right) \cdot \omega, \quad \omega \in \mathbb{S}^2.$$

Therefore, the corresponding support function is given by

$$\hat{h}_D(x_0, R, \omega) = \sup_{x \in D} \left\{ R^2 \left(\frac{x - x_0}{|x - x_0|^2}\right) \cdot \omega + x_0 \cdot \rho \right\}.$$

Theorem 2.7 [Zhou 2010]. *Given $x_0 \in \mathbb{R}^3 \setminus \overline{\Omega}$ and $R > 0$ such that $\overline{\Omega} \subset B(x_0, R)$, there is a zero measure subset Σ of $\mathbb{S}^2$, s.t., when $\omega \in \mathbb{S}^2 \setminus \Sigma$, we have*

$$\hat{h}_D(x_0, R, \omega) = \inf\{t \in \mathbb{R} \mid \lim_{\tau \to \infty} \hat{I}(\tau, t) = 0\}.$$

3. Enclosing inclusions using acoustic and elastic waves

In this section we will consider the enclosure method for the case where the unknown domain is an inclusion by using acoustic and elastic waves. In other words, the obstacle is a penetrable one. In this situation, the reflected solution will satisfy the elliptic equation with discontinuous coefficients. Unlike the case of impenetrable obstacle, the Sobolev embedding theorem is not sufficient to provide us estimates of the reflected solution we need. In the case of acoustic waves, the difficulty was overcome in [Nagayasu et al. 2011] for dimension $n = 2$, using estimates from [Li and Vogelius 2000]. The extension to $n = 3$ was accomplished in [Yoshida 2010]. Sini and Yoshida [2010] then improved the result in [Nagayasu et al. 2011] with the help of Meyers' L^p estimate and the sharp Friedrichs inequality. Kuan [2012] extended Sini and Yoshida's result to elastic waves.

3A. *Acoustic penetrable obstacle.* Here we will review the result in [Nagayasu et al. 2011] for $n = 2$. For $n = 3$, one simply replaces CGO solutions in $n = 2$

by complex spherical waves [Yoshida 2010]. We assume $D \Subset \Omega \subset \mathbb{R}^2$. For technical simplicity, we suppose that both D and Ω have C^2 boundaries. Let $\gamma_D \in C^2(\bar{D})$ satisfy $\gamma_D \geq c_\gamma$ for some positive constant c_γ and $\tilde{\gamma} := 1 + \gamma_D \chi_D$, where χ_D is the characteristic function of D. Let $k > 0$ and consider the steady state acoustic wave equation in Ω with Dirichlet condition

$$\begin{cases} \nabla \cdot (\tilde{\gamma} \nabla v) + k^2 v = 0 & \text{in } \Omega, \\ v = f & \text{on } \partial \Omega. \end{cases} \quad (3\text{-}1)$$

We assume that k^2 is not a Dirichlet eigenvalue of the operator $-\nabla \cdot (\tilde{\gamma} \nabla \cdot)$. Let $\Lambda_D : H^{1/2}(\partial \Omega) \to H^{-1/2}(\partial \Omega)$ be the associated Dirichlet-to-Neumann map. As before, our aim is to reconstruct the shape of D by Λ_D. The key in the enclosure method is the CGO solutions. To construct the CGO solutions to the Helmholtz equation for $n = 2$, we begin with the CGO solutions with polynomial phases to the Laplacian operator, then apply the Vekua transform [1967, page 58].

More precisely, let us define $\eta(x) := c_* \big((x_1 - x_{*,1}) + i(x_2 - x_{*,2}) \big)^N$ as the phase function, where $c_* \in \mathbb{C}$ satisfies $|c_*| = 1$, N is a positive integer, and $x_* = (x_{*,1}, x_{*,2}) \in \mathbb{R}^2 \setminus \bar{\Omega}$. Without loss of generality we may assume that $x_* = 0$ using an appropriate translation. Denote $\eta_R(x) := \operatorname{Re} \eta(x)$ and note that

$$\eta_R(x) = r^N \cos N(\theta - \theta_*) \text{ for } x = r(\cos \theta, \sin \theta) \in \mathbb{R}^2.$$

It is readily seen that $\eta_R(x) > 0$ for all $x \in \Gamma$, where

$$\Gamma := \left\{ r(\cos \theta, \sin \theta) : |\theta - \theta_*| < \frac{\pi}{2N} \right\},$$

i.e., a cone with opening angle π/N.

Given any $h > 0$, $\check{V}_\tau(x) := \exp(\tau \eta(x))$ is a harmonic function. Following [Vekua 1967], we define a map T_k on any harmonic function $\check{V}(x)$ by

$$T_k \check{V}(x) := \check{V}(x) - \int_0^1 \check{V}(tx) \frac{\partial}{\partial t} \big\{ J_0 \big(k|x| \sqrt{1-t} \big) \big\} dt$$

$$= \check{V}(x) - k|x| \int_0^1 \check{V}\big((1-s^2) x \big) J_1(k|x|s) \, ds,$$

where J_m is the Bessel function of the first kind of order m. We now set $V_\tau^\sharp(x) := T_k \check{V}_\tau(x)$. Then $V_\tau^\sharp(x)$ satisfies the Helmholtz equation $\Delta V_\tau^\sharp + k^2 V_\tau^\sharp = 0$ in $\mathbb{R}^2$. One can show that $V_\tau^\sharp$ satisfies the following estimate in Γ:

Lemma 3.1 [Nagayasu et al. 2011]. *We have*

$$V_\tau^\sharp(x) = \exp(\tau \eta(x)) (1 + R_0(x)) \text{ in } \Gamma, \quad (3\text{-}2)$$

where $R_0(x) = R_0(x; \tau)$ satisfies

$$|R_0(x)| \leq \frac{1}{\tau} \frac{k^2|x|^2}{4\eta_R(x)}, \quad \left|\frac{\partial R_0}{\partial x_j}(x)\right| \leq \frac{Nk^2|x|^{N+1}}{4\eta_R(x)} + \frac{1}{\tau} \frac{k^2|x_j|}{2\eta_R(x)} \quad \text{in } \Gamma.$$

Notice that here $V_\tau^\#(x)$ is only defined in $\Gamma \cap \Omega$. We now extend it to the whole domain Ω by using an appropriate cut-off. Let $l_s := \{x \in \Gamma : \eta_R(x) = 1/s\}$ for $s > 0$. For $\varepsilon > 0$ small enough and $t^\# > 0$ large enough, we choose a function $\phi_t \in C^\infty(\mathbb{R}^2)$ satisfying

$$\phi_t(x) = \begin{cases} 1 & \text{for } x \in \overline{\bigcup_{0<s<t+\varepsilon/2} l_s}, \ t \in [0, t^\#], \\ 0 & \text{for } x \in \mathbb{R}^2 \setminus \bigcup_{0<s<t+\varepsilon} l_s, \ t \in [0, t^\#], \end{cases}$$

and

$$|\partial_x^\alpha \phi_t(x)| \leq C_\phi \quad \text{for } |\alpha| \leq 2, \ x \in \Omega, \ t \in [0, t^\#]$$

for some positive constant C_ϕ depending only on Ω, N, $t^\#$ and ε. Next we define the function $V_{t,\tau}$ by

$$V_{t,\tau}(x) := \phi_t(x) \exp\left(-\frac{\tau}{t}\right) V_\tau^\#(x) \text{ for } x \in \overline{\Omega}.$$

Then we know by Lemma 3.1 that the dominant parts of $V_{t,\tau}$ and its derivatives are as follows:

$$V_{t,\tau}(x) = \begin{cases} 0 \text{ for } x \in \Omega \setminus \bigcup_{0<s<t+\varepsilon} l_s, \\ \exp\left(\tau\left(-\frac{1}{t} + \eta(x)\right)\right)(\phi_t(x) + S_0(x)h) \\ \qquad \text{for } x \in \Omega \cap \bigcup_{0<s<t+\varepsilon} l_s, \end{cases} \quad (3\text{-}3)$$

$$\nabla V_{t,\tau}(x) = \begin{cases} 0 \text{ for } x \in \Omega \setminus \bigcup_{0<s<t+\varepsilon} l_s, \\ \tau \exp\left(\tau\left(-\frac{1}{t} + \eta(x)\right)\right)(\phi_t(x)\nabla\eta(x) + \boldsymbol{S}(x)h) \\ \qquad \text{for } x \in \Omega \cap \bigcup_{0<s<t+\varepsilon} l_s \end{cases} \quad (3\text{-}4)$$

for $t \in (0, t^\#]$ and $\tau^{-1} \in (0, 1]$, where $S_0(x) = S_0(x; t, \tau)$ and $\boldsymbol{S}(x) = \boldsymbol{S}(x; t, \tau)$ satisfy

$$|S_0(x)|, |\boldsymbol{S}(x)| \leq C_V \quad \text{for any } x \in \Omega \cap \bigcup_{0<s<t+\varepsilon} l_s, \ t \in (0, t^\#], \ \tau^{-1} \in (0, 1]$$

with a positive constant C_V depending only on Ω, N, $t^\#$, ε and k. It should be remarked that the function $V_{t,\tau}$ does not satisfy the Helmholtz equation in Ω. Nonetheless, if we let $v_{0,t,\tau}$ be the solution to the Helmholtz equation in Ω

with boundary value $f_{t,\tau} := V_{t,\tau}|_{\partial\Omega}$, then the error between $V_{t,\tau}$ and $v_{0,t,\tau}$ is exponentially small

Lemma 3.2. *There exist constants $C_0, C_0' > 0$ and $a > 0$ such that*

$$\|v_{0,t,\tau} - V_{t,\tau}\|_{H^2(\Omega)} \leq \tau C_0' e^{-\tau a_t} \leq C_0 e^{-\tau a}$$

for any $\tau^{-1} \in (0, 1]$, where the constants C_0 and C_0' depend only on $\Omega, k, N, t^\sharp$ and ε; the constant a depends only on $t^\sharp$ and ε; and we set $a_t := 1/t - 1/(t+\varepsilon/2)$.

This lemma can be proved in the same way as Lemma 4.1 in [Uhlmann and Wang 2008].

Now we consider the energy gap

$$I(\tau, t) = \int_{\partial\Omega} (\Lambda_D - \Lambda_\emptyset) f_{t,\tau} \, \overline{f}_{t,\tau} \, dS.$$

It can be shown that

$$I(\tau, t) \leq k^2 \int_\Omega |w_{t,\tau}|^2 \, dx + \int_D \gamma_D |\nabla v_{0,t,\tau}|^2 \, dx, \tag{3-5}$$

$$I(\tau, t) \geq \int_D \frac{\gamma_D}{1+\gamma_D} |\nabla v_{0,t,\tau}|^2 \, dx - k^2 \int_\Omega |w_{t,\tau}|^2 \, dx, \tag{3-6}$$

where $v_{0,t,\tau}$ satisfies the Helmholtz equation in Ω with Dirichlet condition $v_{0,t,\tau}|_{\partial\Omega} = f_{t,\tau}$ and $w_{t,\tau} = v_{t,\tau} - v_{0,t,\tau}$ is the reflected solution, i.e.,

$$\begin{cases} \nabla \cdot (\tilde{\gamma} \nabla w_{t,\tau}) + k^2 w_{t,\tau} = -\nabla \cdot ((\tilde{\gamma} - 1)\nabla v_{0,t,\tau}) & \text{in } \Omega, \\ w_{t,\tau} = 0 & \text{on } \partial\Omega \end{cases} \tag{3-7}$$

(see [Nagayasu et al. 2011, Lemma 4.1]). It is easy to see that

$$\int_\Omega |w_{t,\tau}|^2 \, dx \leq C \int_D |\nabla v_{0,t,\tau}|^2 \, dx.$$

In other words, in view of (3-5), the upper bound of $I(t, \tau)$ solely depends on $\int_D |\nabla v_{0,t,\tau}|^2 \, dx$.

To estimate the lower bound of $I(\tau, t)$, we proceed as above and introduce

$$I_{x_0,\alpha} := \int_{\partial D} |\partial_\nu v_{0,t,\tau}(x)| \, |x - x_0|^\alpha \, dS$$

for any $x_0 \in \Omega$ and $0 < \alpha < 1$. The following estimate is crucial in determining the behavior of $I(\tau, t)$ when the level curve of η_R intersects D.

Lemma 3.3 [Nagayasu et al. 2011, Lemma 3.7]. *For any $x_0 \in \Omega$, $0 < \alpha < 1$ and $2 < q \leq 4$, we have*

$$\int_\Omega |w_{t,\tau}|^2 \, dx \leq C_{q,\alpha} \big(I_{x_0,\alpha}^2 + I_{x_0,\alpha} \|\nabla v_{0,t,\tau}\|_{L^q(D)} + \|v_{0,t,\tau}\|_{L^2(D)}^2\big). \tag{3-8}$$

It should be noted that $w_{t,\tau}$ satisfies an elliptic equation with coefficients having jump interfaces. To get the desired estimate (3-8), we make use of the Hölder estimate of Li and Vogelius [2000] for the this type of equations.

The enclosure method is now based on the following theorem regarding the behavior of $I(\tau, t)$.

Theorem 3.4 [Nagayasu et al. 2011, Theorem 4.1]. *Assume $D \cap \Gamma \neq \varnothing$. Suppose that $\{x \in \Gamma : \eta_R(x) = \Theta_D\} \cap \partial D$ consists only of one point x_0 and the relative curvature (see [Nagayasu et al. 2011] for the definition) to $\eta_R(x) = \Theta_D$ of ∂D at x_0 is not zero. Then there exist positive constants C_1, c_1 and τ_1 such that for any $0 < t \leq t^{\sharp}$ and $\tau \geq \tau_1$ the following holds:*

(I) *if $1/t > \Theta_D$ then*

$$|I(\tau,t)| \leq \begin{cases} C_1 \tau^2 \exp\left(2\tau\left(-\dfrac{1}{t} + \dfrac{1}{t+\varepsilon/2}\right)\right) & \text{if } \Theta_D \leq \dfrac{1}{t+\varepsilon/2}, \\ C_1 \tau^2 \exp\left(2\tau\left(-\dfrac{1}{t} + \Theta_D\right)\right) & \text{if } \dfrac{1}{t+\varepsilon/2} < \Theta_D < \dfrac{1}{t}. \end{cases}$$

(II) *if $1/t \leq \Theta_D$ then*

$$I(\tau,t) \geq c_1 \exp\left(2\tau\left(-\dfrac{1}{t} + \Theta_D\right)\right) \tau^{1/2}.$$

The proof of this theorem relies on estimates we obtained above. Moreover, even though we impose some restriction on the curvature of ∂D at x_0, one can show that the curvature assumption is always satisfied as long as N is large enough for C^2 boundary ∂D.

3B. *An improvement by Sini and Yoshida.* In the enclosure method discussed above (for impenetrable or penetrable obstacles), two conditions are assumed, that is, the level curve of real part of the phase function in CGO solutions touches ∂D at one point and the nonvanishing of the relative curvature at the touching point. These two assumptions are removed in [Sini and Yoshida 2010]. Roughly speaking, the authors use following estimates for the reflected solution w

$$\|w\|_{L^2(\Omega)} \leq C_p \|v\|_{W^{1,p}(D)} \quad \text{with} \quad p < 2 \tag{3-9}$$

for a penetrable obstacle, and

$$\|w\|_{L^2(\Omega \setminus \bar{D})} \leq C_t \|v\|_{H^{-t+\frac{3}{2}}(D)} \quad \text{with} \quad t < 1 \tag{3-10}$$

for an impenetrable obstacle. Here v satisfies the Helmholtz equation in Ω.

The derivation of (3-9) is based on Meyers' theorem [1963] and the sharp Friedrichs inequality, while, the proof of (3-10) relies on layer potential techniques on Sobolev spaces and integral estimates of the p-powers of Green's function. We refer to [Sini and Yoshida 2010] for details. Here we would like to see how (3-9) and (3-10) lead to the characteristic behaviors of the energy gap in the enclosure method. To illustrate the ideas, we follow Sini and Yoshida in considering only CGO solutions with linear phases, i.e.,

$$v(x;\tau,t) = e^{\tau(x\cdot\omega - t) + i\sqrt{\tau^2 + k^2}\, x\cdot\omega^{\perp}}.$$

It is clear that v is a solution of the Helmholtz equation. Denote the energy gap by

$$I(\tau, t) = \int_{\partial\Omega} (\Lambda_D - \Lambda_\varnothing) v \bar{v}\, dS.$$

The following behavior of I can be obtained.

Theorem 3.5 [Sini and Yoshida 2010, Theorem 2.4]. *Let $D \Subset \Omega$ with Lipschitz boundary ∂D. For both the penetrable and impenetrable cases, we have*:

(i) $\lim_{\tau\to\infty} I(\tau, t) = 0 \quad \text{if} \quad t > h_D(\omega)$,

$\liminf_{\tau\to\infty} |I(\tau, h_D(\omega))| = \infty \ (n = 2), \quad \liminf_{\tau\to\infty} |I(\tau, h_D(\omega))| > 0 \ (n = 3),$

$\lim_{\tau\to\infty} |I(\tau, t)| = \infty \quad \text{if} \quad t < h_D(\omega).$

(ii) $$h_D(\omega) - t = \lim_{\tau\to\infty} \frac{\ln |I(\tau, t)|}{2\tau}.$$

To prove Theorem 3.5, it is enough to estimate the lower bound of $I(\tau, t)$ at $t = h_D(\omega)$ for $n = 3$. Let $y \in \partial D \cap \{x\cdot\omega = h_D(\omega)\} := K$. Since K is compact, there exist $y_1, \ldots, y_N \in K$ such that

$$K \subset D_\delta \quad \text{for } \delta > 0 \text{ sufficiently small,}$$

where

$$D_\delta = \bigcup_{j=1}^{N} (D \cap B(y_j, \delta)).$$

It is obvious that $\int_{D\setminus D_\delta} |\nabla^m v|^p\, dx$ is exponentially small in τ for $m = 0, 1$. Therefore, to obtain the behaviors of $\int_D |\nabla^m v|^p\, dx$ in τ, it suffices to study the integrals over D_δ. Using the change of coordinates, it is tedious but not difficult to show that

$$\|v\|_{L^2(D)}^2 \geq C\tau^{-2}, \quad \frac{\|\nabla v\|_{L^2(D)}^2}{\|v\|_{L^2(D)}^2} \geq C\tau^2 \quad (3\text{-}11)$$

and

$$\frac{\|v\|_{L^p(D)}^2}{\|v\|_{L^2(D)}^2} \leq C\tau^{1-2/p}, \quad \frac{\|\nabla v\|_{L^p(D)}^2}{\|v\|_{L^2(D)}^2} \leq C\tau^{3-2/p} \quad (3\text{-}12)$$

with max$\{2-\varepsilon, 6/5\} < p \leq 2$ (see [Sini and Yoshida 2010, pages 6–9]). Using (3-9) we get from (3-12) that

$$\frac{\|w\|^2_{L^p(D)}}{\|v\|^2_{L^2(D)}} \leq C\tau^{3-2/p}. \tag{3-13}$$

Recall that

$$I(\tau, t) \geq \int_D \frac{\gamma_D}{1+\gamma_D} |\nabla v|^2 \, dx - k^2 \int_\Omega |w|^2 \, dx.$$

Thus, combining (3-11) and (3-13) implies that

$$I(\tau, h_D(\omega)) \geq C\tau^2 \|v\|^2_{L^2(D)} \geq C' > 0$$

As for an impenetrable obstacle (sound hard), we recall that

$$-I(\tau, t) \geq \int_D |\nabla v|^2 dx - k^2 \int_{\Omega \setminus \bar{D}} |w|^2 \tag{3-14}$$

(see for example [Ikehata 1999a, Lemma 4.1]). Let $s = \frac{3}{2} - t$, then $\frac{1}{2} < s \leq \frac{3}{2}$ if $0 \leq t < 1$. From (3-10) and (3-14), we have that

$$-I(\tau, t) \geq \int_D |\nabla v|^2 dx - C\|v\|^2_{H^s(D)}.$$

Using the interpolation and Young's inequalities, one can choose appropriate parameters such that

$$-I(\tau, t) \geq C \int_D |\nabla v|^2 dx - C' \int_D |v|^2 dx$$

and thus

$$-I(\tau, h_D(\omega)) > 0$$

follows from (3-11).

3C. *Elastic penetrable obstacles.* Recently, Kuan [2012] extended the enclosure method to the reconstruction of a penetrable obstacle using elastic waves. Her result is in 2 dimensions, but it can be generalized to 3 dimensions without serious difficulties. Consider the elastic waves in $\Omega \subset \mathbb{R}^2$ with smooth boundary $\partial \Omega$

$$\nabla \cdot (\sigma(u)) + k^2 u = 0 \quad \text{in } \Omega, \tag{3-15}$$

where u is the displacement vector and

$$\sigma(u) = \lambda(\nabla \cdot u) I_2 + 2\mu\epsilon(u)$$

is the stress tensor. Here $\epsilon(u) = \frac{1}{2}(\nabla u + (\nabla u)^T)$ denotes the infinitesimal strain tensor. Assume that

$$\lambda = \lambda_0 + \lambda_D \chi_D \quad \text{and} \quad \mu = \mu_0 + \mu_D \chi_D,$$

where D is an open subset of Ω with $\bar{D} \subset \Omega$ and λ_D, μ_D belong to $L^\infty(D)$. Assume that

$$\lambda_0 + \mu_0 > 0, \ \mu_0 > 0, \ \lambda + \mu > 0, \ \mu > 0 \quad \text{in } \Omega.$$

We would like to discuss the reconstruct of the shape of D from boundary measurements in the spirit of enclosure method.

Assume that $-k^2$ is not a Dirichlet eigenvalue of the Lamé operator $\nabla \cdot (\sigma(\cdot))$. Define the Dirichlet-to-Neumann (displacement-to-traction) map

$$\Lambda_D : u|_{\partial\Omega} \to \sigma(u)\nu|_{\partial\Omega}.$$

Let v satisfy the Lamé equation with Lamé coefficients λ_0, μ_0, i.e.,

$$\nabla \cdot (\sigma(v)) + k^2 v = 0 \quad \text{in } \Omega, \tag{3-16}$$

with

$$\sigma(v) = \lambda_0(\nabla \cdot v)I_2 + 2\mu_0 \epsilon(v).$$

Likewise, we assume that $-k^2$ is not a Dirichlet eigenvalue of the free Lamé operator. We then define the corresponding Dirichlet-to-Neumann map

$$\Lambda_\varnothing : v|_{\partial\Omega} \to \sigma(v)\nu|_{\partial\Omega}.$$

Similar as above, in the enclosure method, we need to construct the CGO solutions for the Lamé equation (3-16). For simplicity, we assume that both λ_0 and μ_0 are constants. To construct the CGO solutions in this case, we take advantage of the Helmholtz decomposition and consider two Helmholtz equations

$$\begin{cases} \Delta \varphi + k_1^2 \varphi = 0, \\ \Delta \psi + k_2^2 \psi = 0, \end{cases} \tag{3-17}$$

where

$$k_1 = \left(\frac{k^2}{\lambda_0 + 2\mu_0}\right)^{1/2} \quad \text{and} \quad k_2 = \left(\frac{k^2}{\mu_0}\right)^{1/2}.$$

Then $v = \nabla \varphi + \nabla^\perp \psi$ solves (3-16). Here $\nabla^\perp \psi := (-\partial_2 \psi, \partial_1 \psi)^T$. For (3-17), we can construct the CGO solutions having linear or polynomial phases, which will give us the CGO solutions v for (3-16).

We will not repeat the construction of CGO solutions here. We simply denote $v(\tau, t)$ the CGO solution. Similarly, we define the energy gap

$$I(\tau, t) = \int_{\partial\Omega} (\Lambda_D - \Lambda_\varnothing) f_{\tau,t} \cdot \bar{f}_{\tau,t} dS,$$

where $f_{\tau,t} = v(\tau, t)|_{\partial\Omega}$. Let Γ be the domain where the real part of the phase function of v, denoted by $\rho(x)$, is positive. Let

$$h_D = \begin{cases} \sup_{x \in D \cap \Gamma} \rho(x) & \text{if } D \cap \Gamma \neq \varnothing, \\ 0 & \text{if } D \cap \Gamma = \varnothing. \end{cases}$$

Assume appropriate jump conditions on λ_D and μ_D. Then the following behaviors of $I(\tau, t)$ are obtained in [Kuan 2012].

Theorem 3.6. (i) $\lim_{\tau \to \infty} I(\tau, t) = 0$ *if* $t > h_D$.
(ii) *If* $t = h_D$ *and* $\partial D \in C^{0,\alpha}$, $1/3 < \alpha \leq 1$, *then* $\liminf_{\tau \to \infty} |I(\tau, h_D(\omega))| = \infty$.
(iii) *If* $t < h_D$ *and* $\partial D \in C^0$, *then* $\lim_{\tau \to \infty} |I(\tau, t)| = \infty$.

The proof of Theorem 3.6 is based on the following inequalities for the energy gap:

$$I(\tau, t) \leq \int_D (\lambda_D + \mu_D) |\nabla \cdot v|^2 dx$$
$$+ 2 \int_D \mu_D \left| \epsilon(v) - \tfrac{1}{2}(\nabla \cdot v) I_2 \right|^2 dx + k^2 \|w\|^2_{L^2(\Omega)},$$

$$I(\tau, t) \geq \int_D \frac{(\lambda_0 + \mu_0)(\lambda_D + \mu_D)}{\lambda + \mu} |\nabla \cdot v|^2 dx$$
$$+ 2 \int_D \frac{\mu_0 \mu_D}{\mu} \left| \epsilon(v) - \tfrac{1}{2}(\nabla \cdot v) I_2 \right|^2 dx - k^2 \|w\|^2_{L^2(\Omega)},$$

where w is the reflected solution. The rest of the proof is similar to that in [Sini and Yoshida 2010], which relies on the following L^p estimate:

Lemma 3.7 [Kuan 2012, Lemma 4.2]. *There exist constants* $C > 0$ *and* p_0 *in* $[1, 2)$ *such that, for* $p_0 < p \leq 2$,

$$\|w\|_{L^2(\Omega)} \leq C \|\nabla v\|_{L^p(D)}.$$

Lemma 3.7 can be proved adapting arguments from [Meyers 1963].

4. Open problems

The enclosure method in the electromagnetic waves we discussed in Section 2B is for the case of an impenetrable obstacle. Therefore, it is a legitimate project to study the penetrable case for the electromagnetic waves. However, the tools used in the acoustic waves, i.e., Li–Vogelius type estimates or Meyers type L^p estimates, are not available in the electromagnetic waves. The derivation of these estimates itself is an interesting problem.[1] Another interesting problem is to extend the enclosure method to the plate or shell equations. The distinct feature of these equations is the appearance of the biharmonic operator Δ^2.

Finally, it is desirable to design stable and efficient algorithms for the enclosure method. Attempts of numerical implementations have been made for the conductivity equation [Brühl and Hanke 2000; Ide et al. 2007; Ikehata and Siltanen 2000; Uhlmann and Wang 2008] and for the 2D static elastic equation [Uhlmann et al. 2009]. There are two obvious difficulties. On one hand, the boundary data involves large parameter which gives rise to highly oscillatory functions. On the other hand, a reliable way of numerically determining whether $I(\tau, t)$ decays or blows up as $\tau \to \infty$ is yet to be found.

References

[Arens 2004] T. Arens, "Why linear sampling works", *Inverse Problems* **20**:1 (2004), 163–173. MR 2005b:35036 Zbl 1055.35131

[Arens and Lechleiter 2009] T. Arens and A. Lechleiter, "The linear sampling method revisited", *J. Integral Equations Appl.* **21**:2 (2009), 179–202. MR 2011a:35564 Zbl 1237.65118

[Astala and Päivärinta 2006] K. Astala and L. Päivärinta, "Calderón's inverse conductivity problem in the plane", *Ann. of Math.* (2) **163**:1 (2006), 265–299. MR 2007b:30019 Zbl 1111.35004

[Brühl and Hanke 2000] M. Brühl and M. Hanke, "Numerical implementation of two noniterative methods for locating inclusions by impedance tomography", *Inverse Problems* **16**:4 (2000), 1029–1042. MR 2001d:65080 Zbl 0955.35076

[Cakoni and Colton 2006] F. Cakoni and D. Colton, *Qualitative methods in inverse scattering theory: an introduction*, Springer, Berlin, 2006. MR 2008c:35334 Zbl 1099.78008

[Calderón 1980] A.-P. Calderón, "On an inverse boundary value problem", pp. 65–73 in *Seminar on numerical analysis and its applications to continuum physics* (Rio de Janeiro, 1980), Soc. Brasil. Mat., Rio de Janeiro, 1980. MR 81k:35160

[Colton and Päivärinta 1992] D. Colton and L. Päivärinta, "The uniqueness of a solution to an inverse scattering problem for electromagnetic waves", *Arch. Rational Mech. Anal.* **119**:1 (1992), 59–70. MR 93h:78009 Zbl 0756.35114

[Dos Santos Ferreira et al. 2007] D. Dos Santos Ferreira, C. E. Kenig, J. Sjöstrand, and G. Uhlmann, "Determining a magnetic Schrödinger operator from partial Cauchy data", *Comm. Math. Phys.* **271**:2 (2007), 467–488. MR 2008a:35044 Zbl 1148.35096

[Dos Santos Ferreira et al. 2009] D. Dos Santos Ferreira, C. E. Kenig, M. Salo, and G. Uhlmann, "Limiting Carleman weights and anisotropic inverse problems", *Invent. Math.* **178**:1 (2009), 119–171. MR 2010h:58033 Zbl 1181.35327

[Haberman and Tataru 2011] B. Haberman and D. Tataru, "Uniqueness in Calderón's problem for Lipschitz conductivities", preprint, 2011. arXiv 1108.6068

---

[1] After the completion of this article, the enclosure type method for the Maxwell equations with a penetrable obstacle was established by Manars Kar and Mourad Sini in the paper "Reconstruction of interfaces using CGO solutions for the Maxwell equations".

[Ide et al. 2007] T. Ide, H. Isozaki, S. Nakata, S. Siltanen, and G. Uhlmann, "Probing for electrical inclusions with complex spherical waves", *Comm. Pure Appl. Math.* **60**:10 (2007), 1415–1442. MR 2008j:35186

[Ikehata 1998] M. Ikehata, "Reconstruction of the shape of the inclusion by boundary measurements", *Comm. Partial Differential Equations* **23**:7-8 (1998), 1459–1474. MR 99f:35222 Zbl 0915.35114

[Ikehata 1999a] M. Ikehata, "How to draw a picture of an unknown inclusion from boundary measurements. Two mathematical inversion algorithms", *J. Inverse Ill-Posed Probl.* **7**:3 (1999), 255–271. MR 2000d:35254 Zbl 0928.35207

[Ikehata 1999b] M. Ikehata, "Reconstruction of obstacle from boundary measurements", *Wave Motion* **30**:3 (1999), 205–223. MR 2000i:35214 Zbl 1067.35506

[Ikehata 2000] M. Ikehata, "Reconstruction of the support function for inclusion from boundary measurements", *J. Inverse Ill-Posed Probl.* **8**:4 (2000), 367–378. MR 2002g:35212

[Ikehata and Siltanen 2000] M. Ikehata and S. Siltanen, "Numerical method for finding the convex hull of an inclusion in conductivity from boundary measurements", *Inverse Problems* **16**:4 (2000), 1043–1052. MR 2001g:65164 Zbl 0956.35133

[Isakov 1990] V. Isakov, "On uniqueness in the inverse transmission scattering problem", *Comm. Partial Differential Equations* **15**:11 (1990), 1565–1587. MR 91i:35203 Zbl 0728.35148

[Kenig et al. 2007] C. E. Kenig, J. Sjöstrand, and G. Uhlmann, "The Calderón problem with partial data", *Ann. of Math.* (2) **165**:2 (2007), 567–591. MR 2008k:35498 Zbl 1127.35079

[Kirsch and Grinberg 2008] A. Kirsch and N. Grinberg, *The factorization method for inverse problems*, Oxford Lecture Series in Mathematics and its Applications **36**, Oxford University Press, 2008. MR 2009k:35322 Zbl 1222.35001

[Kuan 2012] R. Kuan, "Reconstruction of penetrable inclusions in elastic waves by boundary measurements", *J. Differential Equations* **252**:2 (2012), 1494–1520. MR 2012j:35451 Zbl 1243.35174

[Li and Vogelius 2000] Y. Y. Li and M. Vogelius, "Gradient estimates for solutions to divergence form elliptic equations with discontinuous coefficients", *Arch. Ration. Mech. Anal.* **153**:2 (2000), 91–151. MR 2001m:35083 Zbl 0958.35060

[Meyers 1963] N. G. Meyers, "An L^pe-estimate for the gradient of solutions of second order elliptic divergence equations", *Ann. Scuola Norm. Sup. Pisa* (3) **17** (1963), 189–206. MR 28 #2328

[Nagayasu et al. 2011] S. Nagayasu, G. Uhlmann, and J.-N. Wang, "Reconstruction of penetrable obstacles in acoustic scattering", *SIAM J. Math. Anal.* **43**:1 (2011), 189–211. MR 2012b:35369 Zbl 1234.35315

[Nakamura and Yoshida 2007] G. Nakamura and K. Yoshida, "Identification of a non-convex obstacle for acoustical scattering", *J. Inverse Ill-Posed Probl.* **15**:6 (2007), 611–624. MR 2009j:35387 Zbl 1126.35103

[Nakamura et al. 2005] G. Nakamura, G. Uhlmann, and J.-N. Wang, "Oscillating-decaying solutions, Runge approximation property for the anisotropic elasticity system and their applications to inverse problems", *J. Math. Pures Appl.* (9) **84**:1 (2005), 21–54. MR 2005i:35289 Zbl 1067.35151

[Ola and Somersalo 1996] P. Ola and E. Somersalo, "Electromagnetic inverse problems and generalized Sommerfeld potentials", *SIAM J. Appl. Math.* **56**:4 (1996), 1129–1145. MR 97b:35194 Zbl 0858.35138

[Ola et al. 1993] P. Ola, L. Päivärinta, and E. Somersalo, "An inverse boundary value problem in electrodynamics", *Duke Math. J.* **70**:3 (1993), 617–653. MR 94i:35196 Zbl 0804.35152

[Potthast 2001] R. Potthast, *Point sources and multipoles in inverse scattering theory*, Research Notes in Mathematics **427**, CRC, Boca Raton, FL, 2001. MR 2002j:35313 Zbl 0985.78016

[Sini and Yoshida 2010] M. Sini and K. Yoshida, "On the reconstruction of interfaces using CGO solutions for the acoustic case", preprint, 2010, Available at www.ricam.oeaw.ac.at/people/page/sini/Sini-Yoshida-preprint.pdf.

[Sylvester and Uhlmann 1987] J. Sylvester and G. Uhlmann, "A global uniqueness theorem for an inverse boundary value problem", *Ann. of Math.* (2) **125**:1 (1987), 153–169. MR 88b:35205 Zbl 0625.35078

[Takuwa et al. 2008] H. Takuwa, G. Uhlmann, and J.-N. Wang, "Complex geometrical optics solutions for anisotropic equations and applications", *J. Inverse Ill-Posed Probl.* **16**:8 (2008), 791–804. MR 2010d:35408 Zbl 1152.35519

[Uhlmann and Wang 2007] G. Uhlmann and J.-N. Wang, "Complex spherical waves for the elasticity system and probing of inclusions", *SIAM J. Math. Anal.* **38**:6 (2007), 1967–1980. MR 2008a:35295 Zbl 1131.35088

[Uhlmann and Wang 2008] G. Uhlmann and J.-N. Wang, "Reconstructing discontinuities using complex geometrical optics solutions", *SIAM J. Appl. Math.* **68**:4 (2008), 1026–1044. MR 2009d:35347 Zbl 1146.35097

[Uhlmann et al. 2009] G. Uhlmann, J.-N. Wang, and C.-T. Wu, "Reconstruction of inclusions in an elastic body", *J. Math. Pures Appl.* (9) **91**:6 (2009), 569–582. MR 2010d:35409 Zbl 1173.35123

[Vekua 1967] I. N. Vekua, *New methods for solving elliptic equations*, Applied Mathematics and Mechanics **1**, North-Holland, Amsterdam, 1967. MR 35 #3243 Zbl 0146.34301

[Yoshida 2010] K. Yoshida, "Reconstruction of a penetrable obstacle by complex spherical waves", *J. Math. Anal. Appl.* **369**:2 (2010), 645–657. MR 2011i:65192 Zbl 1196.35229

[Zhou 2010] T. Zhou, "Reconstructing electromagnetic obstacles by the enclosure method", *Inverse Probl. Imaging* **4**:3 (2010), 547–569. MR 2011f:35366 Zbl 1206.35262

Department of Mathematics, NCTS (Tapei), National Taiwan University, Taipei 106, Taiwan
jnwang@math.ntu.edu.tw

Department of Mathematics. University of California, Irvine, CA 92697, United States
tzhou@math.washington.edu

Multiwave methods via ultrasound

PLAMEN STEFANOV AND GUNTHER UHLMANN

We survey recent results by the authors on multiwave methods where the high-resolution method is ultrasound. We consider the inverse problem of determining a source inside a medium from ultrasound measurements made on the boundary of the medium. Some multiwave medical imaging methods where this is considered are photoacoustic tomography, thermoacoustic tomography, ultrasound modulated tomography, transient elastography and magnetoacoustic tomography. In the case of measurements on the whole boundary, we give an explicit solution in terms of a Neumann series expansion. We give almost necessary and sufficient conditions for uniqueness and stability when the measurements are taken on a part of the boundary. We study the case of a smooth speed and speeds having jump type of singularities. The latter models propagation of acoustic waves in the brain, where the skull has a much larger sound speed than the rest of the brain. In this paper we emphasize a microlocal viewpoint.

1. Introduction

Multiwave imaging methods, also called hybrid methods, attempt to combine the high resolution of one imaging method with the high contrast capabilities of another through a physical principle. One important medical imaging application is breast cancer detection. Ultrasound provides high (submillimeter) resolution, but it suffers from low contrast. On the other hand, many tumors absorb much more energy from electromagnetic waves (in some specific energy bands) than healthy cells. Photoacoustic tomography (PAT) [Wang 2009] consists of exposing tissues to relatively harmless optical radiation that causes temperature increases in the millikelvin range, resulting in the generation of propagating ultrasound waves (the photoacoustic effect). Such ultrasonic waves are readily measurable. The inverse problem then consists of reconstructing the optical properties of the tissue. In thermoacoustic tomography (TAT) — see, e.g., [Kruger et al. 1999] — low frequency microwaves, with wavelengths on the order of 1 m, are sent into the medium. The rationale for using the latter frequencies is that they are

---

Stefanov was partly supported by NSF Grant DMS-0800428. Uhlmann is partly supported by NSF, a Clay Senior Award and a Chancellor Professorship at UC Berkeley.

less absorbed than optical frequencies. In ultrasound modulated tomography (UMT), radiation is sent through the tissues at the same time as a modulating acoustic signal, which changes the local properties of the optical parameters (the acousto-optic effect) in a controlled manner. The objective is then the same as in PAT: to reconstruct the optical properties of the tissues. In both modalities, we seek to combine the large contrast in optical parameters between normal and cancerous tissues with the high (submillimeter) resolution of ultrasound imaging. Transient elastography (TE) [McLaughlin et al. 2010] images the propagation of shear waves using ultrasound. In magnetoacoustic tomography (MAT) [Xu and He 2005] the medium is located in a static magnetic field and a time-varying magnetic field. The time dependent magnetic field induces an eddy current and therefore induce an acoustic wave by the Lorentz force which are measured at the boundary of the medium. PAT, TAT, UMT, TE and MAT offer potential breakthroughs in the clinical application of multiwave methods to early detection of cancer, functional imaging, and molecular imaging among others.

We remark that we are only considering the first step in solving the inverse problem, namely recovering the source term from ultrasound measurements at the boundary. For a review of the results in recovering optical, elastic, electromagnetic and other properties of tissues see [Bal 2013] in this volume. This first step has been studied extensively in the mathematical literature; see, e.g., [Agranovsky et al. 2009; Finch et al. 2004; Finch and Rakesh 2009; Hristova 2009; Hristova et al. 2008; Kuchment and Kunyansky 2008; Patch 2004; Stefanov and Uhlmann 2009b; 2011] and the references there.

The purpose of this survey is to present an approach to the problem allowing us to treat variable and discontinuous sound speeds, and also consider partial data, based on [Stefanov and Uhlmann 2009b; 2011]. This approach is based on microlocal, PDE and functional analysis methods, rather than trying to find explicit closed form formulas for the partial case of a constant speed. We always assume a variable speed. We will actually formulate the problem in anisotropic media modeled by a Riemannian metric g in $\mathbb{R}^n$. Let $c > 0$, $q \geq 0$ be functions, all smooth and real-valued. Assume for convenience that g is Euclidean outside a large compact, and $c - 1 = q = 0$ there.

Let P be the differential operator

$$P = -c^2 \Delta_g + q, \qquad \Delta_g = \frac{1}{\sqrt{\det g}} \frac{\partial}{\partial x^i} g^{ij} \sqrt{\det g} \frac{\partial}{\partial x^j}. \qquad (1\text{-}1)$$

Let u solve the problem

$$\begin{cases} (\partial_t^2 + P)u = 0 & \text{in } (0, T) \times \mathbb{R}^n, \\ u|_{t=0} = f, \\ \partial_t u|_{t=0} = 0, \end{cases} \qquad (1\text{-}2)$$

where $T > 0$ is fixed.

Assume that f is supported in $\overline{\Omega}$, where $\Omega \subset \mathbb{R}^n$ is some smooth bounded domain. The measurements are modeled by the operator

$$\Lambda f := u|_{[0,T] \times \partial \Omega}. \tag{1-3}$$

The problem is to reconstruct the unknown f, knowing c; and if possible, to reconstruct both. The same problem, but with data on a part of $\partial \Omega$ is of great practical interest, as well.

The accepted mathematical model is as described above with g Euclidean, and $q = 0$; see, e.g., [Xu and Wang 2006a; Wang and Wu 2007; Finch et al. 2004]. Including nontrivial g and q does not complicate the problem further, and one can even include a magnetic field [Stefanov and Uhlmann 2009b].

If $T = \infty$, then one can solve a problem with Cauchy data 0 at $t = \infty$ (as a limit), and boundary data $h = \Lambda f$. The zero Cauchy data are justified by local energy decay that holds for nontrapping geometry, for example (actually, it is always true but much weaker and not uniform in general). Then solving the resulting problem backwards recovers f. This is known as *time reversal* or *back-projection*. For a fixed T, one can still do the same thing with an error $\epsilon(T) \to 0$, as $T \to \infty$. In the nontrapping case, n odd, the error is uniform and $\epsilon(T) = O(e^{-T/C})$. There is no good control over C though. Error estimates based on local energy decay can be found in [Hristova 2009]; see also Corollary 4.2. Other reconstruction methods have been used as well (see, e.g., [Hristova et al. 2008] for a discussion) and they all use measurements for all t in the variable coefficients case, i.e., $T = \infty$; and they are only approximate for $T < \infty$ with an error depending on the local energy decay rate. Of course, if n is odd and $P = -\Delta$, any finite $T > \text{diam}(\Omega)$ suffices by Huygens' principle. In the constant-speed case and for Ω of a specific type such as a ball or a box there are explicit closed-form inversion formulas; see [Finch et al. 2004; Xu and Wang 2005; Haltmeier et al. 2004; 2005; Finch et al. 2007] and references therein.

We describe now briefly the contents of this survey. We study what happens when $T < \infty$ is fixed. When the speed is smooth, Tataru's continuation principle [1995; 1999] provides a sharp time T_0 such that there is uniqueness for $T > T_0$ and no uniqueness for $T < T_0$. This time can be characterized as the least time T such that a signal from any point can reach $\partial \Omega$ before that time. For stable recovery, we need something more: from any point *and any direction*, we need the corresponding unit speed geodesic to hit $\partial \Omega$ for time t such that $|t| < T_1/2$. The optimal T_1 with that property is the length of the longest geodesic in $\overline{\Omega}$. Then when $T > T_1/2$, there is stability. In case of data on $[0, T] \times \partial \Omega$, $T > T_1/2$, we present an explicit Neumann series inversion formula. We also analyze the same questions for observations on a part of the boundary. In Section 3 we give an almost necessary and sufficient condition for uniqueness, and in Section 3 we give another almost necessary and sufficient condition for stability. In Proposition 5.1 we characterize Λ as a sum of two Fourier integral operators with canonical relations of graph type. Under the stability assumption, we do not have an explicit inversion anymore but we show that the problem reduces to a Fredholm equation with a trivial kernel.

In Section 6, we discuss a relation between the problems we consider and boundary control.

In Section 7 we give an estimate of the largest time interval for the geodesics to leave the medium which is important for the stability analysis.

In Section 8 we discuss briefly the connection with integral geometry.

In Section 9, we study the case where c is piecewise smooth, with jumps over smooth surfaces. This case is important for applications since in brain imaging, the acoustic speed jumps by a factor of two in the skull. Propagation of singularities is more complicated in this case: a single singularity can reflect and refract when hitting the boundary, then each branch can do the same. etc. Rays tangent to the boundary behave in an even more complicated manner. We present results similar to some of the ones above, under more restrictive assumptions which would allow us to avoid the analysis of the tangent rays. We review thoroughly the construction of geometrical optics solutions in this case.

In the Appendix we review briefly some basic concepts of microlocal analysis used in this survey. This is based mainly on [Stefanov 2012].

We also mention that a numerical method based on the theoretical developments considered here has been developed in [Qian et al. 2011].

We assume throughout the paper that the speed of sound is known. It has been suggested [Jin and Wang 2006] that one can use ultrasound transmission tomography, which measures travel times, to determine the speed of sound. For a numerical algorithm for UTT and also reflection tomography see [Chung et al. 2011]. This algorithm is based on the theoretical work [Stefanov and Uhlmann 1998a].

2. Preliminaries

2A. *Energy spaces.* Let g, $q \geq 0$ and c be in C^∞ first. The operator P is formally self-adjoint with respect to the measure $c^{-2} \, d\,\text{Vol}$, where $d\,\text{Vol}(x) = \sqrt{\det g}\, dx$. Given a domain U, and a function $u(t, x)$, define the energy

$$E_U(t, u) = \int_U \left(|Du|^2 + c^{-2} q |u|^2 + c^{-2} |u_t|^2 \right) d\,\text{Vol},$$

where $D_j = -i\partial/\partial x^j$, $D = (D_1, \ldots, D_n)$, $|Du|^2 = g^{ij}(D_i u)(\overline{D_j u})$. In particular, we define the space $H_D(U)$ to be the completion of $C_0^\infty(U)$ under the

Dirichlet norm

$$\|f\|_{H_D}^2 = \int_U \left(|Du|^2 + c^{-2}q|u|^2\right) d\text{Vol}. \qquad (2\text{-}1)$$

It is easy to see that $H_D(U) \subset H^1(U)$, if U is bounded with smooth boundary, therefore, $H_D(U)$ is topologically equivalent to $H_0^1(U)$. If $U = \mathbb{R}^n$, this is true for $n \geq 3$ only, if $q = 0$. By the finite speed of propagation, the solution with compactly supported Cauchy data always stays in H^1 even when $n = 2$. The energy norm for the Cauchy data (f, h), that we denote by $\|\cdot\|_{\mathcal{H}}$ is then defined by

$$\|(f,h)\|_{\mathcal{H}}^2 = \int_U \left(|Df|^2 + c^{-2}q|f|^2 + c^{-2}|h|^2\right) d\text{Vol}.$$

This defines the energy space

$$\mathcal{H}(U) = H_D(U) \oplus L^2(U).$$

Here and below, $L^2(U) = L^2(U; c^{-2}d\text{Vol})$. Note also that

$$\|f\|_{H_D}^2 = (Pf, f)_{L^2}. \qquad (2\text{-}2)$$

The wave equation then can be written down as the system

$$\boldsymbol{u}_t = \boldsymbol{P}\boldsymbol{u}, \quad \boldsymbol{P} = \begin{pmatrix} 0 & I \\ -P & 0 \end{pmatrix}, \qquad (2\text{-}3)$$

where $\boldsymbol{u} = (u, u_t)$ belongs to the energy space $\mathcal{H}$. The operator $\boldsymbol{P}$ then extends naturally to a skew-selfadjoint operator on $\mathcal{H}$. In this paper, we will deal with either $U = \mathbb{R}^n$ or $U = \Omega$. In the latter case, the definition of $H_D(U)$ reflects Dirichlet boundary conditions.

Assume now that c, $1/c$ and q are in L^∞. Then again, $\boldsymbol{P}$ is a skew-selfadjoint operator on $\mathcal{H}(U)$ (see [Stefanov and Uhlmann 2011]) and the statements above still hold. The important case for applications is $g = \{\delta_{ij}\}$ and $q = 0$.

By [Lasiecka et al. 1986; Katchalov et al. 2001], the operator

$$\Lambda : H_D(\Omega) \to H_{(0)}^1([0, T] \times \partial\Omega)$$

is bounded, where the subscript (0) indicates the subspace of functions vanishing for $t = 0$.

2B. *Finite propagation speed and unique continuation for the wave equation.*
It is well known (see [Taylor 1996, Chapter 8], for example) that the wave equation (2-7) has the finite speed of propagation property: "signals" propagate with speed no greater that 1, in the metric $c^{-2}g$ (or with speed c, in the metric g).

More precisely, if u solves (2-7) and has Cauchy data (f, h) for $t = 0$, then

$$u(t, x) = 0 \quad \text{for} \quad t > \text{dist}(x, \text{supp}(f, h)), \tag{2-4}$$

where "dist" is the distance in the metric $c^{-2}g$. Another way to say this is that any solution of (2-7) at (t_0, x_0) has a domain of dependence given by the characteristic cone

$$\{(t, x); \text{ dist}(x, x_0) \leq |t - t_0|\}. \tag{2-5}$$

The forward part of this cone is given by $t > t_0$, and the backward one by $t < t_0$.

Recall that given two subsets A and B of a metric space, the distance $\text{dist}(A, B)$ is defined by

$$\text{dist}(A, B) = \sup(\text{dist}(a, B); \ a \in A). \tag{2-6}$$

This function is not symmetric in general; the Hausdorff distance is defined as

$$\text{dist}_\text{H}(A, B) = \max\bigl(\text{dist}(A, B), \text{dist}(B, A)\bigr).$$

The finite speed propagation property can then be formulated in the following form: if u has Cauchy data (f, h) at $t = 0$ supported in the set U, then $u(t, x) = 0$ when $\text{dist}(x, U) > |t|$.

We recall next a Holmgren's type of unique continuation theorem for the wave equation $(\partial_t^2 + P)u = 0$ due mainly to Tataru [1995; 1999]. The local version of this theorem states that we have unique continuation across every surface that is not characteristic for P. One of its global versions, presented below, follows from its local version by Holmgren's type of arguments; see also [Katchalov et al. 2001].

Theorem 2.1. *Let P be the differential operator in $\mathbb{R}^n$ defined in* (1-1). *Assume that $u \in H^1_{\text{loc}}$ satisfies*

$$(\partial_t^2 + P)u = 0, \tag{2-7}$$

near the set in (2-8) *and vanishes in a neighborhood of* $[-T, T] \times \{x_0\}$, *with some $T > 0$, $x_0 \in \mathbb{R}^n$. Then*

$$u(t, x) = 0 \quad \text{for } |t| + \text{dist}(x_0, x) < T. \tag{2-8}$$

Proof. If P has analytic coefficients, this is Holmgren's theorem. In the non-analytic coefficients case, a version of this theorem was proved in [Robbiano 1991] with ρ replaced by $K\rho$ with an unspecified constant $K > 0$. It is derived there from a local unique continuation theorem across a surface that is "not too close to being characteristic". Hörmander [1992, Theorem 1 and Corollary 7] showed that one can choose $K = \sqrt{27/23}$, in both the local theorem and the global theorem. Moreover, he showed that K in the global one can be chosen to be the same as the K in the local one. Finally, Tataru [1995; 1999] proved a unique continuation result that implies unique continuation across any noncharacteristic surface. This shows that actually $K = 1$ in Hörmander's work, and the theorem above then follows from [Hörmander 1992, Corollary 7]. □

For the partial data analysis we need a version of that theorem restricted to a bounded (connected) domain Ω. The inconvenience of the theorem above is that it requires u to solve the wave equation in a cone that may not fit in $\mathbb{R} \times \Omega$. The next theorem shows unique continuation of Cauchy data on $\mathbb{R} \times \partial\Omega$ to their domain of influence; see e.g., [Katchalov et al. 2001, Theorem 3.16].

Proposition 2.2. *Let $\Omega \subset \mathbb{R}^n$ be a domain, and let $u \in H^1$ solve the homogeneous wave equation $Pu = 0$ in $[-T, T] \times \Omega$. Assume that u has Cauchy data zero on $[-T, T] \times \Gamma$, where $\Gamma \subset \partial\Omega$ is open. Then $u = 0$ in the domain of influence $\{(t, x) \in [-T, T] \times \Omega;\ \mathrm{dist}(x, \Gamma) < T - |t|\}$.*

One way to derive Proposition 2.2 from the unique continuation theorem is to extend u as zero in a one sided neighborhood of Γ, in the exterior of Ω (by extending g and c there first), and this extension will still be a solution. Then we apply unique continuation along a curve connecting that exterior neighborhood with an arbitrary point x such that $\mathrm{dist}(x, \Gamma) < T$. To make sure that we always stay in some neighborhood of that curve in the x space, we need to apply the unique continuation Theorem 2.1 in small increments. We refer to the proof of [Stefanov and Uhlmann 2011, Theorem 6.1] for similar arguments.

3. Uniqueness for a smooth speed

Uniqueness and reconstruction results in the constant coefficients case based on spherical means have been known for a while; see e.g., the review paper [Kuchment and Kunyansky 2008]. If $P = -c^2(x)\Delta$, and Λf is known on $[0, T] \times \partial\Omega$, Finch and Rakesh [2009] have proved that Λf recovers f uniquely as long as $T > 2T_0$; see the definition below. A uniqueness result when Γ is a part of $\partial\Omega$ in the constant coefficients case is given in [Finch et al. 2004], and we follow the ideas of that proof below. Holmgren's uniqueness theorem for constant coefficients and its analogue for variable ones (see Theorem 2.1) play a central role in the proofs, which suggests possible instability without further assumptions; see also the remark following Theorem 5.2 below.

Stability of the reconstruction when $P = -\Delta$ and $T = \infty$ follows from known reconstruction formulas; see e.g., [Kuchment and Kunyansky 2008]. In the variable coefficients case, stability estimates as $T \to \infty$ based on local energy decay have been established recently in [Hristova 2009]. When T is fixed, there is the general feeling that if one can recover "stably" all singularities, and if there is uniqueness, there must be stability (although this has been viewed from the point of view of integral geometry; see also Section 8). We prove this to be the case in Theorem 5.2, and we use the analysis in [Stefanov and Uhlmann 2009a] as well.

3A. *Data on the whole boundary.* We study first the uniqueness of recovery of f, given Λf. Since this is a linear problem, we just need to study conditions under which Λ has a trivial kernel.

We would like to use the unique continuation Theorem 2.1 but we only know that the solution u to (1-2) vanishes for $x \in \partial\Omega$ and $t \in [0, T]$. For the application of the uniqueness continuation theorem, we need to know that the normal derivative of u on $\partial\Omega$ vanishes, as well. Then we could apply Proposition 2.2. Here, we would use the simple fact that u extends as a solution to the wave equation for $t < 0$ in an even way, since $u_t = 0$ for $t = 0$.

It turns out, that knowing Λf, one can recover the Neumann derivative of the solution at $[0, T] \times \partial\Omega$ as well. This is done by applying the nonlocal exterior Dirichlet-to-Neumann map to Λf; see Lemma 6.1. We will explain now briefly the uniqueness part of this recovery. Suppose that $\Lambda f = 0$ (on $[0, T] \times \Omega$). The function u also solves the wave equation in the exterior of Ω for $0 < t < T$, with vanishing Dirichlet data on $[0, T] \times \partial\Omega$ by assumption. The Cauchy data at $t = 0$ are zero as well, because $\text{supp } f \subset \bar\Omega$. Therefore, $u = 0$ on $[0, T] \times (\mathbb{R}^n \setminus \Omega)$. Take a normal derivative $\partial/\partial\nu$ on $\partial\Omega$ from the exterior, to get $\partial u/\partial\nu = 0$ on $[0, T] \times \partial\Omega$. We can extend those equalities for $t \in [-T, 0]$, as well, because u is an even function of t. By Proposition 2.2, $f(x) = 0$ for $\text{dist}(x, \partial\Omega) < T$. Note that this is a sharp inequality by the finite speed of propagation. To get $f = 0$ for all $x \in \Omega$, we need to take T greater than the critical "uniqueness time"

$$T_0 = \text{dist}(\Omega, \partial\Omega); \tag{3-1}$$

see (2-6).

We have therefore proved the following.

Theorem 3.1. *Let $\Lambda f = 0$ with $f \in H_D(\Omega)$. Then $f(x) = 0$ for $\text{dist}(x, \partial\Omega) < T$. In particular*:

(a) *If $T < T_0$, then $f(x)$ can be arbitrary for $\text{dist}(x, \partial\Omega) > T$.*

(b) *If $T > T_0$, then $f = 0$.*

If we restrict f to a subspace of functions supported in some compact set $K \subset \bar\Omega$, then the theorem above admits an obvious generalization with T_0 replaced by $T_0(K) := \text{dist}(K, \partial\Omega)$. Also, f can be a distribution supported in $\bar\Omega$, and the theorem would still hold.

3B. *Data on a part of $\partial\Omega$.* The case of partial measurements has been discussed in the literature as well; see, e.g., [Kuchment and Kunyansky 2008; Xu et al.

2004; 2009]. One of the motivations is that in breast imaging, for example, measurements are possible only on part of the boundary. Remember that $P = -\Delta$ outside Ω. All geodesics below are related to the metric $c^{-2}g$.

Let $\Gamma \subset \partial\Omega$ be a relatively open subset of $\partial\Omega$. We are interested in what information about f can be obtained when making measurements on sets of the kind

$$\mathcal{G} := \{(t, x);\ x \in \Gamma,\ 0 < t < s(x)\}, \tag{3-2}$$

where s is a fixed continuous function on Γ. This corresponds to measurements taken at each $x \in \Gamma$ for the time interval $0 < t < s(x)$. The special case studied so far is $s(x) \equiv T$, for some $T > 0$; then $\mathcal{G} = [0, T] \times \Gamma$, and this is where our main interest is.

We assume now that the observations are made on $\mathcal{G}$ only, i.e., we assume we are given

$$\Lambda f|_\mathcal{G}, \tag{3-3}$$

where, with some abuse of notation, we denote by Λ the operator in (1-3), with $T = \infty$ that actually can be replaced by any upper bound of the function s.

We study below functions f with support in some fixed compact $\mathcal{K} \subset \bar\Omega$. By the finite speed of propagation, to be able to recover all f supported in $\mathcal{K}$, we want for any $x \in \mathcal{K}$, at least one signal from x to reach $\mathcal{G}$, i.e., we want to have a signal that reaches some $z \in \Gamma$ for $t \leq s(z)$. In other words, we should at least require that

$$\forall x \in \mathcal{K},\ \exists z \in \Gamma \text{ such that } \mathrm{dist}(x, z) < s(z). \tag{3-4}$$

We strengthened slightly the condition by replacing the $\leq$ sign by $<$. In Theorem 3.2 below, we show that this is a sufficient condition, as well.

Another way to formulate this condition is to say that $f = 0$ in the domain of influence

$$\Omega_\mathcal{G} := \{x \in \Omega;\ \exists z \in \Gamma \text{ such that } \mathrm{dist}(x, z) < s(z)\}.$$

We have the following uniqueness result, which in particular generalizes the result in [Finch et al. 2004] to the case of variable coefficients.

Theorem 3.2. *Let $P = -\Delta$ outside Ω and let $\partial\Omega$ be strictly convex. Under the assumption (3-4), if $\Lambda f = 0$ on $\mathcal{G}$ for $f \in H_D(\Omega)$ with $\mathrm{supp}\, f \subset \mathcal{K}$, then $f = 0$.*

As above, we can make this more precise.

Proposition 3.3. *Let $P = -\Delta$ outside Ω and let $\partial\Omega$ be strictly convex. Assume that $\Lambda f = 0$ on $\mathcal{G}$ for some $f \in H_D(\Omega)$ with $\mathrm{supp}\, f \subset \Omega$ that may not satisfy (3-4). Then $f = 0$ in $\Omega_\mathcal{G}$. Moreover, no information about f in $\Omega \setminus \bar\Omega_\mathcal{G}$ is contained in $\Lambda f|_\mathcal{G}$.*

Sketch of the proof. We follow the proof in [Finch et al. 2004], where c is constant everywhere (and g is Euclidean).

The main difficulty in the partial data case is that we do not have the whole Cauchy data on $\mathcal{G}$, and unlike the case of the whole boundary, we cannot recover the Neumann data directly. If we assume for a moment that the Cauchy data on $\mathcal{G}$ vanishes, the unique continuation principle of Theorem 2.1 finishes the proof.

Note first that it is enough to prove the theorem if $\Gamma = U \times \partial\Omega$, where U is a small neighborhood of some $p \in \partial\Omega$, and $\Omega_\mathcal{G}$ given by $\text{dist}(x, p) \leq s(p)$. We fist recover the Neumann data on a part of $\mathbb{R}_+ \times \Gamma$ (smaller than we would want), using a finite domain of dependence result: Proposition 2 of [Finch et al. 2004], which shows, roughly speaking, that the corresponding solution u to the exterior problem with Dirichlet data equal to zero on $[0, T] \times \Gamma$ vanishes in an exterior neighborhood $[0, T_0] \times \{p\}$ (and therefore has zero normal derivative there) only for $T_0 > 0$ such that no signal traveling in the exterior of Ω can reach p for time not exceeding T_0. In other words, if we define a distance function $\text{dist}_e(x, y)$ outside Ω as the infimum of the Euclidean distance of all curves *outside* Ω, connecting x and y, then any time T_1 with that property would not exceed $\text{dist}_e(p, \partial\Omega \setminus \Gamma)$. A critical observation is that if we are not restricted to the exterior of Ω, the (geodesic) distance between p and $\partial\Omega \setminus \Gamma$ is strictly less. Moreover, if are restricted to a set on $\partial\Omega$ where either of those distances has a uniform positive lower bound, then so does the difference. Now, knowing that $u = 0$ near $[0, T_0] \times \{p\}$, we apply unique continuation to conclude that $f(x) = 0$ for $\text{dist}(x, p) < T_0$, *and* to conclude that u has zero Dirichlet data on a larger part than Γ, by the reason explained above. Then we repeat the same argument using the fact that at each step, we improve the maximal distance at which we can get inside by at least a positive constant, independent of the step. □

4. Reconstruction with data on the whole boundary; the modified back-projection

One method to get an approximate solution of the thermoacoustic problem is the following time reversal (back-projection) method. Given h, which eventually will be replaced by Λf, let v_0 solve

$$\begin{cases} (\partial_t^2 + P)v_0 = 0 & \text{in } (0, T) \times \Omega, \\ v_0|_{[0,T] \times \partial\Omega} = h, \\ v_0|_{t=T} = 0, \\ \partial_t v_0|_{t=T} = 0. \end{cases} \qquad (4\text{-}1)$$

Then we define the *back-projection*

$$A_0 h := v_0(0, \cdot) \quad \text{in } \overline{\Omega}.$$

The function $A_0 \Lambda f$ is viewed as a candidate for a reconstructed f. Since h does not necessarily vanish at $t = T$, the compatibility condition of first order may not be satisfied because there might be a possible jump at $\{T\} \times \partial \Omega$. That singularity will propagate back to $t = 0$ and will affect v_0, and then v_0 may not be in the energy space. For this reason, h is usually cut off smoothly near $t = T$, i.e., h is replaced by $\chi(t)h(t, x)$, where $\chi \in C^\infty(\mathbb{R})$, $\chi = 0$ for $t = T$, and $\chi = 1$ in a neighborhood of $(-\infty, T(\Omega))$. See, e.g., [Hristova 2009, Section 2.2].

As we mentioned above, the back-projection v_0 converges to f, as $T \to \infty$; see [Hristova 2009] for rate of convergence estimates based on local energy decay results. In our analysis, T is fixed however.

We will modify this approach in a way that would make the problem Fredholm, and will make the error operator a contraction for certain explicit $T \gg 1$. Given h (that eventually will be replaced by Λf), solve

$$\begin{cases} (\partial_t^2 + P)v = 0 & \text{in } (0, T) \times \Omega, \\ v|_{[0,T] \times \partial \Omega} = h, \\ v|_{t=T} = \phi, \\ \partial_t v|_{t=T} = 0, \end{cases} \quad (4\text{-}2)$$

where ϕ solves the elliptic boundary value problem

$$P\phi = 0, \quad \phi|_{\partial \Omega} = h(T, \cdot). \quad (4\text{-}3)$$

Since P is a positive operator, 0 is not a Dirichlet eigenvalue of P in Ω, and therefore (4-3) is uniquely solvable. Now the initial data at $t = T$ satisfy compatibility conditions of first order (no jump at $\{T\} \times \partial \Omega$). Then we define the modified back-projection

$$Ah := v(0, \cdot) \quad \text{in } \overline{\Omega}. \quad (4\text{-}4)$$

The operator A maps continuously the closed subspace of $H^1([0, T] \times \partial \Omega)$ consisting of functions that vanish at $t = T$ (compatibility condition) to $H^1(\Omega)$; see [Lasiecka et al. 1986]. It also sends the range of Λ to $H_0^1(\Omega) \cong H_D(\Omega)$, as the proof below indicates.

To explain the idea behind this approach, let us assume for a moment that we knew the Cauchy data $[u, u_t]$ on $\{T\} \times \Omega$. Then one could simply solve the mixed problem in $[0, T] \times \Omega$ with that Cauchy data and boundary data Λf. Then that solution at $t = 0$ would recover f. We do not know the Cauchy data $[u, u_t]$ on $\{T\} \times \Omega$, of course, but we know the trace of u (a priori in H^1 for t fixed) on $\{T\} \times \partial \Omega$. The trace of u_t does not make sense because the latter is only in L^2 for $t = T$. The choice of the Cauchy data in (4-2) can then be explained by the following. Among all possible Cauchy data that belong to the "shifted linear

space"
$$\{g = [g_1, g_2] \subset H^1(\Omega) \oplus L^2(\Omega); \ g_1|_{\partial\Omega} = h(T, \cdot)\},$$

(the linear space $\mathcal{H}(\Omega)$ translated by a single element of the set above) we chose the one that minimizes the energy. The "error" will then be minimized. We refer to the proof of Theorem 4.1 for more details.

In the next theorem and everywhere below, $T_1 = T_1(\Omega)$ is the supremum of the lengths of all geodesics of the metric $c^{-2}g$ in $\overline{\Omega}$. Also, $\mathrm{dist}(x, y)$ denotes the distance function in that metric. We then call $(\Omega, c^{-2}g)$ nontrapping, if $T_1 < \infty$. It is easy to see that

$$T_0 \leq T_1/2. \tag{4-5}$$

Theorem 4.1. *Let $(\Omega, c^{-2}g)$ be nontrapping, and let $T > T_1/2$. Then $A\Lambda = \mathrm{Id} - K$, where K is compact in $H_D(\Omega)$, and $\|K\|_{H_D(\Omega)} < 1$. In particular, $\mathrm{Id} - K$ is invertible on $H_D(\Omega)$, and the inverse thermoacoustic problem has an explicit solution of the form*

$$f = \sum_{m=0}^{\infty} K^m A h, \quad h := \Lambda f. \tag{4-6}$$

Sketch of the proof. Let u solve (1-2) with a given $f \in H_D$, and let v be the solution of (4-2) with $h = \Lambda f$. Then $w := u - v$ solves

$$\begin{cases} (\partial_t^2 + P)w = 0 & \text{in } (0, T) \times \Omega, \\ w|_{[0,T] \times \partial\Omega} = 0, \\ w|_{t=T} = u|_{t=T} - \phi, \\ w_t|_{t=T} = u_t|_{t=T}, \end{cases} \tag{4-7}$$

Restrict w to $t = 0$ to get

$$f = A\Lambda f + w(0, \cdot).$$

Therefore, the "error" is given by

$$Kf = w(0, \cdot).$$

First, we show that

$$\|Kf\|_{H_D(\Omega)} \leq \|f\|_{H_D(\Omega)}, \quad \forall f \in H_D(\Omega), \tag{4-8}$$

for any fixed $T > 0$ (not necessarily greater than T_1). Since the Dirichlet boundary condition is energy preserving, it is enough to estimate th energy of $(u^T - \phi, u^T)$, where $u^T := u(T, \cdot)$.

In what follows, $(\cdot, \cdot)_{H_D(\Omega)}$ is the inner product in $H_D(\Omega)$ — see (2-1) — applied to functions that belong to $H^1(\Omega)$ but maybe not to $H_D(\Omega)$ (because

they may not vanish on $\partial\Omega$). By (2-2) and the fact that $u^T = \phi$ on $\partial\Omega$, we get
$$(u^T - \phi, \phi)_{H_D(\Omega)} = 0.$$
Then
$$\|u^T - \phi\|^2_{H_D(\Omega)} = \|u^T\|^2_{H_D(\Omega)} - \|\phi\|^2_{H_D(\Omega)} \le \|u^T\|^2_{H_D(\Omega)}.$$
Therefore, the energy of the initial conditions in (4-7) satisfies the inequality
$$E_\Omega(w, T) = \|u^T - \phi\|^2_{H_D(\Omega)} + \|u^T_t\|^2_{L^2(\Omega)} \le E_\Omega(u, T). \tag{4-9}$$
As mentioned above, the Dirichlet boundary condition is energy preserving, therefore
$$E_\Omega(w, 0) = E_\Omega(w, T) \le E_\Omega(u, T) \le E_{\mathbb{R}^n}(u, T) = E_\Omega(u, 0) = \|f\|^2_{H_D(\Omega)}.$$
This proves (4-8). Note that no condition on $T > 0$ was needed. If $\operatorname{supp} f \subset K$, and $T < \operatorname{dist}(K, \partial\Omega)$, for example, then $K = \operatorname{Id}$, and $A \wedge f = 0$. Then the "error" is 100%, and we have no information about f but (4-8) is still true.

We show next that the inequality above is strict when $T > T_0(\Omega)$:
$$\|Kf\|_{H_D(\Omega)} < \|f\|_{H_D(\Omega)}, \quad f \ne 0. \tag{4-10}$$
Assuming the opposite, we would get for some $f \ne 0$ that all inequalities leading to (4-8) are equalities. In particular,
$$u(T, x) = u_t(T, x) = 0 \quad \text{for } x \notin \Omega.$$
By the finite domain of dependence then
$$u(t, x) = 0 \quad \text{when } \operatorname{dist}(x, \Omega) > |T - t|. \tag{4-11}$$
One the other hand, we also have
$$u(t, x) = 0 \quad \text{when } \operatorname{dist}(x, \Omega) > |t|. \tag{4-12}$$
Therefore,
$$u(t, x) = 0 \quad \text{when } \operatorname{dist}(x, \partial\Omega) > T/2, \ -T/2 \le t \le 3T/2. \tag{4-13}$$
Since u extends to an even function of t that is still a solution of the wave equation, we get that (4-13) actually holds for $|t| < 3T/2$.

We will conclude next by the unique continuation Theorem 2.1 that $u = 0$ on $[0, T] \times \Omega$, and therefore $f = 0$ (see Figure 1). To this end, notice fist that by John's theorem (equivalent to Tataru's unique continuation result [Stefanov and Uhlmann 2009b, Theorem 2] in the Euclidean setting), we get $u = 0$ on $[-T, T] \times \mathbb{R}^n \setminus \Omega$. Fix $x_0 \in \Omega$. Then there is a piecewise smooth curve starting at x_0 in direction either ξ^0 or $-\xi^0$, where ξ^0 is arbitrary and fixed, of length less

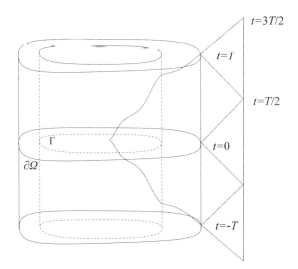

Figure 1

than T that reaches $\partial\Omega$ because $T > T_0$. This means that $\text{dist}(x_0, \mathbb{R}^n \setminus \Omega) < T$. Then by Theorem 2.1, $u(0, \cdot) = 0$ near x_0. Since x_0 was arbitrary, we get $f = 0$. This completes the proof of (4-10).

Finally, we show that $\|K\| < 1$ if $T > T_1/2$ as claimed in the theorem. Indeed, for such T, and $(x, \xi) \in S^*\Omega$, at least one of the rays originating from $(x, \pm\xi)$ leaves $\overline{\Omega}$. Then for any $\varepsilon > 0$, K can be represented as a sum of an operator K_1 with norm not exceeding $1/2 + \varepsilon$, plus a compact one, K_2. The spectrum of K^*K on the interval $((1/2 + \varepsilon)^2, 1]$ then is discrete and consists of eigenvalues only; and 1 cannot be among them, by (4-10). Then

$$\|Kf\|_{H_D(\Omega)} \leq \sqrt{\lambda_1} \|f\|_{H_D(\Omega)}, \quad f \neq 0, \qquad (4\text{-}14)$$

where $\lambda_1 < 1$ is the maximum of $1/2$ and the largest eigenvalue of K^*K greater than $1/2$, if any.

It is worth mentioning that for $T > T_1$, K is compact. □

The proof of Theorem 4.1 provides an estimate of the error in the reconstruction if we use the first term in (4-6) only that is Ah. It is in the spirit of [Hristova 2009] and relates the error to the local energy decay, as can be expected.

Corollary 4.2.

$$\|f - A\Lambda f\|_{H_D(\Omega)} \leq \left(\frac{E_\Omega(u, T)}{E_\Omega(u, 0)}\right)^{\frac{1}{2}} \|f\|_{H_D(\Omega)}, \quad \forall f \in H_D(\Omega), \ f \neq 0,$$

where u is the solution of (1-2).

Note that the $f - A\Lambda f = Kf$, and the corollary actually provides an upper bound for $\|Kf\|$. The estimate above also can be used to estimate the rate of convergence of the Neumann series (4-6) when we have a good control over the uniform local energy decay from time $t = 0$ to time $t = T$.

5. Stability and a microlocal characterization of Λ and the back-projection

Note first that in case of observations on $[0, T] \times \partial\Omega$ with $T > T_1/2$, Theorem 4.1 already implies a Lipschitz stability estimate of the type below. We consider below the partial boundary data case, where Λf is known on $\mathcal{G}$; see (3-2).

If we want that recovery to be stable, we need to be able to recover all singularities of f "in a stable way." By the zero initial velocity condition, each singularity (x, ξ) splits into two parts (see Proposition 5.1 below and the Appendix): one that starts propagating in the direction ξ; and another one propagates in the direction $-\xi$. Moreover, neither one of those singularities vanishes at $t = 0$ (and therefore never vanishes), they actually start with equal amplitudes. For a stable recovery, we need to be able to detect at least one of them, in the spirit of [Stefanov and Uhlmann 2009a], i.e., at least one of them should reach $\mathcal{G}$.

Define $\tau_\pm(x, \xi)$ by the condition

$$\tau_\pm(x, \xi) = \max\big(\tau \geq 0; \, \gamma_{x,\xi}(\pm\tau) \in \overline{\Omega}\big).$$

Based on the arguments above, for a stable recovery we should assume that $\mathcal{G}$ satisfies the condition

$$\forall (x, \xi) \in S^*\mathcal{H}, \, \big(\tau_\sigma(x, \xi), \gamma_{x,\xi}(\tau_\sigma(x, \xi))\big) \in \mathcal{G}$$

for either $\sigma = +$ or $\sigma = -$ (or both). (5-1)

Compared to condition (3-4), this means that for each $x \in \mathcal{H}$ and each unit direction ξ, at least one of the signals from (x, ξ) and $(x, -\xi)$ reaches $\mathcal{G}$. This condition becomes necessary if we replace $\mathcal{G}$ by its closure above; see Remark 5.3. In Theorem 5.2 below, we show that it is also sufficient.

We start with a description of the operator Λ that is of independent interest as well. In the next proposition, we formally choose $T = \infty$. We restrict the result below to functions supported in Ω (the support cannot touch $\partial\Omega$) to avoid the analysis at the boundary, where Λ is of more general class.

Proposition 5.1. $\Lambda = \Lambda_+ + \Lambda_-$, where $\Lambda_\pm : C_0^\infty(\Omega) \to C^\infty((0, \infty) \times \partial\Omega)$ are elliptic Fourier integral operators of zeroth order with canonical relations given by the graphs of the maps

$$(y, \xi) \mapsto \big(\tau_\pm(y, \xi), \, \gamma_{y,\xi}(\pm\tau_\pm(y, \xi)), \, \mp|\xi|, \, \dot{\gamma}'_{y,\xi}(\pm\tau_\pm(y, \xi))\big), \quad (5\text{-}2)$$

where $|\xi|$ is the norm in the metric $c^{-2}g$, and $\dot{\gamma}'$ stands for the tangential projection of $\dot{\gamma}$ on $T\partial\Omega$.

Proof. This statement is well known and follows directly from [Duistermaat 1996], for example. See also the Appendix where microlocal analysis and geometric optics is briefly reviewed. We will give more details that are needed just for the proof of this proposition in order to be able to compute the principal symbol in Theorem 5.2.

We start with a standard geometric optics construction. See Section A.4 in the Appendix.

Fix $x_0 \in \Omega$. In a neighborhood of $(0, x_0)$, the solution to (4-2) is given by

$$u(t, x) = (2\pi)^{-n} \sum_{\sigma = \pm} \int e^{i\phi_\sigma(t,x,\xi)} a_\sigma(x, \xi, t) \hat{f}(\xi) \, d\xi, \qquad (5\text{-}3)$$

modulo smooth terms, where the phase functions $\phi_\pm$ are positively homogeneous of order 1 in ξ and solve the eikonal equations (A-16), (A-17), while $a_\pm$ are classical amplitudes of order 0 solving the corresponding transport equations (A-18). Singularities starting from $(x, \xi) \in \mathrm{WF}(f)$ propagate along geodesics in the phase space issued from (x, ξ), i.e., they stay on the curve $(\gamma_{x,\xi}(t), \dot{\gamma}_{x,\xi}(\sigma t))$ for $\sigma = \pm$. This is consistent with the general propagation of singularities theory for the wave equation because the principal symbol of the wave operator $\tau^2 - c^2|\xi|_g$ has two roots $\tau = \pm c|\xi|_g$.

The construction is valid as long as the eikonal equations are solvable, i.e., along geodesics issued from $(x, \pm\xi)$ that do not have conjugate points. Assume that $\mathrm{WF}(f)$ is supported in a small neighborhood of (x_0, ξ_0) with some $\xi_0 \neq 0$. Assume first that the geodesic from (x_0, ξ_0) with endpoint on $\partial\Omega$ has no conjugate points. We will study the $\sigma = +$ term in (5-3) first. Let ϕ_b, a_b be the restrictions of ϕ_+, a_+, respectively, on $\mathbb{R} \times \partial\Omega$. Then, modulo smooth terms,

$$\Lambda_+ f := u_+(t, x)|_{\mathbb{R} \times \partial\Omega} = (2\pi)^{-n} \int e^{i\phi_\mathrm{b}(t,x,\xi)} a_\mathrm{b}(x, \xi, t) \hat{f}(\xi) \, d\xi, \qquad (5\text{-}4)$$

where u_+ is the $\sigma = +$ term in (5-3). Set $t_0 = \tau_+(x_0, \xi_0)$, $y_0 = \gamma_{x_0,\xi_0}(t_0)$, $\eta_0 = \dot{\gamma}_{x_0,\xi_0}(t_0)$; in other words, (y_0, η_0) is the exit point and direction of the geodesic issued from (x_0, ξ_0) when it reaches $\partial\Omega$. Let $x = (x', x^n)$ be boundary normal coordinates near y_0. Writing $\hat{f}$ in (5-4) as an integral, we see that (5-4) is an oscillating integral with phase function $\Phi = \phi_+(t, x', 0, \xi) - y \cdot \xi$. Then (see [Trèves 1980], for example), the set $\Sigma := \{\Phi_\xi = 0\}$ is given by the equation

$$y = \partial_\xi \phi_+(t, x', 0, \xi)$$

It is well known (see Example 2.1 in [Trèves 1980, VI.2], for example) that this equation implies that $(x', 0)$ is the endpoint of the geodesic issued from

(y, ξ) until it reaches the boundary, and $t = \tau_+(y, \xi)$, i.e., t is the time it takes to reach $\partial \Omega$. In particular, Σ is a manifold of dimension $2n$, parametrized by (y, ξ). Next, the map

$$\Sigma \ni (y, t, x', \xi) \mapsto (y, t, x', -\xi, \partial_t \phi_+, \partial_{x'} \phi_+) \tag{5-5}$$

is smooth of rank $2n$ at any point. This shows that Φ is a nondegenerate phase; see [Trèves 1980, VIII.1], and that $f \mapsto \Lambda_+ f$ is an FIO associated with the Lagrangian given by the right side of (5-5). The canonical relation is then given by

$$C := (y, \xi, t, x', \partial_t \phi_+, \partial_{x'} \phi_+), \quad (y, t, x', \xi) \in \Sigma.$$

Then (5-2) follows from the way ϕ_+ is constructed by the Hamilton-Jacobi theory. The proof in the $\sigma = -$ case is the same.

The proof above was done under the assumption that there are no conjugate points on $\gamma_{y_0, \xi_0}(t)$, $0 \leq t \leq \tau_+(y_0, \xi_0)$. To prove the theorem in the general case, let $t_1 \in (0, \tau_+(y_0, \xi_0))$ be such that there are no conjugate points on that geodesic for $t_1 \leq t \leq \tau_+(y_0, \xi_0)$. Then each of the terms in (5-3) extends to a global elliptic FIO mapping initial data at $t = 0$ to a solution at $t = t_1$; see, e.g., [Duistermaat 1996]. Its canonical relation is the graph of the geodesic flow between those two moments of time (for $\sigma = +$, and with obvious sign changes when $\sigma = -$). We can compose this with the local FIO constructed above, and the result is a well defined elliptic FIO of order 0 with canonical relation (5-2). □

We now consider the situation where Λf is given on a set $\mathcal{G}$ satisfying (5-1). Since $\mathcal{H}$ is compact and $\mathcal{G}$ is closed, one can always choose $\mathcal{G}' \Subset \mathcal{G}$ that still satisfies (5-1). Fix $\chi \in C_0^\infty([0, T] \times \partial \Omega)$ such that $\text{supp}\, \chi \subset \mathcal{G}$ and $\chi = 1$ on $\mathcal{G}'$. The measurements are then modeled by $\chi \Lambda f$, which depends on Λf on $\mathcal{G}$ only.

Choose and fix $T > \sup_\Gamma s$; see (3-2). Let A be the back-projection operator defined in (4-2) and (4-4). Note that A is always applied to $\chi \Lambda$ below, therefore $\phi = 0$ in this case.

Theorem 5.2. *$A\chi\Lambda$ is a zero-order classical pseudodifferential operator* (ΨDO) *in some neighborhood of $\mathcal{H}$ with principal symbol*

$$\tfrac{1}{2} \chi\bigl(\tau_+(x, \xi), \gamma_{x, \xi}(\tau_+(x, \xi))\bigr) + \tfrac{1}{2} \chi\bigl(\tau_+(x, \xi), \gamma_{x, \xi}(\tau_-(x, \xi))\bigr).$$

If $\mathcal{G}$ satisfies (5-1), then

(a) *$A\chi\Lambda$ is elliptic,*

(b) *$A\chi\Lambda$ is a Fredholm operator on $H_D(\mathcal{H})$, and*

(c) *there exists a constant $C > 0$ such that*

$$\|f\|_{H_D(\mathcal{H})} \leq C \|\Lambda f\|_{H^1(\mathcal{G})}. \tag{5-6}$$

Remark 5.3. By [Stefanov and Uhlmann 2009a, Proposition 3], condition (5-1), with $\mathcal{G}$ replaced by its closure, is a necessary condition for stability in any pair of Sobolev spaces. In particular, $c^{-2}g$ has to be nontrapping for stability. Indeed, then the proof below shows that $A\chi\Lambda$ will be a smoothing operator on some nonempty open conic subset of $T^*\mathcal{H} \setminus 0$.

Remark 5.4. Note that $\Lambda : H_D(\mathcal{H}) \to H^1([0, T] \times \partial\Omega)$ is bounded. This follows for example from Proposition 5.1.

Sketch of the proof. To construct a parametrix for $A\chi\Lambda f$, we apply again a geometric optic construction, using the two characteristic roots $\pm c|\xi|_g$. It is enough to assume that $A\chi\Lambda f$ has a wave front set in a conic neighborhood of some point $(t_0, y_0, \tau_0, \xi_0') \in [0, T] \times \partial\Omega$, using the notation above. For simplicity, assume that the eikonal equation is solvable for t in some neighborhood of $[0, T]$. Let $\tau_0 < 0$, for example. We look for a parametrix of the solution of the wave equation (4-2) with zero Cauchy data at $t = T$ and boundary data $\chi\Lambda_+ f$ in the form

$$v(t, x) = (2\pi)^{-n} \int e^{i\phi_+(t,x,\xi)} b(x, \xi, t) \hat{f}(\xi) \, d\xi.$$

Let (x_0, ξ_0) be the intersection of the bicharacteristic issued from $(t_0, y_0, \tau_0, \xi_0')$ with $t = 0$. The choice of that parametrix is justified by the fact that all singularities of that solution must propagate along the geodesics close to γ_{x_0, ξ_0} in the opposite direction, as t decreases because there are no singularities for $t = T$. The critical observation is that the first transport equation for the principal term b_0 of b is a linear ODE along bicharacteristics, and starting from initial data $b_0 = \chi a_0$, where $a_0 = 1/2$, at time $t = 0$, we will get that $b_0(x, \xi)|_{t=0}$ is given by the value of $\chi/2$ at the exit point of $\gamma_{x,\xi}$ on $\partial\Omega$.

This proves the first statement of the theorem.

Parts (a), (b) follows immediately from the ellipticity of $A\chi\Lambda$ that is guaranteed by (5-1).

To prove part (c), note first that the ellipticity of $A\chi\Lambda$ and the mapping property of A (see [Lasiecka et al. 1986]) imply the estimate

$$\|f\|_{H_D(\mathcal{H})} \leq C \left(\|\chi\Lambda f\|_{H^1} + \|f\|_{L^2} \right).$$

By Theorem 3.2, and (5-1), $\chi\Lambda$ is injective on $H_D(\mathcal{H})$. By [Taylor 1981, Proposition V.3.1], one gets estimate (5-6) with a constant $C > 0$ possibly different than the one above. □

6. Relations to boundary control and observability

This problem is closely related but not equivalent to the *observability problem* in boundary control. The observability problem asks the following. Let u solve

$$\begin{cases} (\partial_t^2 + P)u = 0 & \text{in } (0, T) \times \Omega, \\ u|_{(0,T) \times \partial\Omega} = 0, \\ u|_{t=0} = f, \\ \partial_t u|_{t=0} = h, \end{cases} \quad (6\text{-}1)$$

where Ω is a bounded domain with a smooth boundary as above and $T > 0$ is fixed. Comparing this with (1-2), we see that the Cauchy data at $t = 0$ given by (f, h) with h not necessarily zero (which is not essential for the discussion here) but the equation is satisfied for $x \in \Omega$ only and there is a Dirichlet boundary condition for $x \in \partial\Omega$. Then the question is: given $\partial u/\partial \nu$ on $(0, T) \times \Gamma$, with some $\Gamma \subset \Omega$, can we determine (f, h), and therefore, u? One can have Neumann or Robin boundary conditions in (6-1) and measure Dirichlet ones on $(0, T) \times \Gamma$. The essential assumption on a possibly different boundary condition is that the latter defines a well posed problem and the measurement determines the Cauchy data on $(0, T) \times \Gamma$. Physically, and microlocally, the presence of a boundary condition leads to waves that reflect off $\partial\Omega$. In the thermoacoustic case, they do not; actually then there is no boundary for the direct problem. The measurements consist of "half" of the Cauchy data only — the Dirichlet part.

6A. *Measurements on the whole boundary.* If $\Gamma = \partial\Omega$, then the two problems are actually equivalent *in a stable way*. Indeed, we will show here that knowing Λf, one can recover the normal derivative of the solution of (1-2) on $[0, T] \times \partial\Omega$ as well. This is done by applying a nonlocal ΨDO to Λf.

We will define first the outgoing DN map. Given $h \in C_0^\infty([0, \infty) \times \partial\Omega)$, let w solve the exterior mixed problem related to the Euclidean Laplacian:

$$\begin{cases} (\partial_t^2 - \Delta)w = 0 & \text{in } (0, T) \times \mathbb{R}^n \setminus \overline{\Omega}, \\ w|_{[0,T] \times \partial\Omega} = h, \\ w|_{t=0} = 0, \\ \partial_t w|_{t=0} = 0. \end{cases} \quad (6\text{-}2)$$

Then we set

$$Ng = \left.\frac{\partial w}{\partial \nu}\right|_{[0,T] \times \partial\Omega}.$$

By [Lasiecka et al. 1986], for $h \in H^1_{(0)}([0, T] \times \partial\Omega)$, we have

$$[w, w_t] \in C([0, T); \mathcal{H});$$

therefore,

$$N : H^1_{(0)}([0, T] \times \partial\Omega) \to C([0, T] \times H^{\frac{1}{2}}(\partial\Omega))$$

is continuous. Note that the results in [Lasiecka et al. 1986] require the domain to be bounded but by finite domain of dependence we can remove that restriction in our case. We also refer to [Finch et al. 2004, Proposition 2] for a sharp domain of dependence result for exterior problems.

Lemma 6.1. *Let u solve (1-2) with $f \in H_D(\Omega)$ compactly supported in Ω. Assume that $P = -\Delta$ outside Ω. Then for any $T > 0$, Λf determines uniquely u in $[0, T] \times \mathbb{R}^n \setminus \Omega$ and the normal derivative of u on $[0, T] \times \partial \Omega$ as follows:*

(a) *The solution u in $[0, T] \times \mathbb{R}^n \setminus \Omega$ coincides with the solution of (6-2) with $h = \Lambda f$.*

(b) *We have*
$$\left.\frac{\partial w}{\partial \nu}\right|_{[0,T] \times \partial \Omega} = N \Lambda f. \tag{6-3}$$

Proof. Let w be the solution of (6-2) with $g = \Lambda f \in H^1_{(0)}([0, T] \times \partial \Omega)$. Let u be the solution of (1-2). Then $u - w$ solves the unit speed wave equation in $[0, T] \times \mathbb{R}^n \setminus \Omega$ with zero Dirichlet data and zero initial data. Therefore, $u = w$ in $[0, T] \times \mathbb{R}^n \setminus \Omega$. □

The operator N is well known in scattering theory as the outgoing DN map, also called the Neumann operator sometimes. If $\partial \Omega$ is strictly convex, it is a classical ΨDO of order 1 restricted to noncharacteristic codirections (corresponding to either reflecting rays or evanescent waves) and has a more complicated structure near characteristic vectors (corresponding to glancing rays). The range of Λ acting in f with $\operatorname{supp} f \subset \Omega$ can have a wave front set in the hyperbolic region only, corresponding to reflected rays.

Now, knowing Λf, we can recover the whole Cauchy data $(f, N\Lambda f)$ on $(0, T) \times \partial \Omega$. In this case, the observability problem is to recover f from the Cauchy data there as well. One can therefore use all results known in the literature about the observability problem (see [Bardos et al. 1992], for example) to obtain results for the thermoacoustic one. On the other hand, this may not be the best way to do, numerically, at least. Also, the special and in fact the simpler structure of the thermoacoustic solution of (1-2) (no reflected waves) would be ignored if we did so. An essential part of [Bardos et al. 1992] is devoted to the analysis of such reflected waves which do not exist in our case.

6B. *Measurements on a part of the boundary.* When Λ is known restricted to $(0, T) \times \Gamma$, $\Gamma \subset \Omega$, the relation between the two problems is not so straightforward. First, the solution u to (1-2) and that to (6-1) are different as we explained already. In the observability problem, we know u on $[0, T] \times \partial \Omega$ (zero), and $\partial u / \partial \nu$ on the smaller set $(0, T) \times \Gamma$. In the thermoacoustic one, we know that the waves go through $\partial \Omega$, which is equivalent to the hidden boundary condition $\partial u / \partial \nu = Nu$ on $[0, T] \times \partial \Omega$, and we know u on $(0, T) \times \Gamma$. As Theorem 3.2 shows, we can, in a nontrivial way, recover $\partial u / \partial \nu$ on $(0, T) \times \Gamma$. The proof uses unique continuation, which is unstable. Therefore, trying to reduce the thermoacoustic problem to an observability one this way (and no other is known to the authors) goes through a unstable step and will not lead to sharp results because we have showed in Theorem 5.2 that under certain conditions, the recovery is stable.

7. Estimating the uniqueness time T_0 and the stability time T_1

One practical question is how to estimate the times T_0 and T_1 from above, to be certain that the chosen T is large enough for uniqueness or stability.

The max-min Equation (3-1) of T_0 makes it easy to get an upper bound. First, to estimate $\text{dist}(x, \partial\Omega)$ from above for x fixed, we can take any path $[a,b] \ni s \mapsto \gamma(s)$ from x to $\partial\Omega$ and compute the length of that path as $\int_a^b \frac{|\dot\gamma(s)|ds}{c(\gamma(s))}$. Then we take an upper bound with respect to $x \in \Omega$. Let $R > 0$ be such that Ω is contained in the ball $B(0, R)$ and assume that $0 \in \Omega$. Then, for example,

$$T_0 < \max_{|\omega|=1} \int_0^R \frac{dr}{c(r\omega)}.$$

In particular, if $c(x) \geq c_0 = \text{const.}$, we get

$$T_0 < \frac{R}{c_0}.$$

We estimate T_1 now, which (divided by 2) is critical for stability. A possible way to do this is to use a suitable *escape function*, a method well known and used in scattering theory. Consider the Hamiltonian

$$H(x, \xi) = \tfrac{1}{2}c^2(x)g^{ij}(x)\xi_i\xi_j$$

of P on the energy level $\Sigma := \{(x, \xi) \in T^*\overline{\Omega}; \ H = 1/2\}$. Here, g^{ij} are the components of g^{-1}. Let $\psi(x, \xi)$ be a smooth function on $\Omega \times \mathbb{R}^n$ which we regard as $T^*\Omega$ in local coordinates. Assume that for some constant α,

$$X_H \psi \geq \alpha > 0 \quad \text{on } \Sigma, \tag{7-1}$$

where X_H is the Hamiltonian vector field related to H. Relation (7-1) tells us that ψ is strictly increasing along the Hamiltonian flow. Let

$$A = \max_\Sigma |\psi(x, \xi)|.$$

Then any Hamiltonian curve on Σ issued from $T^*\Omega$ will leave Ω for time t such that $\alpha t > 2A$. Thus $T_1 \leq 2A/\alpha$.

For example, assume that g is Euclidean. Then $H = \tfrac{1}{2}c^2|\xi|^2$ and

$$X_H = \sum \left(c^2 \xi_j \frac{\partial}{\partial x^j} - c \frac{\partial c}{\partial x^j}|\xi|^2 \frac{\partial}{\partial \xi_j}\right).$$

Choose $\psi = x \cdot \xi$. Then
$$X_H \psi = c^2 |\xi|^2 \quad |\xi|^2 c x \cdot \partial_x c$$
On the energy level Σ, we have
$$X_H \psi = 1 - c^{-1} x \cdot \partial_x c.$$
Condition (7-1) is then satisfied if
$$x \cdot \partial_x c(x) < c(x) \quad \text{in } \overline{\Omega}. \tag{7-2}$$
In particular, if $c = c(r)$ is radial, condition (7-2) reduces to $r \partial c / \partial r < c$ or $\partial_r (r/c(r)) > 0$. This is the condition imposed by Herglotz [1905] and Wiechert and Zoeppritz [1907] more than a century ago in their solution of the inverse kinematic problem for radial speeds arising in seismology.

We have therefore proved the following.

Proposition 7.1. *Let $0 < c_0 \leq c(x)$ in $\overline{\Omega} \subset \bar{B}(0, R)$. Then*
$$T_0 < R/c_0.$$
Assume that
$$\alpha := \min_{\overline{\Omega}} (1 - c^{-1} x \cdot \partial_x c) > 0.$$
Then $T_1/2 \leq R/(\alpha c_0)$.

To finish the proof it only remains to notice that $|\psi| \leq |x||\xi| \leq R/c(x)$ on the energy level Σ.

8. Multiwave tomography and integral geometry

If $P = -\Delta$, and if n is odd, the solution of the wave equation is given by Kirchhoff's formula and can be expressed in terms of integrals over spheres centered at $\partial \Omega$ with radius t, and their t-derivatives. Then the problem can be formulated as an integral geometry problem — recovering f from integrals over spheres centered at $\partial \Omega$, with radii in $[0, T]$. This point of view has been exploited a lot in the literature. Uniqueness theorems can be proved using analytic microlocal calculus, when the boundary is analytic (a ball, for example). Explicit formulas has been derived when $\partial \Omega$ is a ball. There are also works studying "uniqueness sets" — what configuration of the boundary, not necessarily smooth, provides unique recovery; see [Kuchment and Kunyansky 2008], for example.

One may attempt to apply the same approach in the variable coefficients case; then one has to integrate over geodesic spheres. This has two drawbacks. First, those integrals represent the leading order terms of the solution operator only, not the whole solution. That would still be enough for constructing a parametrix however but not the Neumann series solution in Theorem 4.1. The second problem is that the geodesic spheres become degenerate in presence of caustics. The wave equation viewpoint that we use in this paper is not sensitive to caustics. We still have to require that the metric be nontrapping in some of our theorems. By the remark following Theorem 5.2 however, this is a necessary condition for stability. On the other hand, it is not needed for the uniqueness result as long as (3-4) is satisfied. Also, there is no clear integral geometry approach to uniqueness, except for analytic speeds, that would replace unique continuation. So in this sense, the integral geometry problem is "the wrong approach" when the speed is variable.

9. Brain imaging

In this section, we study the mathematical model of thermoacoustic and photoacoustic tomography when the sound speed has a jump across a smooth surface. This models the change in the sound speed in the skull when trying to image the human brain. This problem was proposed by Lihong Wang at the meeting in Banff on inverse transport and tomography in May, 2010 and it arises in brain imaging [Xu and Wang 2006b; Yang and Wang 2008]. We derive again an explicit inversion formula in the form of a convergent Neumann series under the assumptions that all singularities from the support of the source reach the boundary.

The main difference between the case of a smooth speed c and a discontinuous one with jump type of singularities is the propagation of singularities. In the present case, each ray may split into two parts when it hits the surface Γ where the speed jumps, then each branch may split again, etc. This is illustrated in Figure 2. Each such branch carries a positive fraction of the high frequency energy if there are segments tangent to Γ. The stability condition (9-5) then requires that we can detect at least one of those branches issued from supp f and any direction at time $|t| < T$. Then we also have an explicit inversion in the form of a convergent Neumann series as shown in Theorem 4.1. That reconstruction is based on applying a modified time reversal with a harmonic extension step, and then iterating it. While for a smooth speed, the classical time reversal already provides a parametrix but not necessarily an actual inversion, in the case under consideration the harmonic extension and the iteration are even more important because the first term or the classical time reversal are not parametrices. This has been also numerically observed in [Qian et al. 2011].

We describe the mathematical model now. Let $\Omega \subset \mathbb{R}^n$ be a bounded domain with smooth boundary. Let $\Gamma \subset \Omega$ be a smooth closed, orientable, not necessarily connected surface. Let the sound speed $c(x) > 0$ be smooth up to Γ with a

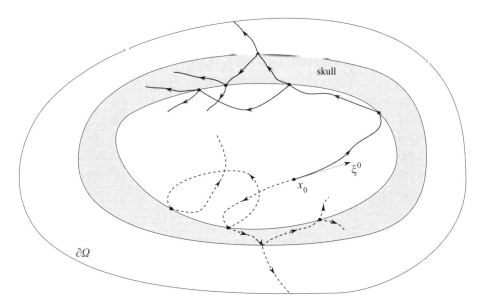

Figure 2. Propagation of singularities for the transmission problem in the skull example. The shaded region represents the skull, and the speed there is higher than in the nonshaded part. The dotted curves represent the propagation of the same singularity but moving with a negative wave speed.

nonzero jump across it. For $x \in \Gamma$, and a fixed orientation of Γ, we introduce the notation

$$c_{\text{int}}(x) = c|_{\Gamma_{\text{int}}}, \quad c_{\text{ext}}(x) = c|_{\Gamma_{\text{ext}}} \tag{9-1}$$

for the limits from the interior and from the exterior of $\Omega \setminus \Gamma$. Our assumption then is that those limits are positive as well, and

$$c_{\text{int}}(x) \neq c_{\text{ext}}(x) \quad \text{for all } x \in \Gamma. \tag{9-2}$$

In the case of brain imaging, the brain is represented by some domain $\Omega_0 \Subset \Omega$. Let Ω_1 be another domain representing the brain and the skull, so that $\Omega_0 \Subset \Omega_1 \Subset \Omega$, and $\overline{\Omega}_1 \setminus \Omega_0$ is the skull; see Figure 2. The measuring devices are then typically placed on a surface encompassing the skull, modeled by $\partial\Omega$ in our case. Then

$$c|_{\Omega_0} < c|_{\Omega_1 \setminus \Omega_0}, \quad c|_{\Omega_1 \setminus \Omega_0} > c|_{\Omega \setminus \Omega_1},$$

with the speed jumping by about a factor of two inside the skull $\overline{\Omega}_1 \setminus \Omega_0$. Another motivation to study this problem is to model the classical case of a smooth speed in the patient's body but account for a possible jump of the speed when the acoustic waves leave the body and enter the liquid surrounding it.

Let u solve the problem

$$\begin{cases} (\partial_t^2 - c^2 \Delta)u = 0 & \text{in } (0, T) \times \mathbb{R}^n, \\ u|_{\Gamma_{\text{int}}} = u|_{\Gamma_{\text{ext}}}, \\ (\partial u/\partial \nu)|_{\Gamma_{\text{int}}} = (\partial u/\partial \nu)|_{\Gamma_{\text{ext}}}, \\ u|_{t=0} = f, \\ \partial_t u|_{t=0} = 0, \end{cases} \quad (9\text{-}3)$$

where $T > 0$ is fixed, $u|_{\Gamma_{\text{int,ext}}}$ is the limit value (the trace) of u on Γ when taking the limit from the exterior and from the interior of Γ, respectively, and f is the source that we want to recover. We similarly define the interior/exterior normal derivatives, and ν is the exterior unit (in the Euclidean metric) normal to Γ.

Assume that f is supported in $\overline{\Omega}$, where $\Omega \subset \mathbb{R}^n$ is some smooth bounded domain. The measurements are modeled by the operator Λf as in (1-3). The problem is to reconstruct the unknown f.

We study the case where f is supported in some compact $\mathcal{H}$ in Ω. In applications, this corresponds to f, that is not necessarily zero outside $\mathcal{H}$ but are known there. By subtracting the known part, we arrive at the formulation that we described above. We also assume that $c = 1$ on $\mathbb{R}^n \setminus \Omega$.

The propagation of singularities for the transmission problem is well understood, at least away from possible gliding rays [Hansen 1984; Taylor 1976; Petkov 1982a; 1982b]. When a singularity traveling along a geodesic hits the interface Γ transversely, there is a reflected ray carrying a singularity, that reflects at Γ according to the usual reflection laws. If the speed on the other side is smaller, there is a transmitted (refracted) ray as well, at an angle satisfying Snell's law; see (9-41). In the opposite case, such a ray exists only if the angle with Γ is above some critical angle; see (9-42). If that angle is smaller than the critical one, there is no transmitted singularity on the other side of Γ. This is known as a full internal reflection. This is what happens in the case of the skull when a ray hits the skull boundary from inside at a small enough angle; see Figure 2. Therefore, the initial ray splits into two parts, or does not split; or it hits the boundary exactly at the critical angle. The latter case is more delicate, and we refer to Section 9B for some discussion on that.

Next, consider the propagation of each branch, if more than one. Each branch may split into two, etc. In the skull example, a ray coming from the interior of the skull hitting the boundary goes to a region with a smaller speed; and therefore there is always a transmitted ray, together with the reflected one. Then a single singularity starting at time $t = 0$ until time $t = T$ in general propagates along a few branches that look like a directed graph. This is true at least under the assumption than none of those branches, including possible transmitted ones, is tangent to the boundary.

Since $u_t|_{t=0} = 0$, singularities from (x_0, ξ^0) start to propagate in the direction ξ^0 and in the negative one $-\xi^0$. If none of the branches reaches $\partial\Omega$ at time T or less, a stable recovery is not possible [Stefanov and Uhlmann 1998b]. In the next subsection, we study the case where the initial data is supported in some compact $\mathcal{H} \subset \Omega \setminus \Gamma$ and for each $(x_0, \xi^0) \in T^*\mathcal{H} \setminus 0$, each ray through it, or through $(x_0, -\xi^0)$ has a branch that reaches $\partial\Omega$ transversely at time less than T. The main idea of the proof is to estimate the energy that each branch carries at high energies. If there is branching into rays not tangent to the boundary, we show that a positive portion of the energy is transmitted, and a positive one is reflected, at high energies. As long as one of these branches reaches the boundary transversely, at a time at which measurements are still done, we can detect that singularity. If we can do that for all singularities originating from $\mathcal{H}$, we have stability. This explains condition (9-5) below. Uniqueness follows from unique continuation results.

Similarly to the case of smooth speed studied above, assuming (9-5), we also get an explicit converging Neumann series formula for reconstructing f; see Theorem 4.1. As in the case of a smooth speed considered in [Stefanov and Uhlmann 2009b] the "error" operator K in (4-6) is a contraction. An essential difference in this case is that K is not necessarily compact. Roughly speaking, Kf corresponds to that part of the high frequency energy that is still held in Ω until time T due to reflected or transmitted signals that have not reached $\partial\Omega$ yet. While the first term only in (4-6) will still recover all singularities of f, it will not recover their strength, in contrast to the situation in [Stefanov and Uhlmann 2009b], where the speed is smooth. Thus one can expect somewhat slower convergence in this case.

9A. Main result.

Let u solve the problem (9-3) where $T > 0$ is fixed. Let $\Lambda f := u|_{[0,T]\times\partial\Omega}$ as in (1-3). The trace Λf is well defined in $C_{(0)}([0, T]; H^{1/2}(\partial\Omega))$, where the subscript (0) indicates that we take the subspace of functions h such that $h = 0$ for $t = 0$. For a discussion of other mapping properties, we refer to [Isakov 2006], when c has no jumps. By finite speed of propagation, one can reduce the analysis of the mapping properties of Λ to that case.

As in the case of a smooth speed, one could use the standard back-projection that would serve as some kind of approximation of the actual solution. We cut off smoothly Λf near $t = T$ to satisfy the compatibility conditions in the next step; and then we solve a backward mixed problem with boundary data the so cut Λf; and Cauchy data $[0, 0]$ at $t = T$. As in the case of a smooth speed (see [Hristova 2009; Stefanov and Uhlmann 2009b]) one can show that such a back-projection would converge to f as $T \to \infty$ at a rate that depends on f and that is at least a slow logarithmic one if one knows a priori that $f \in H^2$; see [Bellassoued 2003].

If $\Gamma = \partial\Omega_0$, where $\Omega_0 \subset \Omega$ is strictly convex, then in the case that the speed outside Ω_0 is faster than the speed inside (then there is full internal reflection), the convergence would be no faster than logarithmic, as suggested by the result in [Popov and Vodev 1999]. In the opposite case, it is exponential if n is odd, and polynomial when n is even [Cardoso et al. 1999]. Our goal in this work is to fix T however.

Consider the modified back-projection described in (4-2)–(4-4). The function Ah with $h = \Lambda f$ can be thought of as the "first approximation" of f. On the other hand, the proof of Theorem 9.1 below shows that it is not even a parametrix, in contrast to the case where c is smooth; see Remark 9.2.

The discussion in the Introduction and in Section 9B indicates that the singularities that we are certain to detect at $\partial\Omega$ lie in the *nontrapped set*

$$\mathcal{U} = \{(x, \xi) \in S^*(\Omega \setminus \Gamma);\ \text{there is a geodesic path issued from either } (x, \xi) \text{ or } (x, -\xi) \text{ at } t = 0, \text{ never tangent to } \Gamma, \text{ and outside } \overline{\Omega} \text{ at time } t = T\}. \quad (9\text{-}4)$$

Actually, $\mathcal{U}$ is the maximal open set with the property that a singularity in $\mathcal{U}$ is visible at $[0, T] \times \partial\Omega$; and what happens at the boundary of that set, that includes for example rays tangent to Γ, will not be important for our analysis. We emphasize here that "visible" means that some positive fraction of the energy and high frequencies can be detected as a singularity of the data; and of course there is a fraction that is reflected; then some trace of it may appear later on $\partial\Omega$, and so on.

One special case is the following. Take a compact set $\mathcal{H} \subset \Omega \setminus \Gamma$ with smooth boundary, and assume that

$$S^*\mathcal{H} \subset \mathcal{U}. \quad (9\text{-}5)$$

In other words, we require that for any $x \in \mathcal{H}$ and any unit $\xi \in S^*_x\mathcal{H}$, at least one of the multi-branched "geodesics" starting from (x, ξ), and from $(x, -\xi)$, at $t = 0$ has a path that hits $\partial\Omega$ for time $t < T$ and satisfies the nontangency assumption of (9-4). Such a set may not even exist for some speeds c.

Example 1. Let $\Omega_0 \subset \Omega$ be two concentric balls, and let c be piecewise constant; more precisely, assume

$$\Omega = B(0, R), \quad \Omega_0 = B(0, R_0), \quad 0 < R_0 < R,$$

and let

$$c = \begin{cases} c_0 < 1 & \text{in } \Omega_0, \\ 1 & \text{in } \mathbb{R}^n \setminus \Omega_0. \end{cases}$$

Then such a set $\mathcal{H}$ always exist and can be taken to be a ball with the same center and small enough radius. Indeed, the requirement then is that all rays starting from $\mathcal{H}$ hit Γ at an angle greater than a critical one $\pi/2 - \alpha_0$; see (9-42).

This can be achieved by choosing $\mathcal{H} = B(0, \rho)$ with $\rho \ll R_0$. An elementary calculation shows that we need to satisfy the inequality $\rho/R_0 < \sin\alpha_0 = c_0$, i.e., it is enough to choose $\rho < c_0 < R_0$. Then there exists T_0 such that (9-5) holds for $T > T_0$, and T_0 is easy to compute. On can also add to $\mathcal{H}$ any compact included in $\{R_0 < |x| < R\}$. In other words, $\mathcal{H}$ can be any compact in Ω not intersecting $\{c_0 R_0 \leq |x| \leq R_0\}$, the zone where the trapped rays lie.

If $c = c_0 > 1$ in Ω_0, then any compact $\mathcal{H}$ in Ω satisfies (9-5). In that case, there is always a transmitted ray leaving Ω_0.

Example 2. This is a simplified version of the skull model. Let $\Omega_0 \subset \Omega_1 \subset \Omega$ be balls such that
$$\Omega = B(0, R), \quad \Omega_0 = B(0, R_0), \quad \Omega_1 = B(0, R_1) \quad 0 < R_0 < R_1 < R,$$
Assume that
$$c|_{\Omega_0} = c_0, \quad c|_{\Omega_1 \setminus \Omega_0} = c_1, \quad c|_{\mathbb{R}^n \setminus \Omega_1} = 1$$
with some constants c_0, c_1 such that $c_0 < c_1$, $c_1 > 1$. Here, c_0 models the acoustic speed in the brain, c_1 is the speed in the skull, and 1 is the acoustic speed in the liquid outside the head. If for a moment we consider Ω_0 and Ω_1 only, we have the configuration of the previous example. If $\mathcal{H} = B(0, \rho)$ with $\rho < (c_0/c_1)R_0$, then $\mathcal{H}$ satisfies (9-5) with an appropriate T. Now, since $c_1 > 1$, rays that hit $\partial\Omega_1$ always have a transmitted part outside Ω_1, and therefore (9-5) is still satisfied in Ω. Rays originating outside Ω_1 are not trapped, therefore, more generally, $\mathcal{H}$ can be any compact in $\Omega \setminus \{c_0 R_0/c_1 \leq |x| \leq R_0\}$.

Let $\Pi_{\mathcal{H}} : H_D(\Omega) \to H_D(\mathcal{H})$ be the orthogonal projection of elements of the former space to the latter (considered as a subspace of $H_D(\Omega)$). It is easy to check that $\Pi_{\mathcal{H}} f = f|_{\mathcal{H}} - P_{\partial\mathcal{H}}(f|_{\partial\mathcal{H}})$, where $P_{\partial\mathcal{H}}$ is the Poisson operator of harmonic extension in $\mathcal{H}$.

Our main result about discontinuous speeds is the following.

Theorem 9.1. *Let $\mathcal{H}$ satisfy (9-5). Then $\Pi_{\mathcal{H}} A_1 \Lambda = \mathrm{Id} - K$ in $H_D(\mathcal{H})$, with $\|K\|_{H_D(\mathcal{H})} < 1$. In particular, $\mathrm{Id} - K$ is invertible on $H_D(\mathcal{H})$, and Λ restricted to $H_D(\mathcal{H})$ has an explicit left inverse of the form*

$$f = \sum_{m=0}^{\infty} K^m \Pi_{\mathcal{H}} Ah, \quad h = \Lambda f. \tag{9-6}$$

Remark 9.2. As discussed in the Introduction, K is not a compact operator as in the case of smooth sound speed. It follows from the proof of the theorem that the least upper bound of its essential spectrum (always less that 1) corresponds to the maximal portion of the high-frequency energy that is still held in Ω at time $t = T$.

Remark 9.3. Consider the case now where $\mathcal{H}$ does not satisfy (9-5). If there is an open set of singularities that does not reach $\partial\Omega$, a stable recovery is impossible [Stefanov and Uhlmann 1998b]. In either case however, a truncated version of the series (9-6) would provide an approximate parametrix that would recover the visible singularities, i.e., those in $\mathcal{U}$. By an approximate parametrix we mean a pseudodifferential operator elliptic in $\mathcal{U}$ with a principal symbol converging to 1 in any compact in that set as the number of the terms in (9-6) increases. This shows that roughly speaking, if a recovery of the singularities is the primary goal, then only those in $\mathcal{U}$ can be recovered in a "stable way", and (9-6) works in that case as well, without the assumption (9-5).

9B. Sketch of the proof; geometric optics.

The proof of Theorem 9.1 is based on a detailed microlocal analysis of the solution of the forward equation (9-3). As we explained above, propagation of singularities is well understood, and we avoided the most delicate cases with our assumptions about $\mathcal{H}$. To prove that the "error operator" K is a contraction however, we show first that it is a contraction up to a compact operator by studying the parametrix first. Then we use a suitable adaptation of the unique continuation property to this setting, combined with arguments similar to those in the smooth case to show that the whole K is a contraction as well. The most essential part of the proof is to show that the parametrix is a contraction. This requires not only to trace the propagation of singularities but to show that each time a ray splits into a reflected and transmitted one (neither one tangent), both rays carry a positive fraction of the energy.

Analysis at the boundary. We will analyze what happens when the geodesic (x_0, ξ^0) issued from (x_0, ξ^0), $x_0 \notin \Gamma$, hits Γ for first time, under some assumptions. Let the open sets Ω_{int}, Ω_{ext}, be the interior and the exterior part of Ω near x_0, according to the orientation of Γ. They only need to be defined near the first contact with Γ. Let us assume that this geodesic hits Γ from Ω_{int}. We will construct here a microlocal representation of the reflected and the transmitted waves near the boundary. We refer to Section A.4 for the geometric optics construction.

Extend $c|_{\Omega_{\text{int}}}$ in a smooth way in a small neighborhood on the other side of Γ, and let u_+ be the solution described above, defined in some neighborhood of that geodesic segment. Since we are only going to use u_+ in the microlocal construction described below, and we will need only the trace of u_+ on $\mathbb{R}_+ \times \Gamma$ near the first contact of the bicharacteristic from (x_0, ξ^0) with Γ, the particular extension of c would not affect the microlocal expansion but may affect the smoothing part.

Set
$$h := u_+|_{\mathbb{R} \times \Gamma}. \qquad (9\text{-}7)$$

Let $(t_1, x_1) \in \mathbb{R}_+ \times \Gamma$ be the point where the geodesic from γ_{x_0, ξ^0} hits Γ for the first time (see Figure 3). We assume that such a t_1 exists. Let ξ^1 be the tangent covector to that geodesic at (t_1, x_1). Assume that ξ^0 is unit covector in the metric $c^{-2} dx^2$, then so is ξ^1 (in the metric $c_{\text{int}}^{-2} dx^2$), i.e., $c_{\text{int}} |\xi| = 1$, where $|\xi|$ is the Euclidean norm. Assume that ξ^1 is transversal to Γ. In view of condition (9-5), this is the case that we need to study.

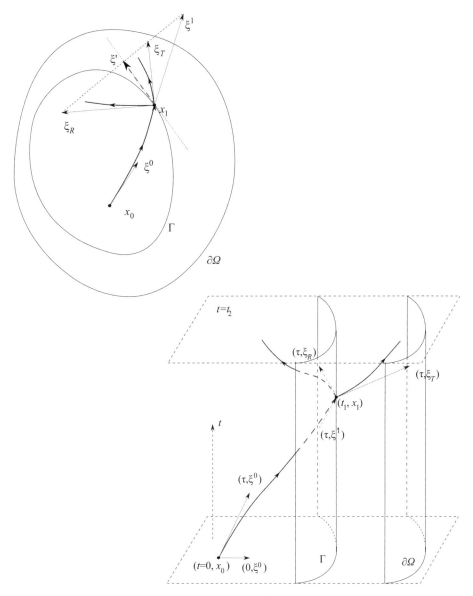

Figure 3. Reflected and the transmitted rays in x-space (top) and in (t, x)-space (bottom).

Standard microlocal arguments show that the map $[f_1, f_2] \mapsto h$ is an elliptic Fourier integral operator (FIO) with a canonical relation that is locally a canonical graph described in Proposition 3 of [Stefanov and Uhlmann 2009b]; see the proof of that proposition for details. That diffeomorphism maps (x_0, ξ^0) into $(t_1, x_1, 1, (\xi^1)')$, where the prime stands for the tangential projection onto $T^*\Gamma$; and that maps extends as a positively homogeneous one of order one with respect to the dual variable. In particular, the dual variable τ to t stays positive. In fact, WF(u) is in the characteristic set $\tau^2 - c^2(x)|\xi|^2 = 0$, and (x, ξ) belongs to some small neighborhood of (x_1, ξ^1). The wave front set WF(h) is given by $(x, \xi') \in T^*\Gamma$, $(x, \xi) \in$ WF(u), where ξ' is the tangential projection of ξ to the boundary. Then (t, x, τ, ξ') is the image of some $(\tilde{x}, \tilde{\xi})$ close to (x_0, ξ^0) under the canonical map above. Here $(\tilde{x}, \tilde{\xi})$ is such that the x-projection $x(s)$ of the bicharacteristic from it hits Γ for the first time at time for the value of s given by $sc(\tilde{x}) = t$. Since $\tau^2 - c_{\text{int}}^2(x)|\xi|^2 = 0$, for the projection ξ' we have $\tau^2 - c_{\text{int}}^2(x)|\xi'|^2 > 0$, where $(x, \xi') \in T^*\Gamma$, and $|\xi'|$ is the norm of the covector ξ' in the metric on Γ induced by the Euclidean one.

The microlocal regions of $T^*(\mathbb{R} \times \Gamma) \ni (t, x, \tau, \xi')$ with respect to the sound speed c_{int}, i.e., in $\overline{\Omega}_{\text{int}}$, are defined as follows:

hyperbolic region: $c_{\text{int}}(x)|\xi'| < \tau$,

glancing manifold: $c_{\text{int}}(x)|\xi'| = \tau$,

elliptic region: $c_{\text{int}}(x)|\xi'| > \tau$.

One has a similar classification of $T^*\Gamma$ with respect to the sound speed c_{ext}. A ray that hits Γ transversely, coming from Ω_{int}, has a tangential projection on $T^*(\mathbb{R} \times \Gamma)$ in the hyperbolic region relative to c_{int}. If $c_{\text{int}} < c_{\text{ext}}$, that projection may belong to any of the three microlocal regions with respect to the speed c_{int}. If $c_{\text{int}} > c_{\text{ext}}$, then that projection is always in the hyperbolic region for c_{ext}. When we have a ray that hits Γ from Ω_{ext}, then those two cases are reversed.

Reflected and transmitted waves. We will analyze the case where $(\xi^1)'$ belongs to the hyperbolic region with respect to both c_{int} and c_{ext}, i.e., we will work with ξ' in a neighborhood of $(\xi^1)'$ satisfying

$$c_{\text{int}}^{-2}\tau^2 - |\xi'|^2 > 0, \quad c_{\text{ext}}^{-2}\tau^2 - |\xi'|^2 > 0. \tag{9-8}$$

The analysis also applies to the case of a ray coming from Ω_{ext}, under the same assumption. We will confirm below in this setting the well known fact that under that condition, such a ray splits into a reflected ray with the same tangential component of the velocity that returns to the interior Ω_{int}, and a transmitted one, again with the same tangential component of the velocity, that propagates in Ω_{ext}. We will also compute the amplitudes and the energy at high frequencies of the corresponding asymptotic solutions.

Choose local coordinates on Γ that we denote by x', and a normal coordinate x^n to Γ such that $x^n > 0$ in Ω_{ext}, and $|x^n|$ is the Euclidean distance to Γ; then $x = (x', x^n)$. We will express the solution u_+ in $\mathbb{R} \times \overline{\Omega}_{\text{int}}$ that we defined above, as well as a reflected solution u_R in the same set; and a transmitted one u_T in $\mathbb{R} \times \overline{\Omega}_{\text{ext}}$, up to smoothing terms in the form

$$u_\sigma = (2\pi)^{-n} \int e^{i\varphi_\sigma(t,x,\tau,\xi')} b_\sigma(t,x,\tau,\xi') \hat{h}(\tau,\xi') \, d\tau \, d\xi', \quad \sigma = +, R, T, \tag{9-9}$$

where $\hat{h} := \int_{\mathbb{R} \times \mathbb{R}^{n-1}} e^{-i(-t\tau + x' \cdot \xi')} h(t, x') dt \, dx'$. We chose to alter the sign of τ so that if $c = 1$, the phase function in (9-9) equals φ_+, i.e., then $\varphi_+ = -t\tau + x \cdot \xi$. The three phase functions φ_+, φ_R, φ_T solve the eikonal equation

$$\partial_t \varphi_\sigma + c(x) |\nabla_x \varphi_\sigma| = 0, \quad \varphi_\sigma|_{x^n=0} = -t\tau + x' \cdot \xi'. \tag{9-10}$$

The right choice of the sign in front of $\partial_t \varphi_+$ (see (A-16)) is the positive one because $\partial_t \varphi_+ = -\tau < 0$ for $x^n = 0$, and that derivative must remain negative near the boundary as well. We see below that $\varphi_{R,T}$ have the same boundary values on $x^n = 0$, therefore they satisfy the same eikonal equation, with the same choice of the sign.

Let now h be a compactly supported distribution on $\mathbb{R} \times \Gamma$ with WF(h) in a small conic neighborhood of $(t_1, x_1, 1, (\xi^1)')$. We will take h as in (9-7) eventually, with u_+ the solution corresponding to initial data f at $t = 0$ but in what follows, h is arbitrary as long as WF(h) has that property, and u_+ is determined through h. We now look for a parametrix

$$\tilde{u} = u_+ + u_R + u_T \tag{9-11}$$

near (t_1, x_1) with u_+, u_R, u_T of the type (9-9), satisfying the wave equation and (9-7). We use the notation for u_+ now for a parametrix in Ω_{int} having singularities that come from the past and hit Γ; i.e., for an outgoing solution. The subscript $+$ is there to remind us that this is related to the positive sound speed $c(x)|\xi|$. Next, u_R is a solution with singularities that are obtained form those of u_+ by reflection; they propagate back to Ω_{int}. It is an outgoing solution in Ω_{int}. And finally, u_T is a solution in Ω_{ext} with singularities that go away from Γ as time increases; hence it is outgoing there. To satisfy the first transmission condition in (9-3), we need to have

$$\varphi_T = \varphi_R = \varphi_+ = -t\tau + x \cdot \xi' \quad \text{for } x^n = 0, \tag{9-12}$$

that explains the same boundary condition in (9-10), and

$$1 + b_R = b_T \quad \text{for } x^n = 0. \tag{9-13}$$

In particular, for the leading terms of the amplitudes we get

$$b_T^{(0)} - b_R^{(0)} = 1 \quad \text{for } x^n = 0. \tag{9-14}$$

To satisfy the second transmission condition, we require

$$i\frac{\partial \varphi_+}{\partial x^n} + \frac{\partial b_+}{\partial x^n} + i\frac{\partial \varphi_R}{\partial x^n} b_R + \frac{\partial b_R}{\partial x^n} = i\frac{\partial \varphi_T}{\partial x^n} b_T + \frac{\partial b_T}{\partial x^n} \quad \text{for } x^n = 0. \tag{9-15}$$

Expanding this in a series of homogeneous in (τ, ξ) terms, we get series of initial conditions for the transport equations that follow. Comparing the leading order terms only, we get

$$\frac{\partial \varphi_T}{\partial x^n} b_T^{(0)} - \frac{\partial \varphi_R}{\partial x^n} b_R^{(0)} = \frac{\partial \varphi_+}{\partial x^n} \quad \text{for } x^n = 0. \tag{9-16}$$

The linear system (9-14), (9-16) for $b_R^{(0)}|_{x^n=0}$, $b_T^{(0)}|_{x^n=0}$ has determinant

$$-\left(\frac{\partial \varphi_T}{\partial x^n} - \frac{\partial \varphi_R}{\partial x^n}\right)\bigg|_{x^n=0}. \tag{9-17}$$

Provided that this determinant is nonzero near x_1, we can solve for $b_R^{(0)}|_{x^n=0}$ and $b_T^{(0)}|_{x^n=0}$. Moreover, the determination of each subsequent term $b_R^{(-j)}|_{x^n=0}$ or $b_T^{(-j)}|_{x^n=0}$ in the asymptotic expansion of $b_R|_{x^n=0}$ and $b_T|_{x^n=0}$ can be found from (9-15) by solving a linear system with the same (nonzero) determinant.

Solving the eikonal equations. As is well known, the eikonal equation (9-10) in any fixed side of $\mathbb{R} \times \Gamma$, near (t_1, x_1), has two solutions. They are determined by a choice of the sign of the normal derivative on $\mathbb{R} \times \Gamma$ and the boundary condition. We will make the choice of the signs according to the desired properties for the singularities of u_+, u_R, u_T. Let $\nabla_{x'}$ denote the tangential gradient on Γ. By (9-12),

$$\nabla_{x'}\varphi_T = \nabla_{x'}\varphi_R = \nabla_{x'}\varphi_+ = \xi', \quad \partial_t \varphi_T = \partial_t \varphi_R = \partial_t \varphi_+ = -\tau \quad \text{for } x^n = 0. \tag{9-18}$$

Using the eikonal equation (9-10) and the boundary condition there, we get

$$\frac{\partial \varphi_+}{\partial t} = -\tau, \quad \frac{\partial \varphi_+}{\partial x^n} = \sqrt{c_{\text{int}}^{-2}\tau^2 - |\xi'|^2} \quad \text{for } x^n = 0. \tag{9-19}$$

We made a sign choice for the square root here based on the required property of u_+ described above. This shows in particular, that the map $h \mapsto \partial u_+/\partial t$ (that is just d/dt), and the interior incoming Dirichlet-to-Neumann map

$$N_{\text{int,in}} : h \mapsto \frac{\partial u_+}{\partial \nu}\bigg|_{\mathbb{R} \times \Gamma}$$

are locally ΨDOs of order 1 with principal symbols given by $-i\tau$, and

$$\sigma_p(N_{\text{int,in}}) = i\frac{\partial\varphi_+}{\partial x^n} = i\sqrt{c_{\text{int}}^{-2}\tau^2 - |\xi'|^2}. \tag{9-20}$$

The notion "interior incoming" is related to the fact that locally, near (t_1, x_1), we are solving a mixed problem in $\mathbb{R} \times \Omega_{\text{int}}$ with lateral boundary value h and zero Cauchy data for $t \gg 0$.

Consider φ_R next. The reflected phase φ_R solves the same eikonal equation, with the same boundary condition, as φ_+. By the eikonal equation (9-10), we must have

$$\frac{\partial\varphi_R}{\partial x^n} = \pm\frac{\partial\varphi_+}{\partial x^n} \quad \text{for } x^n = 0. \tag{9-21}$$

The "+" choice will give us the solution φ_+ for φ_R. We chose the negative sign, that uniquely determines a solution locally, that we call φ_R, i.e.,

$$\frac{\partial\varphi_R}{\partial x^n} = -\frac{\partial\varphi_+}{\partial x^n} \quad \text{for } x^n = 0. \tag{9-22}$$

Therefore, $\nabla\varphi_R$ on the boundary is obtained from $\nabla\varphi_+$ by inverting the sign of the normal derivative. This corresponds to the usual law of reflection. Therefore,

$$\frac{\partial\varphi_R}{\partial t} = -\tau, \quad \frac{\partial\varphi_R}{\partial x^n} = -\sqrt{c_{\text{int}}^{-2}\tau^2 - |\xi'|^2} \quad \text{for } x^n = 0. \tag{9-23}$$

In particular, $\partial u_R/\partial x^n|_{\mathbb{R}\times\Gamma}$ can be obtained from $u_R|_{\mathbb{R}\times\Gamma}$, that we still need to determine, via the interior outgoing Dirichlet-to-Neumann map

$$N_{\text{int,out}} : u_R\Big|_{\mathbb{R}\times\Gamma} \longmapsto \frac{\partial u_R}{\partial x^n}\Big|_{\mathbb{R}\times\Gamma}$$

that is locally a first order ΨDO with principal symbol

$$\sigma_p(N_{\text{int,out}}) = i\frac{\partial\varphi_R}{\partial t} = -i\sqrt{c_{\text{int}}^{-2}\tau^2 - |\xi'|^2}. \tag{9-24}$$

To construct φ_T, we work in $\overline{\Omega}_{\text{ext}}$. We define φ_T as the solution of (9-10) with the following choice of a normal derivative. This time φ_T and φ_+ solve the eikonal equation at different sides of Γ, and c has a jump at Γ. By (9-18),

$$c_{\text{ext}}^2\left(|\xi'|^2 + \left|\frac{\partial\varphi_T}{\partial x^n}\right|^2\right) = \tau^2 \quad \text{for } x^n = 0. \tag{9-25}$$

We solve this equation for $|\partial\varphi_T/\partial x^n|^2$. Under the assumption (9-8), this solution is positive, therefore we can solve for $\partial\varphi_T/\partial x^n$ to get

$$\frac{\partial\varphi_T}{\partial x^n} = \sqrt{c_{\text{ext}}^{-2}\tau^2 - |\xi'|^2} \quad \text{for } x^n = 0. \tag{9-26}$$

The positive sign of the square root is determined by the requirement the singularity to be outgoing. In particular, we get that the exterior outgoing Dirichlet-to-Neumann map

$$N_{\text{ext,out}} : u_T|_{\mathbb{R}\times\Gamma} \longmapsto \frac{\partial u_T}{\partial x^n}\bigg|_{\mathbb{R}\times\Gamma}$$

has principal symbol

$$\sigma_p(N_{\text{ext,out}}) = \mathrm{i}\frac{\partial\varphi_T}{\partial x^n} = \mathrm{i}\sqrt{c_{\text{ext}}^{-2}\tau^2 - |\xi'|^2}. \qquad (9\text{-}27)$$

For future reference, we note the inequality

$$0 \le \frac{\partial\varphi_T}{\partial x^n} \le \gamma \frac{\partial\varphi_+}{\partial x^n}, \quad \text{where } \gamma := \max_{\Gamma} \frac{c_{\text{int}}}{c_{\text{ext}}} < 1. \qquad (9\text{-}28)$$

Amplitude and energy calculations. By (9-23), (9-26), the determinant (9-17) is negative. Solving (9-14) and (9-16) then yields

$$b_T^{(0)} = \frac{2\partial\varphi_+/\partial x^n}{\partial\varphi_+/\partial x^n + \partial\varphi_T/\partial x^n}, \quad b_R^{(0)} = \frac{\partial\varphi_+/\partial x^n - \partial\varphi_T/\partial x^n}{\partial\varphi_+/\partial x^n + \partial\varphi_T/\partial x^n} \quad \text{for } x^n = 0. \qquad (9\text{-}29)$$

As explained below (9-17), we can get initial conditions for the subsequent transport equations, and then solve those transport equation. By (9-12), the maps

$$P_R : h \mapsto u_R|_{\mathbb{R}\times\Gamma}, \quad P_T : h \mapsto u_T|_{\mathbb{R}\times\Gamma} \qquad (9\text{-}30)$$

are ΨDOs of order 0 with principal symbols equal to $b_R^{(0)}$, $b_T^{(0)}$ restricted to $\mathbb{R}\times\Gamma$; see (9-29). We recall (9-7) as well.

We estimate next the amount of energy that is transmitted in Ω_{ext}. We will do it only based on the principal term in our parametrix. That corresponds to an estimate of the solution operator corresponding to transmission, up to compact operators, as we show below.

A quick look at (9-29) (see also (9-14)) shows that $b_T^{(0)} > 1$. This may look strange because we should have only a fraction of the energy transmitted, and the rest is reflected. There is no contradiction however because the energy is not proportional to the amplitude.

Let u solve $(\partial_t^2 - c^2\Delta)u = 0$ in the bounded domain U with smooth boundary for $t' \le t \le t''$ with some $t' < t''$. A direct calculation yields

$$E_U(u(t'')) = E_U(u(t')) + 2\Re \int_{[t',t'']\times\partial U} u_t \frac{\partial \bar{u}}{\partial \nu}\,\mathrm{d}t\,\mathrm{d}S. \qquad (9\text{-}31)$$

We will use this to estimate the energy of u_T in Ω_{ext}. Since the wave front set of u_T is contained in some small neighborhood of the transmitted bicharacteristic, we have smooth data for $t = 0$. Therefore, if $t_2 > t_1$ is fixed closed enough to

t_1, we can apply (9-31) to a large ball minus Ω_{int} to get that modulo a compact operator applied to h,

$$E_{\Omega_{\text{ext}}}(\boldsymbol{u}_T(t_2)) \cong 2\Re \int_{[0,t_2]\times\Gamma} \frac{\partial u_T}{\partial t} \frac{\partial \bar{u}_T}{\partial \nu} \, dt \, dS. \tag{9-32}$$

Therefore,

$$E_{\Omega_{\text{ext}}}(\boldsymbol{u}_T(t_2)) \cong 2\Re(P_t u_T, N_{\text{ext,out}} u_T) = \Re(2 P_T^* N_{\text{ext,out}}^* P_t P_T h, h), \tag{9-33}$$

where $(\cdot, \cdot)$ is the inner product in $\mathbb{R} \times \mathbb{R}^{n-1}$, and $P_t = d/dt$.

Apply similar arguments to u_+ in Ω_{int}. Since the bicharacteristics leave Ω_{int}, we have modulo smoother terms

$$0 \cong E_{\Omega_{\text{int}}}(\boldsymbol{u}_+(0)) + 2\Re \int_{[0,t_2]\times\Gamma} \frac{\partial u_+}{\partial t} \frac{\partial \bar{u}_+}{\partial \nu} \, dt \, dS. \tag{9-34}$$

Similarly we get (see again (9-30))

$$E_{\Omega_{\text{int}}}(\boldsymbol{u}_+(0)) \cong -2\Re(P_t h, N_{\text{int,in}} h) = \Re(2 N_{\text{int,in}}^* P_t h, h). \tag{9-35}$$

For the principal symbols of the operators in (9-33), (9-35) we have

$$\frac{\sigma_p(2 P_T^* N_{\text{ext,out}}^* P_t P_T)}{\sigma_p(2 N_{\text{int,in}}^* P_t)} = \frac{\partial \varphi_T/\partial \nu}{\partial \varphi_+/\partial \nu} (b_T^{(0)})^2 = \frac{4(\partial \varphi_+/\partial \nu)(\partial \varphi_T/\partial \nu)}{(\partial \varphi_+/\partial \nu + \partial \varphi_T/\partial \nu)^2}. \tag{9-36}$$

Denote for a moment $a := \partial \varphi_+/\partial \nu$, $b := \partial \varphi_T/\partial \nu$. Then the quotient above equals $4ab/(a+b)^2 \leq 1$ that confirms that the reflected wave has less energy than the incident one. By (9-28), $0 \leq b \leq \gamma a$, $0 < \gamma < 1$. This easily implies

$$\frac{4ab}{(a+b)^2} \leq \frac{4\gamma}{(1+\gamma)^2} < 1. \tag{9-37}$$

Therefore, the expression in the middle represents an upper bound of the portion of the total energy that gets transmitted in the asymptotic regime when the frequency tends to infinity. To get a lower bound, assume in addition that $b \geq b_0 > 0$ and $a \leq a_0$ for some a_0, b_0, i.e.,

$$0 < b_0 < \frac{\partial \varphi_T}{\partial \nu}, \quad \frac{\partial \varphi_+}{\partial \nu} \leq a_0. \tag{9-38}$$

Then

$$\frac{4ab}{(a+b)^2} \geq \frac{4b_0^2/\gamma}{(1+\gamma)^2 a_0^2} > 0. \tag{9-39}$$

This is a lower bound of the ratio of the high frequency energy that is transmitted. As we can see, if the transmitted ray gets very close to a tangent one, that ratio tends to 0.

So far this is still not a proof of such a statement but just a heuristic argument. For the a precise statement, see [Stefanov and Uhlmann 2011].

Snell's law. Now assume that $(\xi^1)'$ is in the hyperbolic region for c_{int} but not necessarily for c_{ext}. This corresponds to a ray hitting Γ from the interior Ω_{int}. There is no change in solving the eikonal equation for φ_R but a real phase φ_T does not exist if the expression under the square root in (9-26) is negative. This happens when $(\xi^1)'$ is in the elliptic region for c_{ext}. Then there is no transmitted singularity in the parametrix. We analyze this case below. If $c_{\text{int}} > c_{\text{ext}}$, then $(\xi^1)'$ that is in the hyperbolic region for c_{int} by assumption, also falls into the hyperbolic region for the speed c_{ext}, i.e., there is always a transmitted ray. If $c_{\text{int}} < c_{\text{ext}}$, then existence of a transmitted wave depends on where $(\xi^1)'$ belongs with respect to c_{ext}.

Let α be the angle that $\xi^1 = \partial\varphi_+/\partial x^n$ makes with the (co-)normal represented by dx^n, and let β be the angle between the latter and $\xi_T := \partial\varphi_T/\partial x^n$. We have

$$|\xi'| = |\xi^1|\sin\alpha = c_{\text{int}}^{-1}\tau\sin\alpha, \quad |\xi'| = |\xi_T|\sin\beta = c_{\text{ext}}^{-1}\tau\sin\beta \quad (9\text{-}40)$$

By (9-40), we recover Snell's law

$$\frac{\sin\alpha}{\sin\beta} = \frac{c_{\text{int}}}{c_{\text{ext}}}, \quad (9\text{-}41)$$

Assume now that $c_{\text{int}} < c_{\text{ext}}$, which is the case where there might be no transmitted ray. Denote by

$$\alpha_0(x) = \arcsin(c_{\text{int}}(x)/c_{\text{ext}}(x)) \quad (9\text{-}42)$$

the critical angle at any $x \in \Gamma$ that places $(\xi^1)'$ in the glancing manifold with respect to c_{ext}. Then the transmitted wave does not exist when $\alpha > \alpha_0$; more precisely we do not have a real phase function φ_T in that case. It exists, when $\alpha < \alpha_0$. In the critical case $\alpha = \alpha_0$, this construction provides an outgoing ray tangent to Γ that we are not going to analyze.

The full internal reflection case. Assume now that $(\xi^1)'$ is in the elliptic region with respect to c_{ext}, then there is no transmitted singularity, but one can still construct a parametrix for the "evanescent" wave in Ω_{ext}; and there is a reflected ray. This is known as a full internal reflection. We give details below.

We proceed as above with one essential difference. There is no real-valued solution φ_T to the eikonal equation (9-10) outside Ω_0. Similarly to (9-26), we get formally,

$$\frac{\partial\varphi_T}{\partial\nu} = i\sqrt{|\xi'|^2 - c_{\text{ext}}^{-2}\tau^2} \quad \text{for } x^n = 0. \quad (9\text{-}43)$$

The choice of the sign of the square root is dictated by the requirement that the parametrix (9-9) with $\sigma = T$ be exponentially decreasing away from Γ instead of exponentially increasing.

In general, the eikonal equation may not be solvable but one can still construct solutions modulo $O((x^n)^\infty)$. The same applies to the transport equations. One can show that the $O((x^n)^\infty)$ error does not change the properties of u_T to be a parametrix. In particular, in (9-33) in this case one gets

$$E_{\Omega_{\text{ext}}}(\boldsymbol{u}_T(t_2)) \cong 0, \tag{9-44}$$

because the principal term of $\partial \bar{u}_T / \partial \nu$ in (9-32) now is pure imaginary instead of being real. Moreover, u_T is smooth in $\overline{\Omega}_{\text{ext}}$. Therefore, no energy, as far as the principal part only is considered, is transmitted to Ω_{ext}. That does not mean that the solution vanishes there, of course.

Glancing, gliding rays and other cases. We do not analyze the cases where $(\xi^1)'$ is in the glancing manifold with respect one of the speeds. We can do that because the analysis of those cases in not needed because of our assumptions guaranteeing no tangent rays. The analysis there is more delicate, and we refer to [Taylor 1976; Petkov 1982a; 1982b] for more details and examples. We do not analyze either the case where $(\xi^1)'$ is in the elliptic region with respect to either speed.

Justification of the parametrix. Denote by $\boldsymbol{u}_R = [u_R, \partial_t u_R]$, $\boldsymbol{u}_T = [u_T, \partial_t u_T]$ the approximate solutions constructed above, defined for t in some neighborhood of t_2. Then $\boldsymbol{u}_R = V_R h$, $\boldsymbol{u}_T = V_T h$, where $V_{R,T}$ are the FIOs constructed above. Let $\boldsymbol{u}_+$ be the solution of (9-3) defined above, with initial data $\Pi_+ \boldsymbol{f}$ at $t = 0$ having wave front set in a small neighborhood of (x_0, ξ^0). The map $\Lambda_+ : \boldsymbol{f} \mapsto u_+|_{\mathbb{R} \times \Gamma} = h$ is an FIO described in [Stefanov and Uhlmann 2009b]. Then near (t_1, x_1),

$$\boldsymbol{u}_R = V_R \Lambda \boldsymbol{f}, \quad \boldsymbol{u}_T = V_T \Lambda \boldsymbol{f},$$

the former supported in $\mathbb{R} \times \overline{\Omega}_{\text{int}}$, and the later in $\mathbb{R} \times \overline{\Omega}_{\text{ext}}$. So far we had two objects that we denoted by u_+: first, the parametrix of the solution of (9-3) corresponding to the positive sound speed $c(x)|\xi|$; and the parametrix in $\mathbb{R} \times \overline{\Omega}_{\text{int}}$ for the incoming solution corresponding to boundary value h. When $h = \Lambda_+ \boldsymbol{f}$, those two parametrices coincide up to a smooth term, as it is not hard to see (the second one is a back-projection and is discussed in [Stefanov and Uhlmann 2009b], in fact). This justifies the same notation for them that we will keep.

Consider the parametrix $v_p := u_+ + u_R + u_T$. We can always assume that its support is in some small neighborhood of the geodesic that hits $\mathbb{R} \times \Gamma$ at (t_1, x_1) and is tangent to ξ^1 there; and then reflects, and another branch refracts (see Figure 3 on page 300). In particular, then h has t-support near $t = t_1$, let us say that this included in the interval $[t_1-\varepsilon, t_1+\varepsilon]$ with some $\varepsilon > 0$. At $t = t_2 := t_1 + 2\varepsilon$, let x_2 be the position of the reflected ray, and let ξ^2 be its unit co-direction. Then $\mathrm{WF}(u_R(t_2, \cdot))$ is in a small conic neighborhood of (x_2, ξ^2).

Let $v(t, \cdot) = e^{tP}\Pi_+ f$ be the exact solution — see (A-22) — with some fixed choice of the parametrix Q^{-1} in the definition of Π_+, properly supported. Consider $w = v - v_p$ in $[0, t_2] \times \mathbb{R}^n$. It satisfies

$$(\partial_t^2 - c^2\Delta)w|_{[0,t_2]\times\overline{\Omega}_{\mathrm{int}}} \in C^\infty, \qquad (\partial_t^2 - c^2\Delta)w|_{[0,t_2]\times\overline{\Omega}_{\mathrm{ext}}} \in C^\infty, \qquad (9\text{-}45)$$
$$w|_{[0,t_2]\times\Gamma_{\mathrm{ext}}} - w|_{[0,t_2]\times\Gamma_{\mathrm{int}}} \in C^\infty, \quad \frac{\partial w}{\partial\nu}\Big|_{[0,t_2]\times\Gamma_{\mathrm{ext}}} - \frac{\partial w}{\partial\nu}\Big|_{[0,t_2]\times\Gamma_{\mathrm{int}}} \in C^\infty.$$

On the other hand, for $0 \le t \ll 1$, v is smooth. Let $\chi \in C^\infty(\mathbb{R})$ be a function that vanishes in $(-\infty, \delta]$ and equals 1 on $[2\delta, \infty)$, $0 < \delta \ll 1$. Then $\tilde{w} := \chi(t)w(t, x)$ still satisfies (9-45) and also vanishes for $t \le 0$. By [Williams 1992, Theorem 1.36], $\tilde{w}$ is smooth in $[0, t_2] \times \overline{\Omega}_{\mathrm{int}}$, up to the boundary, and is also smooth in $[0, t_2] \times \overline{\Omega}_{\mathrm{ext}}$, up to the boundary. Therefore,

$$v(t, \cdot) = v_p(t, \cdot) + K_t f, \qquad (9\text{-}46)$$

for any $t \in [0, t_2]$, where K_t is a compact operator in $\mathcal{H}$, depending smoothly on t. The operator K_t depends on Q as well. Therefore, the parametrix coincides with the exact solution up to a compact operator that is also smoothing in the sense described above.

This concludes the description of the microlocal part of the proof. The rest of the proof of Theorem 9.1 is as indicated above. Write $A\Lambda = \mathrm{Id} - K$, as in the smooth case. This time K is not compact any more, regardless of how large T is. Based on our assumptions and on what we proved, its essential spectrum is supported in a disk $|z| < C_0 < 1$ in the complex plane; and by unique continuation, we still have (4-10). This situation is similar to the proof of Theorem 4.1; see (4-14). The difference is that in the smooth case, $C_0 = 1/2$, if $T_1/2 < T < T_1$, and $C_0 = 0$, if $T > T_1$, while in the "skull" case, $0 < C_1 < 1$ and we can only make C_1 as small as we want but not zero, as $T \to \infty$, under our assumptions.

Numerical experiments done in [Qian et al. 2011] based on this approach show that one gets very good reconstruction even without restricting the support of f to sets $\mathcal{H}$ satisfying (9-5), i.e., if we allow for invisible singularities. The reconstruction is worse in the trapping region, and trapped conormal singularities are not recovered.

The partial data case for a discontinuous speed, i.e, when we have data on a part of $\partial\Omega$ has not been studied yet. It seems plausible that the methods in [Stefanov and Uhlmann 2009b] for a smooth speed described above can be extended but there are new technical difficulties. Even for a smooth speed however, a convergent series solution is not known. On the other hand, such reconstruction has been tried numerically in [Qian et al. 2011] with success. Under the condition that all singularities issued from supp f are visible, for a smooth speed, the inverse problem reduces to a Fredholm equation with a trivial kernel. For a discontinuous speed of the type we study in this paper, it follows from our analysis that we still get a Fredholm equation but the triviality of the kernel is a more delicate question.

Appendix: Microlocal analysis and geometric optics

One of the fundamental ideas of classical analysis is a thorough study of functions near a point, i.e., locally. Microlocal analysis, loosely speaking, is analysis near points and directions, i.e., in the "phase space".

A.1. Wave front sets.
The phase space in $\mathbb{R}^n$ is the cotangent bundle $T^*\mathbb{R}^n$ that can be identified with $\mathbb{R}^n \times \mathbb{R}^n$. Given a distribution $f \in \mathcal{D}'(\mathbb{R}^n)$, a fundamental object to study is the wave front set $\mathrm{WF}(f) \subset T^*\mathbb{R}^n \setminus 0$ that we define below.

The basic idea goes back to the properties of the Fourier transform. If f is an integrable compactly supported function, one can tell whether f is smooth by looking at the behavior of $\hat{f}(\xi)$ (that is smooth, even analytic) when $|\xi| \to \infty$. It is known that f is smooth if and only if for any N, $|\hat{f}(\xi)| \leq C_N |\xi|^{-N}$ for some C_N. If we localize this requirement to a conic neighborhood V of some $\xi_0 \neq 0$ (V is conic if $\xi \in V \Rightarrow t\xi \in V$, $\forall t > 0$), then we can think of this as a smoothness in the cone V. To localize in the base x variable however, we first have to cut smoothly near a fixed x_0.

We say that $(x_0, \xi_0) \in \mathbb{R}^n \times (\mathbb{R}^n \setminus 0)$ is *not* in the wave front set $\mathrm{WF}(f)$ of $f \in \mathcal{D}'(\mathbb{R}^n)$ if there exists $\phi \in C_0^\infty(\mathbb{R}^n)$ with $\phi(x_0) \neq 0$ such that for any N, there exists C_N so that

$$|\widehat{\phi f}(\xi)| \leq C_N |\xi|^{-N}$$

for ξ in some conic neighborhood of ξ_0. This definition is independent of the choice of ϕ. If $f \in \mathcal{D}'(\Omega)$ with some open $\Omega \subset \mathbb{R}^n$, to define $\mathrm{WF}(f) \subset \Omega \times (\mathbb{R}^n \setminus 0)$, we need to choose $\phi \in C_0^\infty(\Omega)$. Clearly, the wave front set is a closed conic subset of $\mathbb{R}^n \times (\mathbb{R}^n \setminus 0)$. Next, multiplication by a smooth function cannot enlarge the wave front set. The transformation law under coordinate changes is that of covectors making it natural to think of $\mathrm{WF}(f)$ as a subset of $T^*\mathbb{R}^n \setminus 0$, or $T^*\Omega \setminus 0$, respectively.

The wave front set $\mathrm{WF}(f)$ generalizes the notion $\mathrm{singsupp}(f)$ — the complement of the largest open set where f is smooth. The points (x, ξ) in $\mathrm{WF}(f)$ are referred too as *singularities* of f. Its projection onto the base is

$$\mathrm{singsupp}(f) = \{x;\ \exists \xi, (x, \xi) \in \mathrm{WF}(f)\}.$$

Examples. (a) $\mathrm{WF}(\delta) = \{(0, \xi); \xi \neq 0\}$. In other words, the Dirac delta function is singular at $x = 0$ in all directions.

(b) Let $x = (x', x'')$, where $x' = (x_1, \ldots, x_k)$, $x'' = (x_{k+1}, \ldots, x_n)$ with some k. Then $\mathrm{WF}(\delta(x')) = \{(0, x'', \xi', 0), \xi' \neq 0\}$, where $\delta(x')$ is the Dirac delta function on the plane $x' = 0$, defined by $\langle \delta(x'), \phi \rangle = \int \phi(0, x'') \, dx''$. In other words, $\mathrm{WF}(\delta(x'))$ consists of all (co)vectors with a base point on that plane, perpendicular to it.

(c) Let f be a piecewise smooth function that has a nonzero jump across some smooth surface S. Then $\mathrm{WF}(f)$ consists of all (co)vectors at points of S, normal to it. This follows from (a) and a change of variables that flatten S locally.

(d) Let $f = \mathrm{pv}\frac{1}{x} - \pi i \delta(x)$ in $\mathbb{R}$. Then $\mathrm{WF}(f) = \{(0, \xi); \xi > 0\}$.

In example (d) we see a distribution with a wave front set that is not symmetric under the change $\xi \mapsto -\xi$. In fact, wave front sets do not have a special structure except for the requirement to be closed conic sets; given any such set, there is a distribution with a wave front set exactly that set.

Two distributions cannot be multiplied in general. However, if their wave front sets do not intersect, there is a "natural way" to define a product.

A.2. Pseudodifferential operators.

Definitions. We first define the symbol class $S^m(\Omega)$, $m \in \mathbb{R}$, as the set of all smooth functions $p(x, \xi)$, $(x, \xi) \in \Omega \times \mathbb{R}^n$, called symbols, satisfying the following symbol estimates: for any compact $K \subset \Omega$, and any multi-indices α, β, there is a constant $C_{K,\alpha,\beta} > 0$ such that

$$|\partial_\xi^\alpha \partial_x^\beta p(x, \xi)| \leq C_{K,\alpha,\beta}(1 + |\xi|)^{m-|\alpha|} \quad \text{for all } (x, \xi) \in K \times \mathbb{R}^n. \quad \text{(A-1)}$$

More generally, one can define the class $S_{\rho,\delta}^m(\Omega)$ with $0 \leq \rho, \delta \leq 1$ by replacing $m - |\alpha|$ there by $m - \rho|\alpha| + \delta|\beta|$. Then $S^m(\Omega) = S_{1,0}^m(\Omega)$. Often, we omit Ω and simply write S^m. There are other classes in the literature, for example $\Omega = \mathbb{R}^n$, and (A-1) is required to hold for all $x \in \mathbb{R}^n$.

The estimates (A-1) do not provide any control of p when x approaches boundary points of Ω, or ∞.

Given $p \in S^m(\Omega)$, we define the $\Psi\mathrm{DO}$ with symbol p, denoted by $p(x, D)$, by

$$p(x, D) f = (2\pi)^{-n} \int e^{ix \cdot \xi} p(x, \xi) \hat{f}(\xi) \, d\xi, \quad f \in C_0^\infty(\Omega). \quad \text{(A-2)}$$

The definition is inspired by the following. If $P = \sum_{|\alpha| \leq m} a_\alpha(x) D^\alpha$ is a differential operator, where $D = -i\partial$, then using the Fourier inversion formula we can write P as in (A-2) with a symbol $p = \sum_{|\alpha| \leq m} a_\alpha(x) \xi^\alpha$ that is a polynomial in ξ with x-dependent coefficients. The symbol class S^m allows for more general functions. The class of the pseudodifferential operators with symbols in S^m is denoted usually by Ψ^m. The operator P is called a ΨDO if it belongs to Ψ^m for some m. By definition, $S^{-\infty} = \cap_m S^m$, and $\Psi^{-\infty} = \cap_m \Psi^m$.

An important subclass is the set of the *classical symbols* that have an asymptotic expansion of the form

$$p(x,\xi) \sim \sum_{j=0}^{\infty} p_{m-j}(x,\xi), \tag{A-3}$$

where $m \in \mathbb{R}$, and p_{m-j} are smooth and positively homogeneous in ξ of order $m - j$ for $|\xi| > 1$, i.e., $p_{m-j}(x, \lambda\xi) = \lambda^{m-j} p_{m-j}(x, \xi)$ for $|\xi| > 1$, $\lambda > 1$; and the sign $\sim$ means that

$$p(x,\xi) - \sum_{j=0}^{N} p_{m-j}(x,\xi) \in S^{m-N-1} \quad \text{for all } N \geq 0. \tag{A-4}$$

Any ΨDO $p(x, D)$ is continuous from $C_0^\infty(\Omega)$ to $C^\infty(\Omega)$, and can be extended by duality as a continuous map from $\mathcal{E}'(\Omega)$ to $\mathcal{D}'(\Omega)$.

Principal symbol. The principal symbol of a ΨDO given by (A-2) is the equivalence class $S^m(\Omega)/S^{m-1}(\Omega)$, and any its representative is called a principal symbol as well. In case of classical ΨDOs, the convention is to choose the principal symbol to be the first term p_m, that in particular is positively homogeneous in ξ.

Smoothing operators. Those are operators than map continuously $\mathcal{E}'(\Omega)$ into $C^\infty(\Omega)$. They coincide with operators with smooth Schwartz kernels in $\Omega \times \Omega$. They can always be written as ΨDOs with symbols in $S^{-\infty}$, and vice versa — all operators in $\Psi^{-\infty}$ are smoothing. Smoothing operators are viewed in this calculus as negligible and ΨDOs are typically defined modulo smoothing operators, i.e., $A = B$ if and only if $A - B$ is smoothing. Smoothing operators are not "small".

The pseudolocal property. For any ΨDO P and any $f \in \mathcal{E}'(\Omega)$,

$$\operatorname{singsupp}(Pf) \subset \operatorname{singsupp} f. \tag{A-5}$$

In other words, a ΨDO cannot increase the singular support. This property is preserved if we replace singsupp by WF; see (A-11).

Symbols defined by an asymptotic expansion. In many applications, a symbol is defined by consecutively constructing symbols $p_j \in S^{m_j}$, $j = 0, 1, \ldots$, where $m_j \searrow -\infty$, and setting

$$p(x,\xi) \sim \sum_j p_j(x,\xi). \tag{A-6}$$

The series on the right may not converge but we can make it convergent by using our freedom to modify each p_j for ξ in expanding compact sets without changing the large ξ behavior of each term. This extends the Borel idea of constructing a smooth function with prescribed derivatives at a fixed point. The asymptotic (A-6) then is understood in a sense similar to (A-4). This shows that there exists a symbol $p \in S^{m_0}$ satisfying (A-6). That symbol is not unique but the difference of two such symbols is always in $S^{-\infty}$.

Amplitudes. A seemingly larger class of ΨDOs is defined by

$$Af = (2\pi)^{-n} \int e^{i(x-y)\cdot\xi} a(x, y, \xi) f(y) \, dy \, d\xi, \quad f \in C_0^\infty(\Omega), \qquad \text{(A-7)}$$

where the amplitude a satisfies

$$|\partial_\xi^\alpha \partial_x^\beta \partial_y^\gamma a(x, y, \xi)| \leq C_{K,\alpha,\beta,\gamma} (1+|\xi|)^{m-|\alpha|} \quad \text{for all } (x, y, \xi) \in K \times \mathbb{R}^n \qquad \text{(A-8)}$$

for any compact $K \subset \Omega \times \Omega$, and any α, β, γ. In fact, any such ΨDO A is a ΨDO with a symbol $p(x, \xi)$ (independent of y) with the formal asymptotic expansion

$$p(x, \xi) \sim \sum_{\alpha \geq 0} D_\xi^\alpha \partial_y^\alpha a(x, x, \xi).$$

In particular, the principal symbol of that operator can be taken to be $a(x, x, \xi)$.

Transpose and adjoint operators to a ΨDO. The mapping properties of any ΨDO A indicate that it has a well defined transpose A', and a complex adjoint A^* with the same mapping properties. They satisfy

$$\langle Au, v \rangle = \langle u, A'v \rangle, \quad \langle Au, \bar{v} \rangle = \langle u, \overline{A^*v} \rangle \quad \text{for all } u, v \in C_0^\infty$$

where $\langle \cdot, \cdot \rangle$ is the pairing in distribution sense; and in this particular case just an integral of uv. In particular, $A^*u = \overline{A'\bar{u}}$, and if A maps L^2 to L^2 in a bounded way, then A^* is the adjoint of A in L^2 sense.

The transpose and the adjoint are ΨDOs in the same class with amplitudes $a(y, x, -\xi)$ and $\bar{a}(y, x, \xi)$, respectively; and symbols

$$\sum_{\alpha \geq 0} (-1)^{|\alpha|} \frac{1}{\alpha!} (\partial_\xi^\alpha D_x^\alpha p)(x, -\xi), \quad \sum_{\alpha \geq 0} \frac{1}{\alpha!} \partial_\xi^\alpha D_x^\alpha \bar{p}(x, \xi),$$

if $a(x, y, \xi)$ and $p(x, \xi)$ are the amplitude and/or the symbol of that ΨDO. In particular, the principal symbols are $p_0(x, -\xi)$ and $\bar{p}_0(x, \xi)$, respectively, where p_0 is (any representative of) the principal symbol.

Composition of ΨDOs and ΨDOs with properly supported kernels. Given two ΨDOs A and B, their composition may not be defined even if they are smoothing ones because each one maps C_0^∞ to C^∞ but may not preserve the compactness of the support. For example, if $A(x, y)$, and $B(x, y)$ are their Schwartz kernels, the candidate for the kernel of AB given by $\int A(x, z)B(z, y)$ may be a divergent integral. On the other hand, for any ΨDO A, one can find a smoothing correction R such that $A + R$ has properly supported kernel, i.e., the kernel of $A + R$, has a compact intersection with $K \times \Omega$ and $\Omega \times K$ for any compact $K \subset \Omega$. The proof of this uses the fact that the Schwartz kernel of a ΨDO is smooth away from the diagonal $\{x = y\}$ and one can always cut there in a smooth way to make the kernel properly supported at the price of a smoothing error. ΨDOs with properly supported kernels preserve $C_0^\infty(\Omega)$, and also $\mathcal{E}'(\Omega)$, and therefore can be composed in either of those spaces. Moreover, they map $C^\infty(\Omega)$ to itself, and can be extended from $\mathcal{D}'(\Omega)$ to itself. The property of the kernel to be properly supported is often assumed, and it is justified by considering each ΨDO as an equivalence class.

If $A \in \Psi^m(\Omega)$ and $B \in \Psi^k(\Omega)$ are properly supported ΨDOs with symbols a and b, respectively, then AB is again a ΨDO in $\Psi^{m+k}(\Omega)$ and its symbol is given by

$$\sum_{\alpha \geq 0} (-1)^{|\alpha|} \frac{1}{\alpha!} \partial_\xi^\alpha a(x, \xi) D_x^\alpha b(x, \xi).$$

In particular, the principal symbol can be taken to be ab.

Change of variables and ΨDOs on manifolds. Let Ω' be another domain, and let $\phi : \Omega \to \tilde\Omega$ be a diffeomorphism. For any $P \in \Psi^m(\Omega)$, $\tilde P f := (P(f \circ \phi)) \circ \phi^{-1}$ maps $C_0^\infty(\tilde\Omega)$ into $C^\infty(\tilde\Omega)$. It is a ΨDO in $\Psi^m(\tilde\Omega)$ with principal symbol

$$p(\phi^{-1}(y), (\mathrm{d}\phi)' \eta) \tag{A-9}$$

where p is the symbol of P, $\mathrm{d}\phi$ is the Jacobi matrix $\{\partial \phi_i / \partial x_j\}$ evaluated at $x = \phi^{-1}(y)$, and $(\mathrm{d}\phi)'$ stands for the transpose of that matrix. We can also write $(\mathrm{d}\phi)' = ((\mathrm{d}\phi^{-1})^{-1})'$. An asymptotic expansion for the whole symbol can be written down as well.

Relation (A-9) shows that the transformation law under coordinate changes is that of a covector. Therefore, the principal symbol is a correctly defined function on the cotangent bundle $T^*\Omega$. The full symbol is not invariantly defined there in general.

Let M be a smooth manifold, and $A : C_0^\infty(M) \to C^\infty(M)$ be a linear operator. We say that $A \in \Psi^m(M)$, if its kernel is smooth away from the diagonal in $M \times M$, and if in any coordinate chart (A, χ), where $\chi : U \to \Omega \subset \mathbb{R}^n$, we have $(A(u \circ \chi)) \circ \chi^{-1} \in \Psi^m(\Omega)$. As before, the principal symbol of A, defined in any local chart, is an invariantly defined function on T^*M.

Mapping properties in Sobolev spaces. In $\mathbb{R}^n$, Sobolev spaces $H^s(\mathbb{R}^n)$, $s \in \mathbb{R}$, are defined as the completion of $\mathcal{S}'(\mathbb{R}^n)$ in the norm

$$\|f\|^2_{H^s(\mathbb{R}^n)} = \int (1+|\xi|^2)^s |\hat{f}(\xi)|^2 \, d\xi.$$

When s is a nonnegative integer, an equivalent norm is the square root of $\sum_{|\alpha| \le s} \int |\partial^\alpha f(x)|^2 \, dx$. For such s, and a bounded domain Ω, one defines $H^s(\Omega)$ as the completion of $C^\infty(\overline{\Omega})$ using the latter norm with the integral taken in Ω. Sobolev spaces in Ω for other real values of s are defined by different means, including duality or complex interpolation.

Sobolev spaces are also Hilbert spaces.

Any $P \in \Psi^m(\Omega)$ is a continuous map from $H^s_{\text{comp}}(\Omega)$ to $H^{s-m}_{\text{loc}}(\Omega)$. If the symbols estimates (A-1) are satisfied in the whole $\mathbb{R}^n \times \mathbb{R}^n$, then $P : H^s(\mathbb{R}^n) \to H^{s-m}(\mathbb{R}^n)$.

Elliptic ΨDOs and their parametrices. The operator $P \in \Psi^m(\Omega)$ with symbol p is called elliptic of order m, if for any compact $K \subset \Omega$, there exist constants $C > 0$ and $R > 0$ such that

$$C|\xi|^m \le |p(x,\xi)| \quad \text{for } x \in K, \text{ and } |\xi| > R. \tag{A-10}$$

Then the symbol p is called also elliptic of order m. It is enough to require the principal symbol only to be elliptic (of order m). For classical ΨDOs, as in (A-3), the requirement can be written as $p_m(x,\xi) \ne 0$ for $\xi \ne 0$. A fundamental property of elliptic operators is that they have parametrices. In other words, given an elliptic ΨDO P of order m, there exists $Q \in \Psi^{-m}(\Omega)$ such that

$$QP - \text{Id} \in \Psi^{-\infty}, \quad PQ - \text{Id} \in \Psi^{-\infty}.$$

The proof of this is to construct a left parametrix first by choosing a symbol $q_0 = 1/p$, cut off near the possible zeros of p, that form a compact any time when x is restricted to a compact as well. The corresponding ΨDO Q_0 will then satisfy $Q_0 P = \text{Id} + R$, $R \in \Psi^{-1}$. Then we take a ΨDO E with asymptotic expansion $E \sim \text{Id} - R + R^2 - R^3 + \cdots$, that would be the formal Neumann series expansion of $(\text{Id} + R)^{-1}$, if the latter existed. Then EQ_0 is a left parametrix that is also a right parametrix.

An important consequence is the following elliptic regularity statement. If P is elliptic (and properly supported), then

$$\text{singsupp}(PF) = \text{singsupp}(f) \quad \text{for all } f \in \mathcal{D}'(\Omega).$$

In particular, $Pf \in C^\infty$ implies $f \in C^\infty$.

ΨDOs and wave front sets. The microlocal version of the pseudolocal property is given by the following:

$$\mathrm{WF}(Pf) \subset \mathrm{WF}(f) \tag{A-11}$$

for any (properly supported) ΨDO P and $f \in \mathcal{D}'(\Omega)$. In other words, a ΨDO cannot increase the wave front set. If P is elliptic for some m, it follows from the existence of a parametrix that there is equality above, i.e., $\mathrm{WF}(Pf) = \mathrm{WF}(f)$.

We say that the ΨDO P is of order $-\infty$ in the open conic set $U \subset T^*\Omega \setminus 0$, if for any closed conic set $K \subset U$ with a compact projection on the base "x-space", (A-1) is fulfilled for any m. The *essential support* $\mathrm{ES}(P)$, sometimes also called the *microsupport* of P, is defined as the smallest closed conic set on the complement of which the symbol p is of order $-\infty$. Then

$$\mathrm{WF}(Pf) \subset \mathrm{WF}(f) \cap \mathrm{ES}(P).$$

Let P have a homogeneous principal symbol p_m. The characteristic set Char P is defined by

$$\operatorname{Char} P = \{(x,\xi) \in T^*\Omega \setminus 0; \; p_m(x,\xi) = 0\}.$$

Char P can be defined also for general ΨDOs that may not have homogeneous principal symbols. For any ΨDO P, we have

$$\mathrm{WF}(f) \subset \mathrm{WF}(Pf) \cup \operatorname{Char} P \quad \text{for all } f \in \mathcal{E}'(\Omega). \tag{A-12}$$

P is called *microlocally elliptic* in the open conic set U, if (A-10) is satisfied in all compact subsets, similarly to the definition of $\mathrm{ES}(P)$ above. If it has a homogeneous principal symbol p_m, ellipticity is equivalent to $p_m \neq 0$ in U. If P is elliptic in U, then Pf and f have the same wave front set restricted to U, as follows from (A-12) and (A-11).

A.3. The Hamilton flow and propagation of singularities. Let $P \in \Psi^m(M)$ be properly supported, where M is a smooth manifold, and suppose that P has a real homogeneous principal symbol p_m. The Hamiltonian vector field of p_m on $T^*M \setminus 0$ is defined by

$$H_{p_m} = \sum_{j=1}^n \left(\frac{\partial p_m}{\partial x_j} \frac{\partial}{\partial \xi_j} - \frac{\partial p_m}{\partial \xi_j} \frac{\partial}{\partial x_j} \right).$$

The integral curves of H_{p_m} are called *bicharacteristics* of P. Clearly, $H_{p_m} p_m = 0$, thus p_m is constant along each bicharacteristics. The bicharacteristics along which $p_m = 0$ are called *zero bicharacteristics*.

The Hörmander's theorem about propagation of singularities is one of the fundamental results in the theory. It states that if P is an operator as above, and

$Pu = f$ with $u \in \mathcal{D}'(M)$, then
$$\mathrm{WF}(u) \setminus \mathrm{WF}(f) \subset \mathrm{Char}\, P,$$
and is invariant under the flow of H_{p_m}.

An important special case is the wave operator $P = \partial_t^2 - \Delta_g$, where Δ_g is the Laplace Beltrami operator associated with a Riemannian metric g. We may add lower order terms without changing the bicharacteristics. Let (τ, ξ) be the dual variables to (t, x). The principal symbol is $p_2 = -\tau^2 + |\xi|_g^2$, where $|\xi|_g^2 := \sum g^{ij}(x)\xi_i\xi_j$, and $(g^{ij}) = (g_{ij})^{-1}$. The bicharacteristics equations then are $\dot{t} = 0$, $\dot{\tau} = -2\tau$, $\dot{x}^j = 2\sum g^{ij}\xi_i$, $\dot{\xi}_j = -2\partial_{x^j}\sum g^{ij}(x)\xi_i\xi_j$, and they are null one if $\tau^2 = |\xi|_g^2$. Here, $\dot{x} = dx/ds$, etc. The latter two equations are the Hamiltonian curves of $\tilde{H} := \sum g^{ij}(x)\xi_i\xi_j$ and they are known to coincide with the geodesics $(\gamma, \dot{\gamma})$ on TM when identifying vectors and covectors by the metric. They lie on the energy surface $\tilde{H} = \mathrm{const}$. The first two equations imply that τ is a constant, positive or negative, and up to rescaling, one can choose the parameter along the geodesics to be t. That rescaling forces the speed along the geodesic to be 1. The null condition $\tau^2 = |\xi|_g^2$ defines two smooth surfaces away from $(\tau, \xi) = (0, 0)$: $\tau = \pm|\xi|_g$. This corresponds to geodesics starting from x in direction either ξ or $-\xi$. To summarize, for the homogeneous equation $Pu = 0$, we get that each singularity (x, ξ) of the initial conditions at $t = 0$ starts to propagate from x in direction either ξ or $-\xi$ or both (depending on the initial conditions) along the unit speed geodesic. In fact, we get this first for the singularities in $T^*(\mathbb{R}_t \times \mathbb{R}_x^n)$ first, but since they lie in $\mathrm{Char}\, P$, one can see that they project to $T^*\mathbb{R}_x^n$ as singularities again.

A.4. Geometric optics. Geometric optics describes asymptotically the solutions of hyperbolic equations at large frequencies. It also provides a parametrix (a solution up to smooth terms) of the initial value problem for hyperbolic equations. The resulting operators are not ΨDOs anymore; they are actually examples of Fourier integral operators. Geometric optics also studies the large frequency behavior of solutions that reflect from a smooth surface (obstacle scattering) including diffraction; reflect from an edge or a corner; reflect and refract from a surface where the speed jumps (transmission problems).

As an example, consider the acoustic equation

$$(\partial_t^2 - c^2(x)\Delta)u = 0, \quad (t, x) \in \mathbb{R}^n, \tag{A-13}$$

with initial conditions $u(0, x) = f_1(x)$, $u_t(0, x) = f_2$. It is enough to assume first that f_1 and f_2 are in C_0^∞, and extend the resulting solution operator to larger spaces later.

We are looking for a solution of the form
$$u(t,x) = (2\pi)^{-n}$$
$$\times \sum_{\sigma=\pm} \int e^{i\phi_\sigma(t,x,\xi)} \big(a_{1,\sigma}(x,\xi,t)\hat{f}_1(\xi) + |\xi|^{-1} a_{2,\sigma}(x,\xi,t)\hat{f}_2(\xi)\big) d\xi, \quad \text{(A-14)}$$

modulo terms involving smoothing operators of f_1 and f_2. The reason to expect two terms is already clear by the propagation of singularities theorem, and is also justified by the eikonal equation below. Here the phase functions $\phi_\pm$ are positively homogeneous of order 1 in ξ. Next, we seek the amplitudes in the form

$$a_{j,\sigma} \sim \sum_{k=0}^\infty a_{j,\sigma}^{(k)}, \quad \sigma = \pm, \ j = 1,2, \quad \text{(A-15)}$$

where $a_{j,\sigma}^{(k)}$ is homogeneous in ξ of degree $-k$ for large $|\xi|$. To construct such a solution, we plug (A-14) into (A-13) and try to kill all terms in the expansion in homogeneous (in ξ) terms.

Equating the terms of order 2 yields the *eikonal equation*

$$(\partial_t \phi)^2 - c^2(x)|\nabla_x \phi|^2 = 0. \quad \text{(A-16)}$$

Write $f_j = (2\pi)^{-n} \int e^{ix\cdot\xi} \hat{f}_j(\xi)\, d\xi$, $j = 1,2$, to get the following initial conditions for $\phi_\pm$

$$\phi_\pm|_{t=0} = x\cdot\xi. \quad \text{(A-17)}$$

The eikonal equation can be solved by the method of characteristics. First, we determine $\partial_t \phi$ and $\nabla_x \phi$ for $t = 0$. We get $\partial_t \phi|_{t=0} = \mp c(x)|\xi|$, $\nabla_x \phi|_{t=0} = \xi$. This implies existence of two solutions $\phi_\pm$. If $c = 1$, we easily get $\phi_\pm = \mp|\xi|t + x\cdot\xi$. Let for any (z,ξ), $\gamma_{z,\xi}(s)$ be unit speed geodesic through (z,ξ). Then ϕ_+ is constant along the curve $(t, \gamma_{z,\xi}(t))$ that implies that $\phi_+ = z(x,\xi)\cdot\xi$ in any domain in which (t,z) can be chosen to be coordinates. Similarly, ϕ_- is constant along the curve $(t, \gamma_{z,-\xi}(t))$. In general, we cannot solve the eikonal equation globally, for all (t,x). Two geodesics $\gamma_{z,\xi}$ and $\gamma_{w,\xi}$ may intersect, for example, giving a value for $\phi_\pm$ that is not unique. However, we always have a solution in a neighborhood of $t = 0$.

Now equate the first-order terms in the expansion of $(\partial_t^2 - c^2\Delta)u$ to get that the principal terms of the amplitudes must solve the *transport equation*

$$\big((\partial_t \phi_\pm)\partial_t - c^2 \nabla_x \phi_\pm \cdot \nabla_x + C_\pm\big) a_{j,\pm}^{(0)} = 0, \quad \text{(A-18)}$$

with

$$2C_\pm = (\partial_t^2 - c^2\Delta)\phi_\pm.$$

This is an ODE along the vector field $(\partial_t \phi_\pm, c^2 \nabla_x \phi)$, and the integral curves of it coincide with the curves $(t, \gamma_{z, \pm\xi})$. Given an initial condition at $t = 0$, it has a unique solution along the integral curves as long as ϕ is well defined.

Equating terms homogeneous in ξ of lower order we get transport equations for $a_{j,\sigma}^{(k)}$, $j = 1, 2, \ldots$ with the same left-hand side as in (A-18) with a right-hand side determined by $a_{k,\sigma}^{(k-1)}$.

Taking into account the initial conditions, we get

$$a_{1,+} + a_{1,-} = 1, \quad a_{2,+} + a_{2,-} = 0 \quad \text{for } t = 0.$$

This is true in particular for the leading terms $a_{1,\pm}^{(0)}$ and $a_{2,\pm}^{(0)}$. Since $\partial_t \phi_\pm = \mp c(x)|\xi|$ for $t = 0$, and $u_t = f_2$ for $t = 0$, from the leading order term in the expansion of u_t we get

$$a_{1,+}^{(0)} = a_{1,-}^{(0)}, \quad ic(x)(a_{2,-}^{(0)} - a_{2,+}^{(0)}) = 1 \quad \text{for } t = 0.$$

Therefore,

$$a_{1,+}^{(0)} = a_{1,-}^{(0)} = \frac{1}{2}, \quad a_{2,+}^{(0)} = -a_{2,-}^{(0)} = \frac{i}{2c(x)} \quad \text{for } t = 0. \qquad \text{(A-19)}$$

Note that if $c = 1$, then $\phi_\pm = x \cdot \xi \mp t|\xi|$, and $a_{1,+} = a_{1,-} = 1/2$, $a_{2,+} = -a_{2,-} = i/2$. Using those initial conditions, we solve the transport equations for $a_{1,\pm}^{(0)}$ and $a_{2,\pm}^{(0)}$. Similarly, we derive initial conditions for the lower order terms in (A-15) and solve the corresponding transport equations. Then we define $a_{j,\sigma}$ by (A-15) as a symbol.

The so constructed u in (A-14) is a solution only up to smoothing operators applied to (f_1, f_2). Using standard hyperbolic estimates, we show that adding such terms to u, we get an exact solution to (A-13). As mentions above, this construction may fail for t too large, depending on the speed. On the other hand, the solution operator $(f_1, f_2) \mapsto u$ makes sense as a global Fourier integral operator for which this construction is just one if its local representations.

Projections to the positive and the negative wave speeds. The zeros of the principal symbol of the wave operator, in regions where c is smooth, are given by $\tau = \pm c(x)|\xi|$, that we call wave speeds. We constructed above parametrices $u_\pm$ for the corresponding solutions. We will present here a functional analysis point of view that allows us to project the initial data f to data $\Pi_\pm f$ such that, up to smoothing operators, $u_\pm$ corresponds to initial data $\Pi_\pm f$.

Assume that $c(x)$ is extended from the maximal connected component of $\mathbb{R}^n \setminus \Gamma$ containing x_0 to the whole $\mathbb{R}^n$ in a smooth way so that $0 < 1/C \le c(x) \le C$. Let

$$Q = (-c^2 \Delta)^{1/2}, \qquad \text{(A-20)}$$

where the operator in the parentheses is the natural self-adjoint extension of $-c^2\Delta$ to $L^2(\mathbb{R}^n, c^{-2}dx)$, and the square root exists by the functional calculus. Moreover, Q is an elliptic ΨDO of order 1 in any open set; and let Q^{-1} denote a fixed parametrix.

It is well known that the solution to (9-3) can be written as

$$u = \cos(tQ)f_1 + \frac{\sin(tQ)}{Q}f_2, \quad (A\text{-}21)$$

and the latter operator is defined by the functional calculus as $\phi(t, Q)$ with $\phi(t, \cdot) = \sin(t\cdot)/\cdot \in C^\infty$. Based on that, we can write

$$e^{tP} = e^{itQ}\Pi_+ + e^{-itQ}\Pi_-, \quad (A\text{-}22)$$

where

$$\Pi_+ = \frac{1}{2}\begin{pmatrix} 1 & -iQ^{-1} \\ iQ & 1 \end{pmatrix}, \quad \Pi_- = \frac{1}{2}\begin{pmatrix} 1 & iQ^{-1} \\ -iQ & 1 \end{pmatrix}. \quad (A\text{-}23)$$

It is straightforward to see that $\Pi\pm$ are orthogonal projections in $\mathcal{H}$, up to errors of smoothing type. Then given $f \in \mathcal{H}$ supported on Ω, one has $u_\pm = e^{tP}f_\pm$, with $f_\pm := \Pi_\pm f$.

References

[Agranovsky et al. 2009] M. Agranovsky, P. Kuchment, and L. Kunyansky, "On reconstruction formulas and algorithms for the thermoacoustic tomography", Chapter 8, pp. 89–101 in *Photoacoustic imaging and spectroscopy*, edited by L. V. Wang, Optical Science and Engineering **144**, CRC, Boca Raton, FL, 2009. arXiv 0706.1303

[Bal 2013] G. Bal, "Hybrid inverse problems and internal functionals", pp. 325–368 in *Inverse problems and applications: Inside out II*, edited by G. Uhlmann, Publ. Math. Sci. Res. Inst. **60**, Cambridge University Press, New York, 2013.

[Bardos et al. 1992] C. Bardos, G. Lebeau, and J. Rauch, "Sharp sufficient conditions for the observation, control, and stabilization of waves from the boundary", *SIAM J. Control Optim.* **30**:5 (1992), 1024–1065. MR 94b:93067 Zbl 0786.93009

[Bellassoued 2003] M. Bellassoued, "Carleman estimates and distribution of resonances for the transparent obstacle and application to the stabilization", *Asymptot. Anal.* **35**:3-4 (2003), 257–279. MR 2004i:35139 Zbl 1137.35388

[Cardoso et al. 1999] F. Cardoso, G. Popov, and G. Vodev, "Distribution of resonances and local energy decay in the transmission problem, II", *Math. Res. Lett.* **6**:4 (1999), 377–396. MR 2000i:35029 Zbl 0968.35035

[Chung et al. 2011] E. Chung, J. Qian, G. Uhlmann, and H. Zhao, "An adaptive phase space method with application to reflection traveltime tomography", *Inverse Problems* **27**:11 (2011), Art. ID #115002. MR 2851908 Zbl 05989438

[Duistermaat 1996] J. J. Duistermaat, *Fourier integral operators*, Progress in Mathematics **130**, Birkhäuser, Boston, MA, 1996. MR 96m:58245 Zbl 0841.35137

[Finch and Rakesh 2009] D. Finch and Rakesh, "Recovering a function from its spherical mean values in two and three dimensions", Chapter 7, pp. 77–88 in *Photoacoustic imaging and spectroscopy*, edited by L. V. Wang, Optical Science and Engineering **144**, CRC, Boca Raton, FL, 2009.

[Finch et al. 2004] D. Finch, S. K. Patch, and Rakesh, "Determining a function from its mean values over a family of spheres", *SIAM J. Math. Anal.* **35**:5 (2004), 1213–1240. MR 2005b:35290 Zbl 1073.35144

[Finch et al. 2007] D. Finch, M. Haltmeier, and Rakesh, "Inversion of spherical means and the wave equation in even dimensions", *SIAM J. Appl. Math.* **68**:2 (2007), 392–412. MR 2008k:35494 Zbl 1159.35073

[Haltmeier et al. 2004] M. Haltmeier, O. Scherzer, P. Burgholzer, and G. Paltauf, "Thermoacoustic computed tomography with large planar receivers", *Inverse Problems* **20**:5 (2004), 1663–1673. MR 2006a:35311 Zbl 1065.65143

[Haltmeier et al. 2005] M. Haltmeier, T. Schuster, and O. Scherzer, "Filtered backprojection for thermoacoustic computed tomography in spherical geometry", *Math. Methods Appl. Sci.* **28**:16 (2005), 1919–1937. MR 2006d:92023 Zbl 1085.65092

[Hansen 1984] S. Hansen, "Singularities of transmission problems", *Math. Ann.* **268**:2 (1984), 233–253. MR 86a:58107 Zbl 0523.58041

[Herglotz 1905] G. Herglotz, "Über die Elastizität der Erde bei Berücksichtigung ihrer variablen Dichte", *Z. Math. Phys.* **52** (1905), 275–299. JFM 36.1008.02

[Hörmander 1992] L. Hörmander, "A uniqueness theorem for second order hyperbolic differential equations", *Comm. Partial Differential Equations* **17**:5-6 (1992), 699–714. MR 93h:35116 Zbl 0815.35063

[Hristova 2009] Y. Hristova, "Time reversal in thermoacoustic tomography: an error estimate", *Inverse Problems* **25**:5 (2009), Art. ID #055008. MR 2010d:78036 Zbl 1167.35051

[Hristova et al. 2008] Y. Hristova, P. Kuchment, and L. Nguyen, "Reconstruction and time reversal in thermoacoustic tomography in acoustically homogeneous and inhomogeneous media", *Inverse Problems* **24**:5 (2008), Art. ID #055006. MR 2010c:65162 Zbl 1180.35563

[Isakov 2006] V. Isakov, *Inverse problems for partial differential equations*, 2nd ed., Applied Mathematical Sciences **127**, Springer, New York, 2006. MR 2006h:35279 Zbl 1092.35001

[Jin and Wang 2006] X. Jin and L. V. Wang, "Thermoacoustic tomography with correction for acoustic speed variations", *Phys. Med. Biol.* **51**:24 (2006), 6437–6448.

[Katchalov et al. 2001] A. Katchalov, Y. Kurylev, and M. Lassas, *Inverse boundary spectral problems*, Chapman & Hall/CRC Monographs and Surveys in Pure and Applied Mathematics **123**, Chapman & Hall/CRC, Boca Raton, FL, 2001. MR 2003e:58045 Zbl 1037.35098

[Kruger et al. 1999] R. A. Kruger, D. R. Reinecke, and G. A. Kruger, "Thermoacoustic computed tomography: technical considerations", *Med. Phys.* **26**:9 (1999), 1832–1837.

[Kuchment and Kunyansky 2008] P. Kuchment and L. Kunyansky, "Mathematics of thermoacoustic tomography", *European J. Appl. Math.* **19**:2 (2008), 191–224. MR 2009c:92026 Zbl 1185.35327

[Lasiecka et al. 1986] I. Lasiecka, J.-L. Lions, and R. Triggiani, "Nonhomogeneous boundary value problems for second order hyperbolic operators", *J. Math. Pures Appl.* (9) **65**:2 (1986), 149–192. MR 88c:35092

[McLaughlin et al. 2010] J. R. McLaughlin, N. Zhang, and A. Manduca, "Calculating tissue shear modules and pressure by 2D log-elastographic methods", *Inverse Problems* **26**:8 (2010), Art. ID #085007. MR 2658824 Zbl 1195.35301

[Patch 2004] S. K. Patch, "Thermoacoustic tomography: consistency conditions and the partial scan problem", *Phys. Med. Biol.* **49**:11 (2004), 2305–2315.

[Petkov 1982a] V. Petkov, "Inverse scattering problem for transparent obstacles", *Math. Proc. Cambridge Philos. Soc.* **92**:2 (1982), 361–367. MR 84a:35230b Zbl 0504.35073

[Petkov 1982b] V. Petkov, "Propagation of singularities and inverse scattering problem for transparent obstacles", *J. Math. Pures Appl.* (9) **61**:1 (1982), 65–90. MR 84a:35230a Zbl 0487.35055

[Popov and Vodev 1999] G. Popov and G. Vodev, "Resonances near the real axis for transparent obstacles", *Comm. Math. Phys.* **207**:2 (1999), 411–438. MR 2001d:58036 Zbl 0951.35036

[Qian et al. 2011] J. Qian, P. D. Stefanov, G. Uhlmann, and H. Zhao, "An efficient Neumann series-based algorithm for thermoacoustic and photoacoustic tomography with variable sound speed", *SIAM J. Imaging Sci.* **4**:3 (2011), 850–883. MR 2836390 Zbl 06000102

[Robbiano 1991] L. Robbiano, "Théorème d'unicité adapté au contrôle des solutions des problèmes hyperboliques", *Comm. Partial Diff. Equations* **16**:4-5 (1991), 789–800. MR 92j:35002 Zbl 0735.35086

[Stefanov 2012] P. D. Stefanov, "Microlocal analysis methods", preprint, 2012, available at http://www.math.purdue.edu/~stefanov/publications/encyclopedia.pdf.

[Stefanov and Uhlmann 1998a] P. D. Stefanov and G. Uhlmann, "Rigidity for metrics with the same lengths of geodesics", *Math. Res. Lett.* **5**:1 (1998), 83–96. MR 99e:53059 Zbl 0934.53031

[Stefanov and Uhlmann 1998b] P. D. Stefanov and G. Uhlmann, "Stability estimates for the hyperbolic Dirichlet to Neumann map in anisotropic media", *J. Funct. Anal.* **154**:2 (1998), 330–358. MR 99f:35120 Zbl 0915.35066

[Stefanov and Uhlmann 2009a] P. D. Stefanov and G. Uhlmann, "Linearizing non-linear inverse problems and an application to inverse backscattering", *J. Funct. Anal.* **256**:9 (2009), 2842–2866. MR 2010f:47028 Zbl 1169.65049

[Stefanov and Uhlmann 2009b] P. D. Stefanov and G. Uhlmann, "Thermoacoustic tomography with variable sound speed", *Inverse Problems* **25**:7 (2009), Art. ID #075011. MR 2010i:35439 Zbl 1177.35256

[Stefanov and Uhlmann 2011] P. D. Stefanov and G. Uhlmann, "Thermoacoustic tomography arising in brain imaging", *Inverse Problems* **27**:4 (2011), Art. ID #045004. MR 2012f:76113 Zbl 1220.35195

[Tataru 1995] D. Tataru, "Unique continuation for solutions to PDE's; between Hörmander's theorem and Holmgren's theorem", *Comm. Partial Differential Equations* **20**:5-6 (1995), 855–884. MR 96e:35019 Zbl 0846.35021

[Tataru 1999] D. Tataru, "Unique continuation for operators with partially analytic coefficients", *J. Math. Pures Appl.* (9) **78**:5 (1999), 505–521. MR 2000e:35005 Zbl 0936.35038

[Taylor 1976] M. E. Taylor, "Grazing rays and reflection of singularities of solutions to wave equations, II: Systems", *Comm. Pure Appl. Math.* **29**:5 (1976), 463–481. MR 55 #869 Zbl 0335.35059

[Taylor 1981] M. E. Taylor, *Pseudodifferential operators*, Princeton Mathematical Series **34**, Princeton University Press, Princeton, NJ, 1981. MR 82i:35172 Zbl 0453.47026

[Taylor 1996] M. E. Taylor, *Partial differential equations, I: Basic theory*, Applied Mathematical Sciences **115**, Springer, New York, 1996. 2nd ed. in 2011. MR 98b:35002b Zbl 0869.35002

[Trèves 1980] F. Trèves, *Introduction to pseudodifferential and Fourier integral operators, 2: Fourier integral operators*, Plenum, New York, 1980. MR 82i:58068 Zbl 0453.47027

[Wang 2009] L. V. Wang (editor), *Photoacoustic imaging and spectroscopy*, Optical Science and Engineering **144**, CRC, Boca Raton, FL, 2009.

[Wang and Wu 2007] L. V. Wang and H.-I. Wu, *Biomedical optics: principles and imaging*, Wiley-Interscience, Hoboken, NJ, 2007.

[Wiechert and Zoeppritz 1907] E. Wiechert and K. Zoeppritz, "Über Erdbebenwellen", *Nachr. K. Ges. Wiss. Göttingen* **4** (1907), 415–549.

[Williams 1992] M. Williams, "Transmission across a moving interface: necessary and sufficient conditions for (L^2) well-posedness", *Indiana Univ. Math. J.* **41**:2 (1992), 303–338. MR 94c:35119 Zbl 0799.35139

[Xu and He 2005] Y. Xu and B. He, "Magnetic acoustic tomography with magnetic induction (MAT-MI)", *Phys. Med. Biol.* **50**:21 (2005), 5175–5187.

[Xu and Wang 2005] M. Xu and L. V. Wang, "Universal back-projection algorithm for photoacoustic computed tomography", *Phys. Rev. E* **71**:1 (2005), Art. ID #016706.

[Xu and Wang 2006a] M. Xu and L. V. Wang, "Photoacoustic imaging in biomedicine", *Rev. Sci. Instrum.* **77**:4 (2006), Art. ID #041101.

[Xu and Wang 2006b] Y. Xu and L. V. Wang, "Rhesus monkey brain imaging through intact skull with thermoacoustic tomography", *IEEE Trans. Ultrason. Ferroelectr. Freq. Control* **53**:3 (2006), 542–548.

[Xu et al. 2004] Y. Xu, L. V. Wang, P. Kuchment, and G. Ambartsoumian, "Reconstructions in limited view thermoacoustic tomography", *Med. Phys.* **31**:4 (2004), 724–733.

[Xu et al. 2009] Y. Xu, L. V. Wang, P. Kuchment, and G. Ambartsoumian, "Limited view thermoacoustic tomography", Chapter 6, pp. 61–73 in *Photoacoustic imaging and spectroscopy*, edited by L. V. Wang, Optical Science and Engineering **144**, CRC, Boca Raton, FL, 2009.

[Yang and Wang 2008] X. Yang and L. V. Wang, "Monkey brain cortex imaging by photoacoustic tomography", *J. Biomed. Opt.* **13**:4 (2008), Art. ID #044009.

stefanov@math.purdue.edu   *Department of Mathematics, Purdue University, 150 N. University Street, West Lafayette, IN 47907, United States*

gunther@math.washington.edu *Department of Mathematics, University of Washington, Box 354350, Seattle, WA 98195, United States*

Department of Mathematics, University of California, Irvine, CA 92617, United States

Hybrid inverse problems and internal functionals

GUILLAUME BAL

This paper reviews recent results on hybrid inverse problems, which are also called coupled-physics inverse problems of multiwave inverse problems. Inverse problems tend to be most useful in, e.g., medical and geophysical imaging, when they combine high contrast with high resolution. In some settings, a single modality displays either high contrast or high resolution but not both. In favorable situations, physical effects couple one modality with high contrast with another modality with high resolution. The mathematical analysis of such couplings forms the class of hybrid inverse problems.

Hybrid inverse problems typically involve two steps. In a first step, a well-posed problem involving the high-resolution low-contrast modality is solved from knowledge of boundary measurements. In a second step, a quantitative reconstruction of the parameters of interest is performed from knowledge of the point-wise, internal, functionals of the parameters reconstructed during the first step. This paper reviews mathematical techniques that have been developed in recent years to address the second step.

Mathematically, many hybrid inverse problems find interpretations in terms of linear and nonlinear (systems of) equations. In the analysis of such equations, one often needs to verify that qualitative properties of solutions to elliptic linear equations are satisfied, for instance the absence of any critical points. This paper reviews several methods to prove that such qualitative properties hold, including the method based on the construction of complex geometric optics solutions.

1.	Introduction	326
2.	Physical modeling	328
3.	Reconstructions from functionals of u	335
4.	Reconstructions from functionals of ∇u	345
5.	Qualitative properties of forward solutions	355
6.	Conclusions and perspectives	362
	Acknowledgment	364
	References	364

1. Introduction

The success of most medical imaging modalities rests on their high, typically submillimeter, resolution. Computerized tomography (CT), magnetic resonance imaging (MRI), and ultrasound imaging (UI) are typical examples of such modalities. In some situations, these modalities fail to exhibit a sufficient contrast between different types of tissues, whereas other modalities, for example based on the optical, elastic, or electrical properties of these tissues, do display such high contrast. Unfortunately, the latter modalities, such as optical tomography (OT), electrical impedance tomography (EIT) and elastographic imaging (EI), involve a highly smoothing measurement operator and are thus typically low-resolution as stand-alone modalities.

Hybrid inverse problems concern the combination of a high contrast modality with a high resolution modality. By combination, we mean the existence of a physical mechanism that couples these two modalities. Several examples of physical couplings are reviewed in Section 2. A different strategy, consisting of fusing data acquired independently for two or more imaging modalities, is referred to as multimodality imaging and is not considered in this paper. Examples of possible physical couplings include: optics or electromagnetism with ultrasound in photoacoustic tomography (PAT), thermoacoustic tomography (TAT) and in ultrasound modulated optical tomography (UMOT), also called acousto-optic tomography (AOT); electrical currents with ultrasound in ultrasound modulated electrical impedance tomography (UMEIT), also called electroacoustic tomography (EAT); electrical currents with magnetic resonance in magnetic resonance EIT (MREIT) or current density impedance imaging (CDII); and elasticity with ultrasound in transient elastography (TE). Some hybrid modalities have been explored experimentally whereas other hybrid modalities have not been tested yet. Some have received quite a bit of mathematical attention whereas other ones are less well understood. While more references will be given throughout the review, we refer the reader at this point to the recent books [Ammari 2008; Scherzer 2011; Wang and Wu 2007] and their references for general information about practical and theoretical aspects of medical imaging.

Reconstructions in hybrid inverse problems typically involve two steps. In a first step, an inverse problem involving the high-resolution-low-contrast modality needs to be solved. In PAT and TAT for instance, this corresponds to reconstructing the initial condition of a wave equation from available boundary measurements. In UMEIT and UMOT, this corresponds in an idealized setting to inverting a Fourier transform that is reminiscent of the reconstructions performed in MRI. In transient elastography, this essentially corresponds to solving an inverse scattering problem in a time-dependent wave equation. In this review, we assume that this first step has been performed.

Our interest is in the second step of the procedure, which consists of reconstructing the coefficients that display high contrasts from the mappings obtained during the first step. These mappings involve internal functionals of the coefficients of interest. Typically, if γ is a coefficient of interest and u is the solution to a partial differential equation involving γ, then the internal "measurements" obtained in the first step take the form $H(x) = \gamma(x)u^j(x)$ for $j = 1, 2$ or $H(x) = \gamma(x)|\nabla u|^j(x)$ again for $j = 1, 2$.

Several questions can then be raised: are the coefficients, e.g., γ, uniquely characterized by the internal measurements $H(x)$? How stable are the reconstructions? If specific boundary conditions are prescribed at the boundary of the domain of interest, how do the answers to the above questions depend on such boundary conditions? The answers to these questions depend on the physical model of interest. However, there are important common features that we would like to present in this review.

One such feature relates to the stability of the reconstructions. Loosely speaking, an inverse problem is well-posed, or at least not severely ill-posed, when singularities in the coefficients of interest propagate into singularities in the available data. The map reconstructed during step 1 provides local, point-wise, information about the coefficients. Singularities of the coefficient do not need to propagate to the domain's boundary and we thus expect resolution of hybrid modalities to be significantly improved compared to the stand-alone high-contrast-low-resolution modalities. This will be verified in the examples reviewed here.

Another feature is the relationship between hybrid inverse problems and nonlinear partial differential equations. Typically, both the coefficient γ and the solution u are unknown. However, for measurements of the form $H(x) = \gamma(x)u^j(x)$, one can eliminate γ in the equation for u using the expression for $H(x)$. This results in a nonlinear equation for $u(x)$. The resulting nonlinear equations often do not display any of the standard features that are amenable to proofs of uniqueness, such as admitting a variational formulation with a strictly convex functional. The main objective is to obtain uniqueness and stability results for such equations, often in the presence of redundant (overdetermined) information.

A third feature shared by many hybrid inverse problems is that their solution strategies often require that the forward solution u satisfy certain qualitative properties, such as for instance the absence of any critical point (points where $\nabla u = 0$). The derivation of qualitative properties such as lower bounds for the modulus of a gradient is a difficult problem. In two dimensions of space, the fact that critical points of elliptic solutions are necessarily isolated is of great help. In higher dimension, such results no longer hold in general. A framework to obtain the requested qualitative behavior of the elliptic solutions is based on the so-called complex geometric optics (CGO) solutions. Such solutions, when they can be constructed, essentially allow us to treat the unknown coefficients as perturbations of known operators, typically the Laplace operator. Using these solutions, we can construct an open set of boundary conditions for which the requested property is guaranteed. This procedure provides a restricted class of boundary conditions for which the solutions to the hybrid inverse problems are shown to be uniquely and stably determined by the internal measurements. From a practical point of view, these mathematical results confirm the physical intuition that the coupling of high contrast and high resolution modalities indeed provides reconstructions that are robust with respect to errors in the measurements.

The rest of this paper is structured as follows. Section 2 is devoted to the modeling of the hybrid inverse problems and the derivation of the internal measurements for the applications considered in this paper, namely: PAT, TAT, UMEIT, UMOT, TE, CDII. The following two sections present recent results of uniqueness and stability obtained for such hybrid inverse problems: Section 3 focuses on internal functionals of the solution u of the forward problem, whereas Section 4 is concerned with internal functionals of the gradient of the solution ∇u. As we mentioned above, these uniqueness and stability results hinge on the forward solutions u to verify some qualitative properties. Section 5 summarizes some of these properties in the two-dimensional case and presents the derivation of such properties in higher spatial dimensions by means of complex geometric optics (CGO) solutions. Some concluding remarks are proposed in Section 6.

2. Physical modeling

High resolution imaging modalities include ultrasound imaging and magnetic resonance imaging. High contrast modalities include optical tomography, electrical impedance tomography, and elastography. This sections briefly presents four couplings between high-contrast and high-resolution modalities: two different methods to couple ultrasound and optics or (low frequency) electromagnetism in PAT/TAT via the photoacoustic effect and in UMOT/UMEIT via ultrasound modulation; the coupling between ultrasound and elastography in transient elastography; and the coupling between electrical impedance tomography and magnetic resonance imaging in CDII/MREIT.

2A. *The photoacoustic effect.* The photoacoustic effect may be described as follows. A pulse of radiation is sent into a domain of interest. A fraction of the propagating radiation is absorbed by the medium. This generates a thermal expansion, which is the source of ultrasonic waves. Ultrasound then propagates to the boundary of the domain where ultrasonic transducers measure the pressure field. The physical coupling between the absorbed radiation and the emitted sound is called the photoacoustic effect. This is the premise for the medical imaging technique photoacoustic tomography (PAT).

Two types of radiation are typically considered. In optoacoustic tomography (OAT), near-infrared photons, with wavelengths typically between $600nm$ and $900nm$ are used. The reason for this frequency window is that they are not significantly absorbed by water molecules and thus can propagate relatively deep into tissues. OAT is often simply referred to as PAT and we will follow this convention here. In thermoacoustic tomography (TAT), low-frequency microwaves, with wavelengths on the order of $1m$, are sent into the medium. The rationale for using such frequencies is that they are less absorbed than optical frequencies and thus propagate into deeper tissues.

In both PAT and TAT, the first step of an inversion procedure is the reconstruction of the map of absorbed radiation from the ultrasonic measurements. In both applications, the inversion may be recast as the reconstruction of an initial condition of a wave equation from knowledge of ultrasound measurements. Assuming a domain of infinite extension with nonperturbative measurements to simplify the presentation, ultrasound propagation is modeled by the wave equation

$$\frac{1}{c_s^2(x)} \frac{\partial^2 p}{\partial t^2} - \Delta p = 0, \qquad t > 0, \ x \in \mathbb{R}^n, \quad (1)$$
$$p(0, x) = H(x) \quad \text{and} \quad \frac{\partial p}{\partial t}(0, x) = 0, \quad x \in \mathbb{R}^n.$$

Here c_s is the speed of sound (assumed to be known), n is spatial dimension, and $H(x)$ is the ultrasonic signal generated at time $t = 0$. Measurements are then of the form $p(t, x)$ for $t > 0$ and $x \in \partial X$ at the boundary of a domain X where $H(x)$ is supported.

Note that the effect of propagating radiation is modeled as an *initial* condition at $t = 0$. The reason for this stems from the large difference between light speed (roughly $2.3\,10^8 m/s$ in water) and sound speed (roughly $1.5\,10^3 m/s$ in water). When a short pulse of radiation is emitted into the medium, we may assume that it propagates into the medium at a time scale that is very short compared to that of ultrasound. This is a very valid approximation in PAT but is a limiting factor in the (still significantly submillimeter) spatial resolution we expect to obtain in TAT; see [Bal et al. 2010; 2011b], for example.

For additional references to the photoacoustic effect, we refer the reader to the works [Cox et al. 2009a; 2009b; Fisher et al. 2007; Xu and Wang 2006; Xu et al. 2009] and their references. The first step in thermo- and photoacoustics is the reconstruction of the absorbed radiation map $H(x)$ from boundary acoustic wave measurements. There is a vast literature on this inverse source problem in the mathematical and physical literature; we refer the reader to [Ammari et al. 2010; Finch et al. 2004; Haltmeier et al. 2004; Hristova et al. 2008; Kuchment and Kunyansky 2008; Patch and Scherzer 2007; Stefanov and Uhlmann 2009], for example. Serious difficulties may need to be addressed in this first step, such as limited data, spatially varying acoustic sound speed [Ammari et al. 2010; Hristova et al. 2008; Stefanov and Uhlmann 2009], and the effects of acoustic wave attenuation [Kowar and Scherzer 2012]. In this paper, we assume that the absorbed radiation map $H(x)$ has been reconstructed. This provides now *internal* information about the properties of the domain of interest. What we can extract from such information depends on the model of radiation propagation. The resulting inverse problems are called quantitative PAT (QPAT) and quantitative TAT (QTAT) for the different modalities of radiation propagation, respectively.

In the PAT setting with near-infrared photons, arguably the most accurate model for radiation propagation is the radiative transfer equation. We shall not describe this model here and refer the reader to [Bal et al. 2010] for QPAT in this setting and to [Bal 2009] for more general inverse problems for the radiative transfer equation. The models we consider for radiation propagation are as follows.

QPAT modeling. In the diffusive regime, photon (radiation) propagation is modeled by the second-order elliptic equation

$$-\nabla \cdot \gamma(x) \nabla u + \sigma(x) u = 0 \quad \text{in } X$$
$$u = f \quad \text{on } \partial X. \tag{2}$$

To simplify, we assume that Dirichlet conditions are prescribed at the boundary of the domain ∂X. Throughout the paper, we assume that X is a bounded open domain in $\mathbb{R}^n$ with smooth boundary ∂X. The optical coefficients $(\gamma(x), \sigma(x))$ are $\gamma(x)$ the diffusion coefficient and $\sigma(x)$ the absorption coefficient, which are assumed to be bounded from above and below by positive constants.

The information about the coefficients in QPAT takes the following form:

$$H(x) = \Gamma(x) \sigma(x) u(x) \quad \text{a.e. } x \in X. \tag{3}$$

The coefficient $\Gamma(x)$ is the Grüneisen coefficient. It models the strength of the photoacoustic effect, which converts absorption of radiation into emission of ultrasound. The objective of QPAT is to reconstruct (γ, σ, Γ) from knowledge of $H(x)$ in (3) obtained for a given number of illuminations f in (2). This is an example of an internal measurement that is *linear* in the solution $u(x)$ and the absorption coefficient σ. For more on QPAT, see, e.g., [Ren and Bal 2012; Bal and Uhlmann 2010; Cox et al. 2009a; 2009b; Ripoll and Ntziachristos 2005; Zemp 2010] and their references.

QTAT modeling. Low-frequency radiation in QTAT is modeled by the system of Maxwell's equations

$$-\nabla \times \nabla \times E + k^2 E + ik\sigma(x)E = 0 \quad \text{in } X, \qquad (4)$$
$$\nu \times E = f \quad \text{on } \partial X.$$

Here, E is the (time-harmonic) electromagnetic field with fixed wavenumber $k = \frac{\omega}{c}$ where ω is the frequency and c the speed of light. We assume that radiation is controlled by the boundary condition $f(x)$ on ∂X. The unknown coefficient is the conductivity (absorption) coefficient $\sigma(x)$. Setting $\Gamma = 1$ to simplify, the map of absorbed electromagnetic radiation is then of the form

$$H(x) = \sigma(x)|E|^2(x). \qquad (5)$$

The above system of equations may be simplified by modeling radiation by a scalar quantity $u(x)$. In this setting, radiation is modeled by the Helmholtz equation

$$\Delta u + k^2 u + ik\sigma(x)u = 0 \quad \text{in } X \qquad (6)$$
$$u = f \quad \text{on } \partial X,$$

for a given boundary condition $f(x)$. The internal data are then of the form

$$H(x) = \sigma(x)|u|^2(x). \qquad (7)$$

For such models, QTAT then consists of reconstructing $\sigma(x)$ from knowledge of $H(x)$. Note that $H(x)$ is now a *quadratic* quantity in the solutions $E(x)$ or $u(x)$. There are relatively few results on QTAT; see [Bal et al. 2011b; Li et al. 2008].

2B. *The ultrasound modulation effect.* We consider the elliptic equation

$$-\nabla \cdot \gamma(x)\nabla u + \sigma(x)u = 0 \quad \text{in } X, \qquad (8)$$
$$u = f \quad \text{on } \partial X.$$

The objective of ultrasound modulation is to send an acoustic signal through the domain X that modifies the coefficients γ and σ. We assume here that the sound speed is constant and that we are able to generate an acoustic signal that takes the form of the plane wave $p \cos(k \cdot x + \varphi)$ where p is amplitude, k wave-number and φ an additional phase. We assume that the acoustic signal modifies the properties of the diffusion equation and that the effect is small. The coefficients in (8) are thus modified as

$$\gamma_\varepsilon(x) = \gamma(x)(1 + \zeta\varepsilon\mathfrak{c}) + O(\varepsilon^2), \quad \sigma_\varepsilon(x) = \sigma(x)(1 + \eta\varepsilon\mathfrak{c}) + O(\varepsilon^2), \quad (9)$$

where we have defined $\mathfrak{c} = \mathfrak{c}(x) = \cos(k \cdot x + \varphi)$ and where $\varepsilon = p\Gamma$ is the product of the acoustic amplitude $p \in \mathbb{R}$ and a measure $\Gamma > 0$ of the coupling between the acoustic signal and the modulations of the constitutive parameters in (8). We assume that $\varepsilon \ll 1$. The terms in the expansion are characterized by ζ and η and depend on the specific application.

Let u and v be solutions of (8) with fixed boundary conditions f and h, respectively. When the acoustic field is turned on, the coefficients are modified as described in (9) and we denote by u_ε and v_ε the corresponding solution. Note that $u_{-\varepsilon}$ is the solution obtained by changing the sign of p or equivalently by replacing φ by $\varphi + \pi$.

By standard regular perturbation arguments, we find that $u_\varepsilon = u_0 + \varepsilon u_1 + O(\varepsilon^2)$. Multiplying the equation for u_ε by $v_{-\varepsilon}$ and the equation for $v_{-\varepsilon}$ by u_ε, subtracting the results, and using standard integrations by parts, we obtain that

$$\int_X (\gamma_\varepsilon - \gamma_{-\varepsilon})\nabla u_\varepsilon \cdot \nabla v_{-\varepsilon} + (\sigma_\varepsilon - \sigma_{-\varepsilon})u_\varepsilon v_{-\varepsilon} dx = \int_{\partial X} \gamma_{-\varepsilon}\frac{\partial v_{-\varepsilon}}{\partial \nu}u_\varepsilon - \gamma_\varepsilon \frac{\partial u_\varepsilon}{\partial \nu}v_{-\varepsilon} d\sigma. \quad (10)$$

We assume that $\gamma_\varepsilon \partial_\nu u_\varepsilon$ and $\gamma_\varepsilon \partial_\nu v_\varepsilon$ are measured on ∂X, at least on the support of $v_\varepsilon = h$ and $u_\varepsilon = f$, respectively, for several values ε of interest. The above equation also holds if the Dirichlet boundary conditions are replaced by Neumann boundary conditions. Let us define

$$J_\varepsilon := \frac{1}{2}\int_{\partial X} \gamma_{-\varepsilon}\frac{\partial v_{-\varepsilon}}{\partial \nu}u_\varepsilon - \gamma_\varepsilon \frac{\partial u_\varepsilon}{\partial \nu}v_{-\varepsilon}d\sigma = \varepsilon J_1 + \varepsilon^2 J_2 + O(\varepsilon^3). \quad (11)$$

We assume that the real valued functions $J_m = J_m(k, \varphi)$ are known from the physical measurement of the Cauchy data of the form $(u_\varepsilon, \gamma_\varepsilon \partial_\nu u_\varepsilon)$ and $(v_\varepsilon, \gamma_\varepsilon \partial_\nu v_\varepsilon)$ on ∂X.

Equating like powers of ε, we find that at the leading order

$$\int_X [\zeta\gamma(x)\nabla u_0 \cdot \nabla v_0(x) + \eta\sigma(x)u_0 v_0(x)]\cos(k \cdot x + \varphi)\, dx = J_1(k, \varphi). \quad (12)$$

Acquiring this for all $k \in \mathbb{R}^n$ and $\varphi = 0, \frac{\pi}{2}$, this yields after inverse Fourier transform:

$$H[u_0, v_0](x) = \zeta\gamma(x)\nabla u_0 \cdot \nabla v_0(x) + \eta\sigma(x)u_0 v_0(x). \quad (13)$$

In the setting of ultrasound modulated optical tomography (UMOT), the coefficients γ_ε and σ_ε in (9) take the form [Bal and Schotland 2010]

$$\gamma_\varepsilon(x) = \frac{\tilde\gamma_\varepsilon}{c_\varepsilon^{n-1}}(x) \quad \text{and} \quad \sigma_\varepsilon(x) = \frac{\tilde\sigma_\varepsilon}{c_\varepsilon^{n-1}}(x),$$

where $\tilde\sigma_\varepsilon$ is the absorption coefficient, $\tilde\gamma_\varepsilon$ is the diffusion coefficient, c_ε is the light speed, and n is spatial dimension. When the pressure field is turned on, the amount of scatterers and absorbers is modified by compression and dilation. Since the diffusion coefficient is inversely proportional to the scattering coefficient, we find that

$$\tilde\sigma_\varepsilon(x) = \tilde\sigma(x)\bigl(1 + \varepsilon\mathfrak{c}(x)\bigr), \quad \frac{1}{\tilde\gamma_\varepsilon(x)} = \frac{1}{\tilde\gamma(x)}\bigl(1 + \varepsilon\mathfrak{c}(x)\bigr).$$

The pressure field changes the index of refraction of light as follows $c_\varepsilon(x) = c(x)(1 + \psi\varepsilon\mathfrak{c}(x))$, where ψ is a constant (roughly equal to $\frac{1}{3}$ for water). This shows that

$$\zeta = -(1 + (n-1)\psi), \quad \eta = 1 - (n-1)\psi. \tag{14}$$

In the application of ultrasound modulated electrical impedance tomography (UMEIT), $\gamma(x)$ is a conductivity coefficient and $\sigma = 0$. We then have $\gamma_\varepsilon(x) = \gamma(x)(1 + \varepsilon\mathfrak{c}(x))$ with thus $\zeta = 1$ and $\eta = 0$. The objective of UMOT and UMEIT is to reconstruct (part of) the coefficients $(\gamma(x), \sigma(x))$ in the elliptic equation

$$\begin{aligned} -\nabla \cdot \gamma(x)\nabla u + \sigma(x)u &= 0 \quad \text{in } X, \\ u &= f \quad \text{on } \partial X \end{aligned} \tag{15}$$

from measurements of the form

$$H[u_0, v_0](x) = \zeta\gamma(x)\nabla u_0 \cdot \nabla v_0(x) + \eta\sigma(x)u_0 v_0(x), \tag{16}$$

for one or several values of the illumination $f(x)$ on ∂X.

In a simplified version of UMOT (also called acousto-optic tomography; AOT), $\zeta = 0$ and the measurements are quadratic (or bilinear) in the solutions to the elliptic equation. More challenging mathematically is the case $\zeta = 1$ and $\eta = 0$ where the measurements are quadratic (or bilinear) in the *gradients* of the solution. No theoretical results exist to date in the setting where both ζ and η are nonvanishing.

The effect of ultrasound modulation is difficult to observe experimentally as the coupling coefficient Γ above is rather small. For references on ultrasound modulation in different contexts, see [Ammari et al. 2008; Bal 2012; Bal and Schotland 2010; Capdeboscq et al. 2009; Gebauer and Scherzer 2008; Kuchment and Kunyansky 2011; Zhang and Wang 2004]. These references concern the so-called incoherent regime of wave propagation, while the coherent regime, whose mathematical structure is different, is addressed in the physical literature in, e.g., [Atlan et al. 2005; Kempe et al. 1997; Wang 2004].

2C. *Transient elastography.* Transient elastography images the (slow) propagation of shear waves using ultrasound. For more details, see [McLaughlin et al. 2010] and its extended list of references. As shear waves propagate, the resulting displacements can be imaged by ultrafast ultrasound. Consider a scalar approximation of the equations of elasticity

$$\begin{aligned}\nabla \cdot \gamma(x)\nabla u(x,t) &= \rho(x)\partial_{tt}u(x,t), & t \in \mathbb{R},\ x \in X, \\ u(x,t) &= f(x,t), & t \in \mathbb{R},\ x \in \partial X,\end{aligned} \quad (17)$$

where $u(x,t)$ is the (say, downward) displacement, $\gamma(x)$ is one of the Lamé parameters and $\rho(x)$ is density. Using ultrafast ultrasound measurements, the displacement $u(x,t)$ can be imaged. This results in a very simplified model of transient elastography where we aim to reconstruct (γ, ρ) from knowledge of $u(x,t)$; see [McLaughlin et al. 2010] for more complex models. We may slightly generalize the model as follows. Upon taking Fourier transforms in the time domain and accounting for possible dispersive effects of the tissues, we obtain

$$\begin{aligned}\nabla \cdot \gamma(x;\omega)\nabla u(x;\omega) + \omega^2 \rho(x;\omega)u(x;\omega) &= 0, & \omega \in \mathbb{R},\ x \in X, \\ u(x;\omega) &= f(x;\omega), & \omega \in \mathbb{R},\ x \in \partial X.\end{aligned} \quad (18)$$

The inverse transient elastography problem with dispersion effect would then be the reconstruction of $(\gamma(x;\omega), \rho(x;\omega))$ from knowledge of $u(x;\omega)$ corresponding to one or several boundary conditions $f(x;\omega)$ applied at the boundary ∂X. This hybrid inverse problem again involves measurements that are *linear* in the solution u.

2D. *Current density imaging.* Magnetic impedance electrical impedance tomography (MREIT) and current density impedance imaging (CDII) are two modalities aiming to reconstruct the conductivity in an equation using magnetic resonance imaging (MRI). The electrical potential u solves the following elliptic equation

$$\begin{aligned}-\nabla \cdot \gamma(x)\nabla u &= 0 & \text{in } X, \\ u &= f & \text{on } \partial X,\end{aligned} \quad (19)$$

with $\gamma(x)$ the unknown conductivity and f a prescribed voltage at the domain's boundary. The electrical current density $J = -\gamma \nabla u$ satisfies the system of Maxwell's equations

$$\nabla \cdot J = 0, \quad J = \frac{1}{\mu_0}\nabla \times B, \quad x \in X. \quad (20)$$

Here μ_0 is a constant, known, magnetic permeability.

Ideally, the whole field B can be reconstructed from MRI measurements. This provides access to the current density $J(x)$ in the whole domain X. CDI then corresponds to reconstructing γ from knowledge of J. In practice, acquiring B requires rotation of the domain of interest (or of the MRI apparatus) which is not straightforward. MREIT thus assumes knowledge of the third component B_z of the magnetic field for several possible boundary conditions. This provides information about $\gamma(x)$. We do not consider the MREIT inverse problem further and refer the reader to the recent review [Seo and Woo 2011] and its references for additional information.

Several works have considered the problem of the reconstruction of γ in (19) from knowledge of the scalar information $|J|$ rather than the full current J. This inverse problem, referred to as the 1-Laplacian, will be addressed below and compared to the 0-Laplacian that appears in UMEIT and UMOT. For more on MREIT and CDII, we refer the reader to [Kim et al. 2002; Nachman et al. 2007; 2009; 2011] and their references.

3. Reconstructions from functionals of u

In this section, we consider internal measurements $H(x)$ of the form $H(x) = \tau(x)u(x)$ for $\tau(x)$ a function that depends linearly on unknown coefficients such as the diffusion coefficient γ or the absorption coefficient σ in Section 3A and internal measurements $H(x)$ of the form $H(x) = \tau(x)|u(x)|^2$ in Section 3B, where τ again depends linearly on unknown coefficients. Measurements of the first form find applications in quantitative photoacoustic tomography (QPAT) and transient elastography (TE) while measurements of the second form find applications in quantitative thermoacoustic tomography (QTAT) and simplified models of acousto-optics tomography (AOT).

3A. *Reconstructions from linear functionals in u.* Recall the elliptic model for photon propagation in tissues:

$$-\nabla \cdot \gamma(x)\nabla u + \sigma(x)u = 0 \quad \text{in } X, \qquad (21)$$
$$u = f \quad \text{on } \partial X.$$

The information about the coefficients in QPAT takes the form

$$H(x) = \Gamma(x)\sigma(x)u(x) \quad \text{a.e. } x \in X. \qquad (22)$$

The coefficient $\Gamma(x)$ is the Grüneisen coefficient. In many works in QPAT, it is assumed to be constant. We assume here that it is Lipschitz continuous and bounded above and below by positive constants.

Nonunique reconstruction of three coefficients. Let f_1 and f_2 be two Dirichlet conditions on ∂X and u_1 and u_2 be the corresponding solutions to (21). We make the following assumptions:

(i) The coefficients (γ, σ, Γ) are of class $W^{1,\infty}(X)$ and bounded above and below by positive constants. The coefficients (γ, σ, Γ) are known on ∂X.

(ii) The illuminations f_1 and f_2 are positive functions on ∂X and are the traces on ∂X of functions of class $C^3(\bar{X})$.

(iii) the vector field

$$\beta := H_1 \nabla H_2 - H_2 \nabla H_1 = H_1^2 \nabla \frac{H_2}{H_1} = H_1^2 \nabla \frac{u_2}{u_1} = -H_2^2 \nabla \frac{H_1}{H_2} \quad (23)$$

is a vector field in $W^{1,\infty}(X)$ such that $\beta \not\equiv 0$ a.e.

(iii′) Same as (iii) above with

$$|\beta|(x) \geq \alpha_0 > 0 \quad \text{a.e. } x \in X. \quad (24)$$

Beyond the regularity assumptions on (γ, σ, Γ), the domain X, and the boundary conditions f_1 and f_2, the only real assumption we impose is (24). In general, there is no guaranty that the gradient of u_2/u_1 does not vanish. In dimension $d = 2$, a simple condition guarantees that (24) holds. We have the following result [Alessandrini 1986; Nachman et al. 2007]:

Lemma 3.1 [Bal and Ren 2011a]. *Assume that $h = g_2/g_1$ on ∂X is an almost two-to-one function in the sense of* [Nachman et al. 2007], *i.e., a function that is a two-to-one map except possibly at its minimum and at its maximum. Then* (24) *is satisfied.*

In dimension $d \geq 3$, the above result on the (absence of) critical points of elliptic solutions no longer holds. By continuity, we verify that (24) is satisfied for a large class of illuminations when γ is close to a constant and σ is sufficiently small. For arbitrary coefficients (γ, σ) in dimension $d \geq 3$, a proof based on CGO solutions shows that (24) is satisfied for an open set of illuminations; see [Bal and Uhlmann 2010] and Section 5B below. Note also that (24) is a sufficient condition for us to solve the inverse problem of QPAT. In [Alessandrini 1986], a similar problem is addressed in dimension $d = 2$ without assuming a constraint of the form (24).

We first prove a result that provides uniqueness up to a specified transformation.

Theorem 3.2 [Bal and Ren 2011a; Bal and Uhlmann 2010]. *Assume that the hypotheses* (i)–(iii) *hold.*

(a) $H_1(x)$ *and* $H_2(x)$ *uniquely determine the measurement operator*

$$\mathcal{H}: H^{\frac{1}{2}}(\partial X) \to H^1(X),$$

which to f defined on ∂X associates $\mathcal{H}(f) = H$ in X defined by (3).

(b) *The measurement operator $\mathcal{H}$ uniquely determines the two functionals*

$$\chi(x) := \frac{\sqrt{\gamma}}{\Gamma \sigma}(x), \quad q(x) := -\left(\frac{\Delta\sqrt{\gamma}}{\sqrt{\gamma}} + \frac{\sigma}{\gamma}\right)(x). \tag{25}$$

Here Δ is the Laplace operator.

(c) *Knowledge of the two functionals χ and q uniquely determines $H_1(x)$ and $H_2(x)$. In other words, the reconstruction of (γ, σ, Γ) is unique up to transformations that leave (χ, q) invariant.*

The proof of this theorem is given in [Bal and Ren 2011a] under the additional assumption (iii′). The following minor modification allows one to prove the theorem as stated above. The original proof is based on the fact that the equality $\int_X (\rho - 1)^2 |\beta|^2 dx = 0$ implies that $\rho = 1$ a.e. Such a result clearly still holds provided that $\beta \neq 0$ a.e.

Note that the result $\beta \neq 0$ a.e. holds for a very large class of boundary conditions (f_1, f_2). Indeed, β is the solution of (26) below with χ^2 bounded from below by a positive constant. This implies that u_2/u_1 is the solution of an elliptic equation.

Thus, the set $\beta = 0$ corresponds to the set of critical points $\nabla(u_2/u_1) = 0$. When u_2/u_1 is not constant, it is proved that for such elliptic equations, the set of critical points $\nabla(u_2/u_1) = 0$ is of Lebesgue measure zero provided that the coefficients in (21) are sufficiently smooth [Hardt et al. 1999; Robbiano and Salazar 1990]. We thus find that so long that f_1/f_2 is not a constant a.e., then $\beta \neq 0$ a.e. and the two internal functionals (H_1, H_2) uniquely characterize the coefficients (χ, q).

Reconstruction of two coefficients. The above result shows that the unique reconstruction of (γ, σ, Γ) is not possible even from knowledge of the full measurement operator $\mathcal{H}$ defined in Theorem 3.2. Two well-chosen illuminations uniquely determine the functionals (χ, q) and acquiring additional measurements does not provide any new information. However, we can prove that if one coefficient in (γ, σ, Γ) is known, then the other two coefficients are uniquely determined:

Corollary 3.3 [Bal and Ren 2011a]. *Under the hypotheses of Theorem 3.2, let (χ, q) in (25) be known.*

(a) *If Γ is known, then (γ, σ) are uniquely determined.*

(b) *If γ is known, then (σ, Γ) are uniquely determined.*

(c) *If σ is known, then (γ, Γ) are uniquely determined.*

The above uniqueness results are *constructive*. In all cases, we need to solve the following transport equation for χ:

$$-\nabla \cdot (\chi^2 \beta) = 0 \quad \text{in } X, \qquad \chi_{|\partial X} \text{ known on } \partial X, \tag{26}$$

with β the vector field defined in (23). This uniquely defines $\chi > 0$. Then we find that

$$q(x) = -\frac{\Delta(H_1 \chi)}{H_1 \chi} = -\frac{\Delta(H_2 \chi)}{H_2 \chi}. \tag{27}$$

This provides explicit reconstructions for (χ, q) from knowledge of (H_1, H_2) when (24) holds.

In case (b), no further equation needs to be solved. In cases (a) and (c), we need to solve an elliptic equation for $\sqrt{\gamma}$, which is the linear equation

$$(\Delta + q)\sqrt{\gamma} + \frac{1}{\Gamma \chi} = 0 \quad \text{in } X,$$
$$\sqrt{\gamma}_{|\partial X} = \sqrt{\gamma_{|\partial X}} \quad \text{on } \partial X, \tag{28}$$

in (a) and the (uniquely solvable) nonlinear (semilinear) equation

$$\sqrt{\gamma}(\Delta + q)\sqrt{\gamma} + \sigma = 0 \quad \text{in } X,$$
$$\sqrt{\gamma}_{|\partial X} = \sqrt{\gamma_{|\partial X}} \quad \text{on } \partial X, \tag{29}$$

in (c). These inversion formulas were implemented numerically in [Bal and Ren 2011a]. Moreover, reconstructions are known to be Hölder or Lipschitz stable depending on the metric used in the stability estimate. For instance:

Theorem 3.4 [Bal and Ren 2011a]. *Assume that the hypotheses of Theorem 3.2 and (iii') hold. Let $H = (H_1, H_2)$ be the measurements corresponding to the coefficients (γ, σ, Γ) for which hypothesis (iii) holds. Let $\tilde{H} = (\tilde{H}_1, \tilde{H}_2)$ be the measurements corresponding to the same illuminations (f_1, f_2) with another set of coefficients $(\tilde{\gamma}, \tilde{\sigma}, \tilde{\Gamma})$ such that (i) and (ii) still hold. Then we find that*

$$\|\chi - \tilde{\chi}\|_{L^p(X)} \leq C \|H - \tilde{H}\|^{\frac{1}{2}}_{(W^{1,\frac{p}{2}}(X))^2} \quad \text{for all } 2 \leq p < \infty. \tag{30}$$

Let us assume, moreover, that $\gamma(x)$ is of class $C^3(\bar{X})$. Then we have that

$$\|\chi - \tilde{\chi}\|_{L^\infty(X)} \leq C \|H - \tilde{H}\|^{\frac{p}{3(d+p)}}_{(L^{\frac{p}{2}}(X))^2} \quad \text{for all } 2 \leq p < \infty. \tag{31}$$

We may for instance choose $p = 4$ above to measure the noise level in the measurement H in the square integrable norm when noise is described by its power spectrum in the Fourier domain. This shows that reconstructions in QPAT are Hölder stable, unlike the corresponding reconstructions in optical tomography [Bal 2009; Uhlmann 2009].

An application to transient elastography. We can apply the above results to the time-harmonic reconstruction in a simplified model of transient elastography. Let us assume that γ and ρ are unknown functions of $x \in X$ and $\omega \in \mathbb{R}$. Recall that the displacement solves (18). Assuming that $u(x;\omega)$ is known after step 1 of the reconstruction using the ultrasound measurements, then we are in the setting of Theorem 3.2 with $\Gamma\sigma = 1$. Let us then assume that the two illuminations $f_1(x;\omega)$ and $f_2(x;\omega)$ are chosen such that for u_1 and u_2 the corresponding solutions of (18), we have that (24) holds. We have seen a sufficient condition for this to hold in dimension $n = 2$ in Lemma 3.1 and will present other sufficient conditions in Section 5B, which is devoted to CGO solutions in the setting $n \geq 3$. Then, (25) shows that the reconstructed function χ uniquely determines the Lamé parameter $\gamma(x;\omega)$ and that the reconstructed function q then uniquely determines $\omega^2 \rho$ and hence the density parameter $\rho(x;\omega)$. The reconstructions are performed for each frequency ω independently. We may summarize this as follows:

Corollary 3.5. *Under the hypotheses of Theorem 3.2 and the hypotheses described above, let (χ, q) in (25) be known. Then $(\gamma(x;\omega), \rho(x;\omega))$ are uniquely determined by two well-chosen measurements. Moreover, the stability results in Theorem 3.4 hold.*

Alternatively, we may assume that in a given range of frequencies, $\gamma(x)$ and $\rho(x)$ are independent of ω. In such a setting, we expect that one measurement $u(x;\omega)$ for two different frequencies will provide sufficient information to reconstruct $(\gamma(x), \rho(x))$. Assume that $u(x;\omega)$ is known for $\omega = \omega_j$, $j = 1, 2$ and define $0 < \alpha = \omega_2^2 \omega_1^{-2} \neq 1$. Then straightforward calculations show that

$$\nabla \cdot \gamma \beta_\alpha = 0, \quad \beta_\alpha = (u_1 \nabla u_2 - \alpha u_2 \nabla u_1). \tag{32}$$

This provides a transport equation for γ that can be solved stably provided that $|\beta_\alpha| \geq c_0 > 0$, i.e., β_α does not vanish on X. Then, Theorem 3.2 and Theorem 3.4 apply in this setting. Since β_α cannot be written as the ratio of two solutions as in (23) when $\alpha = 1$, the results obtained in Lemma 3.1 do not apply when $\alpha \neq 1$. However, we prove in Section 5B that $|\beta_\alpha| \geq c_0 > 0$ is satisfied for an open set of illuminations constructed by means of CGO solutions for all $\alpha > 0$; see (96) below.

Reconstruction of one coefficient. Let us conclude this section by some comments on the reconstruction of a single coefficient from a measurement linear in u. From an algorithmic point of view, such reconstructions are significantly simpler. Let us consider the framework of Corollary 3.3. When Γ is the only unknown coefficient, then we solve for u in (21) and reconstruct Γ from knowledge of H.

When only σ is unknown, then we solve the elliptic equation for u

$$-\nabla \cdot \gamma \nabla u + \frac{H}{\Gamma} = 0 \quad \text{in } X,$$
$$u = g \quad \text{on } \partial X,$$

and then evaluate $\sigma = H/(\Gamma u)$.

When only γ is unknown with either $H = \sigma u$ in QPAT or with $H = u$ in elastography or in applications to ground water flows [Alessandrini 1986; Richter 1981], then u is known and γ solves the transport equation

$$-\nabla \cdot \gamma \nabla u = S \quad \text{in } X,$$
$$\gamma = \gamma_{|\partial X} \quad \text{on } \partial X,$$

with S known. Provided that the vector field ∇u does not vanish, the above equation admits a unique solution as in (26). The stability results of Theorem 3.4 then apply. Other stability results based on solving the transport equation by the method of characteristics are presented in [Richter 1981]. In two dimensions of space, the constraint that the vector field ∇u does not vanish can be partially removed. Under appropriate conditions on the oscillations of the illumination g on ∂X, stability results are obtained in [Alessandrini 1986] in cases where ∇u is allowed to vanish.

3B. *Reconstructions from quadratic functionals in u.*

Reconstructions under smallness conditions. The TAT and (simplified) AOT problems are examples of a more general class we define as follows. Let $P(x, D)$ be an operator acting on functions defined in $\mathbb{C}^m$ for $m \in \mathbb{N}^*$ an integer and with values in the same space. Consider the equation

$$P(x, D)u = \sigma(x)u, \quad x \in X,$$
$$u = f, \quad x \in \partial X. \tag{33}$$

We assume that the above equation admits a unique weak solution in some Hilbert space $\mathcal{H}_1$ for sufficiently smooth illuminations $f(x)$ on ∂X.

For instance, P could be the Helmholtz operator $ik^{-1}(\Delta + k^2)$ seen in the preceding section with $u \in \mathcal{H}_1 := H^1(X; \mathbb{C})$ and $f \in H^{\frac{1}{2}}(\partial X; \mathbb{C})$. Time-harmonic Maxwell's equations can be put in that framework with m and

$$P(x, D) = \frac{1}{ik}(\nabla \times \nabla \times -k^2). \tag{34}$$

We impose an additional constraint on $P(x, D)$ that the equation $P(x, D)u = f$ on X with $u = 0$ on ∂X admits a unique solution in $\mathcal{H} = L^2(X; \mathbb{C}^m)$. For Maxwell's equations, this constraint is satisfied so long as k^2 is not an internal eigenvalue of the Maxwell operator [Dautray and Lions 1990]. This is expressed by the existence of a constant $\alpha > 0$ such that:

$$(P(x,D)u,u)_{\mathcal{H}} \geq \alpha(u,u)_{\mathcal{H}}. \tag{35}$$

We assume that the conductivity σ is bounded from above by a positive constant:

$$0 < \sigma(x) \leq \sigma_M \quad \text{a.e. } x \in X. \tag{36}$$

We denote by Σ_M the space of functions $\sigma(x)$ such that (36) holds. Measurements are of the form

$$H(x) = \sigma(x)|u|^2,$$

where $|\cdot|$ is the Euclidean norm on $\mathbb{C}^m$. Then we have the following result.

Theorem 3.6 [Bal et al. 2011b]. *Let $\sigma_j \in \Sigma_M$ for $j = 1, 2$. Let u_j be the solution to $P(x,D)u_j = \sigma_j u_j$ in X with $u_j = f$ on ∂X for $j = 1, 2$. Define the internal functionals $H_j(x) = \sigma_j(x)|u_j(x)|^2$ on X.*

Assume that σ_M is sufficiently small that $\sigma_M < \alpha$. Then:

(i) (Uniqueness) *If $H_1 = H_2$ a.e. in X, then $\sigma_1(x) = \sigma_2(x)$ a.e. in X where $H_1 = H_2 > 0$.*

(ii) (Stability) *We have the stability estimate*

$$\|(\sqrt{\sigma_1} - \sqrt{\sigma_2})w_1\|_{\mathcal{H}} \leq C\|(\sqrt{H_1} - \sqrt{H_2})w_2\|_{\mathcal{H}}, \tag{37}$$

for some universal constant C and for positive weights given by

$$\begin{aligned} w_1^2(x) &= \prod_{j=1,2} \frac{|u_j|}{\sqrt{\sigma_j}}(x), \\ w_2(x) &= \frac{1}{\alpha - \sup_{x \in X} \sqrt{\sigma_1 \sigma_2}} \max_{j=1,2} \frac{\sqrt{\sigma_j}}{|u_{j'}|}(x) + \max_{j=1,2} \frac{1}{\sqrt{\sigma_j}}(x). \end{aligned} \tag{38}$$

Here $j' = j'(j)$ is defined as $j'(1) = 2$ and $j'(2) = 1$.

The theorem uses the spectral gap in (35). Some straightforward algebra shows that

$$P(x,D)(u_1 - u_2) = \sqrt{\sigma_1 \sigma_2}\big(|u_2|\hat{u}_1 - |u_1|\hat{u}_2\big) + (\sqrt{H_1} - \sqrt{H_2})\left(\frac{\sqrt{\sigma_1}}{|u_1|} - \frac{\sqrt{\sigma_2}}{|u_2|}\right).$$

Here we have defined $\hat{u} = u/|u|$. Although this does not constitute an equation for $u_1 - u_2$, it turns out that

$$\big||u_2|\hat{u}_1 - |u_1|\hat{u}_2\big| = |u_2 - u_1|.$$

This combined with (35) yields the theorem after some elementary manipulations; see [Bal et al. 2011b].

Reconstructions for the Helmholtz equation. Let us consider the scalar model of TAT. We assume that $\sigma \in H^p(X)$ for $p > n/2$ and construct

$$q(x) = k^2 + ik\sigma(x) \in H^p(X), \quad p > \frac{n}{2}. \tag{39}$$

We assume that $q(x)$ is the restriction to X of the compactly supported function (still called q) $q \in H^p(\mathbb{R}^n)$. The extension is chosen so that $\|q_{|X}\|_{H^p(X)} \leq C\|q\|_{H^p(\mathbb{R}^n)}$ for some constant C independent of q; see [Bal and Uhlmann 2010]. Then (6) may be recast as

$$\begin{aligned}\Delta u + q(x)u &= 0 \quad \text{in } X, \\ u &= f \quad \text{on } \partial X.\end{aligned} \tag{40}$$

The measurements are of the form $H(x) = \sigma(x)|u|^2(x)$.

The inverse problem consists of reconstructing $q(x)$ from knowledge of $H(x)$. Note that $q(x)$ need not be of the form (39). It could be a real-valued potential in a Helmholtz equation as considered in [Triki 2010] with applications in the so-called inverse medium problem. The reconstruction of $q(x)$ in (40) from knowledge of $H(x) = \sigma(x)|u|^2(x)$ has been analyzed in [Bal et al. 2011b; Triki 2010]. We cite two stability results, one global and one local.

Theorem 3.7 [Bal et al. 2011b]. *Let $Y = H^p(X)$ and $Z = H^{p-\frac{1}{2}}(\partial X)$, where $p > n/2$. Let $\mathcal{M}$ be the space of functions in Y with norm bounded by a fixed (arbitrary) $M > 0$. Let σ and $\tilde{\sigma}$ be functions in $\mathcal{M}$. Let $f \in Z$ be a given (complex-valued) illumination and $H(x)$ the measurement given in (7) for a solution u of (6). Define $\tilde{H}(x)$ similarly, with $\tilde{\sigma}$ replacing σ in (7) and (6).*

Then there is an open set of illuminations f in Z such that $H(x) = \tilde{H}(x)$ in Y implies that $\sigma(x) = \tilde{\sigma}(x)$ in Y. Moreover, there exists a constant C independent of σ and $\tilde{\sigma}$ in $\mathcal{M}$ such that

$$\|\sigma - \tilde{\sigma}\|_Y \leq C\|H - \tilde{H}\|_Y. \tag{41}$$

The theorem is written in terms of σ, which is the parameter of interest in TAT. The same result holds if σ is replaced by $q(x)$ in (41). The reconstruction of σ is also constructive as the application of a Banach fixed point theorem. The proof is based on the construction of complex geometric optics solutions that will be presented in Section 5B.

Theorem 3.8 [Triki 2010]. *Let $q(x) \geq c_0 > 0$ be real-valued, positive, bounded on X and such that 0 is not an eigenvalue of $\Delta + q$ with domain equal to*

$H_0^1(X) \cap H_2(\overline{X})$. *Let $\tilde{q}$ satisfy the same hypotheses and let H and $\tilde{H}$ be the corresponding measurements.*

Then there is a constant $\varepsilon > 0$ such that if q and $\tilde{q}$ are ε-close in $L^\infty(X)$ and if f is in an ε-dependent open set of (complex-valued) illuminations, then there is a constant C such that

$$\|q - \tilde{q}\|_{L^2(X)} \leq C \|H - \tilde{H}\|_{L^2(X)}. \tag{42}$$

Theorems 3.7 and 3.8 show that the TAT and the inverse medium problem are stable inverse problems. This is confirmed by the numerical reconstructions in [Bal et al. 2011b]. The first result is more global but requires more regularity of the coefficients. It is based on the use of complex geometric optics CGO solutions to show that an appropriate functional is contracting in the space of continuous functions. The second result is more local in nature (a global uniqueness result is also proved in [Triki 2010]) but requires less smoothness on the coefficient $q(x)$ and provides a stability estimate in the larger space $L^2(X)$. It also uses CGO solutions to show that the norm of a complex-valued solution to an elliptic equation is bounded from below by a positive constant. In both cases, the CGO solutions have traces at the boundary ∂X and the chosen illumination f needs to be chosen in the vicinity of such traces.

The results obtained in Theorem 3.6 under smallness constraints on σ apply for very general illuminations f. The above two results apply for more general (essentially arbitrary) coefficients but require more severe constraints on the illuminations f.

Nonunique reconstruction in the AOT setting. The results above concern the uniqueness of the reconstruction of the potential in a Helmholtz equation when well-chosen complex-valued boundary conditions are imposed. They also show that the reconstruction of $0 < c_0 \leq q(x)$ in $\Delta u + qu = 0$ with real-valued $u = f$ from knowledge of qu^2 is unique. This corresponds to $P(x, D) = -\Delta$. In a simplified version of the acousto-optics problem considered in [Bal and Schotland 2010], it is interesting to look at the problem where $P(x, D) = \Delta$ and where the measurements are given by $H(x) = \sigma(x)u^2(x)$. Here, u is the solution of the elliptic equation $(-\Delta + \sigma)u = 0$ on X with $u = f$ on ∂X. Assuming that f is nonnegative, which is the physically interesting case, we obtain that $|u| = u$ and hence

$$\Delta(u_1 - u_2) = \sqrt{\sigma_1 \sigma_2}(u_2 - u_1) + (\sqrt{H_1} - \sqrt{H_2})\left(\frac{\sigma_1}{\sqrt{H_1}} - \frac{\sigma_2}{\sqrt{H_2}}\right).$$

Therefore, as soon as 0 is not an eigenvalue of $\Delta + \sqrt{\sigma_1 \sigma_2}$, we obtain that $u_1 = u_2$ and hence that $\sigma_1 = \sigma_2$. For σ_0 such that 0 is not an eigenvalue of

$\Delta + \sigma_0$, we find that for σ_1 and σ_2 sufficiently close to σ_0, then $H_1 = H_2$ implies that $\sigma_1 = \sigma_2$ on the support of $H_1 - H_2$.

However, it is shown in [Bal and Ren 2011b] that two different, positive, absorptions σ_j for $j = 1, 2$, may in some cases provide the same measurement $H = \sigma_j u_j^2$ with $\Delta u_j = \sigma_j u_j$ on X with $u_j = f$ on ∂X and in fact $\sigma_1 = \sigma_2$ on ∂X so that these absorptions cannot be distinguished by their traces on ∂X. This counterexample shows that conditions such as the smallness condition in Theorem 3.6 are necessary in general.

More generally, and following [Bal and Ren 2011b], consider an elliptic problem of the form

$$Pu = \sigma u \quad \text{in } X, \qquad u = f \quad \text{on } \partial X, \tag{43}$$

and assume that measurements of the form $H(x) = \sigma(x)u^2(x)$ are available. Here, P is a self-adjoint, nonpositive, elliptic operator, which for concreteness we will take of the form $Pu = \nabla \cdot \gamma(x) \nabla u$ with $\gamma(x)$ known, sufficiently smooth, and bounded above and below by positive constants. We assume $f > 0$ and $\gamma > 0$ so that by the maximum principle, $u > 0$ on X. We also assume enough regularity on ∂X and f so that $u \in C^{2,\beta}(\overline{X})$ for some $\beta > 0$ [Gilbarg and Trudinger 1977].

We observe that

$$uPu = H \quad \text{in } X, \qquad u = f \quad \text{on } \partial X, \tag{44}$$

so that the inverse problem may be recast as a semilinear problem. The nonuniqueness result is derived from [Ambrosetti and Prodi 1972] and in some sense generalizes the observation that $x \mapsto x^2$ admits 0, 1, or 2 (real-valued) solutions depending on the value of x^2. Let us define

$$\phi : C^{2,\beta}(\overline{X}) \to C^{0,\beta}(\overline{X}), \quad u \mapsto \phi(u) = uPu. \tag{45}$$

The singular points of ϕ are calculated from its first-order Fréchet derivative:

$$\phi'(u)v = vPu + uPv. \tag{46}$$

The operator $\phi'(u)$ is not invertible when $\sigma := (Pu)/u$ is such that $P + \lambda \sigma$ admits $\lambda = 1$ as an eigenvalue. Let σ_0 be such that $P + \sigma_0$ is not invertible. We assume that the corresponding eigen-space is one-dimensional and spanned by the eigenvector $\psi > 0$ on ∂X such that $(P + \sigma_0)\psi = 0$ and $\psi = 0$ on ∂X. Let us define u_0 as

$$Pu_0 = \sigma_0 u_0 \quad \text{in } X, \qquad u_0 = f \quad \text{on } \partial X, \qquad \sigma_0 > 0. \tag{47}$$

Moreover, u_0 is a singular point of $\phi(u)$ with $\phi'(u_0)\psi = 0$. Then *define*

$$u_\delta := u_0 + \delta\psi \quad \text{in } X, \qquad \delta \in (-\delta_0, \delta_0), \tag{48}$$

$$\sigma_\delta := \frac{Pu_\delta}{u_\delta} = \sigma_0 \frac{u_0 - \delta\psi}{u_0 + \delta\psi}, \tag{49}$$

$$H_\delta := \sigma_\delta u_\delta^2 = \sigma_0 u_\delta u_{-\delta} = \sigma_0(u_0^2 - \delta^2 \psi^2). \tag{50}$$

We choose δ_0 such that $\sigma_\delta > 0$ a.e. on X for all $\delta \in (-\delta_0, \delta_0)$. Then:

Proposition 3.9 [Bal and Ren 2011b]. *Let u_0 be a singular point and $H_0 = \phi(u_0)$ a critical value of ϕ as above and let ψ be the normalized solution of $\phi'(u_0)\psi = 0$. Let u_δ, σ_δ, and H_δ be defined as in* (48)–(50) *for $0 \neq \delta \in (-\delta_0, \delta_0)$ for δ_0 sufficiently small. Then we verify that:*

$$\sigma_\delta \neq \sigma_{-\delta}, \quad \sigma_\delta > 0, \quad H_\delta = H_{-\delta}, \quad Pu_\delta = \sigma_\delta u_\delta \text{ in } X, \quad u_\delta = f \text{ on } \partial X.$$

This shows the nonuniqueness of the reconstruction of σ from knowledge of $H = \sigma u^2$. Moreover we verify that $\sigma_{\pm\delta}$ agree on ∂X so that this boundary information cannot be used to distinguish between σ_δ and $\sigma_{-\delta}$. The nonuniqueness result is not very restrictive since we have seen that two coefficients, hence one coefficient, may be uniquely reconstructed from *two* well-chosen illuminations in the PAT results. Nonetheless, the above result shows once more that identifiability of the unknown coefficients is not always guaranteed by the availability of internal measurements.

4. Reconstructions from functionals of ∇u

We have seen two models of hybrid inverse problems with measurements involving ∇u. In UMEIT, the measurements are of the form $H(x) = \gamma(x)|\nabla u|^2(x)$ whereas in CDII, they are of the form $H(x) = \gamma(x)|\nabla u|(x)$.

We consider more generally measurements of the form $H(x) = \gamma(x)|\nabla u|^{2-p}$ for u the solution to the elliptic equation

$$\begin{aligned} -\nabla \cdot \gamma(x)\nabla u &= 0 \quad \text{in } X, \\ u &= f \quad \text{on } \partial X. \end{aligned} \tag{51}$$

Since $H(x)$ is linear in $\gamma(x)$, we have formally what appears to be an extension to $p \geq 0$ of the p-Laplacian elliptic equations

$$-\nabla \cdot \frac{H(x)}{|\nabla u|^{2-p}} \nabla u = 0, \tag{52}$$

posed on a bounded, smooth, open domain $X \subset \mathbb{R}^n$, $n \geq 2$, with prescribed Dirichlet conditions, say. When $1 < p < \infty$, the above problem is known to admit a variational formulation with convex functional $J[\nabla u] = \int_X H(x)|\nabla u|^p dx$, which admits a unique minimizer in an appropriate functional setting solution of the above associated Euler Lagrange equation [Evans 1998].

When $p = 1$, the equation becomes degenerate while for $p < 1$, the equation is in fact hyperbolic. We consider the problem of measurements that are quadratic (or bilinear) in ∇u (with applications to UMEIT and UMOT) in the next two sections. In the following section, we consider the case $p = 1$.

4A. *Reconstruction from a single power density measurement.* The presentation follows [Bal 2012]. When $p = 0$ so that measurements are of the form $H(x) = \gamma(x)|\nabla u|^2$, the above 0-Laplacian turns out to be a hyperbolic equation. Anticipating this behavior, we assume the availability of Cauchy data (i.e., u and $\gamma \nu \cdot \nabla u$ with ν the unit outward normal to X) on ∂X rather than simply Dirichlet data. Then (52) with $p = 0$ becomes after some algebra

$$(I - 2\widehat{\nabla u} \otimes \widehat{\nabla u}) : \nabla^2 u + \nabla \ln H \cdot \nabla u = 0 \quad \text{in } X,$$
$$u = f \quad \text{and} \quad \frac{\partial u}{\partial \nu} = j \quad \text{on } \partial X. \tag{53}$$

Here $\widehat{\nabla u} = \nabla u/|\nabla u|$. With

$$g^{ij} = g^{ij}(\nabla u) = -\delta^{ij} + 2(\widehat{\nabla u})_i(\widehat{\nabla u})_j \quad \text{and} \quad k^i = -(\nabla \ln H)_i,$$

the above equation is in coordinates

$$g^{ij}(\nabla u)\partial^2_{ij} u + k^i \partial_i u = 0 \quad \text{in } X,$$
$$u = f \quad \text{and} \quad \frac{\partial u}{\partial \nu} = j \quad \text{on } \partial X. \tag{54}$$

Since g^{ij} is a definite matrix of signature $(1, n-1)$, (54) is a quasilinear strictly hyperbolic equation. The Cauchy data f and j then need to be provided on a spacelike hyper-surface in order for the hyperbolic problem to be well-posed [Hörmander 1983]. This is the main difficulty in solving (54) with redundant Cauchy boundary conditions.

In general, we cannot hope to reconstruct $u(x)$, and hence $\gamma(x)$ on the whole domain X. The reason is that the direction of "time" in the second-order hyperbolic equation is $\widehat{\nabla u}(x)$. The normal $\nu(x)$ at the boundary ∂X will distinguish between the (good) part of ∂X that is "spacelike" and the (bad) part of ∂X that is "timelike". Spacelike surfaces such as $t = 0$ provide stable information to solve the standard wave equation whereas in general it is known that arbitrary singularities can form in a wave equation from information on "timelike" surfaces such as $x = 0$ or $y = 0$ in a three-dimensional setting (where (t, x, y) are local coordinates of X) [Hörmander 1983].

Uniqueness and stability. Let (u, γ) and $(\tilde u, \tilde\gamma)$ be two solutions of the Cauchy problem (54) with measurements (H, f, j) and $(\tilde H, \tilde f, \tilde j)$, where we define the reconstructed conductivities

$$\gamma(x) = \frac{H}{|\nabla u|^2}(x), \quad \tilde\gamma(x) = \frac{\tilde H}{|\nabla \tilde u|^2}(x). \tag{55}$$

Let $v = \tilde u - u$. We find that

$$\nabla \cdot \left(\frac{H}{|\nabla \tilde u|^2 |\nabla u|^2} \big((\nabla u + \nabla \tilde u) \otimes (\nabla u + \nabla \tilde u) - (|\nabla u|^2 + |\nabla \tilde u|^2)I\big) \nabla v \right.$$
$$\left. + \delta H \left(\frac{\nabla \tilde u}{|\nabla \tilde u|^2} + \frac{\nabla u}{|\nabla u|^2} \right) \right) = 0.$$

This equation is recast as

$$\mathfrak{g}^{ij}(x)\partial^2_{ij} v + \mathfrak{k}^i \partial_i v + \partial_i(l^i \delta H) = 0 \quad \text{in } X,$$
$$v = \tilde f - f, \quad \frac{\partial v}{\partial \nu} = \tilde j - j \quad \text{on } \partial X, \tag{56}$$

for appropriate coefficients $\mathfrak{k}^i$ and l^i, where

$$\mathfrak{g}(x) = \frac{H}{|\nabla \tilde u|^2 |\nabla u|^2} \big((\nabla u + \nabla \tilde u) \otimes (\nabla u + \nabla \tilde u) - (|\nabla u|^2 + |\nabla \tilde u|^2)I\big)$$
$$= \alpha(x)\big(e(x) \otimes e(x) - \beta^2(x)(I - e(x) \otimes e(x))\big), \tag{57}$$

with

$$e(x) = \frac{\nabla u + \nabla \tilde u}{|\nabla u + \nabla \tilde u|}(x), \quad \beta^2(x) = \frac{|\nabla u + \nabla \tilde u|^2}{|\nabla u + \nabla \tilde u|^2 - (|\nabla u|^2 + |\nabla \tilde u|^2)}(x), \tag{58}$$

and $\alpha(x)$ is the appropriate (scalar) normalization coefficient. For ∇u and $\nabla \tilde u$ sufficiently close so that $\nabla u \cdot \nabla \tilde u > 0$, then the above linear equation for v is strictly hyperbolic. We define the Lorentzian metric $\mathfrak{h} = \mathfrak{g}^{-1}$ so that $\mathfrak{h}_{ij}$ are the coordinates of the inverse of the matrix $\mathfrak{g}^{ij}$. We denote by $\langle \cdot, \cdot \rangle$ the bilinear product associated to $\mathfrak{h}$ so that $\langle u, v \rangle = \mathfrak{h}_{ij} u^i v^j$ where the two vectors u and v have coordinates u^i and v^i, respectively. We verify that

$$\mathfrak{h}(x) = \frac{1}{\alpha(x)}\left(e(x) \otimes e(x) - \frac{1}{\beta^2(x)}(I - e(x) \otimes e(x))\right). \tag{59}$$

The spacelike part Σ_g of ∂X is given by $\mathfrak{h}(v, v) > 0$, i.e., v is a timelike vector, or equivalently

$$|v(x) \cdot e(x)|^2 > \frac{1}{1 + \beta^2(x)}, \quad x \in \partial X. \tag{60}$$

Above, the dot product is with respect to the standard Euclidean metric and ν is a unit vector for the Euclidean metric, not for the metric $\mathfrak{h}$. Let Σ_1 be an open connected component of Σ_g and let $\mathbb{O} = \cup_{0 < \tau < s} \Sigma_2(\tau)$ be a domain of influence of Σ_1 swept out by the spacelike surfaces $\Sigma_2(\tau)$; see [Bal 2012; Taylor 1996]. Then we have the following local stability result:

Theorem 4.1 (local uniqueness and stability). *Let u and $\tilde{u}$ be two solutions of (54). We assume that $\mathfrak{g}$ constructed in (57) is strictly hyperbolic. Let Σ_1 be an open connected component of Σ_g the spacelike component of ∂X and let $\mathbb{O}$ be a domain of influence of Σ_1 constructed as above. Let us define the energy*

$$E(dv) = \langle dv, v_2 \rangle^2 - \tfrac{1}{2} \langle dv, dv \rangle \langle v_2, v_2 \rangle. \tag{61}$$

Here, dv is the gradient of v in the metric $\mathfrak{h}$ given in coordinates by $\mathfrak{g}^{ij} \partial_j v$. Then

$$\int_{\mathbb{O}} E(dv) \, dx \le C \left(\int_{\Sigma_1} |f - \tilde{f}|^2 + |j - \tilde{j}|^2 \, d\sigma + \int_{\mathbb{O}} |\nabla \delta H|^2 \, dx \right), \tag{62}$$

where dx and $d\sigma$ are the standard measures on $\mathbb{O}$ and Σ_1, respectively.

In the Euclidean metric, let $v_2(x)$ be the unit vector to $x \in \Sigma_2(\tau)$, define $c(x) := v_2(x) \cdot e(x)$ and

$$\theta := \min_{x \in \Sigma_2(\tau)} \left(c^2(x) - \frac{1}{1 + \beta^2(x)} \right). \tag{63}$$

Then

$$\int_{\mathbb{O}} |v^2| + |\nabla v|^2 + (\gamma - \tilde{\gamma})^2 \, dx$$
$$\le \frac{C}{\theta^2} \left(\int_{\Sigma_1} |f - \tilde{f}|^2 + |j - \tilde{j}|^2 \, d\gamma + \int_{\mathbb{O}} |\nabla \delta H|^2 \, dx \right), \tag{64}$$

where γ and $\tilde{\gamma}$ are the conductivities in (55). Provided that $f = \tilde{f}$, $j = \tilde{j}$, and $H = \tilde{H}$, we obtain that $v = 0$ and the uniqueness result $u = \tilde{u}$ and $\gamma = \tilde{\gamma}$.

The proof is based on adapting energy methods for hyperbolic equations as they are summarized in [Taylor 1996]. The energy $E(dv)$ fails to control dv for null-like or spacelike vectors, i.e., $\mathfrak{h}(dv, dv) \le 0$. The parameter θ measures how timelike the vector dv is on the domain of influence $\mathbb{O}$. As $\mathbb{O}$ approaches the boundary of the domain of influence of Σ_g and θ tends to 0, the energy estimates deteriorate as indicated in (64).

Assuming that the errors on the Cauchy data f and j are negligible, we obtain the following stability estimate for the conductivity

$$\|\gamma - \tilde{\gamma}\|_{L^2(\mathbb{O})} \le \frac{C}{\theta} \|H - \tilde{H}\|_{H^1(X)}. \tag{65}$$

Under additional regularity assumptions on γ, for instance assuming that $H \in H^s(X)$ for $s \geq 2$, we find by standard interpolation that

$$\|\gamma - \tilde{\gamma}\|_{L^2(\mathbb{O})} \leq \frac{C}{\theta} \|H - \tilde{H}\|_{L^2(X)}^{1-\frac{1}{s}} \|H + \tilde{H}\|_{H^s(X)}^{\frac{1}{s}}, \tag{66}$$

We thus obtain Hölder-stable reconstructions in the practical setting of square integrable measurement errors. However, stability is *local*. Only on the domain of influence of the spacelike part of the boundary can we obtain a stable reconstruction. This can be done by solving a nonlinear strictly hyperbolic equation analyzed in [Bal 2012] using techniques summarized in [Hörmander 1997].

Global reconstructions. In the preceding result, the main roadblock to global reconstructions was that the domain of influence of the spacelike part of the boundary was a strict subset of X. There is a simple solution to this problem: simply make sure that the whole boundary is a level set of u and that no critical points of u (where $\nabla u = 0$) exist. Then all of X is in the domain of influence of the spacelike part of ∂X, which is the whole of ∂X. This setting can be made possible independent of the conductivity γ in two dimensions of space but not always in higher dimensions.

Let $n = 2$. We assume that X is an open smooth domain diffeomorphic to an annulus with boundary $\partial X = \partial X_0 \cup \partial X_1$. We assume that $f = 0$ on the external boundary ∂X_0 and that $f = 1$ on the internal boundary ∂X_1. The boundary of X is thus composed of two smooth connected components that are different level sets of the solution u. The solution u to (51) is uniquely defined on X. Then we can show:

Proposition 4.2 [Bal 2012]. *We assume that both the geometry of X and $\gamma(x)$ are sufficiently smooth. Then $|\nabla u|$ is bounded from above and below by positive constants. The level sets $\Sigma_c = \{x \in X : u(x) = c\}$ for $0 < c < 1$ are smooth curves that separate X into two disjoint subdomains.*

The proof is based on the fact that critical points of solutions to elliptic equations in two dimensions are isolated [Alessandrini 1986]. The result extends to higher dimensions provided that $|\nabla u|$ does not vanish with exactly the same proof. In the absence of critical points, we thus obtain that $e(x) = \widehat{\nabla u} = v(x)$ so that $v(x)$ is clearly a timelike vector. Then the local results of Theorem 4.1 become global results, which yields the following proposition:

Proposition 4.3. *Let X be the geometry described above in dimension $n \geq 2$ and $u(x)$ the solution to (51). We assume here that both the geometry and $\gamma(x)$ are sufficiently smooth. We also assume that $|\nabla u|$ is bounded from above and below by positive constants. Then the nonlinear (54) admits a unique solution* and the reconstruction of u and of γ is stable in X in the sense described in Theorem 4.1.

In dimensions $n \geq 3$, we cannot guaranty that u does not have any critical point independent of the conductivity. If the conductivity is close to a constant, then by continuity, u does not have any critical point and the above result applies. This proves the result for sufficiently small perturbations of the case $\gamma(x) = \gamma_0$. In the general case, however, we cannot guaranty that ∇u does not vanish and in fact can produce counterexamples (see [Bal 2012]):

Proposition 4.4 [Bal 2012; Briane et al. 2004; Melas 1993]. *There is an example of a smooth conductivity such that u admits critical points.*

So in dimensions $n \geq 3$, we are not guaranteed that the nonlinear equation will remain strictly hyperbolic. What we can do, however, is again to use the notion of complex geometric optics solutions. We have the result:

Theorem 4.5 [Bal 2012]. *Let γ be extended by $\gamma_0 = 1$ on $\mathbb{R}^n \setminus \tilde{X}$, where $\tilde{X}$ is the domain where γ is not known. We assume that γ is smooth on $\mathbb{R}^n$. Let $\gamma(x) - 1$ be supported without loss of generality on the cube $(0, 1) \times (-\frac{1}{2}, \frac{1}{2})^{n-1}$. Define the domain $X = (0, 1) \times B_{n-1}(a)$, where $B_{n-1}(a)$ is the $(n-1)$-dimensional ball of radius a centered at 0 and where a is sufficiently large that the light cone for the Euclidean metric emerging from $B_{n-1}(a)$ strictly includes $\tilde{X}$.*

There exists an open set of illuminations (f_1, f_2) such that if u_1 and u_2 are the corresponding solutions of (51), then $\gamma(x)$ is uniquely determined by the measurements

$$\begin{aligned} H_1(x) &= \gamma(x) |\nabla u_1|^2(x), \\ H_2(x) &= \gamma(x) |\nabla u_2|^2(x), \\ H_3(x) &= \gamma(x) |\nabla (u_1 + u_2)|^2, \end{aligned} \quad (67)$$

with the corresponding Cauchy data (f_1, j_1), (f_2, j_2) and $(f_1 + f_2, j_1 + j_2)$ at $x_1 = 0$.

Let $\tilde{H}_i$ be measurements corresponding to $\tilde{\gamma}$ and let $(\tilde{f}_1, \tilde{j}_1)$ and $(\tilde{f}_2, \tilde{j}_2)$ be the corresponding Cauchy data at $x_1 = 0$. Assume that $\gamma(x) - 1$ and $\tilde{\gamma}(x) - 1$ are smooth and such that their norm in $H^{\frac{n}{2}+3+\varepsilon}(\mathbb{R}^n)$ for some $\varepsilon > 0$ are bounded by M. Then for a constant C that depends on M, we have the global stability result

$$\|\gamma - \tilde{\gamma}\|_{L^2(\tilde{X})} \leq C \left(\|d_C - \tilde{d}_C\|_{(L^2(B_{n-1}(a)))^4} + \sum_{i=1}^{3} \|\nabla H_i - \nabla \tilde{H}_i\|_{L^2(X)} \right). \quad (68)$$

Here, we have defined $d_C = (f_1, j_1, f_2, j_2)$ with $\tilde{d}_C$ being defined similarly.

The three measurements H_i in (67) actually correspond to two physical measurements since H_3 may be determined from the experiments yielding H_1 and H_2, as demonstrated in [Bal 2012; Kuchment and Kunyansky 2011]. The three measurements are constructed so that two independent strictly hyperbolic Lorentzian metrics can be constructed everywhere inside the domain. These metrics are constructed by means of CGO solutions. The boundary conditions f_j have to be close to the traces of the CGO solutions. We thus obtain a global Lipschitz stability result. The price to pay is that the open set of illuminations is not very explicit and may depend on the conductivities one seeks to reconstruct.

For conductivities that are close to a constant, several reconstructions are therefore available. We have seen that geometries of the form of an annulus (with a hole that can be arbitrarily small and arbitrarily close to the boundary where $f = 0$) allowed us to obtain globally stable reconstructions since in such situations, it is relatively easy to avoid the presence of critical points. The method of CGO solutions can be shown to apply for a well-defined set of illuminations since the (harmonic) CGO solutions are explicitly known for the Euclidean metric and of the form $e^{\rho \cdot x}$ for ρ a complex valued vector such that $\rho \cdot \rho = 0$. After linearization in the vicinity of the Euclidean metric, another explicit reconstruction procedure was introduced in [Kuchment and Kunyansky 2011].

4B. Reconstructions from multiple power density measurements. Rather than reconstructing γ from one given measurement of the form $\gamma(x)|\nabla u|^2$, we can instead acquire several measurements of the form

$$H_{ij}(x) = \gamma(x)\nabla u_i(x) \cdot \nabla u_j(x) \quad \text{in } X, \quad 1 \le i, j \le M, \qquad (69)$$

where u_j solves the elliptic problem (51) with f given by f_j for $1 \le j \le M$. The result presented in Theorem 4.5 above provides a positive answer for $M = 2$ when the available internal functionals are augmented by Cauchy data at the boundary of the domain of interest.

Results obtained in [Bal et al. 2011a; Capdeboscq et al. 2009; Monard and Bal 2012] and based on an entirely different procedure and not requiring knowledge of boundary data show that $M = 2\lfloor \frac{n+1}{2} \rfloor$ measurements allow for a global reconstruction of γ, i.e., M for n even and $M + 1$ for n odd. Such results were first obtained in [Capdeboscq et al. 2009] in the case $n = 2$ and have been extended with a slightly different presentation to the cases $n = 2$ and $n = 3$ in [Bal et al. 2011a] while the general case $n \ge 2$ is treated in [Monard and Bal 2012]. Let us assume that $n = 3$ for concreteness. Then $H_{ij} = S_i \cdot S_j$, where we have defined

$$S_j(x) = \sqrt{\gamma(x)}\nabla u_j(x), \quad 1 \le j \le M.$$

Let $S = (S_1, S_2, S_3)$ be a matrix of $n = 3$ column vectors S_j. Then $S^T S = H$, where S^T is the transpose matrix made of the rows given by the S_j. We do not know S or the S_j, but we know its normal matrix $S^T S = H$. Let T be a matrix such that $R = S T^T$ is a rotation-valued field on X. Two examples are $T = H^{-\frac{1}{2}}$ or the lower-triangular T obtained by the Gram–Schmidt procedure. We thus have information on S. We need additional equations to solve for S, or equivalently R, uniquely. The elliptic equation may be written as $\nabla \cdot \sqrt{\gamma} S_j = 0$, or equivalently

$$\nabla \cdot S_j + F \cdot S_j = 0, \quad F = \nabla(\log \sqrt{\gamma}) = \tfrac{1}{2} \nabla \log \gamma. \tag{70}$$

Now, since $\gamma^{-\frac{1}{2}} S_j$ is a gradient, its curl vanishes and we find that

$$\nabla \times S_j - F \times S_j = 0. \tag{71}$$

Here, F is unknown. We first eliminate it from the equations and then find a closed form equation for S or equivalently for R as a field in $SO(n; \mathbb{R})$.

Let T be the aforementioned matrix T, say $T = H^{-\frac{1}{2}}$ with entries t_{ij} for $1 \leq i, j \leq n$. Let t^{ij} be the entries of T^{-1} and define the vector fields

$$V_{ij} := \nabla(t_{ik}) t^{kj}, \quad \text{i.e.,} \quad V_{ij}^l = \partial_l(t_{ij}) t^{kl}, \quad 1 \leq i, j, l \leq n. \tag{72}$$

We then define $R(x) = S(x) T^T(x) \in SO(n; \mathbb{R})$ the matrix whose columns are composed of the column vectors $R_j = S_j T^T$. Then in all dimension $n \geq 2$, we find

Lemma 4.6 [Bal et al. 2011a; Monard and Bal 2012]. *In $n \geq 2$, we have*

$$F = \frac{1}{n}\left(\frac{1}{2} \nabla \log \det H + \sum_{i,j=1}^n \left((V_{ij} + V_{ji}) \cdot R_i\right) R_j\right). \tag{73}$$

The proof in dimension $n = 2, 3$ can be found in [Bal et al. 2011a] and in arbitrary dimension in [Monard and Bal 2012].

Note that the determinant of H needs to be positive on the domain X in order for the above expression for F to make sense. It is, however, difficult to ensure that the determinant of several gradients remains positive and there are in fact counterexamples as shown in [Briane et al. 2004]. Here again, complex geometric optics solutions are useful to control the determinant of gradients of elliptic solutions locally and globally using several solutions. We state a global result in the practical setting $n = 3$.

Let there be $m \geq 3$ solutions of the elliptic equation and assume that there exists an open covering $\mathcal{O} = \{\Omega_k\}_{1 \leq k \leq N}$ ($X \subset \bigcup_{k=1}^N \Omega_i$), a constant $c_0 > 0$

and a function $\tau : [1, N] \ni i \mapsto \tau(i) = (\tau(i)_1, \tau(i)_2, \tau(i)_3) \in [1, m]^3$, such that

$$\inf_{x \in \Omega_i} \det(S_{\tau(i)_1}(x), S_{\tau(i)_2}(x), S_{\tau(i)_3}(x)) \geq c_0, \quad 1 \leq i \leq N. \tag{74}$$

Then we have the following result:

Theorem 4.7 (3D global uniqueness and stability). *Let $X \subset \mathbb{R}^3$ be an open convex bounded set, and let two sets of $m \geq 3$ solutions of (51) generate measurements $(H, \tilde{H})$ whose components belong to $W^{1,\infty}(X)$. Assume that one can define a couple $(\mathbb{O}, \tau)$ such that (74) is satisfied for both sets of solutions S and $\tilde{S}$. Let also $x_0 \in \overline{\Omega}_{i_0} \subset \overline{X}$ and $\gamma(x_0), \tilde{\gamma}(x_0), \{S_{\tau(i_0)_i}(x_0), \tilde{S}_{\tau(i_0)_i}(x_0)\}_{1 \leq i \leq 3}$ be given. Let γ and $\tilde{\gamma}$ be the conductivities corresponding to the measurements H and $\tilde{H}$, respectively. Then we have the following stability estimate:*

$$\|\log \gamma - \log \tilde{\gamma}\|_{W^{1,\infty}(X)} \leq C\big(\epsilon_0 + \|H - \tilde{H}\|_{W^{1,\infty}(X)}\big), \tag{75}$$

where ϵ_0 is the error at the initial point x_0

$$\epsilon_0 = |\log \gamma_0 - \log \tilde{\gamma}_0| + \sum_{i=1}^{3} \|S_{\tau(i_0)_i}(x_0) - \tilde{S}_{\tau(i_0)_i}(x_0)\|.$$

This shows that the reconstruction of γ is stable from such redundant measurements. Moreover, the reconstruction is constructive. Indeed, after eliminating F from the equations for R, we find an equation of the form $\nabla R = G(x, R)$, where $G(x, R)$ is polynomial of degree three in the entries of R. This is a redundant equation whose solution, when it exists, is unique and stable with respect to perturbations in G and the conditions at a given point x_0.

That (74) is satisfied can again be proved by means of complex geometric optics solutions, as is briefly mentioned in Section 5B below; see [Bal et al. 2011a].

4C. Reconstruction from a single current density measurement.

Let us now come back to the 1-Laplacian, which is a degenerate elliptic problem. In many cases, this problem admits multiple admissible solutions [Kim et al. 2002]. The inverse problem then cannot be solved uniquely. In some settings, however, uniqueness can be restored [Kim et al. 2002; Nachman et al. 2007; 2009; 2011].

Recall that the measurements are of the form $H(x) = \gamma |\nabla u|$ so that u solves the following degenerate quasilinear equation

$$\nabla \cdot \frac{H(x)}{|\nabla u|} \nabla u = 0 \quad \text{in } X. \tag{76}$$

Different boundary conditions may then be considered. It is shown in [Kim et al. 2002] that the above equation augmented with Neumann boundary conditions of

the form
$$\frac{H}{|\nabla u|}\frac{\partial u}{\partial \nu} = h \text{ on } \partial X, \quad \int_{\partial X} u d\sigma = 0,$$

admits an infinite number of solutions once it admits a solution, and may also admit no solution at all. One possible strategy is to acquire two measurements of the form $H(x) = \gamma |\nabla u|$ corresponding to two prescribed currents. In this setting, it is shown in [Kim et al. 2002] that (appropriately defined) singularities of γ are uniquely determined by the measurements. We refer the reader to the latter reference for the details.

Alternatively, we may augment the above (76) with Dirichlet data. Then the reconstruction of γ was shown to be uniquely determined in [Nachman et al. 2007; 2009; 2011]. Why Dirichlet conditions help to stabilize the equation may be explained as follows. The 1-Laplace equation (76) may be recast as

$$(I - \widehat{\nabla u} \otimes \widehat{\nabla u}) : \nabla^2 u + \nabla \ln H \cdot \nabla u = 0,$$

following similar calculations to those leading to (53). The only difference is the "2" in front of $\widehat{\nabla u} \otimes \widehat{\nabla u}$ replaced by "1", or more generally $2 - p$ for a p-Laplacian. When $p > 1$, the problem remains strictly elliptic. When $p < 1$, the problem is hyperbolic, and when $p = 1$, it is degenerate in the direction $\widehat{\nabla u}$ and elliptic in the transverse directions. We can therefore modify u so that its level sets remain unchanged and still satisfy the above partial differential equation. This modification can also be performed so that Neumann boundary conditions are not changed. This is the procedure used in [Kim et al. 2002] to show the nonuniqueness of the reconstruction for the 1-Laplacian with Neumann boundary conditions.

Dirichlet conditions, however, are modified by changes in the level sets of u. It turns out that even with Dirichlet conditions, several (viscosity) solutions to (76) may be constructed when $H \equiv 1$; see [Nachman et al. 2007; 2011]. However, such solutions involve vanishing gradients on sets of positive measure.

The right formulation for the CDII inverse problem that allows one to avoid vanishing gradients is to recast (76) as the minimization of the functional

$$F[\nabla v] = \int_X H(x)|\nabla v|dx, \tag{77}$$

over $v \in H^1(X)$ with $v = f$ on ∂X. Note that $F[\nabla v]$ is convex although it is not *strictly* convex. Moreover, let γ be the conductivity and $H = \gamma |\nabla u|$ the corresponding measurement. Let then $v \in H^1(X)$ with $v = f$ on ∂X. Then,

$$F[\nabla v] = \int_X \gamma |\nabla u||\nabla v|dx \geq \int_X \gamma \nabla u \cdot \nabla v dx = \int_{\partial X} \sigma \frac{\partial u}{\partial \nu} f ds = F[\nabla u],$$

by standard integrations by parts. This shows that u minimizes F. We have the following result:

Theorem 4.8 [Nachman et al. 2011]. *Let $(f, H) \in C^{1,\alpha}(\partial X) \times C^{\alpha}(\overline{X})$ with $H = \gamma |\nabla u|$ for some $\gamma \in C^{\alpha}(\bar{X})$. Assume that $H(x) > 0$ a.e. in X. Then the minimization of*

$$\mathrm{argmin}\{F[\nabla v] : v \in W^{1,1}(X) \cap C(\overline{X}), \; v_{|\partial X} = f\}, \tag{78}$$

has a unique solution u_0. Moreover $\sigma_0 = H|\nabla u_0|^{-1}$ is the unique conductivity associated to the measurement $H(x)$.

It is known in two dimensions of space that $H(x) > 0$ is satisfied for a large class of boundary conditions $f(x)$; see Lemma 5.2 in the next section. In three dimensions of space, however, critical points of u may arise as observed earlier in this paper; see [Bal 2012]. The CGO solutions that are analyzed in the following section allow us to show that $H(x) > 0$ holds for an open set of illuminations f at the boundary of the domain ∂X; see (95) below.

Unfortunately, no such results exists for real-valued solutions and constraints such as $H(x) > 0$ in dimension $n \geq 3$ will not hold for a given f independent of the conductivity γ.

Several reconstruction algorithms have been devised in [Kim et al. 2002; Nachman et al. 2007; 2009; 2011], to which we refer for additional details. The numerical simulations presented in these papers show that when uniqueness is guaranteed, then the reconstructions are very high resolution and quite robust with respect to noise in the data, as is expected for general hybrid inverse problems.

5. Qualitative properties of forward solutions

5A. *The case of two spatial dimensions.* Several explicit reconstructions obtained in hybrid inverse problems require that the solutions to the considered elliptic equations satisfy specific qualitative properties such as the absence of any critical point or the positivity of the determinant of gradients of solutions. Such results can be proved in great generality in dimension $n = 2$ but do not always hold in dimension $n \geq 3$.

In dimension $n = 2$, the critical points of u (points x where $\nabla u(x) = 0$) are necessarily isolated as is shown in [Alessandrini 1986]. From this and techniques of quasiconformal mappings that are also restricted to two dimensions of space, we can show the following results.

Lemma 5.1 [Alessandrini and Nesi 2001]. *Let u_1 and u_2 be the solutions of (51) on X simply connected with boundary conditions $f_1 = x_1$ and $f_2 = x_2$ on ∂X, respectively, where $x = (x_1, x_2)$ are Cartesian coordinates on X. Assume* that γ is sufficiently smooth. Then $(x_1, x_2) \mapsto (u_1, u_2)$ from X to its image is a diffeomorphism. In other words, $\det(\nabla u_1, \nabla u_2) > 0$ uniformly on X.

This result is useful in the analysis of UMEIT and UMOT in the case of redundant measurements. It is shown in [Briane et al. 2004] that the appropriate extension of this result is false in dimension $n \geq 3$.

We recall that a function continuous on a simple closed contour is almost two-to-one if it is two-to-one except possibly at its maximum and minimum [Nachman et al. 2007]. Then we have, quite similarly to the result in Lemma 3.1 and Proposition 4.2, which also use the results in [Alessandrini 1986], the following:

Lemma 5.2 [Nachman et al. 2007]. *Let X be a simply connected planar domain and let u be solution of* (51) *with f almost two-to-one and σ sufficiently smooth. Then $|\nabla u|$ is bounded from below by a positive constant on $\bar{X}$. Moreover, the level sets of u are open curves inside X with their two end points on ∂X.*

This shows that for a large class of boundary conditions with one maximum and one minimum, the solution u cannot have any critical point in $\bar{X}$. On an annulus with boundaries equal to level sets of u, we saw in Proposition 4.2 that u had no critical points on X in dimension $n = 2$. This was used to show that the normal vector to the level sets of u always forms a timelike vector for the Lorentzian metric defined in (54).

All these results no longer hold in dimension $n \geq 3$. See [Bal 2012; Briane et al. 2004] for counterexamples. In dimension $n \geq 3$, the required qualitative properties cannot be obtained for a given set of illuminations (boundary conditions) independent of the conductivity. However, for conductivities that are bounded (with an arbitrary bound) in an appropriate norm, there are open sets of illuminations that allow us to obtain the required qualitative properties. One way to construct such solutions is by means of the complex geometric optics solutions that are analyzed in the next section.

5B. *Complex geometric optics solutions.*

CGO solutions and Helmholtz equations. Complex geometrical optics (CGO) solutions allow us to treat the potential q in the equation

$$\begin{aligned}(\Delta + q)u &= 0 \quad \text{in } X, \\ u &= f \quad \text{on } \partial X,\end{aligned} \quad (79)$$

as a perturbation of the leading operator Δ. When $q = 0$, CGO solutions are harmonic solutions defined on $\mathbb{R}^n$ and are of the form

$$u_\rho(x) = e^{\rho \cdot x}, \qquad \rho \in \mathbb{C}^n \text{ such that } \rho \cdot \rho = 0.$$

For $\rho = \rho_r + i\rho_i$ with ρ_r and ρ_i vectors in $\mathbb{R}^n$, this means that $|\rho_r|^2 = |\rho_i|^2$ and $\rho_r \cdot \rho_i = 0$.

When $q \not\equiv 0$, CGO solutions are solutions of the following problem
$$\Delta u_\rho + q u_\rho = 0, \qquad u_\rho \sim e^{\rho \cdot x} \text{ as } |x| \to \infty. \tag{80}$$

More precisely, we say that u_ρ is a solution of the above equation with $\rho \cdot \rho = 0$ and the proper behavior at infinity when it is written as
$$u_\rho(x) = e^{\rho \cdot x}(1 + \psi_\rho(x)), \tag{81}$$

for $\psi_\rho \in L^2_\delta$ a weak solution of
$$\Delta \psi_\rho + 2\rho \cdot \nabla \psi_\rho = -q(1 + \psi_\rho). \tag{82}$$

The space L^2_δ for $\delta \in \mathbb{R}$ is defined as the completion of $C^\infty_0(\mathbb{R}^n)$ with respect to the norm $\|\cdot\|_{L^2_\delta}$ defined as
$$\|u\|_{L^2_\delta} = \left(\int_{\mathbb{R}^n} \langle x \rangle^{2\delta} |u|^2 dx\right)^{\frac{1}{2}}, \qquad \langle x \rangle = (1 + |x|^2)^{\frac{1}{2}}. \tag{83}$$

Let $-1 < \delta < 0$ and $q \in L^2_{\delta+1}$ and $\langle x \rangle q \in L^\infty$. One of the main results in [Sylvester and Uhlmann 1987] is that there exists $\eta = \eta(\delta)$ such that the above problem admits a unique solution with $\psi_\rho \in L^2_\delta$ provided that
$$\|\langle x \rangle q\|_{L^\infty} + 1 \leq \eta |\rho|.$$

Moreover, $\|\psi_\rho\|_{L^2_\delta} \leq C|\rho|^{-1}\|q\|_{L^2_{\delta+1}}$ for some $C = C(\delta)$. In the analysis of many hybrid problems, we need smoother CGO solutions than what was recalled above. We introduce the spaces H^s_δ for $s \geq 0$ as the completion of $C^\infty_0(\mathbb{R}^n)$ with respect to the norm $\|\cdot\|_{H^s_\delta}$ defined as
$$\|u\|_{H^s_\delta} = \left(\int_{\mathbb{R}^n} \langle x \rangle^{2\delta} |(I-\Delta)^{\frac{s}{2}} u|^2 dx\right)^{\frac{1}{2}}. \tag{84}$$

Here $(I-\Delta)^{\frac{s}{2}} u$ is defined as the inverse Fourier transform of $\langle \xi \rangle^s \hat{u}(\xi)$, where $\hat{u}(\xi)$ is the Fourier transform of $u(x)$.

Proposition 5.3 [Bal and Uhlmann 2010]. *Let $-1 < \delta < 0$ and $k \in \mathbb{N}^*$. Let $q \in H^{\frac{n}{2}+k+\varepsilon}_1$ (hence $q \in H^{\frac{n}{2}+k+\varepsilon}_{\delta+1}$) and let ρ be such that*
$$\|q\|_{H^{\frac{n}{2}+k+\varepsilon}_1} + 1 \leq \eta|\rho|. \tag{85}$$

Then ψ_ρ, the unique solution to (82), belongs to $H_\delta^{\frac{n}{2}+k+\varepsilon}$ and

$$|\rho|\|\psi_\rho\|_{H_\delta^{\frac{n}{2}+k+\varepsilon}} \leq C\|q\|_{H_{\delta+1}^{\frac{n}{2}+k+\varepsilon}}, \tag{86}$$

for a constant C that depends on δ and η.

We also want to obtain estimates for ψ_ρ and u_ρ restricted to the bounded domain X. We have the following result.

Corollary 5.4 [Bal and Uhlmann 2010]. *Let us assume the regularity hypotheses of the previous proposition. Then we find that*

$$|\rho|\|\psi_\rho\|_{H^{\frac{n}{2}+k+\varepsilon}(X)} + \|\psi_\rho\|_{H^{\frac{n}{2}+k+1+\varepsilon}(X)} \leq C\|q\|_{H^{\frac{n}{2}+k+\varepsilon}(X)}. \tag{87}$$

These results show that for ρ sufficiently large, ψ_ρ is small compared to 1 in the class $C^k(\bar{X})$ by Sobolev imbedding.

Let $Y = H^p(X)$ and $\mathcal{M}$ the ball in Y of functions with norm bounded by a fixed $M > 0$. Not only do we have that ψ_ρ is small for $|\rho|$ large, but we have the following Lipschitz stability with respect to changes in the potential $q(x)$:

Lemma 5.5 [Bal et al. 2011b]. *Let ψ_ρ be the solution of*

$$\Delta\psi_\rho + 2\rho \cdot \nabla\psi_\rho = -q(1+\psi_\rho), \tag{88}$$

and let $\tilde{\psi}_\rho$ be the solution of the same equation with q replaced by $\tilde{q}$, where $\tilde{q}$ is defined as in (27) with σ replaced by $\tilde{\sigma}$. We assume that q and $\tilde{q}$ are in $\mathcal{M}$. Then there is a constant C such that for all ρ with $|\rho| \geq |\rho_0|$, we have

$$\|\psi_\rho - \tilde{\psi}_\rho\|_Y \leq \frac{C}{|\rho|}\|\sigma - \tilde{\sigma}\|_Y. \tag{89}$$

This is the property used in [Bal et al. 2011b] to show that σ in the TAT problem (6)–(7) solves the equation

$$\sigma(x) = e^{(\rho+\bar{\rho})\cdot x} H(x) - \mathcal{H}_f[\sigma](x) \text{ on } X,$$

where

$$\mathcal{H}_f[\sigma](x) = \sigma\big(\psi_f + \overline{\psi_f} + \psi_f\overline{\psi_f}(x)\big),$$

is a contraction map for f in an open set of illuminations; see [Bal et al. 2011b]. The result in Theorem 3.7 then follows by a Banach fixed point argument.

CGO solutions and elliptic equations. Consider the more general elliptic equation

$$\begin{aligned}-\nabla \cdot \gamma \nabla u + \sigma u &= 0 \quad \text{in } X, \\ u &= f \quad \text{on } \partial X.\end{aligned} \tag{90}$$

Upon defining $v = \sqrt{\gamma}u$, we find that

$$(\Delta + q)v = 0 \text{ in } X, \quad q = -\frac{\Delta\sqrt{\gamma}}{\sqrt{\gamma}} - \frac{\sigma}{\gamma}.$$

In other words, we find CGO solutions for (90) defined on $\mathbb{R}^n$ and of the form

$$u_\rho(x) = \frac{1}{\sqrt{\gamma}} e^{\rho \cdot x}\big(1 + \psi_\rho(x)\big), \tag{91}$$

with $|\rho|\psi_\rho(x)$ bounded uniformly provided that γ and σ are sufficiently smooth coefficients.

5C. *Application of CGO solutions to qualitative properties of elliptic solutions.*

Lower bound for the modulus of complex valued solutions. The above results show that for $|\rho|$ sufficiently large, then $|u_\rho|$ is uniformly bounded from below by a positive constant on compact domains. Note that u_ρ is complex valued and that its real and imaginary parts oscillate very rapidly. Indeed,

$$e^{\rho \cdot x} = e^{\rho_r \cdot x}\big(\cos(\rho_i \cdot x) + i\sin(\rho_i \cdot x)\big),$$

which is rapidly increasing in the direction ρ_r and rapidly oscillating in the direction ρ_i. Nonetheless, on a compact domain such as X, then $|u_\rho|$ is uniformly bounded from below by a positive constant.

Let now $f_\rho = u_{\rho|\partial X}$ the trace of the CGO solution on ∂X. Then for f close to f_ρ and u the solution to, say, (79) or (90), we also obtain that $|u|$ is bounded from below by a positive constant. Such results were used in [Triki 2010].

Lower bound for vector fields.

Theorem 5.6 [Bal and Uhlmann 2010]. *Let u_{ρ_j} for $j = 1, 2$ be CGO solutions with q as above for both ρ_j and $k \geq 1$ and with $c_0^{-1}|\rho_1| \leq |\rho_2| \leq c_0|\rho_1|$ for some $c_0 > 0$. Then we have*

$$\hat{\beta} := \frac{1}{2|\rho_1|} e^{-(\rho_1+\rho_2)\cdot x}\big(u_{\rho_1}\nabla u_{\rho_2} - u_{\rho_2}\nabla u_{\rho_1}\big) = \frac{\rho_1 - \rho_2}{2|\rho_1|} + \hat{h}, \tag{92}$$

where the vector field $\hat{h}$ satisfies the constraint

$$\|\hat{h}\|_{C^k(\bar{X})} \leq \frac{C_0}{|\rho_1|}, \tag{93}$$

for some constant C_0 independent of ρ_j, $j = 1, 2$.

With $\rho_2 = \overline{\rho_1}$ so that $u_{\rho_2} = \overline{u_{\rho_1}}$, the imaginary part of (92) is a vector field that does not vanish on X for $|\rho_1|$ sufficiently large. Moreover, let $u_{\rho_1} = v + iw$

and $u_{\rho_2} = v - iw$ for v and w real-valued functions. Then the imaginary and real parts of (92) are given by

$$\Im\hat\beta = \frac{1}{|\rho_1|} e^{-2\Re\rho_1 \cdot x}(w\nabla v - v\nabla w) = \frac{\Im\rho_1}{|\rho_1|} + \Im\hat h, \quad \Re\hat\beta = 0.$$

Let u_1 and u_2 be solutions of the elliptic problem (79) on X such that $u_1 + iu_2$ on ∂X is close to the trace of u_{ρ_1}. The above result shows that

$$|u_1\nabla u_2 - u_2\nabla u_1| \geq c_0 > 0 \quad \text{in } X.$$

This yields (24) and the result on unique and stable reconstructions in QPAT.

The above derivation may be generalized to the vector field β_α in (32) with applications in elastography. Indeed let us start from (81) with $\rho = \mathbf{k} + i\mathbf{l}$ such that $\mathbf{k}\cdot\mathbf{l} = 0$ and $k := |\mathbf{k}| = |\mathbf{l}|$. Then, using Corollary 5.4, we find that

$$\Re u_\rho = e^{\mathbf{k}\cdot x}(\mathfrak{c} + \varphi_\rho^r), \qquad \Im u_\rho = e^{\mathbf{k}\cdot x}(\mathfrak{s} + \varphi_\rho^i),$$
$$\nabla\Re u_\rho = ke^{\mathbf{k}\cdot x}(\mathfrak{c}\hat{\mathbf{k}} - \mathfrak{s}\hat{\mathbf{l}} + \chi_\rho^r), \quad \nabla\Im u_\rho = ke^{\mathbf{k}\cdot x}(\mathfrak{s}\hat{\mathbf{k}} + \mathfrak{c}\hat{\mathbf{l}} + \chi_\rho^i), \qquad (94)$$

where $\mathfrak{c} = \cos(\mathbf{l}\cdot x)$, $\mathfrak{s} = \sin(\mathbf{l}\cdot x)$, $\hat{\mathbf{k}} = \mathbf{k}/|\mathbf{k}|$, $\hat{\mathbf{l}} = \mathbf{l}/|\mathbf{k}|$, and where $|\rho|\,|\zeta|$ is bounded as indicated in Corollary 5.4 for $\zeta \in \{\varphi_\rho^r, \varphi_\rho^i, \chi_\rho^r, \chi_\rho^i\}$.

Let u_1 on ∂X be close to $\Re u_\rho$. Then we find by continuity that $|\nabla u_1|$ is close to $|\nabla \Re u_\rho|$ so that for k sufficiently large, we find that

$$|\nabla u_1| \geq c_0 > 0 \quad \text{in } X. \qquad (95)$$

This proves that $H(x) = \gamma|\nabla u|$ is bounded from below by a positive constant provided that the boundary condition f is in a well-chosen open set of illuminations.

For the application to elastography, define

$$\beta_\alpha = \Im u_\rho \nabla \Re u_\rho - \alpha \Re u_\rho \nabla \Im u_\rho, \quad \alpha > 0.$$

For $|\rho| > \rho_\alpha$ large enough that $|\zeta| < \dfrac{(\min(1,\alpha))^2}{4(1+\alpha)}$ for $\zeta \in \{\varphi_\rho^r, \varphi_\rho^i, \chi_\rho^r, \chi_\rho^i\}$, we verify using (94) that

$$|\beta_\alpha| \geq ke^{2\mathbf{k}\cdot x}\tfrac{1}{2}\big((\mathfrak{c}\mathfrak{s}(1-\alpha))^2 + (\mathfrak{s}^2 + \alpha\mathfrak{c}^2)\big) \geq ke^{2\mathbf{k}\cdot x}\tfrac{1}{2}(\min(1,\alpha))^2. \quad (96)$$

This provides a lower bound for β_α uniformly on compact sets. For an open set of illuminations (f_1, f_2) close to the traces of $(\Im u_\rho, \Re u_\rho)$ on ∂X, we find by continuity that the vector field $\beta_\alpha = u_1\nabla u_2 - \alpha u_2\nabla u_1$ in (32) also has a norm bounded from below uniformly on X.

Lower bound for determinants. The reconstruction in Theorem 4.7 requires that the determinants in (74) be bounded from below. In specific situations, for instance when the conductivity is close to a given constant, such a determinant is indeed bounded from below by a positive constant for a large class of boundary conditions. However, it has been shown in [Briane et al. 2004] that the determinant of the gradients of three solutions could change signs on a domain with conductivities with large gradient. Unlike what happens in two dimensions of space, it is therefore not possible in general to show that the determinant of gradients of solutions has a given sign. However, using CGO solutions, we can be assured that on given bounded domains, the larger of two determinants is indeed uniformly positive for well-chosen boundary conditions.

Let $u_\rho(x)$ be given by (91) solution of the elliptic problem (90). Upon treating the term ψ_ρ and its derivative as in (94) above and making them arbitrary small by choosing ρ sufficiently large, we find that $\sqrt{\gamma} u_\rho = e^{\rho \cdot x} + \text{l.o.t.}$, so that, to leading order,

$$\sqrt{\gamma} \nabla u_\rho = e^{k \cdot x}(k + il)\big(\cos(l \cdot x) + i \sin(l \cdot x)\big) + \text{l.o.t.}, \quad \rho = k + il.$$

Let $n = 3$ and (e_1, e_2, e_3) a constant orthonormal frame of $\mathbb{R}^3$. It remains to take the real and imaginary parts of the above terms and choose $\hat{k} = e_2$ or $\hat{k} = e_3$ with $\hat{l} = e_1$ to obtain, up to normalization and negligible contributions (for $k = |k|$ sufficiently large), that for

$$\tilde{S}_1 = e_2 \cos kx_1 - e_1 \sin kx_1, \quad \tilde{S}_2 = e_1 \cos kx_1 + e_2 \sin kx_1,$$
$$\tilde{S}_3 = e_3 \cos kx_1 - e_1 \sin kx_1, \quad \tilde{S}_4 = e_2 \cos|k|x_1 + e_3 \sin|k|x_1,$$

we verify that $\det(\tilde{S}_1, \tilde{S}_2, \tilde{S}_3) = -\cos kx_1$ and that $\det(\tilde{S}_1, \tilde{S}_2, \tilde{S}_4) = -\sin kx_1$. Upon changing the sign of S_3 or S_4 if necessary to make both determinants nonnegative, we find that the maximum of these two determinants is always bounded from below by a positive constant uniformly on X. This result is sufficient to prove Theorem 4.7; see [Bal et al. 2011a].

Hyperbolicity of a Lorentzian metric. As a final application of CGO solutions, we mention the proof that a given constant vector field remains a timelike vector of a Lorentzian metric. This finds applications in the proof of Theorem 4.1 in [Bal 2012].

Indeed, let $\hat{k}$ be a given direction in $\mathbb{S}^{n-1}$ and $\rho = ik + k^\perp$ and $u_\rho = e^{\rho \cdot x}$, once again neglecting ψ_ρ. The real and imaginary parts of ∇u_ρ are such that

$$e^{-k^\perp \cdot x} \Im \nabla e^{\rho \cdot x} = |k| \theta(x), \quad e^{-k^\perp \cdot x} \Re \nabla e^{\rho \cdot x} = |k| \theta^\perp(x), \qquad (97)$$

where $\theta(x) = \hat{k} \cos k \cdot x + \hat{k}^\perp \sin k \cdot x$ and $\theta^\perp(x) = -\hat{k} \sin k \cdot x + \hat{k}^\perp \cos k \cdot x$. As usual, $\hat{k} = k/|k|$.

Define the Lorentzian metrics

$$\mathfrak{h}_\theta = 2\theta \otimes \theta - I, \quad \mathfrak{h}_{\theta^\perp} = 2\theta^\perp \otimes \theta^\perp - I.$$

Note that $\theta(x)$ and $\theta^\perp(x)$ oscillate in the plane $(\boldsymbol{k}, \boldsymbol{k}^\perp)$. The given vector $\hat{\boldsymbol{k}}$ thus cannot be a time like vector for one of the Lorentzian metrics for all $x \in X$ (unless X is a domain included in a thin slab). However, in the vicinity of any point x_0, we can construct a linear combination $\psi(x) = \cos\alpha\, \theta(x) + \sin\alpha\, \theta^\perp(x)$ for $\alpha \in [0, 2\pi)$ such that

$\hat{\boldsymbol{k}}$ is a timelike vector for $\mathfrak{h}_\psi = 2\psi \otimes \psi - I$, i.e., $\mathfrak{h}_\psi(\hat{\boldsymbol{k}}, \hat{\boldsymbol{k}}) = 2(\psi \cdot \hat{\boldsymbol{k}})^2 - 1 > 0$,

uniformly for x close to x_0; see [Bal 2012] for more details. When $\theta(x)$ is constructed as $\widehat{\nabla u}$ for u solution to (79) or (90) for boundary conditions f close to the trace of the corresponding CGO solution u_ρ, then the Lorentzian metric $\mathfrak{h}_\psi$ constructed above still verifies that $\hat{\boldsymbol{k}}$ is a timelike vector with $\mathfrak{h}_\psi(\hat{\boldsymbol{k}}, \hat{\boldsymbol{k}})$ uniformly bounded from below by a positive constant locally.

6. Conclusions and perspectives

Research in hybrid inverse problems has been very active in recent years, primarily in the mathematical and medical imaging communities but also in geophysical imaging, see [White 2005] and references on the electrokinetic effect. This review focused on time-independent equations primarily with scalar-valued solutions. We did not consider the body of work done in the setting of time-dependent measurements, which involves different techniques than those presented here; see [McLaughlin et al. 2010] and references. We considered scalar equations with the exception of the system of Maxwell's equations as it appears in thermoacoustic tomography. Very few results exist for systems of equations. The diffusion and conductivity equations considered in this review involve a scalar coefficient γ. The reconstruction of more general tensors remains an open problem.

Compared to boundary value inverse problems, inverse problems with internal measurements enjoy better stability estimates precisely because local information is available. However, the derivation of such stability estimates often requires that specific, qualitative properties of solutions be satisfied, such as for instance the absence of critical points. This imposes constraints on the illuminations (boundary conditions) used to generate the internal data that forms one of the most difficult mathematical questions raised by the hybrid inverse problems.

What are the "optimal" illuminations (boundary conditions) for a given class of unknown parameters and how robust will the reconstructions be when such illuminations are modified are questions that are not fully answered. The theory of complex geometrical optics (CGO) solutions provides a useful tool to address these questions and construct suitable illuminations or prove their existence in several cases of interest. Numerical simulations will presumably be of great help to better understand whether such theoretical predictions are useful or reasonable in practice. Many numerical simulations performed in two dimensions of space confirm the good stability properties predicted by theory [Ammari et al. 2008; Bal and Ren 2011a; Bal et al. 2011b; Capdeboscq et al. 2009; Gebauer and Scherzer 2008; Kuchment and Kunyansky 2011; Nachman et al. 2009]. The two-dimensional setting is special as we saw in Section 5A. Very few simulations have been performed in the theoretically more challenging case of three (or more) dimensions of space. Simulations in [Kuchment and Kunyansky 2011] show very promising three-dimensional reconstructions in the setting of diffusion coefficients that are close to the constant case, which is also understood theoretically since $|\nabla u|$ then does not vanish for a large class of boundary conditions.

The main interest of hybrid inverse problems is that they combine high contrast with high resolution. This translates mathematically into good (Lipschitz or Hölder) stability estimates. Ideally, we would like to reconstruct highly oscillatory coefficients with a minimal influence of the noise in the measurements. Yet, all the results presented in this review paper and the cited references require that the coefficients satisfy some unwanted smoothness properties. To focus on one example for concreteness, the reconstructions in photoacoustic tomography involve the solution of the transport equation (26), which is well-posed provided that the vector field β is sufficiently smooth. Using theories of renormalization, the regularity of such vector fields can be decreased to $W^{1,1}$ or to the BV category [Ambrosio 2004; Bouchut and Crippa 2006; DiPerna and Lions 1989]. Yet, $u_1 \nabla u_2 - u_2 \nabla u_1$ is a priori only in L^2 when γ is arbitrary as a bounded coefficient [Hauray 2003]. The construction of CGO solutions presented in Section 5B also requires sufficient smoothness of the coefficients. How such reconstructions and stability estimates might degrade in the presence of nonsmooth coefficients is quite open. Many similar problems are also open for boundary-value inverse problems [Uhlmann 2009].

Finally, we have assumed in this review that the first step of the hybrid inverse problems had been done accurately. In practice, this may not quite always be so. PAT and TAT require that we solve an inverse source problem for a wave equation, which is a difficult problem in the presence of partial data and variable sound speed and is not entirely understood when realistic absorbing effects are accounted for [Kowar and Scherzer 2012; Stefanov and Uhlmann 2011]. In UMEIT and UMOT, we have assumed in the derivation in Section 2 that standing plane waves could be generated. This is practically difficult to achieve and different (equivalent) mechanisms have been proposed [Ammari et al. 2008; Kuchment and Kunyansky 2011]. In transient elastography, we have assumed that the full (scalar) displacement could be reconstructed as a function of time and space. This is also sometimes an idealized approximation of what can be achieved in practice [McLaughlin et al. 2010]. Finally, we have assumed knowledge of the current $\gamma|\nabla u|$ in CDII, which is also difficult to acquire in practical settings as typically only the z component of the magnetic field B_z can be constructed; see the recent review [Seo and Woo 2011]. The modeling of errors generated during the first step of the procedure and the influence that such errors may have on the reconstructions during the second step of the hybrid inverse problem remain active areas of research.

Acknowledgment

This review and several collaborative efforts that led to papers referenced in the review were initiated during the participation of the author to the program on Inverse Problems and Applications at the Mathematical Sciences Research Institute, Berkeley, California, in the Fall of 2010. I would like to thank the organizers and in particular Gunther Uhlmann for creating a very stimulating research environment at MSRI. Partial funding of this work by the National Science Foundation is also greatly acknowledged.

References

[Alessandrini 1986] G. Alessandrini, "An identification problem for an elliptic equation in two variables", *Ann. Mat. Pura Appl.* (4) **145** (1986), 265–295. MR 88g:35193

[Alessandrini and Nesi 2001] G. Alessandrini and V. Nesi, "Univalent σ-harmonic mappings", *Arch. Ration. Mech. Anal.* **158**:2 (2001), 155–171. MR 2002d:31004 Zbl 0977.31006

[Ambrosetti and Prodi 1972] A. Ambrosetti and G. Prodi, "On the inversion of some differentiable mappings with singularities between Banach spaces", *Ann. Mat. Pura Appl.* (4) **93** (1972), 231–246. MR 47 #9377 Zbl 0288.35020

[Ambrosio 2004] L. Ambrosio, "Transport equation and Cauchy problem for BV vector fields", *Invent. Math.* **158**:2 (2004), 227–260. MR 2005f:35127 Zbl 1075.35087

[Ammari 2008] H. Ammari, *An introduction to mathematics of emerging biomedical imaging*, Mathématiques et applications **62**, Springer, Berlin, 2008. MR 2010j:44002 Zbl 1181.92052

[Ammari et al. 2008] H. Ammari, E. Bonnetier, Y. Capdeboscq, M. Tanter, and M. Fink, "Electrical impedance tomography by elastic deformation", *SIAM J. Appl. Math.* **68**:6 (2008), 1557–1573. MR 2009h:35439 Zbl 1156.35101

[Ammari et al. 2010] H. Ammari, E. Bossy, V. Jugnon, and H. Kang, "Mathematical modeling in photo-acoustic imaging of small absorbers", *SIAM Review* **52**:4 (2010), 677–695.

[Atlan et al. 2005] M. Atlan, B. C. Forget, F. Ramaz, A. C. Boccara, and M. Gross, "Pulsed acousto-optic imaging in dynamic scattering media with heterodyne parallel speckle detection", *Optics Letters* **30**:11 (2005), 1360–1362.

[Bal 2009] G. Bal, "Inverse transport theory and applications", *Inverse Problems* **25** (2009), 053001.

[Bal 2012] G. Bal, "Cauchy problem for ultrasound modulated EIT", preprint, 2012. To appear in *Analysis & PDE*. arXiv 1201.0972

[Bal and Ren 2011a] G. Bal and K. Ren, "Multi-source quantitative photoacoustic tomography in a diffusive regime", *Inverse Problems* **27**:7 (2011), 075003. MR 2012h:65255 Zbl 1225.92024

[Bal and Ren 2011b] G. Bal and K. Ren, "Non-uniqueness results for a hybrid inverse problem", *Contemporary Mathematics* **559** (2011), 29–38.

[Bal and Schotland 2010] G. Bal and J. C. Schotland, "Inverse scattering and acousto-optics imaging", *Phys. Rev. Letters* **104** (2010), 043902.

[Bal and Uhlmann 2010] G. Bal and G. Uhlmann, "Inverse diffusion theory of photoacoustics", *Inverse Problems* **26**:8 (2010), 085010. MR 2658827 Zbl 1197.35311

[Bal et al. 2010] G. Bal, A. Jollivet, and V. Jugnon, "Inverse transport theory of photoacoustics", *Inverse Problems* **26**:2 (2010), 025011. MR 2011b:35552 Zbl 1189.35367

[Bal et al. 2011a] G. Bal, E. Bonnetier, F. Monard, and F. Triki, "Inverse diffusion from knowledge of power densities", preprint, 2011. arXiv 1110.4577

[Bal et al. 2011b] G. Bal, K. Ren, G. Uhlmann, and T. Zhou, "Quantitative thermo-acoustics and related problems", *Inverse Problems* **27**:5 (2011), 055007. MR 2793826 Zbl 1217.35207

[Bouchut and Crippa 2006] F. Bouchut and G. Crippa, "Uniqueness, renormalization, and smooth approximations for linear transport equations", *SIAM J. Math. Anal.* **38**:4 (2006), 1316–1328. MR 2008d:35019 Zbl 1122.35104

[Briane et al. 2004] M. Briane, G. W. Milton, and V. Nesi, "Change of sign of the corrector's determinant for homogenization in three-dimensional conductivity", *Arch. Ration. Mech. Anal.* **173**:1 (2004), 133–150. MR 2005i:49015 Zbl 1118.78009

[Capdeboscq et al. 2009] Y. Capdeboscq, J. Fehrenbach, F. de Gournay, and O. Kavian, "Imaging by modification: numerical reconstruction of local conductivities from corresponding power density measurements", *SIAM J. Imaging Sci.* **2**:4 (2009), 1003–1030. MR 2011c:35611 Zbl 1180.35549

[Cox et al. 2009a] B. T. Cox, S. R. Arridge, and P. C. Beard, "Estimating chromophore distributions from multiwavelength photoacoustic images", *J. Opt. Soc. Am. A* **26** (2009), 443–455.

[Cox et al. 2009b] B. T. Cox, J. G. Laufer, and P. C. Beard, "The challenges for quantitative photoacoustic imaging", article 717713 in *Photons plus ultrasound — imaging and sensing 2009* (*10th Conference on Biomedical Thermoacoustics, Optoacoustics, and Acousto-optics*), edited by A. Oraevsky and L. V. Wang, Proc. of SPIE **7177**, 2009.

[Dautray and Lions 1990] R. Dautray and J.-L. Lions, *Mathematical analysis and numerical methods for science and technology*, vol. 3: *Spectral theory and applications*, Springer, Berlin, 1990. MR 91h:00004a Zbl 0944.47002

[DiPerna and Lions 1989] R. J. DiPerna and P.-L. Lions, "On the Cauchy problem for Boltzmann equations: global existence and weak stability", *Ann. of Math.* (2) **130**:2 (1989), 321–366. MR 90k:82045 Zbl 0698.45010

[Evans 1998] L. C. Evans, *Partial differential equations*, Graduate Studies in Mathematics **19**, American Mathematical Society, Providence, RI, 1998. MR 99e:35001 Zbl 0902.35002

[Finch et al. 2004] D. Finch, S. K. Patch, and Rakesh, "Determining a function from its mean values over a family of spheres", *SIAM J. Math. Anal.* **35**:5 (2004), 1213–1240. MR 2005b:35290 Zbl 1073.35144

[Fisher et al. 2007] A. R. Fisher, A. J. Schissler, and J. C. Schotland, "Photoacoustic effect for multiply scattered light", *Phys. Rev. E* **76** (2007), 036604.

[Gebauer and Scherzer 2008] B. Gebauer and O. Scherzer, "Impedance-acoustic tomography", *SIAM J. Appl. Math.* **69**:2 (2008), 565–576. MR 2009j:35381 Zbl 1159.92027

[Gilbarg and Trudinger 1977] D. Gilbarg and N. S. Trudinger, *Elliptic partial differential equations of second order*, Grundlehren der Math. Wiss. **224**, Springer, Berlin, 1977. MR 57 #13109 Zbl 0361.35003

[Haltmeier et al. 2004] M. Haltmeier, O. Scherzer, P. Burgholzer, and G. Paltauf, "Thermoacoustic computed tomography with large planar receivers", *Inverse Problems* **20**:5 (2004), 1663–1673. MR 2006a:35311 Zbl 1065.65143

[Hardt et al. 1999] R. Hardt, M. Hoffmann-Ostenhof, T. Hoffmann-Ostenhof, and N. Nadirashvili, "Critical sets of solutions to elliptic equations", *J. Differential Geom.* **51**:2 (1999), 359–373. MR 2001g:35052 Zbl 1144.35370

[Hauray 2003] M. Hauray, "On two-dimensional Hamiltonian transport equations with L_{loc}^p coefficients", *Ann. Inst. H. Poincaré Anal. Non Linéaire* **20**:4 (2003), 625–644. MR 2004g:35037 Zbl 1028.35148

[Hörmander 1983] L. Hörmander, *The analysis of linear partial differential operators, II: Differential operators with constant coefficients*, Grundlehren der Math. Wiss. **257**, Springer, Berlin, 1983. MR 85g:35002b Zbl 0521.35002

[Hörmander 1997] L. Hörmander, *Lectures on nonlinear hyperbolic differential equations*, Mathématiques et Applications **26**, Springer, 1997. Zbl 0881.35001

[Hristova et al. 2008] Y. Hristova, P. Kuchment, and L. Nguyen, "Reconstruction and time reversal in thermoacoustic tomography in acoustically homogeneous and inhomogeneous media", *Inverse Problems* **24**:5 (2008), 055006. MR 2010c:65162 Zbl 1180.35563

[Kempe et al. 1997] M. Kempe, M. Larionov, D. Zaslavsky, and A. Z. Genack, "Acousto-optic tomography with multiply scattered light", *J. Opt. Soc. Am. A* **14** (1997), 1151–1158.

[Kim et al. 2002] S. Kim, O. Kwon, J. K. Seo, and J.-R. Yoon, "On a nonlinear partial differential equation arising in magnetic resonance electrical impedance tomography", *SIAM J. Math. Anal.* **34**:3 (2002), 511–526. MR 2004f:35185 Zbl 1055.35142

[Kowar and Scherzer 2012] R. Kowar and O. Scherzer, "Attenuation modes in photoacoustics", pp. 85–130 in *Mathematical modeling in biomedical imaging II*, Lecture Notes in Mathematics **2035**, Springer, New York, 2012. arXiv 1009.4350

[Kuchment and Kunyansky 2008] P. Kuchment and L. Kunyansky, "Mathematics of thermoacoustic tomography", *Euro. J. Appl. Math.* **19** (2008), 191–224.

[Kuchment and Kunyansky 2011] P. Kuchment and L. Kunyansky, "2D and 3D reconstructions in acousto-electric tomography", *Inverse Problems* **27**:5 (2011), 055013. MR 2012b:65166

[Li et al. 2008] C. H. Li, M. Pramanik, G. Ku, and L. V. Wang, "Image distortion in thermoacoustic tomography caused by microwave diffraction", *Phys. Rev. E* **77** (2008), 031923.

[McLaughlin et al. 2010] J. R. McLaughlin, N. Zhang, and A. Manduca, "Calculating tissue shear modules and pressure by 2D log-elastographic methods", *Inverse Problems* **26**:8 (2010), 085007. MR 2658824

[Melas 1993] A. D. Melas, "An example of a harmonic map between Euclidean balls", *Proc. Amer. Math. Soc.* **117**:3 (1993), 857–859. MR 93d:58037 Zbl 0836.54007

[Monard and Bal 2012] F. Monard and G. Bal, "Inverse diffusion problems with redundant internal information", *Inv. Probl. Imaging* **6**:2 (2012), 289–313.

[Nachman et al. 2007] A. Nachman, A. Tamasan, and A. Timonov, "Conductivity imaging with a single measurement of boundary and interior data", *Inverse Problems* **23**:6 (2007), 2551–2563. MR 2009k:35325 Zbl 1126.35102

[Nachman et al. 2009] A. Nachman, A. Tamasan, and A. Timonov, "Recovering the conductivity from a single measurement of interior data", *Inverse Problems* **25**:3 (2009), 035014. MR 2010g:35340 Zbl 1173.35736

[Nachman et al. 2011] A. Nachman, A. Tamasan, and A. Timonov, "Current density impedance imaging", pp. 135–149 in *Tomography and inverse transport theory. International workshop on mathematical methods in emerging modalities of medical imaging* (Banff, 2009/2010), edited by G. Bal et al., American Mathematical Society, 2011. Zbl 1243.78033

[Patch and Scherzer 2007] S. K. Patch and O. Scherzer, "Guest editors' introduction: Photo- and thermo-acoustic imaging", *Inverse Problems* **23**:6 (2007), S1–S10. MR 2440994

[Ren and Bal 2012] K. Ren and G. Bal, "On multi-spectral quantitative photoacoustic tomography", *Inverse Problems* **28** (2012), 025010.

[Richter 1981] G. R. Richter, "An inverse problem for the steady state diffusion equation", *SIAM J. Appl. Math.* **41**:2 (1981), 210–221. MR 82m:35143 Zbl 0501.35075

[Ripoll and Ntziachristos 2005] J. Ripoll and V. Ntziachristos, "Quantitative point source photoacoustic inversion formulas for scattering and absorbing medium", *Phys. Rev. E* **71** (2005), 031912.

[Robbiano and Salazar 1990] L. Robbiano and J. Salazar, "Dimension de Hausdorff et capacité des points singuliers d'une solution d'un opérateur elliptique", *Bull. Sci. Math.* **114**:3 (1990), 329–336. MR 91g:35084 Zbl 0713.35025

[Scherzer 2011] O. Scherzer, *Handbook of mathematical methods in imaging*, Springer, New York, 2011.

[Seo and Woo 2011] J. K. Seo and E. J. Woo, "Magnetic resonance electrical impedance tomography (MREIT)", *SIAM Rev.* **53**:1 (2011), 40–68. MR 2012d:35412 Zbl 1210.35293

[Stefanov and Uhlmann 2009] P. Stefanov and G. Uhlmann, "Thermoacoustic tomography with variable sound speed", *Inverse Problems* **25** (2009), 075011.

[Stefanov and Uhlmann 2011] P. Stefanov and G. Uhlmann, "Thermoacoustic tomography arising in brain imaging", *Inverse Problems* **27** (2011), 045004.

[Sylvester and Uhlmann 1987] J. Sylvester and G. Uhlmann, "A global uniqueness theorem for an inverse boundary value problem", *Ann. of Math.* (2) **125**:1 (1987), 153–169. MR 88b:35205 Zbl 0625.35078

[Taylor 1996] M. E. Taylor, *Partial differential equations, I: Basic theory*, Applied Mathematical Sciences **115**, Springer, New York, 1996. MR 98b:35002b Zbl 1206.35002

[Triki 2010] F. Triki, "Uniqueness and stability for the inverse medium problem with internal data", *Inverse Problems* **26**:9 (2010), 095014. MR 2011h:35322 Zbl 1200.35333

[Uhlmann 2009] G. Uhlmann, "Calderón's problem and electrical impedance tomography", *Inverse Problems* **25** (2009), 123011.

[Wang 2004] L. V. Wang, "Ultrasound-mediated biophotonic imaging: a review of acousto-optical tomography and photo-acoustic tomography", *Journal of Disease Markers* **19** (2004), 123–138.

[Wang and Wu 2007] L. V. Wang and H. Wu, *Biomedical optics: principles and imaging*, Wiley, New York, 2007.

[White 2005] B. S. White, "Asymptotic theory of electroseismic prospecting", *SIAM J. Appl. Math.* **65**:4 (2005), 1443–1462. MR 2006e:86014 Zbl 1073.86005

[Xu and Wang 2006] M. Xu and L. V. Wang, "Photoacoustic imaging in biomedicine", *Rev. Sci. Instr.* **77** (2006), 041101.

[Xu et al. 2009] Y. Xu, L. Wang, P. Kuchment, and G. Ambartsoumian, "Limited view thermoacoustic tomography", pp. 61–73 (Chapter 6) in *Photoacoustic imaging and spectroscopy*, edited by L. H. Wang, CRC Press, 2009.

[Zemp 2010] R. J. Zemp, "Quantitative photoacoustic tomography with multiple optical sources", *Applied Optics* **49** (2010), 3566–3572.

[Zhang and Wang 2004] H. Zhang and L. V. Wang, "Acousto-electric tomography", article 145 in *Photons plus ultrasound – imaging and sensing 2009 (10th Conference on Biomedical Thermoacoustics, Optoacoustics, and Acousto-optics)*, Proc. SPIE **5320**, 2004.

gb2030@columbia.edu *Department of Applied Physics and Applied Mathematics, Columbia University, New York, NY 10027, United States*

Inverse problems for connections

GABRIEL P. PATERNAIN

We discuss various recent results related to the inverse problem of determining a unitary connection from its parallel transport along geodesics.

1. Introduction

Let (M, g) be a compact oriented Riemannian manifold with smooth boundary, and let $SM = \{(x, v) \in TM \; ; \; |v| = 1\}$ be the unit tangent bundle with canonical projection $\pi : SM \to M$. The geodesics going from ∂M into M can be parametrized by the set $\partial_+(SM) = \{(x, v) \in SM \; ; \; x \in \partial M, \langle v, \nu \rangle \leq 0\}$, where ν is the outer unit normal vector to ∂M. For any $(x, v) \in SM$ we let $t \mapsto \gamma(t, x, v)$ be the geodesic starting from x in direction v. We assume that (M, g) is nontrapping, which means that the time $\tau(x, v)$ when the geodesic $\gamma(t, x, v)$ exits M is finite for each $(x, v) \in SM$. The scattering relation

$$\alpha = \alpha_g : \partial_+(SM) \to \partial_-(SM)$$

maps a starting point and direction of a geodesic to the end point and direction, where $\partial_-(SM) = \{(x, v) \in SM \; ; \; x \in \partial M, \langle v, \nu \rangle \geq 0\}$.

Suppose now that $E \to M$ is a Hermitian vector bundle of rank n over M and ∇ is a unitary connection on E. Associated with ∇ there is the following additional piece of scattering data: given $(x, v) \in \partial_+(SM)$, let $P(x, v) = P_\nabla(x, v) : E(x) \to E(\pi \circ \alpha(x, v))$ denote the parallel transport along the geodesic $\gamma(t, x, v)$. This map is a linear isometry and the main inverse problem we wish to discuss here is the following:

Question. Does P determine ∇?

The first observation is that the problem has a natural gauge equivalence. Let ψ be a gauge transformation, that is, a smooth section of the bundle of automorphisms $\mathrm{Aut}\,E$. The set of all these sections naturally forms a group (known as the gauge group) which acts on the space of unitary connections by the rule

$$(\psi^*\nabla)s := \psi\nabla(\psi^{-1}s)$$

where s is any smooth section of E. If in addition $\psi|_{\partial M} = \mathrm{Id}$, then it is a simple exercise to check that
$$P_\nabla = P_{\psi^*\nabla}.$$

Thus we can rephrase the question above more precisely as follows:

Question I (manifolds with boundary). Let ∇_1 and ∇_2 be two unitary connections with $P_{\nabla_1} = P_{\nabla_2}$. Does there exist a gauge transformation ψ with $\psi|_{\partial M} = \mathrm{Id}$ and $\psi^*\nabla_1 = \nabla_2$?

There is a version of this question which makes sense also for closed manifolds, that is, $\partial M = \varnothing$. Let $\gamma : [0, T] \to M$ be a closed geodesic and let $P_\nabla(\gamma) : E(\gamma(0)) \to E(\gamma(0))$ be the parallel transport along γ.

Question II (closed manifolds). Let ∇_1 and ∇_2 be two unitary connections and suppose there is a connection ∇ gauge equivalent to ∇_1 such that $P_\nabla(\gamma) = P_{\nabla_2}(\gamma)$ for every closed geodesic γ. Are ∇_1 and ∇_2 gauge equivalent?

A connection ∇ is said to be *transparent* if $P_\nabla(\gamma) = \mathrm{Id}$ for all closed geodesics γ. Understanding the set of transparent connections modulo gauge is an important special case of Question II.

To make further progress on Questions I and II we need to impose some conditions on the manifold (M, g).

In the case of manifolds with boundary a typical hypothesis is that of *simplicity*. A compact Riemannian manifold with boundary is said to be *simple* if for any point $x \in M$ the exponential map $\exp_x$ is a diffeomorphism onto M, and if the boundary is strictly convex. The notion of simplicity arises naturally in the context of the boundary rigidity problem [Michel 1981]. For the case of closed manifolds there are two reasonable disjoint options. One is to assume that (M, g) is a Zoll manifold, that is, a Riemannian manifold all of whose geodesics are closed, but we shall not really discuss this case in any detail here. The other is to assume that the geodesic flow is *Anosov*. Recall that the geodesic flow ϕ_t is Anosov if there is a continuous splitting $TSM = E^0 \oplus E^u \oplus E^s$, where E^0 is the flow direction, and there are constants $C > 0$ and $0 < \rho < 1 < \eta$ such that for all $t > 0$ we have
$$\|d\phi_{-t}|_{E^u}\| \leq C\,\eta^{-t} \quad \text{and} \quad \|d\phi_t|_{E^s}\| \leq C\,\rho^t.$$

It is very well known that the geodesic flow of a closed negatively curved Riemannian manifold is a contact Anosov flow [Katok and Hasselblatt 1995]. The Anosov property automatically implies that the manifold is free of conjugate points [Klingenberg 1974; Anosov 1985; Mañé 1987]. Simple manifolds are also free of conjugate points (this follows directly from the definition) and both conditions (simplicity and Anosov) are open conditions on the metric. It is remarkable that similar results exist in both situations.

It is easy to see from the definition that a simple manifold must be diffeomorphic to a ball in $\mathbb{R}^n$. Therefore any bundle over such M is necessarily trivial. For most of this paper we shall consider Questions I and II only for the case of trivial bundles; this will make the presentation clearer without removing substantial content.

Question I arises naturally when considering the hyperbolic Dirichlet-to-Neumann map associated to the Schrödinger equation with a connection. It was shown in [Finch and Uhlmann 2001] that when the metric is Euclidean, the scattering data for a connection can be determined from the hyperbolic Dirichlet-to-Neumann map. A similar result holds true on simple Riemannian manifolds: a combination of the methods in [Finch and Uhlmann 2001; Uhlmann 2004] shows that the hyperbolic Dirichlet-to-Neumann map for a connection determines the scattering data P_∇.

2. Elementary background on connections

Consider the trivial bundle $M \times \mathbb{C}^n$. For us a connection A will be a complex $n \times n$ matrix whose entries are smooth 1-forms on M. Another way to think of A is to regard it as a smooth map $A : TM \to \mathbb{C}^{n \times n}$ which is linear in $v \in T_x M$ for each $x \in M$.

Very often in physics and geometry one considers *unitary* or *Hermitian* connections. This means that the range of A is restricted to skew-Hermitian matrices. In other words, if we denote by $\mathfrak{u}(n)$ the Lie algebra of the unitary group $U(n)$, we have a smooth map $A : TM \to \mathfrak{u}(n)$ which is linear in the velocities. There is yet another equivalent way to phrase this. The connection A induces a covariant derivative d_A on sections $s \in C^\infty(M, \mathbb{C}^n)$ by setting $d_A s = ds + As$. Then A being Hermitian or unitary is equivalent to requiring compatibility with the standard Hermitian inner product of $\mathbb{C}^n$ in the sense that

$$d\langle s_1, s_2 \rangle = \langle d_A s_1, s_2 \rangle + \langle s_1, d_A s_2 \rangle$$

for any pair of functions s_1, s_2.

Given two unitary connections A and B we shall say that A and B are gauge equivalent if there exists a smooth map $u : M \to U(n)$ such that

$$B = u^{-1} du + u^{-1} A u. \tag{1}$$

It is an easy exercise to check that this definition coincides with the one given in the previous section if we set $\psi = u^{-1}$.

The *curvature* of the connection is the 2-form F_A with values in $u(n)$ given by
$$F_A := dA + A \wedge A.$$

If A and B are related by (1) then:
$$F_B = u^{-1} F_A u.$$

Given a smooth curve $\gamma : [a, b] \to M$, the *parallel transport* along γ is obtained by solving the linear differential equation in $\mathbb{C}^n$:
$$\begin{cases} \dot{s} + A(\gamma(t), \dot{\gamma}(t))s = 0, \\ s(a) = w \in \mathbb{C}^n. \end{cases} \quad (2)$$

The isometry $P_A(\gamma) : \mathbb{C}^n \to \mathbb{C}^n$ is defined as $P_A(\gamma)(w) := s(b)$. We may also consider the fundamental unitary matrix solution $U : [a, b] \to U(n)$ of (2). It solves
$$\begin{cases} \dot{U} + A(\gamma(t), \dot{\gamma}(t))U = 0, \\ U(a) = \mathrm{Id}. \end{cases} \quad (3)$$

Clearly $P_A(\gamma)(w) = U(b)w$.

3. The transport equation and the attenuated ray transform

Consider now the case of a compact simple Riemannian manifold. We would like to pack the information provided by (3) along every geodesic into one PDE in SM. For this we consider the vector field X associated with the geodesic flow ϕ_t and we look at the unique solution $U_A : SM \to U(n)$ of
$$\begin{cases} X(U_A) + A(x, v)U_A = 0, \ (x, v) \in SM \\ U_A|_{\partial_+(SM)} = \mathrm{Id} \end{cases} \quad (4)$$

The scattering data of the connection A is now the map $C_A : \partial_-(SM) \to U(n)$ defined as $C_A := U_A|_{\partial_-(SM)}$.

We can now rephrase Question I as follows:

Question I (manifolds with boundary). Let A and B be two unitary connections with $C_A = C_B$. Does there exist a smooth map $U : M \to U(n)$ with $U|_{\partial M} = \mathrm{Id}$ and $B = U^{-1}dU + U^{-1}AU$?

Suppose $C_A = C_B$ and define $U := U_A(U_B)^{-1} : SM \to U(n)$. One easily checks that U satisfies:
$$\begin{cases} XU + AU - UB = 0, \\ U|_{\partial(SM)} = \mathrm{Id}. \end{cases}$$

If we show that U is in fact smooth *and* it only depends on the base point $x \in M$ we would have an answer to Question I, since the equation above reduces to $dU + AU - UB = 0$ and $U|_{\partial M} = \mathrm{Id}$ which is exactly gauge equivalence. Showing that U only depends of x is not an easy task and it often is the crux of the matter in these type of problems. To tackle this issue we will rephrase the problem in terms of an *attenuated ray transform*.

Consider $W := U - \mathrm{Id} : SM \to \mathbb{C}^{n \times n}$, where as before $\mathbb{C}^{n \times n}$ stands for the set of all $n \times n$ complex matrices. Clearly W satisfies

$$XW + AW - WB = B - A, \tag{5}$$
$$W|_{\partial(SM)} = 0. \tag{6}$$

We introduce a new connection $\hat{A}$ on the trivial bundle $M \times \mathbb{C}^{n \times n}$ as follows: given a matrix $R \in \mathbb{C}^{n \times n}$ we define $\hat{A}(R) := AR - RB$. One easily checks that $\hat{A}$ is Hermitian if A and B are. Then equations (5) and (6) are of the form:

$$\begin{cases} Xu + Au = -f, \\ u|_{\partial(SM)} = 0. \end{cases}$$

where A is a unitary connection, $f : SM \to \mathbb{C}^N$ is a smooth function linear in the velocities, $u : SM \to \mathbb{C}^N$ is a function that we would like to prove smooth and only dependent on $x \in M$ and $N = n \times n$. As we will see shortly this amounts to understanding which functions f linear in the velocities are in the kernel of the attenuated ray transform of the connection A.

First recall that in the scalar case, the attenuated ray transform $I_a f$ of a function $f \in C^\infty(SM, \mathbb{C})$ with attenuation coefficient $a \in C^\infty(SM, \mathbb{C})$ can be defined as the integral

$$I_a f(x, v) := \int_0^{\tau(x,v)} f(\varphi_t(x, v)) \exp\left[\int_0^t a(\varphi_s(x, v))\, ds\right] dt,$$
$$(x, v) \in \partial_+(SM).$$

Alternatively, we may set $I_a f := u|_{\partial_+(SM)}$ where u is the unique solution of the *transport equation*

$$Xu + au = -f \text{ in } SM, \quad u|_{\partial_-(SM)} = 0.$$

The last definition generalizes without difficulty to the case of connections. Assume that A is a unitary connection and let $f \in C^\infty(SM, \mathbb{C}^n)$ be a vector valued function. Consider the following transport equation for $u : SM \to \mathbb{C}^n$,

$$Xu + Au = -f \text{ in } SM, \quad u|_{\partial_-(SM)} = 0.$$

On a fixed geodesic the transport equation becomes a linear ODE with zero initial condition, and therefore this equation has a unique solution $u = u^f$.

Definition 3.1. The attenuated ray transform of $f \in C^\infty(SM, \mathbb{C}^n)$ is given by
$$I_A f := u^f |_{\partial_+(SM)}.$$

We note that I_A acting on sums of 0-forms and 1-forms always has a nontrivial kernel, since
$$I_A(dp + Ap) = 0 \text{ for any } p \in C^\infty(M, \mathbb{C}^n) \text{ with } p|_{\partial M} = 0.$$

Thus from the ray transform $I_A f$ one only expects to recover f up to an element having this form.

The transform I_A also has an integral representation. Consider the unique matrix solution $U_A : SM \to U(n)$ from above. Then it is easy to check that
$$I_A f(x, v) = \int_0^{\tau(x,v)} U_A^{-1}(\phi_t(x, v)) f(\phi_t(x, v)) \, dt.$$

We are now in a position to state the next main question:

Question III (kernel of I_A). Let (M, g) be a compact simple Riemannian manifold and let A be a unitary connection. Assume that $f : SM \to \mathbb{C}^n$ is a smooth function of the form $F(x) + \alpha_j(x)v^j$, where $F : M \to \mathbb{C}^n$ is a smooth function and α is a $\mathbb{C}^n$-valued 1-form. If $I_A(f) = 0$, is it true that $F = 0$ and $\alpha = d_A p = dp + Ap$, where $p : M \to \mathbb{C}^n$ is a smooth function with $p|_{\partial M} = 0$?

As explained above a positive answer to Question III gives a positive answer to Question I. The next recent result provides a full answer to Question III in the two-dimensional case:

Theorem 3.2 [Paternain et al. 2011a]. *Let M be a compact simple surface. Assume that $f : SM \to \mathbb{C}^n$ is a smooth function of the form $F(x) + \alpha_j(x)v^j$, where $F : M \to \mathbb{C}^n$ is a smooth function and α is a $\mathbb{C}^n$-valued 1-form. Let also $A : TM \to \mathfrak{u}(n)$ be a unitary connection. If $I_A(f) = 0$, then $F = 0$ and $\alpha = d_A p$, where $p : M \to \mathbb{C}^n$ is a smooth function with $p|_{\partial M} = 0$.*

Let us explicitly state the positive answer to Question I in the case of surfaces:

Theorem 3.3 [Paternain et al. 2011a]. *Assume M is a compact simple surface and let A and B be two unitary connections. Then $C_A = C_B$ implies that there exists a smooth $U : M \to U(n)$ such that*
$$U|_{\partial M} = \text{Id} \quad \text{and} \quad B = U^{-1}dU + U^{-1}AU.$$

We will provide a sketch of the proof of Theorem 3.2 in the next section, but first we survey some prior results on this topic.

In the case of Euclidean space with the Euclidean metric the attenuated ray transform is the basis of the medical imaging technology of SPECT and has been extensively studied; see [Finch 2003] for a review. We remark that in connection with injectivity results for ray transforms, there is great interest in reconstruction procedures and inversion formulas. For the attenuated ray transform in $\mathbb{R}^2$ with Euclidean metric and scalar attenuation function, an explicit inversion formula was proved by R. Novikov [2002a]. A related formula also including 1-form attenuations appears in [Boman and Strömberg 2004], inversion formulas for matrix attenuations in Euclidean space are given in [Eskin 2004; Novikov 2002b], and the case of hyperbolic space $\mathbb{H}^2$ is considered in [Bal 2005].

In our general geometric setting an essential contribution is made in the paper [Salo and Uhlmann 2011] in which it is was shown that the attenuated ray transform is injective in the scalar case with $a \in C^\infty(M, \mathbb{C})$ for simple two dimensional manifolds. This paper also contains the proof of existence of holomorphic integrating factors of a for arbitrary simple surfaces; a result that extends to the case when a is a 1-form and that will be crucial in the proof of Theorem 3.2.

Various versions of Theorem 3.3 have been proved in the literature. Sharafutdinov [2000] proves the theorem assuming that the connections are C^1 close to another connection with small curvature (but in any dimension). In the case of domains in the Euclidean plane the theorem was proved by Finch and Uhlmann [2001] assuming that the connections have small curvature and by G. Eskin [2004] in general. R. Novikov [2002b] considers the case of connections which are not compactly supported (but with suitable decay conditions at infinity) and establishes local uniqueness of the trivial connection and gives examples in which global uniqueness fails (existence of "ghosts"). His examples are based on a remarkable connection between the Bogomolny equation in Minkowski $(2 + 1)$-space and the scattering data associated with the transport equation considered above. As explained in [Ward 1988] (see also [Dunajski 2010, Section 8.2.1]), certain soliton solutions A have the property that when restricted to space-like planes the scattering data is trivial. In this way one obtains connections in $\mathbb{R}^2$ with the property of having trivial scattering data but which are not gauge equivalent to the trivial connection. Of course these pairs are not compactly supported in $\mathbb{R}^2$ but they have a suitable decay at infinity. Motivated by this L. Mason obtained a full classification of $U(n)$ transparent connections for the round metric on S^2 (unpublished) using methods from twistor theory as in [Mason 2006].

4. Sketch of proof of Theorem 3.2

Let (M, g) be a compact oriented two dimensional Riemannian manifold with smooth boundary ∂M. As before SM will denote the unit circle bundle which is a compact 3-manifold with boundary given by $\partial(SM) = \{(x, v) \in SM : x \in \partial M\}$. Since M is assumed oriented there is a circle action on the fibers of SM with infinitesimal generator V called the *vertical vector field*. It is possible to complete the pair X, V to a global frame of $T(SM)$ by considering the vector field $X_\perp := [X, V]$. There are two additional structure equations given by $X = [V, X_\perp]$ and $[X, X_\perp] = -KV$ where K is the Gaussian curvature of the surface. Using this frame we can define a Riemannian metric on SM by declaring $\{X, X_\perp, V\}$ to be an orthonormal basis and the volume form of this metric will be denoted by $d\Sigma^3$. The fact that $\{X, X_\perp, V\}$ are orthonormal together with the commutator formulas implies that the Lie derivative of $d\Sigma^3$ along the three vector fields vanishes.

Given functions $u, v : SM \to \mathbb{C}^n$ we consider the inner product

$$(u, v) = \int_{SM} \langle u, v \rangle_{\mathbb{C}^n} \, d\Sigma^3.$$

Since $X, X_\perp, V$ are volume preserving we have $(Vu, v) = -(u, Vv)$ for $u, v \in C^\infty(SM, \mathbb{C}^n)$, and if additionally $u|_{\partial(SM)} = 0$ or $v|_{\partial(SM)} = 0$ then also $(Xu, v) = -(u, Xv)$ and $(X_\perp u, v) = -(u, X_\perp v)$.

The space $L^2(SM, \mathbb{C}^n)$ decomposes orthogonally as a direct sum

$$L^2(SM, \mathbb{C}^n) = \bigoplus_{k \in \mathbb{Z}} H_k$$

where H_k is the eigenspace of $-iV$ corresponding to the eigenvalue k. A function $u \in L^2(SM, \mathbb{C}^n)$ has a Fourier series expansion

$$u = \sum_{k=-\infty}^{\infty} u_k,$$

where $u_k \in H_k$. Let $\Omega_k = C^\infty(SM, \mathbb{C}^n) \cap H_k$.

An important ingredient is the fiberwise *Hilbert transform* $\mathcal{H}$. This can be introduced in various ways (see [Pestov and Uhlmann 2005; Salo and Uhlmann 2011]), but perhaps the most informative approach is to indicate that it acts fiberwise and for $u_k \in \Omega_k$,

$$\mathcal{H}(u_k) = -\mathrm{sgn}(k)\, i u_k$$

where we use the convention $\mathrm{sgn}(0) = 0$. Moreover, $\mathcal{H}(u) = \sum_k \mathcal{H}(u_k)$. Observe that

$$(\mathrm{Id} + i\mathcal{H})u = u_0 + 2\sum_{k=1}^{\infty} u_k \quad \text{and} \quad (\mathrm{Id} - i\mathcal{H})u = u_0 + 2\sum_{k=-\infty}^{-1} u_k.$$

Definition 4.1. A function $u : SM \to \mathbb{C}^n$ is said to be holomorphic if

$$(\mathrm{Id} - i\mathcal{H})u = u_0.$$

Equivalently, u is holomorphic if $u_k = 0$ for all $k < 0$. Similarly, u is said to be antiholomorphic if $(\mathrm{Id} + i\mathcal{H})u = u_0$ which is equivalent to saying that $u_k = 0$ for all $k > 0$.

As in previous works the following commutator formula from [Pestov and Uhlmann 2005] will come into play:

$$[\mathcal{H}, X]u = X_\perp u_0 + (X_\perp u)_0, \quad u \in C^\infty(SM, \mathbb{C}^n). \tag{7}$$

We will give a proof of this formula in Lemma 6.7.

It is easy to extend this bracket relation so that it includes a connection A. We often think of A as a function restricted to SM. We also think of A as acting on smooth functions $u \in C^\infty(SM, \mathbb{C}^n)$ by multiplication. Note that $V(A)$ is a new function on SM which can be identified with the restriction of $-\star A$ to SM, so we will simply write $V(A) = -\star A$. Here $\star$ denotes the Hodge star operator of the metric g.

Lemma 4.2. *For any smooth function u we have*

$$[\mathcal{H}, X + A]u = (X_\perp + \star A)(u_0) + \{(X_\perp + \star A)(u)\}_0.$$

The proof makes use of this regularity result:

Proposition 4.3 [Paternain et al. 2011a, Proposition 5.2]. *Let $f : SM \to \mathbb{C}^n$ be smooth with $I_A(f) = 0$. Then $u^f : SM \to \mathbb{C}^n$ is smooth.*

The next proposition will provide the holomorphic integrating factors in the scalar case.

Proposition 4.4 [Paternain et al. 2011a, Theorem 4.1]. *Let (M, g) be a simple two-dimensional manifold and $f \in C^\infty(SM, \mathbb{C})$. The following conditions are equivalent.*

(a) *There exist a holomorphic $w \in C^\infty(SM, \mathbb{C})$ and an antiholomorphic $\tilde{w} \in C^\infty(SM, \mathbb{C})$ such that $Xw = X\tilde{w} = -f$.*

(b) *$f(x, v) = F(x) + \alpha_j(x)v^j$ where F is a smooth function on M and α is a 1-form.*

The existence of holomorphic and antiholomorphic solutions for the case $\alpha = 0$ was first proved in [Salo and Uhlmann 2011], but here we will need the case in which $F = 0$, but α is nonzero.

Another key ingredient is an energy identity or a "Pestov type identity", which generalizes the standard Pestov identity [Sharafutdinov 1994] to the case where a connection is present. There are several predecessors for this formula [Vertgeĭm 1991; Sharafutdinov 2000] and its use for simple surfaces is in the spirit of [Sharafutdinov and Uhlmann 2000; Dairbekov and Paternain 2007]. Recall that the curvature F_A of the connection A is defined as $F_A = dA + A \wedge A$ and $\star F_A$ is a function $\star F_A : M \to \mathfrak{u}(n)$.

Lemma 4.5 (energy identity). *If $u : SM \to \mathbb{C}^n$ is a smooth function such that $u|_{\partial(SM)} = 0$, then*

$$\|(X+A)Vu\|^2 - (KVu, Vu) - (\star F_A u, Vu) = \|V(X+A)(u)\|^2 - \|(X+A)u\|^2.$$

Remark 4.6. The same energy identity holds true for *closed* surfaces.

To use the energy identity we need to control the signs of various terms. The first easy observation is the following:

Lemma 4.7. *Assume $(X + A)u = F(x) + \alpha_j(x)v^j$, where $F : M \to \mathbb{C}^n$ is a smooth function and α is a $\mathbb{C}^n$-valued 1-form. Then*

$$\|V(X + A)u\|^2 - \|(X + A)u\|^2 = -\|F\|^2 \leq 0.$$

Proof. It suffices to note the identities

$$\|V(X + A)u\|^2 = \|V\alpha\|^2 = \|\alpha\|^2 \quad \text{and} \quad \|F + \alpha\|^2 = \|\alpha\|^2 + \|F\|^2. \qquad \square$$

Next we have the following lemma due to the absence of conjugate points on simple surfaces (compare with [Dairbekov and Paternain 2007, Theorem 4.4]):

Lemma 4.8. *Let M be a compact simple surface. If $u : SM \to \mathbb{C}^n$ is a smooth function such that $u|_{\partial(SM)} = 0$, then*

$$\|(X + A)Vu\|^2 - (KVu, Vu) \geq 0.$$

Proof. Consider a smooth function $a : SM \to \mathbb{R}$ which solves the Riccati equation $X(a) + a^2 + K = 0$. These exist by the absence of conjugate points (see for example [Sharafutdinov 1999, Theorem 6.2.1] or the proof of Lemma 4.1 in [Sharafutdinov and Uhlmann 2000]). Set for simplicity $\psi = V(u)$. Clearly $\psi|_{\partial(SM)} = 0$.

Using that A is skew-Hermitian, we compute

$$|(X+A)(\psi) - a\psi|^2_{\mathbb{C}^n} = |(X+A)(\psi)|^2_{\mathbb{C}^n} - 2\Re\langle (X+A)(\psi), a\psi \rangle_{\mathbb{C}^n} + a^2|\psi|^2_{\mathbb{C}^n}$$
$$= |(X+A)(\psi)|^2_{\mathbb{C}^n} - 2a\Re\langle X(\psi), \psi \rangle_{\mathbb{C}^n} + a^2|\psi|^2_{\mathbb{C}^n}.$$

Using the Riccati equation we have
$$X(a|\psi|^2) = (-a^2 - K)|\psi|^2 + 2a\Re\langle X(\psi), \psi\rangle_{\mathbb{C}^n};$$
thus
$$|(X+A)(\psi) - a\psi|_{\mathbb{C}^n}^2 = |(X+A)(\psi)|_{\mathbb{C}^n}^2 - K|\psi|_{\mathbb{C}^n}^2 - X(a|\psi|_{\mathbb{C}^n}^2).$$
Integrating this equality with respect to $d\Sigma^3$ and using that ψ vanishes on $\partial(SM)$ we obtain
$$\|(X+A)(\psi)\|^2 - (K\psi, \psi) = \|(X+A)(\psi) - a\psi\|^2 \geq 0. \qquad \square$$

Theorem 4.9. *Let $f : SM \to \mathbb{C}^n$ be a smooth function. Suppose $u : SM \to \mathbb{C}^n$ satisfies*
$$\begin{cases} Xu + Au = -f, \\ u|_{\partial(SM)} = 0. \end{cases}$$
Then if $f_k = 0$ for all $k \leq -2$ and $i \star F_A(x)$ is a negative definite Hermitian matrix for all $x \in M$, the function u must be holomorphic. Moreover, if $f_k = 0$ for all $k \geq 2$ and $i \star F_A(x)$ is a positive definite Hermitian matrix for all $x \in M$, the function u must be antiholomorphic.

Proof. Let us assume that $f_k = 0$ for $k \leq -2$ and $i \star F_A$ is a negative definite Hermitian matrix; the proof of the other claim is similar.

We need to show that $(\mathrm{Id} - i\mathcal{H})u$ only depends on x. We apply $X + A$ to it and use Lemma 4.2 together with $(\mathrm{Id} - i\mathcal{H})f = f_0 + 2f_{-1}$ to derive

$$\begin{aligned}(X+A)&[(\mathrm{Id}-i\mathcal{H})u] \\
&= -f - i(X+A)(\mathcal{H}u) \\
&= -f - i(\mathcal{H}((X+A)(u)) - (X_\perp + \star A)(u_0) - \{(X_\perp + \star A)(u)\}_0) \\
&= -(\mathrm{Id} - i\mathcal{H})(f) + i(X_\perp + \star A)(u_0) + i\{(X_\perp + \star A)(u)\}_0 \\
&= -f_0 - 2f_{-1} + i(X_\perp + \star A)(u_0) + i\{(X_\perp + \star A)(u)\}_0 \\
&= F(x) + \alpha_x(v),\end{aligned}$$

where $F : M \to \mathbb{C}^n$ and α is a $\mathbb{C}^n$-valued 1-form. Now we are in good shape to use the energy identity from Lemma 4.5. We will apply it to
$$v = (\mathrm{Id} - i\mathcal{H})u = u_0 + 2 \sum_{k=-\infty}^{-1} u_k.$$
We know from Lemma 4.7 that its right-hand side is ≤ 0 and using Lemma 4.8 we deduce
$$(\star F_A v, Vv) \geq 0.$$

On the other hand,

$$(\star F_A v, Vv) = -4 \sum_{k=-\infty}^{-1} k(i \star F_A u_k, u_k)$$

and since $i \star F_A$ is negative definite this forces $u_k = 0$ for all $k < 0$. □

Outline of proof of Theorem 3.2. Consider the area form ω_g of the metric g. Since M is a disk there exists a smooth 1-form φ such that $\omega_g = d\varphi$. Given $s \in \mathbb{R}$, consider the Hermitian connection

$$A_s := A - is\varphi \, \mathrm{Id}.$$

Clearly its curvature is given by

$$F_{A_s} = F_A - is\omega_g \mathrm{Id};$$

therefore

$$i \star F_{A_s} = i \star F_A + s\mathrm{Id},$$

from which we see that there exists $s_0 > 0$ such that for $s > s_0$, $i \star F_{A_s}$ is positive definite and for $s < -s_0$, $i \star F_{A_s}$ is negative definite.

Since $I_A(f) = 0$, Proposition 4.3 implies that there is a *smooth* $u : SM \to \mathbb{C}^n$ such that $(X + A)(u) = -f$ and $u|_{\partial(SM)} = 0$ (to abbreviate the notation we write u instead of u^f).

Let e^{sw} be an integrating factor of $-is\varphi$. In other words $w : SM \to \mathbb{C}$ satisfies $X(w) = i\varphi$. By Proposition 4.4 we know we can choose w to be holomorphic or antiholomorphic. Observe now that $u_s := e^{sw}u$ satisfies $u_s|_{\partial(SM)} = 0$ and solves

$$(X + A_s)(u_s) = -e^{sw} f.$$

Choose w to be holomorphic. Since $f = F(x) + \alpha_j(x)v^j$, the function $e^{sw} f$ has the property that its Fourier coefficients $(e^{sw} f)_k$ vanish for $k \leq -2$. Choose s such that $s < -s_0$ so that $i \star F_{A_s}$ is negative definite. Then Theorem 4.9 implies that u_s is holomorphic and thus $u = e^{-sw} u_s$ is also holomorphic.

Choosing w antiholomorphic and $s > s_0$ we show similarly that u is antiholomorphic. This implies that $u = u_0$ which together with $(X + A)u = -f$, gives $d_A u_0 = -f$. If we set $p = -u_0$ we see right away that $F \equiv 0$ and $\alpha = d_A p$ as desired. □

5. Applications to tensor tomography

In this section we explain how the ideas of the previous section can be used to tackle a well-known inverse problem which is a priori unrelated with unitary connections.

We consider the geodesic ray transform acting on symmetric m-tensor fields on M. When the metric is Euclidean and $m = 0$ this transform reduces to the usual X-ray transform obtained by integrating functions along straight lines. More generally, given a symmetric (covariant) m-tensor field

$$f = f_{i_1 \cdots i_m} \, dx^{i_1} \otimes \cdots \otimes dx^{i_m}$$

on M, we define the corresponding function on SM by

$$f(x, v) = f_{i_1 \cdots i_m} v^{i_1} \cdots v^{i_m}.$$

The ray transform of f is defined by

$$If(x, v) = \int_0^{\tau(x,v)} f(\phi_t(x, v)) \, dt, \quad (x, v) \in \partial_+(SM),$$

where ϕ_t denotes the geodesic flow of the Riemannian metric g. If h is a symmetric $(m-1)$-tensor field, its inner derivative dh is a symmetric m-tensor field defined by $dh = \sigma \nabla h$, where σ denotes symmetrization and ∇ is the Levi-Civita connection. A direct calculation in local coordinates shows that

$$dh(x, v) = Xh(x, v),$$

where X as before is the geodesic vector field associated with ϕ_t. If additionally $h|_{\partial M} = 0$, then one clearly has $I(dh) = 0$. The ray transform on symmetric m-tensors is said to be s-injective if these are the only elements in the kernel. The terminology arises from the fact that any tensor field f may be written uniquely as $f = f^s + dh$, where f^s is a symmetric m-tensor with zero divergence and h is an $(m-1)$-tensor with $h|_{\partial M} = 0$ (see [Sharafutdinov 1994]). The tensor fields f^s and dh are called respectively the *solenoidal* and *potential* parts of the tensor f. Saying that I is s-injective is saying precisely that I is injective on the set of solenoidal tensors.

The next result shows that the ray transform on simple surfaces is s-injective for tensors of any rank. This settles a long standing question in the two-dimensional case [Pestov and Sharafutdinov 1988; Sharafutdinov 1994, Problem 1.1.2].

Theorem 5.1 [Paternain et al. 2011b]. *Let (M, g) be a simple surface and let $m \geq 0$. If f is a smooth symmetric m-tensor field on M which satisfies $If = 0$, then $f = dh$ for some smooth symmetric $(m-1)$-tensor field h on M with $h|_{\partial M} = 0$. (If $m = 0$, then $f = 0$.)*

It is not the objective of this article to discuss the vast literature on the tensor tomography problem for simple manifolds. Instead we refer the reader to [Sharafutdinov 1994] and to the references in [Paternain et al. 2011b] and we limit ourselves to supplying a proof of Theorem 5.1 based on the ideas of the previous section. The proof reduces to proving the next result. We say that $f \in C^\infty(SM, \mathbb{C})$ has degree m if $f_k = 0$ for $|k| \geq m+1$ and $m \geq 0$ is the smallest nonnegative integer with that property.

Proposition 5.2. *Let (M, g) be a simple surface. Assume that $u \in C^\infty(SM, \mathbb{C})$ satisfies $Xu = -f$ in SM with $u|_{\partial(SM)} = 0$. If $f \in C^\infty(SM, \mathbb{C})$ has degree $m \geq 1$, then u has degree $m - 1$. If f has degree 0, then $u = 0$.*

Proof of Theorem 5.1. Let f be a symmetric m-tensor field on SM and suppose that $If = 0$. We write

$$u(x, v) := \int_0^{\tau(x,v)} f(\phi_t(x, v)), \quad (x, v) \in SM.$$

Then $u|_{\partial(SM)} = 0$, and also $u \in C^\infty(SM)$ by Proposition 4.3.

Now f has degree m, and u satisfies $Xu = -f$ in SM with $u|_{\partial(SM)} = 0$. Proposition 5.2 implies that u has degree $m - 1$ (and $u = 0$ if $m = 0$). We let $h := -u$. It is not hard to see that h gives rise to a symmetric $(m-1)$-tensor still denoted by h. Since $X(h) = f$, this implies that dh and f agree when restricted to SM and thus $dh = f$. This proves the theorem. $\square$

Proposition 5.2 is in turn an immediate consequence of the next two results.

Proposition 5.3. *Let (M, g) be a simple surface. Assume that $u \in C^\infty(SM, \mathbb{C})$ satisfies $Xu = -f$ in SM with $u|_{\partial(SM)} = 0$. If $m \geq 0$ and if $f \in C^\infty(SM, \mathbb{C})$ is such that $f_k = 0$ for $k \leq -m-1$, then $u_k = 0$ for $k \leq -m$.*

Proposition 5.4. *Let (M, g) be a simple surface. Assume that $u \in C^\infty(SM, \mathbb{C})$ satisfies $Xu = -f$ in SM with $u|_{\partial(SM)} = 0$. If $m \geq 0$ and if $f \in C^\infty(SM, \mathbb{C})$ is such that $f_k = 0$ for $k \geq m+1$, then $u_k = 0$ for $k \geq m$.*

We will only prove Proposition 5.3, the proof of the other result being completely analogous. We shall need the following result from [Salo and Uhlmann 2011, Proposition 5.1]:

Proposition 5.5. *Let (M, g) be a simple surface and let f be a smooth holomorphic (antiholomorphic) function on SM. Suppose $u \in C^\infty(SM, \mathbb{C})$ satisfies*

$$Xu = -f \text{ in } SM, \quad u|_{\partial(SM)} = 0.$$

Then u is holomorphic (antiholomorphic) and $u_0 = 0$.

Proof of Proposition 5.3. Suppose that u is a smooth solution of $Xu = -f$ in SM where $f_k = 0$ for $k \leq -m-1$ and $u|_{\partial(SM)} = 0$. We choose a nonvanishing function $r \in \Omega_m$ and define the 1-form

$$A := -r^{-1}Xr.$$

Then ru solves the problem

$$(X + A)(ru) = -rf \text{ in } SM, \quad ru|_{\partial(SM)} = 0.$$

Note that rf is a holomorphic function. Next we employ a holomorphic integrating factor: by Proposition 4.4 there exists a holomorphic $w \in C^\infty(SM, \mathbb{C})$ with $Xw = A$. The function $e^w ru$ then satisfies

$$X(e^w ru) = -e^w rf \text{ in } SM, \quad e^w ru|_{\partial(SM)} = 0.$$

The right-hand side, $e^w rf$, is holomorphic. Now Proposition 5.5 implies that the solution $e^w ru$ is also holomorphic and $(e^w ru)_0 = 0$. Looking at Fourier coefficients shows that $(ru)_k = 0$ for $k \leq 0$, and therefore $u_k = 0$ for $k \leq -m$ as required. $\square$

Finally, let us explain the choice of r and A in the proof in more detail. Since M is a disk we can consider global isothermal coordinates (x, y) on M such that the metric can be written as $ds^2 = e^{2\lambda}(dx^2 + dy^2)$ where λ is a smooth real-valued function of (x, y). This gives coordinates (x, y, θ) on SM where θ is the angle between a unit vector v and $\partial/\partial x$. Then Ω_m consists of all functions $a(x, y)e^{im\theta}$ where $a \in C^\infty(M, \mathbb{C})$. We choose the specific nonvanishing function

$$r(x, y, \theta) := e^{im\theta}.$$

In the (x, y, θ) coordinates the geodesic vector field X is given by:

$$X = e^{-\lambda}\left(\cos\theta \frac{\partial}{\partial x} + \sin\theta \frac{\partial}{\partial y} + \left(-\frac{\partial\lambda}{\partial x}\sin\theta + \frac{\partial\lambda}{\partial y}\cos\theta\right)\frac{\partial}{\partial \theta}\right). \quad (8)$$

The connection $A = -Xr/r$ has the form

$$A = ime^{-\lambda}\left(-\frac{\partial\lambda}{\partial y}\cos\theta + \frac{\partial\lambda}{\partial x}\sin\theta\right) = im\left(-\frac{\partial\lambda}{\partial y}dx + \frac{\partial\lambda}{\partial x}dy\right).$$

Here as usual we identify A with $A(x, v)$ where $(x, v) \in SM$. This shows that the connection A is essentially the Levi-Civita connection of the metric g on the tensor power bundle $TM^{\otimes m}$, and since $(X + A)r = 0$ we have that r corresponds to a section of the pull-back bundle $\pi^*(TM^{\otimes m})$ whose covariant derivative along the geodesic vector field vanishes (here $\pi : SM \to M$ is the standard projection).

A second proof of Proposition 5.4 in the same spirit may be found in [Paternain et al. 2011b].

6. Closed manifolds

In this section we will discuss Question II, but before embarking into that we need some preliminary discussion on cocycles with values in a Lie group over a flow ϕ_t.

Let N be a closed manifold and $\phi_t : N \to N$ a smooth flow with infinitesimal generator X. Let G be a compact Lie group; for our purposes it is enough to think of G as a compact matrix group like $U(n)$.

Definition 6.1. A G-valued cocycle over the flow ϕ_t is a map $C : N \times \mathbb{R} \to G$ that satisfies
$$C(x, t+s) = C(\phi_t x, s) \, C(x, t)$$
for all $x \in N$ and $s, t \in \mathbb{R}$.

In this paper the cocycles will always be smooth. In this case C is determined by its infinitesimal generator $B : N \to \mathfrak{g}$ given by
$$B(x) := -\left.\frac{d}{dt}\right|_{t=0} C(x, t).$$
The cocycle can be recovered from B as the unique solution to
$$\frac{d}{dt} C(x, t) = -dR_{C(x,t)}(B(\phi_t x)), \quad C(x, 0) = \mathrm{Id},$$
where R_g is right translation by $g \in G$. We will indistinctly use the word "cocycle" for C or its infinitesimal generator B.

Definition 6.2. The cocycle C is said to be *cohomologically trivial* if there exists a smooth function $u : N \to G$ such that
$$C(x, t) = u(\phi_t x) u(x)^{-1}$$
for all $x \in N$ and $t \in \mathbb{R}$.

Observe that the condition of being cohomologically trivial can be equivalently expressed in terms of the infinitesimal generator B of the cocycle by saying that there exists a smooth function $u : N \to G$ that satisfies the equation
$$d_x u(X(x)) + d_{\mathrm{Id}} R_{u(x)}(B(x)) = 0$$
for all $x \in N$. If G is a matrix group we can write this more succinctly as
$$Xu + Bu = 0$$
where it is understood that differentiation and multiplication is in the set of matrices.

Definition 6.3. A cocycle C is said to satisfy the *periodic orbit obstruction condition* if $C(x, T) = \mathrm{Id}$ whenever $\phi_T x = x$.

Obviously a cohomologically trivial cocycle satisfies the periodic orbit obstruction condition. The converse turns out to be true for transitive Anosov flows: this is one of the celebrated Livšic theorems [Livšic 1971; 1972; Niţică and Török 1998].

Theorem 6.4 (smooth Livšic periodic data theorem). *Suppose ϕ_t is a smooth transitive Anosov flow. Let C be a smooth cocycle such that $C(x, T) = \mathrm{Id}$ whenever $\phi_T x = x$. Then C is cohomologically trivial.*

Given two G-valued cocycles C_1 and C_2 we shall say that they are *cohomologous* (or X-cohomologous) if there is a smooth function $u : N \to G$ such that

$$C_1(x, t) = u(\phi_t x) C_2(x, t) u(x)^{-1}$$

for all $x \in N$ and $t \in \mathbb{R}$. Clearly if C_1 and C_2 are cohomologous, $C_1(x, T) = u(x) C_2(x, T) u(x)^{-1}$, whenever $\phi_T x = x$. An extension of the Livšic theorem due to W. Parry [1999] together with the regularity result from [Niţică and Török 1998] gives the following extension of Theorem 6.4:

Theorem 6.5 (smooth Livšic periodic data theorem for two cocycles). *Suppose ϕ_t is a smooth transitive Anosov flow. Let C_1 and C_2 be two smooth cocycles such that there is a Hölder continuous function $u : N \to G$ for which $C_1(x, T) = u(x) C_2(x, T) u(x)^{-1}$ whenever $\phi_T x = x$. Then C and D are cohomologous.*

Observe that if G is a matrix group then two cocycles C_1 and C_2 are cohomologous if and only if their infinitesimal generators B_1 and B_2 are related by a smooth function $u : N \to G$ such that

$$Xu + B_1 u - u B_2 = 0$$

or equivalently

$$B_2 = u^{-1} Xu + u^{-1} B_1 u.$$

Note the formal similarity of this equation with the one that defines gauge equivalent connections. One could take the viewpoint that the main question raised in this paper is to decide when it is possible to go from cohomology defined by the operator X to cohomology defined by d in the geometric situation when X is the geodesic vector field. Let us be a bit more precise about this.

Let (M, g) be a closed Riemannian manifold with unit tangent bundle SM and projection $\pi : SM \to M$. The geodesic flow ϕ_t acts on SM with infinitesimal generator X.

Consider the trivial bundle $M \times \mathbb{C}^n$ and let $\mathcal{A}$ stand for the set of all unitary connections. Given $A \in \mathcal{A}$, we have a pull-back connection π^*A on the bundle $SM \times \mathbb{C}^n$ and we denote by $\pi^*\mathcal{A}$ the set of all such connections.

Each connection A gives rise to a cocycle over the geodesic flow whose generator is $\pi^*A(X) : SM \to \mathfrak{u}(n)$. Note that $\pi^*A(X)(x,v) = A(x,v)$, in words, $\pi^*A(X)$ is the restriction of $A : TM \to \mathfrak{u}(n)$ to SM.

The cocycle C associated with this generator is nothing but parallel transport along geodesics, so that $C : SM \times \mathbb{R} \to U(n)$ solves

$$\frac{d}{dt}C(x,v,t) + A(\phi_t(x,v))C(x,v,t) = 0, \quad C(x,v,0) = \mathrm{Id}.$$

On the set $\pi^*\mathcal{A}$ we impose the equivalence relation $\sim X$ of being X-cohomologous and on $\mathcal{A}$ we have the equivalence relation $\sim$ given by gauge equivalence. There is a natural map induced by π:

$$\mathcal{A}/\sim \; \mapsto \; \pi^*\mathcal{A}/\sim X. \qquad (9)$$

Suppose now we have two connections A_1 and A_2 as in Question II and the geodesic flow is Anosov. Then Theorem 6.5 implies that $\pi^*A_1(X)$ and $\pi^*A_2(X)$ are cohomologous cocycles, that is, there is a smooth map $u : SM \to U(n)$ such that on SM we have the *cohomological equation*

$$A_2 = u^{-1}Xu + u^{-1}A_1 u. \qquad (10)$$

This is the main dynamical input, that allows the passage from closed geodesics to X-cohomology. What is left is the geometric problem of deciding if the map in (9) is injective. Suppose for a moment that for some reason we can show that $u(x,v)$ only depends on x. Then (10) means exactly that A_1 and A_2 are gauge equivalent since $Xu = du$. Thus understanding the dependence of u in the velocities is crucial and this often can be achieved using Pestov type identities and/or Fourier analysis as in the sketch of proof of Theorem 3.2 before. Let us see a good example of this in the simplest possible case in which $n = 1$. Since $U(1) = S^1$ is abelian we can reduce (10) to the cohomologically trivial case

$$Xu + Au = 0$$

where $A = A_1 - A_2$ and $u : SM \to S^1$. Write $A = i\theta$, where θ is an ordinary real-valued 1-form. Then

$$du(X) + i\theta u = 0. \qquad (11)$$

The function u gives rise to a real-valued closed 1-form in SM given by $\varphi := \frac{du}{iu}$. Since $\pi^* : H^1(M,\mathbb{R}) \to H^1(SM,\mathbb{R})$ is an isomorphism when M is different

from the 2-torus, there exists a closed 1-form ω in M and a smooth function $f : SM \to \mathbb{R}$ such that
$$\varphi = \pi^*\omega + df.$$
(It is easy to see that if ϕ_t is Anosov, then M cannot be a 2-torus since for example, $\pi_1(M)$ must grow exponentially.) When this equality is applied to X and combined with (11) one obtains
$$-\theta_x(v) - \omega_x(v) = df(X(x,v)) = X(f)(x,v)$$
for all $(x,v) \in SM$. It is known that this implies that $\theta + \omega$ is exact and that f only depends on x. This was proved by V. Guillemin and D. Kazhdan [1980a] for surfaces of negative curvature, by C. Croke and Sharafutdinov [1998] for arbitrary manifolds of negative curvature and by N.S. Dairbekov and Sharafutdinov [2003] for manifolds whose geodesic flow is Anosov. It follows easily now that u only depends on x and hence A_1 and A_2 must be gauge equivalent and thus for $n = 1$ we have a full answer. Before going further let us explain why if we have a smooth solution u to the cohomological equation
$$Xu = \theta$$
where θ is a 1-form, then θ is exact and u only depends on x. We can see this for dim $M = 2$ using the energy identity from Lemma 4.5 for the case $A = 0$. Since θ is a 1-form, the right-hand side is zero as in Lemma 4.7; thus
$$\|XVu\|^2 - (KVu, Vu) = 0. \tag{12}$$
If the flow is Anosov there are two solutions $r^{s,u}$ of the Riccati equation $X(r) + r^2 + K = 0$. These solutions are related to the stable and unstable bundles as follows: $-X_\perp + r^{s,u} V \in E^{s,u}$ and $r^s - r^u$ never vanishes; for an account of these results we refer to [Paternain 1999]. Hence using the proof of Lemma 4.8 and (12) we deduce:
$$XVu - r^{s,u} Vu = 0$$
from which it follows that $(r^s - r^u)Vu = 0$ and thus $Vu = 0$. This shows that u only depends on x and therefore θ is exact.

Now that we have a better understanding of the abelian case $n = 1$, let us go back to the general equation (10). As in Section 3 we can introduce a new unitary connection $\hat{A}$ on the trivial bundle $M \times \mathbb{C}^{n \times n}$ as follows: given a matrix $R \in \mathbb{C}^{n \times n}$ we define $\hat{A}(R) := AR - RB$. Then (10) is the form
$$Xu + Au = 0$$
at the price of course, of increasing the rank of our trivial vector bundle. Note that $F_{\hat{A}}(R) = F_A R - R F_B$.

This suggests that in general we should study the following problem on closed manifolds. Given a unitary connection A on $M \times \mathbb{C}^n$ and $f : SM \to \mathbb{C}^n$ a smooth function of the form $F(x) + \alpha_j(x)v^j$, where $F : M \to \mathbb{C}^n$ is a smooth function and α is a $\mathbb{C}^n$-valued 1-form, describe the set of smooth solutions $u : SM \to \mathbb{C}^n$ to the equation

$$Xu + Au = -f. \tag{13}$$

Unfortunately we know very little about (13) in the general Anosov case. However for closed surfaces of negative curvature we have the following fundamental result which should be regarded as an extension of [Guillemin and Kazhdan 1980a, Theorem 3.6].

The Fourier analysis that we set up in Section 4 works equally well in the case of closed oriented surfaces. Given $u \in C^\infty(SM, \mathbb{C}^n)$, we write $u = \sum_{m \in \mathbb{Z}} u_m$, where $u_m \in \Omega_m$. We will say that u has degree N, if N is the smallest nonnegative integer such that $u_m = 0$ for all m with $|m| \geq N+1$.

Theorem 6.6 [Paternain 2009, Theorem 5.1]. *If M is a closed surface of negative curvature and $f : SM \to \mathbb{C}^n$ has finite degree, then any smooth solution u of $Xu + Au = -f$ has finite degree.*

Below we shall sketch the proof of this theorem, but first we need some preliminaries. As in [Guillemin and Kazhdan 1980a] we introduce the first-order elliptic operators

$$\eta_+, \eta_- : C^\infty(SM, \mathbb{C}^n) \to C^\infty(SM, \mathbb{C}^n)$$

given by

$$\eta_+ := (X + iX_\perp)/2, \quad \eta_- := (X - iX_\perp)/2.$$

Clearly $X = \eta_+ + \eta_-$. We have

$$\eta_+ : \Omega_m \to \Omega_{m+1}, \quad \eta_- : \Omega_m \to \Omega_{m-1}, \quad (\eta_+)^* = -\eta_-.$$

Before going further, let us use these operators to give a short proof of the bracket relation (7):

Lemma 6.7. *The following formula holds*:

$$[\mathcal{H}, X]u = X_\perp u_0 + (X_\perp u)_0, \quad u \in C^\infty(SM, \mathbb{C}^n).$$

Proof. It suffices to show that

$$[\mathrm{Id} + i\mathcal{H}, X]u = iX_\perp u_0 + i(X_\perp u)_0.$$

Since $X = \eta_+ + \eta_-$ we need to compute $[\mathrm{Id}+i\mathcal{H}, \eta_\pm]$, so let us find $[\mathrm{Id}+i\mathcal{H}, \eta_+]u$, where $u = \sum_k u_k$. Recall that $(\mathrm{Id}+i\mathcal{H})u = u_0 + 2\sum_{k\geq 1} u_k$. We find:

$$(\mathrm{Id}+i\mathcal{H})\eta_+ u = \eta_+ u_{-1} + 2\sum_{k\geq 0} \eta_+ u_k,$$

$$\eta_+(\mathrm{Id}+i\mathcal{H})u = \eta_+ u_0 + 2\sum_{k\geq 1} \eta_+ u_k.$$

Thus

$$[\mathrm{Id}+i\mathcal{H}, \eta_+]u = \eta_+ u_{-1} + \eta_+ u_0.$$

Similarly we find

$$[\mathrm{Id}+i\mathcal{H}, \eta_-]u = -\eta_- u_0 - \eta_- u_1.$$

Therefore using that $iX_\perp = \eta_+ - \eta_-$ we obtain

$$[\mathrm{Id}+i\mathcal{H}, X]u = iX_\perp u_0 + i(X_\perp u)_0$$

as desired. $\square$

To deal with the equation $Xu + Au = -f$, we introduce the "twisted" operators

$$\mu_+ := \eta_+ + A_1, \quad \mu_- := \eta_- + A_{-1},$$

where $A = A_{-1} + A_1$ and

$$A_1 := \frac{A - iV(A)}{2} \in \Omega_1, \quad A_{-1} := \frac{A + iV(A)}{2} \in \Omega_{-1}.$$

This decomposition corresponds precisely with the usual decomposition of $\mathfrak{u}(n)$-valued 1-forms on a surface:

$$\Omega^1(M, \mathfrak{u}(n)) \otimes \mathbb{C} = \Omega^{1,0}(M, \mathfrak{u}(n)) \oplus \Omega^{0,1}(M, \mathfrak{u}(n)),$$

where $\star = -i$ on $\Omega^{1,0}$ and $\star = i$ on $\Omega^{0,1}$ (here $\star$ is the Hodge star operator of the metric).

We also have

$$\mu_+ : \Omega_m \to \Omega_{m+1}, \quad \mu_- : \Omega_m \to \Omega_{m-1}, \quad (\mu_+)^* = -\mu_-.$$

The equation $Xu + Au = -f$ is now $\mu_+(u) + \mu_-(u) = -f$.

Sketch of proof of Theorem 6.6. We shall use the following equality proved in [Paternain 2009, Corollary 4.4]. Given $u \in C^\infty(SM, \mathbb{C}^n)$ we have

$$\|\mu_+ u\|^2 = \|\mu_- u\|^2 + \frac{i}{2}((KVu, u) + (\star F_A u, u)),$$

where K is the Gaussian curvature of the metric and F_A is the curvature of A. This L^2 identity is a close relative of the identity in Lemma 4.5. For $u_m \in \Omega_m$ we have

$$\|\mu_+ u_m\|^2 = \|\mu_- u_m\|^2 + \frac{1}{2}\big((i \star F_A - mK \text{ Id})u_m, u_m\big).$$

Hence if $K < 0$, there exist a constant $c > 0$ and a positive integer ℓ such that

$$\|\mu_+ u_m\|^2 \geq \|\mu_- u_m\|^2 + c\|u_m\|^2 \tag{14}$$

for all $m \geq \ell$. Projecting the equation $Xu + Au = -f$ onto Ω_m-components we obtain

$$\mu_+(u_{m-1}) + \mu_-(u_{m+1}) = -f_m \tag{15}$$

for all $m \in \mathbb{Z}$. Since f has finite degree, combining (15) and (14) we obtain

$$\|\mu_+(u_{m+1})\| \geq \|\mu_+(u_{m-1})\| \tag{16}$$

for all m sufficiently large. Since the function u is smooth, $\mu_+(u_m)$ must tend to zero in the L^2-topology as $m \to \infty$. It follows from (16) that $\mu_+(u_m) = 0$ for all m sufficiently large. However, (14) implies that μ_+ is injective for m large enough and thus $u_m = 0$ for all m large enough.

A similar argument shows that $u_m = 0$ for all m sufficiently large and negative thus concluding that u has finite degree as desired. □

A glance at the proof shows that we can obtain the same finiteness result under the following weaker hypothesis: $K \leq 0$ and the support of $\star F_A$ is contained in the region where $K < 0$. A more careful inspection shows the following:

Corollary 6.8. *Suppose that the Hermitian matrix $\pm i \star F_A(x) - K(x)$ Id is positive definite for all $x \in M$ and that f has degree N. Then, any solution u of $Xu + Au = -f$ must have degree $N - 1$. If $N = 0$, then $f = 0$ and $u = u(x)$ with $d_A u = 0$.*

Let us apply these ideas to show that for closed negatively curved surfaces, the map (9) is locally injective at flat connections. If A and B are two connections with sufficiently small curvatures, then $F_{\hat{A}}$ will be small enough so that the hypothesis of Corollary 6.8 is satisfied. Hence the map u solving (10) depends only on $x \in M$ and A and B must be gauge equivalent. Putting everything together we have shown:

Theorem 6.9. *Let M be a closed negatively curved surface. There is $\varepsilon > 0$ such that if A and B are two connections as in Question II with $\|F_A\|_{C^0}, \|F_B\|_{C^0} < \varepsilon$, then A and B are gauge equivalent.*

When A is the trivial connection, this is essentially [Paternain 2009, Theorem A].

Let us see an easy example which shows that the result in the theorem fails for $n = 2$ without assumptions on the smallness of the curvature. The tangent bundle of an orientable Riemannian surface M is naturally a Hermitian line bundle. It is certainly not trivial in general, but it carries the Levi-Civita connection which is easily seen to be transparent. Indeed, the parallel transport along a closed geodesic γ must fix $\dot{\gamma}(0)$ and consequently any vector orthogonal to it since the parallel transport is an isometry and the surface is orientable. The Levi-Civita connection on T^*M is also transparent and thus we obtain a transparent unitary connection on $TM \oplus T^*M$. But $TM \oplus T^*M$ has zero first Chern class and thus it is unitarily equivalent to the trivial bundle $M \times \mathbb{C}^2$. In this way we obtain a transparent connection on $M \times \mathbb{C}^2$ which in general is not equivalent to the trivial connection (it is nonflat if the Gaussian curvature is not identically zero). Taking higher tensor powers of TM and T^*M and adding them we obtain more examples of transparent connections all arising from the Levi-Civita connection. It turns out that these are not the only examples, but the failure of the uniqueness can be fully understood at least in some important cases. This will be the content of the next section, but before that, we would like to take another look at Theorem 6.6 in the abelian case $n = 1$.

When $n = 1$, $A = i\theta$ and the set Ω_m can be identified with the set of smooth sections of $\kappa^{\otimes m}$, where κ is the canonical line bundle of M. In this case, well known results on the theory of Riemann surfaces imply that μ_- is surjective for $m \geq 2$ (see for example [Duistermaat 1972]) since μ_- is essentially a $\bar{\partial}_A$-operator (we are assuming here that M has genus ≥ 2); see (25) below. It follows that μ_+ is injective for $m \geq 1$. Hence if f has degree N and u has finite degree and solves $Xu + i\theta u = -f$, then u must have degree $N - 1$. Thus in the abelian case we have:

Theorem 6.10. *Suppose that M is a closed surface of negative curvature and $n = 1$. If f has degree N, then any solution u of $Xu + i\theta u = -f$ must have degree $N - 1$. If $N = 0$, then $f = 0$ and $u = u(x)$ with $du + i\theta u = 0$.*

When $n \geq 2$, the operators μ_+ could have nontrivial kernels for $m \geq 1$, and this precisely gives room for the existence of transparent connections.

7. Transparent connections

We start with some motivation for the constructions in this section. How can we construct a cohomologically trivial connection on $M \times \mathbb{C}^2$? Let us suppose that we start with the simplest possible nontrivial u. This would be a smooth map $u : SM \to \mathrm{SU}(2)$ such that $u = u_{-1} + u_1$. We would need $A = -X(u)u^{-1}$ to be a connection, thus its Fourier expansion should have only terms of degree ± 1.

Writing $X = \eta_+ + \eta_-$ we discover that A is a connection if and only if

$$\eta_+(u_1)u^*_{-1} = \eta_-(u_{-1})u^*_1 = 0.$$

In fact $\eta_-(u_{-1})u^*_1 = 0$ implies $\eta_+(u_1)u^*_{-1} = 0$ and vice versa. This can be seen simply by conjugating each relation. So we need to ensure that:

$$\eta_-(u_{-1})u^*_1 = 0. \tag{17}$$

What does this mean? Since u is unitary we have $u_{-1}u^*_1 = 0$ from which we see that $L := \mathrm{Ker}(u_{-1}) = \mathrm{Im}(u^*_1)$ is a line subbundle of $\mathbb{C}^2$. Now (17) can be rewritten as

$$u_{-1}\eta_-(u^*_1) = 0$$

so if we pick $0 \neq \xi \in \mathbb{C}^2$, then $s := u^*_1\xi \in L$ and the equation above says $\eta_-(s) \in L$. In (22) below we will write an equation for η_- in local coordinates which shows that it is essentially a $\bar{\partial}$-operator and hence (17) is saying that L must be a *holomorphic* line bundle. But there is an ample supply of these: it is equivalent to providing a meromorphic function on M. Now we can ask, given a holomorphic line bundle L can we find a function $u : SM \to \mathrm{SU}(2)$ such that $u = u_{-1} + u_1$ and $\mathrm{Ker}(u_{-1}) = L$? We will see below that this is indeed the case, but here is one way to think about it. Given the line bundle L consider the unique map $f : M \to \mathfrak{su}(2)$ with $\det f = 1$ (so that it hits the unit sphere in $\mathfrak{su}(2)$) such that L is the eigenspace of f corresponding to the eigenvalue i. The map u in local coordinates is now

$$u(x, \theta) = \cos\theta \; \mathrm{Id} + \sin\theta \; f(x).$$

Thus for every meromorphic function we obtain a cohomologically trivial connection. Are these all? Not quite, there are many more in which u has higher-order dependence on velocities as we will see below.

We now provide details and we begin by giving a general classification result for cohomologically trivial connections on any surface.

As before let M be an oriented surface with a Riemannian metric and let SM be its unit tangent bundle. Let

$$\mathcal{A} := \{A : SM \to \mathfrak{u}(n) : V^2(A) = -A\}.$$

The set $\mathcal{A}$ is identified with the set of all unitary connections on the trivial bundle $M \times \mathbb{C}^n$. Indeed, a function A satisfying $V^2(A) + A = 0$ extends to a function on TM depending linearly on the velocities.

Recall from the previous section that A is said to be cohomologically trivial if there exists a smooth $u : SM \to U(n)$ such that $C(x, v, t) = u(\phi_t(x, v))u(x, v)^{-1}$.

Differentiating with respect to t and setting $t = 0$ this is equivalent to

$$Xu + Au = 0. \qquad (18)$$

Let $\mathcal{A}_0$ be the set of all cohomologically trivial connections, that is, the set of all $A \in \mathcal{A}$ such that there exists $u : SM \to U(n)$ for which (18) holds.

Given a vector field W in SM, let G_W be the set of all $u : SM \to U(n)$ such that $W(u) = 0$, that is, first integrals of W. Note that G_V is nothing but the group of gauge transformations of the trivial bundle $M \times \mathbb{C}^n$.

We wish to understand $\mathcal{A}_0/G_V$. Now let $\mathcal{D}$ be the set of all $f : SM \to \mathfrak{u}(n)$ such that

$$-X_\perp(f) + VX(f) = [X(f), f]$$

and there is $u : SM \to U(n)$ such that $f = u^{-1}V(u)$. It is easy to check that G_X acts on $\mathcal{D}$ by $f \mapsto a^{-1}fa + a^{-1}V(a)$ where $a \in G_X$.

Theorem 7.1. *There is a one-to-one correspondence between $\mathcal{A}_0/G_V$ and $\mathcal{D}/G_X$.*

Proof. Forward direction: a cohomologically trivial connection A comes with a u such that $Xu + Au = 0$. If we set $f := u^{-1}V(u)$, then $f \in \mathcal{D}$, that is, f satisfies the PDE $-X_\perp(f) + VX(f) = [X(f), f]$. This a calculation (see [Paternain 2009, Theorem B] for details), but for the reader's convenience we explain the geometric origin of this equation. Using u we may define a connection on SM gauge equivalent to π^*A by setting $B := u^{-1}du + u^{-1}\pi^*Au$, where $\pi : SM \to M$ is the foot-point projection. Since π^*A is the pull-back of a connection on M, the curvature F_B of B must vanish when one of the entries is the vertical vector field V. The PDE $-X_\perp(f) + VX(f) = [X(f), f]$ arises by combining the two equations $F_B(X, V) = F_B(X_\perp, V) = 0$ with $B(X) = 0$.

Backward direction: Given f with $uf = V(u)$, set $A := -X(u)u^{-1}$. Then $A \in \mathcal{A}_0$, that is, $V^2(A) = -A$; again this is a calculation done fully in Theorem B in [Paternain 2009].

Now there are two ambiguities here. Going forward, we may change u as long as we solve $Xu + Au = 0$. This changes f by the action of G_X. Going backwards we may change u as long as $uf = V(u)$, this changes A by a gauge transformation, that is, an element in G_V. $\square$

Note that if the geodesic flow is transitive (i.e., there is a dense orbit) the only first integrals are the constants and thus $G_X = U(n)$ acts simply by conjugation. If M is closed and of negative curvature, the geodesic flow is Anosov and therefore transitive.

The fact that the PDE describing cohomologically trivial connections arises from zero curvature conditions is an indication of the "integrable" nature of the problem at hand. The existence of a Bäcklund transformation that we will introduce shortly is another typical feature of integrable systems. Note that the space $\mathcal{D}/G_X$ is in some sense simpler and larger when the underlying geodesic flow is more complicated, that is, when it is transitive G_X reduces to $U(n)$.

The Bäcklund transformation. For the remainder of this section we restrict to the case in which the structure group is SU(2). This is the simplest nontrivial case.

Suppose there is a smooth map $b : SM \to \mathrm{SU}(2)$ such that $f := b^{-1}V(b)$ solves the PDE:
$$-X_\perp(f) + VX(f) = [X(f), f]. \tag{19}$$

Then, by Theorem 7.1, $A := -X(b)b^{-1}$ defines a cohomologically trivial connection on M and $-\star A = V(A) = -bX(f)b^{-1} + X_\perp(b)b^{-1}$.

Lemma 7.2. *Let $g : M \to \mathfrak{su}(2)$ be a smooth map with $\det g = 1$ (i.e., $g^2 = -\mathrm{Id}$). Then there exists $a : SM \to \mathrm{SU}(2)$ such that $g = a^{-1}V(a)$.*

Proof. Let $L(x)$ and $U(x)$ be the eigenspaces corresponding respectively to the eigenvalues i and $-i$ of $g(x)$. We have an orthogonal decomposition $\mathbb{C}^2 = L(x) \oplus U(x)$ for every $x \in M$. Consider sections
$$\alpha \in \Omega^{1,0}(M, \mathbb{C}) \quad \text{and} \quad \beta \in \Omega^{1,0}(M, \mathrm{Hom}(L, U)) = \Omega^{1,0}(M, L^*U)$$
such that $|\alpha|^2 + |\beta|^2 = 1$. Such a pair of sections always exists; for example, we can choose a section $\tilde{\beta}$ with a finite number of isolated zeros and then choose a $\tilde{\alpha}$ that does not vanish on the zeros of $\tilde{\beta}$. Then we set $\alpha := \tilde{\alpha}/(|\tilde{\alpha}|^2 + |\tilde{\beta}|^2)^{1/2}$ and $\beta := \tilde{\beta}/(|\tilde{\alpha}|^2 + |\tilde{\beta}|^2)^{1/2}$. Note that
$$\bar{\alpha} \in \Omega^{0,1}(M, \mathbb{C}) \quad \text{and} \quad \beta^* \in \Omega^{0,1}(M, \mathrm{Hom}(U, L)) = \Omega^{0,1}(M, U^*L).$$

Using the orthogonal decomposition we define $a : SM \to \mathrm{SU}(2)$ by
$$a(x, v) = \begin{pmatrix} \alpha(x, v) & \beta^*(x, v) \\ -\beta(x, v) & \bar{\alpha}(x, v) \end{pmatrix}.$$

Clearly $a = a_{-1} + a_1$, where
$$a_1 = \begin{pmatrix} \alpha & 0 \\ -\beta & 0 \end{pmatrix} \quad \text{and} \quad a_{-1} = \begin{pmatrix} 0 & \beta^* \\ 0 & \bar{\alpha} \end{pmatrix}.$$

It is straightforward to check that $ag = V(a)$. $\square$

Remark 7.3. There is an alternative proof of this lemma along the following lines. Consider an open set U in M over which the circle fibration $\pi : SM \to M$ trivializes as $U \times S^1$, where $S^1 = \mathbb{R}/2\pi\mathbb{Z}$. In this trivialization $V = \partial/\partial\theta$ and any solution to $ag = V(a)$ has the form $a_U := r_U(x)(\cos\theta\,\mathrm{Id} + \sin\theta\,g(x))$, where $r_U : U \to SU(2)$ is smooth. Consider another set U' which trivializes $\pi : SM \to M$ and which intersects U. We obtain a transition function $\psi_{UU'} : U \cap U' \to S^1$. The functions a_U can be glued to define a global function $a : SM \to SU(2)$ as long as

$$r_U(x)(\cos\theta\, \text{Id} + \sin\theta\, g(x)) = r_{U'}(x)(\cos\theta'\, \text{Id} + \sin\theta'\, g(x))$$

where $\theta = \theta' + \psi_{UU'}(x)$ and $x \in U \cap U'$. Hence to have a globally defined a we need to show the existence of smooth functions $r_U : U \to SU(2)$ such that

$$\varphi_{UU'}(x) := \cos(\psi_{UU'}(x))\text{Id} + \cos(\psi_{UU'}(x))g(x) = (r_U(x))^{-1} r_{U'}(x).$$

The key observation is that $\varphi_{UU'}$ defines an $SU(2)$-cocycle in the sense of principal bundles. Indeed, the cocycle property $\varphi_{UU''}(x) = \varphi_{UU'}(x)\,\varphi_{U'U''}(x)$ follows right away from the fact that $\psi_{UU'}$ is an S^1-cocycle. But an $SU(2)$-bundle over a surface is trivial. The existence of the functions $r_U : U \to SU(2)$ follows.

Note that by construction, $\text{Ker}\, a_{\pm 1}$ coincides with the $\mp i$ eigenspace of g.

Now let $u := ab : SM \to SU(2)$ and let $F := (ab)^{-1} V(ab) = b^{-1} g\, b + f$.

Question. When does F satisfy (19)?

If it does, then it defines (via Theorem 7.1) a new cohomologically trivial connection given by

$$A_F = -X(ab)(ab)^{-1} = -X(a)a^{-1} + aAa^{-1},$$

where A is the cohomologically trivial connection associated to f.

Recall that the connection A defines a covariant derivative $d_A g = dg + [A, g]$.

Lemma 7.4. *F satisfies (19) if and only if*

$$-\star d_A g = (d_A g)\, g. \tag{20}$$

Proof. Starting with $F = b^{-1} g\, b + f$ and using that $A = -X(b)b^{-1} = bX(b^{-1})$ we compute

$$X(F) = b^{-1}\left([A, g] + X(g)\right) b + X(f).$$

Similarly, using $X_\perp(b) = -(\star A)b + bX(f)$ we find

$$X_\perp(F) = b^{-1}\left([\star A, g] + X_\perp(g)\right) b - [X(f), b^{-1} g\, b] + X_\perp(f).$$

Now we compute $VX(F)$; here we use that $V(g) = 0$. We obtain

$$VX(F) = [b^{-1}([A, g] + X(g))b,\, f] + b^{-1}\left([-\star A, g] + VX(g)\right) b + VX(f).$$

The last term we need for (19) is

$$[X(F), F] = b^{-1}[[A, g] + X(g), g]b + [b^{-1}([A, g] + X(g))b, f]$$
$$+ [X(f), b^{-1}g\,b] + [X(f), f].$$

Since f satisfies (19) we see that F satisfies (19) if and only if

$$-X_\perp(g) + VX(g) - 2[\star A, g] = [[A, g] + X(g), g].$$

Since g depends only on the base point and $g^2 = -\mathrm{Id}$ we can rewrite this as

$$-2\star(dg + [A, g]) = [dg + [A, g], g] = 2(dg + [A, g])\,g.$$

Thus F satisfies (19) if and only if

$$-\star d_A g = (d_A g)\,g,$$

as claimed. $\square$

We will now rephrase (20) in terms of holomorphic line bundles. Recall that the connection A induces a holomorphic structure on the trivial bundle $M \times \mathbb{C}^2$ and on the endomorphism bundle $M \times \mathbb{C}^{2\times 2}$. We have an operator $\bar{\partial}_A = (d_A - i\star d_A)/2 = \bar{\partial} + [A_{-1}, \cdot]$ acting on sections $f : M \to \mathbb{C}^{2\times 2}$.

Set $\pi := (\mathrm{Id} - ig)/2$ and $\pi^\perp = (\mathrm{Id} + ig)/2$ so that $\pi + \pi^\perp = \mathrm{Id}$. Let $L(x)$ be as above the eigenspace corresponding to the eigenvalue i of $g(x)$. Note that π is the Hermitian orthogonal projection over $L(x) = \mathrm{Image}(\pi(x))$.

Lemma 7.5. *Let $g : M \to \mathfrak{su}(2)$ be a smooth map with $\det g = 1$. The following conditions are equivalent.*

(1) $-\star d_A g = (d_A g)g.$

(2) L is a $\bar{\partial}_A$-holomorphic line bundle.

(3) $\pi^\perp \bar{\partial}_A \pi = 0.$

Proof. Suppose that (1) holds. Apply $\star$ to obtain $d_A g = (\star d_A g)\,g$. Thus

$$d_A g - i \star d_A g = i(d_A g - i\star d_A g)g.$$

In other words $\bar{\partial}_A g = i(\bar{\partial}_A g)\,g = -ig(\bar{\partial}_A g)$ (recall that $g^2 = -\mathrm{Id}$). Since $\pi = (\mathrm{Id} - ig)/2$, then $\bar{\partial}_A g = -ig(\bar{\partial}_A g)$ is equivalent to $\pi^\perp \bar{\partial}_A \pi = 0$ which is (3).

Using the condition $\pi^2 = \pi$, we see that $\pi^\perp \bar{\partial}_A \pi = 0$ is equivalent to $(\bar{\partial}_A \pi)\pi = 0$. The line bundle L is holomorphic if and only if given a local section ξ of L, then $\bar{\partial}_A \xi \in L$. Using that $\pi \xi = \xi$ we see that $\bar{\partial}_A \xi \in L$ if and only if $(\bar{\partial}_A \pi)\xi = 0$. Clearly, this happens if and only if $(\bar{\partial}_A \pi)\pi = 0$ and thus (2) holds if and only if (3) holds. $\square$

The next theorem summarizes the Bäcklund transformation that we just introduced and it follows directly from Lemmas 7.4 and 7.5 and Theorem 7.1.

Theorem 7.6. *Let A be a cohomologically trivial connection and let L be a holomorphic line subbundle of the trivial bundle $M \times \mathbb{C}^2$ with respect to the complex structure induced by A. Define a map $g : M \to \mathfrak{su}(2)$ with $\det g = 1$ by declaring L to be its eigenspace with eigenvalue i. Consider $a : SM \to \mathrm{SU}(2)$ with $g = a^{-1}V(a)$ as given by Lemma 7.2. Then*
$$A_F := -X(a)a^{-1} + aAa^{-1}$$
defines a cohomologically trivial connection.

Definition 7.7. Let A be a cohomologically trivial connection. Given a map $g : M \to \mathfrak{su}(2)$ with $\det g = 1$ and $-\star d_A g = (d_A g)g$, let $a : SM \to \mathrm{SU}(2)$ be any smooth map with $ag = V(a)$. Then the *Bäcklund transformation* of the connection A with respect to the pair (g, a) is:
$$\mathcal{B}_{g,a}(A) := -X(a)a^{-1} + aAa^{-1}.$$

By Theorem 7.6, $\mathcal{B}_{g,a}(A)$ is a new cohomologically trivial connection.

Remark 7.8. Note that if the geodesic flow is transitive, two solutions u, w of $Xu + Au = 0$ are related by $u = wg$ where g is a constant unitary matrix, because $X(w^{-1}u) = 0$. Thus the degrees of u and w are the same. We can then talk about the "degree" of a cohomologically trivial connection as the degree of any solution of $Xu + Au = 0$.

Remark 7.9. If we let $q := aga^{-1}$, then a simple calculation shows that $V(q) = 0$ and $d_{A_F} q = a(d_A g)a^{-1}$. Moreover, $\star d_{A_F} q = (d_{A_F} q)q$ which means that $-q$ satisfies (20) with respect to A_F. Hence if we run the Bäcklund transformation on A_F with $g' := -q$ and $a' := a^{-1}$ we recover A (note that $a'g' = V(a')$). In other words $\mathcal{B}_{-q,a^{-1}}(\mathcal{B}_{g,a}(A)) = A$. Thus the Bäcklund transformation described in Theorem 7.6 has a natural "inverse".

If we start, for example, with the trivial connection $A = 0$ (which is obviously cohomologically trivial), then a map $g : M \to \mathfrak{su}(2)$ with $\det g = 1$ and $-\star dg = (dg)g$ can be identified with a meromorphic function. The connections of degree one $A_F = -X(a)a^{-1}$ given by Theorem 7.6 were first found in [Paternain 2009] and coincide with the ones described at the beginning of the section. In the next subsection we will show that any cohomologically trivial connection such that the associated u has a finite Fourier series can be built up by successive applications of the transformation described in Theorem 7.6, provided that the geodesic flow is transitive. This will provide a full classification of transparent $\mathrm{SU}(2)$-connections over negatively curved surfaces.

The classification result. Let A be a transparent connection with $A = -X(b)b^{-1}$ and $f = b^{-1}V(b)$, where $b : SM \to \mathrm{SU}(2)$.

We first make some remarks concerning the SU(2)-structure. Let $j : \mathbb{C}^2 \to \mathbb{C}^2$ be the antilinear map given by

$$j(z_1, z_2) = (-\bar{z}_2, \bar{z}_1).$$

If we think of a matrix $a \in \mathrm{SU}(2)$ as a linear map $a : \mathbb{C}^2 \to \mathbb{C}^2$, then $ja = aj$. This implies that given $b : SM \to \mathrm{SU}(2)$ with $b = \sum_{k \in \mathbb{Z}} b_k$, then $jb_k = b_{-k}j$ for all $k \in \mathbb{Z}$.

Assumption. Suppose b has a finite Fourier expansion, that is, $b = \sum_{k=-N}^{k=N} b_k$, where $N \geq 1$. By Theorem 6.6 we know that this holds if M has negative curvature.

Let us assume also that N is the degree of b, so both b_N and $b_{-N} = -jb_N j$ are nonzero.

The unitary condition $bb^* = b^*b = \mathrm{Id}$ implies that $b_N b^*_{-N} = b^*_{-N} b_N = 0$. These relations imply that the rank of b_{-N} and b_N is at most one and equals one on an open set, which, as we will see shortly, must be all of M except for perhaps a finite number of points. But first we need some preliminaries.

Consider isothermal coordinates (x, y) on M such that the metric can be written as $ds^2 = e^{2\lambda}(dx^2 + dy^2)$, where λ is a smooth real-valued function of (x, y). This gives coordinates (x, y, θ) on SM, where θ is the angle between a unit vector v and $\partial/\partial x$. In these coordinates X is given by (8) and $X_\perp$ by:

$$X_\perp = -e^{-\lambda}\left(-\sin\theta \frac{\partial}{\partial x} + \cos\theta \frac{\partial}{\partial y} - \left(\frac{\partial \lambda}{\partial x}\cos\theta + \frac{\partial \lambda}{\partial y}\sin\theta\right)\frac{\partial}{\partial \theta}\right). \quad (21)$$

Consider $u \in \Omega_m$ and write it locally as $u(x, y, \theta) = h(x, y)e^{im\theta}$. Using (8) and (21) a straightforward calculation shows that

$$\eta_-(u) = e^{-(1+m)\lambda}\bar{\partial}(he^{m\lambda})e^{i(m-1)\theta}, \quad (22)$$

where $\bar{\partial} = \frac{1}{2}(\partial/\partial x + i\,\partial/\partial y)$. In order to write μ_- suppose that $A(x, y, \theta) = a(x, y)\cos\theta + b(x, y)\sin\theta$. If we also write $A = A_x dx + A_y dy$, then $A_x = ae^\lambda$ and $A_y = be^\lambda$. Let $A_{\bar{z}} := \frac{1}{2}(A_x + iA_y)$. Using the definition of A_{-1} we derive

$$A_{-1} = \tfrac{1}{2}(a + ib)e^{-i\theta} = A_{\bar{z}}d\bar{z}. \quad (23)$$

Putting this together with (22) we obtain

$$\mu_-(u) = e^{-(1+m)\lambda}\left(\bar{\partial}(he^{m\lambda}) + A_{\bar{z}}he^{m\lambda}\right)e^{i(m-1)\theta}. \quad (24)$$

Note that Ω_m can be identified with the set of smooth sections of the bundle $(M \times \mathbb{M}_2(\mathbb{C})) \otimes \kappa^{\otimes m}$ where κ is the canonical line bundle. The identification takes

$u = he^{im\theta}$ into $he^{m\lambda}(dz)^m$ ($m \geq 0$) and $u = he^{-im\theta} \in \Omega_{-m}$ into $he^{m\lambda}(d\bar{z})^m$. The second equality in (23) should be understood using this identification.

Consider now a fixed vector $\xi \in \mathbb{C}^2$ such that $s(x,v) := b_{-N}(x,v)\xi \in \mathbb{C}^2$ is not zero identically. Clearly s can be seen as a section of $(M \times \mathbb{C}^2) \otimes \kappa^{\otimes -N}$. We may write b_{-N} and s in local isothermal coordinates as $b_{-N} = he^{-iN\theta}$ and $s = e^{N\lambda}h\xi(d\bar{z})^N$.

Lemma 7.10. *The local section $e^{-2N\lambda}s$ is $\bar{\partial}_A$-holomorphic.*

Proof. Using the operators $\mu_\pm$ we can write $X(b) + Ab = 0$ as

$$\mu_+(b_{k-1}) + \mu_-(b_{k+1}) = 0$$

for all k. This gives $\mu_+(b_N) = \mu_-(b_{-N}) = 0$. But $\mu_-(b_{-N}) = 0$ is saying that $e^{-2N\lambda}s$ is $\bar{\partial}_A$-holomorphic. Indeed, using (24), we see that $\mu_-(b_{-N}) = 0$ implies

$$\bar{\partial}(he^{-N\lambda}) + A_{\bar{z}}he^{-N\lambda} = 0$$

which in turn implies

$$\bar{\partial}(e^{-N\lambda}h\xi) + A_{\bar{z}}e^{-N\lambda}h\xi = 0.$$

This equation says that $e^{-2N\lambda}s = e^{-N\lambda}h\xi(d\bar{z})^N$ is $\bar{\partial}_A$-holomorphic. □

The section s spans a line bundle L over M which by the previous lemma is $\bar{\partial}_A$-holomorphic. The section s may have zeros, but at a zero z_0, the line bundle extends holomorphically. Indeed, in a neighborhood of z_0 we may write $e^{-2N\lambda(z)}s(z) = (z-z_0)^k w(z)$, where w is a local holomorphic section with $w(z_0) \neq 0$. The section w spans a holomorphic line subbundle which coincides with the one spanned by s off z_0. Therefore L is a $\bar{\partial}_A$-holomorphic line bundle that contains the image of b_{-N} (and $U = jL$ is an antiholomorphic line bundle that contains the image of b_N). We summarize this in a lemma:

Lemma 7.11. *The line bundle L determined by the image of b_{-N} is $\bar{\partial}_A$-holomorphic.*

We will now use the line bundle L to construct an appropriate $g : M \to \mathfrak{su}(2)$ such that when we run the Bäcklund transformation from the previous subsection we obtain a cohomologically trivial connection of degree $\leq N-1$. But first we need the following lemma. Recall that a matrix-valued function f is said to be *odd* if $f(x,v) = -f(x,-v)$ and *even* if $f(x,v) = f(x,-v)$.

Lemma 7.12. *Assume that the geodesic flow is transitive and let $b : SM \to \mathrm{SU}(2)$ solve $X(b) + Ab = 0$. Then b is either even or odd.*

Proof. Write $b = b_o + b_e$ where b_o is odd and b_e is even. Since the operator $(X + A)$ maps even to odd and odd to even, the equation $X(b) + Ab = 0$ decouples as
$$X(b_o) + Ab_o = 0,$$
$$X(b_e) + Ab_e = 0.$$

A calculation using these equations shows that $X(b_o^* b_o)$, $X(b_e^* b_e)$, $X(b_o^* b_e)$ all vanish. Since the geodesic flow is transitive, these matrices are all constant. Moreover, since $b_o^* b_e$ is odd it must be zero. On the other hand $jb = bj$ implies that $jb_o = b_o j$ and $jb_e = b_e j$, which in turn implies that both b_o and b_e cannot have rank 1. Putting all this together, we see that either b_o or b_e must vanish identically. $\square$

Suppose the geodesic flow is transitive. By Lemma 7.12, $b = b_{-N} + d + b_N$, where d has degree $\leq N - 2$. We now seek $a : SM \to \mathrm{SU}(2)$ of degree one such that $u := ab$ has degree $\leq N - 1$. For this we need $a_1 b_N = a_{-1} b_{-N} = 0$. We take a map $g : M \to \mathfrak{su}(2)$ with $\det g = 1$ such that its i eigenspace is L and its $-i$ eigenspace is U. By Lemmas 7.5 and 7.11, $-\star d_A g = (d_A g) g$. The construction of a with $ag = V(a)$ from Lemma 7.2 is precisely such that the kernel of a_{-1} is L and the kernel of a_1 is U, so the needed relations to lower the degree hold.

Finally by Theorem 7.6, u gives rise to a cohomologically trivial connection $-X(u)u^{-1}$. Combining this with Theorem 6.6 we have arrived at the main result of this section:

Theorem 7.13. *Let M be a closed orientable surface of negative curvature. Then any transparent* $\mathrm{SU}(2)$-*connection can be obtained by successive applications of Bäcklund transformations as described in Theorem 7.6.*

We finish this section with some remarks on the operators $\mu_\pm$. Let $\Gamma(M, \kappa^{\otimes m})$ denote the space of smooth sections of the m-th tensor power of the canonical line bundle κ. Locally its elements have the form $w(z)dz^m$ for $m \geq 0$ and $w(z)d\bar{z}^{-m}$ for $m \leq 0$. Given a metric g on M, there is map
$$\varphi_g : \Gamma(M, \kappa^{\otimes m}) \to \Omega_m$$

given by restriction to SM. This map is a complex linear isomorphism. Let us check what this map looks like in isothermal coordinates. An element of $\Gamma(M, \kappa^{\otimes m})$ is locally of the form $w(z)dz^m$. Consider a tangent vector $\dot{z} = \dot{x}_1 + i\dot{x}_2$. It has norm 1 in the metric g if and only if $e^{i\theta} = e^\lambda \dot{z}$. Hence the restriction of $w(z)dz^m$ to SM is
$$w(z)e^{-m\lambda}e^{im\theta}$$

as indicated above. Moreover there is also a restriction map

$$\psi_g : \Gamma(M, \kappa^{\otimes m} \otimes \bar{\kappa}) \to \Omega_{m-1}$$

which is an isomorphism. The restriction of $w(z)dz^m \otimes d\bar{z}$ to SM is

$$w(z)e^{-(m+1)\lambda}e^{i(m-1)\theta},$$

because $e^{-i\theta} = e^{\lambda}\bar{z}$.

Given any holomorphic bundle ξ over M, there is a $\bar{\partial}$-operator defined on:

$$\bar{\partial} : \Gamma(M, \xi) \to \Gamma(M, \xi \otimes \bar{\kappa}).$$

In particular we can take $\xi = \kappa^{\otimes m}$ (or $\kappa^{\otimes m} \otimes \mathbb{C}^n$). Combining this with (22) we get the following commutative diagram:

$$\begin{array}{ccc} \Gamma(M, \kappa^{\otimes m}) & \xrightarrow{\varphi_g} & \Omega_m \\ \bar{\partial} \downarrow & & \downarrow \eta_- \\ \Gamma(M, \kappa^{\otimes m} \otimes \bar{\kappa}) & \xrightarrow{\psi_g} & \Omega_{m-1} \end{array}$$

In other words:

$$\eta_- = \psi_g \bar{\partial} \varphi_g^{-1}.$$

This equation exhibits explicitly the relation of η_- with the metric. More generally, if we let $\bar{\partial}_A := \bar{\partial} + A_{\bar{z}}$, then (24) shows that

$$\mu_- = \psi_g \bar{\partial}_A \varphi_g^{-1}. \tag{25}$$

In particular we see from (25) that the injectivity and surjectivity properties of μ_- only depend on the conformal class of the metric and are the same as those of $\bar{\partial}_A$. Also, the index of μ_- may be computed using Riemann–Roch; see [McDuff and Salamon 2004, Appendix C]. If $\xi = \kappa^{\otimes m} \otimes \mathbb{C}^n$ and g denotes the genus of M, then

$$\text{index}(\mu_-) = n(1-g) + c_1(\xi) = (g-1)n(2m-1).$$

For the abelian case $n = 1$, it is a classical result that $\bar{\partial}_A$ is surjective if $g \geq 2$ and $m \geq 2$.

8. Higgs fields

Virtually everything that we have said above extends when a Higgs field is present. For us, a Higgs field is a smooth matrix-valued function $\Phi : M \to \mathbb{C}^{n \times n}$. Often in gauge theories, the structure group is $U(n)$ and the field Φ is required to take values in $\mathfrak{u}(n)$. We call a Higgs field $\Phi : M \to \mathfrak{u}(n)$ a skew-Hermitian Higgs field. The pairs (A, Φ) often appear in the so-called Yang–Mills–Higgs theories. A good example of this is the Bogomolny equation in Minkowski $(2+1)$-space given by $d_A \Phi = \star F_A$. Here d_A stands for the covariant derivative induced on endomorphism $d_A \Phi = d\Phi + [A, \Phi]$, $F_A = dA + A \wedge A$ is the curvature of A and $\star$ is the Hodge star operator of Minkowski space. The Bogomolny equation appears as a reduction of the self-dual Yang–Mills equation in $(2+2)$-space and has been object of intense study in the literature of solitons and integrable systems; see for instance [Dunajski 2010; Manton and Sutcliffe 2004, Chapter 8; Hitchin et al. 1999, Chapter 4; Mason and Woodhouse 1996].

To include the Higgs field in the discussions above, we consider the following transport equation for $u : SM \to \mathbb{C}^n$,

$$Xu + Au + \Phi u = -f \quad \text{in } SM, \quad u|_{\partial_-(SM)} = 0.$$

As before, on a fixed geodesic the transport equation becomes a linear system of ODEs with zero initial condition, and therefore this equation has a unique solution $u = u^f$.

Definition 8.1. The geodesic ray transform of $f \in C^\infty(SM, \mathbb{C}^n)$ with attenuation determined by the pair (A, Φ) is given by

$$I_{A,\Phi} f := u^f|_{\partial_+(SM)}.$$

Obviously $I_A = I_{A,0}$ when $\Phi = 0$. The following extension of Theorem 3.2 holds:

Theorem 8.2 [Paternain et al. 2011a]. *Let M be a compact simple surface. Assume that $f : SM \to \mathbb{C}^n$ is a smooth function of the form $F(x) + \alpha_j(x)v^j$, where $F : M \to \mathbb{C}^n$ is a smooth function and α is a $\mathbb{C}^n$-valued 1-form. Let also $A : SM \to \mathfrak{u}(n)$ be a unitary connection and $\Phi : M \to \mathfrak{u}(n)$ a skew-Hermitian matrix function. If $I_{A,\Phi}(f) = 0$, then $F = \Phi p$ and $\alpha = d_A p$, where $p : M \to \mathbb{C}^n$ is a smooth function with $p|_{\partial M} = 0$.*

The introduction of the Higgs field complicates matters from a technical point of view: more terms appear in the Pestov identity, and these need to be carefully controlled, we refer the reader to [Paternain et al. 2011a] for details.

Given a pair (A, Φ) one can also associate to it scattering data. We look at the unique solution $U_{A,\Phi} : SM \to U(n)$ of

$$\begin{cases} X(U_{A,\Phi}) + (A(x,v) + \Phi(x))U_{A,\Phi} = 0, & (x,v) \in SM, \\ U_{A,\Phi}|_{\partial_+(SM)} = \mathrm{Id}. \end{cases}$$

The scattering data of the pair (A, Φ) is now the map $C_{A,\Phi} : \partial_-(SM) \to U(n)$ defined as $C_{A,\Phi} := U_{A,\Phi}|_{\partial_-(SM)}$.

Using Theorem 8.2 we can derive the following result just as we have done for the proof of Theorem 3.3.

Theorem 8.3 [Paternain et al. 2011a]. *Assume M is a compact simple surface, let A and B be two Hermitian connections, and let Φ and Ψ be two skew-Hermitian Higgs fields. Then $C_{A,\Phi} = C_{B,\Psi}$ implies that there exists a smooth $U : M \to U(n)$ such that $U|_{\partial M} = \mathrm{Id}$ and $B = U^{-1}dU + U^{-1}AU$, $\Psi = U^{-1}\Phi U$.*

A Higgs field can also be included for the case of closed manifolds. A classification of SO(3)-transparent pairs (A, Φ) for surfaces of negative curvature may be found in [Paternain 2012].

9. Arbitrary bundles

In this section we briefly discuss the case of closed surfaces and arbitrary (not necessarily trivial) bundles. We begin with some generalities.

Suppose E is a rank n Hermitian vector bundle over a closed manifold N and $\phi_t : N \to N$ is a smooth transitive Anosov flow.

Definition 9.1. A cocycle over ϕ_t is an action of $\mathbb{R}$ by bundle automorphisms which covers ϕ_t. In other words, for each $(x,t) \in N \times \mathbb{R}$, we have a unitary map $C(x,t) : E_x \to E_{\phi_t x}$ such that $C(x, t+s) = C(\phi_t x, s) \, C(x,t)$.

If E admits a unitary trivialization $f : E \to N \times \mathbb{C}^n$, then

$$f \, C(x,t) \, f^{-1}(x,a) = (\phi_t x, D(x,t)a),$$

where $D : N \times \mathbb{R} \to U(n)$ is a cocycle as in Definition 6.1.

Let E^* denote the dual vector bundle to E. If E carries a Hermitian metric h, we have a conjugate isomorphism $\ell_h : E \to E^*$, which induces a Hermitian metric h^* on E^*. Given a cocycle C on E, $C^* := \ell_h \, C \, \ell_h^{-1}$ is a cocycle on (E^*, h^*).

Proposition 9.2. *Let E be a Hermitian vector bundle over N such that $E \oplus E^*$ is a trivial vector bundle. Let C be a smooth cocycle on E such that $C(x,T) = \mathrm{Id}$ whenever $\phi_T x = x$. Then E is a trivial vector bundle.*

Proof. As explained above, the cocycle C on E induces a cocycle C^* on E^*. On the trivial vector bundle $E \oplus E^*$ we consider the cocycle $C \oplus C^*$. Clearly $C \oplus C^*(x,T) = \mathrm{Id}$ every time that $\phi_T x = x$. Choose a unitary trivialization $f : E \oplus E^* \to N \times \mathbb{C}^{2n}$ and write

$$f \, C \oplus C^*(x,t) \, f^{-1}(x,a) = (\phi_t x, D(x,t)a).$$

By Theorem 6.4, there exists a smooth function $u : N \to U(2n)$ such that $D(x,t) = u(\phi_t x)u^{-1}(x)$. Since ϕ_t is a transitive flow, we may choose $x_0 \in N$

with a dense orbit and without loss of generality we may suppose that $u(x_0) = \mathrm{Id}$. Let
$$\{e_1(x_0), \ldots, e_n(x_0)\}$$
be a unitary frame at E_{x_0}. Write $f(x_0, e_i(x_0)) = (x_0, a_i)$, where $a_i \in \mathbb{C}^{2n}$. Let
$$e_i(x) := f^{-1}(x, u(x)a_i).$$
Clearly at every $x \in N$, $\{e_1(x), \ldots, e_n(x)\}$ is a smooth unitary n-frame of $E_x \oplus E_x^*$. We claim that in fact $e_i(x) \in E_x$ for all $x \in N$. This, of course, implies the triviality of E. Note that
$$e_i(\phi_t x_0) = f^{-1}(\phi_t x_0, u(\phi_t x_0)a_i)$$
$$= f^{-1}(\phi_t x_0, D(x_0, t)a_i) = C \oplus C^*(x_0, t)e_i(x_0).$$
But $e_i(x_0) \in E_{x_0}$, thus $e_i(\phi_t x_0) \in E_{\phi_t x_0}$. It follows that $e_i(x) \in E_x$ for a dense set of points in N. By continuity of e_i, $e_i(x) \in E_x$ for all $x \in N$. □

Remark 9.3. The hypothesis of $E \oplus E^*$ being trivial is not needed in Proposition 9.2. Ralf Spatzier has informed me that it is possible to adapt the proof of the usual Livšic periodic data theorem to show directly that E is trivial. However, this weaker version is all that we will need below.

Let M be a closed orientable surface. In this case, complex vector bundles E over M are classified topologically by the first Chern class $c_1(E) \in H^2(M, \mathbb{Z}) = \mathbb{Z}$. Since $c_1(E^*) = -c_1(E)$ and c_1 is additive with respect to direct sums, we see that $E \oplus E^*$ is the trivial bundle and therefore we will be able to apply Proposition 9.2. In fact we will show:

Theorem 9.4 [Paternain 2009]. *Let M be a closed orientable Riemannian surface of genus g whose geodesic flow is Anosov. A complex vector bundle E over M admits a transparent connection if and only if $2 - 2g$ divides $c_1(E)$.*

Proof. Suppose E admits a transparent connection. As explained above we may apply Proposition 9.2 to deduce that $\pi^* E$ is a trivial bundle and since $c_1(\pi^* E) = \pi^* c_1(E)$ we conclude that $\pi^* c_1(E) = 0$. Consider now the Gysin sequence of the unit circle bundle $\pi : SM \to M$:

$$0 \to H^1(M, \mathbb{Z}) \xrightarrow{\pi^*} H^1(SM, \mathbb{Z})$$
$$\xrightarrow{0} H^0(M, \mathbb{Z}) \xrightarrow{\times(2-2g)} H^2(M, \mathbb{Z}) \xrightarrow{\pi^*} H^2(SM, \mathbb{Z}) \to \cdots.$$

We see that $\pi^* c_1(E) = 0$ if and only if $c_1(E)$ is in the image of the map $H^0(M, \mathbb{Z}) \to H^2(M, \mathbb{Z})$ given by cup product with the Euler class of the unit circle bundle. Equivalently, $2 - 2g$ must divide $c_1(E)$.

Let κ be the canonical line bundle of M. We can think of κ as the cotangent bundle to M; it has $c_1(\kappa) = 2g - 2$. The tensor powers κ^s of κ (positive and negative) generate all possible line bundles with first Chern class divisible by $2 - 2g$ and they all carry the unitary connection induced by the Levi-Civita connection of the Riemannian metric on M. All these connections are clearly transparent. Topologically, all complex vector bundles over M whose first Chern class is divisible by $2 - 2g$ are of the form $\kappa^s \oplus \varepsilon$, where ε is the trivial vector bundle. Since the trivial connection on the trivial bundle is obviously transparent, it follows that every complex vector bundle whose first Chern class is divisible by $2 - 2g$ admits a transparent connection. □

A similar argument shows that if E is a Hermitian *line* bundle with a transparent connection and dim $M \geq 3$, then E must be trivial and the connection is gauge equivalent to the trivial connection.

10. Open problems

To organize the discussion we will divide the set of open questions into the two cases: compact simple M and closed manifolds with Anosov geodesic flow.

Compact simple manifolds with boundary.

(1) The most important problem here is to decide if Theorem 8.2 (or Theorem 3.2) holds when dim $M \geq 3$. This will automatically extend Theorem 8.3 to any dimension.

(2) Of equal importance is the tensor tomography problem in dimension ≥ 3. In other words, does Theorem 5.1 extend to any dimension? This problem is explicitly stated in [Sharafutdinov 1994, Problem 1.1.2] and it has been solved by Pestov and Sharafutdinov [1988] for negatively curved manifolds and then by Sharafutdinov [1994] under a weaker curvature condition. It is also known that if "ghosts" exist, they must be regular: on a simple Riemannian manifold, every L^2 solenoidal tensor field belonging to the kernel of the ray transform is C^∞ smooth [Sharafutdinov et al. 2005].

(3) We have only considered unitary connections and skew-Hermitian Higgs fields, mostly because these are the most relevant in physics, but the problems addressed here make sense for any structure group. In particular, does Theorem 8.2 extend to the case of $GL(n, \mathbb{C})$?

(4) The proof of Theorem 3.2 uses in an essential way the existence of holomorphic integrating factors from Proposition 4.4 for scalar 1-forms and

carefully avoids the question of existence of holomorphic integrating factors for matrix valued 1-forms. In other words, suppose A is a $GL(n,\mathbb{C})$-connection with $n \geq 2$. Does there exist a smooth fiberwise holomorphic map $R: SM \to GL(n,\mathbb{C})$ such that $XR + AR = 0$ on SM?

(5) Are there versions of Theorems 8.2 and Theorem 8.3 when the set of geodesics of a simple surface is replaced by another set of distinguished curves? I would expect a positive answer for magnetic geodesics in view of the work in [Dairbekov et al. 2007].

Closed manifolds with Anosov geodesic flow. Here, the lack of answers is more pronounced, even for surfaces, but this is reasonable as one expects this setting to be harder. As we have seen, the appearance of ghosts (nontrivial transparent connections) has to do with the different holomorphic structures that one can have on a complex vector bundle over the surface. For simple surfaces this does not appear because there is essentially only one $\bar{\partial}_A$ operator on a disk.

(1) Perhaps one of the most important questions for surfaces is whether in Theorem 6.6 one can replace "negative curvature" by "Anosov geodesic flow". This question is of great interest even when $A = 0$.

(2) Does Theorem 6.9 extend to higher dimensions? I would expect a positive answer based on the Fourier analysis displayed in [Guillemin and Kazhdan 1980b]. There is virtually nothing known on transparent connections in dim $M \geq 3$ as the next question shows.

(3) Are there nontrivial transparent connections on $M \times \mathbb{C}^2$, where M is a closed hyperbolic 3-manifold?

(4) Classify transparent $U(n)$-connections (and pairs) over a negatively curved surface using the ideas displayed in Section 7 for SU(2).

(5) Let M be a surface with an Anosov geodesic flow and suppose there is a smooth $u: SM \to \mathbb{R}$ such that $Xu = f$, where f arises from a symmetric m-tensor. Must f be potential? (The tensor tomography problem for an Anosov surface). The proof given in this paper for simple surfaces does not extend since we do not have the analogue of holomorphic integrating factors from Proposition 4.4. The best result available for 2-tensors appears in [Sharafutdinov and Uhlmann 2000] where a positive answer is given assuming in addition that the surface is free of focal points. A solution of this problem for the case of symmetric 2-tensors would give right away infinitesimal spectral rigidity for Anosov surfaces [Guillemin and Kazhdan 1980a].

Acknowledgements

I am very grateful to Gareth Ainsworth, Will Merry, Mikko Salo and the referee for several comments and corrections on an earlier draft.

References

[Anosov 1985] D. V. Anosov, "Geodesic flows that satisfy the U-condition (**Y**)", *Trudy Mat. Inst. Steklov.* **167** (1985), 3–24, 276. In Russian; translated in *Proc. Steklov Inst. of Math.* **167** (1985) 3–24. MR 87f:58132

[Bal 2005] G. Bal, "Ray transforms in hyperbolic geometry", *J. Math. Pures Appl.* (9) **84**:10 (2005), 1362–1392. MR 2007a:53140 Zbl 1099.44002

[Boman and Strömberg 2004] J. Boman and J.-O. Strömberg, "Novikov's inversion formula for the attenuated Radon transform—a new approach", *J. Geom. Anal.* **14**:2 (2004), 185–198. MR 2005d:44002

[Croke and Sharafutdinov 1998] C. B. Croke and V. A. Sharafutdinov, "Spectral rigidity of a compact negatively curved manifold", *Topology* **37**:6 (1998), 1265–1273. MR 99e:58191 Zbl 0936.58013

[Dairbekov and Paternain 2007] N. S. Dairbekov and G. P. Paternain, "Entropy production in Gaussian thermostats", *Comm. Math. Phys.* **269**:2 (2007), 533–543. MR 2007m:37071 Zbl 1113.37013

[Dairbekov and Sharafutdinov 2003] N. S. Dairbekov and V. A. Sharafutdinov, "Some problems of integral geometry on Anosov manifolds", *Ergodic Theory Dynam. Systems* **23**:1 (2003), 59–74. MR 2005b:58053 Zbl 1140.58302

[Dairbekov et al. 2007] N. S. Dairbekov, G. P. Paternain, P. Stefanov, and G. Uhlmann, "The boundary rigidity problem in the presence of a magnetic field", *Adv. Math.* **216**:2 (2007), 535–609. MR 2008m:37107 Zbl 1131.53047

[Duistermaat 1972] J. J. Duistermaat, "On first order elliptic equations for sections of complex line bundles", *Compositio Math.* **25** (1972), 237–243. MR 51 #3718 Zbl 0243.58003

[Dunajski 2010] M. Dunajski, *Solitons, instantons, and twistors*, Oxford Graduate Texts in Mathematics **19**, Oxford University Press, Oxford, 2010. MR 2011b:53046 Zbl 1197.35209

[Eskin 2004] G. Eskin, "On non-abelian Radon transform", *Russ. J. Math. Phys.* **11**:4 (2004), 391–408. MR 2007d:43004 Zbl 1186.43009

[Finch 2003] D. V. Finch, "The attenuated x-ray transform: recent developments", pp. 47–66 in *Inside out: inverse problems and applications*, edited by G. Uhlmann, Math. Sci. Res. Inst. Publ. **47**, Cambridge Univ. Press, 2003. MR 2005a:44002 Zbl 1083.44500

[Finch and Uhlmann 2001] D. Finch and G. Uhlmann, "The x-ray transform for a non-abelian connection in two dimensions", *Inverse Problems* **17**:4 (2001), 695–701. MR 2003b:35210 Zbl 1004.53055

[Guillemin and Kazhdan 1980a] V. Guillemin and D. Kazhdan, "Some inverse spectral results for negatively curved 2-manifolds", *Topology* **19**:3 (1980), 301–312. MR 81j:58082 Zbl 0465.58027

[Guillemin and Kazhdan 1980b] V. Guillemin and D. Kazhdan, "Some inverse spectral results for negatively curved n-manifolds", pp. 153–180 in *Geometry of the Laplace operator*, edited by R. Osserman and A. Weinstein, Proc. Sympos. Pure Math. **36**, Amer. Math. Soc., Providence, R.I., 1980. MR 81i:58048 Zbl 0456.58031

[Hitchin et al. 1999] N. J. Hitchin, G. B. Segal, and R. S. Ward, *Integrable systems: Twistors, loop groups, and Riemann surfaces*, Oxford Graduate Texts in Mathematics **4**, Oxford Univ. Press, New York, 1999. MR 2000g:37003 Zbl 1082.37501

[Katok and Hasselblatt 1995] A. Katok and B. Hasselblatt, *Introduction to the modern theory of dynamical systems*, Encyclopedia of Mathematics and its Applications **54**, Cambridge University Press, 1995. MR 96c:58055 Zbl 0878.58020

[Klingenberg 1974] W. Klingenberg, "Riemannian manifolds with geodesic flow of Anosov type", *Ann. of Math.* (2) **99** (1974), 1–13. MR 51 #14149 Zbl 0272.53025

[Livšic 1971] A. N. Livšic, "Certain properties of the homology of Y-systems", *Mat. Zametki* **10** (1971), 555–564. MR 45 #2746

[Livšic 1972] A. N. Livšic, "Cohomology of dynamical systems", *Izv. Akad. Nauk SSSR Ser. Mat.* **36** (1972), 1296–1320. In Russian; translated in *Math. USSR Izv.*, **6**:6 (1972), 1278–1301. MR 48 #12606

[Mañé 1987] R. Mañé, "On a theorem of Klingenberg", pp. 319–345 in *Dynamical systems and bifurcation theory* (Rio de Janeiro, 1985), edited by M. I. Camacho et al., Pitman Res. Notes Math. Ser. **160**, Longman Sci. Tech., Harlow, 1987. MR 88k:58129 Zbl 0633.58022

[Manton and Sutcliffe 2004] N. Manton and P. Sutcliffe, *Topological solitons*, Cambridge University Press, 2004. MR 2006d:58020 Zbl 1100.37044

[Mason 2006] L. J. Mason, "Global anti-self-dual Yang–Mills fields in split signature and their scattering", *J. Reine Angew. Math.* **597** (2006), 105–133. MR 2008a:53018 Zbl 1110.53019

[Mason and Woodhouse 1996] L. J. Mason and N. M. J. Woodhouse, *Integrability, self-duality, and twistor theory*, London Mathematical Society Monographs (N.S.) **15**, Oxford Univ. Press, New York, 1996. MR 98f:58002

[McDuff and Salamon 2004] D. McDuff and D. Salamon, *J-holomorphic curves and symplectic topology*, American Mathematical Society Colloquium Publications **52**, American Mathematical Society, Providence, RI, 2004. MR 2004m:53154 Zbl 1064.53051

[Michel 1981] R. Michel, "Sur la rigidité imposée par la longueur des géodésiques", *Invent. Math.* **65**:1 (1981), 71–83. MR 83d:58021 Zbl 0471.53030

[Niţică and Török 1998] V. Niţică and A. Török, "Regularity of the transfer map for cohomologous cocycles", *Ergodic Theory Dynam. Systems* **18**:5 (1998), 1187–1209. MR 2000m:37030 Zbl 0918.58057

[Novikov 2002a] R. G. Novikov, "An inversion formula for the attenuated X-ray transformation", *Ark. Mat.* **40**:1 (2002), 145–167. MR 2003k:44004 Zbl 1036.53056

[Novikov 2002b] R. G. Novikov, "On determination of a gauge field on $\mathbb{R}^d$ from its non-abelian Radon transform along oriented straight lines", *J. Inst. Math. Jussieu* **1**:4 (2002), 559–629. MR 2004b:53132 Zbl 1072.53023

[Parry 1999] W. Parry, "The Livšic periodic point theorem for non-abelian cocycles", *Ergodic Theory Dynam. Systems* **19**:3 (1999), 687–701. MR 2000d:37019

[Paternain 1999] G. P. Paternain, *Geodesic flows*, Progress in Mathematics **180**, Birkhäuser, Boston, MA, 1999. MR 2000h:53108 Zbl 0930.53001

[Paternain 2009] G. P. Paternain, "Transparent connections over negatively curved surfaces", *J. Mod. Dyn.* **3**:2 (2009), 311–333. MR 2010d:53031 Zbl 1185.53024

[Paternain 2012] G. P. Paternain, "Transparent pairs", *J. Geom. Anal.* **22**:4 (2012), 1211–1235. MR 2965367

[Paternain et al. 2011a] G. Paternain, M. Salo, and G. Uhlmann, "The attenuated ray transform for connections and Higgs fields", preprint, 2011. To appear in *Geom. Funct. Anal.* arXiv 1108.1118

[Paternain et al. 2011b] G. Paternain, M. Salo, and G. Uhlmann, "Tensor tomography on surfaces", preprint, 2011. arXiv 1109.0505

[Pestov and Sharafutdinov 1988] L. N. Pestov and V. A. Sharafutdinov, "Integral geometry of tensor fields on a manifold of negative curvature", *Sibirsk. Mat. Zh.* **29**:3 (1988), 114–130, 221. In Russian; translated in *Siberian Math. J.* **29** (1988), 427–441. MR 89k:53066 Zbl 0675.53048

[Pestov and Uhlmann 2005] L. Pestov and G. Uhlmann, "Two dimensional compact simple Riemannian manifolds are boundary distance rigid", *Ann. of Math.* (2) **161**:2 (2005), 1093–1110. MR 2006c:53038 Zbl 1076.53044

[Salo and Uhlmann 2011] M. Salo and G. Uhlmann, "The attenuated ray transform on simple surfaces", *J. Differential Geom.* **88**:1 (2011), 161–187. MR 2819758 Zbl 1238.53058

[Sharafutdinov 1994] V. A. Sharafutdinov, *Integral geometry of tensor fields*, Inverse and Ill-posed Problems Series **1**, VSP, Utrecht, 1994. MR 97h:53077 Zbl 0883.53004

[Sharafutdinov 1999] V. A. Sharafutdinov, "Ray transform on Riemannian manifolds: eight lectures on integral geometry", preprint, 1999, Available at http://goo.gl/oFup3.

[Sharafutdinov 2000] V. A. Sharafutdinov, "On the inverse problem of determining a connection on a vector bundle", *J. Inverse Ill-Posed Probl.* **8**:1 (2000), 51–88. MR 2001h:53037 Zbl 0959.53011

[Sharafutdinov and Uhlmann 2000] V. Sharafutdinov and G. Uhlmann, "On deformation boundary rigidity and spectral rigidity of Riemannian surfaces with no focal points", *J. Differential Geom.* **56**:1 (2000), 93–110. MR 2002i:53056 Zbl 1065.53039

[Sharafutdinov et al. 2005] V. Sharafutdinov, M. Skokan, and G. Uhlmann, "Regularity of ghosts in tensor tomography", *J. Geom. Anal.* **15**:3 (2005), 499–542. MR 2006m:58034

[Uhlmann 2004] G. Uhlmann, "The Cauchy data and the scattering relation", pp. 263–287 in *Geometric methods in inverse problems and PDE control*, edited by C. B. Croke et al., IMA Vol. Math. Appl. **137**, Springer, New York, 2004. MR 2006f:58032 Zbl 1061.35175

[Vertgeĭm 1991] L. B. Vertgeĭm, "Integral geometry with a matrix weight and a nonlinear problem of the reconstruction of matrices", *Dokl. Akad. Nauk SSSR* **319**:3 (1991), 531–534. In Russian; translated in *Sov. Math.-Dokl.* **44** (1992) 132–135. MR 93a:53061

[Ward 1988] R. S. Ward, "Soliton solutions in an integrable chiral model in 2 + 1 dimensions", *J. Math. Phys.* **29**:2 (1988), 386–389. MR 89h:81126 Zbl 0644.58038

g.p.paternain@dpmms.cam.ac.uk   Department of Pure Mathematics and Mathematical Statistics, University of Cambridge, Cambridge CB3 0WB, United Kingdom

Elastic-wave inverse scattering based on reverse time migration with active and passive source reflection data

VALERIY BRYTIK, MAARTEN V. DE HOOP
AND ROBERT D. VAN DER HILST

We develop a comprehensive theory and microlocal analysis of reverse-time imaging — also referred to as reverse-time migration or RTM — for the anisotropic elastic wave equation based on the single scattering approximation. We consider a configuration representative of the seismic inverse scattering problem. In this configuration, we have an interior (point) body-force source that generates elastic waves, which scatter off discontinuities in the properties of earth's materials (anisotropic stiffness, density), and are observed at receivers on the earth's surface. The receivers detect all the components of displacement. We introduce (i) an anisotropic elastic-wave RTM inverse scattering transform, and for the case of mode conversions (ii) a microlocally equivalent formulation avoiding knowledge of the source via the introduction of so-called array receiver functions. These allow a seamless integration of passive source and active source approaches to inverse scattering.

1. Introduction

We develop a program and analysis for elastic wave-equation inverse scattering, based on the single scattering approximation, from two interrelated points of view, known in the seismic imaging literature as "receiver functions" (passive source) and "reverse-time migration" (active source).

We consider an interior (point) body-force source that generates elastic waves, which scatter off discontinuities in the properties of earth's materials (anisotropic stiffness, density), and which are observed at receivers on the earth's surface. The receivers detect all the components of displacement. We decompose the medium into a smooth background model and a singular contrast and assume the single scattering or Born approximation. The inverse scattering problem concerns the reconstruction of the contrast given a background model.

Keywords: elastic wave equation, inverse scattering, receiver function, microlocal analysis.

In this paper, we extend the original reverse-time imaging or migration (RTM) procedure for scalar waves [Whitmore 1983; McMechan 1983; Baysal et al. 1983] to elastic waves. We generalize the analysis developed in [Op 't Root et al. 2012] for inverse scattering based on RTM for scalar waves; part of this analysis contains elements of the original integral formulation of [Schneider 1978] and the inverse scattering integral equation of [Bojarski 1982]. Elastic-wave RTM has recently become a subject of considerable interest. The current developments have been mostly limited to approaches based on certain polarized qP-wave approximations [Sun and McMechan 2001; Zhang et al. 2007; Jones et al. 2007; Lu et al. 2009; Fletcher et al. 2009a; 2009b; Fowler et al. 2010]. In our framework, the RTM imaging condition is connected to a decomposition into polarizations (for an implementation of such a decomposition in quasihomogeneous media, see [Yan and Sava 2007; 2008]).

We develop a comprehensive theory and microlocal analysis of reverse-time imaging for the anisotropic elastic wave equation. We construct a transform that yields inverse scattering up to the contrast-source radiation patterns and which naturally removes the "smooth artifacts" discussed in [Yoon et al. 2004; Mulder and Plessix 2004; Fletcher et al. 2005; Xie and Wu 2006; Guitton et al. 2007]. Our work is based on results presented in [de Hoop and de Hoop 2000; Stolk and de Hoop 2002] while assuming a common-source data acquisition. The main results are: (i) the introduction of an (anisotropic) elastic-wave RTM inverse scattering transform, and (ii) the reformulation of (i) using mode-converted wave constituents removing the knowledge of the source while introducing the notion of array receiver functions, which generalize the notion of receiver functions in planarly layered media. Under the assumption of absence of source caustics (the generation of caustics between the source and scattering points), the RTM inverse scattering transform defines a Fourier integral operator the propagation of singularities of which is described by a canonical graph. The array receiver functions provide a seamless integration of passive source and active source approaches to inverse scattering.

A key application concerns the reconstruction of discontinuities in Earth's upper mantle, such as the Moho (the crust-mantle interface) and the 660 discontinuity (the discontinuity at an approximate depth of 660 km marking the lower boundary of the upper mantle transition zone). In Figure 1 we illustrate the propagation of singularities associated with certain body-wave reflections off and mode conversion at a conormal singularity (a piece of smooth interface) in the transition zone.

Over the past decades, converted seismic waves have been extensively used in global seismology to identify discontinuities in earth's crust, lithosphere-asthenosphere boundary, and mantle transition zone. The method commonly

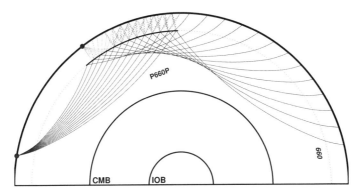

Figure 1. Propagation of singularities (body-wave phases) in Earth's mantle (for illustration purposes we use here a spherically symmetric isotropic model). The dots indicate the locations of seismic events. (We note the absence of source-wave caustics; for underside reflections a caustic is generated between scattering points and receiver networks.) A singular coefficient perturbation is indicated by a curved line segment (representing the 660 discontinuity). To image and characterize this perturbation, we use topside (*P*-wave) reflections (green ray segments), underside (*P*- or *S*-wave) reflections (blue ray segments) and (*P*-to-*S*) mode conversions (red squiggly lines). We note the possibility, with limited regions data acquisition, to globally illuminate singularities in Earth's transition zone. (CMB: core-mantle boundary; ICB: inner-core boundary.)

used has been the one of receiver functions, which were introduced and developed in [Vinnik 1977] and [Langston 1979]. In this method, essentially, the converted (scattered) *S*-wave observation is deconvolved (in time) with the corresponding incident *P*-wave observation at each available receiver, and assumes a planarly layered earth model. Various refinements have been developed for arrays of receivers. We mention binning according to common-conversion points [Dueker and Sheehan 1997] and diffraction stacking [Revenaugh 1995]. An analysis of (imaging with) receiver functions starting from plane-wave single scattering has been given in [Rydberg and Weber 2000]. (Plane-wave) Kirchhoff migration for mode-converted waves was considered in [Bostock 1999] and [Poppeliers and Pavlis 2003], while its extension to wave-form inversion was developed in [Frederiksen and Revenaugh 2004]. Receiver functions, however, being bilinear in the data (through cross correlation in time), do not fit a description directly in terms of Kirchhoff migration, being linear in the data. We resolve this issue by making precise under which limiting assumptions receiver function imaging is equivalent with (Kirchhoff-style) RTM via the synthesis of source plane waves.

The outline of the paper is as follows. In the next section, we summarize our inverse scattering procedures both for a known source and an unknown source. In Section 3 we discuss various aspects of the parametrix construction for the elastic wave equation, as well as the WKBJ approximation. In Section 4, we introduce the single scattering approximation and the notion of continued scattered field. In Section 5, we discuss and analyze reverse-time continuation from the boundary. In Section 6 we present the inverse scattering analysis, and in Section 7 we construct array receiver functions. We also discuss how receiver functions that are commonly used in global seismology can be recovered from array receiver functions in flat, planarly layered earth models using the WKBJ approximation. In Section 8 we discuss applications in global seismology and conclude with some final remarks.

2. Reverse-time migration based inverse scattering

2A. *Elastic waves.* The propagation and scattering of seismic waves is governed by the elastic wave equation, which is written in the form

$$P_{il} u_l = 0, \qquad (2\text{-}1)$$

$$u_l|_{t=0} = 0, \quad \partial_t u_l|_{t=0} = h_l, \qquad (2\text{-}2)$$

where

$$u_l = \sqrt{\rho}\,(\text{displacement})_l, \qquad (2\text{-}3)$$

and

$$P_{il} = \delta_{il} \frac{\partial^2}{\partial t^2} + A_{il}, \quad A_{il} = -\frac{\partial}{\partial x_j} \frac{c_{ijkl}(x)}{\rho(x)} \frac{\partial}{\partial x_k} + \text{l.o.t.}, \qquad (2\text{-}4)$$

where l.o.t. stands for lower-order terms. Here, $x \in \mathbb{R}^n$ and the subscripts $i, j, k, l \in \{1, \ldots, n\}$; $c_{ijkl} = c_{ijkl}(x)$ denotes the stiffness tensor and $\rho = \rho(x)$ the density of mass. The system of partial differential equations is assumed to be of principal type. It supports different wave types (modes). System (2-1) is real, time reversal invariant, and its solutions satisfy reciprocity.

We decompose the medium into a smooth background model and a singular contrast, and assume that the contrast is supported in a bounded subset X of $\mathbb{R}^n$.

Polarizations. We consider here propagation in the background model which has smoothly varying coefficients. Decoupling of the modes is then accomplished by diagonalizing the system. We describe how the system (2-1) can be decoupled by transforming it with appropriate matrix-valued pseudodifferential operators, $Q(x, D_x)_{iM}$, $D_x = -i\partial/\partial x$; see [Taylor 1975; Ivrii 1979; Dencker 1982]. Since the time derivative in P_{il} is already in diagonal form, it remains only to diagonalize its spatial part, $A_{il}(x, D_x)$. The goal becomes finding Q_{iM} and A_M such that

$$Q(x, D_x)^{-1}_{Mi} A_{il}(x, D_x) Q(x, D_x)_{lN} = \text{diag}(A_M(x, D_x); M = 1, \ldots, n)_{MN}. \tag{2-5}$$

The indices M, N denote the mode of propagation. Then

$$u_M = Q(x, D_x)^{-1}_{Mi} u_i, \quad h_M = Q(x, D_x)^{-1}_{Mi} h_i \tag{2-6}$$

satisfy the uncoupled equations

$$P_M(x, D_x, D_t) u_M = 0, \tag{2-7}$$

$$u_M|_{t=0} = 0, \quad \partial_t u_M|_{t=0} = h_M, \tag{2-8}$$

where $D_t = -i\partial/\partial t$, and in which

$$P_M(x, D_x, D_t) = \partial_t^2 + A_M(x, D_x).$$

Because of the properties of stiffness related to (i) the conservation of angular momentum, (ii) the properties of the strain-energy function, and (iii) the positivity of strain energy, subject to the adiabatic and isothermal conditions, the principal symbol $A^{\text{prin}}_{il}(x, \xi)$ of $A_{il}(x, D_x)$ is a positive, symmetric matrix. Hence, it can be diagonalized by an orthogonal matrix. On the level of principal symbols, composition of pseudodifferential operators reduces to multiplication. Therefore, we let $Q^{\text{prin}}_{iM}(x, \xi)$ be this orthogonal matrix, and we let $A^{\text{prin}}_M(x, \xi)$ be the eigenvalues of $A^{\text{prin}}_{il}(x, \xi)$, so that

$$Q^{\text{prin}}_{Mi}(x, \xi)^{-1} A^{\text{prin}}_{il}(x, \xi) Q^{\text{prin}}_{lN}(x, \xi) = \text{diag}(A^{\text{prin}}_M(x, \xi))_{MN}. \tag{2-9}$$

The principal symbol $Q^{\text{prin}}_{iM}(x, \xi)$ is the matrix that has as its columns the orthonormalized polarization vectors associated with the modes of propagation. If the $A^{\text{prin}}_M(x, \xi)$ are all different, A_{il} can be diagonalized with a unitary operator, that is, $Q(x, D_x)^{-1} = Q(x, D_x)^*$; see Appendix.

We introduce $B_M(x, D_x) = \sqrt{A_M(x, D_x)}$. Furthermore, we introduce boundary normal coordinates, $x = (x', x_n)$; that is, $x' = (x_1, \ldots, x_{n-1})$ and $x_n = 0$ defines the boundary. We also write $z = x_n$ and $\zeta = \xi_n$. We let Σ denote a bounded open subset of the boundary where the receivers are placed. In Section 3C, we also introduce operators $C_\mu(x', z, D_{x'}, D_t)$, the principal parts of the symbols, $C_\mu(x', z, \xi', \tau)$, of which are the solutions for ζ of

$$A^{\text{prin}}_M(x', z, \xi', \zeta) = \tau^2. \tag{2-10}$$

2B. *A known source.* We introduce the polarized Green's function, G_N, to be the causal solution of the equation

$$P_N(x, D_x, D_t) G_N(x, \tilde{x}, t) = \delta_{\tilde{x}}(x) \delta_0(t); \qquad (2\text{-}11)$$

G_N is identified as the source or incident field.

We let d_{MN} denote the N-to-M converted data, which, for a given source at $\tilde{x}$, are observed on $\Sigma \times (0, T)$. We introduce the reverse-time continued field, v_r, as the anticausal solution of the equation

$$[\partial_t^2 + A_M(x, D_x)] v_r(x, t) = \delta(x_n) \mathcal{N}_M(x', D_{x'}, D_t) \Psi_{\mu, \Sigma}(x', t, D_{x'}, D_t) d_{MN}(x', t; \tilde{x}). \qquad (2\text{-}12)$$

Here,

$$\mathcal{N}_M(x', D_{x'}, D_t) = -2\mathrm{i} D_t \frac{\partial B_M^{\text{prin}}}{\partial \xi_n}(x', 0, D_t^{-1} D_{x'}, C_\mu(0, x', D_t^{-1} D_{x'}, 1)), \qquad (2\text{-}13)$$

and $\Psi_{\mu, \Sigma}$ is a pseudodifferential cutoff, which removes grazing rays. We define first-order partial differential and pseudodifferential operators $\Xi(x, D_x, D_t)$ and $\Theta(x, D_x, D_t)$ with (principal) symbols

$$\Xi_0(x, \xi, \tau) = \tau, \qquad \Xi_j(x, \xi, \tau) = \xi_j$$

and

$$\Theta_0(x, \xi, \tau) = \tau, \qquad \Theta_j(x, \xi, \tau) = \tau \frac{\partial B_M^{\text{prin}}}{\partial \xi_j}(x, \xi).$$

We then define the operator, H_{MN}, as

$$(H_{MN} d_{MN})_{ijkl}(x) =$$

$$-\frac{1}{2\pi} \int \frac{2\Omega(\tau)}{\mathrm{i}\tau |\hat{G}_N(x, \tilde{x}, \tau)|^2} \sum_{p=0}^{n} \left(\frac{\partial}{\partial x_k} Q(x, D_x)_{lN} \, \Xi_p(x, D_x, \tau) \overline{\hat{G}_N(x, \tilde{x}, \tau)} \right)$$

$$\times \left(\frac{\partial}{\partial x_j} Q(x, D_x)_{iM} \, \Theta_p(x, D_x, \tau) \hat{v}_r(x, \tau) \right) d\tau \qquad (2\text{-}14)$$

for imaging the contrast in stiffness tensor in X, and similarly for the density contrast (to be indexed by a subscript $_0$) upon replacing $\partial/\partial x_j$ and $\partial/\partial x_k$ by $\mathrm{i}\tau$ and the subscript l by i. In this expression, $\hat{}$ denotes the Fourier transform in time. The Fourier multiplier $\Omega(\tau)$ is a smooth function which is zero in a neighborhood of $\tau = 0$. In Theorem 6.2 we present the inverse scattering properties of operator H_{MN}.

2C. *An unknown source and array receiver functions.* In the case of mode conversions ($M \neq N$) one may observe separately the source field; the incident field data are represented by d_N. We then introduce the reverse-time continued field, $w_{\tilde{x};r}$, as the anticausal solution of the equation

$$[\partial_t^2 + A_N(x, D_x)] w_{\tilde{x};r}(x, t)$$
$$= \delta(x_n) \mathcal{N}_N(x', D_{x'}, D_t) \Psi_{\nu,\Sigma}(x', t, D_{x'}, D_t) d_N(x', t; \tilde{x}). \quad (2\text{-}15)$$

We replace $\hat{G}_N(x, \tilde{x}, \tau)$ by $\hat{w}_{\tilde{x};r}(x, \tau)$ in (2-14). In Lemma 7.1, we obtain an operator which is bilinear in the data in as much as it acts on array receiver functions, which we define as

Definition 2.1. For $M \neq N$, the array receiver function (ARF), R_{MN}, is defined as the bilinear map, $d \to R_{MN}$, with

$$(R_{MN}(d(\cdot,\cdot;\tilde{x})))(r, t, r')$$
$$= \int (\mathcal{N}_M \Psi_{\mu,\Sigma} d_{MN})(r, t' + t; \tilde{x}) (\mathcal{N}_N \Psi_{\nu,\Sigma} d_N)(r', t'; \tilde{x}) \, dt' \quad (2\text{-}16)$$

(no sums over N, M).

3. Parametrix construction

Having assumed that P_{il} is of principal type, the multiplicities of the eigenvalues $A_M^{\text{prin}}(x, \xi)$ are constant, whence the principal symbol $Q_{iM}^{\text{prin}}(x, \xi)$ depends smoothly on (x, ξ) and microlocally Equation (2-9) carries over to an operator equation. Taylor [1975] has shown that if this condition is satisfied, then decoupling can be accomplished to all orders.

The second-order equations (2-7) inherit the symmetries of the original system, such as time-reversal invariance and reciprocity. Time-reversal invariance follows because the operators $Q_{iM}(x, D_x)$, $A_M(x, D_x)$ can be chosen in such a way that $Q_{iM}(x, \xi) = -\overline{Q_{iM}(x, -\xi)}$, $A_M(x, \xi) = \overline{A_M(x, \xi)}$. Then Q_{iM}, A_M are real-valued. Reciprocity for the causal Green's function $G_{ij}(x, x_0, t)$ means that $G_{ij}(x, x_0, t) = G_{ji}(x_0, x, t)$. Such a relationship also holds (modulo smoothing operators) for the Green's function $G_M(x, x_0, t)$ associated with (2-7).

Remark 3.1. In the isotropic case, for $n = 3$, the symbol matrix $A_{il}^{\text{prin}}(x, \xi)$ attains the form

$$\rho A_{il}^{\text{prin}}(x, \xi) = \begin{pmatrix} (\lambda+\mu)\xi_1^2 + \mu|\xi|^2 & (\lambda+\mu)\xi_1\xi_2 & (\lambda+\mu)\xi_1\xi_3 \\ (\lambda+\mu)\xi_1\xi_2 & (\lambda+\mu)\xi_2^2 + \mu|\xi|^2 & (\lambda+\mu)\xi_2\xi_3 \\ (\lambda+\mu)\xi_1\xi_3 & (\lambda+\mu)\xi_2\xi_3 & (\lambda+\mu)\xi_3^2 + \mu|\xi|^2 \end{pmatrix},$$

where $\lambda = \lambda(x)$ and $\mu = \mu(x)$ denote the Lamé parameters. We find that

$$\tilde{Q}^{\text{prin}} = \tilde{Q}^{\text{prin}}(\xi) = \begin{pmatrix} | & | & | \\ \tilde{Q}_P & \tilde{Q}_{SV} & \tilde{Q}_{SH} \\ | & | & | \end{pmatrix},$$

which is independent of x and where

$$\tilde{Q}_P = \begin{pmatrix} \xi_1 \\ \xi_2 \\ \xi_3 \end{pmatrix}, \quad \tilde{Q}_{SH} = n \times \tilde{Q}_P = \begin{pmatrix} -\xi_2 \\ \xi_1 \\ 0 \end{pmatrix}, \quad \tilde{Q}_{SV} = \tilde{Q}_P \times \tilde{Q}_{SH} = \begin{pmatrix} -\xi_1 \xi_3 \\ -\xi_2 \xi_3 \\ \xi_1^2 + \xi_2^2 \end{pmatrix},$$

with $n = (0, 0, 1)^t$, diagonalizes $A_{il}^{\text{prin}}(x, \xi)$:

$$\text{diag}(\rho A_M^{\text{prin}}(x, \xi); \ M = 1, \ldots, n) = \begin{pmatrix} (\lambda + 2\mu)|\xi|^2 & 0 & 0 \\ 0 & \mu|\xi|^2 & 0 \\ 0 & 0 & \mu|\xi|^2 \end{pmatrix}.$$

The polarizations are identified as P, SV and SH. Upon normalizing the columns of $\tilde{Q}^{\text{prin}}$, we obtain the unitary symbol matrix, Q^{prin}, with

$$(Q^{\text{prin}})^{-1} = (Q^{\text{prin}})^* = \begin{pmatrix} \dfrac{\xi_1}{|\xi|} & \dfrac{\xi_2}{|\xi|} & \dfrac{\xi_3}{|\xi|} \\ \dfrac{-\xi_1 \xi_3}{(\xi_1^2 + \xi_2^2)^{1/2}|\xi|} & \dfrac{-\xi_2 \xi_3}{(\xi_1^2 + \xi_2^2)^{1/2}|\xi|} & \dfrac{(\xi_1^2 + \xi_2^2)^{1/2}}{|\xi|} \\ \dfrac{-\xi_2}{(\xi_1^2 + \xi_2^2)^{1/2}} & \dfrac{\xi_1}{(\xi_1^2 + \xi_2^2)^{1/2}} & 0 \end{pmatrix}.$$

We note that $\tilde{Q}_{SV}$ and $\tilde{Q}_{SH}$ are zero if $\xi \parallel n$. This reflects the fact that it is not possible to construct a nonvanishing continuous tangent vector field on S^2 (the Euler characteristic of S^2 is nonvanishing).

With the projections onto P and S, it follows that

$$Q_{i1}^{\text{prin}}(Q^{\text{prin}})^*_{1j} u_j = \left(-\nabla(-\Delta^{-1}(\nabla \cdot (u_1 \ u_2 \ u_3)^T))\right)_i,$$

and

$$[Q_{i2}^{\text{prin}}(Q^{\text{prin}})^*_{2j} + Q_{i3}^{\text{prin}}(Q^{\text{prin}})^*_{3j}] u_j = \left(\nabla \times (-\Delta^{-1}(\nabla \times (u_1 \ u_2 \ u_3)^T))\right)_i,$$

in accordance with the Helmholtz decomposition of u. Here superscript T denotes transposition.

3A. *A particular oscillatory integral representation.* To evaluate the parametrix, we use the first-order system for u_M that is equivalent to (2-7),

$$\frac{\partial}{\partial t}\begin{pmatrix} u_M \\ \frac{\partial u_M}{\partial t} \end{pmatrix} = \begin{pmatrix} 0 & 1 \\ -A_M(x,D_x) & 0 \end{pmatrix}\begin{pmatrix} u_M \\ \partial u_M/\partial t \end{pmatrix}. \tag{3-1}$$

This system can be decoupled also, namely, by the matrix-valued pseudodifferential operators

$$V_M(x,D_x) = \begin{pmatrix} 1 & 1 \\ -iB_M(x,D_x) & iB_M(x,D_x) \end{pmatrix},$$

$$\Lambda_M(x,D_x) = \tfrac{1}{2}\begin{pmatrix} 1 & iB_M(x,D_x)^{-1} \\ 1 & -iB_M(x,D_x)^{-1} \end{pmatrix},$$

where $B_M(x,D_x) = \sqrt{A_M(x,D_x)}$ is a pseudodifferential operator of order 1 that exists because $A_M(x,D_x)$ is positive definite. The principal symbol of $B_M(x,D_x)$ is given by $B_M^{\text{prin}}(x,\xi) = \sqrt{A_M^{\text{prin}}(x,\xi)}$. (In the isotropic case — see Remark 3.1 — we have $B_P^{\text{prin}}(x,\xi) = \rho^{-1}(\lambda+2\mu)(x)|\xi|$ and $B_{SV}^{\text{prin}}(x,\xi) = B_{SH}^{\text{prin}}(x,\xi) = \rho^{-1}\mu(x)|\xi|$.) Then

$$u_{M,\pm} = \tfrac{1}{2}u_M \pm \tfrac{1}{2}iB_M(x,D_x)^{-1}\frac{\partial u_M}{\partial t} \tag{3-2}$$

satisfy the two first-order ("half wave") equations

$$P_{M,\pm}(x,D_x,D_t)u_{M,\pm} = 0, \tag{3-3}$$

where

$$P_{M,\pm}(x,D_x,D_t) = \partial_t \pm iB_M(x,D_x), \quad P_{M,+}P_{M,-} = P_M, \tag{3-4}$$

supplemented with the initial conditions

$$u_{M,\pm}|_{t=0} = h_{M,\pm}, \quad h_{M,\pm} = \pm\tfrac{1}{2}iB_M(x,D_x)^{-1}h_M. \tag{3-5}$$

We construct operators $S_{M,\pm}(t)$ that solve the initial value problem (3-3), (3-5): $u_{M,\pm}(y,t) = (S_{M,\pm}(t)h_{M,\pm})(y)$; then

$$u_M(y,t) = ([S_{M,+}(t) - S_{M,-}(t)]\tfrac{1}{2}iB_M^{-1}h_M)(y).$$

The operators $S_{M,\pm}(t)$ are Fourier integral operators. Their construction is well known; see for example [Duistermaat 1996, Chapter 5]. Microlocally, the solution operator associated with (3-1) can be written in the form

$$S_M(t) = V_M\begin{pmatrix} S_{M,+}(t) & 0 \\ 0 & S_{M,-}(t) \end{pmatrix}\Lambda_M;$$

in this notation, $S_{M,12}(t) = ([S_{M,+}(t) - S_{M,-}(t)]\tfrac{1}{2}iB_M^{-1}$.

For the later analysis, we introduce the operators $S_M(t,s)$ and $S_{M,\pm}(t,s)$: $S_M(t,s)$ solves the problem

$$P_M(x, D_x, D_t) S_M(\cdot, s) = 0,$$
$$S_M(\cdot, s)|_{t=s} = 0, \quad \partial_t S_M(\cdot, s)|_{t=s} = \mathrm{Id},$$

so that the solution of

$$P_M(x, D_x, D_t) u_M = f_M, \quad u_M(t < 0) = 0,$$

is given by

$$u_M(y, t) = \int_0^t P_1 S_M(t, s) \begin{pmatrix} 0 \\ f_M(\cdot, s) \end{pmatrix}(y) \, ds$$
$$= \iint G_M(y, x, t-s) f_M(x, s) \, dx \, ds,$$

where we identified the causal Green's function $G_M(y, x, t-s)$. Here, P_1 is the projection onto the first component. Likewise, $S_{M,+}(t,s)$ solves (for $t \in \mathbb{R}$) the problem

$$P_{M,+}(x, D_x, D_t) S_{M,+}(\cdot, s) = 0,$$
$$S_{M,+}(\cdot, s)|_{t=s} = \mathrm{Id},$$

so that the causal solution of

$$P_{M,+}(x, D_x, D_t) u_{M,+} = f_{M,+}$$

is given by

$$u_{M,+}(y, t) = \int_{-\infty}^t (S_{M,+}(t, s) f_{M,+}(\cdot, s))(y) \, ds$$
$$= \iint G_{M,+}(y, x, t-s) f_{M,+}(x, s) \, dx \, ds,$$

while the anticausal solution is given by

$$u_{M,+}(y, t) = -\int_t^\infty (S_{M,+}(t, s) f_{M,+}(\cdot, s))(y) \, ds$$
$$= \iint G_{M,+}(y, x, s-t) f_{M,+}(x, s) \, dx \, ds.$$

A similar construction holds with $+$ replaced by $-$.

For sufficiently small t (in the absence of conjugate points), we obtain the oscillatory integral representation

$$(S_{M,\pm}(t) h_{M,\pm})(y)$$
$$= (2\pi)^{-n} \iint a_{M,\pm}(y,t,\xi) \exp(i\phi_{M,\pm}(y,t,x,\xi)) h_{M,\pm}(x) \, dx \, d\xi, \quad (3\text{-}6)$$

where

$$\phi_{M,\pm}(y,t,x,\xi) = \alpha_{M,\pm}(y,t,\xi) - \langle \xi, x \rangle. \quad (3\text{-}7)$$

We note that $\alpha_{M,-}(y,t,\xi) = -\alpha_{M,+}(y,t,-\xi)$. Singularities are propagated along the bicharacteristics, that are determined by Hamilton's equations generated by the principal symbol $\pm B_M^{\text{prin}}(x,\xi)$

$$\frac{dy^t}{dt} = \pm \frac{\partial B_M^{\text{prin}}(y^t, \eta^t)}{\partial \eta}, \quad \frac{d\eta^t}{dt} = \mp \frac{\partial B_M^{\text{prin}}(y^t, \eta^t)}{\partial y}. \quad (3\text{-}8)$$

(In the seismological literature, one refers to "ray tracing".) We denote the solution of (3-8) with the $+$ sign and initial values (x, ξ) at $t = 0$ by

$$(y_M^t(x,\xi), \eta_M^t(x,\xi)) = \Phi_M^t(x,\xi).$$

The solution with the $-$ sign is found upon reversing the time direction and is given by $(y_M^{-t}(x,\xi), \eta_M^{-t}(x,\xi))$. Away from conjugate points, y_M^t and ξ determine η_M^t and x; we write $x = x_M^t(y,\xi)$. Then

$$\alpha_{M,+}(y,t,\xi) = \langle \xi, x_M^t(y,\xi) \rangle.$$

To highest order,

$$a_{M,+}(y,t,\xi) = \left| \frac{\partial(y_M^t)}{\partial(x)} \right|_{\xi, x = x_M^t(y,\xi)}^{-1/2}. \quad (3\text{-}9)$$

We consider the perturbations of (y_M^t, η_M^t) with respect to the initial conditions (x, ξ),

$$W_M^t(x,\xi) = \begin{pmatrix} W_{M,1}^t(x,\xi) & W_{M,2}^t(x,\xi) \\ W_{M,3}^t(x,\xi) & W_{M,4}^t(x,\xi) \end{pmatrix}$$
$$= \begin{pmatrix} \partial_x y_M^t(x,\xi) & \partial_\xi y_M^t(x,\xi) \\ \partial_x \eta_M^t(x,\xi) & \partial_\xi \eta_M^t(x,\xi) \end{pmatrix}. \quad (3\text{-}10)$$

This matrix solves the (linearized) Hamilton–Jacobi equations,

$$\frac{dW^t}{dt}(x,\xi) = \begin{pmatrix} \partial_{\eta y} B_M^{\text{prin}}(y^t, \eta^t) & \partial_{\eta \eta} B_M^{\text{prin}}(y^t, \eta^t) \\ -\partial_{yy} B_M^{\text{prin}}(y^t, \eta^t) & -\partial_{yn} B_M^{\text{prin}}(y^t, \eta^t) \end{pmatrix} W^t(x,\xi), \quad (3\text{-}11)$$

subject to initial conditions $W^{t=0} = I$. We note that away from conjugate points, the submatrix $W^t_{M,1}$ is invertible. Because

$$x^t_M = \frac{\partial \alpha_{M,+}}{\partial \xi}, \quad \eta^t_M = \frac{\partial \alpha_{M,+}}{\partial y},$$

integration of (3-11) along (y^t, η^t) yields

$$\frac{\partial^2 \alpha_{M,+}}{\partial y \partial \xi}(y^t_M(x,\xi),t,\xi) = (W^t_{M,1}(x,\xi))^{-1}, \qquad (3\text{-}12)$$

$$\frac{\partial^2 \alpha_{M,+}}{\partial \xi^2}(y^t_M(x,\xi),t,\xi) = (W^t_{M,1}(x,\xi))^{-1} W^t_{M,2}(x,\xi), \qquad (3\text{-}13)$$

$$\frac{\partial^2 \alpha_{M,+}}{\partial y^2}(y^t_M(x,\xi),t,\xi) = W^t_{M,3}(x,\xi)(W^t_{M,1}(x,\xi))^{-1}, \qquad (3\text{-}14)$$

which we evaluate at $x = x^t_M(y,\xi)$. It follows that

$$a_{M,+}(y,t,\xi) = \left| \det W^t_{M,1}|_{x=x^t_M(y,\xi),\xi} \right|^{-1/2}.$$

The amplitude of $S_{M,+}(t) \frac{1}{2} i B_M^{-1}$, then becomes

$$\tilde{a}_{M,+}(y,t,\xi) = a_{M,+}(y,t,\xi) \tfrac{1}{2} i B_M^{\text{prin}}(x^t_M(y,\xi),\xi)^{-1} \qquad (3\text{-}15)$$

to leading order. The amplitude $a_{M,-}$ follows from time reversal:

$$a_{M,-}(y,t,\xi) = \overline{a_{M,+}(y,t,-\xi)}.$$

3B. *Absence of caustics: The source field.* In the absence of caustics, we can change phase variables in the oscillatory integral representation of G_N according to

$$G_{N,+}(y,x,t) = (2\pi)^{-1} \int (2\pi)^{-n} \int a_{N,+}(y,t',\xi) \exp(i\phi_{N,+}(y,t',x,\xi)) \, d\xi$$
$$\times \exp(i\tau(t-t')) \, dt' \, d\tau$$
$$= (2\pi)^{-1} \int a'_{N,+}(y,x,\tau) \exp(i\tau(t-T_N(y,x))) \, d\tau.$$

We find the leading-order contribution to $a'_{N,+} = \mathcal{A}_{N,+}$ by applying the method of stationary phase in the variables (ξ,t'):

$$\frac{\partial \alpha_{N,+}}{\partial \xi}(y,t',\xi) = x, \qquad (3\text{-}16)$$

$$\frac{\partial \alpha_{N,+}}{\partial t'}(y,t',\xi) = \tau, \qquad (3\text{-}17)$$

at $\xi = \xi(y, x, \tau)$, $t' = t'(y, x, \tau) = T_N(y, x)$; $\xi(y, x, \tau)$ is homogeneous of degree 1 in τ, whence $\partial \xi / \partial \tau = \tau^{-1} \xi$. With the matrix product

$$\begin{pmatrix} W_{N,1}^t & 0 \\ 0 & 1 \end{pmatrix}\bigg|_{t=T_N(y,x), \xi=\xi(y,x,\tau)} \overbrace{\begin{pmatrix} \frac{\partial^2 \alpha_{N,+}}{\partial \xi^2} & \frac{\partial^2 \alpha_{N,+}}{\partial \xi \partial t} \\ \frac{\partial^2 \alpha_{N,+}}{\partial t \partial \xi} & \frac{\partial^2 \alpha_{N,+}}{\partial t^2} \end{pmatrix}}^{\Delta(y,t,\xi)}\bigg|_{t=T_N(y,x), \xi=\xi(y,x,\tau)}$$

$$= \begin{pmatrix} W_{M,2}^t & \frac{\partial y_N^t}{\partial t} \\ \frac{\partial \tau}{\partial \xi} & \frac{\partial \tau}{\partial t} \end{pmatrix}\bigg|_{t=T_N(y,x), \xi=\xi(y,x,\tau)},$$

we find that

$$|\mathcal{A}_{N,+}(y, x, \tau)| = (2\pi)^{-n} a_{N,+}(y, T_N(y, x), \xi(y, x, \tau))$$
$$\times (2\pi)^{(n+1)/2} \big|\det \Delta(y, T_N(y, x), \xi(y, x, \tau))\big|^{-1/2}$$
$$= (2\pi)^{-(n-1)/2} \bigg|\det \frac{\partial(x, \xi, t)}{\partial(y, x, \tau)}\bigg|^{1/2}. \tag{3-18}$$

Furthermore, $\phi_{N,+}(y, T_N(y, x), x, \xi(y, x, \tau)) = 0$. Thus the source field can be written in the form

$$G_N(x, \tilde{x}, t) = (2\pi)^{-1} \int a'_N(x, \tilde{x}, \tau) \exp(i\tau(t - T_N(x, \tilde{x}))) \, d\tau. \tag{3-19}$$

Here, $\tilde{x}$ is the source location and T_N is the travel time satisfying the eikonal equation

$$B_N(x, -\partial_x T_N(x, \tilde{x})) = -1; \tag{3-20}$$

to highest order, $a'_N = \mathcal{A}_N$ with

$$|\mathcal{A}_N(x, \tilde{x}, \tau)| = |\mathcal{A}_{N,\pm}(x, \tilde{x}, \tau)| \frac{1}{2|\tau|}. \tag{3-21}$$

We introduce

$$n_{\tilde{x}}(x) = \frac{\partial_x T_N(x, \tilde{x})}{|\partial_x T_N(x, \tilde{x})|},$$

and, using (3-20) and the homogeneity of B_N, we can write

$$\partial_x T_N(x, \tilde{x}) = \frac{1}{B_N(x, n_{\tilde{x}}(x))} n_{\tilde{x}}(x). \tag{3-22}$$

With a point-body force, $f_k = e_k \delta_{\tilde{x}} \delta_0$, the polarized source field is modeled by

$$\int G_N(x, \tilde{x}', t) \mathfrak{Q}_{Nk'}^{-1}(\tilde{x}', \tilde{x}) e_{k'} \, d\tilde{x}' \quad \text{(no sum over } N\text{)}.$$

To simplify the analysis, we will consider a polarized source, $f_k = \mathfrak{Q}_{kN}(\cdot, \tilde{x}) \delta_0$, where $\mathfrak{Q}$ denotes the kernel of Q. Then the source field reduces to $G_N(x, \tilde{x}, t)$.

We also denote the source field as $w_{\tilde{x}}(x,t)$, and use the time-decomposed wavefields

$$\begin{pmatrix} w_{\tilde{x};+} \\ w_{\tilde{x};-} \end{pmatrix} = \Lambda_N \begin{pmatrix} w_{\tilde{x}} \\ \partial_t w_{\tilde{x}} \end{pmatrix}.$$

We suppress the subscript N in $w_{\tilde{x}}$.

3C. Flat, smoothly layered media.
Here, we make use of results in [Woodhouse 1974; Garmany 1983; 1988; Fryer and Frazer 1984; 1987; Singh and Chapman 1988]. We introduce coordinates $x = (x', z)$ if $x_n = z$ is the (depth) coordinate normal to the surface, and write $c_{jk;il} = (c_{jk})_{il} = c_{ijkl}$. We consider the displacement, $\rho^{-1/2} u_i$, and the traction, $\sum_{k,l=1}^{n} c_{nk;il} \partial(\rho^{-1/2} u_l)/\partial x_k$, and form

$$W = \begin{pmatrix} \rho^{-1/2} u_i \\ \sum_{k,l=1}^{n} c_{nk;il} \dfrac{\partial(\rho^{-1/2} u_l)}{\partial x_k} \end{pmatrix}, \quad F = \begin{pmatrix} 0 \\ f_i \end{pmatrix}, \quad i = 1,\ldots,n. \quad (3\text{-}23)$$

The elastic wave equation—see (2-1)–(2-4)—can then be rewritten as the system of equations,

$$\frac{\partial W_a}{\partial z} = i \sum_{b=1}^{2n} C_{ab}(x', z, D_{x'}, D_t) W_b + F_a, \quad (3\text{-}24)$$

with

$C_{ab}(x', z, D_{x'}, D_t)$

$$= -i \begin{pmatrix} -\sum_{q=1}^{n-1}\sum_{j=1}^{n}(c_{nn})_{ij}^{-1} c_{nq;jl} \dfrac{\partial}{\partial x_q} & (c_{nn})_{il}^{-1} \\ -\sum_{p,q=1}^{n-1} \dfrac{\partial}{\partial x_p} b_{pq;il} \dfrac{\partial}{\partial x_q} + \rho \delta_{il} \dfrac{\partial^2}{\partial t^2} & -\sum_{p=1}^{n-1} \dfrac{\partial}{\partial x_p} c_{pn;ij}(c_{nn})_{jl}^{-1} \end{pmatrix}_{ab},$$

$$i,l = 1,\ldots,n, \quad (3\text{-}25)$$

where $b_{pq;il} = c_{pq;il} - \sum_{j,k=1}^{n} c_{pn;ij}(c_{nn})_{jk}^{-1} c_{nq;kl}$. Diagonalizing the system, microlocally, involves

$$C_{ab}(x', z, D_{x'}, D_t) = \sum_{\mu,\nu=1}^{2n} L(x', z, D_{x'}, D_t)_{a\mu}$$

$$\times \operatorname{diag}(C_\mu(x', z, D_{x'}, D_t); \mu = 1, \ldots, 2n)_{\mu\nu} L(x', z, D_{x'}, D_t)_{\nu b}^{-1}; \quad (3\text{-}26)$$

the principal parts of the symbols $C_\mu(x', z, \xi', \tau)$ are the solutions for ζ of (2-10).

In smoothly layered media one can Fourier transform (3-24) with respect to x' and t and obtain a system of ordinary differential equations for

$$\tilde{W}(z) = \tilde{W}(\xi', z, \tau) = \int W(x', z, t) \exp\left(-i\left(\sum_{j=1}^{n-1} \xi_j x_j + \tau t\right)\right) dx' dt,$$

namely

$$\frac{\partial \tilde{W}_a}{\partial z} = i \sum_{b=1}^{2n} C_{ab}(z, \xi', \tau) \tilde{W}_b + \tilde{F}_a. \tag{3-27}$$

We choose the C_μ such that the homogeneity property

$$C_\mu(z, \xi', \tau) = \tau C_\mu(z, \tau^{-1}\xi', 1)$$

extends to $\tau < 0$. We have

$$L_{a\mu}(z, \xi', \tau)$$

$$= \begin{pmatrix} Q_{iM(\mu)}(z, (\xi', C_\mu(z, \xi', \tau))) \\ \sum_{k,l=1}^{n} c_{nk;il}(-i)(\xi', C_\mu(z, \xi', \tau))_k Q_{lM(\mu)}(z, (\xi', C_\mu(z, \xi', \tau))) \end{pmatrix}_{a\mu}, \tag{3-28}$$

with inverse

$$L^{-1}(z, \xi', \tau) = N(z, \xi', \tau) L^t(z, \xi', \tau) J, \quad \text{where } J = \begin{pmatrix} 0 & I_n \\ I_n & 0 \end{pmatrix}. \tag{3-29}$$

Here, $N(z, \xi', \tau)$ is a diagonal normalization matrix, $\mathrm{diag}(N_\mu(z, \xi', \tau))_{\mu\nu}$. It follows that

$$N_\mu(z, \xi', \tau)^{-1} = \sum_{i=1}^{n} Q_{iM(\mu)}(z, (\xi', C_\mu(z, \xi', \tau)))$$

$$\times \sum_{k,l=1}^{n} (c_{nk;il} + c_{nk;li})(-i)(\xi', C_\mu(z, \xi', \tau))_k Q_{lM(\mu)}(z, (\xi', C_\mu(z, \xi', \tau))). \tag{3-30}$$

The index mapping $\mu \to M(\mu)$ assigns the appropriate mode to the depth component of the wave vector.

We cast (3-27) into an equivalent initial value problem. Let $\tilde{W}_{ab}(z, z_0)$ be the solution to

$$\frac{\partial \tilde{W}_a}{\partial z} = i \sum_{b=1}^{2n} C_{ab}(z, \xi', \tau) \tilde{W}_b, \quad \tilde{W}(z_0) = I_{2n}.$$

Then $\tilde{W}_a(z) = \int_{z_0}^z \sum_{b=1}^{2n} \tilde{W}_{ab}(z,z_0) \tilde{F}_b(z_0)\, dz_0$ solves (3-27). We introduce

$$\dot{W} = \begin{pmatrix} \rho^{-1/2} u_i \\ \sum_{k,l=1}^n c_{nk;il} D_t^{-1} \dfrac{\partial(\rho^{-1/2} u_l)}{\partial x_k} \end{pmatrix}, \quad \dot{F} = \begin{pmatrix} 0 \\ D_t^{-1} f_i \end{pmatrix}; \quad (3\text{-}31)$$

with $\xi' = \tau p'$, we make the identification $\tilde{W}(p', z, \tau) = \dot{W}(\tau p', z, \tau)$, whereas

$$\frac{\partial \tilde{W}_a}{\partial z} = i\tau \sum_{b=1}^{2n} C_{ab}(z, p', 1)\tilde{W}_b + \tilde{F}_b.$$

In the WKBJ approximation, in the absence of turning rays (the characteristics are nowhere horizontal), we have

$$\tilde{W}_{ab}(z, z_0)$$

$$\approx \sum_{\mu=1}^{2n} L_{a\mu}(z, p', 1) Y_\mu(z, p', 1) \exp\left(i\tau \int_{z_0}^z C_\mu(\bar{z}, p', 1)\, d\bar{z}\right)$$

$$\times Y_\mu(z_0, p', 1)^{-1} L^{-1}_{\mu b}(z_0, p', 1)$$

$$= \sum_{\mu=1}^{2n} L_{a\mu}(z, p', 1) Y_\mu(z, p', 1) \exp\left(i\tau \int_{z_0}^z C_\mu(\bar{z}, p', 1)\, d\bar{z}\right)$$

$$\times Y_\mu(z_0, p', 1)(L^t(z_0, p', 1)J)_{\mu b}.$$

Here, $Y_\mu(z, p', 1) = [N_\mu(z, p', 1)]^{1/2}$. We identify the "vertical" travel time

$$\tau_\mu(z, z_0, p') = -\int_{z_0}^z C_\mu(\bar{z}, p', 1)\, d\bar{z}. \quad (3\text{-}32)$$

To obtain the tensor G_{ij}, we substitute a δ source for f_i, yielding $J\tilde{F} = \begin{pmatrix} I_n \\ 0 \end{pmatrix} \delta(\cdot - z_0)$:

$$G_{ij}(x', z, x'_0, z_0, t - t_0)$$

$$\approx \sum_{\mu=1}^{2n}{}' \frac{1}{(2\pi)^n} \iint Q_{iM(\mu)}(z, (p', C_\mu(z, p', 1))) Y_\mu(z, p', 1)$$

$$\times \exp\left(i\tau\left(-\tau_\mu(z, z_0, p') + \sum_{l=1}^{n-1} p'_l(x' - x'_0)_l + t - t_0\right)\right)$$

$$\times Y_\mu(z_0, p', 1) Q^t_{M(\mu)j}(z_0, (p', C_\mu(z, p', 1)))\, dp' |\tau|^{n-1}\, d\tau; \quad (3\text{-}33)$$

which values of μ contribute depends on whether $z > z_0$ ("downgoing") or $z < z_0$ ("upgoing"). The (negative) values of the components of p' associated with the ray connecting (z_0, x'_0) with (z, x') is the solution of the equation

$$\partial_{p'} \tau_\mu(z, z_0, p') = x' - x'_0.$$

4. Continued scattered field

Here, we introduce and analyze the scattered field. To this end, we consider the contrast formulation, in which the total value of the medium parameters ρ, c_{ijkl} is written as the sum of a smooth background component $\rho(x), c_{ijkl}(x)$ and a singular perturbation $\delta\rho(x), \delta c_{ijkl}(x)$, namely $\rho + \delta\rho, c_{ijkl} + \delta c_{ijkl}$; we assume that $\delta\rho, \delta c_{ijkl} \in \mathcal{E}'(X)$ with X a compact subset of $\mathbb{R}^n$. This decomposition induces a perturbation of P_{il} (cf. (2-4)),

$$\delta P_{il} = \delta_{il} \frac{\delta\rho(x)}{\rho(x)} \frac{\partial^2}{\partial t^2} - \frac{\partial}{\partial x_j} \frac{\delta c_{ijkl}(x)}{\rho(x)} \frac{\partial}{\partial x_k}.$$

The first-order perturbation, δG_{il}, of the (causal) kernel G_{il} of the solution operator admits the representation

$$\delta G_{jk}(y, \tilde{x}, t)$$
$$= -\int_0^t \int_X G_{ji}(y, x, t-t') \delta P_{il}(x, D_x, D_{t'}) G_{lk}(x, \tilde{x}, t') \, dx \, dt', \quad (4\text{-}1)$$

which is the Born approximation. Here, $\tilde{x}$ denotes a source location as before, and x a scattering point location. We restrict our time window (of observation) to $(0, T)$ for some $0 < T < \infty$.

We introduce the MN contribution, δG_{MN}, to δG_{jk} as follows:

$$\int \delta G_{jk}(y, \tilde{x}, t - \tilde{t}) f_k(\tilde{x}, \tilde{t}) \, d\tilde{x} \, d\tilde{t}$$
$$= Q(y, D_y)_{jM} \int \delta G_{MN}(y, \tilde{x}, t - \tilde{t}) (Q(\tilde{x}, D_{\tilde{x}})_{Nk}^{-1} f_k)(\tilde{x}, \tilde{t}) \, d\tilde{x} \, d\tilde{t}. \quad (4\text{-}2)$$

We apply reciprocity in (y, x) to the integrand of the right-hand side and obtain

$$\delta G_{MN}(y, \tilde{x}, t)$$
$$= -\int_0^t \int_X (Q(x, D_x)^{-1})_{iM}^* G_M(x, y, t-t')$$
$$\times \delta P_{il}(x, D_x, D_{t'}) Q(x, D_x)_{lN} G_N(x, \tilde{x}, t') \, dx \, dt'$$
$$= -\int_0^t \int_X (Q(x, D_x)^{-1})_{iM}^* G_M(x, y, t-t') \frac{\partial}{\partial(t', x_j)} \left(\delta_{il} \frac{\delta\rho(x)}{\rho(x)}, -\frac{\delta c_{ijkl}(x)}{\rho(x)} \right)$$
$$\times \frac{\partial}{\partial(t', x_k)} Q(x, D_x)_{lN} G_N(x, \tilde{x}, t') \, dx \, dt'$$

$$= \int_0^t \int_X \frac{\partial}{\partial(t, x_j)} (Q(x, D_x)^{-1})^*_{iM} G_M(x, y, t - t') \left(\delta_{il} \frac{\delta \rho(x)}{\rho(x)}, -\frac{\delta c_{ijkl}(x)}{\rho(x)} \right)$$

$$\times \frac{\partial}{\partial(t', x_k)} Q(x, D_x)_{lN} G_N(x, \tilde{x}, t') \, dx \, dt'$$

$$= \int_X \left(\int_0^t \frac{\partial}{\partial(t', x_j)} Q(x, D_x)_{iM} G_M(x, y, t - t') \right.$$

$$\left. \times \frac{\partial}{\partial(t', x_k)} Q(x, D_x)_{lN} G_N(x, \tilde{x}, t') \, dt' \right)$$

$$\times \left(\delta_{il} \frac{\delta \rho(x)}{\rho(x)}, -\frac{\delta c_{ijkl}(x)}{\rho(x)} \right) \, dx \quad (4\text{-}3)$$

upon integration by parts. Reciprocity implies that

$$G_M(x, y, t - t') = G_M(y, x, t - t') \quad \text{and} \quad G_N(x, \tilde{x}, t') = G_N(\tilde{x}, x, t').$$

Also, $\delta G_{MN}(x, \tilde{x}, t)$ is the solution to the initial value problem

$$P_M(x, D_x, D_t)v = Q(x, D_x)^{-1}_{Mi} \delta P_{il}(x, D_x, D_t) Q(x, D_x)_{lN} G_N(x, \tilde{x}, t), \quad (4\text{-}4)$$

$$v|_{t=0} = 0, \quad \partial_t v|_{t=0} = 0. \quad (4\text{-}5)$$

The continued scattered field, v_h, is defined as the solution to a final value problem such that the Cauchy data at $t = T_1$ coincide with the Cauchy data of the scattered field:

$$P_M(x, D_x, D_t)v_h = 0 \quad (4\text{-}6)$$

$$v_h|_{t=T_1} = v|_{t=T_1}, \quad \partial_t v_h|_{t=T_1} = \partial_t v|_{t=T_1}. \quad (4\text{-}7)$$

We assume that the contributions from the scattered field entirely come to pass within the time interval $[T_0, T_1]$; $T_1 < T$. Then, for $t \geq T_1$, $v_h = v$, but these fields differ from one another for $t < T_1$. The corresponding the time-decomposed wavefields are given by

$$\begin{pmatrix} v_{h,+} \\ v_{h,-} \end{pmatrix} = \Lambda_M \begin{pmatrix} v_h \\ \partial_t v_h \end{pmatrix}.$$

We suppress the subscripts M, N in v and v_h.

The single scattering operator, $F(t)$, is defined by the map

$$\left(\frac{\delta \rho}{\rho}, -\frac{\delta c_{ijkl}}{\rho} \right) \mapsto \begin{pmatrix} v_h \\ \partial_t v_h \end{pmatrix}.$$

We decompose $F(t)$ into operators $F_\pm(t)$ mapping the pair on the left to $v_{h,\pm}(\cdot, t)$. We carry out the analysis for a small time interval in the neighborhood of a point in the scattering region, X. Let $\{\chi_i\}_{i \in \mathcal{I}}$ be a finite partition of unity.

The time interval $[t_{0\iota}, t_{1\iota}]$ satisfies $T_N(\mathrm{supp}(\chi_\iota), \tilde{x}) \subset [t_{0\iota}, t_{1\iota}]$. Then

$$F_+(t) = \sum_{\iota \in \mathcal{J}} S_{M,+}(t - t_{1\iota}) F_+(t_{1\iota}) \chi_\iota,$$

and similarly for $F_-(t)$. We construct an oscillatory integral representation for the kernel of $F_+(t_{1\iota})\chi_\iota$ using the representations developed in Section 3, which is enabled by the partition of unity. We omit the subscript ι below.

From the source field we get an amplitude contribution

$$\mathcal{A}_N(x, \tilde{x}, \tau) Q_{lN}^{\mathrm{prin}}(x, -\tau \partial_x T_N(x, \tilde{x})) \, i\tau \, (1, -\partial_{x_k} T_N(x, \tilde{x}))$$

to highest order, and from the solution operator we get an amplitude contribution

$$\tilde{a}_{M,+}(y, t_1 - T_N(x, \tilde{x}), \xi) \, Q_{iM}^{\mathrm{prin}}(x, -\xi) \, i(\tau, -\xi_j)$$

to highest order; here

$$\tau = \partial_t \alpha_{M,+}(y, t_1 - T_N(x, \tilde{x}), \xi). \qquad (4\text{-}8)$$

We introduce the radiation patterns $(w_{MN;0}, w_{MN;ijkl})$ as

$$w_{MN;0}(y, t_1, x, \xi) = -Q_{iM}^{\mathrm{prin}}(x, -\xi) Q_{lN}^{\mathrm{prin}}(x, -\tau \partial_x T_N(x, \tilde{x})) \tau^2, \qquad (4\text{-}9)$$

$$w_{MN;ijkl}(y, t_1, x, \xi) = Q_{iM}^{\mathrm{prin}}(x, -\xi) Q_{lN}^{\mathrm{prin}}(x, -\tau \partial_x T_N(x, \tilde{x})) \xi_j \tau \partial_{x_k} T_N(x, \tilde{x}), \qquad (4\text{-}10)$$

again subject to the substitution (4-8).

Then

$$v_{h,+}(y, t_1) =$$

$$(2\pi)^{-n} \iint_X A_{F,MN}(y, t_1, x, \xi)$$

$$\times \left(w_{MN;0}(y, t_1, x, \xi) \frac{\delta \rho(x)}{\rho(x)} + w_{MN;ijkl}(y, t_1, x, \xi) \frac{\delta c_{ijkl}(x)}{\rho(x)} \right) \chi(x)$$

$$\times \exp(i\varphi_{MN}(y, t_1, x, \xi)) \, \mathrm{d}x \, \mathrm{d}\xi \qquad (4\text{-}11)$$

modulo lower-order terms in amplitude, where

$$A_{F,MN}(y, t_1, x, \xi)$$
$$= \tilde{a}_{M,+}(y, t_1 - T_N(x, \tilde{x}), \xi) \mathcal{A}_N(x, \tilde{x}, \partial_t \alpha_{M,+}(y, t_1 - T_N(x, \tilde{x}), \xi)), \qquad (4\text{-}12)$$

and

$$\varphi_{MN}(y, t_1, x, \xi) = \alpha_{M,+}(y, t_1 - T_N(x, \tilde{x}), \xi) - \langle \xi, x \rangle. \qquad (4\text{-}13)$$

We obtain a similar representation for $v_{h,-}$: $v_{h,-}(y, t_1) = \overline{v_{h,+}(y, t_1)}$. In the above $y \in X$.

$F_+(t_1)\chi$ is a Fourier integral operator if direct source waves are excluded. Lemma 4.1 below implies that the phase function, φ_{MN}, is nondegenerate. The canonical relation, $\Lambda_{MN}^{F_+}$, of $F_+(t_1)\chi$ is obtained as follows. The stationary point set associated with φ_{MN} contains (y, x, ξ) satisfying

$$\partial_\xi \alpha_{M,+}(y, t_1 - T_N(x, \tilde{x}), \xi) = x, \quad x \in \operatorname{supp} \chi; \qquad (4\text{-}14)$$

$F_+(t_1)\chi$ propagates singularities from (x, ζ) with

$$\zeta = \xi + \partial_t \alpha_{M,+}(y, t_1 - T_N(x, \tilde{x}), \xi) \partial_x T_N(x, \tilde{x}), \qquad (4\text{-}15)$$

to (y, t_1, η, τ) with

$$\eta = \partial_y \alpha_{M,+}(y, t_1 - T_N(x, \tilde{x}), \xi). \qquad (4\text{-}16)$$

We note that

$$\partial_t \alpha_{M,+}(y, t_1 - T_N(x, \tilde{x}), \xi) = -B_M^{\mathrm{prin}}(x, \xi).$$

Thus we can write

$$\partial_t \alpha_{M,+}(y, t_1 - T_N(x, \tilde{x}), \xi) \partial_x T_N(x, \tilde{x}) = -\frac{B_M^{\mathrm{prin}}(x, \xi)}{B_N^{\mathrm{prin}}(x, n_{\tilde{x}}(x))} n_{\tilde{x}}(x), \qquad (4\text{-}17)$$

cf. (3-22); hence,

$$\zeta = \xi - \frac{B_M^{\mathrm{prin}}(x, \xi)}{B_N^{\mathrm{prin}}(x, n_{\tilde{x}}(x))} n_{\tilde{x}}(x). \qquad (4\text{-}18)$$

For $F_-(t_1)\chi$ we get the relationship $\zeta = \xi + \dfrac{B_M^{\mathrm{prin}}(x, \xi)}{B_N^{\mathrm{prin}}(x, n_{\tilde{x}}(x))} n_{\tilde{x}}(x)$. Then

$$\Lambda_{MN}^{F_+} = \Big\{(y, t, \eta, \tau; x, \zeta) \,\Big|\, (y, \eta) \in (T^* X \setminus 0) \setminus V_{\tilde{x},t},\ t \in \mathbb{R},\ \tau = -B_M^{\mathrm{prin}}(y, \eta),$$
$$(x, \xi) = \Phi_M^{T_N(x,\tilde{x})-t}(y, \eta),\ \zeta = \xi - \frac{B_M^{\mathrm{prin}}(x, \xi)}{B_N^{\mathrm{prin}}(x, n_{\tilde{x}}(x))} n_{\tilde{x}}(x),\ x \in X \Big\}. \qquad (4\text{-}19)$$

Here, we replaced t_1 by t using that this canonical relation naturally extends to the canonical relation of $F_+(t)$ through Φ_M. In the above, $V_{\tilde{x},t}$ signifies the (conic neighborhood of a) set on which φ_{MN} is not nondegenerate.

Lemma 4.1. *The phase function φ_{MN} is nondegenerate if*

$$\partial_x T_N(x, \tilde{x}) \cdot \partial_\xi B_M^{\mathrm{prin}}(x, \xi) \neq 1.$$

Proof. Because $\partial_\xi \partial_x \varphi_{MN} = \partial_\xi \zeta$ on the stationary point set of φ_{MN}, we need to establish whether the Jacobian, $|\partial_\xi \zeta|$, is singular. Using (4-18) we find that

$$|\partial_\xi \zeta| = \left|\det\left(I - \partial_\xi B_M^{\text{prin}}(x,\xi) \otimes \frac{1}{B_N^{\text{prin}}(x, n_{\tilde{x}}(x))} n_{\tilde{x}}(x)\right)\right|. \quad (4\text{-}20)$$

Hence,

$$|\partial_\xi \zeta| = \left|1 - n_{\tilde{x}}(x) \cdot \frac{1}{B_N^{\text{prin}}(x, n_{\tilde{x}}(x))} \partial_\xi B_M^{\text{prin}}(x,\xi)\right|$$

$$= \left|1 - \partial_x T_N(x, \tilde{x}) \cdot \partial_\xi B_M^{\text{prin}}(x, \xi)\right|, \quad (4\text{-}21)$$

from which the statement follows. $\square$

Hence, for $F_+(t)$ to be a Fourier integral operator, we need to invoke the assumption which excludes scattering such that $|\partial_\xi \zeta|$ is singular. The homogeneity of B_M^{prin} implies that $\xi \cdot \partial_\xi B_M^{\text{prin}}(x, \xi) = B_M^{\text{prin}}(x, \xi)$, from which it is clear that $|\partial_\xi \zeta|$ is singular if

$$\frac{1}{B_M^{\text{prin}}(x, \xi)} \xi = \partial_x T_N(x, \tilde{x}).$$

If $N \neq M$, the assumption is generically satisfied; if $N = M$, this excludes scattering over π, hence the reference to this assumption as the *absence of direct source waves*. In this case, $V_{\tilde{x},t}$ is a conic neighborhood of $\Xi_{\tilde{x},t}$. We introduce a t-family of pseudodifferential cutoffs, $\pi_+(t) = \pi_+(t)(y, D_y)$. For some t_c, the symbol of $\pi_+(t_c)$ vanishes on a conic neighborhood of $\Xi_{\tilde{x},t_c}$; we then set $\pi_+(t) = S_{M,+}(t - t_c)\pi_+(t_c)S_{M,+}(t_c - t)$. It follows that $\pi_+(t) F_+(t)$ is a Fourier integral operator with canonical relation given by (4-19). A similar analysis can be carried out for $F_-(t)$.

In the further analysis we will focus on the conversion where N corresponds with qP and M corresponds with qSV, in particular with a view to developing array receiver functions.

5. Reverse-time continuation from the boundary

We consider solutions to the homogeneous polarized wave equation,

$$P_M(x, D_x, D_t) w = 0.$$

We use boundary normal coordinates. We denote the restriction of w to Σ by $R_\Sigma w$, where Σ is a bounded open subset of the boundary as before. We let w_r be an anticausal solution to

$$[\partial_t^2 + A_M(x, D_x)] w_r(x, t)$$
$$= \delta(x_n) \mathcal{N}_M(x', D_{x'}, D_t) \Psi_{\mu,\Sigma}(x', t, D_{x'}, D_t)(R_\Sigma w)(x', t), \quad (5\text{-}1)$$

where $\mathcal{N}_M(x', D_{x'}, D_t)$ was defined in (2-13), and $\Psi_{\mu,\Sigma}$ is a pseudodifferential cutoff, which removes grazing rays; that is, its symbol vanishes where

$$\frac{\partial B_M^{\text{prin}}}{\partial \xi_n}(x', 0, \tau^{-1}\xi', C_\mu(0, x', \tau^{-1}\xi', 1)) = 0.$$

In this formulation, elements in the wavefront set satisfying

$$C_\mu^{\text{prin}}(0, x', \tau^{-1}\xi', 1) = 0$$

need to be removed as well; moreover, the cutoff is designed to remove direct source waves. An alternative representation of the (principal) symbol of $\mathcal{N}_M$ is obtained using the identity

$$2\mathrm{i} B_M^{\text{prin}}(x, \xi) \frac{\partial B_M^{\text{prin}}}{\partial \xi_n}(x, \xi) = c_{nk;il}(x)\,\mathrm{i}\xi_k\, Q_{iM}^{\text{prin}}(x, \xi) Q_{lM}^{\text{prin}}(x, \xi) \quad (5\text{-}2)$$

(no sums over M) which appears in the relevant representation theorems.

We assume that X is contained in $\{x_n > 0\}$ and let $X_t = X \times \{t\} \subset \mathbb{R}_x^n \times \mathbb{R}_t$. We revisit the bicharacteristic flow

$$(x, \xi) \to ((y_M^t)'(x, \xi), t, (\eta_M^t)'(x, \xi), -B_M^{\text{prin}}(x, \xi))$$

from $T^* X_0 \setminus 0 \to T^* \Sigma \setminus 0$ (cf. (3-8)), and introduce pseudodifferential illumination operators with principal symbols $\Psi_{X_s, +}$ defined by

$$\Psi_{X_s, +}(x, \xi) = \Psi_{\mu, \Sigma}\big((y_M^{t-s})'(x, \xi), t, (\eta_M^{t-s})'(x, \xi), -B_M^{\text{prin}}(x, \xi)\big)$$

if there exists t such that $y_M^{t-s}(x, \xi) \in \Sigma$, and $\Psi_{X_s, +}(x, \xi) = 0$ otherwise. Similarly, $\Psi_{X_0, -}$ is obtained by using

$$(x, \xi) \mapsto \big((y_M^{-t})'(x, \xi), -t, (\eta_M^{-t})'(x, \xi), B_M^{\text{prin}}(x, \xi)\big).$$

We assume that bicharacteristics which illuminate X intersect Σ only once, with $\mathrm{d}(y_M^t)_n/\mathrm{d}t < 0$.

Theorem 5.1. *The reverse-time continued field and the original field are related as*

$$\chi_n\, w_{r,+}(\,\cdot\,, t) = \chi_n\, [\Pi_+(t) w_+(\,\cdot\,, t) + R_{+-}(t) w_-(t)], \quad (5\text{-}3)$$

$$\chi_n\, w_{r,-}(\,\cdot\,, t) = \chi_n\, [\Pi_-(t) w_-(\,\cdot\,, t) + R_{-+}(t) w_+(t)], \quad (5\text{-}4)$$

where the $\Pi_\pm(t)$ are pseudodifferential operators of order zero with principal symbols

$$\Pi_+(t)(x, \xi) = \Psi_{X_t, +}(x, \xi), \quad (5\text{-}5)$$

$$\Pi_-(t)(x, \xi) = \Psi_{X_t, -}(x, \xi), \quad (5\text{-}6)$$

$R_{+-}(t)$ and $R_{-+}(t)$ are regularizing operators, and χ_n is a smooth cutoff supported in $x_n > 0$.

This theorem applies to the continued scattered field, $w_{\pm} = v_{h,\pm}$; then we write $w_{r,\pm} = v_{r,\pm}$. It also applies to the source field, $w_{\pm} = w_{\tilde{x};\pm}$ (replacing M by N in the above); then we write $w_{r,\pm} = w_{\tilde{x};r,\pm}$.

Proof. The anticausal solution of

$$[\partial_t + iB_M(x, D_x)] w_+ = f_+$$

is given by

$$S_{M,+}(t, \cdot) f_+ = -\int_t^{\infty} S_{M,+}(t,s) f_+(\cdot, s) \, ds.$$

The restriction operator, R_{Σ}, gives $R_{\Sigma} w(y', t) = w(y', 0, t)$ for $(y', t) \in \Sigma$, while

$$R_{\Sigma}^* g(y, t) = \delta(y_n) g(y', t),$$

for functions g defined on $\Sigma \times \mathbb{R}_t$. We use the notation

$$g_{M,\Sigma}(y', t) = \mathcal{N}_M \Psi_{\mu, \Sigma} (R_{\Sigma} w)(y', t).$$

For a given time, $t = t_c$, we study the maps $(w_+(\cdot, t_c), w_-(\cdot, t_c)) \mapsto g_{M,\Sigma}$, using that $w(\cdot, t) = S_{M,+}(t, t_c) w_+(\cdot, t_c) + S_{M,-}(t, t_c) w_-(\cdot, t_c)$ microlocally, and $g_{M,\Sigma} \mapsto \chi_n R_{t_c} w_{r,\pm}$, where R_{t_c} is the restriction to $t = t_c$, and their composition. For simplicity of notation we set $t_c = 0$. We proceed with the assumption that $\Psi_{\mu,\Sigma}(y', t, \eta', \tau)$ is supported in $t \in [0, t_1]$ with t_1 such that we can use the particular oscillatory integral representation (3-6)–(3-7) for the kernel of the parametrix

The solution operator $S_{M,+}(t, 0)$ has canonical relation

$$\{(y_M^t(x, \xi), t, \eta_M^t(x, \xi), -B_M(x, \xi); x, \xi)\};$$

the restriction operator R_{Σ} has canonical relation

$$\{(y', t, \eta', \tau; y', 0, t, \eta', \eta_n, \tau)\}.$$

The composition of these canonical relations is transversal because grazing rays have been removed. Hence, the operator $\mathcal{N}_M \Psi_{\mu,\Sigma} R_{\Sigma} S_{M,+}(\cdot, 0)$ is a Fourier integral operator. Its canonical relation is a subset (determined by $\Psi_{\mu,\Sigma}$) of

$$\{((y_M^t)'(x, \xi), t, (\eta_M^t)'(x, \xi), -B_M(x, \xi); x, \xi) \mid y_M^t(x, \xi) \in \Sigma\}$$

and is the graph of an invertible transformation. The kernel of this Fourier integral operator admits an oscillatory integral representation with amplitude

$$a^{(\text{fwd})}(y',t,x,\xi) = -2i\tau \frac{\partial B_M^{\text{prin}}}{\partial \eta_n}(y',0,\tau^{-1}\eta',C_\mu(0,y',\tau^{-1}\eta',1))$$

$$\times \Psi_{\mu,\Sigma}(y',t,\eta',\tau) \left| \frac{\partial(y_M^t)}{\partial(x)} \right|_{\xi,x=x_M^t(y',0,\xi)}^{-1/2} \quad (5\text{-}7)$$

mod S^0, subject to the substitutions

$$\eta' = \partial_{y'}\alpha_{M,+}(y',0,t,\xi),$$
$$\tau = \partial_t \alpha_{M,+}(y',0,t,\xi) = -B_M(x,\xi). \quad (5\text{-}8)$$

Then

$$g_{M,\Sigma,+}(y',t) = (2\pi)^{-n} \iint_X a^{(\text{fwd})}(y',t,x,\xi)$$
$$\times \exp(i(\alpha_{M,+}(y',0,t,\xi) - \langle \xi, x \rangle))w_+(x,0)\,dx\,d\xi. \quad (5\text{-}9)$$

We introduce a pseudodifferential cutoff,

$$\tilde{\Psi}_{\mu,\Sigma} = \tilde{\Psi}_{\mu,\Sigma}(y',t,D_{y'},D_t),$$

which removes grazing rays, such that

$$\tilde{\Psi}_{\mu,\Sigma}\Psi_{\mu,\Sigma} = \Psi_{\mu,\Sigma}.$$

Using the decoupling procedure, $\Lambda_M \begin{pmatrix} 0 \\ g_{M,\Sigma} \end{pmatrix}$, we find that

$$\chi_n R_0 w_{r,+} = \chi_n S_{M,+}(0,\cdot) \tfrac{1}{2} i B_M^{-1} R_\Sigma^* \tilde{\Psi}_{\mu,\Sigma} g_{M,\Sigma}.$$

The operator $\chi_n S_{M,+}(0,\cdot)\tfrac{1}{2}iB_M^{-1}R_\Sigma^*\tilde{\Psi}_{\mu,\Sigma}$ is a Fourier integral operator, the canonical relation of which is a subset of

$$\{(z,\zeta;(y_M^t)'(z,\zeta),t,(\eta_M^t)'(z,\zeta),-B_M(z,\zeta)) \mid (y_M^t)_n(z,\zeta) = 0\}.$$

The kernel of this Fourier integral operator admits an oscillatory integral representation with amplitude

$$a^{(\text{bkd})}(y',t,z,\zeta)$$

$$= \chi_n(z_n) \left| \frac{\partial(y_M^t)}{\partial(x)} \right|_{\zeta,x=x_M^t(y',0,\zeta)}^{-1/2} \tfrac{1}{2}i\tau^{-1}\tilde{\Psi}_{\mu,\Sigma}(y',t,\eta',\tau) \quad (5\text{-}10)$$

ELASTIC-WAVE INVERSE SCATTERING BASED ON REVERSE TIME MIGRATION 435

mod S^{-2}, subject to the substitutions (5-8). Then

$$\chi_n R_0 w_{r,+}(z) = (2\pi)^{-n} \iiint a^{(\text{bkd})}(y',t,z,\zeta)$$
$$\times \exp\bigl(i(-\alpha_{M,+}(y',0,t,\zeta) + \langle \zeta, z \rangle)\bigr) g_{M,\Sigma}(y',t) \,dy'\,dt\,d\zeta. \quad (5\text{-}11)$$

We now consider the composition

$$\chi_n S_{M,+}(0,\cdot) \tfrac{1}{2} i B_M^{-1} R_\Sigma^* \tilde{\Psi}_{\mu,\Sigma} \mathcal{N}_M \Psi_{\mu,\Sigma} R_\Sigma S_{M,+}(\cdot,0).$$

Considering the composition of canonical relations, it follows immediately that this is a pseudodifferential operator. We construct the following representation:

$$(2\pi)^{-n} \iiint a^{(\text{bkd})}(y',t,z,\zeta) a^{(\text{fwd})}(y',t,x,\xi)$$
$$\times \exp\bigl(i(-\alpha_{M,+}(y',0,t,\zeta) + \alpha_{M,+}(y',0,t,\xi) + \langle \zeta, z \rangle - \langle \xi, x \rangle)\bigr) dy' \,dt\,d\zeta$$
$$= \sigma(z,x,\xi) \exp\bigl(i\langle \xi, z-x \rangle\bigr). \quad (5\text{-}12)$$

We write

$$-\alpha_{M,+}(y',0,t,\zeta) + \alpha_{M,+}(y',0,t,\xi) = \langle \xi - \zeta, X(y',t,\zeta,\xi) \rangle,$$
$$X(y',t,\zeta,\xi) = \int_0^1 \partial_\xi \alpha_{M,+}(y',0,t,\zeta + s(\xi - \zeta)) \,ds, \quad (5\text{-}13)$$

and change variables of integration, $(y',t) \to X$. The phase is stationary if $y' = (y_M^t)'(x,\xi)$ and $(y_M^t)_n(x,\xi) = 0$, and $\zeta = \xi$; we have

$$X(y',t,\xi,\xi) = \partial_\xi \alpha_{M,+}(y',0,t,\xi) = x_M^t(y',0,\xi).$$

Using the absence of grazing rays, the relevant Jacobian can be written in the form

$$\left| \frac{\partial(X)}{\partial(y',t)} \right|_{\zeta=\xi} = \left| \frac{\partial(y_M^t)}{\partial(x)} \right|^{-1}_{\xi,x=x_M^t(y',0,\xi)} \left| \frac{\partial(y_M^t)_n}{\partial t} \right|_{\xi,x=x_M^t(y',0,\xi)}, \quad (5\text{-}14)$$

where

$$\frac{\partial(y_M^t)_n}{\partial t} = \frac{\partial B_M^{\text{prin}}(y_M^t,\eta_M^t)}{\partial \eta_n},$$

and

$$y_M^t(x_M^t(y',0,\xi),\xi) = (y',0),$$
$$\eta_M^t(x_M^t(y',0,\xi),\xi) = \partial_y \alpha_{M,+}(y',0,t,\xi).$$

Applying the method of stationary phase, we find the principal symbol of the composition under consideration:

$$\sigma(x,x,\xi) =$$

$$-2i\tau \frac{\partial B_M^{\text{prin}}}{\partial \eta_n}(y',0,\tau^{-1}\eta',C_\mu(0,y',\eta',1))\Psi_{\mu,\Sigma}(y',t,\eta',\tau)$$

$$\times \left|\frac{\partial(y_M^t)}{\partial(x)}\right|_{\xi,x=x_M^t(y',0,\xi)}^{-1/2} \chi_n(x_n) \left|\frac{\partial(y_M^t)}{\partial(x)}\right|_{\xi,x=x_M^t(y',0,\xi)}^{-1/2}$$

$$\times \tfrac{1}{2}i\tau^{-1}\tilde{\Psi}_{\mu,\Sigma}(y',t,\eta',\tau)\left|\frac{\partial(y_M^t)}{\partial(x)}\right|_{\xi,x=x_M^t(y',0,\xi)}$$

$$\times \left(\frac{\partial B_M^{\text{prin}}}{\partial \eta_n}(y',0,\tau^{-1}\eta',C_\mu(0,y',\tau^{-1}\eta',1))\right)^{-1}\bigg|_{\substack{y'=(y_M^t)'(x,\xi)\\ \eta'=(\eta_M^t)'(x,\xi)\\ \tau=-B_M(x,\xi)}}$$

$$= \chi_n(x_n)\Psi_{X_0,+}(x,\xi), \quad (5\text{-}15)$$

using that t is determined by $(y_M^t)_n(x,\xi) = 0$.

We extend the proof to longer times. Using a partition of unity we can decompose $\Psi_{\mu,\Sigma}$ into terms, covering time intervals $[s, s+t_1]$ $(s > 0)$, say. It is sufficient to prove the result for each term. For this, we simply change the time variable from t to $t-s$ in the above. We then use the semigroup property, microlocally, of $S_{M,+}(t,s)$. □

6. Inverse scattering: common source

Here, we develop the inverse scattering with the goal to reconstruct the singular medium perturbation given observations of the scattered field on part of the surface and the background medium. We assume that bicharacteristics which enter the region $x_n < 0$ do not return to the region $x_n \geq 0$. As mentioned before, we invoke an additional hypothesis:

Assumption 6.1 (Bolker condition). No caustics form between the source and scattering points in mode N.

Essentially, we assume the absence of multipathing in the characteristics or rays associated with the source wave field. The reflection data, d_{MN}, are modeled by $R_\Sigma v$, cf. (4-4)–(4-5). We substitute d_{MN} for $R_\Sigma w$ in (5-1) when w_r is identified with v_r, and consider the operator, H_{MN}, defined in (2-14); its canonical relation is illustrated in Figure 2.

Theorem 6.2. *Let H_{MN} be the transform defined in (2-14) and let w_{MN} be as defined in (4-9)–(4-10). With Assumption 6.1, the following holds true:*

$$H_{MN}P_1 F(\cdot) = w_{MN,+}\mathcal{R}_+ w_{MN,+}^T + w_{MN,-}\mathcal{R}_- w_{MN,-}^T,$$

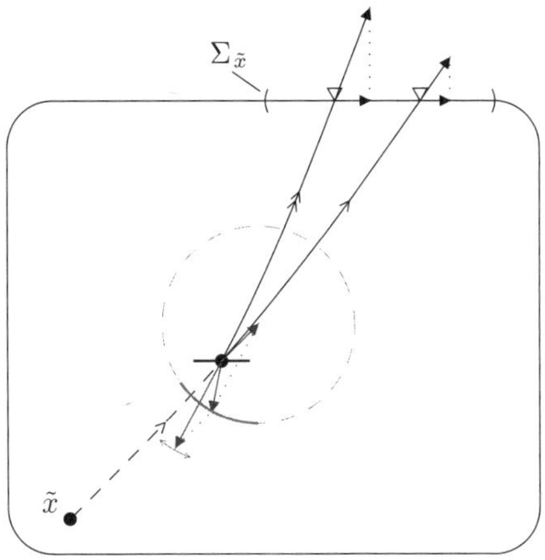

Figure 2. Illustration of the canonical relation of RTM-based inverse scattering. The receivers are contained in the set $\Sigma_{\tilde{x}}$ (the array). The ray with single arrow corresponds with the source field, which may also be observed at the boundary; the ray with double arrows corresponds with the scattered field. The covectors at the scattering point illustrate the construction of $\theta(x, x, \xi)$ (isotropic case).

where $\mathcal{R}_\pm$ are pseudodifferential operators of order zero with principal symbols given by

$$\mathcal{R}_\pm^{\text{prin}}(z, \zeta) = \Pi_+(T_N(z, \tilde{x}))(z, \xi(\pm\zeta)),$$

and $w_{MN,\pm}$ are pseudodifferential operators with principal symbols given by

$$w_{MN,\pm}(z, \zeta) = w_{MN}(z, T_N(z, \tilde{x}), z, \xi(\pm\zeta)).$$

Here, the map $\zeta \to \xi$ is given in (4-18).

Proof. We first carry out the analysis for symbols up to leading orders. Let w denote a wave field in $\mathcal{E}'(X \times \mathbb{R})$. We introduce the "reverse-time migration" imaging condition through the operator K, with

$$Kw(z) = w(z, T_N(z, \tilde{x})).$$

We define the pseudodifferential operator L by

$$Lw(y, t) = \mathcal{A}_N(y, \tilde{x}, D_t)^{-1} 2\mathrm{i} D_t$$
$$\sum_{p=0}^{n} \left(\frac{\partial T_N}{\partial y_k}(y, \tilde{x}) \, Q(y, \partial_y T_N(y, \tilde{x}))_{lN} \, \Xi_p(y, -\partial_y T_N(y, \tilde{x}), 1) \right)$$
$$\times \left(D_{y_j} \, Q(y, D_y)_{iM} \, \Theta_p(y, D_y, D_t) w(y, t) \right);$$

hence $KL v_r$ is an asymptotic approximation of $H_{MN} d_{MN}$.

We consider negative frequencies, identify $v_{r,+}(y,t)$ with

$$\Pi_+(t) F_+(t) \left(\frac{\delta \rho}{\rho}, -\frac{\delta c}{\rho} \right)(y),$$

and analyze the composition $L \Pi_+(\cdot) F_+(\cdot) \chi$. The composition $\Pi_+(t) F_+(t) \chi$ is a Fourier integral operator with a phase function inherited from $F_+(t) \chi$. To highest order, its amplitude is given by

$$\Pi_+(t_1)(y, \partial_y \varphi_{MN}) A_{F,MN}(y, t_1, x, \xi) w_{MN}^T(y, t_1, x, \xi);$$

see (4-12). Also, $L \Pi_+(\cdot) F_+(\cdot) \chi$ is a Fourier integral operator with leading-order amplitude

$$A_{L\Pi F, MN}(y, t_1, x, \xi) = 2i\, w_{MN}(y, t_1, y, \partial_y \alpha_{M,+})$$
$$\times \sum_{p=0}^{n} \Xi_p(y, -\partial_y T_N(y, \tilde{x}), 1) \Theta_p(y, \partial_y \alpha_{M,+}, \partial_t \alpha_{M,+})$$
$$\times \Pi_+(t_1)(y, \partial_y \alpha_{M,+}) \tilde{a}_{M,+}(y, t_1 - T_N(x, \tilde{x}), \xi)$$
$$\times \frac{\mathcal{A}_N(x, \tilde{x}, \partial_t \alpha_{M,+})}{\mathcal{A}_N(y, \tilde{x}, \partial_t \alpha_{M,+})} w_{MN}^T(y, t_1, x, \xi), \qquad (6\text{-}1)$$

in which the argument of $\alpha_{M,+}$ is $(y, t_1 - T_N(x, \tilde{x}), \xi)$.

The local phase function of the oscillatory integral representation of the kernel of $K L \Pi_+(\cdot) F_+(\cdot) \chi$ is obtained by setting $t_1 = T_N(z, \tilde{x})$ in (4-13):

$$\alpha_{M,+}(z, T_N(z, \tilde{x}) - T_N(x, \tilde{x}), \xi) - \langle \xi, x \rangle.$$

This phase is stationary at points (z, x, ξ) for which

$$\partial_\xi \alpha_{M,+}(z, T_N(z, \tilde{x}) - T_N(x, \tilde{x}), \xi) = x.$$

The stationarity condition implies that the bicharacteristic with initial condition (x, ξ) arrives at z after a time lapse $T_N(z, \tilde{x}) - T_N(x, \tilde{x})$. Then, however, $\tilde{x}$, x and z would lie on the same characteristic. Having excluded such (direct source) characteristics, we must have $z = x$. We note that then

$$\zeta = \partial_x \alpha_{M,+}(x, 0, \xi) = \xi.$$

We write

$$\alpha_{M,+}(z, T_N(z, \tilde{x}) - T_N(x, \tilde{x}), \xi) - \langle \xi, x \rangle = \langle \theta(z, x, \xi), z - x \rangle, \qquad (6\text{-}2)$$

ELASTIC-WAVE INVERSE SCATTERING BASED ON REVERSE TIME MIGRATION   439

with

$$\theta(z, x, \xi) = -\int_0^1 [\partial_x \alpha_{M,+}(z, T_N(z, \tilde{x}) - T_N(z + \mu(x-z), \tilde{x}), \xi) - \xi] d\mu$$

$$= \xi + \int_0^1 \partial_t \alpha_{M,+}(z, T_N(z, \tilde{x}) - T_N(z + \mu(x-z), \tilde{x}), \xi)$$
$$\times \partial_x T_N(z + \mu(x-z), \tilde{x}) d\mu. \quad (6\text{-}3)$$

We introduce the point $\check{x} = x_M^{T_N(z,\tilde{x}) - T_N(z+\mu(x-z),\tilde{x})}(z, \xi)$, that is,

$$\check{x} = \partial_\xi \alpha_{M,+}(z, T_N(z, \tilde{x}) - T_N(z + \mu(x-z), \tilde{x}), \xi),$$

with the property that the bicharacteristic with initial condition $(\check{x}, \xi)$ reaches z at time $T_N(z, \tilde{x}) - T_N(z + \mu(x-z), \tilde{x})$. Then

$$\partial_t \alpha_{M,+}\big(z, T_N(z, \tilde{x}) - T_N(z + \mu(x-z), \tilde{x}), \xi\big) \partial_x T_N(z + \mu(x-z), \tilde{x})$$

$$= -\frac{B_M^{\text{prin}}(\check{x}, \xi)}{B_N^{\text{prin}}(z + \mu(x-z), n_{\tilde{x}}(z + \mu(x-z)))} n_{\tilde{x}}(z + \mu(x-z)) \quad (6\text{-}4)$$

cf. (4-17), and

$$\theta(z, x, \xi) = \xi - \int_0^1 B_M^{\text{prin}}(\check{x}, \xi) \gamma_{\tilde{x}}(z, x) d\mu,$$
$$\gamma_{\tilde{x}}(z, x) = \frac{1}{B_N^{\text{prin}}(z+\mu(x-z), n_{\tilde{x}}(z+\mu(x-z)))} n_{\tilde{x}}(z + \mu(x-z)), \quad (6\text{-}5)$$

so that

$$|\partial_\xi \theta(z, x, \xi)| = \left|\det\left(I - \int_0^1 \partial_\xi B_M^{\text{prin}}(\check{x}, \xi) \otimes \gamma_{\tilde{x}}(z, x) d\mu\right)\right|; \quad (6\text{-}6)$$

$\theta(z, x, -\xi) = -\theta(z, x, \xi)$. We note that

$$\gamma_{\tilde{x}}(z, z) = \frac{1}{B_N^{\text{prin}}(z, n_{\tilde{x}}(z))} n_{\tilde{x}}(z) = \partial_x T_N(z, \tilde{x}),$$

while at $x = z$, $\check{x}$ can be replaced by z, and

$$\theta(z, z, \xi) = \xi - \frac{B_M^{\text{prin}}(z, \xi)}{B_N^{\text{prin}}(z, n_{\tilde{x}}(z))} n_{\tilde{x}}(z),$$

$$|\partial_\xi \theta(z, z, \xi)| = |1 - \partial_\xi B_M^{\text{prin}}(z, \xi) \cdot \gamma_{\tilde{x}}(z, z)|,$$

defining a mapping $\xi \to \theta(z, z, \xi)$, which is invertible; see Figure 2 and also Lemma 4.1. Thus the Schwartz kernel of $KL\Pi_+(\cdot) F_+(\cdot) \chi$ can be written in the form

$$(2\pi)^{-n} \int A_{L\Pi F,MN}(z, T_N(z, \tilde{x}), z, \xi(\theta))$$
$$\times \left|\partial_\xi \theta(z, z, \xi(\theta))\right|^{-1} \exp(i\langle\theta, z - x\rangle) \, d\theta \, \chi(x).$$

We evaluate the principal symbol. We have

$$\sum_{p=0}^{n} \Xi_p(z, -\partial_y T_N(z, \tilde{x}), 1) \Theta_p(z, \partial_y \alpha_{M,+}, \partial_t \alpha_{M,+})$$
$$= \partial_t \alpha_{M,+} (1 - \partial_\xi B_M^{\text{prin}}(z, \partial_y \alpha_{M,+}) \cdot \partial_y T_N(z, \tilde{x})), \quad (6\text{-}7)$$

using that the argument of $\alpha_{M,+}$ is $(z, T_N(z, \tilde{x}) - T_N(x, \tilde{x}), \xi)$; at $x = z$ we have

$$\partial_t \alpha_{M,+}(z, 0, \xi) = -B_M^{\text{prin}}(z, \xi), \quad \partial_y \alpha_{M,+}(z, 0, \xi) = \xi, \quad \partial_y T_N(z, \tilde{x}) = \gamma_{\tilde{x}}(z, z),$$

by (4-17), whence this expression reduces to $-B_M^{\text{prin}}(z, \xi) \left|\partial_\xi \theta(z, z, \xi)\right|$. Furthermore, $\tilde{a}_{M,+}(z, 0, \xi) = \frac{1}{2} i B_M^{\text{prin}}(z, \xi)^{-1}$. We obtain

$$A_{L\Pi F,MN}(z, T_N(z, \tilde{x}), z, \xi(\theta)) \left|\partial_\xi \theta(x, x, \xi(\theta))\right|^{-1}$$
$$= w_{MN}(z, T_N(z, \tilde{x}), z, \xi(\theta)) \, \Pi_+(T_N(z, \tilde{x}))(z, \xi(\theta)) \, w_{MN}^T(z, T_N(z, \tilde{x}), z, \xi(\theta)).$$

Combining the negative with the positive frequency contributions yields a point symmetry of the domain of θ integration; we obtain a pseudodifferential operator with (real-valued) principal symbol

$$w_{MN}(z, T_N(z, \tilde{x}), z, \xi(\theta)) \, \Pi_+(T_N(z, \tilde{x}))(z, \xi(\theta)) \, w_{MN}^T(z, T_N(z, \tilde{x}), z, \xi(\theta))$$
$$+ w_{MN}(z, T_N(z, \tilde{x}), z, \xi(-\theta)) \, \Pi_+(T_N(z, \tilde{x}))(z, \xi(-\theta)) \, w_{MN}^T(z, T_N(z, \tilde{x}), z, \xi(-\theta)),$$

from which the statement follows. $\square$

We note that the principal symbol matrix representing the spatial resolution and contrast source radiation patterns (for a fixed source) has rank 1.

7. Array receiver functions

In this section, we assume we also observe the source field and focus on converted waves ($M \neq N$); in fact, we assume that N corresponds with qP. Thus we remove the knowledge of the source. We generalize the notion of receiver functions used in the seismological literature; in the last subsection of this section we will explain under which conditions receiver functions can be obtained from the generalization introduced here. The incident data, d_N, are modeled by $R_\Sigma w_{\tilde{x}}$; see Figure 3.

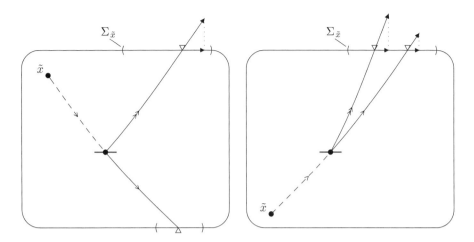

Figure 3. Array receiver functions: Detecting the incident field, and scattered field. Left: distinct arrays. Right: single array (teleseismic situation). The available set Σ may dependent on $\tilde{x}$. The ray with single arrow corresponds with the source field; the ray with double arrows corresponds with the (converted) scattered field. Knowledge of the source is eliminated.

7A. *Inverse scattering: Reverse-time continued source wave field.* We obtain $w_{\tilde{x};r}$ by substituting d_N for $R_\Sigma w$ in (5-1) with M replaced by N. We note that the equation which d_N satisfies is homogeneous in the relevant time interval. Applying Theorem 5.1, we obtain

$$\chi_n \, w_{\tilde{x};r,\pm}(\cdot,t) = \chi_n \, \Pi_{N,\pm}(t) w_{\tilde{x};\pm}(\cdot,t),$$

microlocally. We apply $\mathsf{P}_1 V_N$ to $w_{\tilde{x};r,\pm}$, which we use to replace G_N in the operator H_{MN} of (2-14).

In Theorem 6.2, $\mathcal{R}_\pm$ is affected by $\Pi_{N,\pm}(t)$ in a natural way. Following the propagation of singularities, it becomes clear that the singularities in the source field are recovered at x_0 only if the ray connecting $\tilde{x}$ with x_0 intersects the boundary at a point in $\Sigma_{\tilde{x}}$, see Figure 4. We reemphasize that we admit the formation of caustics between receivers and scattering points.

Without knowledge of $\tilde{x}$, the factor $1/|\hat{G}_N(\cdot,\tilde{x},\tau)|^2$ cannot be evaluated. Instead, we consider

$$-\frac{1}{2\pi} \int \frac{2\Omega(\tau)}{i\tau} \sum_{p=0}^{n} \left(\frac{\partial}{\partial x_k} Q(x,D_x)_{lN} \, \Xi_p(x,D_x,\tau) \overline{\hat{w}_{\tilde{x};r}(x,\tau)} \right)$$
$$\times \left(\frac{\partial}{\partial x_j} Q(x,D_x)_{iM} \, \Theta_p(x,D_x,\tau) \hat{v}_r(x,\tau) \right) d\tau.$$

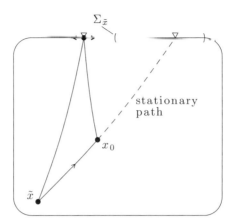

Figure 4. Source wave field: Reverse-time continuation and propagation of singularities. Reconstruction of singular perturbations can be accomplished at points (x_0) which can be connected to $\tilde{x}$ with a ray which intersects $\Sigma_{\tilde{x}}$.

We then adjust $\Omega(\tau)$ by a multiplication with $(\mathcal{A}_N(\cdot,\cdot,1)/\mathcal{A}_N(\cdot,\cdot,\tau))^2$ such that the result yields a (partial) reconstruction up to the factor

$$\left[|\mathcal{T}|^{-1} \int_{\mathcal{T}} |\hat{w}_{\tilde{x};r}(x,\tau)|^2 \left(\frac{\mathcal{A}_N(\cdot,\cdot,1)}{\mathcal{A}_N(\cdot,\cdot,\tau)} \right)^2 d\tau \right], \quad (7\text{-}1)$$

cf. (2-14). Here, $\mathcal{T}$ is the bandwidth of the data.

7B. *Cross correlation formulation.* We reformulate the inverse scattering procedure outlined in the previous subsection in terms of a single Fourier integral operator. To achieve this, we introduce array receiver functions; see Definition 2.1. The observational assumption is that $d_{MN} = d_M$, whence $\mathsf{R}_{MN}(d(\cdot,\cdot;\tilde{x}))$ can be obtained from the multicomponent data. (In receiver functions, correlation, or deconvolution, is considered only for $r' = r$.) We introduce an operator K_{MN} by identifying $(K_{MN}\mathsf{R}_{MN})_{ijkl}(x)$ with the transformation introduced in the previous subsection:

$$(K_{MN}\mathsf{R}_{MN})_{ijkl}(x) = -\int 2\Omega(D_t)(iD_t)^{-1} \sum_{p=0}^{n} \int_{-\infty}^{t} H(-t')$$

$$\times \left[\frac{\partial}{\partial x_k} Q(x,D_x)_{lN} \,\Xi_p(x,D_x,D_t)(S_{N,+}(t',0) - S_{N,-}(t',0)) \tfrac{1}{2}iB_N^{-1} R_\Sigma^* \tilde{\Psi}_{\nu,\Sigma} \right]_{(r')}$$

$$\times \left[\frac{\partial}{\partial x_j} Q(x,D_x)_{iM} \,\Theta_p(x,D_x,D_{t_0})(S_{M,+}(t',t) - S_{M,-}(t',t)) \tfrac{1}{2}iB_M^{-1} R_\Sigma^* \tilde{\Psi}_{\mu,\Sigma} \right]_{(r)} dt'$$

$$\times \mathsf{R}_{MN}(\overset{r}{\cdot},\overset{r'}{\cdot},t) \, dt.$$

ELASTIC-WAVE INVERSE SCATTERING BASED ON REVERSE TIME MIGRATION   443

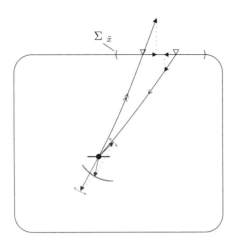

Figure 5. The canonical relation, Λ_{MN}^K, of K_{MN} (isotropic case). This is an adaption of the canonical relation illustrated in Figure 2, reflecting the cross correlation in the definition of array receiver functions.

Lemma 7.1. *Let* $M \neq N$, $Z = \Sigma_{\tilde{x}} \times \Sigma_{\tilde{x}} \times (0, T)$. $K_{MN} : \mathcal{E}'(Z) \to \mathcal{D}'(X)$ *is a Fourier integral operator with canonical relation*

$$\Lambda_{MN}^K = \{(x, \hat{\xi} - \tilde{\xi}; (y_M^{\hat{t}})'(x, \hat{\xi}), \hat{t} - \tilde{t}, (y_N^{\tilde{t}})'(x, -\tilde{\xi}), (\eta_M^{\hat{t}})'(x, \hat{\xi}), \tau, -(\eta_N^{\tilde{t}})'(x, -\tilde{\xi})) \mid $$
$$B_M(x, \hat{\xi}) = B_N(x, \tilde{\xi}) = \mp \tau, \; (y_M^{\hat{t}})_n(x, \hat{\xi}) = 0, \; (y_N^{\tilde{t}})_n(x, -\tilde{\xi}) = 0\}.$$

The canonical relation Λ_{MN}^K is illustrated in Figure 5.

7C. *Flat, translationally invariant models: propagation of singularities, receiver functions.* In view of translational invariance, (4-3) attains the form

$$\delta G_{MN}(\hat{x}, \tilde{x}, t) = \int_{[0,Z]} \left(\int_0^t \int \frac{\partial}{\partial (t_0, x_{0,j})} Q(x_0, D_{x_0})_{iM} G_M(x_0, \hat{x}, t - t_0) \right.$$
$$\left. \times \frac{\partial}{\partial (t_0, x_{0,k})} Q(x_0, D_{x_0})_{lN} G_N(x_0, \tilde{x}, t_0) \, dx_0' \, dt_0 \right)$$
$$\times \left(\delta_{il} \frac{\delta \rho(z_0)}{\rho(z_0)}, -\frac{\delta c_{ijkl}(z_0)}{\rho(z_0)} \right) dz_0, \quad (7\text{-}2)$$

writing $x_0 = (x_0', z_0)$ as before. Upon restriction to $\hat{x} = (r, 0)$, writing $\tilde{x} = s$, the expression in between braces on the right-hand side defines the kernel,

$\mathcal{F}^0_{MN;ijkl}(r,t;z_0)$ say, of a single scattering operator F^0_{MN}:

$$\mathcal{F}^0_{MN;ijkl}(r,t;z_0) = \int\int_0^t\int \frac{\partial}{\partial(t_0,x_{0,j})} Q(x_0,D_{x_0})_{iM} G_M(x_0,r,0,t-t_0)$$

$$\times \frac{\partial}{\partial(t_0,x_{0,k})} Q(x_0,D_{x_0})_{lN} G_N(x_0,\tilde{x}',t_0)\,\mathrm{d}x'_0\,\mathrm{d}t_0\, \mathcal{D}^{-1}_{Nk'}(\tilde{x}',s) e_{k'}\,\mathrm{d}\tilde{x}'. \quad (7\text{-}3)$$

The associated imaging operator, $(F^0_{MN})^*$, maps the (conversion) data to an image as a function of z_0 (and s).

We introduce so-called midpoint-offset coordinates $r' = m - h$ and $r = m + h$ (so $\mathrm{d}r\,\mathrm{d}r' = 2\,\mathrm{d}m\,\mathrm{d}h$) and find that

$$(F^0_{MN})^* d_{MN}(\cdot,\cdot;s) = K^0_{MN}(\mathsf{R}_{MN}(d(\cdot,\cdot;s)))(\cdot,\cdot,\cdot),$$

where K^0_{MN} is an operator with kernel

$$\mathcal{K}^0_{ijkl;MN}(z_0;m+h,t,m-h)$$

$$= 2\int_{-\infty}^t H(-t_0)\int \frac{\partial}{\partial x_{0,j}} Q(x_0,D_{x_0})_{iM} G_M(x'_0,z_0,m+h,0,t-t_0)$$

$$\times \frac{\partial}{\partial x_{0,k}} Q(x_0,D_{x_0})_{lN} G_N(x'_0,z_0,m-h,0,-t_0)\,\mathrm{d}x'_0\,\mathrm{d}t_0. \quad (7\text{-}4)$$

To study the propagation of singularities by this operator, we substitute the WKBJ approximations for G_M and G_N (cf. (3-33)) in this expression.

K^0_{MN}, *imaging*. We focus on the propagation of singularities and, hence, the relevant phase functions; the amplitudes follow from standard stationary phase arguments. The WKBJ phase function associated with $\mathcal{K}^0_{ijkl;MN}$ becomes

$$\tau\left[-\tau_\mu(0,z_0,\hat{p}) + \tau_\nu(0,z_0,\hat{p}') + \sum_{j=1}^{n-1}(\hat{p}-\hat{p}')_j(m-x'_0)_j - \sum_{j=1}^{n-1}(\hat{p}+\hat{p}')_j h_j + t\right].$$

Carrying out the integrations over x'_0 and $\hat{p}'$ leads to

$$\mathcal{K}^0_{ijkl;MN}(z_0;m+h,t,m-h) \approx \dot{\mathcal{K}}^0_{ijkl;MN}(z_0;h,t),$$

which admits an integral representation with WKBJ phase function

$$\tau\left[-\tau_\mu(0,z_0,\hat{p}) + \tau_\nu(0,z_0,\hat{p}) - 2\sum_{j=1}^{n-1}\hat{p}_j h_j + t\right].$$

We get

$$K^0_{MN}(\mathsf{R}_{MN}(d(\cdot,\cdot;s)))(\cdot,\cdot,\cdot) \approx \dot{K}^0_{MN}(\mathsf{R}^0_{MN}(d(\cdot,\cdot;s)))(\cdot,\cdot,\cdot),$$

where

$$(R^0_{MN}(d(\,\cdot\,,\,\cdot\,;s)))(h,t) = \int (R_{MN}(d(\,\cdot\,,\,\cdot\,;s)))(m+h,t,m-h)\,\mathrm{d}m. \quad (7\text{-}5)$$

Applying the method of stationary phase to the integral representation for $\dot{\mathcal{H}}^0_{ijkl;MN}(z_0;h,t)$ in $\hat{p}$, yields stationary points $\hat{p} = \hat{p}^0(z_0,h)$ satisfying

$$-\left[\frac{\partial \tau_\mu(0,z_0,\hat{p})}{\partial \hat{p}} - \frac{\partial \tau_\nu(0,z_0,\hat{p})}{\partial \hat{p}}\right] = 2h, \quad (7\text{-}6)$$

revealing the propagation of singularities: This equation defines a pair of rays sharing the same horizontal slowness $\hat{p}$, originating at (image) depth z_0, and reaching the acquisition surface at

$$r = \frac{\partial \tau_\mu(0,z_0,\hat{p})}{\partial \hat{p}} \quad \text{and} \quad r' = \frac{\partial \tau_\nu(0,z_0,\hat{p})}{\partial \hat{p}},$$

respectively; in the imaging point of view, the rays intersect at depth z_0, whence $r' - r = 2h$. The corresponding differential travel time is given by

$$\tau_\mu(0,z_0,\hat{p}^0(z_0,h)) - \tau_\nu(0,z_0,\hat{p}^0(z_0,h)) + 2\sum_{j=1}^{n-1}\hat{p}^0_j(z_0,h)h_j$$

(we note that $\hat{p}^0_j(z_0,h)$ is the negative of the usual geometric ray parameter in view of our Fourier transform convention). The geometry is illustrated in Figure 6 (pair of solid rays).

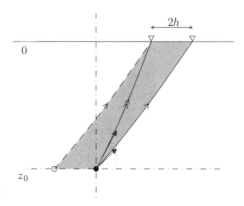

Figure 6. Propagation of singularities by K^0_{MN}. The double arrows relate to (scattered) mode μ, while the single arrow relates to (incident) mode ν. Translational invariance yields alternative ray pairs (dashed) for which the phase of $\mathcal{H}^0_{ijkl;MN}$ is stationary.

R^0_{MN}, *modeling.* Using (7-2), substituting the WKBJ approximations for G_M and G_N in $(R^0_{MN}(d(\,\cdot\,,\cdot\,,s)))(h,t)$, and carrying out the integrations over x'_0 and $\tilde{p}'$, we obtain a linear integral operator representation, acting on

$$\left(\delta_{il}\frac{\delta\rho(z'_0)}{\rho(z'_0)}, -\frac{\delta c_{ijkl}(z'_0)}{\rho(z'_0)}\right),$$

for the integrand in (7-5): For fixed $s = (s', z_s)$, $z_s > z_0$, the WKBJ phase function of its kernel representation is

$$\tau\left[-\tau_\mu(0, z'_0, \hat{p}) - \tau_\nu(z'_0, z_s, \hat{p}) + \tau_\nu(0, z_s, \tilde{p})\right.$$
$$\left. + \sum_{j=1}^{n-1}\hat{p}_j(m - h - s')_j - \sum_{j=1}^{n-1}\tilde{p}_j(m + h - s')_j + t\right].$$

Carrying out the integrations over m and then $\hat{p}$ leads to

$$\int R_{MN}(d(\,\cdot\,,\cdot\,;s))(m + h, t, m - h)\,dm \approx \dot{R}^0_{MN}(d(\,\cdot\,,\cdot\,;s))(h, t),$$

which admits an integral representation with WKBJ phase function

$$\tau\left[-\tau_\mu(0, z'_0, \tilde{p}) + \tau_\nu(0, z'_0, \tilde{p}) - 2\sum_{j=1}^{n-1}\tilde{p}_j h_j + t\right];$$

the integration over $\tilde{p}$ signifies a "plane-wave" decomposition of the source, while the explicit dependence on (s', z_s) has disappeared. Thus

$$\dot{K}^0_{MN}\big(R^0_{MN}(d(\,\cdot\,,\cdot\,;s))(\,\cdot\,,\cdot\,)\big) \approx \dot{K}^0_{MN}\big(\dot{R}^0_{MN}(d(\,\cdot\,,\cdot\,;s))(\,\cdot\,,\cdot\,)\big),$$

yielding a resolution analysis in depth. The stationary phase analysis in, and following (7-6) applies, leading to the introduction of $\tilde{p}^0(z'_0, h)$ and associated ray geometry if there is a nonvanishing contrast (horizontal reflector) at depth z'_0.

For the singularities to appear in $(R^0_{MN}(d(\,\cdot\,,\cdot\,;s)))(h, t)$, it is necessary that $\tilde{p}^0(z'_0, h)$ coincides with the stationary value, $\tilde{p}_s$ say, of $\tilde{p}$ associated with the WKBJ approximation of the *incident* field (d_N), determined by s and $m + h$.

Receiver functions, plane-wave synthesis. Let $\tilde{d}(\,\cdot\,,\cdot\,;\tilde{p}_s)$ denote the frequency-domain data obtained by synthesizing a source *plane wave* with parameter $\tilde{p}_s$ (as in plane-wave Kirchhoff migration) including an appropriate amplitude weighting function derived from the WKBJ approximation (here, we depart from the single source acquisition). We correlate these data using (2-16) subjected to a Fourier transform in time, with $d(\,\cdot\,,\cdot\,;s)$ replaced by $\tilde{d}(\,\cdot\,,\cdot\,;\tilde{p}_s)$. We obtain

$$\widehat{R}_{MN}(\tilde{d}(\,\cdot\,,\cdot\,;\tilde{p}_s))(m + h, \tau, m - h),$$

with the property

$$\dot{\mathsf{R}}^0_{MN}(d(\cdot,\cdot;s))(h,t)$$
$$\sim \int \widehat{\mathsf{R}}_{MN}(\tilde{d}(\cdot,\cdot;\tilde{p}_s))(m_0,\tau,m_0) \exp\left(i\tau\left[-2\sum_{j=1}^{n-1}\tilde{p}_{sj}h_j + t\right]\right) d\tilde{p}_s \, d\tau$$

for any $(m_0, 0)$ contained in Σ_s. The quantity $\widehat{\mathsf{R}}_{MN}(\tilde{d}(\cdot,\cdot;\tilde{p}_s))(m_0,\tau,m_0)$ is what seismologists call a receiver function. The phase shift and receiver function are illustrated by the dashed ray (single array) paired with the ray indicated by double arrows in Figure 6.

8. Applications in global seismology

We presented an approach to elastic-wave inverse scattering of reflection seismic data generated by an active source or a passive source, focused on mode conversions, and based on reverse-time migration (RTM). We introduced array receiver functions (ARFs), a generalization of the notion of receiver functions, which can be used for inverse scattering of passive source data with a resolution comparable to RTM. In principle, the ARFs can be used also for imaging with certain (mode-converted) multiple scattered waves. Microlocally, RTM and generalized Radon transform (GRT) based inverse scattering are the same, the key restriction being the absence of source caustics; however, typically, the GRT is obtained as the composition of the parametrix of a normal operator with the imaging (that is, adjoint of the single scattering) operator. We note that the implementation of our RTM-based inverse scattering does not involve any ray-geometrical computation.

Our RTM-based inverse scattering transform defines a Fourier integral operator the propagation of singularities of which is described by a canonical graph. Thus it directly admits expansions into wave packets or curvelets, accommodates partial reconstruction as developed in [de Hoop et al. 2009], and associated algorithms can be applied. In practice, one may carry out the addition of terms in the inverse scattering transform adaptively. Also, the polarized-wave equation formulation is well-suited for a frequency-domain implementation of the type presented in [Wang et al. 2010].

A key application of the analysis presented in this paper concerns the detection, mapping, and characterization of interfaces in Earth's upper mantle (see Figure 1). The analysis allows one to integrate contributions from different body-wave phases (for example, underside reflections beneath oceans and mode conversions, that is, ARFs beneath continents), accommodating the fact that earthquakes are highly unevenly distributed and that the data are inherently restricted to parts of the earth's surface. Concerning ARFs, we elaborate on the feasibility of

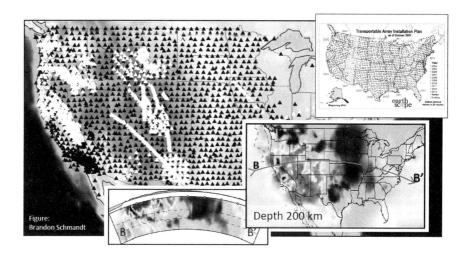

Figure 7. USArray (the seismology component of EarthScope, http://www.earthscope.org); black triangles indicate locations of permanent sensors and white triangles indicate more densely space temporary arrays. Insets: Transportable array installation plan (2004-2013), map of P-wave speed variations relative to a spherically symmetric model obtained with linearized transmission tomography at 200 km depth, and vertical mantle section to 600 km depth depicting in blue the seismically fast slab of subducted Gorda/Farallon lithosphere beneath the western States. Such a model serves as a background model in ARFs and RTM based inverse scattering. Generated by Scott Burdick.

inverse scattering beneath the North American continent using available data from USArray; see Figure 7, which also shows a recently obtained (isotropic) model, which can be used as a background model for application of the ARFs and RTM based inverse scattering presented here. The grid spacing of ~ 70 km of the three-component broadband seismograph stations that constitute the TA component of USArray is too sparse for ARF imaging of the crust mantle interface (near $35-40$ km depth) but is adequate for the imaging of upper mantle discontinuities. The images that we can produce will aid in better constraining the lateral variations in temperature and composition (including melt, volatile content) and the geological processes that produced them.

We end with a numerical example using modeled data designed for the detection of a piecewise smooth reflector. The (isotropic) model is depicted in Figure 9, top left. The finite bandwidth data, for a source indicated by an asterisk in Figure 9, bottom left, are shown in Figure 8. The smooth background model, which could be obtained from, for instance, tomography, is illustrated in Figure 9, top right. Applying the procedure outlined in Section 7A to P-to-S conversions yields the results shown in Figure 9, bottom; the bottom left figure is obtained with data generated by a single source, while the bottom right figure is obtained with data generated by a sparse set of sources.

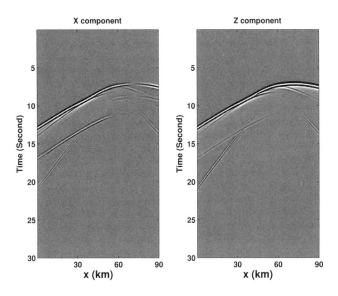

Figure 8. Array receiver functions ($n = 2$). Synthetic data generated in the model depicted in Figure 9 top, left; left: u_1; right: u_2. $\Sigma_{\tilde{x}}$ coincides with the top boundary.

Appendix: Diagonalization of A_{il} with a unitary operator

If the $A_M^{\text{prin}}(x, \xi)$ are all different, A_{il} can be diagonalized with a unitary operator, that is, $Q(x, D_x)^{-1} = Q(x, D_x)^*$. We write $\tilde{Q}_{lN}^0(x, D_x) = Q_{lN}^{\text{prin}}(x, D_x)$; then $(\tilde{Q}^0)^*_{Mi}(x, D_x)$ has principal symbol $(Q^{\text{prin}})^t_{Mi}(x, \xi)$. We have

$$(\tilde{Q}^0)^*_{Mi}(x, D_x) \tilde{Q}^0_{iN}(x, D_x) = \delta_{MN} + R^0_{MN}(x, D_x)$$

where $R^0_{MN}(x, D_x)$ is self adjoint and of order -1. Then $Q^0_{iN}(x, D_x) = (\tilde{Q}^0 (I + R^0)^{-1/2})_{iN}(x, D_x)$ is, microlocally, unitary. We write

$$A_1^0(x, D_x) = \tfrac{1}{2}[A_1^{\text{prin}}(x, D_x) + (A_1^{\text{prin}})^*(x, D_x)],$$
$$A_2^0(x, D_x) = \text{diag}(\tfrac{1}{2}[A_M^{\text{prin}}(x, D_x) + (A_M^{\text{prin}})^*(x, D_x)]; M = 2, \ldots, n),$$

so that

$$(Q^0)^* A Q^0 = \begin{pmatrix} A_1^0 & 0 \\ 0 & A_2^0 \end{pmatrix} + \begin{pmatrix} B_{11} & B_{12} \\ B_{21} & B_{22} \end{pmatrix},$$

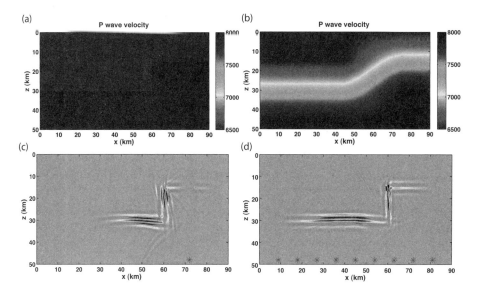

Figure 9. Array receiver functions ($n = 2$); (a) model (P-wave speed); (b) smooth background model (P-wave speed); (c) image reconstruction using data from a single source (location indicated by an asterisk); (d) image reconstruction using data from a sparse set of sources (locations indicated by asterisks). We note the effect of the illumination operators. Generated by Xuefeng Shang.

where B_{11} and the elements of B_{12} (a $1 \times (n-1)$ matrix), B_{21} (a $(n-1) \times 1$ matrix of pseudodifferential operators), and B_{22} (a $(n-1) \times (n_1)$ matrix of pseudodifferential operators), are of order 1; B must be self adjoint, whence $B_{12} = B_{21}^*$.

Next, we seek an operator, $\widetilde{Q}_{lN}^1(x, D_x) = \delta_{lN} + r_{lN}^1(x, D_x)$, assuming that

$$r^1 = \begin{pmatrix} 0 & -(r_{21}^1)^* \\ r_{21}^1 & 0 \end{pmatrix},$$

whence $(r^1)^* = -r^1$, such that

$$\left((\widetilde{Q}^1)^* \left(\begin{pmatrix} A_1^0 & 0 \\ 0 & A_2^0 \end{pmatrix} + \begin{pmatrix} B_{11} & B_{12} \\ B_{21} & B_{22} \end{pmatrix}\right) \widetilde{Q}^1\right)_{21} = 0,$$

$$\left((\widetilde{Q}^1)^* \left(\begin{pmatrix} A_1^0 & 0 \\ 0 & A_2^0 \end{pmatrix} + \begin{pmatrix} B_{11} & B_{12} \\ B_{21} & B_{22} \end{pmatrix}\right) \widetilde{Q}^1\right)_{12} = 0,$$

modulo terms of order 0. This holds true if

$$r_{21}^1 A_1^0 - A_2^0 r_{21}^1 = B_{21};$$

r_{21}^1 must be of order -1. Up to principal parts, this is a matrix equation for the symbol of r_{21}^1, given the principal symbols of A_1^0 and A_2^0; we note that the principals of A_1^{prin} and A_1^0, and of $\text{diag}(A_M^{\text{prin}}; M = 2, \ldots, n)$ and A_2^0, coincide. Because the eigenvalues of $A_2^0(x, \xi)$ all differ from $A_1^0(x, \xi)$, it follows that this system of algebraic equations has a solution. With the solution we form the unitary operator $Q_{iN}^1(x, D_x) = ((I + r^1)(I - (r^1)^2)^{-1/2})_{iN}(x, D_x)$. Then

$$(Q^1)^*(Q^0)^* A Q^0 Q^1 = \begin{pmatrix} A_1^1 & 0 \\ 0 & A_2^1 \end{pmatrix} + \begin{pmatrix} C_{11} & C_{12} \\ C_{21} & C_{22} \end{pmatrix},$$

where A_1^1 and A_2^1 are self adjoint. C_{11} and the elements of C_{12} (a $1 \times (n-1)$ matrix of pseudodifferential operators), C_{21} (a $(n-1) \times 1$ matrix of pseudodifferential operators), and C_{22} (a $(n-1) \times (n_1)$ matrix of pseudodifferential operators), are of order 0; C must be self adjoint, whence $C_{12} = C_{21}^*$. This procedure is continued to find $Q^0 Q^1 \cdots Q^k$ which is microlocally unitary and brings A in block diagonal form modulo terms of order $1-k$. Next, we repeat the procedure for A_2^k.

Acknowledgments

The authors would like to thank Xuefeng Shang for generating the numerical results and figures presented in Section 8. This research was supported in part under NSF CSEDI grant EAR 0724644, and in part by the members of the Geo-Mathematical Imaging Group at Purdue University.)

References

[Baysal et al. 1983] K. Baysal, D. D. Kosloff, and J. W. C. Sherwood, "Reverse time migration", *Geophysics* **48**:11 (1983), 1514–1524.

[Bojarski 1982] N. N. Bojarski, "A survey of the near-field far-field inverse scattering inverse source integral equation", *IEEE Trans. Antennas Propag.* **30**:5 (1982), 975–979. MR 83j:78016 Zbl 0947.78540

[Bostock 1999] M. G. Bostock, "Seismic imaging of lithospheric discontinuities and continental evolution", *Lithos* **48** (1999), 1–16.

[Dencker 1982] N. Dencker, "On the propagation of polarization sets for systems of real principal type", *J. Funct. Anal.* **46**:3 (1982), 351–372. MR 84c:58081 Zbl 0487.58028

[Dueker and Sheehan 1997] K. G. Dueker and A. F. Sheehan, "Mantle discontinuity structure from midpoint stacks of converted P to S waves across the Yellowstone hotspot track", *J. Geophys. Res.* **102** (1997), 8313–8327.

[Duistermaat 1996] J. J. Duistermaat, *Fourier integral operators*, Progress in Mathematics **130**, Birkhäuser, Boston, 1996. MR 96m:58245 Zbl 0841.35137

[Fletcher et al. 2005] R. P. Fletcher, P. J. Fowler, P. Kitchenside, and U. Albertin, "Suppressing artifacts in prestack reverse time migration", *Soc. Explor. Geophys. Expand. Abstr.* **24** (2005), 2049–2052.

[Fletcher et al. 2009a] R. P. Fletcher, X. Du, and P. J. Fowler, "Reverse time migration in tilted transversely isotropic (TTI) media", *Geophysics* **74**:6 (2009), 179–187.

[Fletcher et al. 2009b] R. P. Fletcher, X. Du, and P. J. Fowler, "Stabilizing acoustic reverse-time migration in TTI media", *Soc. Explor. Geophys. Expand. Abstr.* **28** (2009), 2985–2989.

[Fowler et al. 2010] P. J. Fowler, X. Du, and R. P. Fletcher, "Coupled equations for reverse time migration in transversely isotropic media", *Geophysics* **75**:1 (2010), 11–22.

[Frederiksen and Revenaugh 2004] A. W. Frederiksen and J. Revenaugh, "Lithospheric imaging via teleseismic scattering tomography", *Geophys. J. Int.* **159** (2004), 978–990.

[Fryer and Frazer 1984] G. J. Fryer and L. N. Frazer, "Seismic waves in stratified anisotropic media", *Geophys. J. R. Astr. Soc.* **78** (1984), 691–710. Zbl 0556.73092

[Fryer and Frazer 1987] G. J. Fryer and L. N. Frazer, "Seismic waves in stratified anisotropic media, II: Elastodynamic eigensolutions for some anisotropic systems", *Geophys. J. R. Astr. Soc.* **91** (1987), 73–101.

[Garmany 1983] J. Garmany, "Some properties of elastodynamic eigensolutions in stratified media", *Geophys. J. R. Astr. Soc.* **75**:2 (1983), 565–569. Zbl 0525.73029

[Garmany 1988] J. Garmany, "Seismograms in stratified anisotropic media, I: WKBJ theory", *Geophys. J. Int.* **92**:3 (1988), 379–389.

[Guitton et al. 2007] A. Guitton, B. Kaelin, and B. Biondi, "Least-square attenuation of reverse-time migration artifacts", *Geophysics* **72**:1 (2007), 19–23.

[de Hoop and de Hoop 2000] M. V. de Hoop and A. T. de Hoop, "Wave-field reciprocity and optimization in remote sensing", *R. Soc. Lond. Proc. Ser. A Math. Phys. Eng. Sci.* **456**:1995 (2000), 641–682. MR 2001j:86014 Zbl 0974.76075

[de Hoop et al. 2009] M. V. de Hoop, H. Smith, G. Uhlmann, and R. D. van der Hilst, "Seismic imaging with the generalized Radon transform: a curvelet transform perspective", *Inverse Problems* **25**:2 (2009), Art. ID #025005. MR 2010d:65377 Zbl 1157.86002

[Ivrii 1979] V. Y. Ivrii, "Wave fronts of solutions of symmetric pseudodifferential systems", *Sibirsk. Mat. Zh.* **20**:3 (1979), 557–578. In Russian; translated in *Siberian Math. J.* **20**:3 (1979), 390–405. MR 81a:35100 Zbl 0453.35091

[Jones et al. 2007] I. F. Jones, M. C. Goodwin, I. D. Berranger, H. Zhou, and P. A. Farmer, "Application of anisotropic 3D reverse time migration to complex North Sea imaging", *Soc. Explor. Geophys. Expand. Abstr.* **26** (2007), 2140–2144.

[Langston 1979] C. A. Langston, "Structure under Mount Rainier, Washington, inferred from teleseismic body waves", *J. Geophys. Res.* **84** (1979), 4749–4762.

[Lu et al. 2009] R. Lu, P. Traynin, and J. E. Anderson, "Comparison of elastic and acoustic reverse-time migration on the synthetic elastic Marmousi-II OBC dataset", *Soc. Explor. Geophys. Expand. Abstr.* **28** (2009), 2799–2803.

[McMechan 1983] G. A. McMechan, "Migration by extrapolation of time-dependent boundary values", *Geophys. Prosp.* **31**:3 (1983), 413–420.

[Mulder and Plessix 2004] W. A. Mulder and R. E. Plessix, "A comparison between one-way and two-way wave-equation migration", *Geophysics* **69**:6 (2004), 1491–1504.

[Poppeliers and Pavlis 2003] C. Poppeliers and G. L. Pavlis, "Three-dimensional, prestack, plane wave migration of teleseismic P-to-S converted phases, 1: Theory", *J. Geophys. Res.* **108** (2003), Art. ID #2112.

[Revenaugh 1995] J. Revenaugh, "A scattered-wave image of subduction beneath the transverse ranges", *Science* **268**:5219 (1995), 1888–1892.

[Op 't Root et al. 2012] T. J. P. M. Op 't Root, C. C. Stolk, and M. V. de Hoop, "Linearized inverse scattering based on seismic reverse time migration", *J. Math. Pures Appl.* (9) **98**:2 (2012), 211–238. MR 2944376 Zbl 06064167

[Rydberg and Weber 2000] T. Rydberg and M. Weber, "Receiver function arrays: a reflection seismic approach", *Geophys. J. Int.* **141** (2000), 1–11.

[Schneider 1978] W. A. Schneider, "Integral formulation for migration in two and three dimensions", *Geophysics* **43**:1 (1978), 49–76.

[Singh and Chapman 1988] S. C. Singh and C. H. Chapman, "WKBJ seismogram theory in anisotropic media", *J. Acoust. Soc. Am.* **84**:2 (1988), 732–741.

[Stolk and de Hoop 2002] C. C. Stolk and M. V. de Hoop, "Microlocal analysis of seismic inverse scattering in anisotropic elastic media", *Comm. Pure Appl. Math.* **55**:3 (2002), 261–301. MR 2002i:74029 Zbl 1018.86002

[Sun and McMechan 2001] R. Sun and G. A. McMechan, "Scalar reverse-time depth migration of prestack elastic seismic data", *Geophysics* **66**:5 (2001), 1519–1527.

[Taylor 1975] M. E. Taylor, "Reflection of singularities of solutions to systems of differential equations", *Comm. Pure Appl. Math.* **28**:4 (1975), 457–478. MR 58 #22994 Zbl 0332.35058

[Vinnik 1977] L. Vinnik, "Detection of waves converted from P to SV in the mantle", *Phys. Earth Planet. Inter.* **15** (1977), 39–45.

[Wang et al. 2010] S. Wang, M. V. de Hoop, and J. Xia, "Acoustic inverse scattering via Helmholtz operator factorization and optimization", *J. Comput. Phys.* **229**:22 (2010), 8445–8462. MR 2011f:76155 Zbl 1201.65193

[Whitmore 1983] D. Whitmore, "Iterative depth migration by backward time propagation", *Soc. Explor. Geophys. Expand. Abstr.* **2** (1983), 382–385.

[Woodhouse 1974] J. H. Woodhouse, "Surface waves in a laterally varying layered structure", *Geophys. J. R. Astr. Soc.* **37**:3 (1974), 461–490. Zbl 0297.73028

[Xie and Wu 2006] X.-B. Xie and R.-S. Wu, "A depth migration method based on the full-wave reverse-time calculation and local one-way propagation", *Soc. Explor. Geophys. Expand. Abstr.* **25** (2006), 2333–2336.

[Yan and Sava 2007] J. Yan and P. Sava, "Elastic wavefield imaging with scalar and vector potentials", *Soc. Explor. Geophys. Expand. Abstr.* **26** (2007), 2150–2154.

[Yan and Sava 2008] J. Yan and P. Sava, "Elastic wavefield separation for VTI media", *Soc. Explor. Geophys. Expand. Abstr.* **27** (2008), 2191–2195.

[Yoon et al. 2004] K. Yoon, K. J. Marfurt, and W. Starr, "Challenges in reverse-time migration", *Soc. Explor. Geophys. Expand. Abstr.* **23** (2004), 1057–1060.

[Zhang et al. 2007] Y. Zhang, J. Sun, and S. Gray, "Reverse-time migration: amplitude and implementation issues", *Soc. Explor. Geophys. Expand. Abstr.* **26** (2007), 2145–2149.

vbrytik@math.purdue.edu	*Department of Mathematics, Purdue University, West Lafayette, IN 47907, United States*
mdehoop@purdue.edu	*Department of Mathematics, Purdue University, West Lafayette, IN 47907, United States*
hilst@mit.edu	*Department of Earth, Atmospheric and Planetary Sciences, Massachusetts Institute of Technology, Cambridge, MA 02139, United States*

Inverse problems in spectral geometry

KIRIL DATCHEV AND HAMID HEZARI

In this survey we review positive inverse spectral and inverse resonant results for the following kinds of problems: Laplacians on bounded domains, Laplace–Beltrami operators on compact manifolds, Schrödinger operators, Laplacians on exterior domains, and Laplacians on manifolds which are hyperbolic near infinity.

1. Introduction

Marc Kac [1966], in a famous paper, raised the following question: Let $\Omega \subset \mathbb{R}^2$ be a bounded domain and let

$$0 \le \lambda_0 < \lambda_1 \le \lambda_2 \le \cdots$$

be the eigenvalues of the nonnegative Euclidean Laplacian Δ_Ω with either Dirichlet or Neumann boundary conditions. Is Ω determined up to isometries from the sequence $\lambda_0, \lambda_1, \ldots$? We can ask the same question about bounded domains in $\mathbb{R}^n$, and below we will discuss other generalizations as well. Physically, one motivation for this problem is identifying distant physical objects, such as stars or atoms, from the light or sound they emit. These inverse spectral problems, as some engineers have recently proposed in [Reuter 2007; Reuter et al. 2007; 2009; Peinecke et al. 2007], may also have interesting applications in shape-matching, copyright and medical shape analysis.

The only domains in $\mathbb{R}^n$ known to be spectrally distinguishable from all other domains are balls. It is not even known whether or not ellipses are spectrally rigid, i.e., whether or not any continuous family of domains containing an ellipse and having the same spectrum as that ellipse is necessarily trivial. We can go further and ask the same question about a compact Riemannian manifold (M, g) (with or without boundary): can we determine (M, g) up to isometries from the spectrum of the Laplace–Beltrami operator Δ_g? Or in general, what can we hear from the spectrum? For example, can we hear the area (volume in higher

---

The first author is partially supported by a National Science Foundation postdoctoral fellowship, and the second author is partially supported by the National Science Foundation under grant DMS-0969745. The authors are grateful for the hospitality of the Mathematical Sciences Research Institute, where part of this research was carried out.

dimensions or in the case of Riemannian manifolds) or the perimeter of the domain? For the sake of brevity we only mention the historical background for the case of domains.

In 1910, Lorentz gave a series of physics lectures in Göttingen, and he conjectured that the asymptotics of the counting function of the eigenvalues are given by

$$N(\lambda) = \sharp\{\lambda_j; \lambda_j \leq \lambda\} = \frac{\text{Area}(\Omega)}{2\pi}\lambda + O(\sqrt{\lambda}).$$

This asymptotic in particular implies that Area(Ω) is a spectral invariant. Hilbert thought this conjecture would not be proven in his lifetime, but less than two years later Hermann Weyl proved it using the theory of integral equations taught to him by Hilbert. Pleijel [1954] proved that one knows the perimeter of Ω, and Kac [1966] rephrased these results in terms of asymptotics of the heat trace

$$\text{Tr}\, e^{-t\Delta_\Omega} \sim t^{-1} \sum_{j=0}^{\infty} a_j t^{j/2}, \quad t \to 0^+,$$

where the first coefficient a_0 gives the area and the second coefficient gives the perimeter. McKean and Singer [1967] proved Pleijel's conjecture that the Euler characteristic $\chi(\Omega)$ is also a spectral invariant (this is in fact given by a_2) and hence the number of holes is known. Gordon, Webb and Wolpert [1992] found examples of pairs of distinct plane domains with the same spectrum. However, their examples were nonconvex and nonsmooth, and it remains an open question to prove that convex domains are determined by the spectrum (although there are higher-dimensional counterexamples for this in [Gordon and Webb 1994]) or that smooth domains are determined by the spectrum.

In this survey we review positive inverse spectral and inverse resonant results for the following kinds of problems: Laplacians on bounded domains, Laplace–Beltrami operators on compact manifolds, Schrödinger operators, Laplacians on exterior domains, and Laplacians on manifolds which are hyperbolic near infinity. We also recommend the survey [Zelditch 2004b]. For negative results (counterexamples) we refer the reader to the surveys [Gordon 2000; Gordon, Perry, Schueth 2005].

In the next two sections of the paper we review uniqueness results for radial problems (Section 2), and for real analytic and symmetric problems (Section 3). In the first case the object to be identified satisfies very strong assumptions (radialness includes full symmetry as well as analyticity) but it is identified in a broad class of objects. In this case the first few heat invariants, together with an isoperimetric or isoperimetric-type inequality, often suffice. In the second case the assumptions on the object to be identified are somewhat weaker (only analyticity and finitely many reflection symmetries are assumed) but the identification is only within a class of objects which also satisfies the same assumptions, and generic nondegeneracy assumptions are also needed. These proofs are based on wave trace invariants corresponding to a single nondegenerate simple periodic orbit and its iterations.

In Section 4 we consider rigidity and local uniqueness results, where it is shown in the first case that isospectral deformations of a given object are necessarily trivial, and in the second case that a given object is determined by its spectrum among objects which are nearby in a suitable sense. Here the objects to be determined are more general than in the cases considered in Section 2, but less general than those in Section 3: they are ellipses, spheres, flat manifolds (which have completely integrable dynamics), and manifolds of constant negative curvature (which have chaotic dynamics). The proofs use these special features of the classical dynamics.

In Section 5 we consider compactness results, where it is shown that certain isospectral families are compact in a suitable topology. These proofs are based on heat trace invariants and on the determinant of the Laplacian, and much more general assumptions are possible than in the previous cases.

Finally, in Section 6 we review the trace invariants used for the positive results in the previous sections, and give examples of their limitations, that is to say examples of objects which have the same trace invariants but which are not isospectral. At this point we also discuss the history of these invariants, going back to the seminal paper [Selberg 1956].

We end the introduction by presenting the four basic settings we consider in this survey:

1.1. Dirichlet and Neumann Laplacians on bounded domains in $\mathbb{R}^n$. Let Ω be a bounded open set with piecewise smooth boundary. Let Δ_Ω be the nonnegative Laplacian on Ω with Dirichlet or Neumann boundary conditions. Let

$$\text{spec}(\Delta_\Omega) = (\lambda_j)_{j=0}^\infty, \quad \lambda_0 < \lambda_1 \leq \lambda_2 \leq \cdots$$

be the eigenvalues included according to multiplicity, and u_j the corresponding eigenfunctions, that is to say

$$\Delta_\Omega u_j = \lambda_j u_j.$$

Recall that $\lambda_0 > 0$ in the Dirichlet case and $\lambda_0 = 0$ in the Neumann case.

1.2. Laplace–Beltrami operators on compact manifolds. Let (M, g) be a compact Riemannian manifold without boundary. Let $\Delta_g = -\text{div}_g \, \text{grad}_g$ be the nonnegative Laplace–Beltrami operator on M, which we also call the Laplacian

for short. Let
$$\operatorname{spec}(\Delta_g) = (\lambda_j)_{j=0}^\infty, \quad 0 = \lambda_0 < \lambda_1 \leq \lambda_2 \leq \cdots$$
be the eigenvalues included according to multiplicity, and u_j the corresponding eigenfunctions, that is to say
$$\Delta_g u_j = \lambda_j u_j.$$

1.3. *Nonsemiclassical and semiclassical Schrödinger operators on $\mathbb{R}^n$.* Let

$$V \in C^\infty(\mathbb{R}^n; \mathbb{R}), \quad \lim_{|x| \to \infty} V(x) = \infty, \tag{1-1}$$

and let Δ be the nonnegative Laplacian on $\mathbb{R}^n$. Let
$$P_{V,h} = h^2 \Delta + V, \quad h > 0,$$
$$P_V = P_{V,1}.$$

We call P_V the nonsemiclassical Schrödinger operator associated to V, and $P_{V,h}$ the semiclassical operator. For any $h > 0$, the spectrum of $P_{V,h}$ on $\mathbb{R}^n$ is discrete, and we write it as
$$\operatorname{spec}(P_{V,h}) = (\lambda_j)_{j=0}^\infty, \quad \lambda_0 < \lambda_1 \leq \lambda_2 \leq \cdots.$$
The eigenvalues λ_j depend on h, but we do not include this in the notation. We denote by u_j the corresponding eigenfunctions (which also depend on h), so that
$$P_{V,h} u_j = \lambda_j u_j.$$

1.4. *Resonance problems for obstacle and potential scattering.* In this section we discuss problems where the spectrum consists of a half line of essential spectrum, together with possibly finitely many eigenvalues. In such settings the spectrum contains limited information, but one can often define resonances, which supplement the discrete spectral data and contain more information.

Obstacle scattering in $\mathbb{R}^n$. Let $O \subset \mathbb{R}^n$ be a bounded open set with smooth boundary, let $\Omega = \mathbb{R}^n \setminus \overline{O}$, and suppose that Ω is connected. Let Δ_Ω be the nonnegative Dirichlet or Neumann Laplacian on Ω. Then the spectrum of Δ_Ω is continuous and equal to $[0, \infty)$, and so it contains no (further) information about Ω. One way to reformulate the inverse spectral problem in this case is in terms of *resonances*, which are defined as follows. Introduce a new spectral parameter $z = \sqrt{\lambda}$, with $\sqrt{\ }$ taken so as to map $\mathbb{C} \setminus [0, \infty)$ to the upper half-plane. As $\operatorname{Im} z \to 0^+$, z^2 approaches $[0, \infty)$ and the resolvent $(\Delta_\Omega - z^2)^{-1}$ has no limit as a map $L^2(\Omega) \to L^2(\Omega)$. However, if we restrict the domain of the resolvent and expand the range it is possible not only to take the limit but also to take a meromorphic continuation to a larger set. More precisely the resolvent

$$(\Delta_\Omega - z^2)^{-1} : L^2_{\text{comp}} \to L^2_{\text{loc}},$$

(where L^2_{comp} denotes compactly supported L^2 functions and L^2_{loc} denotes functions which are locally L^2) continues meromorphically as an operator-valued function of z from $\{\text{Im } z > 0\}$ to $\mathbb{C}$ when n is odd and to the Riemann surface of $\log z$ when n is even. Resonances are defined to be the poles of this continuation of the resolvent. Let $\text{res}(\Delta_\Omega)$ denote the set of resonances, included according to multiplicity. See for example [Melrose 1995; Sjöstrand 2002; Zworski 2011] for more information.

Potential scattering in $\mathbb{R}^n$. Let $P_{V,h}$ be as before, but instead of (1-1) assume $V \in C_0^\infty(\mathbb{R}^n)$. Then the continuous spectrum of $P_{V,h}$ is equal to $[0, \infty)$, but if V is not everywhere nonnegative then $P_{V,h}$ may have finitely many negative eigenvalues. In either case, the resolvent

$$(P_{V,h} - z^2)^{-1} : L^2_{\text{comp}} \to L^2_{\text{loc}},$$

has a meromorphic continuation from $\{\text{Im } z > 0\}$ to $\mathbb{C}$ when n is odd and to the Riemann surface of $\log z$ when n is even, and resonances are defined to be the poles of this continuation. Let $\text{res}(P_{V,h})$ denote the set of resonances, included according to multiplicity. Again, see for example [Melrose 1995; Sjöstrand 2002; Zworski 2011] for more information.

Scattering on asymptotically hyperbolic manifolds. The problem of determining a noncompact manifold from the scattering resonances of the associated Laplace–Beltrami is in general a much more difficult one, but some progress has been made in the asymptotically hyperbolic setting. Meromorphic continuation of the resolvent was established in [Mazzeo and Melrose 1987], and a wave trace formula in the case of surfaces with exact hyperbolic ends was found by Guillopé and Zworski [1997], which has led to some compactness results: see Section 5.4.

2. The radial case

In this case one makes a strong assumption (radial symmetry) on the object to be spectrally determined (whether it is an open set in $\mathbb{R}^n$, a compact manifold, or a potential) but makes almost no assumption on the class of objects within which it is determined. The methods involved use the first few heat invariants, and in many cases the isoperimetric inequality or an isoperimetric-type inequality.

2.1. *Bounded domains in $\mathbb{R}^n$.* The oldest inverse spectral results are for radial problems. If $\Omega \subset \mathbb{R}^n$ is a bounded open set with smooth boundary, then the spectrum of the Dirichlet (or Neumann) Laplacian on Ω agrees with the spectrum on the unit ball if and only if Ω is a translation of this ball. This can be proved in many ways; one way is to use heat trace invariants. These are defined to be the coefficients of the asymptotic expansion of the heat trace as $t \to 0^+$:

$$\sum_{j=0}^{\infty} e^{-t\lambda_j} = \operatorname{Tr} e^{-t\Delta_\Omega} \sim t^{-n/2} \sum_{j=0}^{\infty} a_j t^{j/2}, \qquad (2\text{-}1)$$

where in both the Dirichlet and the Neumann case a_0 is a universal constant times $\operatorname{vol}(\Omega)$, and a_1 is a universal constant times $\operatorname{vol}(\partial\Omega)$. The left-hand side is clearly determined by the spectrum, and so the conclusion follows from the isoperimetric inequality.

2.2. Compact manifolds. Let (M, g) be a smooth Riemannian manifold of dimension n without boundary. If $n \leq 6$, then the spectrum of the Laplacian on M agrees with the spectrum on S^n (equipped with the round metric) if and only if M is isometric to S^n. This was proved in [Tanno 1973; 1980] using the first four coefficients, a_0, a_1, a_2, a_3, of the heat trace expansion, which in this case takes the form

$$\sum_{j=0}^{\infty} e^{-t\lambda_j} = \operatorname{Tr} e^{-t\Delta_g} \sim t^{-n/2} \sum_{j=0}^{\infty} a_j t^j.$$

In higher dimensions the analogous result is not known. Zelditch [1996] proved that if the multiplicities m_k of the *distinct* eigenvalues $0 = E_0 < E_1 < E_2 < \cdots$ of the Laplacian on M obey the asymptotic $m_k = ak^{n-1} + O(k^{n-2})$, for some $a > 0$ as $k \to \infty$ (this is the asymptotic behavior for the multiplicities of the eigenvalues of the sphere), then (M, g) is a Zoll manifold, that is to say a manifold on which all geodesics are periodic with the same period.

2.3. Schrödinger operators. In general it is impossible to determine a potential V from the spectrum of the nonsemiclassical Schrödinger operator $\Delta + V$. For example, McKean and Trubowitz [1981] found an infinite-dimensional family of potentials in $C^\infty(\mathbb{R})$ which are isospectral with the harmonic oscillator $V(x) = x^2$.

However, analogous uniqueness results to those above were proved in [Datchev, Hezari, and Ventura 2011], where it is shown that radial, monotonic potentials in $\mathbb{R}^n$ (such as for example the harmonic oscillator) are determined by the spectrum of the associated *semiclassical* Schrödinger operator among all potentials with discrete spectrum. The approach is based in part on that of [Colin de Verdière 2011] and [Guillemin and Wang 2009] (see also [Guillemin and Sternberg 2010, §10.6]), where a one-dimensional version of the result is proved. Colin de Verdière and Guillemin–Wang show that an even function (or a suitable noneven function) is determined by its spectrum within the class of functions monotonic away from 0.

The method of proof is similar to that used to prove spectral uniqueness of balls in $\mathbb{R}^n$ as discussed in Section 2.1 above. Namely, we use the first two trace invariants, this time of the semiclassical trace formula of Helffer and Robert [1983], together with the isoperimetric inequality. We show that if V, V_0 are as in (1-1), if $V_0(x) = R(|x|)$ where $R(0) = 0$ and $R'(r) > 0$ for $r > 0$, and if $\text{spec}(P_{V,h}) = \text{spec}(P_{V_0,h})$ up to order[1] $o(h^2)$ for $h \in \{h_j\}_{j=0}^{\infty}$ with $h_j \to 0^+$, then $V(x) = V_0(x - x_0)$ for some $x_0 \in \mathbb{R}^n$.

The semiclassical trace formula we use is

$$\text{Tr}(f(P_{V,h})) = \tag{2-2}$$
$$\frac{1}{(2\pi h)^n} \left(\int_{\mathbb{R}^{2n}} f(|\xi|^2 + V) \, dx \, d\xi + \frac{h^2}{12} \int_{\mathbb{R}^{2n}} |\nabla V|^2 f^{(3)}(|\xi|^2 + V) \, dx \, d\xi + \mathcal{O}(h^4) \right),$$

where $f \in C_0^{\infty}(\mathbb{R})$.

Because the spectrum of $P_{V,h}$ is known up to order $o(h^2)$ we obtain from (2-2) the two trace invariants

$$\int_{\{|\xi|^2 + V(x) < \lambda\}} dx \, d\xi, \quad \int_{\{|\xi|^2 + V(x) < \lambda\}} |\nabla V(x)|^2 \, dx \, d\xi, \tag{2-3}$$

for each λ. It follows in particular that V is nonnegative. By integrating in the ξ variable, we rewrite these invariants as follows:

$$\int_{\{V(x) < \lambda\}} (\lambda - V)^{n/2} dx, \quad \int_{\{V(x) < \lambda\}} |\nabla V(x)|^2 (\lambda - V)^{n/2} dx. \tag{2-4}$$

Using the coarea formula we rewrite the invariants in (2-4) as

$$\int_0^{\lambda} \left(\int_{\{V=s, \nabla V \neq 0\}} \frac{(\lambda - V)^{n/2}}{|\nabla V|} dS \right) ds,$$
$$\int_0^{\lambda} \left(\int_{\{V=s\}} |\nabla V| (\lambda - V)^{n/2} dS \right) ds.$$

Using the fact that $V = s$ in the inner integrand, the factor of $(\lambda - V)^{n/2} = (\lambda - s)^{n/2}$ can be taken out of the surface integral, leaving

$$\int_0^{\lambda} (\lambda - s)^{n/2} I_1(s) \, ds, \quad \int_0^{\lambda} (\lambda - s)^{n/2} I_2(s) \, ds, \tag{2-5}$$

---

[1]The implicit rate of convergence here must be uniform on $[0, \lambda_0]$ for each $\lambda_0 > 0$.

where

$$I_1(s) = \int_{\{V=s, \nabla V \neq 0\}} \frac{1}{|\nabla V|} dS, \quad I_2(s) = \int_{\{V=s\}} |\nabla V| dS. \qquad (2\text{-}6)$$

We denote the integrals (2-5) by $A_{1+n/2}(I_1)(\lambda)$ and $A_{1+n/2}(I_1)(\lambda)$. These are Abel fractional integrals of I_1 and I_2 (see for example [Zelditch 1998a, §5.2] and [Guillemin and Sternberg 2010, (10.45)]), and they can be inverted by applying $A_{1+n/2}$, using the formula

$$\frac{1}{\Gamma(\alpha)} A_\alpha \circ \frac{1}{\Gamma(\beta)} A_\beta = \frac{1}{\Gamma(\alpha+\beta)} A_{\alpha+\beta}, \qquad (2\text{-}7)$$

and differentiating $n+1$ times. From this we conclude that the functions I_1 and I_2 in (2-6) are spectral invariants for every $s > 0$.

Integrating I_1 and using the coarea formula again we find that the volumes of the sets $\{V < s\}$ are spectral invariants:

$$\int_0^s I_1(s') \, ds' = \int_0^s \int_{\{V=s', \nabla V \neq 0\}} \frac{1}{|\nabla V|} dS \, ds' = \int_{\{V<s\}} 1 \, dx. \qquad (2\text{-}8)$$

From Cauchy–Schwarz and the fact that I_1 and I_2 are spectral invariants we obtain

$$\left(\int_{\{V=s\}} 1 \, dS\right)^2 \leq \int_{\{V=s\}} \frac{1}{|\nabla V|} dS \int_{\{V=s\}} |\nabla V| \, dS$$
$$= \int_{\{R=s\}} \frac{1}{R'} dS \int_{\{R=s\}} R' dS, \qquad (2\text{-}9)$$

when s is not a critical value of V, and thus, by Sard's theorem, for almost every $s \in (0, \lambda_0)$. On the other hand, using the invariants obtained in (2-8) and the fact that the sets $\{R < s\}$ are balls, by the isoperimetric inequality we find

$$\int_{\{R=s\}} 1 \, dS \leq \int_{\{V=s\}} 1 \, dS. \qquad (2\text{-}10)$$

However,

$$\left(\int_{\{R=s\}} 1 \, dS\right)^2 = \int_{\{R=s\}} \frac{1}{R'} dS \int_{\{R=s\}} R' dS,$$

because $1/R'$ and R' are constant on $\{R = s\}$. Consequently

$$\int_{\{R=s\}} 1 \, dS = \int_{\{V=s\}} 1 \, dS,$$

and so $\{V = s\}$ is a sphere for almost every s, because only spheres extremize the isoperimetric inequality. Moreover,

$$\left(\int_{\{V=s\}} 1 \, dS\right)^2 = \int_{\{V=s\}} \frac{1}{|\nabla V|} dS \int_{\{V=s\}} |\nabla V| \, dS,$$

and so $|\nabla V|^{-1}$ and $|\nabla V|$ are proportional on the surface $\{V = s\}$ for almost every s, again by Cauchy–Schwarz. Using (2-9) to determine the constant of proportionality, we find that

$$|\nabla V|^2 = R'(R^{-1}(s))^2 = (R^{-1})'(s)^{-2} \stackrel{\text{def}}{=} F(s)$$

on $\{V = s\}$. In other words

$$|\nabla V|^2 = F(V), \qquad (2\text{-}11)$$

for all $x \in V^{-1}(s)$ for almost all s. However, because $F(V) \neq 0$ when $V \neq 0$, it follows by continuity that this equation holds for all $x \in V^{-1}((0, \infty))$.

We solve this equation by restricting it to flowlines of ∇V, with initial conditions taken on a fixed level set $\{V = s_0\}$, and conclude that, the level surfaces are not only spheres (as follows from (2-10)) but are moreover spheres with a common center. Hence, up to a translation, V is radial. Since the volumes (2-8) are spectral invariants, it follows that $V(x) = R(|x|)$.

2.4. Resonance problems. We first mention briefly some results for inverse problems for resonances for the nonsemiclassical Schrödinger problem when $n = 1$. Zworski [2001] proved that a compactly supported even potential $V \in L^1(\mathbb{R})$ is determined from the resonances of P_V among other such potentials, and Korotyaev [2005] showed that a potential which is not necessarily even is determined by some additional scattering data.

Analogous results to those discussed in Section 2.1 hold in the case of obstacle scattering. Hassell and Zworski [1999] showed that a ball is determined by its Dirichlet resonances among all compact obstacles in $\mathbb{R}^3$. Christiansen [2008] extended this result to multiple balls, to higher odd dimensions, and to Neumann resonances. As in the other results discussed above, the proofs use two trace invariants and isoperimetric-type inequalities, although the invariants and inequalities are different here. There is also a large literature of inverse scattering results where data other than the resonances are used. A typical datum here is the *scattering phase*: see for example [Melrose 1995, §4.1].

In [Datchev and Hezari 2012] we prove the analogue for resonances of the result in the previous section for semiclassical Schrödinger operators with discrete spectrum. Let $n \geq 1$ be odd, and let $V_0, V \in C_0^\infty(\mathbb{R}^n; [0, \infty))$. Suppose $V_0(x) = R(|x|)$, and $R'(r)$ vanishes only at $r = 0$ and whenever $R(r) = 0$, and suppose that $\mathrm{res}(P_{V_0,h}) = \mathrm{res}(P_{V,h})$, up to order[2] $o(h^2)$, for $h \in \{h_j\}_{j=1}^\infty$ for some sequence $h_j \to 0$. Then there exists $x_0 \in \mathbb{R}^n$ such that $V(x) = V_0(x - x_0)$.

Our proof is, as before, based on recovering and analyzing first two integral invariants of the Helffer–Robert semiclassical trace formula [1983, Proposition 5.3] (see also [Guillemin and Sternberg 2010, §10.5]):

$$\mathrm{Tr}(f(P_{V,h}) - f(P_{0,h}))$$
$$= \frac{1}{(2\pi h)^n} \left(\int_{\mathbb{R}^{2n}} f(|\xi|^2 + V) - f(|\xi|^2) \, dx \, d\xi \right.$$
$$\left. + \frac{h^2}{12} \int_{\mathbb{R}^{2n}} |\nabla V|^2 f^{(3)}(|\xi|^2 + V) \, dx \, d\xi + \mathcal{O}(h^4) \right). \quad (2\text{-}12)$$

To express the left-hand side of (2-12) in terms of the resonances of $P_{V,h}$, we use Melrose's Poisson formula [Melrose 1982], an extension of the formula of Bardos, Guillot, and Ralston [1982]:

$$2\,\mathrm{Tr}\bigl(\cos(t\sqrt{P_{V,h}}) - \cos(t\sqrt{P_{0,h}})\bigr) = \sum_{\lambda \in \mathrm{res}(P_{V,h})} e^{-i|t|\lambda}, \quad t \neq 0, \quad (2\text{-}13)$$

where equality is in the sense of distributions on $\mathbb{R} \setminus 0$.

From (2-13), it follows that if

$$\hat{g} \in C_0^\infty(\mathbb{R} \setminus 0) \text{ is even}, \quad (2\text{-}14)$$

then

$$\mathrm{Tr}\bigl(g(\sqrt{-h^2\Delta + V}) - g(\sqrt{-h^2\Delta})\bigr) = \frac{1}{4\pi} \sum_{\lambda \in \mathrm{res}(P_{V,h})} \int_\mathbb{R} e^{-i|t|\lambda} \hat{g}(t) \, dt. \quad (2\text{-}15)$$

Now setting the right-hand sides of (2-15) and (2-12) equal and taking $h \to 0$, we find that

$$\int_{\mathbb{R}^{2n}} f(|\xi|^2 + V) - f(|\xi|^2) \, dx \, d\xi, \quad \int_{\mathbb{R}^{2n}} |\nabla V|^2 f^{(3)}(|\xi|^2 + V) \, dx \, d\xi \quad (2\text{-}16)$$

are resonant invariants (i.e., are determined by knowledge of the resonances up to $o(h^2)$) provided that $f(\tau^2) = g(\tau)$ for all τ and for some g as in (2-14). Taylor expanding, we write the first invariant as

$$\sum_{k=1}^m \frac{1}{k!} \int_{\mathbb{R}^n} f^{(k)}(|\xi|^2) \, d\xi \int_{\mathbb{R}^n} V(x)^k \, dx$$
$$+ \int_{\mathbb{R}^{2n}} \frac{V(x)^{m+1}}{m!} \int_0^1 (1-t)^m f^{(m+1)}(|\xi|^2 + tV(x)) \, dt \, dx \, d\xi.$$

---

[2]The implicit rate of convergence here must be uniform on the disk of radius λ_0 for each $\lambda_0 > 0$.

Replacing f by f_λ, where $f_\lambda(\tau) = f(\tau/\lambda)$ (note that $g_\lambda(\tau) = f_\lambda(\tau^2)$ satisfies (2-14)) gives

$$\sum_{k=1}^{m} \lambda^{n/2-k} \frac{1}{k!} \int_{\mathbb{R}^n} f^{(k)}(|\xi|^2) \, d\xi \int_{\mathbb{R}^n} V(x)^k \, dx + \mathcal{O}(\lambda^{n/2-m-1})$$

Taking $\lambda \to \infty$ and $m \to \infty$ we obtain the invariants

$$\int_{\mathbb{R}^n} f^{(k)}(|\xi|^2) \, d\xi \int_{\mathbb{R}^n} V(x)^k \, dx,$$

for every $k \geq 1$.

In [Datchev and Hezari 2012, Lemma 2.1] it is shown that there exists g satisfying (2-14) such that if $f(\tau^2) = g(\tau)$, then $\int_{\mathbb{R}^n} f^{(k)}(|\xi|^2) \, d\xi \neq 0$, provided $k \geq n$.

This shows that

$$\int_{\mathbb{R}^n} V(x)^k \, dx = \int_{\mathbb{R}^n} V_0(x)^k \, dx \tag{2-17}$$

for every $k \geq n$, and a similar analysis of the second invariant of (2-16) proves that

$$\int_{\mathbb{R}^n} V(x)^k |\nabla V(x)|^2 \, dx = \int_{\mathbb{R}^n} V_0(x)^k |\nabla V_0(x)|^2 \, dx \tag{2-18}$$

for every $k \geq n$.

We rewrite the invariant (2-17) using $V_* dx$, the pushforward of Lebesgue measure by V, as

$$\int_{\mathbb{R}^n} V(x)^k \, dx = \int_{\mathbb{R}} s^k (V_* dx)_s = i^k \widehat{V_* dx}^{(k)}(0). \tag{2-19}$$

Since V and V_0 are both bounded functions, the pushforward measures are compactly supported and hence have entire Fourier transforms, and we conclude that

$$V_* dx = V_{0*} dx + \sum_{k=0}^{n-1} c_k \delta_0^{(k)} = V_{0*} dx + c_0 \delta_0.$$

For the first equality we used the invariants (2-19), and for the second the fact that $V_* dx$ is a measure. In other words

$$\mathrm{vol}(\{V > \lambda\}) = \mathrm{vol}(\{V_0 > \lambda\})$$

whenever $\lambda > 0$. Moreover, this shows that $V_* dx$ is absolutely continuous on $(0, \infty)$, and so by Sard's lemma the critical set of V is Lebesgue-null on

$V^{-1}((0,\infty))$. As a result we may use the coarea formula[3] to write

$$V_* dx = \int_{\{V=s\}} |\nabla V|^{-1} dS \, ds \quad \text{on } (0,\infty)$$

and to conclude that

$$\int_{\{V=s\}} |\nabla V|^{-1} dS = \int_{\{V_0=s\}} |\nabla V_0|^{-1} dS$$

for almost every $s > 0$. Similarly, rewriting the invariants (2-18) as

$$\int_{\mathbb{R}^n} V(x)^k |\nabla V(x)|^2 dx = \int_{\mathbb{R}} s^k \int_{\{V=s\}} |\nabla V| \, dS \, ds,$$

we find that

$$\int_{\{V=s\}} |\nabla V| \, dS = \int_{\{V_0=s\}} |\nabla V_0| \, dS \quad s > 0.$$

From this point on the proof proceeds as in the previous section.

To our knowledge it is not known whether such results hold in even dimensions. The higher-dimensional results discussed above all rely on the Poisson formula (2-13) which is only valid for odd dimensions. A similar formula is also true in the obstacle case [Melrose 1983b], although slightly more care is needed in the definition of $\cos(t\sqrt{\Delta_\Omega}) - \cos(t\sqrt{\Delta_{\mathbb{R}^n}})$ because the two operators act on different spaces. poles of R_V. When $n = 1$ a stronger trace formula, valid for all $t \in \mathbb{R}$, is known: see for example [Zworski 1997, page 3]. When n is even, because the meromorphic continuation of the resolvent is not to $\mathbb{C}$ but to the Riemann surface of the logarithm, Poisson formulæ for resonances are more complicated and contain error terms: see [Sjöstrand 1997; Zworski 1998]. A proof based on Sjöstrand's local trace formula [1997] would be of particular interest, firstly because this formula applies in all dimensions and to a very general class of operators, and also because it uses only resonances in a sector (and in certain versions, as in [Bony 2002], resonances in a strip) around the real axis. This would strengthen the known results in odd dimensions as well as proving results in even dimensions, as one would only have to assume that these resonances agreed and not that all resonances do.

3. The real analytic and symmetric case

In this case uniqueness results about nonradial objects are obtained, so the assumptions on the object to be determined are weaker. However, the assumptions on the class of objects within which it is determined are much stronger – in fact

---

[3]If $n=1$ we put $\int_{\{V=s\}} |\nabla V|^{-1} dS = \sum_{x \in V^{-1}(s)} |V'(x)|^{-1}$.

they are the same as the assumptions on the object to be determined. The two main assumptions are analyticity and symmetry. In each case wave invariants are used which are microlocalized near certain periodic orbits, as opposed to the nonmicrolocal heat invariants of the previous section.

3.1. Bounded domains in $\mathbb{R}^n$. Here the main tool is the following result from [Guillemin and Melrose 1979b]. When $\Omega \subset \mathbb{R}^n$ is a bounded, open set with smooth boundary, they prove that $\text{Tr}(\cos(t\sqrt{\Delta_\Omega}))$ is a tempered distribution in $\mathbb{R}$ with the property

$$\text{sing supp } \text{Tr}(\cos(t\sqrt{\Delta_\Omega})) \subset \{0\} \cup \overline{\text{Lsp}(\Omega)},$$

where Lsp denotes the length spectrum, that is to say the lengths of periodic billiard orbits in Ω. Moreover, they show that if $T \in \text{Lsp}(\Omega)$ is of simple length[4] and γ_T is nondegenerate,[5] then for t sufficiently near T we have

$$\text{Tr} \cos(t\sqrt{\Delta_\Omega})$$
$$= \text{Re}\left[i^{\sigma_T} \frac{T^\sharp}{\sqrt{|\det(I - P_T)|}} (t - T + i0)^{-1} \right.$$
$$\left. \times \left(1 + \sum_{j=1}^{\infty} a_j (t-T)^j \log(t - T + i0)\right) \right] + S(t), \quad (3\text{-}1)$$

where S is smooth near T. Here $T^\sharp$ is the primitive length of γ_T, which is the length of γ_T without retracing, and σ_T is the Maslov index of γ_T (which can be defined geometrically but which appears here as the signature of the Hessian in the stationary phase expansion of the wave trace). The coefficients a_j are known as *wave invariants*. Viewing the boundary locally as the graph of a function f, they are polynomials in the Taylor coefficients of f at the reflection points of γ_T. In general there is no explicit formula, but they were computed in [Zelditch 2009] in the special case discussed below.

We now assume that $n = 2$, with coordinates (x, y), and make these further assumptions:

(1) Ω is simply connected, symmetric about the x-axis, and $\partial \Omega$ is analytic on $\{y \neq 0\}$.

(2) There is a nondegenerate vertical bouncing ball orbit γ of length T such that both T and $2T$ are simple lengths in $\text{Lsp}(\Omega)$.

(3) The endpoints of γ are not critical points of the curvature of $\partial \Omega$.

---

[4]This means that only one periodic orbit (up to time reversal), γ_T, has length T.

[5] This means that γ_T is transversal to the boundary and P_T the linearized Poincaré map of γ_T, which is the derivative of the first return map, does not have eigenvalue 1.

We recall that a bouncing ball orbit is a 2-link periodic trajectory of the billiard flow, i.e., a reversible periodic billiard trajectory that bounces back and forth along a line segment orthogonal to the boundary at both endpoints. Without loss of generality we may assume that the bouncing ball orbit in assumption (2) is on the y-axis.

Zelditch [2009] proved that if Ω and Ω' both satisfy these assumptions, and if $\mathrm{spec}(\Delta_\Omega) = \mathrm{spec}(\Delta_{\Omega'})$ (for either Dirichlet or Neumann boundary conditions), then $\Omega = \Omega'$ up to a reflection about the y-axis. This improves a previous result in [Zelditch 2000], where an additional symmetry assumption is needed, which in turn improves upon [Colin de Verdière 1984], where rigidity is proved in the class of analytic domains with two reflection symmetries. Under the above assumptions, for $\varepsilon > 0$ sufficiently small, there exists a real analytic function $f : (-\varepsilon, \varepsilon) \to \mathbb{R}$ such that

$$\Omega \cap \{|x| < \varepsilon\} = \{(x, y) : |x| < \varepsilon, |y| < f(x)\}.$$

To prove the theorem it is enough to show that the Taylor coefficients of f at 0 are determined by $\mathrm{spec}(\Omega)$ (up to possibly replacing $f(x)$ by $f(-x)$). Zelditch does this by writing a formula for the coefficients a_j of (3-1), which are determined by $\mathrm{spec}(\Omega)$, applied to γ and to γ^2 (the iteration of γ):

$$a_j(\gamma^r) = A_j(r) f^{(2j+2)}(0) + B_j(r) f^{(2j+1)}(0) f^{(3)}(0)$$
$$+ \big[\text{terms containing } f^{(k)}(0) \text{ only for } k \leq j\big]. \quad (3\text{-}2)$$

Here $A_j(r)$ and $B_j(r)$ are spectral invariants which are determined by the first term of (3-1). One can show that $(A_j(1), B_j(1))$ as a vector is linearly independent from $(A_j(2), B_j(2))$. Hence, by an inductive argument, if $f^{(3)}(0) \neq 0$, all the coefficients are determined (up to a sign ambiguity for $f^{(3)}(0)$, which corresponds to reflection about the y-axis). The condition $f^{(3)}(0) \neq 0$ is equivalent to assumption (3) above, and Zelditch [2009, §6.9] outlined a possible proof in the case where $f^{(3)}(0) = 0$.

In [Hezari and Zelditch 2010], it is proved that bounded analytic domains $\Omega \subset \mathbb{R}^n$ with $\pm$ reflection symmetries across all coordinate axes, and with one axis height fixed (and also satisfying some generic nondegeneracy conditions) are spectrally determined among other such domains. This inverse result gives a higher-dimensional analogue of the result discussed above from [Zelditch 2009], but with n axes of symmetry rather than $n-1$. To our knowledge, it is the first positive higher-dimensional inverse spectral result for Euclidean domains which is not restricted to balls. The proof is based as before on (3-1) and on formulas for the $a_j(\gamma^r)$, but there are additional algebraic and combinatorial complications coming from the fact that Taylor coefficients must be recovered corresponding to all possible combinations of partial derivatives. These complications are very similar to those that arise for higher-dimensional semiclassical Schrödinger operators discussed below in Section 3.3.

3.2. Compact manifolds. To our knowledge all uniqueness results in this category are about surfaces of revolution. Bérard [1976] and Gurarie [1995] have shown that the joint spectrum of Δ_g and $\partial/\partial\theta$ (the generator of rotations) of a smooth surface of revolution determines the metric among smooth surfaces of revolution, by reducing the problem to a semiclassical Schrödinger operator in one dimension.

Brüning and Heintze [1984] showed that the spectrum of Δ_g alone determines the metric of a smooth surface of revolution with an up-down symmetry among such surfaces. They proved that the spectrum of Δ_g determines the S^1-invariant spectrum (but not necessarily the full joint spectrum), allowing them to apply [Marchenko 1952, Theorem 2.3.2] for one-dimensional Schrödinger operators.

Zelditch [1998a] proved that a convex analytic surface of revolution satisfying a nondegeneracy condition and a simplicity condition is determined uniquely by the spectrum among all such surfaces. He used analyticity and convexity to show that the spectrum determines the full joint spectrum of Δ_g and $\partial/\partial\theta$, reducing the problem to a semiclassical Schrödinger operator in one dimension.

3.3. Schrödinger operators. When $n = 1$, Marchenko [1952] showed that an even potential is determined by the spectrum of the associated nonsemiclassical Schrödinger operator among all even potentials. More specifically, Marchenko's Theorem 2.3.2 states that a Schrödinger operator on $[0, \infty)$ is determined by knowledge of both the Dirichlet and the Neumann spectrum. The result for even potentials on $\mathbb{R}$ follows from the result on $[0, \infty)$ as follows: Let $V \in C^\infty(\mathbb{R})$ obey $\lim_{|x|\to\infty} V(x) = \infty$. If u_j is the eigenfunction of $-\frac{d^2}{dx^2} + V$ corresponding to the eigenvalue λ_j, then u_j has exactly j zeros and they are all simple; see [Berezin and Shubin 1991, Chapter 2, Theorem 3.5]. If V is even then every eigenfunction is either odd or even and this result shows that the parity of u_j is the same as the parity of j. In particular $(\lambda'_j)_{j=0}^\infty$ with $\lambda'_j = \lambda_{2j}$ is the spectrum of

$$-\frac{d^2}{dx^2} + V \text{ on } L^2([0, \infty)) \text{ with Neumann boundary condition,}$$

and $(\lambda''_j)_{j=0}^\infty$ with $\lambda''_j = \lambda_{2j+1}$ is the spectrum of

$$-\frac{d^2}{dx^2} + V \text{ on } L^2([0, \infty)) \text{ with Dirichlet boundary condition.}$$

This reduces the problem on $\mathbb{R}$ to the result of Marchenko.

However, noneven potentials may have the same spectrum: indeed, McKean and Trubowitz [1981] constructed an infinite-dimensional family of potentials having the same spectrum as the one-dimensional harmonic oscillator $V(x) = x^2$.

Guillemin and Uribe [2007] considered potentials V in $\mathbb{R}^n$ which are analytic and even in all variables, which have a unique global minimum $V(0) = 0$, which obey $\liminf_{|x| \to \infty} V(x) > 0$, and such that the square roots of the eigenvalues of Hess $V(0)$ are linearly independent over $\mathbb{Q}$. They showed that such potentials are determined by their low lying semiclassical eigenvalues, that is to say by $\text{spec}(P_{V,h}) \cap [0, \varepsilon]$ for any $\varepsilon > 0$. In [Hezari 2009], the second author removed the symmetry assumption in the case $n = 1$ but assumed $V'''(0) \neq 0$, and for $n \geq 2$ he replaced the symmetry assumption by the assumption that $V(x) = f(x_1^2, \ldots x_n^2) + x_n^3 g(x_1^2, \ldots, x_n^2)$. Another proof of this result is given in [Colin de Verdière and Guillemin 2011; Colin de Verdière 2011] for the case $n = 1$, and in [Guillemin and Uribe 2011] in the higher-dimensional case.

The proofs in these last three works and in [Guillemin and Uribe 2007] are based on quantum Birkhoff normal forms, a quantum version of the Birkhoff normal forms of classical mechanics. In the classical case, one constructs a symplectomorphism which puts a Hamiltonian function into a canonical form in a neighborhood of a periodic orbit. In the quantum case, one constructs a Fourier integral operator associated to this symplectomorphism which puts a pseudodifferential operator which is a quantization of this Hamiltonian into a canonical form, microlocally near the periodic orbit. Quantum Birkhoff normal forms were developed by Sjöstrand [1992] for semiclassical Schrödinger operators near a global minimum of the potential. Guillemin [1996] and Zelditch [1997; 1998b] put the Laplace–Beltrami operator on a compact Riemannian manifold into a quantum Birkhoff normal form. General semiclassical Schrödinger operators on a manifold at nondegenerate energy levels were studied in [Sjöstrand and Zworski 2002; Iantchenko, Sjöstrand, and Zworski 2002].

The proof in [Hezari 2009] (that the Taylor coefficients of the potential at the bottom of the well are determined by the low-lying eigenvalues) is based on Schrödinger trace invariants. These are coefficients of the expansion

$$\text{Tr}\big(e^{-itP_{V,h}/h}\chi(P_{V,h})\big) = \sum_{j=0}^{\infty} a_j(t) h^j, \quad h \to 0^+,$$

where $\chi \in C_0^\infty(\mathbb{R})$ is 1 near 0 and is supported in a sufficiently small neighborhood of 0. The coefficients a_j in dimension $n = 1$ have exactly the form (3-2) (and in higher dimensions they have the same form as the higher-dimensional coefficients of the wave trace on a bounded domain) and hence, once this fact is established, the remainder of the uniqueness proof is the same for both problems.

3.4. Resonance problems.
The case of an analytic obstacle with two mutually symmetric connected components is treated in [Zelditch 2004a] by using the singularities of the wave trace generated by the bouncing ball between the two components. Zworski [2007] gave a general method for reducing inverse problems for resonances on a noncompact space to corresponding inverse problems for spectra on a compact space.

Iantchenko [2008] considered potentials V in $\mathbb{R}^n$ which are analytic and even in all variables, which have a unique global maximum at $V(0) = E$, which extend holomorphically to a sector around the real axis and obey $\liminf_{|x|\to\infty} V(x) = 0$ in that sector, and such that the square roots of the eigenvalues of Hess $V(0)$ are linearly independent over $\mathbb{Q}$. He used the quantum Birkhoff normal form method of [Guillemin and Uribe 2007] to recover the Taylor coefficients of the potential at the maximum and to show that potentials V in this class are determined by the resonances in a small neighborhood of E.

4. Rigidity and local uniqueness results

In this section we consider results which show nonexistence of nontrivial isospectral deformations.

4.1. Bounded domains in $\mathbb{R}^n$.
Marvizi and Melrose [1982] introduced new invariants for strictly convex bounded domains $\Omega \subset \mathbb{R}^2$ based on the length spectrum, associated with the boundary. They show that, for $m \in \mathbb{N}$ fixed,

$\sup\{L(\gamma): \gamma$ is a periodic billiard orbit with m rotations and n reflections$\}$

$$\sim mL(\partial\Omega) + \sum_{k=1}^{\infty} c_{k,m} n^{-2k}, \quad n \to \infty, \quad (4\text{-}1)$$

where L denotes the length. Then they introduce the following *noncoincidence condition* on Ω, which holds for a dense open family (in the C^∞ topology) of strictly convex domains: suppose there exists $\varepsilon > 0$ such that if γ is a closed orbit with $L(\partial\Omega) - \varepsilon < L(\gamma) < L(\partial\Omega)$, then γ consists of one rotation. They show that under this condition, the coefficients $c_{k,m}$ are spectral invariants, and they use the invariants $c_{1,1}$ and $c_{2,1}$ to construct a two-parameter family of planar domains which are locally spectrally unique (meaning that each domain has a neighborhood in the C^∞ topology within which it is determined by its spectrum). The two-parameter family consists of domains defined by elliptic integrals, and which resemble, but are not, ellipses.

Guillemin and Melrose [1979a] considered the Laplacian on an ellipse Ω given by $x^2/a + y^2/b = 1$, with $a > b > 0$, and with boundary condition

$$\partial u/\partial n = Ku \quad \text{on } \partial\Omega, \quad (4\text{-}2)$$

where $K \in C^\infty(\partial\Omega)$ and is even in both x and y. They showed K is determined by $\text{spec}(\Delta_{\Omega,K})$, where $\Delta_{\Omega,K}$ is the Laplacian on Ω with boundary condition (4-2).

To explain their method, let us introduce some terminology. For $T > 0$ the length of a periodic orbit, the fixed point set of T, denoted by Y_T, is the set of $(q, \eta) \in B^*\partial\Omega$, the coball bundle of $\partial\Omega$, such that the billiard orbit corresponding to the initial condition (q, η) is periodic and has length T. For a more general domain there will often be only one periodic orbit of length T (up to time reversal), but an ellipse, because of the complete integrability of its billiard flow, always has one or several one-parameter families of such orbits. Guillemin and Melrose proved that, in the case of the ellipse, for any T which is the length of a periodic orbit such that $L(\partial\Omega) - T > 0$ is sufficiently small, Y_T has one connected component Γ (up to time reversal). This connected component is necessarily a curve which is invariant under the billiard map. Moreover, they showed that the asymptotic expansion of

$$\text{Tr}\big(\cos(t\sqrt{\Delta_{\Omega,K}})\big) - \text{Tr}\big(\cos(t\sqrt{\Delta_{\Omega,0}})\big)$$

in fractional powers of $t - T$ has leading coefficient

$$\int_\Gamma \frac{K}{\sqrt{1-\eta^2}} d\mu_\Gamma. \tag{4-3}$$

Here μ_Γ is the Leray measure on Γ. Under the symmetry assumptions, K is determined from a sequence of such integrals for T_j with T_j tending to $L(\partial\Omega)$ from below.

In [Hezari and Zelditch 2010], it is proved that an ellipse is infinitesimally spectrally rigid among C^∞ domains with the symmetries of the ellipse. This means that if Ω_0 is an ellipse, and if ρ_ϵ is a smooth one-parameter family of smooth functions on $\partial\Omega_0$ which are even in x and y, and if Ω_ϵ is a domain whose boundary is defined by

$$\partial\Omega_\epsilon = \{z + \rho_\epsilon(z)n_z : z \in \partial\Omega_0\},$$

and if $\text{spec}(\Omega_0) = \text{spec}(\Omega_\epsilon)$ for $\epsilon \in [0, \epsilon_0)$, then the Taylor expansion of ρ_ϵ vanishes at $\epsilon = 0$. In particular, if ρ depends on ϵ analytically, the deformation is constant. The proof uses Hadamard's variational formula for the wave trace:

$$\frac{d}{d\epsilon}\bigg|_{\epsilon=0} \text{Tr}(\cos(t\sqrt{\Delta_{\Omega_\epsilon}})) = \frac{t}{2} \int_{\partial\Omega_0} \partial_{n_1}\partial_{n_2} S_{\Omega_0}(t,z,z) \bigg(\frac{d}{d\epsilon}\bigg|_{\epsilon=0} \rho_\epsilon(z)\bigg) dz, \tag{4-4}$$

where ∂_{n_1} and ∂_{n_2} denote normal derivatives in the first and second variables respectively, S_{Ω_0} is the kernel of $\sin(t\sqrt{\Delta_{\Omega_0}})/\sqrt{\Delta_{\Omega_0}}$. They then use (4-4) to prove that for any T in the length spectrum of Ω_0, the leading order singularity of the wave trace variation is,

$$\frac{d}{d\epsilon}\bigg|_{\epsilon=0} \operatorname{Tr}(\cos(t\sqrt{\Delta_{\Omega_\epsilon}})) \sim$$
$$\frac{t}{2}\operatorname{Re}\{(\sum_{\Gamma\subset Y_T} C_\Gamma \int_\Gamma \left(\frac{d}{d\epsilon}\bigg|_{\epsilon=0}\rho_\epsilon\right)\sqrt{1-|\eta|^2}d\mu_\Gamma)(t-T+i0)^{-\frac{5}{2}}\},$$
(4-5)

modulo lower order singularities, where the sum is over the connected components Γ of the set Y_T of periodic points of the billiard map on $B^*\partial\Omega_0$ (and its powers) of length T, and where $d\mu_\Gamma$ is as in (4-3). As before, if $L(\partial\Omega_0)-T>0$ is sufficiently small, there is only one connected component and the sum has only one term. For an isospectral deformation, the left-hand side of (4-5) vanishes, and hence the integrals

$$\int_\Gamma \left(\frac{d}{d\epsilon}\bigg|_{\epsilon=0}\rho_\epsilon\right)\sqrt{1-|\eta|^2}d\mu_\Gamma$$

vanish when $L(\partial\Omega_0)-T>0$ sufficiently small. From this point on proceeding as in [Guillemin and Melrose 1979a] above one can show that

$$\frac{d}{d\epsilon}\bigg|_{\epsilon=0}\rho_\epsilon = 0,$$

and reparametrizing the variation one can show that all Taylor coefficients of the variation are 0. In [Hezari and Zelditch 2010] it is shown that expansions of the form (4-4) and (4-5) hold more generally and in higher dimensions; indeed (4-4) holds for any C^1 variation of any bounded domain, and a version of (4-5) holds whenever the fixed point sets Y_T are clean. These formulas may be useful for example in a possible proof of spectral rigidity of ellipsoids.

4.2. Compact manifolds. Tanno [1980] used heat trace invariants to show local spectral uniqueness of spheres in all dimensions. This means that there is a C^∞ neighborhood of the round metric on the sphere within which this metric is spectrally determined. Kuwabara [1980] did this for compact flat manifolds and Sharafutdinov [2009] for compact manifolds of constant negative curvature.

Guillemin and Kazhdan [1980a; 1980b] proved that a negatively curved compact manifold (M,g) with simple length spectrum is spectrally rigid if its sectional curvatures satisfy the pinching condition that for every $x \in M$ there is $A(x) > 0$ such that $|K/A+1| < 1/n$, where K is any sectional curvature at x (note that the pinching condition is satisfied for all negatively curved surfaces because in that case there is only one sectional curvature at each point x and we may take $A(x) = -K$). Spectrally rigid here means if g_ϵ is a smooth family of metrics on M with $g_0 = g$ and with $\text{spec}(\Delta_{g_\epsilon}) = \text{spec}(\Delta_g)$, then (M, g_ϵ) is isometric to (M, g) for every ϵ. They further used a similar method of proof to establish a spectral uniqueness result for Schrödinger operators on these manifolds. The pinching condition was relaxed in [Maung 1986] and removed in [Croke and Sharafutdinov 1998], and the result was extended to Anosov surfaces with no focal points in [Sharafutdinov and Uhlmann 2000].

4.3. Schrödinger operators.
In work in progress, the second author considers anisotropic harmonic oscillators: $V(x) = a_1^2 x_1^2 + \cdots + a_n^2 x_n^2$, where the a_j are linearly independent over $\mathbb{Q}$. It is shown that if $V_\epsilon(x)$ is a smooth deformation of $V(x)$ within the class of C^∞ functions which are even in each x_j, and if $\text{spec}(P_{V,1}) = \text{spec}(P_{V_\epsilon,1})$ for all $\epsilon \in [0, \epsilon)$, then the deformation is flat at $\epsilon = 0$, just as in the infinitesimal rigidity result for the ellipse in Section 4.1.

5. Compactness results

5.1. Bounded domains in $\mathbb{R}^n$.
Melrose [1983a] used heat trace invariants and Sobolev embedding to prove compactness of isospectral sets of domains $\Omega \subset \mathbb{R}^2$ in the sense of the C^∞ topology on the curvature functions in $C^\infty(\partial\Omega)$. This result allows the possibility of a sequence of isospectral domains whose curvatures converge but which "pinch off" in such a way that the limit object is not a domain, but Melrose [1996] showed that this possibility can be ruled out using the fact that the singularity of the wave trace at $t = 0$ is isolated.

Osgood, Phillips, and Sarnak [1989] gave another approach to this problem based on the determinant of the Laplacian. This is defined via the analytic continuation of the zeta function

$$Z(s) = \sum_{j=1}^{\infty} \lambda_j^{-s}, \quad \det \Delta_\Omega = e^{-Z'(0)}.$$

They consider the domain Ω as the image of the unit disk D under a conformal map F, with $e^{2\phi} g_0$ the induced metric on D, where $\phi = \log |F'|$ is a harmonic function. Thus ϕ is determined by its boundary values, and the topology of [Osgood, Phillips, and Sarnak 1989] is the C^∞ topology on $\phi|_{\partial D}$, and in this case pinching degenerations are ruled out automatically. Hassell and Zelditch [1999] gave a nice review of these results and an application of these methods to the compactness problem for isophasal obstacles in $\mathbb{R}^2$.

To our knowledge there is no compactness result in higher dimensions.

5.2. Compact manifolds.
Osgood, Phillips, and Sarnak [1988a; 1988b] extended their determinant methods to the case of surfaces and prove that the set of isospectral metrics on a given Riemannian surface is sequentially compact

in the C^∞ topology, up to isometry. Further compactness results for isospectral metrics in a given conformal class on a three-dimensional manifold appeared in [Chang and Yang 1989; Brooks, Perry, and Yang 1989]. Brooks, Perry, and Petersen [1992] proved compactness for isospectral families of Riemannian manifolds provided that either the sectional curvatures are all negative or there is a uniform lower bound on the Ricci curvatures. Zhou [1997] showed that on a given manifold, the family of isospectral Riemannian metrics with uniformly bounded curvature is compact, with no restriction on the dimension.

5.3. *Schrödinger operators.* Brüning [1984] considered Schrödinger operators $\Delta_g + V$ on a compact Riemannian manifold (M, g), where $V \in C^\infty(M)$, and proves that if the dimension $n \leq 3$, then any set of isospectral potentials is compact. In higher dimensions he proved the same result under the additional condition that the H^s norm of V for some $s > 3(n/2) - 2$ is known to be bounded by some constant C. Donnelly [2005] improved this condition to $s > (n/2) - 2$, and derived alternative compactness criteria: he shows that isospectral families of nonnegative potentials are compact in dimensions $n \leq 9$. If one considers instead $\Delta_g + \gamma V$, he shows that a family of potentials which is isospectral for more than $(n/2) - 1$ different values of γ is compact. In particular, this implies compactness of families which are isospectral for the semiclassical problem $h^2 \Delta_g + V$.

5.4. *Resonance problems.* Let (X_0, g_0) be a conformally compact surface that is hyperbolic (has constant curvature) outside a given compact set $K_0 \subset X_0$. This means that, if K_0 is taken sufficiently large, then $X_0 \setminus K_0$ is a finite disjoint union of funnel ends, which is to say ends of the form

$$(0, \infty)_r \times S^1_\theta, \quad dr^2 + \ell^2 \cosh^2(r) \, d\theta^2, \tag{5-1}$$

where $\ell \neq 0$ may vary between the funnels. Then the continuous spectrum of Δ_{g_0} is given by $[1/4, \infty)$, and the point spectrum is either empty or finite and contained in $(0, 1/4)$ (and there is no other spectrum). If we introduce the spectral parameter $z = \sqrt{\lambda - 1/4}$, where $\sqrt{}$ is taken to map $\mathbb{C} \setminus [0, \infty)$ to the upper half-plane, then the resolvent $(\Delta_g - 1/4 - z^2)^{-1}$ continues meromorphically from $\{\text{Im } z > 0\}$ to $\mathbb{C}$ as an operator $L^2_{\text{comp}} \to L^2_{\text{loc}}$. This meromorphic continuation can be proved by writing a parametrix in terms of the resolvent of the Laplacian on the ends (5-1), which in this case can be written explicitly in terms of special functions: see [Mazzeo and Melrose 1987] for the general construction, and [Guillopé and Zworski 1995, §5] for a simpler version in this case.

Borthwick and Perry [2011] used a Poisson formula for resonances due to Guillopé and Zworski [1997] and a heat trace expansion to show that the set of surfaces which are isoresonant with (X_0, g_0) and for which there is a compact set $K \subset X$ such that $(X_0 \setminus K_0, g_0)$ is isometric to $(X \setminus K, g)$ is compact in the C^∞ topology, improving a previous result of Borthwick, Judge, and Perry [2003]. They also proved related but weaker results in higher dimensions.

6. Trace invariants and their limitations

6.1. Bounded domains in $\mathbb{R}^n$.
For Δ_Ω with $\Omega \subset \mathbb{R}^n$ a bounded smooth domain we have seen two kinds of trace invariants. The first are heat trace invariants, which are the coefficients a_j of the expansion

$$\operatorname{Tr} e^{-t\Delta_\Omega} \sim t^{-n/2} \sum_{j=0}^\infty a_j t^{j/2}, \quad t \to 0^+,$$

are given by integrals along the boundary of polynomials in the curvature and its derivatives. These are equivalent to the invariants obtained from coefficients of the expansion of the wave trace $\operatorname{Tr} \cos(t\sqrt{\Delta_\Omega})$ at $t = 0$.

The other kind are wave trace invariants obtained from coefficients of the expansion of the wave trace at the length of a periodic billiard orbit, always assumed to be nondegenerate and usually assumed to be simple. In this case the formula, as already mentioned in (3-1), is

$$\operatorname{Tr} \cos(t\sqrt{\Delta_\Omega})$$
$$= \operatorname{Re}\left[i^{\sigma_T} \frac{T^\sharp}{\sqrt{\det(I - P_T)}} (t - T + i0)^{-1} \right.$$
$$\left. \times \left(1 + \sum_{j=1}^\infty b_j (t-T)^j \log(t-T+i0)\right) \right] + S(t), \quad (6\text{-}1)$$

where γ_T is the simple periodic orbit of length T, and where the coefficients b_j are polynomials in the Taylor coefficients at the reflection points of γ_T of the function of which the boundary is a graph. Because of this requirement on the periodic orbit, positive inverse results of the kind described above, which are based on the wave trace, always require generic assumptions such as nondegeneracy and simple length spectrum. Although there has been some work on the degenerate case, such as [Popov 1998], it does not seem to have led yet to uniqueness, rigidity, or compactness results. However, Marvizi and Melrose [1982] obtained information from invariants at lengths approaching the length of $\partial\Omega$ (see Section 4.1 above for more information).

Another limitation comes from the fact that domains can have the same trace invariants without being isospectral. That is to say, we can construct Ω and Ω' such that $\operatorname{Tr}(\cos(t\sqrt{\Delta_\Omega})) - \operatorname{Tr}(\cos(t\sqrt{\Delta_{\Omega'}})) \in C^\infty(\mathbb{R})$ (recall that the wave trace invariants are the coefficients in the expansion of the wave trace near a singularity, as in (6-1)), but $\mathrm{spec}(\Delta_\Omega) \neq \mathrm{spec}(\Delta_{\Omega'})$. This was done by Fulling and Kuchment [2005], following a conjecture of Zelditch [2004b], where the following types of domains are considered (these were first introduced by Penrose to study the illumination problem, and then shown by Lifshits to be examples of nonisometric domains with the same length spectrum):

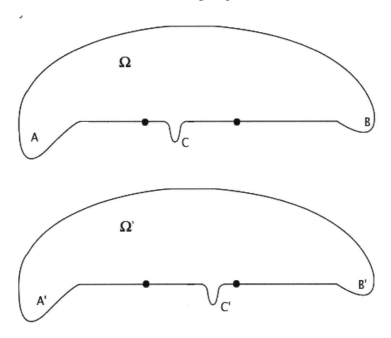

Figure 1. Two domains Ω, Ω' with $\mathrm{Tr}(\cos(t\sqrt{\Delta_\Omega})) - \mathrm{Tr}(\cos(t\sqrt{\Delta_{\Omega'}}))$ in $C^\infty(\mathbb{R})$ but $\mathrm{spec}(\Delta_\Omega) \neq \mathrm{spec}(\Delta_{\Omega'})$.

These two domains are obtained by taking a semiellipse and adding two asymmetric bumps A, B and A', B', with $A = A'$ and $B = B'$, such that the foci are left unperturbed (as in the figure). Then one adds bumps C and C', the small bumps in the middle which are in between the foci, such that $C \neq C'$ but C and C' are reflections of one another. These two domains are not isometric but have the same heat invariants, because heat invariants are given by integrals along the boundary of polynomials in the curvature and its derivatives – indeed we have freedom to 'slide' C back and forth along the boundary without changing any heat invariants, although this is not the case for the wave trace invariants.

We now show that $\mathrm{Tr}(\cos(t\sqrt{\Delta_\Omega})) - \mathrm{Tr}(\cos(t\sqrt{\Delta_{\Omega'}})) \in C^\infty(\mathbb{R})$. This is because of the following separation of the phase spaces[6] $B^*\partial\Omega$ and $B^*\partial\Omega'$ into

---

[6]Recall that $B^*\partial\Omega$ is the ball bundle, the fibers of which are intervals $[-1, 1]$.

two disconnected rooms each, which are invariant under the billiard maps of the domains, and which we denote R_1, R_2, R'_1, and R'_2, and which have the property that $\overline{R_1 \cup R_2} = B^*\partial\Omega$ and $\overline{R'_1 \cup R'_2} = B^*\partial\Omega'$. These are defined as follows: R_1 is the set of points in $B^*\partial\Omega$ whose billiard flowout intersects the part of the boundary strictly in between the two foci, R_2 is the set of points in $B^*\partial\Omega$ whose billiard flowout intersects the part of the boundary which is strictly outside the two foci but on the axis of the ellipse or below (and similarly for R'_1 and R'_2). These two sets are disjoint because billiards in an ellipse which intersect the major axis in between the two foci once do so always. Now we make the generic assumption that no trajectory which passes through the two foci in the initial semiellipse is periodic. Because R_1 is isometric to R'_1, and R_2 is isometric to R'_2, we have $\operatorname{Tr}(\cos(t\sqrt{\Delta_\Omega})) - \operatorname{Tr}(\cos(t\sqrt{\Delta_{\Omega'}})) \in C^\infty(\mathbb{R})$. This is because the singularities of $\operatorname{Tr}(\cos(t\sqrt{\Delta_\Omega}))$ occur at $t = T$, where T is the length of a periodic orbit, and only depend on the structure of $B^*\partial\Omega$ in an arbitrarily small neighborhood of the orbits of length T.

To show that $\operatorname{spec}(\Delta_\Omega) \neq \operatorname{spec}(\Delta_{\Omega'})$, Fulling and Kuchment use a perturbation argument based on Hadamard's variational formula for the ground state to show that, for suitably chosen small C, the ground states are not the same.

6.2. Compact manifolds. As we have already mentioned, heat trace invariants can be defined for compact manifolds (M, g) as well. In the boundaryless case the expansion takes the form

$$\operatorname{Tr} e^{-t\Delta_g} \sim t^{-n/2} \sum_{j=0}^\infty a_j t^j, \quad t \to 0^+,$$

where the a_j are given by integrals on M of polynomials in the curvature and its derivatives. Half powers of t appear only when there is a boundary, as in the case of domains considered above. Once again, these invariants are equivalent to the invariants obtained from coefficients of the expansion of the wave trace $\operatorname{Tr}\cos(t\sqrt{\Delta_g})$ at $t = 0$. In analogy with the example given in the previous section, we can construct manifolds which are not isometric but which have the same heat invariants by taking a sphere, adding two disjoint bumps, and moving them around. For suitable choices of bumps it should be possible to make the length spectra nonequal, as a result of which the manifolds will be nonisospectral. It seems to be an open problem, however, to find an example of two manifolds (M, g) and (M', g') which are nonisospectral but which have $\operatorname{Tr}(\cos(t\sqrt{\Delta_g})) - \operatorname{Tr}(\cos(t\sqrt{\Delta_{g'}})) \in C^\infty(\mathbb{R})$, that is to say which have identical wave trace invariants.

In this setting the wave trace expansion was established by Duistermaat and Guillemin [1975], building off of previous work by Colin de Verdière [1973] and Chazarain [1974]. It is a generalization of Selberg's Poisson formula [1956] to an arbitrary compact boundaryless Riemannian manifold. For T the length of a simple nondegenerate periodic geodesic γ_T, it takes the form

$$\operatorname{Tr} e^{it\sqrt{\Delta_g}}$$
$$= i^{\sigma_T} \frac{T^\sharp}{\sqrt{|\det(I-P_T)|}} (t-T+i0)^{-1} \left(1 + \sum_{j=1}^{\infty} b_j(t-T)^j \log(t-T+i0) \right) + S(t),$$

where $S(t)$ is smooth near T. Using quantum Birkhoff normal forms, Zelditch [1998b] showed these coefficients b_j to be integrals of polynomials in the metric and its derivatives along γ_T. See also [Zelditch 1999] for a more detailed survey on wave invariants. Because of this very local nature of these invariants, to prove uniqueness results one must either assume analyticity (as is done in the results discussed above) or find a way to combine information from many different orbits (no one seems to have been able to do this so far).

6.3. *Schrödinger operators.* In the setting of semiclassical Schrödinger operators the analogue of the Duistermaat–Guillemin wave trace is the Gutzwiller trace formula near the length T of a periodic trajectory of the Hamiltonian vector field H_p in $p^{-1}(E)$, where $p(x,\xi) = |\xi|^2 + V(x)$:

$$\operatorname{Tr} e^{-it(P_{V,h}-E)/h} \chi(P_{V,h}) \sim \sum_\gamma i^{\sigma_\gamma} \frac{e^{iS_\gamma/h}}{\sqrt{|\det(I-P_\gamma)|}} \sum_{j=0}^{\infty} a_{j,\gamma} h^j,$$
$$a_{0,\gamma} = \delta_0(t-T),$$

for t near T, where $\chi \in C_0^\infty(\mathbb{R})$ has $\chi = 1$ near E. Here the sum in γ is over periodic trajectories in $p^{-1}(0)$ of length T and the $a_{j,\gamma}$ are distributions whose singular support is contained in $\{T\}$. This formula goes back to [Gutzwiller 1971], and was proved in various degrees of generality and with various methods from [Guillemin and Uribe 1989] and [Combescure, Ralston, and Robert 1999] (see also this last paper for further history and references). This formula is also valid for more general pseudodifferential operators of real principal type, so long as E is a regular value of the principal symbol p and so long as the periodic trajectories in $p^{-1}(E)$ are nondegenerate (so that the determinants in the denominator are nonzero). In particular it also applies on manifolds. Iantchenko, Sjöstrand, and Zworski used it in [Iantchenko, Sjöstrand, and Zworski 2002] to recover quantum and classical Birkhoff normal forms of semiclassical classical Schrödinger operators, at nondegenerate periodic orbits. When the energy level is degenerate, the Gutzwiller trace formula becomes more complicated: see Section 3.3 for a discussion of the case where p has a unique global minimum at

E, and see for example [Brummelhuis, Paul, and Uribe 1995] and [Khuat-Duy 1997] for other cases.

Colin de Verdière [2011] gave an example of a pair of potentials $V \not\equiv V' \in C^\infty(\mathbb{R})$ such that $\operatorname{spec}(P_{V,h}) = \operatorname{spec}(P_{V',h})$ up to $\mathcal{O}(h^\infty)$, so that in particular all semiclassical trace invariants for these two potentials agree. He conjectured, however, that the spectra are not equal. In [Guillemin and Hezari 2012], two potentials are constructed that have different ground states and hence different spectra, although the spectra still agree up to $\mathcal{O}(h^\infty)$. These potentials are perturbations of the harmonic oscillator analogous to the perturbations of the semiellipse discussed in Section 6.1.

References

[Bardos, Guillot, and Ralston 1982] C. Bardos, J.-C. Guillot, and J. Ralston, "La relation de Poisson pour l'équation des ondes dans un ouvert non borné: application à la théorie de la diffusion", *Comm. Partial Differential Equations* **7**:8 (1982), 905–958. MR 84d:35120 Zbl 0496.35067

[Bérard 1976] P. Bérard, "Quelques remarques sur les surfaces de révolution dans $\mathbf{R}^3$", *C. R. Acad. Sci. Paris Sér. A-B* **282**:3 (1976), A159–A161. MR 52 #15319

[Berezin and Shubin 1991] F. A. Berezin and M. A. Shubin, *The Schrödinger equation*, Mathematics and its Applications **66**, Kluwer Academic Publishers Group, Dordrecht, 1991. MR 93i:81001 Zbl 0749.35001

[Bony 2002] J.-F. Bony, "Minoration du nombre de résonances engendrées par une trajectoire fermée", *Comm. Partial Differential Equations* **27**:5-6 (2002), 1021–1078. MR 2003e:35229 Zbl 1213.35332

[Borthwick and Perry 2011] D. Borthwick and P. A. Perry, "Inverse scattering results for manifolds hyperbolic near infinity", *J. Geom. Anal.* **21**:2 (2011), 305–333. MR 2012i:58023 Zbl 1229.58024

[Borthwick, Judge, and Perry 2003] D. Borthwick, C. Judge, and P. A. Perry, "Determinants of Laplacians and isopolar metrics on surfaces of infinite area", *Duke Math. J.* **118**:1 (2003), 61–102. MR 2004c:58067 Zbl 1040.58013

[Brooks, Perry, and Petersen 1992] R. Brooks, P. Perry, and P. Petersen, V, "Compactness and finiteness theorems for isospectral manifolds", *J. Reine Angew. Math.* **426** (1992), 67–89. MR 93f:53034 Zbl 0737.53038

[Brooks, Perry, and Yang 1989] R. Brooks, P. Perry, and P. Yang, "Isospectral sets of conformally equivalent metrics", *Duke Math. J.* **58**:1 (1989), 131–150. MR 90i:58193 Zbl 0667.53037

[Brummelhuis, Paul, and Uribe 1995] R. Brummelhuis, T. Paul, and A. Uribe, "Spectral estimates around a critical level", *Duke Math. J.* **78**:3 (1995), 477–530. MR 96k:58220

[Brüning 1984] J. Brüning, "On the compactness of isospectral potentials", *Comm. Partial Differential Equations* **9**:7 (1984), 687–698. MR 85h:58170 Zbl 0547.58039

[Brüning and Heintze 1984] J. Brüning and E. Heintze, "Spektrale Starrheit gewisser Drehflächen", *Math. Ann.* **269**:1 (1984), 95–101. MR 85i:58114 Zbl 0553.53028

[Chang and Yang 1989] S.-Y. A. Chang and P. C. Yang, "Compactness of isospectral conformal metrics on S^3", *Comment. Math. Helv.* **64**:3 (1989), 363–374. MR 90c:58181 Zbl 0679.53038

[Chazarain 1974] J. Chazarain, "Formule de Poisson pour les variétés riemanniennes", *Invent. Math.* **24** (1974), 65–82. MR 49 #8062 Zbl 0281.35028

[Christiansen 2008] T. J. Christiansen, "Resonances and balls in obstacle scattering with Neumann boundary conditions", *Inverse Probl. Imaging* **2**:3 (2008), 335–340. MR 2009f:35245 Zbl 1180.35400

[Colin de Verdière 1973] Y. Colin de Verdière, "Spectre du laplacien et longueurs des géodésiques périodiques. II", *Compositio Math.* **27**:2 (1973), 159–184. MR 1557068 Zbl 0281.53036

[Colin de Verdière 1984] Y. Colin de Verdière, "Sur les longueurs des trajectoires périodiques d'un billard", pp. 122–139 in *Séminaire sud-rhodanien en géométrie III: Géométrie symplectique et de contact: autour du théorème de Poincaré–Birkhoff* (Lyon, 1983), edited by P. Dazord and N. Desolneux-Moulis, Hermann, Paris, 1984. MR 86a:58078 Zbl 0599.58039

[Colin de Verdière 2011] Y. Colin de Verdière, "A semi-classical inverse problem, II: reconstruction of the potential", pp. 97–119 in *Geometric aspects of analysis and mechanics*, edited by E. P. van den Ban and J. A. C. Kolk, Progr. Math. **292**, Birkhäuser/Springer, New York, 2011. MR 2012e:34022

[Colin de Verdière and Guillemin 2011] Y. Colin de Verdière and V. Guillemin, "A semi-classical inverse problem, I: Taylor expansions", pp. 81–95 in *Geometric aspects of analysis and mechanics*, edited by E. P. van den Ban and J. A. C. Kolk, Progr. Math. **292**, Birkhäuser/Springer, New York, 2011. MR 2012e:34021

[Combescure, Ralston, and Robert 1999] M. Combescure, J. Ralston, and D. Robert, "A proof of the Gutzwiller semiclassical trace formula using coherent states decomposition", *Comm. Math. Phys.* **202**:2 (1999), 463–480. MR 2000k:81089 Zbl 0939.58031

[Croke and Sharafutdinov 1998] C. B. Croke and V. A. Sharafutdinov, "Spectral rigidity of a compact negatively curved manifold", *Topology* **37**:6 (1998), 1265–1273. MR 99e:58191 Zbl 0936.58013

[Datchev and Hezari 2012] K. Datchev and H. Hezari, "Resonant uniqueness of radial semiclassical Schrödinger operators", *Appl. Math. Res. Express* **2012**:1 (2012), 105–113. MR 2900148 Zbl 1239.35102

[Datchev, Hezari, and Ventura 2011] K. Datchev, H. Hezari, and I. Ventura, "Spectral uniqueness of radial semiclassical Schrödinger operators", *Math. Res. Lett.* **18**:3 (2011), 521–529. MR 2012f:35573 Zbl 1241.35217

[Donnelly 2005] H. Donnelly, "Compactness of isospectral potentials", *Trans. Amer. Math. Soc.* **357**:5 (2005), 1717–1730. MR 2006d:58032 Zbl 1062.58033

[Duistermaat and Guillemin 1975] J. J. Duistermaat and V. W. Guillemin, "The spectrum of positive elliptic operators and periodic bicharacteristics", *Invent. Math.* **29**:1 (1975), 39–79. MR 53 #9307 Zbl 0307.35071

[Fulling and Kuchment 2005] S. A. Fulling and P. Kuchment, "Coincidence of length spectra does not imply isospectrality", *Inverse Problems* **21**:4 (2005), 1391–1395. MR 2006e:58049 Zbl 1083.35077

[Gordon 2000] C. S. Gordon, "Survey of isospectral manifolds", pp. 747–778 in *Handbook of differential geometry*, vol. I, edited by F. J. E. Dillen and L. C. A. Verstraelen, North-Holland, Amsterdam, 2000. MR 2000m:58057 Zbl 0959.58039

[Gordon and Webb 1994] C. S. Gordon and D. L. Webb, "Isospectral convex domains in Euclidean space", *Math. Res. Lett.* **1**:5 (1994), 539–545. MR 95g:58246 Zbl 0840.58046

[Gordon, Perry, Schueth 2005] C. Gordon, P. Perry, and D. Schueth, "Isospectral and isoscattering manifolds: a survey of techniques and examples", pp. 157–179 in *Geometry, spectral theory, groups, and dynamics* (Haifa, 2003/2004), edited by M. Entov et al., Contemp. Math. **387**, Amer. Math. Soc., Providence, RI, 2005. MR 2006k:58058 Zbl 1087.58021

[Gordon, Webb, and Wolpert 1992] C. Gordon, D. Webb, and S. Wolpert, "Isospectral plane domains and surfaces via Riemannian orbifolds", *Invent. Math.* **110**:1 (1992), 1–22. MR 93h:58172 Zbl 0778.58068

[Guillemin 1996] V. Guillemin, "Wave-trace invariants", *Duke Math. J.* **83**:2 (1996), 287–352. MR 97f:58131 Zbl 0858.58051

[Guillemin and Hezari 2012] V. Guillemin and H. Hezari, "A Fulling–Kuchment theorem for the 1D harmonic oscillator", *Inverse Problems* **28**:4 (2012), 045009. MR 2903284 Zbl 1238.35073

[Guillemin and Kazhdan 1980a] V. Guillemin and D. Kazhdan, "Some inverse spectral results for negatively curved 2-manifolds", *Topology* **19**:3 (1980), 301–312. MR 81j:58082 Zbl 0465.58027

[Guillemin and Kazhdan 1980b] V. Guillemin and D. Kazhdan, "Some inverse spectral results for negatively curved n-manifolds", pp. 153–180 in *Geometry of the Laplace operator* (Honolulu, 1979), edited by R. Osserman and A. Weinstein, Proc. Sympos. Pure Math. **36**, Amer. Math. Soc., Providence, R.I., 1980. MR 81i:58048 Zbl 0456.58031

[Guillemin and Melrose 1979a] V. Guillemin and R. Melrose, "An inverse spectral result for elliptical regions in $\mathbf{R}^2$", *Adv. in Math.* **32**:2 (1979), 128–148. MR 80f:35104 Zbl 0415.35062

[Guillemin and Melrose 1979b] V. Guillemin and R. Melrose, "The Poisson summation formula for manifolds with boundary", *Adv. in Math.* **32**:3 (1979), 204–232. MR 80j:58066 Zbl 0421.35082

[Guillemin and Sternberg 2010] V. Guillemin and S. Sternberg, "Semi-classical analysis", lecture notes, 2010, Available at http://math.mit.edu/~vwg/semiclassGuilleminSternberg.pdf.

[Guillemin and Uribe 1989] V. Guillemin and A. Uribe, "Circular symmetry and the trace formula", *Invent. Math.* **96**:2 (1989), 385–423. MR 90e:58159 Zbl 0686.58040

[Guillemin and Uribe 2007] V. Guillemin and A. Uribe, "Some inverse spectral results for semiclassical Schrödinger operators", *Math. Res. Lett.* **14**:4 (2007), 623–632. MR 2008h:35041 Zbl 1138.35006

[Guillemin and Uribe 2011] V. Guillemin and A. Uribe, "Some inverse spectral results for the two-dimensional Schrödinger operator", pp. 319–328 in *Geometry and analysis* (Cambridge, 2008), vol. 1, edited by L. Ji, Adv. Lect. Math. **17**, Int. Press, Somerville, MA, 2011. MR 2012k:35085

[Guillemin and Wang 2009] V. Guillemin and Z. Wang, "Semiclassical spectral invariants for Schrödinger operators", preprint, 2009. arXiv 0905.0919

[Guillopé and Zworski 1995] L. Guillopé and M. Zworski, "Upper bounds on the number of resonances for non-compact Riemann surfaces", *J. Funct. Anal.* **129**:2 (1995), 364–389. MR 96b:58116 Zbl 0841.58063

[Guillopé and Zworski 1997] L. Guillopé and M. Zworski, "Scattering asymptotics for Riemann surfaces", *Ann. of Math.* (2) **145**:3 (1997), 597–660. MR 98g:58181 Zbl 0898.58054

[Gurarie 1995] D. Gurarie, "Semiclassical eigenvalues and shape problems on surfaces of revolution", *J. Math. Phys.* **36**:4 (1995), 1934–1944. MR 96k:58228 Zbl 0826.58041

[Gutzwiller 1971] M. Gutzwiller, "Periodic orbits and classical quantization conditions", *J. Math. Phys.* **12**:3 (1971), 343–358.

[Hassell and Zelditch 1999] A. Hassell and S. Zelditch, "Determinants of Laplacians in exterior domains", *Internat. Math. Res. Notices* **1999**:18 (1999), 971–1004. MR 2001a:58045 Zbl 0941.58020

[Hassell and Zworski 1999] A. Hassell and M. Zworski, "Resonant rigidity of S^2", *J. Funct. Anal.* **169**:2 (1999), 604–609. MR 2001g:58051 Zbl 0954.35125

[Helffer and Robert 1983] B. Helffer and D. Robert, "Calcul fonctionnel par la transformation de Mellin et opérateurs admissibles", *J. Funct. Anal.* **53**:3 (1983), 246–268. MR 85i:47052 Zbl 0524.35103

[Hezari 2009] H. Hezari, "Inverse spectral problems for Schrödinger operators", *Comm. Math. Phys.* **288**:3 (2009), 1061–1088. MR 2010c:81103 Zbl 1170.81042

[Hezari and Zelditch 2010] H. Hezari and S. Zelditch, "Inverse spectral problem for analytic $(\mathbb{Z}/2\mathbb{Z})^n$-symmetric domains in $\mathbb{R}^n$", *Geom. Funct. Anal.* **20**:1 (2010), 160–191. MR 2011i:58054 Zbl 1226.35055

[Iantchenko 2008] A. Iantchenko, "An inverse problem for trapping point resonances", *Lett. Math. Phys.* **86**:2-3 (2008), 151–157. MR 2009m:35538 Zbl 1179.35346

[Iantchenko, Sjöstrand, and Zworski 2002] A. Iantchenko, J. Sjöstrand, and M. Zworski, "Birkhoff normal forms in semi-classical inverse problems", *Math. Res. Lett.* **9**:2-3 (2002), 337–362. MR 2003f:35284 Zbl 01804060

[Kac 1966] M. Kac, "Can one hear the shape of a drum?", *Amer. Math. Monthly* **73**:4, part II (1966), 1–23. MR 34 #1121 Zbl 0139.05603

[Khuat-Duy 1997] D. Khuat-Duy, "A semi-classical trace formula for Schrödinger operators in the case of a critical energy level", *J. Funct. Anal.* **146**:2 (1997), 299–351. MR 1451995 (98j:35134)

[Korotyaev 2005] E. Korotyaev, "Inverse resonance scattering on the real line", *Inverse Problems* **21**:1 (2005), 325–341. MR 2006a:34033 Zbl 1074.34081

[Kuwabara 1980] R. Kuwabara, "On the characterization of flat metrics by the spectrum", *Comment. Math. Helv.* **55**:3 (1980), 427–444. MR 82k:58095a Zbl 0448.58027

[Marchenko 1952] V. A. Marchenko, "Some questions of the theory of one-dimensional linear differential operators of the second order, I", *Trudy Moskov. Mat. Obšč.* **1** (1952), 327–420. In Russian. MR 15,315b

[Marvizi and Melrose 1982] S. Marvizi and R. Melrose, "Spectral invariants of convex planar regions", *J. Differential Geom.* **17**:3 (1982), 475–502. MR 85d:58084 Zbl 0492.53033

[Maung 1986] M.-O. Maung, "Spectral rigidity for manifolds with negative curvature operator", pp. 99–103 in *Nonlinear problems in geometry* (Mobile, AL), edited by D. M. DeTurck, Contemp. Math. **51**, Amer. Math. Soc., Providence, RI, 1986. MR 87k:58278 Zbl 0591.53041

[Mazzeo and Melrose 1987] R. R. Mazzeo and R. B. Melrose, "Meromorphic extension of the resolvent on complete spaces with asymptotically constant negative curvature", *J. Funct. Anal.* **75**:2 (1987), 260–310. MR 89c:58133 Zbl 0636.58034

[McKean and Singer 1967] H. P. McKean, Jr. and I. M. Singer, "Curvature and the eigenvalues of the Laplacian", *J. Differential Geometry* **1**:1 (1967), 43–69. MR 36 #828 Zbl 0198.44301

[McKean and Trubowitz 1981] H. P. McKean and E. Trubowitz, "The spectral class of the quantum-mechanical harmonic oscillator", *Comm. Math. Phys.* **82**:4 (1981), 471–495. MR 83e:34034

[Melrose 1982] R. Melrose, "Scattering theory and the trace of the wave group", *J. Funct. Anal.* **45**:1 (1982), 29–40. MR 83j:35128 Zbl 0525.47007

[Melrose 1983a] R. Melrose, "Isospectral sets of drumheads are compact in C^∞", preprint, 1983, Available at http://www-math.mit.edu/~rbm/papers/isospectral/isospectral.pdf.

[Melrose 1983b] R. Melrose, "Polynomial bound on the number of scattering poles", *J. Funct. Anal.* **53**:3 (1983), 287–303. MR 85k:35180 Zbl 0535.35067

[Melrose 1995] R. B. Melrose, *Geometric scattering theory*, Cambridge University Press, 1995. MR 96k:35129 Zbl 0849.58071

[Melrose 1996] R. Melrose, "The inverse spectral problem for planar domains", pp. 137–160 in *Instructional Workshop on Analysis and Geometry* (Canberra, 1995), vol. I, edited by T. Cranny and J. Hutchinson, Proc. Centre Math. Appl. **34**, Austral. Nat. Univ., Canberra, 1996. MR 98a:58164

[Osgood, Phillips, and Sarnak 1988a] B. Osgood, R. Phillips, and P. Sarnak, "Extremals of determinants of Laplacians", *J. Funct. Anal.* **80**:1 (1988), 148–211. MR 90d:58159 Zbl 0653.53022

[Osgood, Phillips, and Sarnak 1988b] B. Osgood, R. Phillips, and P. Sarnak, "Compact isospectral sets of surfaces", *J. Funct. Anal.* **80**:1 (1988), 212–234. MR 90d:58160 Zbl 0653.53021

[Osgood, Phillips, and Sarnak 1989] B. Osgood, R. Phillips, and P. Sarnak, "Moduli space, heights and isospectral sets of plane domains", *Ann. of Math.* (2) **129**:2 (1989), 293–362. MR 91a:58196 Zbl 0677.58045

[Peinecke et al. 2007] N. Peinecke, F.-E. Wolter, and M. Reuter, "Laplace spectra as fingerprints for image recognition", *Computer-Aided Design* **39** (2007), 460–476.

[Pleijel 1954] Å. Pleijel, "A study of certain Green's functions with applications in the theory of vibrating membranes", *Ark. Mat.* **2** (1954), 553–569. MR 15,798g Zbl 0055.08801

[Popov 1998] G. Popov, "On the contribution of degenerate periodic trajectories to the wave-trace", *Comm. Math. Phys.* **196**:2 (1998), 363–383. MR 2000e:58040 Zbl 0924.58100

[Reuter 2007] M. Reuter, "Can one hear shape?", pp. 1011101–1011102 in *PAMM Proceedings of GAMM07 and ICIAM07*, vol. 1, Wiley, 2007.

[Reuter et al. 2007] M. Reuter, F.-E. Wolter, and N. Peinecke, "Laplace–Beltrami spectra as 'Shape-DNA' of surfaces and solids", *Computer-Aided Design* **38** (2007), 342–366.

[Reuter et al. 2009] M. Reuter, F.-E. Wolter, M. Shenton, and M. Niethammer, "Laplace–Beltrami eigenvalues and topological features of eigenfunctions for statistical shape analysis", *Computer-Aided Design* **41** (2009), 739–755.

[Selberg 1956] A. Selberg, "Harmonic analysis and discontinuous groups in weakly symmetric Riemannian spaces with applications to Dirichlet series", *J. Indian Math. Soc.* (N.S.) **20** (1956), 47–87. MR 19,531g Zbl 0072.08201

[Sharafutdinov 2009] V. A. Sharafutdinov, "Local audibility of a hyperbolic metric", *Sibirsk. Mat. Zh.* **50**:5 (2009), 1176–1194. In Russian; translated in *Sib. Math. J.* **50**:5, 929–944, 2009. MR 2011d:58083 Zbl 1224.58023

[Sharafutdinov and Uhlmann 2000] V. Sharafutdinov and G. Uhlmann, "On deformation boundary rigidity and spectral rigidity of Riemannian surfaces with no focal points", *J. Differential Geom.* **56**:1 (2000), 93–110. MR 2002i:53056 Zbl 1065.53039

[Sjöstrand 1992] J. Sjöstrand, "Semi-excited states in nondegenerate potential wells", *Asymptotic Anal.* **6**:1 (1992), 29–43. MR 93m:35052 Zbl 0782.35050

[Sjöstrand 1997] J. Sjöstrand, "A trace formula and review of some estimates for resonances", pp. 377–437 in *Microlocal analysis and spectral theory* (Lucca, 1996), edited by L. Rodino, NATO Adv. Sci. Inst. Ser. C Math. Phys. Sci. **490**, Kluwer Acad. Publ., Dordrecht, 1997. MR 99e:47064 Zbl 0877.35090

[Sjöstrand 2002] J. Sjöstrand, "Lectures on resonances", lecture notes, 2002, Available at http://www.math.polytechnique.fr/~sjoestrand/CoursgbgWeb.pdf.

[Sjöstrand and Zworski 2002] J. Sjöstrand and M. Zworski, "Quantum monodromy and semi-classical trace formulae", *J. Math. Pures Appl.* (9) **81**:1 (2002), 1–33. MR 2004g:58035 Zbl 1038.58033

[Tanno 1973] S. Tanno, "Eigenvalues of the Laplacian of Riemannian manifolds", *Tôhoku Math. J.* (2) **25** (1973), 391–403. MR 48 #12405 Zbl 0266.53033

[Tanno 1980] S. Tanno, "A characterization of the canonical spheres by the spectrum", *Math. Z.* **175**:3 (1980), 267–274. MR 82g:58093 Zbl 0431.53037

[Zelditch 1996] S. Zelditch, "Maximally degenerate Laplacians", *Ann. Inst. Fourier (Grenoble)* **46**:2 (1996), 547–587. MR 97h:58171 Zbl 0853.58101

[Zelditch 1997] S. Zelditch, "Wave invariants at elliptic closed geodesics", *Geom. Funct. Anal.* **7**:1 (1997), 145–213. MR 98f:58191 Zbl 0876.58010

[Zelditch 1998a] S. Zelditch, "The inverse spectral problem for surfaces of revolution", *J. Differential Geom.* **49**:2 (1998), 207–264. MR 99k:58188 Zbl 0938.58027

[Zelditch 1998b] S. Zelditch, "Wave invariants for non-degenerate closed geodesics", *Geom. Funct. Anal.* **8**:1 (1998), 179–217. MR 98m:58136 Zbl 0908.58022

[Zelditch 1999] S. Zelditch, "Lectures on wave invariants", pp. 284–328 in *Spectral theory and geometry* (Edinburgh, 1998), edited by B. Davies and Y. Safarov, London Math. Soc. Lecture Note Ser. **273**, Cambridge Univ. Press, Cambridge, 1999. MR 2001a:58044 Zbl 0946.58024

[Zelditch 2000] S. Zelditch, "Spectral determination of analytic bi-axisymmetric plane domains", *Geom. Funct. Anal.* **10**:3 (2000), 628–677. MR 2001k:58064 Zbl 0961.58012

[Zelditch 2004a] S. Zelditch, "Inverse resonance problem for $\mathbb{Z}_2$-symmetric analytic obstacles in the plane", pp. 289–321 in *Geometric methods in inverse problems and PDE control*, edited by C. B. Croke et al., IMA Vol. Math. Appl. **137**, Springer, New York, 2004. MR 2007h:58055a Zbl 1061.35176

[Zelditch 2004b] S. Zelditch, "The inverse spectral problem", pp. 401–467 in *Surveys in differential geometry*, vol. 9, International Press, Somerville, MA, 2004. MR 2007c:58053 Zbl 1061.58029

[Zelditch 2009] S. Zelditch, "Inverse spectral problem for analytic domains, II: $\mathbb{Z}_2$-symmetric domains", *Ann. of Math.* (2) **170**:1 (2009), 205–269. MR 2010i:58036

[Zhou 1997] G. Zhou, "Compactness of isospectral compact manifolds with bounded curvatures", *Pacific J. Math.* **181**:1 (1997), 187–200. MR 98m:58140 Zbl 0885.58092

[Zworski 1997] M. Zworski, "Poisson formulae for resonances", exposé XIII in *Séminaire sur les Équations aux Dérivées Partielles, 1996–1997*, École Polytech., Palaiseau, 1997. MR 98j:35036

[Zworski 1998] M. Zworski, "Poisson formula for resonances in even dimensions", *Asian J. Math.* **2**:3 (1998), 609–617. MR 2001g:47091 Zbl 0932.47004

[Zworski 2001] M. Zworski, "A remark on isopolar potentials", *SIAM J. Math. Anal.* **32**:6 (2001), 1324–1326. MR 1856251 Zbl 0988.34067

[Zworski 2007] M. Zworski, "A remark on: 'Inverse resonance problem for $\mathbb{Z}_2$-symmetric analytic obstacles in the plane' by S. Zelditch", *Inverse Probl. Imaging* **1**:1 (2007), 225–227. MR 2007h:58055b

[Zworski 2011] M. Zworski, "Lectures on scattering resonances", Lecture notes, 2011, Available at http://math.berkeley.edu/~zworski/res.pdf.

datchev@math.mit.edu *Mathematics Department, Massachusetts Institute of Technology, Cambridge, MA 02139, United States*

hezari@math.mit.edu *Mathematics Department, Massachusetts Institute of Technology, Cambridge, MA 02139, United States*

Microlocal analysis of asymptotically hyperbolic spaces and high-energy resolvent estimates

ANDRÁS VASY

In this paper we describe a new method for analyzing the Laplacian on asymptotically hyperbolic spaces, which was introduced by the author in 2010. This new method in particular constructs the analytic continuation of the resolvent for even metrics (in the sense of Guillarmou), and gives high-energy estimates in strips. The key idea is an extension across the boundary for a problem obtained from the Laplacian shifted by the spectral parameter. The extended problem is nonelliptic — indeed, on the other side it is related to the Klein–Gordon equation on an asymptotically de Sitter space — but nonetheless it can be analyzed by methods of Fredholm theory. This method is a special case of a more general approach to the analysis of PDEs which includes, for instance, Kerr–de Sitter- and Minkowski-type spaces. The present paper is self-contained, and deals with asymptotically hyperbolic spaces without burdening the reader with material only needed for the analysis of the Lorentzian problems considered in earlier work by the author.

1. Introduction

In this paper we describe a new method for analyzing the Laplacian on asymptotically hyperbolic, or conformally compact, spaces, which was introduced in [Vasy 2010a]. This new method in particular constructs the analytic continuation of the resolvent for even metrics (in the sense of [Guillarmou 2005]), and gives high-energy estimates in strips. The key idea is an extension across the boundary for a problem obtained from the Laplacian shifted by the spectral parameter. The extended problem is nonelliptic — indeed, on the other side it is related to the Klein–Gordon equation on an asymptotically de Sitter space — but nonetheless it can be analyzed by methods of Fredholm theory. In [Vasy 2010a] these methods, with some additional ingredients, were used to analyze the wave equation on

---

The author gratefully acknowledges partial support from the NSF under grant number DMS-0801226 and from a Chambers Fellowship at Stanford University, as well as the hospitality of MSRI in Berkeley in Fall 2010.

Kerr–de Sitter space-times; the present setting is described there as the simplest application of the tools introduced. The purpose of the present paper is to give a self-contained treatment of conformally compact spaces, without burdening the reader with the additional machinery required for the Kerr–de Sitter analysis.

We start by recalling the definition of manifolds with *even* conformally compact metrics. These are Riemannian metrics g_0 on the interior of an n-dimensional compact manifold with boundary X_0 such that near the boundary Y, with a product decomposition nearby and a defining function x, they are of the form

$$g_0 = \frac{dx^2 + h}{x^2},$$

where h is a family of metrics on $Y = \partial X_0$ depending on x in an even manner, that is, only even powers of x show up in the Taylor series. (There is a much more natural way to phrase the evenness condition, see [Guillarmou 2005, Definition 1.2].) We also write $X_{0,\text{even}}$ for the manifold X_0 when the smooth structure has been changed so that x^2 is a boundary defining function; thus, a smooth function on X_0 is even if and only if it is smooth when regarded as a function on $X_{0,\text{even}}$. The analytic continuation of the resolvent in this category, but without the evenness condition, was obtained in [Mazzeo and Melrose 1987] (similar results appear in [Agmon 1986; Perry 1987; 1989] in the restricted setting of hyperbolic quotients), with the possibility of some essential singularities at pure imaginary half-integers noticed by Borthwick and Perry [2002]. Guillarmou [2005] showed that for even metrics the latter do not exist, but generically they do exist for noneven metrics, by a more careful analysis utilizing the work of Graham and Zworski [2003]. Further, if the manifold is actually asymptotic to hyperbolic space (note that hyperbolic space is of this form in view of the Poincaré model), Melrose, Sá Barreto and Vasy [Melrose et al. 2011] proved high-energy resolvent estimates in strips around the real axis via a parametrix construction; these are exactly the estimates that allow expansions for solutions of the wave equation in terms of resonances. Estimates just on the real axis were obtained in [Cardoso and Vodev 2002; Vodev 2004] for more general conformal infinities. One implication of our methods is a generalization of these results: we allow general conformal infinities, and obtain estimates in arbitrary strips.

In the sequel $\dot{\mathcal{C}}^\infty(X_0)$ denotes "Schwartz functions" on X_0, that is, $\mathcal{C}^\infty$ functions vanishing with all derivatives at ∂X_0, and $\mathcal{C}^{-\infty}(X_0)$ is the dual space of "tempered distributions" (these spaces are naturally identified for X_0 and $X_{0,\text{even}}$), while $H^s(X_{0,\text{even}})$ is the standard Sobolev space on $X_{0,\text{even}}$ (corresponding to extension across the boundary, see, for example, [Hörmander 1985a, Appendix B], where these are denoted by $\bar{H}^s(X^\circ_{0,\text{even}})$). For instance,

$\|u\|^2_{H^1(X_{0,\text{even}})} = \|u\|^2_{L^2(X_{0,\text{even}})} + \|du\|^2_{L^2(X_{0,\text{even}})}$, with the norms taken with respect to any *smooth* Riemannian metric on $X_{0,\text{even}}$ (all choices yield equivalent norms by compactness). Here we point out that while $x^2 g_0$ is a smooth nondegenerate section of the pull-back of $T^* X_0$ to $X_{0,\text{even}}$ (which essentially means that it is a smooth, in $X_{0,\text{even}}$, nondegenerate linear combination of dx and dy_j in local coordinates), as $\mu = x^2$ means $d\mu = 2x\, dx$, it is actually not a smooth section of $T^* X_{0,\text{even}}$. However, $x^{n+1} |dg_0|$ is a smooth nondegenerate density, so $L^2(X_{0,\text{even}})$ (up to norm equivalence) is the L^2 space given by the density $x^{n+1}|dg_0|$, i.e., it is the space $x^{-(n+1)/2} L^2_{g_0}(X_0)$, which means that

$$\|x^{-(n+1)/2} u\|_{L^2(X_{0,\text{even}})} \sim \|u\|_{L^2_{g_0}(X_0)}.$$

Further, in local coordinates (μ, y), using $2\partial_\mu = x^{-1}\partial_x$, the $H^1(X_{0,\text{even}})$ norm of u is equivalent to

$$\|u\|^2_{L^2(X_{0,\text{even}})} + \|x^{-1}\partial_x u\|^2_{L^2(X_{0,\text{even}})} + \sum_{j=1}^{n-1} \|\partial_{y_j} u\|^2_{L^2(X_{0,\text{even}})}.$$

We also let $H^s_h(X_{0,\text{even}})$ be the standard semiclassical Sobolev space, that is, for h bounded away from 0 this is equipped with a norm equivalent to the standard fixed (h-independent) norm on $H^s(X_{0,\text{even}})$, but the uniform behavior as $h \to 0$ is different; for example, locally the $H^1_h(X)$ norm is given by

$$\|u\|^2_{H^1_h} = \sum_j \|h D_j u\|^2_{L^2} + \|u\|^2_{L^2};$$

see [Dimassi and Sjöstrand 1999; Evans and Zworski 2010]. Thus, in (1-1), for $s = 1$ (which is possible when $C < \frac{1}{2}$, that is, if one only considers the continuation into a small strip beyond the continuous spectrum),

$$s = 1 \implies \|u\|_{H^{s-1}_{|\sigma|^{-1}}(X_{0,\text{even}})} = \|u\|_{L^2(X_{0,\text{even}})}$$

and $\|u\|^2_{H^s_{|\sigma|^{-1}}(X_{0,\text{even}})} = \|u\|^2_{L^2(X_{0,\text{even}})} + |\sigma|^{-2} \|du\|^2_{L^2(X_{0,\text{even}})}$,

with the norms taken with respect to any smooth Riemannian metric on $X_{0,\text{even}}$.

Theorem. (See Theorem 5.1 for the full statement.) *Suppose that X_0 is an n-dimensional manifold with boundary Y with an even Riemannian conformally compact metric g_0. Then the inverse of*

$$\Delta_{g_0} - \left(\frac{n-1}{2}\right)^2 - \sigma^2,$$

written as $\mathcal{R}(\sigma) : L^2 \to L^2$, has a meromorphic continuation from $\operatorname{Im}\sigma \gg 0$ to $\mathbb{C}$,
$$\mathcal{R}(\sigma) : \dot{\mathcal{C}}^\infty(X_0) \to \mathcal{C}^{-\infty}(X_0),$$
with poles with finite rank residues. If in addition (X_0, g_0) is nontrapping, then nontrapping estimates hold in every strip $-C < \operatorname{Im}\sigma < C_+$, $|\operatorname{Re}\sigma| \gg 0$: for $s > \frac{1}{2} + C$,

$$\left\| x^{-(n-1)/2+\imath\sigma} \mathcal{R}(\sigma) f \right\|_{H^s_{|\sigma|^{-1}}(X_{0,\text{even}})}$$
$$\leq \tilde{C}|\sigma|^{-1} \left\| x^{-(n+3)/2+\imath\sigma} f \right\|_{H^{s-1}_{|\sigma|^{-1}}(X_{0,\text{even}})}. \quad (1\text{-}1)$$

If f has compact support in X_0°, the $s-1$ norm on f can be replaced by the $s-2$ norm. For suitable $\delta_0 > 0$, the estimates are valid in regions $-C < \operatorname{Im}\sigma < \delta_0|\operatorname{Re}\sigma|$ if the multipliers $x^{\imath\sigma}$ are slightly adjusted.

Further, as stated in Theorem 5.1, the resolvent is *semiclassically outgoing* with a loss of h^{-1}, in the sense of recent results of Datchev and Vasy [2011; 2010]. This means that for mild trapping (where, in a strip near the spectrum, one has polynomially bounded resolvent for a compactly localized version of the trapped model) one obtains resolvent bounds of the same kind as for the above-mentioned trapped models, and lossless estimates microlocally away from the trapping. In particular, one obtains logarithmic losses compared to nontrapping on the spectrum for hyperbolic trapping in the sense of [Wunsch and Zworski 2011, Section 1.2], and polynomial losses in strips, since for the compactly localized model this was shown in [Wunsch and Zworski 2011].

Our method is to change the smooth structure, replacing x by $\mu = x^2$, conjugate the operator by an appropriate weight as well as remove a vanishing factor of μ, and show that the new operator continues smoothly and nondegenerately (in an appropriate sense) across $\mu = 0$, that is, Y, to a (nonelliptic) problem which we can analyze utilizing by now almost standard tools of microlocal analysis. These steps are reflected in the form of the estimate (1-1); μ shows up in the use of evenness, conjugation due to the presence of $x^{-(n+1)/2+\imath\sigma}$, and the two halves of the vanishing factor of μ being removed in $x^{\pm 1}$ on the left and right sides.

While it might seem somewhat ad hoc, this construction in fact has origins in wave propagation in Lorentzian spaces of one dimension higher, i.e., $(n+1)$-dimensional: either Minkowski space, or de Sitter space blown up at a point at future infinity. Namely in both cases the wave equation (and the Klein–Gordon equation in de Sitter space) is a totally characteristic, or b-, PDE, and after a Mellin transform this gives a PDE on the sphere at infinity in the Minkowski case, and on the front face of the blow-up in the de Sitter setting. These are exactly the PDE arising by the process described in the previous paragraph, with the original manifold X_0 lying in the interior of the light cone in Minkowski space (so there are two copies, at future and past infinity) and in the interior of the backward light cone from the blow-up point in the de Sitter case; see [Vasy 2010a] for more detail. This relationship, restricted to the X_0-region, was exploited in [Vasy 2010b, Section 7], where the work of Mazzeo and Melrose was used to construct the Poisson operator on asymptotically de Sitter spaces. Conceptually the main novelty here is that we work directly with the extended problem, which turns out to simplify the analysis of Mazzeo and Melrose in many ways and give a new explanation for Guillarmou's results as well as yield high-energy estimates.

We briefly describe this extended operator, P_σ. It has radial points at the conormal bundle $N^*Y \setminus o$ of Y in the sense of microlocal analysis, i.e., the Hamilton vector field is radial at these points — it is a multiple of the generator of dilations of the fibers of the cotangent bundle there. However, tools exist to deal with these, going back to Melrose's geometric treatment of scattering theory on asymptotically Euclidean spaces [Melrose 1994]. Note that $N^*Y \setminus o$ consists of two components, Λ_+, resp. Λ_-, and in $S^*X = (T^*X \setminus o)/\mathbb{R}^+$ the images, L_+, resp. L_-, of these are sources, resp. sinks, for the Hamilton flow. At $L_\pm$ one has choices regarding the direction one wants to propagate estimates (into or out of the radial points), which directly correspond to working with strong or weak Sobolev spaces. For the present problem, the relevant choice is propagating estimates *away from* the radial points, thus working with the "good" Sobolev spaces (which can be taken to have order as positive as one wishes; there is a minimum amount of regularity imposed by our choice of propagation direction — see the requirement $s > \frac{1}{2} + C$ above (1-1)). All other points are either elliptic, or real principal type. It remains to either deal with the noncompactness of the "far end" of the n-dimensional de Sitter space — or instead, as is indeed more convenient when one wants to deal with more singular geometries, adding complex absorbing potentials, in the spirit of [Nonnenmacher and Zworski 2009; Wunsch and Zworski 2011]. In fact, the complex absorption could be replaced by adding a space-like boundary [Vasy 2010a], but for many microlocal purposes complex absorption is more desirable, hence we follow the latter method. However, crucially, these complex absorbing techniques (or the addition of a space-like boundary) already enter in the nonsemiclassical problem in our case, as we are in a nonelliptic setting.

One can reverse the direction of the argument and analyze the wave equation on an n-dimensional even asymptotically de Sitter space X_0' by extending it across the boundary, much like the Riemannian conformally compact space X_0 is extended in this approach. Then, performing microlocal propagation in the opposite direction, which amounts to working with the adjoint operators that we already need in order to prove existence of solutions for the Riemannian spaces, we obtain existence, uniqueness and structure results for asymptotically de Sitter spaces, recovering a large part of the results of [Vasy 2010b]. Here we only briefly indicate this method of analysis in Remark 5.3.

In other words, we establish a Riemannian–Lorentzian duality, that will have counterparts both in the pseudo-Riemannian setting of higher signature and in higher rank symmetric spaces, though in the latter the analysis might become more complicated. Note that asymptotically hyperbolic and de Sitter spaces are not connected by a "complex rotation" (in the sense of an actual deformation); they are smooth continuations of each other in the sense we just discussed.

To emphasize the simplicity of our method, we list all of the microlocal techniques (which are relevant both in the classical and in the semiclassical setting) that we use on a *compact manifold without boundary*; in all cases *only microlocal Sobolev estimates* matter (not parametrices, etc.):

(i) Microlocal elliptic regularity.

(ii) Real principal type propagation of singularities.

(iii) *Rough* analysis at a Lagrangian invariant under the Hamilton flow which roughly behaves like a collection of radial points, though the internal structure does not matter, in the spirit of [Melrose 1994, Section 9].

(iv) Complex absorbing "potentials" in the spirit of [Nonnenmacher and Zworski 2009; Wunsch and Zworski 2011].

These are almost "off the shelf" in terms of modern microlocal analysis, and thus our approach, from a microlocal perspective, is quite simple. We use these to show that on the continuation across the boundary of the conformally compact space we have a Fredholm problem, on a perhaps slightly exotic function space, which however is (perhaps apart from the complex absorption) the simplest possible coisotropic function space based on a Sobolev space, with order dictated by the radial points. Also, we propagate the estimates along bicharacteristics in different directions depending on the component $\Sigma_\pm$ of the characteristic set under consideration; correspondingly the sign of the complex absorbing "potential" will vary with $\Sigma_\pm$, which is perhaps slightly unusual. However, this is completely parallel to solving the standard Cauchy, or forward, problem for the wave equation, where one propagates estimates in *opposite* directions relative to the Hamilton vector field in the two components of the characteristic set.

The complex absorption we use modifies the operator P_σ outside $X_{0,\text{even}}$. However, while $(P_\sigma - \imath Q_\sigma)^{-1}$ depends on Q_σ, its behavior on $X_{0,\text{even}}$, and even near $X_{0,\text{even}}$, is independent of this choice; see the proof of Section 5 for a detailed explanation. In particular, although $(P_\sigma - \imath Q_\sigma)^{-1}$ may have resonances other than those of $\mathcal{R}(\sigma)$, the resonant states of these additional resonances are supported outside $X_{0,\text{even}}$, hence do not affect the singular behavior of the resolvent in $X_{0,\text{even}}$.

While the results are stated for the scalar equation, analogous results hold for operators on natural vector bundles, such as the Laplacian on differential forms. This is so because the results work if the principal symbol of the extended problem is scalar with the demanded properties, and the principal symbol of $\frac{1}{2i}(P_\sigma - P_\sigma^*)$ is either scalar at the 'radial sets', or instead satisfies appropriate estimates (as an endomorphism of the pull-back of the vector bundle to the cotangent bundle) at this location; see Remark 3.1. The only change in terms of results on asymptotically hyperbolic spaces is that the threshold $(n-1)^2/4$ is shifted; in terms of the explicit conjugation of Section 5 this is so because of the change in the first order term in (3-2).

In Section 3 we describe in detail the setup of conformally compact spaces and the extension across the boundary. Then in Section 4 we describe the in detail the necessary microlocal analysis for the extended operator. Finally, in Section 5 we translate these results back to asymptotically hyperbolic spaces.

2. Notation

We start by briefly recalling the basic pseudodifferential objects, in part to establish notation. As a general reference for microlocal analysis, we refer to [Hörmander 1985a; 1985b], while for semiclassical analysis, we refer to [Dimassi and Sjöstrand 1999; Evans and Zworski 2010].

First, $S^k(\mathbb{R}^p; \mathbb{R}^\ell)$ is the set of $\mathcal{C}^\infty$ functions on $\mathbb{R}^p_z \times \mathbb{R}^\ell_\zeta$ satisfying uniform bounds
$$|D_z^\alpha D_\zeta^\beta a| \leq C_{\alpha\beta} \langle \zeta \rangle^{k-|\beta|}, \quad \alpha \in \mathbb{N}^p, \ \beta \in \mathbb{N}^\ell.$$

If $O \subset \mathbb{R}^p$ and $\Gamma \subset \mathbb{R}^\ell_\zeta$ are open, we define $S^k(O; \Gamma)$ by requiring these estimates to hold only for $z \in O$ and $\zeta \in \Gamma$. (We could instead require uniform estimates on compact subsets; this makes no difference here.) The class of classical (or one-step polyhomogeneous) symbols is the subset $S^k_{\text{cl}}(\mathbb{R}^p; \mathbb{R}^\ell)$ of $S^k(\mathbb{R}^p; \mathbb{R}^\ell)$ consisting of symbols possessing an asymptotic expansion
$$a(z, r\omega) \sim \sum a_j(z, \omega) r^{k-j}, \tag{2-1}$$

where $a_j \in \mathcal{C}^\infty(\mathbb{R}^p \times \mathbb{S}^{\ell-1})$. Then on $\mathbb{R}^n_z$, pseudodifferential operators $A \in \Psi^k(\mathbb{R}^n)$ are of the form
$$A = \text{Op}(a); \ (\text{Op}(a)u)(z) = (2\pi)^{-n} \int_{\mathbb{R}^n} e^{i(z-z')\cdot\zeta} a(z, \zeta) u(z') \, d\zeta \, dz',$$
$$u \in \mathcal{S}(\mathbb{R}^n), \ a \in S^k(\mathbb{R}^n; \mathbb{R}^n);$$

understood as an oscillatory integral. Classical pseudodifferential operators, $A \in \Psi_{cl}^k(\mathbb{R}^n)$, form the subset where a is a classical symbol. The principal symbol $\sigma_k(A)$ of $A \in \Psi^k(\mathbb{R}^n)$ is the equivalence class of a in $S^k(\mathbb{R}^n; \mathbb{R}^n)/S^{k-1}(\mathbb{R}^n; \mathbb{R}^n)$, denoted by $[a]$. For classical a, one can instead regard $a_0(z, \omega)r^k$ as the principal symbol; it is a $\mathcal{C}^\infty$ function on $\mathbb{R}^n \times (\mathbb{R}^n \setminus \{0\})$, which is homogeneous of degree k with respect to the $\mathbb{R}^+$-action given by dilations in the second factor, $\mathbb{R}^n \setminus \{0\}$. The principal symbol is multiplicative, that is, $\sigma_{k+k'}(AB) = \sigma_k(A)\sigma_{k'}(B)$. Moreover, the principal symbol of a commutator is given by the Poisson bracket (or equivalently by the Hamilton vector field): $\sigma_{k+k'-1}(\imath[A, B]) = \mathsf{H}_{\sigma_k(A)}\sigma_{k'}(B)$, with $\mathsf{H}_a = \sum_{j=1}^n ((\partial_{\xi_j}a)\partial_{z_j} - (\partial_{z_j}a)\partial_{\xi_j})$. Note that for a homogeneous of order k, H_a is homogeneous of order $k-1$.

There are two very important properties: nondegeneracy (called ellipticity) and extreme degeneracy (captured by the operator wave front set) of an operator. One says that A is elliptic at $\alpha \in \mathbb{R}^n \times (\mathbb{R}^n \setminus \{0\})$ if there exists an open cone Γ (conic with respect to the $\mathbb{R}^+$-action on $\mathbb{R}^n \setminus o$) around α and $R > 0$, $C > 0$ such that $|a(x, \xi)| \geq C|\xi|^k$ for $|\xi| > R$, $(x, \xi) \in \Gamma$, where $[a] = \sigma_k(A)$. If A is classical, and a is taken to be homogeneous, this just amounts to $a(\alpha) \neq 0$.

On the other hand, for $A = \mathrm{Op}(a)$ and $\alpha \in \mathbb{R}^n \times (\mathbb{R}^n \setminus o)$ one says that $\alpha \notin \mathrm{WF}'(A)$ if there exists an open cone Γ around α such that $a|_\Gamma \in S^{-\infty}(\Gamma)$, that is, $a|_\Gamma$ is rapidly decreasing, with all derivatives, as $|\xi| \to \infty$, $(x, \xi) \in \Gamma$. Note that both the elliptic set $\mathrm{ell}(A)$ of A (i.e., the set of points where A is elliptic) and $\mathrm{WF}'(A)$ are conic.

Differential operators on $\mathbb{R}^n$ form the subset of $\Psi(\mathbb{R}^n)$ in which a is polynomial in the second factor, $\mathbb{R}^n_\zeta$, so locally

$$A = \sum_{|\alpha| \leq k} a_\alpha(z) D_z^\alpha, \qquad \sigma_k(A) = \sum_{|\alpha|=k} a_\alpha(z)\zeta^\alpha.$$

If X is a manifold, one can transfer these definitions to X by localization and requiring that the Schwartz kernels are $\mathcal{C}^\infty$ densities away from the diagonal in $X^2 = X \times X$; then $\sigma_k(A)$ is in $S^k(T^*X)/S^{k-1}(T^*X)$, resp. $S^k_{\mathrm{hom}}(T^*X \setminus o)$ when $A \in \Psi^k(X)$, resp. $A \in \Psi_{cl}^k(X)$; here o is the zero section, and hom stands for symbols homogeneous with respect to the $\mathbb{R}^+$ action. If A is a differential operator, then the classical (i.e., homogeneous) version of the principal symbol is a homogeneous polynomial in the fibers of the cotangent bundle of degree k. The notions of $\mathrm{ell}(A)$ and $\mathrm{WF}'(A)$ extend to give conic subsets of $T^*X \setminus o$; equivalently they are subsets of the cosphere bundle $S^*X = (T^*X \setminus o)/\mathbb{R}^+$. We can also work with operators depending on a parameter $\lambda \in O$ by replacing $a \in S^k(\mathbb{R}^n; \mathbb{R}^n)$ by $a \in S^k(\mathbb{R}^n \times O; \mathbb{R}^n)$, with $\mathrm{Op}(a_\lambda) \in \Psi^k(\mathbb{R}^n)$ smoothly dependent on $\lambda \in O$. In the case of differential operators, a_α would simply depend smoothly on the parameter λ.

We next consider the semiclassical operator algebra. We adopt the convention that $\hbar$ denotes semiclassical objects, while h is the actual semiclassical parameter. This algebra, $\Psi_\hbar(\mathbb{R}^n)$, is given by

$$A_h = \mathrm{Op}_\hbar(a); \ \mathrm{Op}_\hbar(a)u(z) = (2\pi h)^{-n} \int_{\mathbb{R}^n} e^{i(z-z')\cdot\zeta/h} a(z,\zeta,h)\, u(z')\, d\zeta\, dz',$$

$$u \in \mathscr{S}(\mathbb{R}^n),\ a \in \mathscr{C}^\infty([0,1)_h; S^k(\mathbb{R}^n; \mathbb{R}^n_\zeta));$$

its classical subalgebra, $\Psi_{\hbar,\mathrm{cl}}(\mathbb{R}^n)$, corresponds to $a \in \mathscr{C}^\infty([0,1)_h; S^k_{\mathrm{cl}}(\mathbb{R}^n; \mathbb{R}^n_\zeta))$. The semiclassical principal symbol is now $\sigma_{\hbar,k}(A) = a|_{h=0} \in S^k(\mathbb{R}^n \times \mathbb{R}^n)$. In the setting of a general manifold X, $\mathbb{R}^n \times \mathbb{R}^n$ is replaced by T^*X. Correspondingly, $\mathrm{WF}'_\hbar(A)$ and $\mathrm{ell}_\hbar(A)$ are subsets of T^*X. We can again add an extra parameter $\lambda \in O$, so $a \in \mathscr{C}^\infty([0,1)_h; S^k(\mathbb{R}^n \times O; \mathbb{R}^n_\zeta))$; then in the invariant setting the principal symbol is $a|_{h=0} \in S^k(T^*X \times O)$.

Differential operators now take the form

$$A_{h,\lambda} = \sum_{|\alpha|\le k} a_\alpha(z,\lambda;h)(hD_z)^\alpha. \tag{2-2}$$

Such a family has two principal symbols, the standard one (but taking into account the semiclassical degeneration, that is, based on $(hD_z)^\alpha$ rather than D_z^α), which depends on h and is homogeneous, and the semiclassical one, which is at $h=0$, and is not homogeneous:

$$\sigma_k(A_{h,\lambda}) = \sum_{|\alpha|=k} a_\alpha(z,\lambda;h)\zeta^\alpha,$$

$$\sigma_\hbar(A_{h,\lambda}) = \sum_{|\alpha|\le k} a_\alpha(z,\lambda;0)\zeta^\alpha.$$

However, the restriction of $\sigma_k(A_{h,\lambda})$ to $h=0$ is the principal symbol of $\sigma_\hbar(A_{h,\lambda})$. In the special case in which $\sigma_k(A_{h,\lambda})$ is independent of h (which is true in the setting considered below), one can simply regard the usual principal symbol as the principal part of the semiclassical symbol.

This is a convenient place to recall from [Melrose 1994] that it is often useful to consider the radial compactification of the fibers of the cotangent bundle to balls (or hemispheres, in Melrose's exposition). Thus, one adds a sphere at infinity to the fiber T_q^*X of T^*X over each $q \in X$. This sphere is naturally identified with S_q^*X, and we obtain compact fibers $\overline{T}_q^*X$ with boundary S_q^*X, with the smooth structure near S_q^*X arising from reciprocal polar coordinates $(\tilde{\rho},\omega) = (r^{-1},\omega)$ for $\tilde{\rho} > 0$, but extending to $\tilde{\rho} = 0$, and with S_q^*X given by

$\tilde{\rho} = 0$. Thus, with $X = \mathbb{R}^n$ the classical expansion (2-1) becomes

$$a(z, \tilde{\rho}, \omega) \sim \tilde{\rho}^{-k} \sum a_j(z, \omega) \tilde{\rho}^j,$$

where $a_j \in \mathscr{C}^\infty(\mathbb{R}^p \times \mathbb{S}^{\ell-1})$, so in particular for $k = 0$, this is simply the Taylor series expansion at S^*X of a function smooth up to $S^*X = \partial \overline{T}^*X$. In the semiclassical context then one considers $\overline{T}^*X \times [0, 1)$, and notes that "classical" semiclassical operators of order 0 are given locally by $\mathrm{Op}_\hbar(a)$ with a extending to be smooth up to the boundaries of this space, with semiclassical symbol given by restriction to $\overline{T}^*X \times \{0\}$, and standard symbol given by restriction to $S^*X \times [0, 1)$. Thus, the claim regarding the limit of the semiclassical symbol at infinity is simply a matching statement of the two symbols at the corner $S^*X \times \{0\}$ in this compactified picture.

Finally, we recall that if $P = \sum_{|\alpha| \leq k} a_\alpha(z) D_z^\alpha$ is an order k differential operator, then the behavior of $P - \lambda$ as $\lambda \to \infty$ can be converted to a semiclassical problem by considering

$$P_{\hbar,\sigma} = h^k(P - \lambda) = \sum_{|\alpha| \leq k} h^{k-|\alpha|} a_\alpha(z)(hD_z)^\alpha - \sigma,$$

where $\sigma = h^k \lambda$. Here there is freedom in choosing h, for example, $h = |\lambda|^{1/k}$, in which case $|\sigma| = 1$, but it is often useful to leave some flexibility in the choice so that $h \sim |\lambda|^{1/k}$ only, and thus σ is in a compact subset of $\mathbb{C}$ disjoint from 0. Note that

$$\sigma_\hbar(P_{\hbar,\sigma}) = \sum_{|\alpha|=k} a_\alpha(z) \zeta^\alpha - \sigma.$$

If we do not want to explicitly multiply by h^k, we write the full high-energy principal symbol of $P - \lambda$ as

$$\sigma_{\mathrm{full}}(P_\lambda) = \sum_{|\alpha|=k} a_\alpha(z) \zeta^\alpha - \lambda.$$

More generally, if $P(\lambda) = \sum_{|\alpha|+|\beta| \leq k} a_\alpha(z) \lambda^\beta D_z^\alpha$ is an order k differential operator depending on a large parameter λ, we let

$$\sigma_{\mathrm{full}}(P(\lambda)) = \sum_{|\alpha|+|\beta|=k} a_\alpha(z) \lambda^\beta \zeta^\alpha$$

be the full large-parameter symbol. With $\lambda = h^{-1}\sigma$,

$$P_{\hbar,\sigma} = h^k P(\lambda) = \sum_{|\alpha|+|\beta| \leq k} h^{k-|\alpha|-|\beta|} a_\alpha(z) \sigma^\beta (hD_z)^\alpha$$

is a semiclassical differential operator with semiclassical symbol

$$\sigma_\hbar(P_{\hbar,\sigma}) = \sum_{|\alpha|+|\beta|=k} a_\alpha(z)\sigma^\beta \zeta^\alpha.$$

Note that the full large-parameter symbol and the semiclassical symbol are "the same", that is, they are simply related to each other.

3. Conformally compact spaces

3A. *From the Laplacian to the extended operator.*
Suppose that g_0 is an even asymptotically hyperbolic metric on X_0, with $\dim X_0 = n$. Then we may choose a product decomposition near the boundary such that

$$g_0 = \frac{dx^2 + h}{x^2} \tag{3-1}$$

there, where h is an even family of metrics; it is convenient to take x to be a globally defined boundary defining function. Then the dual metric is

$$G_0 = x^2(\partial_x^2 + H),$$

with H the dual metric family of h (depending on x as a parameter), and

$$|dg_0| = \sqrt{|\det g_0|}\, dx\, dy = x^{-n}\sqrt{|\det h|}\, dx\, dy,$$

so

$$\Delta_{g_0} = (xD_x)^2 + \iota(n-1+x^2\gamma)(xD_x) + x^2\Delta_h, \tag{3-2}$$

with γ even, and Δ_h the x-dependent family of Laplacians of h on Y.

We show now that if we change the smooth structure on X_0 by declaring that only even functions of x are smooth, that is, introducing $\mu = x^2$ as the boundary defining function, then after a suitable conjugation and division by a vanishing factor the resulting operator smoothly and nondegenerately continues across the boundary, that is, continues to $X_{-\delta_0} = (-\delta_0, 0)_\mu \times Y \sqcup X_{0,\mathrm{even}}$, where $X_{0,\mathrm{even}}$ is the manifold X_0 with the new smooth structure.

First, changing to coordinates (μ, y), $\mu = x^2$, we obtain

$$\Delta_{g_0} = 4(\mu D_\mu)^2 + 2\iota(n-1+\mu\gamma)(\mu D_\mu) + \mu\Delta_h, \tag{3-3}$$

Now we conjugate by $\mu^{-\iota\sigma/2+(n+1)/4}$ to obtain

$$\mu^{\iota\sigma/2-(n+1)/4}\left(\Delta_{g_0} - (n-1)^2/4 - \sigma^2\right)\mu^{-\iota\sigma/2+(n+1)/4}$$
$$= 4(\mu D_\mu - \sigma/2 - \iota(n+1)/4)^2 + 2\iota(n-1+\mu\gamma)(\mu D_\mu - \sigma/2 - \iota(n+1)/4)$$
$$+ \mu\Delta_h - (n-1)^2/4 - \sigma^2$$

$$= 4(\mu D_\mu)^2 - 4\sigma(\mu D_\mu) + \mu \Delta_h - 4\iota(\mu D_\mu) + 2\iota\sigma - 1$$
$$+ 2\iota\mu\gamma(\mu D_\mu - \sigma/2 - \iota(n+1)/4).$$

Next we multiply by $\mu^{-1/2}$ from both sides to obtain

$$\mu^{-1/2}\mu^{\iota\sigma/2-(n+1)/4}(\Delta_{g_0} - (n-1)^2/4 - \sigma^2)\mu^{-\iota\sigma/2+(n+1)/4}\mu^{-1/2}$$
$$= 4\mu D_\mu^2 - \mu^{-1} - 4\sigma D_\mu - 2\iota\sigma\mu^{-1} + \Delta_h - 4\iota D_\mu + 2\mu^{-1} + 2\iota\sigma\mu^{-1} - \mu^{-1}$$
$$+ 2\iota\gamma(\mu D_\mu - \sigma/2 - \iota(n-1)/4)$$
$$= 4\mu D_\mu^2 - 4\sigma D_\mu + \Delta_h - 4\iota D_\mu + 2\iota\gamma(\mu D_\mu - \sigma/2 - \iota(n-1)/4). \quad (3\text{-}4)$$

This operator is in $\mathrm{Diff}^2(X_{0,\text{even}})$, and now it continues smoothly across the boundary, by extending h and γ in an arbitrary smooth manner. This form suffices for analyzing the problem for σ in a compact set, or indeed for σ going to infinity in a strip near the reals. However, it is convenient to modify it as we would like the resulting operator to be semiclassically elliptic when σ is away from the reals. We achieve this via conjugation by a smooth function, with exponent depending on σ. The latter would make no difference even semiclassically in the real regime as it is conjugation by an elliptic semiclassical FIO. However, in the nonreal regime (where we would like ellipticity) it does matter; the present operator is not semiclassically elliptic at the zero section. So finally we conjugate by $(1 + \mu)^{\iota\sigma/4}$ to obtain

$$P_\sigma = 4(1 + a_1)\mu D_\mu^2 - 4(1 + a_2)\sigma D_\mu - (1 + a_3)\sigma^2 + \Delta_h$$
$$- 4\iota D_\mu + b_1\mu D_\mu + b_2\sigma + c_1, \quad (3\text{-}5)$$

with a_j smooth, real, vanishing at $\mu = 0$, b_j and c_1 smooth. In fact, we have $a_1 \equiv 0$, but it is sometimes convenient to have more flexibility in the form of the operator since this means that we do not need to start from the relatively rigid form (3-2).

Writing covectors as $\xi\, d\mu + \eta\, dy$, the principal symbol of $P_\sigma \in \mathrm{Diff}^2(X_{-\delta_0})$, including in the high-energy sense ($\sigma \to \infty$), is

$$p_{\text{full}} = 4(1 + a_1)\mu\xi^2 - 4(1 + a_2)\sigma\xi - (1 + a_3)\sigma^2 + |\eta|^2_{\mu,y}, \quad (3\text{-}6)$$

and is real for σ real. The Hamilton vector field is

$$H_{p_{\text{full}}} = 4(2(1 + a_1)\mu\xi - (1 + a_2)\sigma)\partial_\mu + \tilde{\mathsf{H}}_{|\eta|^2_{\mu,y}}$$
$$- \left(4\left(1 + a_1 + \mu\frac{\partial a_1}{\partial\mu}\right)\xi^2 - 4\frac{\partial a_2}{\partial\mu}\sigma\xi + \frac{\partial a_3}{\partial\mu}\sigma^2 + \frac{\partial|\eta|^2_{\mu,y}}{\partial\mu}\right)\partial_\xi$$
$$- \left(4\frac{\partial a_1}{\partial y}\mu\xi^2 - 4\frac{\partial a_2}{\partial y}\sigma\xi - \frac{\partial a_3}{\partial y}\sigma^2\right)\partial_\eta, \quad (3\text{-}7)$$

where $\tilde{\mathsf{H}}$ indicates that this is the Hamilton vector field in T^*Y, that is, with μ considered a parameter. Correspondingly, the standard, "classical", principal symbol is

$$p = \sigma_2(P_\sigma) = 4(1+a_1)\mu\xi^2 + |\eta|^2_{\mu,y}, \tag{3-8}$$

which is real, independent of σ, while the Hamilton vector field is

$$\mathsf{H}_p = 8(1+a_1)\mu\xi\partial_\mu + \tilde{\mathsf{H}}_{|\eta|^2_{\mu,y}}$$
$$- \left(4\left(1+a_1+\mu\frac{\partial a_1}{\partial \mu}\right)\xi^2 + \frac{\partial |\eta|^2_{\mu,y}}{\partial \mu}\right)\partial_\xi - 4\frac{\partial a_1}{\partial y}\mu\xi^2\partial_\eta. \tag{3-9}$$

It is useful to keep in mind that as $\Delta_{g_0} - \sigma^2 - (n-1)^2/4$ is formally self-adjoint relative to the metric density $|dg_0|$ for σ real, so the same holds for $\mu^{-1/2}(\Delta_{g_0} - \sigma^2 - (n-1)^2/4)\mu^{-1/2}$ (as μ is real), and indeed for its conjugate by $\mu^{-i\sigma/2}(1+\mu)^{i\sigma/4}$ for σ real since this is merely unitary conjugation. As for f real, A formally self-adjoint relative to $|dg_0|$, $f^{-1}Af$ is formally self-adjoint relative to $f^2|dg_0|$, we then deduce that for σ real, P_σ is formally self-adjoint relative to

$$\mu^{(n+1)/2}|dg_0| = \tfrac{1}{2}|dh|\,|d\mu|,$$

as $x^{-n}\,dx = \tfrac{1}{2}\mu^{-(n+1)/2}\,d\mu$. Note that $\mu^{(n+1)/2}|dg_0|$ thus extends to a $\mathcal{C}^\infty$ density to $X_{-\delta_0}$, and we deduce that with respect to the extended density, $\sigma_1(\frac{1}{2i}(P_\sigma - P_\sigma^*))|_{\mu\geq 0}$ vanishes when $\sigma \in \mathbb{R}$. Since in general $P_\sigma - P_{\mathrm{Re}\,\sigma}$ differs from $-4i(1+a_2)\operatorname{Im}\sigma D_\mu$ by a zeroth order operator, we conclude that

$$\sigma_1\left(\frac{1}{2i}(P_\sigma - P_\sigma^*)\right)\bigg|_{\mu=0} = -4(\operatorname{Im}\sigma)\xi. \tag{3-10}$$

We still need to check that μ can be appropriately chosen in the interior away from the region of validity of the product decomposition (3-1) (where we had no requirements so far on μ). This only matters for semiclassical purposes, and (being smooth and nonzero in the interior) the factor $\mu^{-1/2}$ multiplying from both sides does not affect any of the relevant properties (semiclassical ellipticity and possible nontrapping properties), so can be ignored — the same is true for σ-independent powers of μ.

Thus, near $\mu = 0$, but μ bounded away from 0, the only semiclassically nontrivial action we have done was to conjugate the operator by $e^{-i\sigma\phi}$ where $e^\phi = \mu^{1/2}(1+\mu)^{-1/4}$; we need to extend ϕ into the interior. But the semiclassical principal symbol of the conjugated operator is, with $\sigma = z/h$,

$$(\zeta - z\,d\phi, \zeta - z\,d\phi)_{G_0} - z^2 = |\zeta|^2_{G_0} - 2z(\zeta, d\phi)_{G_0} - (1 - |d\phi|^2_{G_0})z^2. \tag{3-11}$$

For z nonreal this is elliptic if $|d\phi|_{G_0} < 1$. Indeed, if (3-11) vanishes then from the vanishing imaginary part we get

$$2 \operatorname{Im} z ((\zeta, d\phi)_{G_0} + (1 - |d\phi|_{G_0}^2) \operatorname{Re} z) = 0, \qquad (3\text{-}12)$$

and then the real part is

$$|\zeta|_{G_0}^2 - 2 \operatorname{Re} z (\zeta, d\phi)_{G_0} - (1 - |d\phi|_{G_0}^2)((\operatorname{Re} z)^2 - (\operatorname{Im} z)^2)$$
$$= |\zeta|_{G_0}^2 + (1 - |d\phi|_{G_0}^2)((\operatorname{Re} z)^2 + (\operatorname{Im} z)^2), \qquad (3\text{-}13)$$

which cannot vanish if $|d\phi|_{G_0} < 1$. But, reading off the dual metric from the principal symbol of (3-3),

$$\tfrac{1}{4}|d(\log \mu - \tfrac{1}{2}\log(1+\mu))|_{G_0}^2 = \left(1 - \frac{\mu}{2(1+\mu)}\right)^2 < 1$$

for $\mu > 0$, with a strict bound as long as μ is bounded away from 0. Correspondingly, $\mu^{1/2}(1+\mu)^{-1/4}$ can be extended to a function e^ϕ on all of X_0 so that semiclassical ellipticity for z away from the reals is preserved, and we may even require that ϕ is constant on a fixed (but arbitrarily large) compact subset of X_0°. Then, after conjugation by $e^{-\imath \sigma \phi}$,

$$P_{h,z} = e^{\imath z \phi / h} \mu^{-(n+1)/4 - 1/2} (h^2 \Delta_{g_0} - z) \mu^{(n+1)/4 - 1/2} e^{-\imath z \phi / h} \qquad (3\text{-}14)$$

is semiclassically elliptic in $\mu > 0$ (as well as in $\mu \leq 0$, μ near 0, where this is already guaranteed), as desired.

Remark 3.1. We have not considered vector bundles over X_0. However, for instance for the Laplacian on the differential form bundles it is straightforward to check that slightly changing the power of μ in the conjugation the resulting operator extends smoothly across ∂X_0, has scalar principal symbol of the form (3-6), and the principal symbol of $\frac{1}{2\imath}(P_\sigma - P_\sigma^*)$, which plays a role below, is also as in the scalar setting, so all the results in fact go through.

3B. *Local dynamics near the radial set.* Let

$$N^* S \setminus o = \Lambda_+ \cup \Lambda_-, \quad \Lambda_\pm = N^* S \cap \{\pm \xi > 0\}, \quad S = \{\mu = 0\};$$

thus $S \subset X_{-\delta_0}$ can be identified with $Y = \partial X_0 (= \partial X_{0, \text{even}})$. Note that $p = 0$ at $\Lambda_\pm$ and H_p is radial there since

$$N^* S = \{(\mu, y, \xi, \eta) : \mu = 0, \eta = 0\},$$

so

$$H_p|_{N^* S} = -4\xi^2 \partial_\xi.$$

This corresponds to $dp = 4\xi^2\, d\mu$ at N^*S, so the characteristic set $\Sigma = \{p=0\}$ is smooth at N^*S.

Let $L_\pm$ be the image of $\Lambda_\pm$ in $S^*X_{-\delta_0}$. Next we analyze the Hamilton flow at $\Lambda_\pm$. First,

$$\mathsf{H}_p |\eta|^2_{\mu,y} = 8(1+a_1)\mu\xi\partial_\mu|\eta|^2_{\mu,y} - 4\frac{\partial a_1}{\partial y}\mu\xi^2 \cdot_h \eta \tag{3-15}$$

and

$$\mathsf{H}_p \mu = 8(1+a_1)\xi\mu. \tag{3-16}$$

In terms of linearizing the flow at N^*S, p and μ are equivalent as $dp = 4\xi^2\, d\mu$ there, so one can simply use $\hat p = p/|\xi|^2$ (which is homogeneous of degree 0, like μ), in place of μ. Finally,

$$\mathsf{H}_p|\xi| = -4\,\mathrm{sgn}(\xi) + b, \tag{3-17}$$

with b vanishing at $\Lambda_\pm$.

It is convenient to rehomogenize (3-15) in terms of $\hat\eta = \eta/|\xi|$. This can be phrased more invariantly by working with $S^*X_{-\delta_0} = (T^*X_{-\delta_0} \setminus o)/\mathbb{R}^+$, briefly discussed in Section 2. Let $L_\pm$ be the image of $\Lambda_\pm$ in $S^*X_{-\delta_0}$. Homogeneous degree zero functions on $T^*X_{-\delta_0} \setminus o$, such as $\hat p$, can be regarded as functions on $S^*X_{-\delta_0}$. For semiclassical purposes, it is best to consider $S^*X_{-\delta_0}$ as the boundary at fiber infinity of the fiber-radial compactification $\overline{T}^*X_{-\delta_0}$ of $T^*X_{-\delta_0}$, also discussed in Section 2. Then at fiber infinity near N^*S, we can take $(|\xi|^{-1}, \hat\eta)$ as (projective, rather than polar) coordinates on the fibers of the

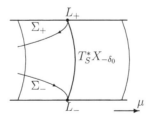

Figure 1. The cotangent bundle of $X_{-\delta_0}$ near $S = \{\mu = 0\}$. It is drawn in a fiber-radially compactified view. The boundary of the fiber compactification is the cosphere bundle $S^*X_{-\delta_0}$; it is the surface of the cylinder shown. $\Sigma_\pm$ are the components of the (classical) characteristic set containing $L_\pm$. They lie in $\mu \leq 0$, only meeting $S_S^*X_{-\delta_0}$ at $L_\pm$. Semiclassically, that is, in the interior of $\overline{T}^*X_{-\delta_0}$, for $z = h^{-1}\sigma > 0$, only the component of the semiclassical characteristic set containing L_+ can enter $\mu > 0$. This is reversed for $z < 0$.

cotangent bundle, with $\tilde{\rho} = |\xi|^{-1}$ defining $S^*X_{-\delta_0}$ in $\overline{T}^*X_{-\delta_0}$. Then $W = |\xi|^{-1}H_p$ is a $\mathcal{C}^\infty$ vector field in this region and

$$|\xi|^{-1}H_p|\hat{\eta}|^2_{\mu,y} = 2|\hat{\eta}|^2_{\mu,y}H_p|\xi|^{-1} + |\xi|^{-3}H_p|\eta|^2_{\mu,y} = 8(\operatorname{sgn}\xi)|\hat{\eta}|^2 + \tilde{a}, \quad (3\text{-}18)$$

where $\tilde{a}$ vanishes cubically at N^*S. In similar notation we have

$$H_p\tilde{\rho} = 4\operatorname{sgn}(\xi) + \tilde{a}', \quad \tilde{\rho} = |\xi|^{-1}, \quad (3\text{-}19)$$

and

$$|\xi|^{-1}H_p\mu = 8(\operatorname{sgn}\xi)\mu + \tilde{a}'', \quad (3\text{-}20)$$

with $\tilde{a}'$ smooth (indeed, homogeneous degree zero without the compactification) vanishing at N^*S, and $\tilde{a}''$ is also smooth, vanishing quadratically at N^*S. As the vanishing of $\hat{\eta}$, $|\xi|^{-1}$ and μ defines ∂N^*S, we conclude that $L_- = \partial\Lambda_-$ is a sink, while $L_+ = \partial\Lambda_+$ is a source, in the sense that all nearby bicharacteristics (in fact, including semiclassical (null)bicharacteristics, since $H_p|\xi|^{-1}$ contains the additional information needed; see (3-29)) converge to $L_\pm$ as the parameter along the bicharacteristic goes to $\mp\infty$. In particular, the quadratic defining function of $L_\pm$ given by

$$\rho_0 = \hat{p} + \hat{p}^2, \text{ where } \hat{p} = |\xi|^{-2}p, \ \hat{\hat{p}} = |\hat{\eta}|^2,$$

satisfies

$$(\operatorname{sgn}\xi)W\rho_0 \geq 8\rho_0 + \mathcal{O}(\rho_0^{3/2}). \quad (3\text{-}21)$$

We also need information on the principal symbol of $\frac{1}{2\imath}(P_\sigma - P_\sigma^*)$ at the radial points. At $L_\pm$ this is given by

$$\sigma_1\left(\frac{1}{2\imath}(P_\sigma - P_\sigma^*)\right)|_{N^*S} = -(4\operatorname{sgn}(\xi))\operatorname{Im}\sigma|\xi|; \quad (3\text{-}22)$$

here $(4\operatorname{sgn}(\xi))$ is pulled out due to (3-19), namely its size relative to $H_p|\xi|^{-1}$ matters. This corresponds to the fact that $(\mu \pm \imath 0)^{\imath\sigma}$, which are Lagrangian distributions associated to $\Lambda_\pm$, solve the PDE (3-5) modulo an error that is two orders lower than what one might a priori expect, that is, $P_\sigma(\mu \pm \imath 0)^{\imath\sigma} \in (\mu \pm \imath 0)^{\imath\sigma}\mathcal{C}^\infty(X_{-\delta_0})$. Note that P_σ is second order, so one should lose two orders a priori, that is, get an element of $(\mu \pm \imath 0)^{\imath\sigma-2}\mathcal{C}^\infty(X_{-\delta_0})$; the characteristic nature of $\Lambda_\pm$ reduces the loss to 1, and the particular choice of exponent eliminates the loss. This has much in common with $e^{\imath\lambda/x}x^{(n-1)/2}$ being an approximate solution in asymptotically Euclidean scattering; see [Melrose 1994].

3C. *Global behavior of the characteristic set.* By (3-8), points with $\xi = 0$ cannot lie in the characteristic set. Thus, with

$$\Sigma_\pm = \Sigma \cap \{\pm\xi > 0\},$$

$\Sigma = \Sigma_+ \cup \Sigma_-$ and $\Lambda_\pm \subset \Sigma_\pm$. Further, the characteristic set lies in $\mu \leq 0$, and intersects $\mu = 0$ only in $\Lambda_\pm$.

Moreover, as $H_p \mu = 8(1 + a_1)\xi\mu$ and $\xi \neq 0$ on Σ, and μ only vanishes at $\Lambda_+ \cup \Lambda_-$ there, for $\epsilon_0 > 0$ sufficiently small the $\mathscr{C}^\infty$ function μ provides a negative global escape function on $\mu \geq -\epsilon_0$ which is decreasing on Σ_+, increasing on Σ_-. Correspondingly, bicharacteristics in Σ_- travel from $\mu = -\epsilon_0$ to L_-, while in Σ_+ they travel from L_+ to $\mu = -\epsilon_0$.

3D. *High energy, or semiclassical, asymptotics.*

We are also interested in the high-energy behavior, as $|\sigma| \to \infty$. For the associated semiclassical problem one obtains a family of operators

$$P_{\hbar,z} = h^2 P_{h^{-1}z},$$

with $h = |\sigma|^{-1}$, and z corresponding to $\sigma/|\sigma|$ in the unit circle in $\mathbb{C}$. Then the semiclassical principal symbol $p_{\hbar,z}$ of $P_{\hbar,z}$ is a function on $T^* X_{-\delta_0}$, whose asymptotics at fiber infinity of $T^* X_{-\delta_0}$ is given by the classical principal symbol p. We are interested in $\operatorname{Im} \sigma \geq -C$, which in semiclassical notation corresponds to $\operatorname{Im} z \geq -Ch$. It is sometimes convenient to think of $p_{\hbar,z}$, and its rescaled Hamilton vector field, as objects on $\overline{T}^* X_{-\delta_0}$. Thus,

$$p_{\hbar,z} = \sigma_{2,\hbar}(P_{\hbar,z}) = 4(1+a_1)\mu\xi^2 - 4(1+a_2)z\xi - (1+a_3)z^2 + |\eta|^2_{\mu,y}, \quad (3\text{-}23)$$

so

$$\operatorname{Im} p_{\hbar,z} = -2\operatorname{Im} z (2(1+a_2)\xi + (1+a_3)\operatorname{Re} z). \quad (3\text{-}24)$$

In particular, for z nonreal, $\operatorname{Im} p_{\hbar,z} = 0$ implies $2(1+a_2)\xi + (1+a_3)\operatorname{Re} z = 0$, so

$$\operatorname{Re} p_{\hbar,z} = \big((1+a_1)(1+a_3)^2(1+a_2)^{-2}\mu + (1+2a_2)(1+a_3)\big)(\operatorname{Re} z)^2$$
$$+ (1+a_3)(\operatorname{Im} z)^2 + |\eta|^2_{\mu,y} > 0 \quad (3\text{-}25)$$

near $\mu = 0$; in other words, $p_{\hbar,z}$ is semiclassically elliptic on $T^* X_{-\delta_0}$, but *not* at fiber infinity, i.e., at $S^* X_{-\delta_0}$ (standard ellipticity is lost only in $\mu \leq 0$, of course). In $\mu > 0$ we have semiclassical ellipticity (and automatically classical ellipticity) by our choice of ϕ following (3-11). Explicitly, if we introduce for instance

$$(\mu, y, \nu, \hat{\eta}), \quad \nu = |\xi|^{-1}, \ \hat{\eta} = \eta/|\xi|, \quad (3\text{-}26)$$

as valid projective coordinates in a (large!) neighborhood of $L_\pm$ in $\overline{T}^* X_{-\delta_0}$, then

$$\nu^2 p_{\hbar,z} = 4(1+a_1)\mu - 4(1+a_2)(\operatorname{sgn}\xi)z\nu - (1+a_3)z^2\nu^2 + |\hat{\eta}|^2_{y,\mu},$$

so
$$v^2 \operatorname{Im} p_{\hbar,z} = -4(1+a_2)(\operatorname{sgn}\xi)\nu \operatorname{Im} z - 2(1+a_3)v^2 \operatorname{Re} z \operatorname{Im} z,$$

which automatically vanishes at $\nu = 0$, that is, at $S^*X_{-\delta_0}$. Thus, for σ large and pure imaginary, the semiclassical problem adds no complexity to the "classical" quantum problem, but of course it does not simplify it. In fact, we need somewhat more information at the characteristic set, which is thus at $\nu = 0$ when $\operatorname{Im} z$ is bounded away from 0:

$$\begin{aligned}\nu \text{ small, } \operatorname{Im} z \geq 0 &\Rightarrow (\operatorname{sgn}\xi)\operatorname{Im} p_{\hbar,z} \leq 0 \Rightarrow \pm \operatorname{Im} p_{\hbar,z} \leq 0 \text{ near } \Sigma_{\hbar,\pm},\\ \nu \text{ small, } \operatorname{Im} z \leq 0 &\Rightarrow (\operatorname{sgn}\xi)\operatorname{Im} p_{\hbar,z} \geq 0 \Rightarrow \pm \operatorname{Im} p_{\hbar,z} \geq 0 \text{ near } \Sigma_{\hbar,\pm},\end{aligned} \quad (3\text{-}27)$$

which, as we recall in Section 4, means that for $P_{\hbar,z}$ with $\operatorname{Im} z > 0$ one can propagate estimates forwards along the bicharacteristics where $\xi > 0$ (in particular, away from L_+, as the latter is a source) and backwards where $\xi < 0$ (in particular, away from L_-, as the latter is a sink), while for $P_{\hbar,z}^*$ the directions are reversed since its semiclassical symbol is $\overline{p_{\hbar,z}}$. The directions are also reversed if $\operatorname{Im} z$ switches sign. This is important because it gives invertibility for $z = \imath$ (corresponding to $\operatorname{Im}\sigma$ large positive, that is, the physical half-plane), but does not give invertibility for $z = -\imath$ negative.

We now return to the claim that even semiclassically, for z almost real (i.e., when z is not bounded away from the reals; we are not fixing z as we let h vary!), when the operator is not semiclassically elliptic on $T^*X_{-\delta_0}$ as mentioned above, the characteristic set can be divided into two components $\Sigma_{\hbar,\pm}$, with $L_\pm$ in different components. The vanishing of the factor following $\operatorname{Im} z$ in (3-24) gives a hypersurface that separates $\Sigma_\hbar$ into two parts. Indeed, this is the hypersurface given by
$$2(1+a_2)\xi + (1+a_3)\operatorname{Re} z = 0,$$
on which, by (3-25), $\operatorname{Re} p_{\hbar,z}$ cannot vanish, so
$$\Sigma_\hbar = \Sigma_{\hbar,+} \cup \Sigma_{\hbar,-}, \quad \Sigma_{\hbar,\pm} = \Sigma_\hbar \cap \{\pm(2(1+a_2)\xi + (1+a_3)\operatorname{Re} z) > 0\}.$$

Farther in $\mu > 0$, the hypersurface is given, due to (3-12), by
$$(\zeta, d\phi)_{G_0} + (1 - |d\phi|^2_{G_0})\operatorname{Re} z = 0,$$
and on it, by (3-13), the real part is $|\zeta|^2_{G_0} + (1 - |d\phi|^2_{G_0})((\operatorname{Re} z)^2 + (\operatorname{Im} z)^2) > 0$; correspondingly
$$\Sigma_\hbar = \Sigma_{\hbar,+} \cup \Sigma_{\hbar,-}, \quad \Sigma_{\hbar,\pm} = \Sigma_\hbar \cap \{\pm((\zeta, d\phi)_{G_0} + (1 - |d\phi|^2_{G_0})\operatorname{Re} z) > 0\}.$$

In fact, more generally, the real part is

$$|\zeta|^2_{G_0} - 2\operatorname{Re} z(\zeta, d\phi)_{G_0} - (1 - |d\phi|^2_{G_0})((\operatorname{Re} z)^2 - (\operatorname{Im} z)^2)$$
$$= |\zeta|^2_{G_0} - 2\operatorname{Re} z((\zeta, d\phi)_{G_0} + (1 - |d\phi|^2_{G_0})\operatorname{Re} z)$$
$$+ (1 - |d\phi|^2_{G_0})((\operatorname{Re} z)^2 + (\operatorname{Im} z)^2),$$

so for $\pm \operatorname{Re} z > 0$, $\mp((\zeta, d\phi)_{G_0} + (1 - |d\phi|^2_{G_0})\operatorname{Re} z) > 0$ implies that $p_{\hbar,z}$ does not vanish. Correspondingly, only one of the two components of $\Sigma_{\hbar,\pm}$ enters $\mu > 0$: for $\operatorname{Re} z > 0$, it is $\Sigma_{\hbar,+}$, while for $\operatorname{Re} z < 0$, it is $\Sigma_{\hbar,-}$.

We finally need more information about the global semiclassical dynamics.

Lemma 3.2. *There exists $\epsilon_0 > 0$ such that the following holds. All semiclassical null-bicharacteristics in $(\Sigma_{\hbar,+} \setminus L_+) \cap \{-\epsilon_0 \leq \mu \leq \epsilon_0\}$ go to either L_+ or to $\mu = \epsilon_0$ in the backward direction and to $\mu = \epsilon_0$ or $\mu = -\epsilon_0$ in the forward direction, while all semiclassical null-bicharacteristics in $(\Sigma_{\hbar,-} \setminus L_-) \cap \{-\epsilon_0 \leq \mu \leq \epsilon_0\}$ go to L_- or $\mu = \epsilon_0$ in the forward direction and to $\mu = \epsilon_0$ or $\mu = -\epsilon_0$ in the backward direction.*

For $\operatorname{Re} z > 0$, only $\Sigma_{\hbar,+}$ enters $\mu > 0$, so the $\mu = \epsilon_0$ possibility only applies to $\Sigma_{\hbar,+}$ then, while for $\operatorname{Re} z < 0$, the analogous remark applies to $\Sigma_{\hbar,-}$.

Proof. We assume that $\operatorname{Re} z > 0$ for the sake of definiteness. Observe that the semiclassical Hamilton vector field is

$$\mathsf{H}_{p_{\hbar,z}} = 4(2(1+a_1)\mu\xi - (1+a_2)z)\partial_\mu + \tilde{\mathsf{H}}_{|\eta|^2_{\mu,y}}$$
$$- \left(4(1+a_1+\mu\frac{\partial a_1}{\partial \mu})\xi^2 - 4\frac{\partial a_2}{\partial \mu}z\xi + \frac{\partial a_3}{\partial \mu}z^2 + \frac{\partial |\eta|^2_{\mu,y}}{\partial \mu}\right)\partial_\xi$$
$$- \left(4\frac{\partial a_1}{\partial y}\mu\xi^2 - 4\frac{\partial a_2}{\partial y}z\xi - \frac{\partial a_3}{\partial y}z^2\right)\partial_\eta; \qquad (3\text{-}28)$$

here we are concerned about z real. Near $S^*X_{-\delta_0} = \partial \overline{T}^*X_{-\delta_0}$, using the coordinates (3-26) (which are valid near the characteristic set), we have

$$\mathsf{W}_\hbar = \nu \mathsf{H}_{p_{\hbar,z}} =$$
$$4(2(1+a_1)\mu(\operatorname{sgn}\xi) - (1+a_2)z\nu)\partial_\mu + \nu\tilde{\mathsf{H}}_{|\hat{\eta}|^2_{\mu,y}}$$
$$+ (\operatorname{sgn}\xi)\left(4(1+a_1+\mu\frac{\partial a_1}{\partial \mu}) - 4\frac{\partial a_2}{\partial \mu}z(\operatorname{sgn}\xi)\nu + \frac{\partial a_3}{\partial \mu}z^2\nu^2 + \frac{\partial |\hat{\eta}|^2_{\mu,y}}{\partial \mu}\right)(\nu\partial_\nu + \hat{\eta}\partial_{\hat{\eta}})$$
$$- \left(4\frac{\partial a_1}{\partial y}\mu - 4(\operatorname{sgn}\xi)\frac{\partial a_2}{\partial y}z\nu - \frac{\partial a_3}{\partial y}z^2\nu^2\right)\partial_{\hat{\eta}}, \qquad (3\text{-}29)$$

with

$$\nu\tilde{\mathsf{H}}_{|\hat{\eta}|^2_{\mu,y}} = \sum_{ij} H_{ij}\hat{\eta}_i \partial_{y_j} - \sum_{ijk} \frac{\partial H_{ij}}{\partial y_k}\hat{\eta}_i\hat{\eta}_j\partial_{\hat{\eta}_k}$$

smooth. Thus, $W_\hbar$ is a smooth vector field on the compactified cotangent bundle, $\overline{T}^* X_{-\delta_0}$ which is tangent to its boundary, $S^* X_{-\delta_0}$, and $W_\hbar - W = \nu W^\sharp$ (with W considered as a homogeneous degree zero vector field) with $W^\sharp$ smooth and tangent to $S^* X_{-\delta_0}$. In particular, by (3-19) and (3-21), using that $\tilde\rho^2 + \rho_0$ is a quadratic defining function of $L_\pm$,

$$(\operatorname{sgn} \xi) W_\hbar (\tilde\rho^2 + \rho_0) \geq 8(\tilde\rho^2 + \rho_0) - \mathbb{O}((\tilde\rho^2 + \rho_0)^{3/2})$$

shows that there is $\epsilon_1 > 0$ such that in $\tilde\rho^2 + \rho_0 \leq \epsilon_1$, $\xi > 0$, $\tilde\rho^2 + \rho_0$ is strictly increasing along the Hamilton flow except at L_+, while in $\tilde\rho^2 + \rho_0 \leq \epsilon_1$, $\xi < 0$, $\tilde\rho^2 + \rho_0$ is strictly decreasing along the Hamilton flow except at L_-. Indeed, all null-bicharacteristics in this neighborhood of $L_\pm$ except the constant ones at $L_\pm$ tend to $L_\pm$ in one direction and to $\tilde\rho^2 + \rho_0 = \epsilon_1$ in the other direction.

Choosing $\epsilon_0' > 0$ sufficiently small, the characteristic set in

$$\overline{T}^* X_{-\delta_0} \cap \{-\epsilon_0' \leq \mu \leq \epsilon_0'\}$$

is disjoint from $S^* X_{-\delta_0} \setminus \{\tilde\rho^2 + \rho_0 \leq \epsilon_1\}$, and indeed only contains points in $\Sigma_{\hbar,+}$ as $\operatorname{Re} z > 0$. Since $\mathsf{H}_{p_{\hbar,z}} \mu = 4(2(1+a_1)\mu\xi - (1+a_2)z)$, it is negative on $\overline{T}^*_{\{\mu=0\}} X_{-\delta_0} \setminus S^* X_{-\delta_0}$. In particular, there is a neighborhood U of $\mu = 0$ in $\Sigma_{\hbar,+} \setminus S^* X_{-\delta_0}$ on which the same sign is preserved; since the characteristic set in $\overline{T}^* X_{-\delta_0} \setminus \{\tilde\rho^2 + \rho_0 < \epsilon_1\}$ is compact, and is indeed a subset of $T^* X_{-\delta_0} \setminus \{\tilde\rho^2 + \rho_0 < \epsilon_1\}$, we deduce that $|\mu|$ is bounded below on $\Sigma \setminus (U \cup \{\tilde\rho^2 + \rho_0 < \epsilon_1\})$, say $|\mu| \geq \epsilon_0'' > 0$ there, so with $\epsilon_0 = \min(\epsilon_0', \epsilon_0'')$, $\mathsf{H}_{p_{\hbar,z}} \mu < 0$ on

$$\Sigma_{\hbar,+} \cap \{-\epsilon_0 \leq \mu \leq \epsilon_0\} \setminus \{\tilde\rho^2 + \rho_0^2 < \epsilon_1\}.$$

As $\mathsf{H}_{p_{\hbar,z}} \mu < 0$ at $\mu = 0$, bicharacteristics can only cross $\mu = 0$ in the outward direction.

Thus, if γ is a bicharacteristic in $\Sigma_{\hbar,+}$, there are two possibilities. If γ is disjoint from $\{\tilde\rho^2 + \rho_0 < \epsilon_1\}$, it has to go to $\mu = \epsilon_0$ in the backward direction and to $\mu = -\epsilon_0$ in the forward direction. If γ has a point in $\{\tilde\rho^2 + \rho_0 < \epsilon_1\}$, then it has to go to L_+ in the backward direction and to $\tilde\rho^2 + \rho_0 = \epsilon_1$ in the forward direction; if $|\mu| \geq \epsilon_0$ by the time $\tilde\rho^2 + \rho_0 = \epsilon_1$ is reached, the result is proved, and otherwise $\mathsf{H}_{p_{\hbar,z}} \mu < 0$ in $\tilde\rho^2 + \rho_0 \geq \epsilon_1$, $|\mu| \leq \epsilon_0$, shows that the bicharacteristic goes to $\mu = -\epsilon_0$ in the forward direction.

If γ is a bicharacteristic in $\Sigma_{\hbar,-}$, only the second possibility exists, and the bicharacteristic cannot leave $\{\tilde\rho^2 + \rho_0 < \epsilon_1\}$ in $|\mu| \leq \epsilon_0$, so it reaches $\mu = -\epsilon_0$ in the backward direction (as the characteristic set is in $\mu \leq 0$). □

If we assume that g_0 is a nontrapping metric, that is, bicharacteristics of g_0 in $T^* X_0^\circ \setminus o$ tend to ∂X_0 in both the forward and the backward directions, then $\mu = \epsilon_0$ can be excluded from the statement of the lemma, and the above argument gives the following stronger conclusion: for sufficiently small $\epsilon_0 > 0$, and for $\operatorname{Re} z > 0$, any bicharacteristic in $\Sigma_{\hbar,+}$ in $-\epsilon_0 \leq \mu$ has to go to L_+ in the backward direction, and to $\mu = -\epsilon_0$ in the forward direction (with the exception of the constant bicharacteristics at L_+), while in $\Sigma_{\hbar,-}$, all bicharacteristics in $-\epsilon_0 \leq \mu$ lie in $-\epsilon_0 \leq \mu \leq 0$, and go to L_- in the forward direction and to $\mu = -\epsilon_0$ in the backward direction (with the exception of the constant bicharacteristics at L_-).

In fact, for applications, it is also useful to remark that for sufficiently small $\epsilon_0 > 0$, and for $\alpha \in T^*X_0$,

$$0 < \mu(\alpha) < \epsilon_0, \ p_{\hbar,z}(\alpha) = 0 \text{ and } (\mathsf{H}_{p_{\hbar,z}}\mu)(\alpha) = 0 \Rightarrow (\mathsf{H}^2_{p_{\hbar,z}}\mu)(\alpha) < 0. \quad (3\text{-}30)$$

Indeed, as $\mathsf{H}_{p_{\hbar,z}}\mu = 4(2(1+a_1)\mu\xi - (1+a_2)z)$, the hypotheses imply that $z = 2(1+a_1)(1+a_2)^{-1}\mu\xi$ and

$$0 = p_{\hbar,z}$$
$$= 4(1+a_1)\mu\xi^2 - 8(1+a_1)\mu\xi^2 - 4(1+a_1)^2(1+a_2)^{-2}(1+a_3)\mu^2\xi^2 + |\eta|^2_{\mu,y}$$
$$= -4(1+a_1)\mu\xi^2 - 4(1+a_1)^2(1+a_2)^{-2}(1+a_3)\mu^2\xi^2 + |\eta|^2_{\mu,y},$$

so $|\eta|^2_{\mu,y} = 4(1+b)\mu\xi^2$, with b vanishing at $\mu = 0$. Thus, at points where $\mathsf{H}_{p_{\hbar,z}}\mu$ vanishes, writing $a_j = \mu\tilde{a}_j$,

$$\mathsf{H}^2_{p_{\hbar,z}}\mu = 8(1+a_1)\mu\mathsf{H}_{p_{\hbar,z}}\xi + 8\mu^2\xi\mathsf{H}_{p_{\hbar,z}}\tilde{a}_1 - 4z\mu\mathsf{H}_{p_{\hbar,z}}\tilde{a}_2$$
$$= 8(1+a_1)\mu\mathsf{H}_{p_{\hbar,z}}\xi + \mathbb{O}(\mu^2\xi^2). \quad (3\text{-}31)$$

Now

$$\mathsf{H}_{p_{\hbar,z}}\xi = -\left(4\left(1+a_1+\mu\frac{\partial a_1}{\partial \mu}\right)\xi^2 - 4\frac{\partial a_2}{\partial \mu}z\xi + \frac{\partial a_3}{\partial \mu}z^2 + \frac{\partial|\eta|^2_{\mu,y}}{\partial \mu}\right).$$

Since $z\xi$ is $\mathbb{O}(\mu\xi^2)$ due to $\mathsf{H}_{p_{\hbar,z}}\mu = 0$, z^2 is $\mathbb{O}(\mu^2\xi^2)$ for the same reason, and $|\eta|^2$ and $\partial_\mu|\eta|^2$ are $\mathbb{O}(\mu\xi^2)$ due to $p_{\hbar,z} = 0$, we deduce that $\mathsf{H}_{p_{\hbar,z}}\xi < 0$ for sufficiently small $|\mu|$, so (3-31) implies (3-30). Thus, μ can be used for gluing constructions as in [Datchev and Vasy 2011].

3E. Complex absorption.

The final step of fitting P_σ into our general microlocal framework is moving the problem to a compact manifold, and adding a complex absorbing second order operator. We thus consider a compact manifold without boundary X for which $X_{\mu_0} = \{\mu > \mu_0\}$, $\mu_0 = -\epsilon_0 < 0$, with $\epsilon_0 > 0$ as above, is identified as an open subset with smooth boundary; it is convenient to take X to be the double of X_{μ_0}, so there are two copies of $X_{0,\text{even}}$ in X.

In the case of hyperbolic space, this doubling process can be realized from the perspective of $(n+1)$-dimensional Minkowski space. Then, as mentioned in the introduction, the Poincaré model shows up in two copies, namely in the interior of the future and past light cone inside the sphere at infinity, while de Sitter space as the "equatorial belt", that is, the exterior of the light cone at the sphere at infinity. One can take the Minkowski equatorial plane, $t = 0$, as $\mu = \mu_0$, and place the complex absorption there, thereby decoupling the future and past hemispheres. See [Vasy 2010a] for more detail.

It is convenient to separate the "classical" (i.e., quantum!) and "semiclassical" problems, for in the former setting trapping for g_0 does not matter, while in the latter it does.

We then introduce a "complex absorption" operator $Q_\sigma \in \Psi_{\text{cl}}^2(X)$ with real principal symbol q supported in, say, $\mu < -\epsilon_1$, with the Schwartz kernel also supported in the corresponding region (i.e., in both factors on the product space this condition holds on the support) such that $p \pm iq$ is elliptic near ∂X_{μ_0}, that is, near $\mu = \mu_0$, and which satisfies that $\pm q \geq 0$ near $\Sigma_\pm$. This can easily be done since $\Sigma_\pm$ are disjoint, and away from these p is elliptic, hence so is $p \pm iq$ regardless of the choice of q; we simply need to make q to have support sufficiently close to $\Sigma_\pm$, elliptic on $\Sigma_\pm$ at $\mu = -\epsilon_0$, with the appropriate sign near $\Sigma_\pm$. Having done this, we extend p and q to X in such a way that $p \pm iq$ are elliptic near ∂X_{μ_0}; the region we added is thus irrelevant at the level of bicharacteristic dynamics (of p) in so far as it is decoupled from the dynamics in X_0, and indeed also for analysis as we see shortly (in so far as we have two essentially decoupled copies of the same problem). This is accomplished, for instance, by using the doubling construction to define p on $X \setminus X_{\mu_0}$ (in a smooth fashion at ∂X_{μ_0}, as can be easily arranged; the holomorphic dependence of P_σ on σ is still easily preserved), and then, noting that the characteristic set of p still has two connected components, making q elliptic on the characteristic set of p near ∂X_{μ_0}, with the same sign in each component as near ∂X_{μ_0}. (An alternative would be to make q elliptic on the characteristic set of p near $X \setminus X_{\mu_0}$; it is just slightly more complicated to write down such a q when the high-energy behavior

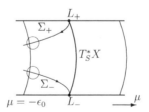

Figure 2. The cotangent bundle near $S = \{\mu = 0\}$. It is drawn in a fiber-radially compactified view, as in Figure 1. The circles on the left show the support of q; it has opposite signs on the two disks corresponding to the opposite directions of propagation relative to the Hamilton vector field.

is taken into account. With the present choice, due to the doubling, there are essentially two copies of the problem on X_0: the original, and the one from the doubling.) Finally we take Q_σ be any operator with principal symbol q with Schwartz kernel satisfying the desired support conditions and which depends on σ holomorphically. We may choose Q_σ to be independent of σ so Q_σ is indeed holomorphic; in this case we may further replace it by $\frac{1}{2}(Q_\sigma + Q_\sigma^*)$ if self-adjointness is desired.

In view of Section 3C we have arranged the following. For $\alpha \in S^*X \cap \Sigma$, let $\gamma_+(\alpha)$, resp. $\gamma_-(\alpha)$ denote the image of the forward, resp. backward, half-bicharacteristic of p from α. We write $\gamma_\pm(\alpha) \to L_\pm$ (and say $\gamma_\pm(\alpha)$ tends to $L_\pm$) if given any neighborhood O of $L_\pm$, $\gamma_\pm(\alpha) \cap O \neq \varnothing$; by the source/sink property this implies that the points on the curve are in O for sufficiently large (in absolute value) parameter values. Then, with $\mathrm{ell}(Q_\sigma)$ denoting the elliptic set of Q_σ,

$$\begin{aligned} \alpha \in \Sigma_- \setminus L_- &\Rightarrow \gamma_+(\alpha) \to L_- \text{ and } \gamma_-(\alpha) \cap \mathrm{ell}(Q_\sigma) \neq \varnothing, \\ \alpha \in \Sigma_+ \setminus L_+ &\Rightarrow \gamma_-(\alpha) \to L_+ \text{ and } \gamma_+(\alpha) \cap \mathrm{ell}(Q_\sigma) \neq \varnothing. \end{aligned} \quad (3\text{-}32)$$

That is, all forward and backward half-(null)bicharacteristics of P_σ either enter the elliptic set of Q_σ, or go to $\Lambda_\pm$, that is, $L_\pm$ in S^*X. The point of the arrangements regarding Q_σ and the flow is that we are able to propagate estimates forward near where $q \geq 0$, backward near where $q \leq 0$, so by our hypotheses we can always propagate estimates for $P_\sigma - \imath Q_\sigma$ from $\Lambda_\pm$ towards the elliptic set of Q_σ. On the other hand, for $P_\sigma^* + \imath Q_\sigma^*$, we can propagate estimates from the elliptic set of Q_σ towards $\Lambda_\pm$. This behavior of $P_\sigma - \imath Q_\sigma$ vs. $P_\sigma^* + \imath Q_\sigma^*$ is important for duality reasons.

An alternative to the complex absorption would be simply adding a boundary at $\mu = \mu_0$; this is easy to do since this is a space-like hypersurface, but this is slightly unpleasant from the point of view of microlocal analysis as one has to work on a manifold with boundary (though as mentioned this is easily done; see [Vasy 2010a]).

For the semiclassical problem, when z is almost real (namely when $\mathrm{Im}\, z$ is bounded away from 0 we only need to make sure we do not mess up the semiclassical ellipticity in $T^* X_{-\delta_0}$) we need to increase the requirements on Q_σ, and what we need to do depends on whether g_0 is nontrapping.

If g_0 is nontrapping, we choose Q_σ such that $h^2 Q_{h^{-1}z} \in \Psi_{\hbar,\mathrm{cl}}^2(X)$ with semiclassical principal symbol $q_{\hbar,z}$, and in addition to the above requirement for the classical symbol, we need semiclassical ellipticity near $\mu = \mu_0$, that is, that $p_{\hbar,z} - \imath q_{\hbar,z}$ and its complex conjugate are elliptic near ∂X_{μ_0}, that is, near $\mu = \mu_0$, and which satisfies that for z real $\pm q_{\hbar,z} \geq 0$ on $\Sigma_{\hbar,\pm}$. Again, we extend P_σ and Q_σ to X in such a way that $p - \imath q$ and $p_{\hbar,z} - \imath q_{\hbar,z}$ (and thus their complex conjugates) are elliptic near ∂X_{μ_0}; the region we added is thus irrelevant. This is straightforward to arrange if one ignores that one wants Q_σ to be holomorphic: one easily constructs a function $q_{\hbar,z}$ on T^*X (taking into account the disjointness of $\Sigma_{\hbar,\pm}$), and defines $Q_{h^{-1}z}$ to be h^{-2} times the semiclassical quantization of $q_{\hbar,z}$ (or any other operator with the same semiclassical and standard principal symbols). Indeed, for our purposes this would suffice since we want high-energy estimates for the analytic continuation resolvent on the original space X_0 (which we will know exists by the nonsemiclassical argument), and as we shall see, the resolvent is given by the same formula in terms of $(P_\sigma - \imath Q_\sigma)^{-1}$ independently whether Q_σ is holomorphic in σ (as long as it satisfies the other properties), so there is no need to ensure the holomorphy of Q_σ. However, it is instructive to have an example of a holomorphic family Q_σ in a strip at least: in view of (3-24) we can take (with $C > 0$)

$$q_{h,z} = 2(2(1+a_2)\xi + (1+a_3)z)(\xi^2 + |\eta|^2 + z^2 + C^2 h^2)^{1/2} \chi(\mu),$$

where $\chi \geq 0$ is supported near μ_0; the corresponding full symbol is

$$\sigma_{\text{full}}(Q_\sigma) = 2(2(1+a_2)\xi + (1+a_3)\sigma)(\xi^2 + |\eta|^2 + \sigma^2)^{1/2} \chi(\mu),$$

and Q_σ is taken as a quantization of this full symbol. Here the square root is defined on $\mathbb{C} \setminus [0, -\infty)$, with real part of the result being positive, and correspondingly $q_{h,z}$ is defined away from $h^{-1}z \in \pm \imath [C, +\infty)$. Note that $\xi^2 + |\eta|^2 + \sigma^2$ is an elliptic symbol in $(\xi, \eta, \operatorname{Re}\sigma, \operatorname{Im}\sigma)$ as long as $|\operatorname{Im}\sigma| < C'|\operatorname{Re}\sigma|$, so the corresponding statement also holds for its square root. While $q_{h,z}$ is only holomorphic away from $h^{-1}z \in \pm \imath [C, +\infty)$, the full (and indeed the semiclassical and standard principal) symbols are actually holomorphic in cones near infinity, and indeed, for example, via convolutions by the Fourier transform of a compactly supported function can be extended to be holomorphic in $\mathbb{C}$, but this is of no importance here.

If g_0 is trapping, we need to add complex absorption inside X_0 as well, at $\mu = \epsilon_0$, so we relax the requirement that Q_σ is supported in $\mu < -\epsilon_0/2$ to support in $|\mu| > \epsilon_0/2$, but we require in addition to the other classical requirements that $p_{\hbar,z} - \imath q_{\hbar,z}$ and its complex conjugate are elliptic near $\mu = \pm\epsilon_0$, and which satisfies that $\pm q_{\hbar,z} \geq 0$ on $\Sigma_{\hbar,\pm}$. This can be achieved as above for μ near μ_0. Again, we extend P_σ and Q_σ to X in such a way that $p - \imath q$ and $p_{\hbar,z} - \imath q_{\hbar,z}$ (and thus their complex conjugates) are elliptic near ∂X_{μ_0}.

In either of these semiclassical cases we have arranged that for sufficiently small $\delta_0 > 0$, $p_{\hbar,z} - \imath q_{\hbar,z}$ and its complex conjugate are *semiclassically nontrapping* for $|\operatorname{Im} z| < \delta_0$, namely the bicharacteristics from any point in $\Sigma_\hbar \setminus (L_+ \cup L_-)$ flow to $\text{ell}(q_{\hbar,z}) \cup L_-$ (i.e., either enter $\text{ell}(q_{\hbar,z})$ at some finite time, or tend to L_-) in the forward direction, and to $\text{ell}(q_{\hbar,z}) \cup L_+$ in the backward direction. Here

$\delta_0 > 0$ arises from the particularly simple choice of $q_{\hbar,z}$ for which semiclassical ellipticity is easy to check for $\operatorname{Im} z > 0$ (bounded away from 0) and small; a more careful analysis would give a specific value of δ_0, and a more careful choice of $q_{\hbar,z}$ would give a better result.

4. Microlocal analysis

4A. *Elliptic and real principal type points.* First, recall the basic elliptic and real principal type regularity results. Let $\operatorname{WF}^s(u)$ denote the H^s wave front set of a distribution $u \in \mathscr{C}^{-\infty}(X)$, that is, $\alpha \notin \operatorname{WF}^s(u)$ if there exists $A \in \Psi^0(X)$ elliptic at α such that $Au \in H^s$. Elliptic regularity states that

$$P_\sigma - \imath Q_\sigma \text{ elliptic at } \alpha, \ \alpha \notin \operatorname{WF}^{s-2}((P_\sigma - \imath Q_\sigma)u) \Rightarrow \alpha \notin \operatorname{WF}^s(u).$$

In particular, if $(P_\sigma - \imath Q_\sigma)u \in H^{s-2}$ and $p - \imath q$ is elliptic at α then $\alpha \notin \operatorname{WF}^s(u)$. Analogous conclusions apply to $P_\sigma^* + \imath Q_\sigma^*$; since both p and q are real, $p - \imath q$ is elliptic if and only if $p + \imath q$ is.

We also have real principal type propagation, in the usual form valid outside $\operatorname{supp} q$:

$$\operatorname{WF}^s(u) \setminus (\operatorname{WF}^{s-1}((P_\sigma - \imath Q_\sigma)u) \cup \operatorname{supp} q)$$

is a union of maximally extended bicharacteristics of H_p in the characteristic set $\Sigma = \{p = 0\}$ of P_σ. Putting it differently,

$$\alpha \notin \operatorname{WF}^s(u) \cup \operatorname{WF}^{s-1}((P_\sigma - \imath Q_\sigma)u) \cup \operatorname{supp} q \Rightarrow \tilde\gamma(\alpha) \cap \operatorname{WF}^s(u) = \varnothing,$$

where $\tilde\gamma(\alpha)$ is the component of the bicharacteristic $\gamma(\alpha)$ of p in the complement of $\operatorname{WF}^{s-1}((P_\sigma - \imath Q_\sigma)u) \cup \operatorname{supp} q$. If $(P_\sigma - \imath Q_\sigma)u \in H^{s-1}$, then the condition $\operatorname{WF}^{s-1}((P_\sigma - \imath Q_\sigma)u) = \varnothing$ can be dropped from all statements above; if $q = 0$ one can thus replace $\tilde\gamma$ by γ.

In general, the result does not hold for nonzero q. However, it holds in one direction (backward/forward) of propagation along H_p if q has the correct sign. Thus, let $\tilde\gamma_\pm(\alpha)$ be a forward $(+)$ or backward $(-)$ bicharacteristic from α, defined on an interval I. If $\pm q \geq 0$ on a neighborhood of $\tilde\gamma_\pm(\alpha)$ (i.e., $q \geq 0$ on a neighborhood of $\tilde\gamma_+(\alpha)$, or $q \leq 0$ on a neighborhood of $\tilde\gamma_-(\alpha)$) then (for the corresponding sign)

$$\alpha \notin \operatorname{WF}^s(u) \text{ and } \operatorname{WF}^{s-1}((P_\sigma - \imath Q_\sigma)u) \cap \tilde\gamma_\pm(\alpha) = \varnothing \Rightarrow \tilde\gamma_\pm(\alpha) \cap \operatorname{WF}^s(u) = \varnothing,$$

that is, one can propagate regularity forward if $q \geq 0$, backward if $q \leq 0$. A proof of this claim that is completely analogous to Hörmander's positive commutator proof in the real principal type setting can easily be given: see [Nonnenmacher and Zworski 2009; Datchev and Vasy 2011] in the semiclassical setting; the changes are minor in the "classical" setting. Note that at points where $q \neq 0$, just

$\alpha \notin \mathrm{WF}^{s-1}((P_\sigma - \imath Q_\sigma)u)$ implies $\alpha \notin \mathrm{WF}^{s+1}(u)$ (stronger than stated above), but at points with $q = 0$ such an elliptic estimate is unavailable (unless P_σ is elliptic).

As $P_\sigma^* + \imath Q_\sigma^*$ has symbol $p + \imath q$, one can propagate regularity in the opposite direction as compared to $P_\sigma - \imath Q_\sigma$. Thus, if $\mp q \geq 0$ on a neighborhood of $\tilde{\gamma}_\pm(\alpha)$ (i.e., $q \leq 0$ on a neighborhood of $\tilde{\gamma}_+(\alpha)$, or $q \geq 0$ on a neighborhood of $\tilde{\gamma}_-(\alpha)$) then (for the corresponding sign)

$$\alpha \notin \mathrm{WF}^s(u) \text{ and } \mathrm{WF}^{s-1}((P_\sigma^* + \imath Q_\sigma^*)u) \cap \tilde{\gamma}_\pm(\alpha) = \varnothing \Rightarrow \tilde{\gamma}_\pm(\alpha) \cap \mathrm{WF}^s(u) = \varnothing.$$

4B. *Analysis near* $\Lambda_\pm$. The last ingredient in the classical setting is an analogue of Melrose's regularity result at radial sets which have the same features as ours. Although it is not stated in this generality in Melrose's paper [1994], the proof is easily adapted. Thus, the results are:

At $\Lambda_\pm$, for $s \geq m > \frac{1}{2} - \mathrm{Im}\,\sigma$, we can propagate estimates *away* from $\Lambda_\pm$:

Proposition 4.1. *Suppose* $s \geq m > \frac{1}{2} - \mathrm{Im}\,\sigma$, *and* $\mathrm{WF}^m(u) \cap \Lambda_\pm = \varnothing$. *Then*

$$\Lambda_\pm \cap \mathrm{WF}^{s-1}(P_\sigma u) = \varnothing \Rightarrow \Lambda_\pm \cap \mathrm{WF}^s(u) = \varnothing.$$

This is completely analogous to Melrose's estimates in asymptotically Euclidean scattering theory at the radial sets [Melrose 1994, Section 9]. Note that the H^s regularity of u at $\Lambda_\pm$ is "free" in the sense that we do not need to impose H^s assumptions on u anywhere; merely H^m at $\Lambda_\pm$ does the job; of course, on $P_\sigma u$ one must make the H^{s-1} assumption, that is, the loss of one derivative compared to the elliptic setting. At the cost of changing regularity, one can propagate estimate *towards* $\Lambda_\pm$. Keeping in mind that taking P_σ^* in place of P_σ, principal symbol of $\frac{1}{2\imath}(P_\sigma - P_\sigma^*)$ switches sign, we have the following:

Proposition 4.2. *For* $s < \frac{1}{2} + \mathrm{Im}\,\sigma$, *and* O *a neighborhood of* $\Lambda_\pm$,

$$\mathrm{WF}^s(u) \cap (O \setminus \Lambda_\pm) = \varnothing, \ \mathrm{WF}^{s-1}(P_\sigma^* u) \cap \Lambda_\pm = \varnothing \Rightarrow \mathrm{WF}^s(u) \cap \Lambda_\pm = \varnothing.$$

Proof of Propositions 4.1 and 4.2. The proof is a positive commutator estimate. Consider commutants $C_\epsilon^* C_\epsilon$ with $C_\epsilon \in \Psi^{s-1/2-\delta}(X)$ for $\epsilon > 0$, uniformly bounded in $\Psi^{s-1/2}(X)$ as $\epsilon \to 0$; with the ϵ-dependence used to regularize the argument. More precisely, let

$$c = \phi(\rho_0)\tilde{\rho}^{-s+1/2}, \quad c_\epsilon = c(1 + \epsilon\tilde{\rho}^{-1})^{-\delta},$$

where $\phi \in \mathcal{C}_c^\infty(\mathbb{R})$ is identically 1 near 0, $\phi' \leq 0$ and ϕ is supported sufficiently close to 0 so that

$$\rho_0 \in \mathrm{supp}\,d\phi \Rightarrow \pm\tilde{\rho}\,\mathsf{H}_p\rho_0 > 0; \tag{4-1}$$

such ϕ exists by (3-21). To avoid using the sharp Gårding inequality, we choose ϕ so that $\sqrt{-\phi\phi'}$ is $\mathscr{C}^\infty$. Note that the sign of $H_p\tilde{\rho}^{-s+1/2}$ depends on the sign of $-s+1/2$ which explains the difference between $s > 1/2$ and $s < 1/2$ in Propositions 4.1–4.2 when there are no other contributions to the threshold value of s. The contribution of the principal symbol of $\frac{1}{2i}(P_\sigma - P_\sigma^*)$, however, shifts the critical value $1/2$.

Now let $C \in \Psi^{s-1/2}(X)$ have principal symbol c, and have $\mathrm{WF}'(C) \subset \mathrm{supp}\,\phi \circ \rho_0$, and let $C_\epsilon = CS_\epsilon$, $S_\epsilon \in \Psi^{-\delta}(X)$ uniformly bounded in $\Psi^0(X)$ for $\epsilon > 0$, converging to Id in $\Psi^{\delta'}(X)$ for $\delta' > 0$ as $\epsilon \to 0$, with principal symbol $(1+\epsilon\tilde{\rho}^{-1})^{-\delta}$. Thus, the principal symbol of C_ϵ is c_ϵ.

First, consider Proposition 4.1. Then

$$\sigma_{2s}(\iota(P_\sigma^* C_\epsilon^* C_\epsilon - C_\epsilon^* C_\epsilon P_\sigma)) = \sigma_1(\iota(P_\sigma^* - P_\sigma))c_\epsilon^2 + 2c_\epsilon H_p c_\epsilon \qquad (4\text{-}2)$$
$$= \pm 8\left(-\mathrm{Im}\,\sigma\phi + \left(-s+\tfrac{1}{2}\right)\phi \pm \tfrac{1}{4}(\tilde{\rho}H_p\rho_0)\phi' + \delta\frac{\epsilon}{\tilde{\rho}+\epsilon}\phi\right)\phi\tilde{\rho}^{-2s}(1+\epsilon\tilde{\rho}^{-1})^{-2\delta},$$

so

$$\pm\sigma_{2s}(\iota(P_\sigma^* C_\epsilon^* C_\epsilon - C_\epsilon^* C_\epsilon P_\sigma)) \leq -8\left(s-\tfrac{1}{2}+\mathrm{Im}\,\sigma-\delta\right)\tilde{\rho}^{-2s}(1+\epsilon\tilde{\rho}^{-1})^{-2\delta}\phi^2$$
$$+ 2(\pm\tilde{\rho}H_p\rho_0)\tilde{\rho}^{-2s}(1+\epsilon\tilde{\rho}^{-1})^{-2\delta}\phi'\phi. \qquad (4\text{-}3)$$

Here the first term on the right-hand side is negative if $s - 1/2 + \mathrm{Im}\,\sigma - \delta > 0$ and this is the same sign as that of ϕ' term; the presence of δ (needed for the regularization) is the reason for the appearance of m in the estimate. Thus,

$$\pm\iota(P_\sigma^* C_\epsilon^* C_\epsilon - C_\epsilon^* C_\epsilon P_\sigma) = -S_\epsilon^*(B^*B + B_1^*B_1 + B_{2,\epsilon}^*B_{2,\epsilon})S_\epsilon + F_\epsilon,$$

with $B, B_1, B_{2,\epsilon} \in \Psi^s(X)$, $B_{2,\epsilon}$ uniformly bounded in $\Psi^s(X)$ as $\epsilon \to 0$, F_ϵ uniformly bounded in $\Psi^{2s-1}(X)$, and $\sigma_s(B)$ an elliptic multiple of $\phi(\rho_0)\tilde{\rho}^{-s}$. Computing the pairing, using an extra regularization (insert a regularizer $\Lambda_r \in \Psi^{-1}(X)$, uniformly bounded in $\Psi^0(X)$, converging to Id in $\Psi^\delta(X)$ to justify integration by parts, and use that $[\Lambda_r, P_\sigma^*]$ is uniformly bounded in $\Psi^1(X)$, converging to 0 strongly, cf. [Vasy 2000, Lemma 17.1] and its use in [Vasy 2000, Lemma 17.2]) yields

$$\langle\iota(P_\sigma^* C_\epsilon^* C_\epsilon - C_\epsilon^* C_\epsilon P_\sigma)u, u\rangle = \langle\iota C_\epsilon^* C_\epsilon u, P_\sigma u\rangle - \langle\iota P_\sigma u, C_\epsilon^* C_\epsilon u\rangle.$$

Using Cauchy–Schwartz on the right-hand side, a standard functional analytic argument (see, for instance, Melrose [1994, Proof of Proposition 7 and Section 9]) gives an estimate for Bu, showing u is in H^s on the elliptic set of B, provided u is microlocally in $H^{s-\delta}$. A standard inductive argument, starting with $s - \delta = m$ and improving regularity by $\leq \tfrac{1}{2}$ in each step proves Proposition 4.1.

For Proposition 4.2, when applied to P_σ in place of P_σ^* (so the assumption is $s < (1-\mathrm{Im}\,\sigma)/2$), the argument is similar, but we want to change the overall sign of the terms in (4-2) corresponding to the first term on the right-hand side of (4-3), that is, we want it to be positive. This is satisfied if $s - 1/2 + \operatorname{Im} \sigma < 0$ since the regularizer now contributes the correct sign. On the other hand, ϕ' now has the wrong sign, so one needs to make an assumption on $\operatorname{supp} d\phi$; one can arrange that this is in $O \setminus \Lambda$ by making ϕ have sufficiently small support, but identically 1 near 0. Since the details are standard — see [Melrose 1994, Section 9] — we leave them to the reader. When interchanging P_σ and P_σ^*, we need to take into account the switch of the sign of the principal symbol of $\frac{1}{2i}(P_\sigma - P_\sigma^*)$, which causes the sign change in front of $\operatorname{Im} \sigma$ in the statement of the proposition. □

4C. *Global estimates.* For our Fredholm results, we actually need estimates. However, these can be easily obtained from regularity results as in, for example, [Hörmander 1985b, Proof of Theorem 26.1.7] by the closed graph theorem. It should be noted that of course one really proved versions of the relevant estimates when proving regularity, but the closed graph theorem provides a particularly simple way of combining these (though it comes at the cost of using a theorem which in principle is unnecessary).

So suppose $s \geq m > \frac{1}{2} - \operatorname{Im} \sigma$, $u \in H^m$ and $(P_\sigma - \imath Q_\sigma)u \in H^{s-1}$. The above results give that, first, $\operatorname{WF}^s(u)$ (indeed, $\operatorname{WF}^{s+1}(u)$) is disjoint from the elliptic set of $P_\sigma - \imath Q_\sigma$. Next $\Lambda_\pm$ is disjoint from $\operatorname{WF}^s(u)$, hence so is a neighborhood of $\Lambda_\pm$ as the complement of the wave front set is open. Thus by propagation of singularities and (3-32), taking into account the sign of q along $\Sigma_\pm$, $\operatorname{WF}^s(u) \cap \Sigma_\pm = \varnothing$. Now, by the regularity result, the inclusion map

$$\mathcal{X}_s = \{u \in H^m : (P_\sigma - \imath Q_\sigma)u \in H^{s-1}\} \to H^m,$$

in fact maps to H^s.

Note that $\mathcal{X}_s$ is complete with the norm $\|u\|_{\mathcal{X}_s}^2 = \|u\|_{H^m}^2 + \|(P_\sigma - \imath Q_\sigma)u\|_{H^{s-1}}^2$. Indeed, if $\{u_j\}_{j=1}^\infty$ is Cauchy in $\mathcal{X}_s$ we have $u_j \to u$ in H^m and

$$(P_\sigma - \imath Q_\sigma)u_j \to v \in H^{s-1}.$$

By the first convergence, $(P_\sigma - \imath Q_\sigma)u_j \to (P_\sigma - \imath Q_\sigma)u$ in H^{m-2}, thus, as $s-1 \geq m-2$, $(P_\sigma - \imath Q_\sigma)u_j \to v$ in H^{m-2} shows $(P_\sigma - \imath Q_\sigma)u = v \in H^{s-1}$, and thus, $(P_\sigma - \imath Q_\sigma)u_j \to (P_\sigma - \imath Q_\sigma)u$ in H^{s-1}, so $u_j \to u$ in $\mathcal{X}_s$.

The graph of the inclusion map, considered as a subset of $\mathcal{X}_s \times H^s$ is closed, for $(u_j, u_j) \to (u, v) \in \mathcal{X}_s \times H^s$ implies in particular $u_j \to u$ and $u_j \to v$ in H^m, so $u = v \in \mathcal{X}_s \cap H^s$. Correspondingly, by the closed graph theorem, the inclusion map is continuous, that is,

$$\|u\|_{H^s} \leq C(\|(P_\sigma - \imath Q_\sigma)u\|_{H^{s-1}} + \|u\|_{H^m}), \quad u \in \mathcal{X}_s. \tag{4-4}$$

This estimate implies that $\text{Ker}(P_\sigma - \imath Q_\sigma)$ in H^s is finite dimensional since elements of this kernel lie in $\mathcal{Y}_s$, and since on the unit ball of this closed subspace of H^s (for $P_\sigma - \imath Q_\sigma : H^s \to H^{s-2}$ is continuous), $\|u\|_{H^s} \leq C\|u\|_{H^m}$, and the inclusion $H^s \to H^m$ is compact. Further, elements of $\text{Ker}(P_\sigma - \imath Q_\sigma)$ are in $\mathscr{C}^\infty(X)$ by our regularity result, and thus this space is independent of the choice of s.

On the other hand, for the adjoint operator $P_\sigma^* + \imath Q_\sigma^*$, if $s' < \frac{1}{2} + \text{Im}\,\sigma$ (recall that replacing P_σ by its adjoint switches the sign of the principal symbol of $\frac{1}{2\imath}(P_\sigma - P_\sigma^*)$), $u \in H^{-N}$ and $(P_\sigma^* + \imath Q_\sigma^*)u \in H^{s'-1}$ then first $\text{WF}^{s'}(u)$ (indeed, $\text{WF}^{s'+1}(u)$) is disjoint from the elliptic set of $P_\sigma^* + \imath Q_\sigma^*$. Next, by propagation of singularities and (3-32), taking into account the sign of q along $\Sigma_\pm$, namely the sign of the imaginary part of the principal symbol switched by taking the adjoints, $\text{WF}^{s'}(u) \cap (\Sigma_\pm \setminus \Lambda_\pm) = \varnothing$. Finally, by the result at the radial points $\Lambda_\pm$ is disjoint from $\text{WF}^{s'}(u)$. Thus, the inclusion map

$$\mathcal{W}_{s'} = \{u \in H^{-N} : (P_\sigma^* + \imath Q_\sigma^*)u \in H^{s'-1}\} \to H^{-N},$$

in fact maps to $H^{s'}$. We deduce, as above, by the closed graph theorem, that

$$\|u\|_{H^{s'}} \leq C(\|(P_\sigma^* + \imath Q_\sigma^*)u\|_{H^{s'-1}} + \|u\|_{H^{-N}}), \quad u \in \mathcal{W}_{s'}. \qquad (4\text{-}5)$$

As above, this estimate implies that $\text{Ker}(P_\sigma^* + \imath Q_\sigma^*)$ in $H^{s'}$ is finite dimensional. Indeed, by our regularity results (elliptic regularity, propagation of singularities, and then regularity at the radial set) elements of $\text{Ker}(P_\sigma^* + \imath Q_\sigma^*)$ have wave front set in $\Lambda_+ \cup \Lambda_-$ and lie in $\cap_{s'<1/2+\text{Im}\,\sigma} H^{s'}$.

The dual of H^s for $s > \frac{1}{2} - \text{Im}\,\sigma$, is $H^{-s} = H^{s'-1}$, $s' = 1-s$, so $s' < \frac{1}{2} + \text{Im}\,\sigma$ in this case, while the dual of H^{s-1}, $s > \frac{1}{2} - \text{Im}\,\sigma$, is $H^{1-s} = H^{s'}$, with $s' = 1-s < \frac{1}{2} + \text{Im}\,\sigma$ again. Thus, the spaces (apart from the residual spaces H^m and H^{-N}, into which the inclusion is compact) in the left, resp. right, side of (4-5), are exactly the duals of those on the right, resp. left, side of (4-4). Thus, by a standard functional analytic argument [Hörmander 1985b, Proof of Theorem 26.1.7], namely dualization and using the compactness of the inclusion $H^{s'} \to H^{-N}$ for $s' > -N$, (4-5) gives the H^s-solvability, $s = 1 - s'$ (i.e., we demand $u \in H^s$), of

$$(P_\sigma - \imath Q_\sigma)u = f, \quad s > \tfrac{1}{2} - \text{Im}\,\sigma,$$

for f in the annihilator (in $H^{s-1} = (H^{s'})^*$ with duality induced by the L^2 inner product) of the finite dimensional subspace $\text{Ker}(P_\sigma^* + \imath Q_\sigma^*)$ of $H^{s'}$.

Recall from [Hörmander 1985b, Proof of Theorem 26.1.7] that this argument has two parts: first for any complementary subspace V of $\text{Ker}(P_\sigma^* + \imath Q_\sigma^*)$ in $H^{s'}$ (i.e., V is closed, $V \cap \text{Ker}(P_\sigma^* + \imath Q_\sigma^*) = \{0\}$, and $V + \text{Ker}(P_\sigma^* + \imath Q_\sigma^*) = H^{s'}$, for example, V is the $H^{s'}$ orthocomplement of $\mathrm{Ker}(P_\sigma^* + \imath Q_\sigma^*))$, one can drop $\|u\|_{H^{-N}}$ from the right-hand side of (4-5) when $u \in V \cap \mathcal{W}_{s'}$ at the cost of replacing C by a larger constant C'. Indeed, if no C' existed, one would have a sequence $u_j \in V \cap \mathcal{W}_{s'}$ such that $\|u_j\|_{H^{s'}} = 1$ and $\|(P_\sigma^* + \imath Q_\sigma^*)u_j\|_{H^{s'-1}} \to 0$, so $(P_\sigma^* + \imath Q_\sigma^*)u_j \to 0$ in $H^{s'-1}$. By weak compactness of the $H^{s'}$ unit ball, there is a weakly convergent subsequence u_{j_ℓ} converging to some $u \in H^{s'}$, by the closedness (which implies weak closedness) of V, $u \in V$, so

$$(P_\sigma^* + \imath Q_\sigma^*) u_{j_\ell} \to (P_\sigma^* + \imath Q_\sigma^*) u$$

weakly in $H^{s'-2}$, and thus $(P_\sigma^* + \imath Q_\sigma^*)u = 0$ so $u \in V \cap \mathrm{Ker}(P_\sigma^* + \imath Q_\sigma^*) = \{0\}$. On the other hand, by compactness of the inclusion $H^{s'} \to H^{-N}$, $u_{j_\ell} \to u$ strongly in H^{-N}, so $\{u_{j_\ell}\}$ is Cauchy in H^{-N}, hence from (4-5), it is Cauchy in $H^{s'}$, so it converges to u strongly in $H^{s'}$ and hence $\|u\|_{H^{s'}} = 1$. This contradicts $u = 0$, completing the proof of

$$\|u\|_{H^{s'}} \leq C' \|(P_\sigma^* + \imath Q_\sigma^*)u\|_{H^{s'-1}}, \quad u \in V \cap \mathcal{W}_{s'}. \tag{4-6}$$

Thus, with $s' = 1-s$, and for f in the annihilator (in H^{s-1}, via the L^2-pairing) of $\mathrm{Ker}(P_\sigma^* + \imath Q_\sigma^*) \subset H^{s'}$, and for $v \in V \cap \mathcal{W}_{s'}$,

$$|\langle f, v\rangle| \leq \|f\|_{H^{s-1}} \|v\|_{H^{s'}} \leq C' \|f\|_{H^{s-1}} \|(P_\sigma^* + \imath Q_\sigma^*)v\|_{H^{s'-1}}.$$

As adding an element of $\mathrm{Ker}(P_\sigma^* + \imath Q_\sigma^*)$ to v does not change either side, the inequality holds for all $v \in \mathcal{W}_{s'} \subset H^{s'}$. Thus, the conjugate-linear map $(P_\sigma^* + \imath Q_\sigma^*)v \mapsto \langle f, v\rangle$, $v \in \mathcal{W}_{s'}$, which is well-defined, is continuous from $\mathrm{Ran}_{\mathcal{W}_{s'}}(P_\sigma^* + \imath Q_\sigma^*) \subset H^{s'-1}$ to $\mathbb{C}$, and by the Hahn–Banach theorem can be extended to a continuous conjugate linear functional ℓ on $H^{s'-1} = (H^s)^*$, so there exists $u \in H^s$ such that $\langle u, \phi\rangle = \ell(\phi)$ for $\phi \in H^{s'-1}$. In particular, when $\phi = (P_\sigma^* + \imath Q_\sigma^*)\psi$, $\psi \in \mathcal{C}^\infty(X) \subset \mathcal{W}_{s'}$,

$$\langle u, (P_\sigma^* + \imath Q_\sigma^*)\psi\rangle = \ell(\phi) = \langle f, \psi\rangle,$$

so $(P_\sigma - \imath Q_\sigma)u = f$ as claimed.

In order to set up Fredholm theory, let $\tilde{P}$ be any operator with principal symbol $p - \imath q$; for example, $\tilde{P}$ is $P_{\sigma_0} - \imath Q_{\sigma_0}$ for some σ_0. Then consider

$$\mathcal{X}^s = \{u \in H^s : \tilde{P}u \in H^{s-1}\}, \quad \mathcal{Y}^s = H^{s-1}, \tag{4-7}$$

with

$$\|u\|_{\mathcal{X}^s}^2 = \|u\|_{H^s}^2 + \|\tilde{P}u\|_{H^{s-1}}^2.$$

Note that the $\mathcal{Z}_s$-norm is equivalent to the $\mathcal{X}^s$-norm, and $\mathcal{Z}_s = \mathcal{X}^s$, by (4-4) and the preceding discussion. Note that $\mathcal{X}^s$ only depends on the principal symbol of $\tilde{P}$. Moreover, $\mathcal{C}^\infty(X)$ is dense in $\mathcal{X}^s$; this follows by considering $R_\epsilon \in \Psi^{-\infty}(X)$,

$\epsilon > 0$, such that $R_\epsilon \to \mathrm{Id}$ in $\Psi^\delta(X)$ for $\delta > 0$, R_ϵ uniformly bounded in $\Psi^0(X)$; thus $R_\epsilon \to \mathrm{Id}$ strongly (but not in the operator norm topology) on H^s and H^{s-1}. Then for $u \in \mathcal{X}^s$, $R_\epsilon u \in \mathcal{C}^\infty(X)$ for $\epsilon > 0$, $R_\epsilon u \to u$ in H^s and $\tilde{P} R_\epsilon u = R_\epsilon \tilde{P} u + [\tilde{P}, R_\epsilon] u$, so the first term on the right converges to $\tilde{P} u$ in H^{s-1}, while $[\tilde{P}, R_\epsilon]$ is uniformly bounded in $\Psi^1(X)$, converging to 0 in $\Psi^{1+\delta}(X)$ for $\delta > 0$, so converging to 0 strongly as a map $H^s \to H^{s-1}$. Thus, $[\tilde{P}, R_\epsilon] u \to 0$ in H^{s-1}, and we conclude that $R_\epsilon u \to u$ in $\mathcal{X}^s$. (In fact, $\mathcal{X}^s$ is a first-order coisotropic space, more general function spaces of this nature are discussed in [Melrose et al. 2009, Appendix A].)

With these preliminaries,

$$P_\sigma - \imath Q_\sigma : \mathcal{X}^s \to \mathcal{Y}^s$$

is bounded for each σ with $s \geq m > \frac{1}{2} - \mathrm{Im}\,\sigma$, and is an analytic family of bounded operators in this half-plane of σ's. Further, it is Fredholm for each σ: the kernel in $\mathcal{X}^s$ is finite dimensional, and it surjects onto the annihilator in H^{s-1} of the (finite dimensional) kernel of $P_\sigma^* + \imath Q_\sigma^*$ in H^{1-s}, which thus has finite codimension, and is closed, since for f in this space there exists $u \in H^s$ with $(P_\sigma - \imath Q_\sigma) u = f$, and thus $u \in \mathcal{X}^s$. Restating this as a theorem:

Theorem 4.3. *Let P_σ, Q_σ be as above, and $\mathcal{X}^s$, $\mathcal{Y}^s$ as in (4-7). Then*

$$P_\sigma - \imath Q_\sigma : \mathcal{X}^s \to \mathcal{Y}^s$$

is an analytic family of Fredholm operators on

$$\mathbb{C}_s = \{\sigma \in \mathbb{C} : \mathrm{Im}\,\sigma > \tfrac{1}{2} - s\}. \tag{4-8}$$

Thus, analytic Fredholm theory applies, giving meromorphy of the inverse provided the inverse exists for a particular value of σ.

Remark 4.4. Note that the Fredholm property means that $P_\sigma^* + \imath Q_\sigma^*$ is also Fredholm on the dual spaces; this can also be seen directly from the estimates. The analogue of this remark also applies to the semiclassical discussion below.

4D. *Semiclassical estimates.* There are semiclassical estimates completely analogous to those in the classical setting; we again phrase these as wave front set statements. Let $H_\hbar^s$ denote the semiclassical Sobolev space of order s, that is, as a function space this is the space of functions $(u_\hbar)_{\hbar \in I}$, $I \subset (0, 1]_\hbar$ with values in the standard Sobolev space H^s, with $A_\hbar u_\hbar$ bounded in L^2 for an elliptic, semiclassically elliptic, operator $A_\hbar \in \Psi_\hbar^s(X)$. (Note that $u_\hbar$ need not be defined for all $\hbar \in (0, 1]$; we suppress I from the notation.) Let

$$\mathrm{WF}_\hbar^{s,r}(u) \subset \partial(\overline{T}^* X \times [0, 1)_\hbar) = S^* X \times [0, 1)_\hbar \cup T^* X \times \{0\}_\hbar$$

denote the semiclassical wave front set of a polynomially bounded family of distributions, that is, $u = (u_h)_{h \in I}$, $I \subset (0, 1]$, satisfying u_h is uniformly bounded in $h^{-N} H_h^{-N}$ for some N. This is defined as follows: we say that $\alpha \notin \mathrm{WF}_h^{s,r}(u)$ if there exists $A \in \Psi_h^0(X)$ elliptic at α such that $Au \in h^r H_h^s$. Note that, in view of the description of the symbols in Section 2, ellipticity at α means the ellipticity of $\sigma_h(A_h)$ if $\alpha \in T^*X \times \{0\}$, that of $\sigma_0(A_h)$ if $\alpha \in S^*X \times (0, 1)$, and that of either (and thus both, in view of the compatibility of these symbols) of these when $\alpha \in S^*X \times \{0\}$. The semiclassical wave front set captures global estimates: if u is polynomially bounded and $\mathrm{WF}_h^{s,r}(u) = \varnothing$, then $u \in h^r H_h^s$.

Elliptic regularity states that

$$P_{h,z} - \imath Q_{h,z} \text{ elliptic at } \alpha, \ \alpha \notin \mathrm{WF}_h^{s-2,0}((P_{h,z} - \imath Q_{h,z})u) \Rightarrow \alpha \notin \mathrm{WF}_h^{s,0}(u).$$

Thus, $(P_{h,z} - \imath Q_{h,z}) \in H_h^{s-2}$ and $p_{h,z} - \imath q_{h,z}$ is elliptic at α then $\alpha \notin \mathrm{WF}_h^s(u)$.

We also have real principal type propagation:

$$\mathrm{WF}_h^{s,-1}(u) \setminus (\mathrm{WF}_h^{s-1,0}((P_{h,z} - \imath Q_{h,z})u) \cup \operatorname{supp} q_{h,z})$$

is a union of maximally extended bicharacteristics of H_p in the characteristic set $\Sigma_{h,z} = \{p_{h,z} = 0\}$ of $P_{h,z}$. Put differently,

$$\alpha \notin \mathrm{WF}^{s,-1}(u) \cup \mathrm{WF}^{s-1,0}((P_{h,z} - \imath Q_{h,z})u) \cup \operatorname{supp} q_{h,z}$$
$$\Rightarrow \tilde{\gamma}(\alpha) \cap \mathrm{WF}^{s,-1}(u) = \varnothing,$$

where $\tilde{\gamma}(\alpha)$ is the component of the bicharacteristic $\gamma(\alpha)$ of $p_{h,z}$ in the complement of $\mathrm{WF}^{s-1,0}((P_{h,z} - \imath Q_{h,z})u) \cup \operatorname{supp} q_{h,z}$. If $(P_{h,z} - \imath Q_{h,z})u \in H^{s-1}$, then $\mathrm{WF}^{s-1,0}((P_{h,z} - \imath Q_{h,z})u) = \varnothing$ can be dropped from all statements above; if $q_{h,z} = 0$ one can thus replace $\tilde{\gamma}$ by γ.

In general, the result does not hold for nonzero $q_{h,z}$. However, it holds in one direction (backward/forward) of propagation along $\mathsf{H}_{p_{h,z}}$ if q_{h_z} has the correct sign. Thus, with $\tilde{\gamma}_\pm(\alpha)$ a forward (+) or backward (−) bicharacteristic from α defined on an interval, if $\pm q_{h,z} \geq 0$ on a neighborhood of $\tilde{\gamma}_\pm(\alpha)$ then

$$\alpha \notin \mathrm{WF}_h^{s,-1}(u) \text{ and } \tilde{\gamma}_\pm(\alpha) \cap \mathrm{WF}_h^{s-1,0}((P_{h,z} - \imath Q_{h,z})u) = \varnothing$$
$$\Rightarrow \tilde{\gamma}_\pm(\alpha) \cap \mathrm{WF}_h^{s,-1}(u) = \varnothing,$$

that is, one can propagate regularity forward if $q_{h,z} \geq 0$, backward if $q_{h,z} \leq 0$; see [Nonnenmacher and Zworski 2009; Datchev and Vasy 2011]. Again, for $P_{h,z}^* + \imath Q_{h,z}^*$ the directions are reversed, that is, one can propagate regularity forward if $q_{h,z} \leq 0$, backward if $q_{h,z} \geq 0$.

A semiclassical version of Melrose's regularity result was proved in [Vasy and Zworski 2000] in the asymptotically Euclidean setting. We need a more general result, which is an easy adaptation:

Proposition 4.5. *Suppose* $s \geq m > \frac{1}{2} - \operatorname{Im}\sigma$, *and* $\operatorname{WF}_\hbar^{m,-N}(u) \cap L_\pm = \emptyset$ *for some* N. *Then*
$$L_\pm \cap \operatorname{WF}^{s-1,0}(P_{\hbar,z}u) = \emptyset \Rightarrow L_\pm \cap \operatorname{WF}^{s,-1}(u) = \emptyset.$$

Again, at the cost of changing regularity, one can propagate estimate *towards* $L_\pm$.

Proposition 4.6. *For* $s < \frac{1}{2} + \operatorname{Im}\sigma$ *and* O *a neighborhood of* $L_\pm$,
$$\operatorname{WF}_\hbar^{s,-1}(u) \cap (O \setminus L_\pm) = \emptyset, \ \operatorname{WF}_\hbar^{s-1,0}(P_{\hbar,z}^*u) \cap L_\pm = \emptyset$$
$$\Rightarrow \operatorname{WF}_\hbar^{s,-1}(u) \cap L_\pm = \emptyset.$$

Proof. We just need to localize in $\tilde{\rho}$ in addition to ρ_0; such a localization in the classical setting is implied by working on S^*X or with homogeneous symbols. We achieve this by modifying the localizer ϕ in the commutant constructed in the proof of Propositions 4.1 and 4.2. As already remarked, the proof is much like at radial points in semiclassical scattering on asymptotically Euclidean spaces, studied in [Vasy and Zworski 2000], but we need to be more careful about localization in ρ_0 and $\tilde{\rho}$ as we are assuming less about the structure.

First, note that $L_\pm$ is defined by $\tilde{\rho} = 0$, $\rho_0 = 0$, so $\tilde{\rho}^2 + \rho_0$ is a quadratic defining function of $L_\pm$. Thus, let $\phi \in \mathcal{C}_c^\infty(\mathbb{R})$ be identically 1 near 0, $\phi' \leq 0$ and ϕ supported sufficiently close to 0 so that
$$\tilde{\rho}^2 + \rho_0 \in \operatorname{supp} d\phi \Rightarrow \pm \tilde{\rho}(H_p \rho_0 + 2\tilde{\rho} H_p \tilde{\rho}) > 0$$
and
$$\tilde{\rho}^2 + \rho_0 \in \operatorname{supp} \phi \Rightarrow \pm \tilde{\rho} H_p \tilde{\rho} > 0.$$

Such a ϕ exists by (3-19) and (3-21) as
$$\pm \tilde{\rho}(H_p \rho_0 + 2\tilde{\rho} H_p \tilde{\rho}) \geq 8\rho_0 + 8\tilde{\rho}^2 - \mathcal{O}((\tilde{\rho}^2 + \rho_0)^{3/2}).$$

Then let c be given by
$$c = \phi(\rho_0 + \tilde{\rho}^2)\tilde{\rho}^{-s+1/2}, \quad c_\epsilon = c(1 + \epsilon \tilde{\rho}^{-1})^{-\delta}.$$

The rest of the proof proceeds exactly as for Propositions 4.1 and 4.2, except one writes $P_{\hbar,z} = P_{\hbar,z,\operatorname{Re}} + \imath P_{\hbar,z,\operatorname{Im}}$ with the two summands being symmetric, resp. antisymmetric, thus with principal symbol $\operatorname{Re} p_{\hbar,z}$ and $\imath \operatorname{Im} p_{\hbar,z}$, one computes the commutator with $P_{\hbar,z,\operatorname{Re}}$ which involves $\operatorname{Re} p_{\hbar,z}$, obtains an extra term from

$P_{h,z,\mathrm{Im}}$ which has the correct sign as in the case of complex absorption in view of (3-27); see [Datchev and Vasy 2011, Lemma 5.1] for its treatment. □

Suppose now that $p_{\hbar,z}$ is semiclassically nontrapping, as discussed at the end of Section 3. Suppose again that $s \geq m > \frac{1}{2} - \mathrm{Im}\,\sigma$, $h^N u_h$ is bounded in $H_\hbar^m$ and $(P_{h,z} - \imath Q_{h,z})u_h \in H_\hbar^{s-1}$. The above results give that, first, $\mathrm{WF}_\hbar^{s,-1}(u)$ (indeed, $\mathrm{WF}_\hbar^{s+1,0}(u)$) is disjoint from the elliptic set of $P_{h,z} - \imath Q_{h,z}$. Next we see that $L_\pm$ is disjoint from $\mathrm{WF}_\hbar^{s,-1}(u)$, hence so is a neighborhood of $L_\pm$. Thus by propagation of singularities and the semiclassically nontrapping property, taking into account the sign of q along $\Sigma_{\hbar,\pm}$, $\mathrm{WF}_\hbar^{s,-1}(u) \cap \Sigma_{\hbar,\pm} = \varnothing$. In summary, $\mathrm{WF}_\hbar^{s,-1}(u) = \varnothing$, i.e., hu_h is bounded in $H_\hbar^s$, i.e.,

$$h^N u_h \text{ bounded in } H_\hbar^m, \ (P_{h,z} - \imath Q_{h,z})u_h \in H_\hbar^{s-1} \Rightarrow hu_h \in H_\hbar^s. \qquad (4\text{-}9)$$

Now suppose that for a decreasing sequence $h_j \to 0$, $w_h \in \mathrm{Ker}(P_{h,z} - \imath Q_{h,z})$ and $\|w_h\|_{H_\hbar^s} = 1$. Then for any N, $u_h = h^{-N} w_h$ satisfies the above hypotheses, and we deduce that hu_h is uniformly bounded in $H_\hbar^s$, that is, $h^{-N+1} w_h$ is uniformly bounded in $H_\hbar^s$. But for $N > 1$ this contradicts that $\|w_h\|_{H_\hbar^s} = 1$, so such a sequence h_j does not exist. Therefore $\mathrm{Ker}(P_{h,z} - \imath Q_{h,z}) = \{0\}$ for sufficiently small h.

Using semiclassical propagation of singularities in the reverse direction, much as we did in the previous section, we deduce that $\mathrm{Ker}(P_{h,z}^* + \imath Q_{h,z}^*) = \{0\}$ for h sufficiently small. Since $P_{h,z} - \imath Q_{h,z} : \mathcal{X}^s \to \mathcal{Y}^s$ is Fredholm, we deduce immediately that there exists h_0 such that it is invertible for $h < h_0$.

In order to obtain uniform estimates for $(P_{h,z} - \imath Q_{h,z})^{-1}$ as $h \to 0$, it is convenient to "renormalize" the problem to make the function spaces (and their norms) independent of h so that one can use the uniform boundedness principle. (Again, this could have been avoided if we had just stated the estimates uniformly in u as well, much like the closed graph theorem could have been avoided in the previous section.) So for $r \in \mathbb{R}$ let $\Lambda_\hbar^r \in \Psi_\hbar^r$ be elliptic and invertible, and let

$$P_{h,z}^s - \imath Q_{h,z}^s = \Lambda_\hbar^{s-1}(P_{h,z} - \imath Q_{h,z})\Lambda_\hbar^s.$$

Then, with $\tilde{P} = P_{h_0,z_0}^s \in \Psi^1(X)$, for instance, independent of h,

$$\mathcal{X} = \{u \in L^2 : \tilde{P}u \in L^2\}, \ \mathcal{Y} = L^2,$$

$P_{h,z}^s - \imath Q_{h,z}^s : \mathcal{X} \to \mathcal{Y}$ is invertible for $h < h_0$ by the above observations. Let $j : \mathcal{X} \to \mathcal{Z} = L^2$ be the inclusion map. Then

$$j \circ h(P_{h,z}^s - \imath Q_{h,z}^s)^{-1} : \mathcal{Y} \to \mathcal{Z}$$

is continuous for each $h < h_0$.

We claim that for each (nonzero) $f \in \mathcal{Y}$, $\{\|h(P^s_{h,z} - \imath Q^s_{h,z})^{-1} f\|_{L^2} : h < h_0\}$ is bounded. Indeed, let $v_h = h(P^s_{h,z} - \imath Q^s_{h,z})^{-1} f$, so we need to show that v_h is bounded. Suppose first that hv_h is not bounded, so consider a sequence h_j with $h_j \|v_{h_j}\|_{L^2} \geq 1$. Then let $u_h = h^{-2} v_h / \|v_h\|_{L^2}$, $h \in \{h_j : j \in \mathbb{N}\}$, so $h^2 u_h$ is bounded in L^2, so u_h is in particular polynomially bounded in L^2. Also, $(P^s_{h,z} - \imath Q^s_{h,z}) u_h = h^{-1} f / \|v_h\|_{L^2}$ is bounded in L^2 as $\|v_h\| \geq h^{-1}$. Thus, by (4-9), hu_h is bounded in L^2, that is, $h^{-1} v_h / \|v_h\|_{L^2}$ is bounded, which is a contradiction, showing that hv_h is bounded. Thus, introducing a new u_h, namely $u_h = h^{-1} v_h$, u_h is polynomially bounded, and $(P^s_{h,z} - \imath Q^s_{h,z}) u_h = f$ is bounded, so, by (4-9), $hu_h = v_h$ is bounded as claimed.

Thus, by the uniform boundedness principle, $j \circ h(P^s_{h,z} - \imath Q^s_{h,z})^{-1}$ is equicontinuous. Undoing the transformation, we deduce that

$$\|(P_{h,z} - \imath Q_{h,z})^{-1} f\|_{H^s_h} \leq C h^{-1} \|f\|_{H^{s-1}_h},$$

which is exactly the high-energy estimate we were after.

Our arguments were under the assumption of semiclassical nontrapping. As discussed in Sections 3D and 3E, this always holds in sectors $\delta |\mathrm{Re}\,\sigma| < \mathrm{Im}\,\sigma < \delta_0 |\mathrm{Re}\,\sigma|$ (with Q_σ supported in $\mu < 0$!) since $P_{h,z} - \imath Q_{h,z}$ is actually semiclassically elliptic then. In particular this gives the meromorphy of $P_\sigma - \imath Q_\sigma$ by giving invertibility of large σ in such a sector. Rephrasing in the large parameter notation, using σ instead of h,

Theorem 4.7. *Let P_σ, Q_σ, $\mathbb{C}_s$ be as above, and $\mathcal{X}^s$, $\mathcal{Y}^s$ as in (4-7). Then, for $\sigma \in \mathbb{C}_s$,*

$$P_\sigma - \imath Q_\sigma : \mathcal{X}^s \to \mathcal{Y}^s$$

has a meromorphic inverse

$$R(\sigma) : \mathcal{Y}^s \to \mathcal{X}^s.$$

Moreover, there is $\delta_0 > 0$ such that for all $\delta \in (0, \delta_0)$ there is $\sigma_0 > 0$ such that $R(\sigma)$ is invertible in

$$\{\sigma : \delta |\mathrm{Re}\,\sigma| < \mathrm{Im}\,\sigma < \delta_0 |\mathrm{Re}\,\sigma|, |\mathrm{Re}\,\sigma| > \sigma_0\},$$

and nontrapping estimates hold:

$$\|R(\sigma) f\|_{H^s_{|\sigma|^{-1}}} \leq C' |\sigma|^{-1} \|f\|_{H^{s-1}_{|\sigma|^{-1}}}.$$

If the metric g_0 is nontrapping then $p_{\hbar,z} - \imath q_{\hbar,z}$ and its complex conjugate are semiclassically nontrapping by Section 3D, so the high-energy estimates are then applicable in half-planes $\mathrm{Im}\,\sigma < -C$, that is, half-planes $\mathrm{Im}\,z \geq -Ch$. The same holds for trapping g_0 provided that we add a complex absorbing operator near the trapping, as discussed in Section 3E.

Translated into the classical setting, this gives:

Theorem 4.8. *Let P_σ, Q_σ, $\mathbb{C}_s$, $\delta_0 > 0$ be as above, in particular semiclassically nontrapping, and $\mathcal{X}^s$, $\mathcal{Y}^s$ as in (4-7). Let $C > 0$. Then there exists σ_0 such that*

$$R(\sigma) : \mathcal{Y}^s \to \mathcal{X}^s,$$

is holomorphic in $\{\sigma : -C < \operatorname{Im}\sigma < \delta_0|\operatorname{Re}\sigma|, |\operatorname{Re}\sigma| > \sigma_0\}$, assumed to be a subset of $\mathbb{C}_s$, and nontrapping estimates

$$\|R(\sigma)f\|_{H^s_{|\sigma|^{-1}}} \leq C'|\sigma|^{-1}\|f\|_{H^{s-1}_{|\sigma|^{-1}}}$$

hold. For $s = 1$ this states that for $|\operatorname{Re}\sigma| > \sigma_0$, $\operatorname{Im}\sigma > -C$,

$$\|R(\sigma)f\|^2_{L^2} + |\sigma|^{-2}\|dR(\sigma)\|^2_{L^2} \leq C''|\sigma|^{-2}\|f\|^2_{L^2}.$$

While we stated just the global results here, one also has microlocal estimates for the solution. In particular we have the following, stated in the semiclassical language, as immediate from the estimates used to derive from the Fredholm property:

Theorem 4.9. *Let P_σ, Q_σ, $\mathbb{C}_s$ be as above, in particular semiclassically nontrapping, and $\mathcal{X}^s$, $\mathcal{Y}^s$ as in (4-7).*

For $\operatorname{Re} z > 0$ and $s' > s$, the resolvent $R_{h,z}$ is semiclassically outgoing with a loss of h^{-1} in the following sense. Let $\alpha \in \overline{T}^*X \cap \Sigma_{\hbar,\pm}$ and let γ_- and γ_+ be the backward and forward bicharacteristic from α, respectively.

If $\operatorname{WF}_{\hbar}^{s'-1,0}(f) \cap \overline{\gamma_{\mp}} = \varnothing$ (where the upper sign in $\mp$ corresponds to the upper sign in $\pm$ in the previous paragraph), then $\alpha \notin \operatorname{WF}_{\hbar}^{s',-1}(R_{h,z}f)$.

In fact, for any $s' \in \mathbb{R}$, the resolvent $R_{h,z}$ extends to $f \in H^{s'}_{\hbar}(X)$, with nontrapping bounds, provided that $\operatorname{WF}_{\hbar}^{s,0}(f) \cap (L_+ \cup L_-) = \varnothing$. The semiclassically outgoing with a loss of h^{-1} result holds for such f and s' as well.

Proof. The only part that is not immediate by what has been discussed is the last claim. This follows immediately, however, by microlocal solvability in arbitrary ordered Sobolev spaces away from the radial points (i.e., solvability modulo $\mathscr{C}^\infty$, with semiclassical estimates), combined with our preceding results to deal with this smooth remainder plus the contribution near $L_+ \cup L_-$, which are assumed to be in $H^s_\hbar(X)$. □

This result is needed for gluing constructions as in [Datchev and Vasy 2011], namely polynomially bounded trapping with appropriate microlocal geometry can be glued to our resolvent. Furthermore, it gives nontrapping estimates microlocally away from the trapped set provided the overall (trapped) resolvent is polynomially bounded, as shown in [Datchev and Vasy 2010].

5. Results in the conformally compact setting

We now state our results in the original conformally compact setting. Without the nontrapping estimate, these are a special case of a result of Mazzeo and Melrose [1987], with improvements by Guillarmou [2005], with "special" meaning that evenness is assumed. If the space is asymptotic to actual hyperbolic space, the nontrapping estimate is a slightly stronger version of the estimate of [Melrose et al. 2011], where it is shown by a parametrix construction; here conformal infinity can have arbitrary geometry. The point is thus that first, we do not need the machinery of the zero calculus here, second, we do have nontrapping high-energy estimates in general (and without a parametrix construction), and third, we add the semiclassically outgoing property which is useful for resolvent gluing, including for proving nontrapping bounds microlocally away from trapping, provided the latter is mild, as shown in [Datchev and Vasy 2010; 2011].

Theorem 5.1. *Suppose that (X_0, g_0) is an n-dimensional manifold with boundary with an even conformally compact metric and boundary defining function x. Let $X_{0,\text{even}}$ denote the even version of X_0, that is, with the boundary defining function replaced by its square with respect to a decomposition in which g_0 is even. Then the inverse of*

$$\Delta_{g_0} - \left(\frac{n-1}{2}\right)^2 - \sigma^2,$$

written as $\mathcal{R}(\sigma) : L^2 \to L^2$, has a meromorphic continuation from $\operatorname{Im} \sigma \gg 0$ to $\mathbb{C}$,

$$\mathcal{R}(\sigma) : \dot{\mathcal{C}}^\infty(X_0) \to \mathcal{C}^{-\infty}(X_0),$$

with poles with finite rank residues. If in addition (X_0, g_0) is nontrapping, then, with ϕ as in Section 3A, and for suitable $\delta_0 > 0$, nontrapping estimates hold in every region $-C < \operatorname{Im} \sigma < \delta_0 |\operatorname{Re} \sigma|$, $|\operatorname{Re} \sigma| \gg 0$: for $s > \frac{1}{2} + C$,

$$\|x^{-(n-1)/2} e^{\iota\sigma\phi} \mathcal{R}(\sigma) f\|_{H^s_{|\sigma|^{-1}}(X_{0,\text{even}})} \leq \tilde{C} |\sigma|^{-1} \|x^{-(n+3)/2} e^{\iota\sigma\phi} f\|_{H^{s-1}_{|\sigma|^{-1}}(X_{0,\text{even}})}. \quad (5\text{-}1)$$

If f is supported in X_0°, the $s-1$ norm on f can be replaced by the $s-2$ norm.

Furthermore, for $\operatorname{Re} z > 0$, $\operatorname{Im} z = \mathcal{O}(h)$, the resolvent $\mathcal{R}(h^{-1}z)$ is semiclassically outgoing with a loss of h^{-1} in the sense that if f has compact support in X_0°, $\alpha \in T^*X$ is in the semiclassical characteristic set and if $\operatorname{WF}_h^{s-1,0}(f)$ is disjoint from the backward bicharacteristic from α, then

$$\alpha \notin \operatorname{WF}_h^{s,-1}(\mathcal{R}(h^{-1}z) f).$$

We remark that although in order to go through without changes, our methods require the evenness property, it is not hard to deduce more restricted results without this. Essentially one would have operators with coefficients that have a conormal singularity at the event horizon; as long as this is sufficiently mild relative to what is required for the analysis, it does not affect the results. The problems arise for the analytic continuation, when one needs strong function spaces (H^s with s large); these are not preserved when one multiplies by the singular coefficients.

Proof. All of the results of Section 4 apply.

By self-adjointness and positivity of Δ_{g_0} and as $\dot{\mathcal{C}}^\infty(X_0)$ is in its domain,

$$\left(\Delta_{g_0} - \sigma^2 - \left(\frac{n-1}{2}\right)^2\right)u = f \in \dot{\mathcal{C}}^\infty(X_0)$$

has a unique solution $u = \mathcal{R}(\sigma)f \in L^2(X_0, |dg_0|)$ when $\operatorname{Im}\sigma \gg 0$. On the other hand, let ϕ be as in Section 3A, so $e^\phi = \mu^{1/2}(1+\mu)^{-1/4}$ near $\mu = 0$ (so $e^\phi \sim x$ there), $\tilde{f}_0 = e^{i\sigma\phi}x^{-(n+1)/2}x^{-1}f$ in $\mu \geq 0$, and $\tilde{f}_0$ still vanishes to infinite order at $\mu = 0$. Let $\tilde{f}$ be an arbitrary smooth extension of $\tilde{f}_0$ to the compact manifold X on which $P_\sigma - \iota Q_\sigma$ is defined. Let $\tilde{u} = (P_\sigma - \iota Q_\sigma)^{-1}\tilde{f}$, with $(P_\sigma - \iota Q_\sigma)^{-1}$ given by our results in Section 4; this satisfies $(P_\sigma - \iota Q_\sigma)\tilde{u} = \tilde{f}$ and $\tilde{u} \in \mathcal{C}^\infty(X)$. Thus, $u' = e^{-i\sigma\phi}x^{(n+1)/2}x^{-1}\tilde{u}|_{\mu>0}$ satisfies $u' \in x^{(n-1)/2}e^{-i\sigma\phi}\mathcal{C}^\infty(X_0)$, and

$$\left(\Delta_{g_0} - \sigma^2 - \left(\frac{n-1}{2}\right)^2\right)u' = f,$$

by (3-5) and (3-14) (as Q_σ is supported in $\mu < 0$). Since $u' \in L^2(X_0, |dg_0|)$ for $\operatorname{Im}\sigma > 0$, by the aforementioned uniqueness, $u = u'$.

To make the extension from $X_{0,\text{even}}$ to X more systematic, let

$$E_s : H^s(X_{0,\text{even}}) \to H^s(X)$$

be a continuous extension operator, $R_s : H^s(X) \to H^s(X_{0,\text{even}})$ the restriction map. Then, as we have just seen, for $f \in \dot{\mathcal{C}}^\infty(X_0)$,

$$\mathcal{R}(\sigma)f = e^{-i\sigma\phi}x^{(n+1)/2}x^{-1}R_s(P_\sigma - \iota Q_\sigma)^{-1}E_{s-1}e^{i\sigma\phi}x^{-(n+1)/2}x^{-1}f. \quad (5\text{-}2)$$

While, for the sake of simplicity, Q_σ is constructed in Section 3E in such a manner that it is not holomorphic in all of $\operatorname{Im}\sigma > -C$ due to a cut in the upper half plane, this cut can be moved outside any fixed compact subset, so taking into account that $\mathcal{R}(\sigma)$ is independent of the choice of Q_σ, the theorem follows immediately from the results of Section 4. □

Our argument proves that every pole of $\mathcal{R}(\sigma)$ is a pole of $(P_\sigma - \iota Q_\sigma)^{-1}$ (for otherwise (5-2) would show $\mathcal{R}(\sigma)$ does not have a pole either), but it is possible for $(P_\sigma - \imath Q_\sigma)^{-1}$ to have poles which are not poles of $\mathcal{R}(\sigma)$. However, in the latter case, the Laurent coefficients of $(P_\sigma - \imath Q_\sigma)^{-1}$ would be annihilated by multiplication by R_s from the left, that is, the resonant states (which are smooth) would be supported in $\mu \leq 0$, in particular vanish to infinite order at $\mu = 0$.

In fact, a stronger statement can be made: by a calculation completely analogous to what we just performed, we can easily see that in $\mu < 0$, P_σ is a conjugate (times a power of μ) of a Klein–Gordon-type operator on n-dimensional de Sitter space with $\mu = 0$ being the boundary (i.e., where time goes to infinity). Thus, if σ is not a pole of $\mathcal{R}(\sigma)$ and $(P_\sigma - \imath Q_\sigma)\tilde{u} = 0$ then one would have a solution u of this Klein–Gordon-type equation near $\mu = 0$, that is, infinity, that rapidly vanishes at infinity. It is shown in [Vasy 2010b, Proposition 5.3] by a Carleman-type estimate that this cannot happen; although there $\sigma^2 \in \mathbb{R}$ is assumed, the argument given there goes through almost verbatim in general. Thus, if Q_σ is supported in $\mu < c$, where $c < 0$, then $\tilde{u}$ is also supported in $\mu < c$. This argument can be iterated for Laurent coefficients of higher order poles; their range (which is finite dimensional) contains only functions supported in $\mu < c$.

Remark 5.2. We now return to our previous remarks regarding the fact that our solution disallows the conormal singularities $(\mu \pm i0)^{\imath\sigma}$ from the perspective of conformally compact spaces of dimension n. Recalling that $\mu = x^2$, the two indicial roots on these spaces correspond to the asymptotics $\mu^{\pm\imath\sigma/2+(n-1)/4}$ in $\mu > 0$. Thus for the operator

$$\mu^{-1/2}\mu^{\imath\sigma/2-(n+1)/4}\left(\Delta_{g_0} - (n-1)^2/4 - \sigma^2\right)\mu^{-\imath\sigma/2+(n+1)/4}\mu^{-1/2},$$

or indeed P_σ, they correspond to

$$\left(\mu^{-\imath\sigma/2+(n+1)/4}\mu^{-1/2}\right)^{-1}\mu^{\pm\imath\sigma/2+(n-1)/4} = \mu^{\imath\sigma/2\pm\imath\sigma/2}.$$

Here the indicial root $\mu^0 = 1$ corresponds to the smooth solutions we construct for P_σ, while $\mu^{\imath\sigma}$ corresponds to the conormal behavior we rule out. Back to the original Laplacian, thus, $\mu^{-\imath\sigma/2+(n-1)/4}$ is the allowed asymptotics and $\mu^{\imath\sigma/2+(n-1)/4}$ is the disallowed one. Notice that $\operatorname{Re} \imath\sigma = -\operatorname{Im}\sigma$, so the disallowed solution is growing at $\mu = 0$ relative to the allowed one, as expected in the physical half plane, and the behavior reverses when $\operatorname{Im}\sigma < 0$. Thus, in the original asymptotically hyperbolic picture one has to distinguish two different rates of growths, whose relative size changes. On the other hand, in our approach, we rule out the singular solution and allow the nonsingular (smooth one), so there is no change in behavior at all for the analytic continuation.

Remark 5.3. For *even* asymptotically de Sitter metrics on an n-dimensional manifold X'_0 with boundary, the methods for asymptotically hyperbolic spaces work, except $P_\sigma - \imath Q_\sigma$ and $P_\sigma^* + \imath Q_\sigma^*$ switch roles, which does not affect

Fredholm properties; see Remark 4.4. Again, evenness means that we may choose a product decomposition near the boundary such that

$$g_0 = \frac{dx^2 - h}{x^2} \tag{5-3}$$

there, where h is an even family of Riemannian metrics; as above, we take x to be a globally defined boundary defining function. Then with $\tilde{\mu} = x^2$, so $\tilde{\mu} > 0$ is the Lorentzian region, $\bar{\sigma}$ in place of σ (recalling that our aim is to get to $P_\sigma^* + \imath Q_\sigma^*$) the above calculations for $\square_{g_0} - (n-1)^2/4 - \bar{\sigma}^2$ in place of $\Delta_{g_0} - (n-1)^2/4 - \sigma^2$ leading to (3-4) all go through with μ replaced by $\tilde{\mu}$, σ replaced by $\bar{\sigma}$ and Δ_h replaced by $-\Delta_h$. Letting $\mu = -\tilde{\mu}$, and conjugating by $(1+\mu)^{\imath\bar{\sigma}/4}$ as above, yields

$$-4\mu D_\mu^2 + 4\bar{\sigma} D_\mu + \bar{\sigma}^2 - \Delta_h + 4\imath D_\mu + 2\imath\gamma(\mu D_\mu - \bar{\sigma}/2 - \imath(n-1)/4), \tag{5-4}$$

modulo terms that can be absorbed into the error terms in operators in the class (3-5), that is, this is indeed of the form $P_\sigma^* + \imath Q_\sigma^*$ in the framework of Section 3E, at least near $\tilde{\mu} = 0$. If now X_0' is extended to a manifold without boundary in such a way that in $\tilde{\mu} < 0$, that is, $\mu > 0$, one has a classically elliptic, semiclassically nontrapping problem, then all the results of Section 4 are applicable.

Acknowledgements

I am very grateful to Maciej Zworski, Richard Melrose, Semyon Dyatlov, Gunther Uhlmann, Jared Wunsch, Rafe Mazzeo, Kiril Datchev, Colin Guillarmou and Dean Baskin for very helpful discussions, careful reading of versions of this manuscript as well as [Vasy 2010a] (with special thanks to Semyon Dyatlov in this regard; Dyatlov noticed an incomplete argument in an earlier version of this paper), and for their enthusiasm for this project, as well as to participants in my Topics in Partial Differential Equations class at Stanford University in Winter Quarter 2011, where this material was covered, for their questions and comments.

References

[Agmon 1986] S. Agmon, "Spectral theory of Schrödinger operators on Euclidean and on non-Euclidean spaces", *Comm. Pure Appl. Math.* **39**:S, suppl. (1986), S3–S16. MR 88c:35113 Zbl 0431.93059

[Borthwick and Perry 2002] D. Borthwick and P. Perry, "Scattering poles for asymptotically hyperbolic manifolds", *Trans. Amer. Math. Soc.* **354**:3 (2002), 1215–1231. MR 2003g:58046 Zbl 1009.58021

[Cardoso and Vodev 2002] F. Cardoso and G. Vodev, "Uniform estimates of the resolvent of the Laplace–Beltrami operator on infinite volume Riemannian manifolds, II", *Ann. Henri Poincaré* **3**:4 (2002), 673–691. MR 2003j:58054 Zbl 1021.58016

[Datchev and Vasy 2010] K. Datchev and A. Vasy, "Propagation through trapped sets and semiclassical resolvent estimates", preprint, 2010. To appear in *Annales de l'Institut Fourier*. arXiv 1010.2190

[Datchev and Vasy 2011] K. Datchev and A. Vasy, "Gluing semiclassical resolvent estimates via propagation of singularities", *Int. Math. Res. Notices* (2011), rnr255.

[Dimassi and Sjöstrand 1999] M. Dimassi and J. Sjöstrand, *Spectral asymptotics in the semi-classical limit*, London Mathematical Society Lecture Note Series **268**, Cambridge Univ. Press, 1999. MR 2001b:35237 Zbl 0926.35002

[Evans and Zworski 2010] L. C. Evans and M. Zworski, "Lectures on semiclassical analysis", preprint, 2010, Available at http://math.berkeley.edu/~evans/semiclassical.pdf.

[Graham and Zworski 2003] C. R. Graham and M. Zworski, "Scattering matrix in conformal geometry", *Invent. Math.* **152**:1 (2003), 89–118. MR 2004c:58064 Zbl 1030.58022

[Guillarmou 2005] C. Guillarmou, "Meromorphic properties of the resolvent on asymptotically hyperbolic manifolds", *Duke Math. J.* **129**:1 (2005), 1–37. MR 2006k:58051 Zbl 1099.58011

[Hörmander 1985a] L. Hörmander, *The analysis of linear partial differential operators, III: Pseudodifferential operators*, Grundlehren der Mathematischen Wissenschaften [Fundamental Principles of Mathematical Sciences] **274**, Springer, Berlin, 1985. MR 87d:35002b

[Hörmander 1985b] L. Hörmander, *The analysis of linear partial differential operators, IV: Fourier integral operators*, Grundlehren der Mathematischen Wissenschaften [Fundamental Principles of Mathematical Sciences] **275**, Springer, Berlin, 1985. MR 87d:35002b

[Mazzeo and Melrose 1987] R. R. Mazzeo and R. B. Melrose, "Meromorphic extension of the resolvent on complete spaces with asymptotically constant negative curvature", *J. Funct. Anal.* **75**:2 (1987), 260–310. MR 89c:58133 Zbl 0636.58034

[Melrose 1994] R. B. Melrose, "Spectral and scattering theory for the Laplacian on asymptotically Euclidian spaces", pp. 85–130 in *Spectral and scattering theory* (Sanda, 1992), edited by M. Ikawa, Lecture Notes in Pure and Appl. Math. **161**, Dekker, New York, 1994. MR 95k:58168

[Melrose et al. 2009] R. B. Melrose, A. Vasy, and J. Wunsch, "Diffraction of singularities for the wave equation on manifolds with corners", 2009. To appear in *Astérisque*. arXiv 0903.3208

[Melrose et al. 2011] R. Melrose, A. S. Barreto, and A. Vasy, "Analytic continuation and semiclassical resolvent estimates on asymptotically hyperbolic spaces", preprint, 2011. arXiv 1103.3507

[Nonnenmacher and Zworski 2009] S. Nonnenmacher and M. Zworski, "Quantum decay rates in chaotic scattering", *Acta Math.* **203**:2 (2009), 149–233. MR 2011c:58063 Zbl 1226.35061

[Perry 1987] P. A. Perry, "The Laplace operator on a hyperbolic manifold, I. Spectral and scattering theory", *J. Funct. Anal.* **75**:1 (1987), 161–187. MR 88m:58191

[Perry 1989] P. A. Perry, "The Laplace operator on a hyperbolic manifold. II. Eisenstein series and the scattering matrix", *J. Reine Angew. Math.* **398** (1989), 67–91. MR 90g:58138

[Vasy 2000] A. Vasy, *Propagation of singularities in three-body scattering*, Astérisque **262**, Société Mathématique de France, Paris, 2000. MR 2002e:35183 Zbl 0941.35001

[Vasy 2010a] A. Vasy, "Microlocal analysis of asymptotically hyperbolic and Kerr–de Sitter spaces", preprint, 2010. arXiv 1012.4391

[Vasy 2010b] A. Vasy, "The wave equation on asymptotically de Sitter-like spaces", *Adv. Math.* **223**:1 (2010), 49–97. MR 2011i:58046 Zbl 1191.35064

[Vasy and Zworski 2000] A. Vasy and M. Zworski, "Semiclassical estimates in asymptotically Euclidean scattering", *Comm. Math. Phys.* **212**:1 (2000), 205–217. MR 2002b:58047 Zbl 0955.58023

[Vodev 2004] G. Vodev, "Local energy decay of solutions to the wave equation for nontrapping metrics", *Ark. Mat.* **42**:2 (2004), 379–397. MR 2005k:58057 Zbl 1061.58024

[Wunsch and Zworski 2011] J. Wunsch and M. Zworski, "Resolvent estimates for normally hyperbolic trapped sets", *Ann. Henri Poincaré* **12**:7 (2011), 1349–1385. MR 2846671 Zbl 1228.81170

andras@math.stanford.edu *Department of Mathematics, Stanford University, Palo Alto, CA 94305-2125, United States*

Transmission eigenvalues in inverse scattering theory

FIORALBA CAKONI AND HOUSSEM HADDAR

In the past few years transmission eigenvalues have become an important area of research in inverse scattering theory with active research being undertaken in many parts of the world. Transmission eigenvalues appear in the study of scattering by inhomogeneous media and are closely related to non-scattering waves. Such eigenvalues provide information about material properties of the scattering media and can be determined from scattering data. Hence they can play an important role in a variety of inverse problems in target identification and nondestructive testing. The transmission eigenvalue problem is a non-selfadjoint and nonlinear eigenvalue problem that is not covered by the standard theory of eigenvalue problems for elliptic operators.

This article provides a comprehensive review of the state-of-the art theoretical results on the transmission eigenvalue problem including a discussion on fundamental questions such as existence and discreteness of transmission eigenvalues as well as Faber–Krahn type inequalities relating the first eigenvalue to material properties of inhomogeneous media. We begin our presentation by showing how the transmission eigenvalue problem appears in scattering theory and how transmission eigenvalues are determined from scattering data. Then we discuss the simple case of spherically stratified media where it is possible to obtain partial results on inverse spectral problems. In the case of more general inhomogeneous media we discuss the transmission eigenvalue problem for various types of media employing different mathematical techniques. We conclude our presentation with a list of open problems that in our opinion merit investigation.

1. Introduction

The interior transmission problem arises in inverse scattering theory for inhomogeneous media. It is a boundary value problem for a coupled set of equations defined on the support of the scattering object and was first introduced by Colton and

---

Cakoni's research was supported in part by the AFOSR under Grant FA 9550-11-1-0189 and NSF Grant DMS-1106972. The paper was written while she was a CNRS visiting researcher at CMAP, École Polytechnique, Palaiseau. The financial support of CNRS is greatly acknowledged.

Monk [1988] and Kirsch [1986]. Of particular interest is the eigenvalue problem associated with this boundary value problem, referred to as the transmission eigenvalue problem and, more specifically, the corresponding eigenvalues which are called transmission eigenvalues. The transmission eigenvalue problem is a nonlinear and nonselfadjoint eigenvalue problem that is not covered by the standard theory of eigenvalue problems for elliptic equations. For a long time research on the transmission eigenvalue problem mainly focused on showing that transmission eigenvalues form at most a discrete set and we refer the reader to [Colton et al. 2007] for the state of the art on this question up to 2007. From a practical point of view the question of discreteness was important to answer, since sampling methods for reconstructing the support of an inhomogeneous medium [Cakoni and Colton 2006; Kirsch and Grinberg 2008] fail if the interrogating frequency corresponds to a transmission eigenvalue. On the other hand, due to the nonselfadjointness of the transmission eigenvalue problem, the existence of transmission eigenvalues for nonspherically stratified media remained open for more than 20 years until Sylvester and Päivärinta [2008] showed the existence of at least one transmission eigenvalue provided that the contrast in the medium is large enough. The story of the existence of transmission eigenvalues was completed by Cakoni, Gintides and Haddar [Cakoni et al. 2010e] where the existence of an infinite set of transmission eigenvalue was proven only under the assumption that the contrast in the medium does not change sign and is bounded away from zero. In addition, estimates on the first transmission eigenvalue were provided. It was then showed by Cakoni, Colton and Haddar [Cakoni et al. 2010c] that transmission eigenvalues could be determined from the scattering data and since they provide information about material properties of the scattering object can play an important role in a variety of problems in target identification.

Since [Päivärinta and Sylvester 2008] appeared, the interest in transmission eigenvalues has increased, resulting in a number of important advancements in this area (throughout this paper the reader can find specific references from the vast available literature on the subject). Arguably, the transmission eigenvalue problem is one of today's central research subjects in inverse scattering theory with many open problems and potential applications. This survey aims to present the state of the art of research on the transmission eigenvalue problem focusing on three main topics, namely the discreteness of transmission eigenvalues, the existence of transmission eigenvalues and estimates on transmission eigenvalues, in particular, Faber–Krahn type inequalities. We begin our presentation by showing how transmission eigenvalue problem appears in scattering theory and how transmission eigenvalues are determined from the scattering data. Then we discuss the simple case of a spherically stratified medium where it is possible to obtain explicit expressions for transmission eigenvalues based on the theory of entire functions. In this case it is also possible to obtain a partial solution to the inverse spectral problem for transmission eigenvalues. We then proceed to discuss the general case of nonspherically stratified inhomogeneous media. As representative of the transmission eigenvalue problem we consider the scalar case for two types of problems namely the physical parameters of the inhomogeneous medium are represented by a function appearing only in the lower-order term of the partial differential equation, or the physical parameters of the inhomogeneous medium are presented by a (possibly matrix-valued) function in the main differential operator. Each of these problems employs different type of mathematical techniques. We conclude our presentation with a list of open problems that in our opinion merit investigation.

2. Transmission eigenvalues and the scattering problem

To understand how transmission eigenvalues appear in inverse scattering theory we consider the direct scattering problem for an inhomogeneous medium of bounded support. More specifically, we assume that the support $D \subset \mathbb{R}^d$, $d = 2, 3$ of the inhomogeneous medium is a bounded connected region with piece-wise smooth boundary ∂D. We denote by ν the outward normal vector ν to the boundary ∂D. The physical parameters in the medium are represented by a $d \times d$ matrix valued function A with $L^\infty(D)$ entries and by a bounded function $n \in L^\infty(D)$. From physical consideration we assume that A is a symmetric matrix such that $\overline{\xi} \cdot \Im(A(x))\xi \leq 0$ for all $\xi \in \mathbb{C}^d$ and $\Im(n(x)) \geq 0$ for almost all $x \in D$. The scattering problem for an incident wave u^i which is assumed to satisfy the Helmholtz equation $\Delta u^i + k^2 u^i = 0$ in $\mathbb{R}^d$ (possibly except for a point outside D in the case of point source incident fields) reads: Find the total field $u := u^i + u^s$ that satisfies

$$\Delta u + k^2 u = 0 \quad \text{in } \mathbb{R}^d \setminus \overline{D}, \tag{1}$$

$$\nabla \cdot A(x) \nabla u + k^2 n(x) u = 0 \quad \text{in } D, \tag{2}$$

$$u^+ = u^- \quad \text{on } \partial D, \tag{3}$$

$$\left(\frac{\partial u}{\partial \nu}\right)^+ = \left(\frac{\partial u}{\partial \nu_A}\right)^- \quad \text{on } \partial D, \tag{4}$$

$$\lim_{r \to \infty} r^{\frac{d-1}{2}} \left(\frac{\partial u^s}{\partial r} - iku^s\right) = 0, \tag{5}$$

where $k > 0$ is the wave number, $r = |x|$, u^s is the scattered field and the Sommerfeld radiation condition (5) is assumed to hold uniformly in $\hat{x} = x/|x|$. Here for a generic function f we denote $f^\pm = \lim_{h \to 0} f(x \pm h\nu)$ for $h > 0$

and $x \in \partial D$ and
$$\frac{\partial u}{\partial \nu_A} := \nu \cdot A(x)\nabla u, \quad x \in \partial D.$$

It is well-known that this problem has a unique solution $u \in H^1_{\text{loc}}(\mathbb{R}^d)$ provided that $\bar{\xi} \cdot \Re(A(x))\xi \geq \alpha |\xi|^2 > 0$ for all $\xi \in \mathbb{C}^d$ and almost all $x \in D$. The direct scattering problem in $\mathbb{R}^3$ models for example the scattering of time harmonic acoustic waves of frequency ω by an inhomogeneous medium with spatially varying sound speed and density and $k = \omega/c_0$ where c_0 is the background sound speed. In $\mathbb{R}^2$, (1)–(5) could be considered as the mathematical model of the scattering of time harmonic electromagnetic waves of frequency ω by an infinitely long cylinder such that either the magnetic field or the electric field is polarized parallel to the axis of the cylinder. Here D is the cross section of the cylinder where A and n are related to relative electric permittivity and magnetic permeability in the medium and $k = \omega/\sqrt{\epsilon_0 \mu_0}$ where ϵ_0 and μ_0 are the constant electric permittivity and magnetic permeability of the background, respectively [Colton and Kress 1998].

The transmission eigenvalue problem is related to nonscattering incident fields. Indeed, if u^i is such that $u^s = 0$ then $w := u|_D$ and $v := u^i|_D$ satisfy the following homogeneous problem:

$$\nabla \cdot A(x)\nabla w + k^2 n w = 0 \quad \text{in } D, \tag{6}$$

$$\Delta v + k^2 v = 0 \quad \text{in } D, \tag{7}$$

$$w = v \quad \text{on } \partial D, \tag{8}$$

$$\frac{\partial w}{\partial \nu_A} = \frac{\partial v}{\partial \nu} \quad \text{on } \partial D. \tag{9}$$

Conversely, if (6)–(9) has a nontrivial solution w and v and v can be extended outside D as a solution to the Helmholtz equation, then if this extended v is considered as the incident field the corresponding scattered field is $u^s = 0$. As will be seen later in this paper, there are values of k for which under some assumptions on A and n, the homogeneous problem (6)–(9) has nontrivial solutions. The homogeneous problem (6)–(9) is referred to as the *transmission eigenvalue problem*, whereas the values of k for which the transmission eigenvalue problem has nontrivial solutions are called *transmission eigenvalues*. (In next sections we will give a more rigorous definition of the transmission eigenvalue problem and corresponding eigenvalues.) As will be shown in the following sections, under further assumptions on the functions A and n, (6)–(9) satisfies the Fredholm property for $w \in H^1(D)$, $v \in H^1(D)$ if $A \neq I$ and for $w \in L^2(D)$, $v \in L^2(D)$ such that $w - v \in H^2(D)$ if $A = I$.

Even at a transmission eigenvalue, it is not possible in general to construct an incident wave that does not scatter. This is because, in general it is not possible to extend v outside D in such away that the extended v satisfies the Helmholtz equation in all of $\mathbb{R}^d$. Nevertheless, it is already known [Colton and Kress 2001; Colton and Sleeman 2001; Weck 2004], that solutions to the Helmholtz equation in D can be approximated by entire solutions in appropriate norms. In particular let $\mathcal{X}(D) := H^1(D)$ if $A \neq I$ and $\mathcal{X}(D) := L^2(D)$ if $A = I$. Then if v_g is a Herglotz wave function defined by

$$v_g(x) := \int_\Omega g(d) e^{ikx \cdot d} \, ds(d), \qquad g \in L^2(\Omega), \; x \in \mathbb{R}^d, \; d = 2, 3 \quad (10)$$

where Ω is the unit $(d-1)$-sphere $\Omega := \{x \in \mathbb{R}^d : |x| = 1\}$ and k is a transmission eigenvalue with the corresponding nontrivial solution v, w, then for a given $\epsilon > 0$, there is a v_{g_ϵ} that approximates v with discrepancy ϵ in the $\mathcal{X}(D)$-norm and the scattered field corresponding to this v_{g_ϵ} as incident field is roughly speaking ϵ-small.

The above analysis suggests that it possible to determine the transmission eigenvalues from the scattering data. To fix our ideas let us assume that the incident field is a plane wave given by $u^i := e^{ikx \cdot d}$, where $d \in \Omega$ is the incident direction. The corresponding scattered field has the asymptotic behavior [Colton and Kress 1998]

$$u^s(x) = e^{ikr} r^{-\frac{d-1}{2}} u_\infty(\hat{x}, d, k) + O\left(r^{-\frac{d+1}{2}}\right) \qquad \text{in } \mathbb{R}^d, \; d = 2, 3. \quad (11)$$

as $r \to \infty$ uniformly in $\hat{x} = x/r$, $r = |x|$ where u_∞ is known as the *far field pattern* which is a function of the observation direction $\hat{x} \in \Omega$ and also depends on the incident direction d and the wave number k. We can now define the *far field operator* $F_k : L^2(\Omega) \to L^2(\Omega)$ by

$$(F_k g)(\hat{x}) := \int_\Omega u_\infty(\hat{x}, d, k) g(d) \, ds(d). \quad (12)$$

Note that the far field operator $F := F_k$ is related to the scattering operator S defined in [Lax and Phillips 1967] by

$$S = I + \frac{ik}{2\pi} F \text{ in } \mathbb{R}^3 \quad \text{and} \quad S = I + \frac{ik}{\sqrt{2\pi k}} F \text{ in } \mathbb{R}^2.$$

To characterize the injectivity of the far field operator we first observe that by linearity $(Fg)(\cdot)$ is the far field pattern corresponding to the scattered field due to the Herglotz wave function (10) with kernel g as incident field. Thus the above discussion on nonscattering incident waves together with the fact that the

L^2-adjoint F^* of F is given by

$$(F^*g)(\hat{x}) = \overline{(Fh)(-\hat{x})},$$

with $h(d) := \overline{g(-d)}$, yield the following theorem:

Theorem 2.1 [Cakoni and Colton 2006; 1998]. *The far field operator $F : L^2(\Omega) \to L^2(\Omega)$ corresponding to the scattering problem (1)–(5) is injective and has dense range if and only if k^2 is not a transmission eigenvalue of (6)–(9) such that the function v of the corresponding nontrivial solution to (6)–(9) has the form of a Herglotz wave function (10).*

Note that the relation between the far field operator and scattering operator says that the far field operator F not being injective is equivalent to the scattering operator S having one as an eigenvalue.

Next we show that it is possible to determine the real transmission eigenvalues from the scattering data. To fix our ideas we consider far field scattering data, i.e., we assume a knowledge of $u_\infty(\hat{x}, d, k)$ for $\hat{x}, d \in \Omega$ and $k \in \mathbb{R}_+$ which implies a knowledge of the far field operator $F := F_k$ for a range of wave numbers k. Thus we can introduce the far field equation

$$(Fg)(\hat{x}) = \Phi_\infty(\hat{x}, z) \tag{13}$$

where $\Phi_\infty(\hat{x}, z)$ is the far field pattern of the fundamental solution $\Phi(x, z)$ of the Helmholtz equation given by

$$\Phi(x, z) := \frac{e^{ik|x-z|}}{4\pi|x-z|} \quad \text{in } \mathbb{R}^3,$$

$$\Phi(x, z) := \frac{i}{4} H_0^{(1)}(k|x-z|) \quad \text{in } \mathbb{R}^2, \tag{14}$$

and $H_0^{(1)}$ is the Hankel function of order zero. By a linearity argument, using Rellich's lemma and the denseness of the Herglotz wave functions in the space of $\mathcal{X}(D)$-solutions to the Helmholtz equation, it is easy to prove the following result (see, e.g., [Cakoni and Colton 2006]).

Theorem 2.2. *Assume that $z \in D$ and k is not a transmission eigenvalue. Then for any given $\epsilon > 0$ there exists $g_{z,\epsilon}$ such that*

$$\|Fg_{z,\epsilon} - \Phi_\infty(\cdot, z)\|_{L^2(\Omega)}^2 < \epsilon$$

and the corresponding Herglotz wave function $v_{g_{z,\epsilon}}$ satisfies

$$\lim_{\epsilon \to 0} \|v_{g_{z,\epsilon}}\|_{\mathcal{X}(D)} = \|v_z\|_{\mathcal{X}(D)}$$

where (w_z, v_z) is the unique solution of the nonhomogeneous interior transmission problem

$$\nabla \cdot A(x)\nabla w_z + k^2 n w_z = 0 \quad \text{in } D, \tag{15}$$

$$\Delta v_z + k^2 v_z = 0 \quad \text{in } D, \tag{16}$$

$$w_z - v_z = \Phi(\cdot, z) \quad \text{on } \partial D, \tag{17}$$

$$\frac{\partial w_z}{\partial \nu_A} - \frac{\partial v_z}{\partial \nu} = \frac{\partial \Phi(\cdot, z)}{\partial \nu} \quad \text{on } \partial D. \tag{18}$$

On the other hand, if k is a transmission eigenvalue, again by linearity argument and applying the Fredholm alternative to the interior transmission problem (15)–(18) it is possible to show the following theorem:

Theorem 2.3. *Assume k is a transmission eigenvalue, and for a given $\epsilon > 0$ let $g_{z,\epsilon}$ be such that*

$$\|Fg_{z,\epsilon} - \Phi_\infty(\cdot, z)\|_{L^2(\Omega)}^2 \leq \epsilon \tag{19}$$

with $v_{g_{z,\epsilon}}$ the corresponding Herglotz wave function. Then, for all $z \in D$, except for a possibly nowhere dense subset, $\|v_{g_{z,\epsilon}}\|_{\mathcal{H}(D)}$ can not be bounded as $\epsilon \to 0$.

For a proof of Theorem 2.3 for the case of $A = I$ we refer the reader to [Cakoni et al. 2010c]. Theorem 2.2 and Theorem 2.3, roughly speaking, state that if D is known and $\|v_{g_{z,\epsilon}}\|_{X(D)}$ is plotted against k for a range of wave numbers $[k_0, k_1]$, the transmission eigenvalues should appear as peaks in the graph. We remark that for some special situations (e.g., if D is a disk centered at the origin, $A = I$, $z = 0$ and n constant) $g_{z,\epsilon}$ satisfying (19) may not exist. However it is reasonable to assume that (19) always holds for the noisy far field operator F^δ given by

$$(F^\delta g)(\hat{x}) := \int_\Omega u_\infty^\delta(\hat{x}, d, k) g(d) \, ds(d),$$

where $u_\infty^\delta(\hat{x}, d, k)$ denotes the noisy measurement with noise level $\delta > 0$ (see the Appendix in [Cakoni et al. 2010c]). Nevertheless, in practice, we have access only to the noisy far field operator F_δ. Due to the ill-posedness of the far field equation (note that F is a compact operator), one looks for the Tikhonov regularized solution $g_{z,\alpha}^\delta$ of the far field equation defined as the unique minimizer of the Tikhonov functional [Colton and Kress 1998]

$$\|F^\delta g - \Phi_\infty(\cdot, z)\|_{L^2(\Omega)}^2 + \alpha \|g\|_{L^2(\Omega)}^2$$

where the positive number $\alpha := \alpha(\delta)$ is the Tikhonov regularization parameter satisfying $\alpha(\delta) \to 0$ as $\delta \to 0$. In [Arens and Lechleiter 2009] and [Arens 2004] it is proven for the case of $A = I$ that Theorem 2.2 is also valid if the approximate solution $g_{z,\epsilon}$ is replaced by the regularized solution $g_{z,\alpha}^\delta$ and the noise level tends to zero. We remark that since the proof of such result relies on the validity of the factorization method (i.e., if F is normal, see [Kirsch and Grinberg 2008] for details), in general for many scattering problems, Theorem 2.2 can only be proven for the approximate solution to the far field equation. On the other hand, Theorem 2.3 remains valid for the regularized solution $g_{z,\alpha}^\delta$ as the noise level $\delta \to 0$ (see [Cakoni et al. 2010c] for the proof).

3. The transmission eigenvalue problem for isotropic media

We start our discussion of the transmission eigenvalue problem with the case of isotropic media, i.e., when $A = I$. The *transmission eigenvalue problem* corresponding to the scattering problem for isotropic media reads: Find $v \in L^2(D)$ and $w \in L^2(D)$ such that $w - v \in H^2(D)$ satisfying

$$\Delta w + k^2 n(x) w = 0 \quad \text{in } D, \tag{20}$$

$$\Delta v + k^2 v = 0 \quad \text{in } D, \tag{21}$$

$$w = v \quad \text{on } \partial D, \tag{22}$$

$$\frac{\partial w}{\partial \nu} = \frac{\partial v}{\partial \nu} \quad \text{on } \partial D. \tag{23}$$

As will become clear later, the above function spaces provide the appropriate framework for the study of this eigenvalue problem which turns out to be nonselfadjoint. Note that since the difference between two equations in D occurs in the lower-order term and only Cauchy data for the difference is available, it is not possible to have any control on the regularity of each field w and v and assuming (20) and (21) in the $L^2(D)$ (distributional) sense is the best one can hope. Let us define

$$H_0^2(D) := \left\{ u \in H^2(D) : \text{ such that } u = 0 \text{ and } \frac{\partial u}{\partial \nu} = 0 \text{ on } \partial D \right\}.$$

Definition 3.1. Values of $k \in \mathbb{C}$ for which (20)–(23) has nontrivial solution $v \in L^2(D)$ and $w \in L^2(D)$ such that $w - v \in H_0^2(D)$ are called *transmission eigenvalues*.

Note that if $n(x) \equiv 1$ every $k \in \mathbb{C}$ is a transmission eigenvalue, since in this trivial case there is no inhomogeneity and any incident field does not scatterer.

3A. *Spherically stratified media.* To shed light into the structure of the eigenvalue problem (20)–(23), we start our discussion with the special case of a spherically stratified medium where D is a ball of radius a and $n(x) := n(r)$ is spherically stratified. It is possible to obtain explicit formulas for the solution of this problem by separation of variables and using tools from the theory of entire functions. This allows the possibility to obtain sharper results than are currently available for the general nonspherically stratified case. In particular, it is possible to solve the inverse spectral problem for transmission eigenvalues, prove that complex transmission eigenvalues can exist for nonabsorbing media and show that real transmission eigenvalues may exist under some conditions for the case of absorbing media, all of which problems are still open in the general case.

Throughout this section we assume that $\Im(n(r)) = 0$ and (unless otherwise specified). Setting $B := \{x \in \mathbb{R}^3 : |x| < a\}$ the transmission eigenvalue problem for spherically stratified medium is:

$$\Delta w + k^2 n(r) w = 0 \quad \text{in } B, \tag{24}$$

$$\Delta v + k^2 v = 0 \quad \text{in } B, \tag{25}$$

$$w = v \quad \text{in } \partial B, \tag{26}$$

$$\frac{\partial w}{\partial r} = \frac{\partial v}{\partial r} \quad \text{on } \partial B. \tag{27}$$

Let us assume that $n(r) \in C^2[0, a]$ (unless otherwise specified). The main concern here is to show the existence of real and complex transmission eigenvalues and solve the inverse spectral problem. To this end, introducing spherical coordinates (r, θ, φ) we look for solutions of (24)–(27) in the form

$$v(r, \theta) = a_\ell j_\ell(kr) P_\ell(\cos \theta) \quad \text{and} \quad w(r, \theta) = b_\ell y_\ell(r) P_\ell(\cos \theta)$$

where P_ℓ is a Legendre's polynomial, j_ℓ is a spherical Bessel function, a_ℓ and b_ℓ are constants and y_ℓ is a solution of

$$y'' + \frac{2}{r} y' + \left(k^2 n(r) - \frac{\ell(\ell+1)}{r^2} \right) y_\ell = 0$$

for $r > 0$ such that $y_\ell(r)$ behaves like $j_\ell(kr)$ as $r \to 0$, i.e.,

$$\lim_{r \to 0} r^{-\ell} y_\ell(r) = \frac{\sqrt{\pi} k^\ell}{2^{\ell+1} \Gamma(\ell + 3/2)}.$$

From [Colton and Kress 1983, pp. 261–264], in particular Theorem 9.9, we can deduce that k is a (possibly complex) transmission eigenvalue if and only if

$$d_\ell(k) = \det \begin{pmatrix} y_\ell(a) & -j_\ell(ka) \\ y'_\ell(a) & -k j'_\ell(ka) \end{pmatrix} = 0. \tag{28}$$

Setting $m := 1 - n$, from [Colton 1979] (see also [Cakoni et al. 2010a]) we can represent $y_\ell(r)$ in the form

$$y_\ell(r) = j_\ell(kr) + \int_0^r G(r,s,k) j_\ell(ks) ds \qquad (29)$$

where $G(r,s,k)$ satisfies the Goursat problem

$$r^2 \left[\frac{\partial^2 G}{\partial r^2} + \frac{2}{r}\frac{\partial G}{\partial r} + k^2 n(r) G \right] = s^2 \left[\frac{\partial^2 G}{\partial s^2} + \frac{2}{s}\frac{\partial G}{\partial s} + k^2 G \right] \qquad (30)$$

$$G(r,r,k) = \frac{k^2}{2r} \int_0^r \rho m(\rho) d\rho, \qquad G(r,s,k) = O\big((rs)^{1/2}\big). \qquad (31)$$

It is shown in [Colton 1979] that (30)–(31) can be solved by iteration and the solution G is an even function of k and an entire function of exponential type satisfying

$$G(r,s,k) = \frac{k^2}{2\sqrt{rs}} \int_0^{\sqrt{rs}} \rho m(\rho) d\rho \big(1 + O(k^2)\big). \qquad (32)$$

Hence for fixed $r > 0$, y_ℓ and spherical Bessel functions are entire function of k of finite type and bounded for k on the positive real axis, and thus $d_\ell(k)$ also has this property. Furthermore, by the series expansion of j_ℓ [Colton and Kress 1998], we see that $d_\ell(k)$ is an even function of k and $d_\ell(0) = 0$. Consequently, if $d_\ell(k)$ does not have a countably infinite number of zeros it must be identically zero. It is easy to show now that $d_\ell(k)$ is not identically zero for every ℓ unless $n(r)$ is identically equal to 1. Indeed, assume that $d_\ell(k)$ is identically zero for every nonnegative integer ℓ. Noticing that $j_\ell(kr) Y_\ell^m(\hat{x})$ is a Herglotz wave function, it follows from the proof of Theorem 8.16 in [Colton and Kress 1998] that

$$\int_0^a j_\ell(k\rho) y_\ell(\rho) \rho^2 m(\rho) d\rho = 0$$

for all k where $m(r) := 1 - n(r)$. Hence, using the Taylor series expansion of $j_\ell(k\rho)$ and (29) we see that

$$\int_0^a \rho^{2\ell+2} m(\rho) d\rho = 0 \qquad (33)$$

for all nonnegative integers ℓ. By Müntz's theorem [Davis 1963], we now have $m(r) = 0$, i.e., $n(r) = 1$. Note that from (33) it is easy to see that none of the integrals (33) can become zero if $m(r) \geq 0$ or $m(r) \leq 0$ (not identically zero) which implies that in these cases the transmission eigenvalues form a discrete set as a countable union of countably many zeros of $d_\ell(k)$. Nothing can be said about discreteness of transmission eigenvalues if our only assumption is that

$n(r)$ is not identically equal to one. However, if B is a ball in $\mathbb{R}^3$, $n \in C^2[0, a]$ and $n(a) \neq 1$, transmission eigenvalues form at most discrete set and there exist infinitely many transmission eigenvalues corresponding to spherically symmetric eigenfunctions.

Theorem 3.1. *Assume that $n \in C^2[0, a]$, $\Im(n(r)) = 0$ and either $n(a) \neq 1$ or $n(a) = 1$ and $\frac{1}{a}\int_0^a \sqrt{n(\rho)}\,d\rho \neq 1$. Then there exists an infinite discrete set of transmission eigenvalues for (24)–(27) with spherically symmetric eigenfunctions. Furthermore the set of all transmission eigenvalues is discrete.*

Proof. To show existence, we restrict ourselves to spherically symmetric solutions to (24)–(27), and look for solutions of the form.

$$v(r) = a_0 j_0(kr) \quad \text{and} \quad w(r) = b_0 \frac{y(r)}{r}$$

where

$$y'' + k^2 n(r) y = 0, \quad y(0) = 0, \quad y'(0) = 1.$$

Using the Liouville transformation

$$z(\xi) := [n(r)]^{\frac{1}{4}} y(r) \quad \text{where} \quad \xi(r) := \int_0^r [n(\rho)]^{\frac{1}{2}}\, d\rho$$

we arrive at the following initial value problem for $z(\xi)$:

$$z'' + [k^2 - p(\xi)]z = 0, \quad z(0) = 0, \quad z'(0) = [n(0)]^{-\frac{1}{4}}, \tag{34}$$

where

$$p(\xi) := \frac{n''(r)}{4[n(r)]^2} - \frac{5}{16}\frac{[n'(r)]^2}{[n(r)]^3}.$$

Now exactly in the same way as in [Colton and Kress 1998; Colton et al. 2007], by writing (34) as a Volterra integral equation and using the methods of successive approximations, we obtain the following asymptotic behavior for y:

$$y(r) = \frac{1}{k\,[n(0)\,n(r)]^{1/4}} \sin\left(k \int_0^r [n(\rho)]^{1/2}\,d\rho\right) + \mathcal{O}\left(\frac{1}{k^2}\right)$$

and

$$y'(r) = \left[\frac{n(r)}{n(0)}\right]^{1/4} \cos\left(k \int_0^r [n(\rho)]^{1/2}\,d\rho\right) + \mathcal{O}\left(\frac{1}{k}\right),$$

uniformly on $[0, a]$. Applying the boundary conditions (26), (27) on ∂B, we see that a nontrivial solution to (24)–(27) exists if and only if

$$d_0(k) = \det \begin{pmatrix} \dfrac{y(a)}{a} & -j_0(ka) \\ \dfrac{d}{dr}\left(\dfrac{y(r)}{r}\right)_{r=a} & -k\, j_0'(ka) \end{pmatrix} = 0. \tag{35}$$

Since $j_0(kr) = \sin kr/kr$, from the above asymptotic behavior of $y(r)$ we have that

$$d_0(k) = \frac{1}{ka^2}[A\sin(k\delta a)\cos(ka) - B\cos(k\delta a)\sin(ka)] + \mathbb{O}\left(\frac{1}{k^2}\right) \tag{36}$$

where

$$\delta = \frac{1}{a}\int_0^a \sqrt{n(\rho)}\, d\rho, \qquad A = \frac{1}{[n(0)n(a)]^{1/4}}, \qquad B = \left[\frac{n(a)}{n(0)}\right]^{1/4}.$$

If $n(a) = 1$, since $\delta \neq 1$ the first term in (36) is a periodic function if δ is rational and almost-periodic [Colton et al. 2007] if δ is irrational, and in either case takes both positive and negative values. This means that for large enough k, $d_0(k)$ has infinitely many real zeros which proves the existence of infinitely many real transmission eigenvalues. Now if $n(a) \neq 1$ then $A \neq B$ and the above argument holds independent of the value of δ.

Concerning the discreteness of transmission eigenvalues, we first observe that similar asymptotic expression to (36) holds for all the determinants $d_\ell(k)$ [Colton and Kress 1998]. Hence the above argument shows that $d_\ell(k) \neq 0$ and hence they have countably many zeros, which shows that transmission eigenvalues are discrete. $\square$

Next we are interested in the inverse spectral problem for the transmission eigenvalue problem (24)–(27). The question we ask is under what conditions do transmission eigenvalues uniquely determine $n(r)$. This question was partially answered in [McLaughlin and Polyakov 1994; McLaughlin et al. 1994] under restrictive assumptions on $n(r)$ and the nature of the spectrum. The inverse spectral problem for the general case is solved in [Cakoni et al. 2010a], provided that all transmission eigenvalues are given, which we briefly sketch in the following:

Theorem 3.2. *Assume that $n \in C^2[0, +\infty)$, $\Im(n(r)) = 0$ and $n(r) > 1$ or $n(r) < 1$ for $r < a$, $0 < n(r) = 1$ for $r > a$. If $n(0)$ is given then $n(r)$ is uniquely determined from a knowledge of the transmission eigenvalues and their multiplicity as a zero of $d_\ell(k)$.*

Proof. We return to the determinant (28) and observe that $d_\ell(k)$ has the asymptotic behavior [Colton and Kress 1998]

$$d_\ell(k) = \frac{1}{a^2 k \, [n(0)]^{\ell/2+1/4}} \sin k \left(a - \int_0^a [n(r)]^{1/2} dr \right) + O\left(\frac{\ln k}{k^2}\right). \quad (37)$$

We first compute the coefficient $c_{2\ell+2}$ of the term $k^{2\ell+2}$ in its Hadamard factorization expression [Davis 1963]. A short computation using (28), (29), and the order estimate

$$j_\ell(kr) = \frac{\sqrt{\pi}(kr)^\ell}{2^{\ell+1}\Gamma(\ell+3/2)} \left(1 + O(k^2 r^2)\right) \quad (38)$$

shows that

$$c_{2\ell+2} \left[\frac{2^{\ell+1}\Gamma(\ell+3/2)}{\sqrt{\pi} a^{(\ell-1)/2}}\right]^2 = a \int_0^a \frac{d}{dr} \left(\frac{1}{2\sqrt{rs}} \int_0^{\sqrt{rs}} \rho m(\rho) \, d\rho\right)_{r=a} s^\ell \, ds$$

$$- \ell \int_0^a \frac{1}{2\sqrt{as}} \int_0^{\sqrt{as}} \rho m(\rho) \, d\rho \, s^\ell \, ds + \frac{a^\ell}{2} \int_0^a \rho m(\rho) \, d\rho. \quad (39)$$

After a rather tedious calculation involving a change of variables and interchange of orders of integration, the identity (39) remarkably simplifies to

$$c_{2\ell+2} = \frac{\pi a^2}{2^{\ell+1}\Gamma(\ell+3/2)} \int_0^a \rho^{2\ell+2} m(\rho) \, d\rho. \quad (40)$$

We note that $j_\ell(r)$ is odd if ℓ is odd and even if ℓ is even. Hence, since G is an even function of k, we have that $d_\ell(k)$ is an even function of k. Furthermore, since both G and j_ℓ are entire function of k of exponential type, so is $d_\ell(k)$. From the asymptotic behavior of $d_\ell(k)$ for $k \to \infty$, that is, (37), we see that the rank of $d_\ell(k)$ is one and hence by Hadamard's factorization theorem [Davis 1963],

$$d_\ell(k) = k^{2\ell+2} e^{a_\ell k + b_\ell} \prod_{\substack{n=-\infty \\ n \neq 0}}^{\infty} \left(1 - \frac{k}{k_{n\ell}}\right) e^{k/k_{n\ell}},$$

where a_ℓ, b_ℓ are constants or, since d_ℓ is even,

$$d_\ell(k) = k^{2\ell+2} c_{2\ell+2} \prod_{n=1}^{\infty} \left(1 - \frac{k^2}{k_{n\ell}^2}\right), \quad (41)$$

where $c_{2\ell+2}$ is a constant given by (40) and $k_{n\ell}$ are zeros in the right half-plane (possibly complex). In particular, $k_{n\ell}$ are the (possibly complex) *transmission eigenvalues* in the right half-plane. Thus, if the transmission eigenvalues are known, so is

$$\frac{d_\ell(k)}{c_{2\ell+2}} = k^{2\ell+2} \prod_{n=1}^{\infty}\left(1 - \frac{k^2}{k_{n\ell}^2}\right),$$

as well as from (37) a nonzero constant γ_ℓ independent of k such that

$$\frac{d_\ell(k)}{c_{2\ell+2}} = \frac{\gamma_\ell}{a^2 k} \sin k\left(a - \int_0^a [n(r)]^{1/2}\,dr\right) + O\left(\frac{\ln k}{k^2}\right),$$

that is,

$$\frac{1}{c_{2\ell+2}[n(0)]^{\ell/2+1/4}} = \gamma_\ell.$$

From (40) we now have

$$\int_0^a \rho^{2\ell+2} m(\rho)\,d\rho = \frac{(2^{\ell+1}\Gamma(\ell+3/2))^2}{[n(0)]^{\ell/2+1/4}\,\gamma_\ell\pi a^2}.$$

If $n(0)$ is given then $m(\rho)$ is uniquely determined by Müntz's theorem [Davis 1963]. □

It was shown in [Aktosun et al. 2011] that in the case when $0 < n(r) < 1$ the eigenvalues corresponding to spherically symmetric eigenfunctions, that is, the zeros of $d_0(kr)$ (together with their multiplicity) uniquely determine $n(r)$. Specifically:

Theorem 3.3 [Aktosun et al. 2011]. *Assume that $n(r) \in C^1(0, a)$ such that $n'(r) \in L^2(0, a)$, $\Im(n(r)) = 0$ and $\frac{1}{a}\int_0^a \sqrt{n(\rho)}d\rho < 1$. Then $n(r)$ is uniquely determined from a knowledge of k_{n0} and its multiplicity as a zero of $d_0(k)$.*

The argument used in the proof refers back to the classic inverse Sturm–Liouville problem and it breaks down if $n(r) > 1$.

As we have just showed, for a spherically symmetric index of refraction the real and complex transmission eigenvalues uniquely determine the index of refraction up to a normalizing constant. From Theorem 3.1 we also know that real transmission eigenvalues exist. This raises the question as to whether or not complex transmission eigenvalues can exist. The following simple example in $\mathbb{R}^2$ shows that in general complex transmission eigenvalues can exist [Cakoni et al. 2010a].

Example of existence of complex transmission eigenvalues. Consider the interior transmission problem (20) and (21) where D is a disk of radius one in $\mathbb{R}^2$ and constant index of refraction $n \neq 1$. We will show that if n is sufficiently small there exist complex transmission eigenvalues in this particular case. To this end we note that k is a transmission eigenvalue provided

$$d_0(k) = k\left(J_1(k)J_0(k\sqrt{n}) - \sqrt{n}J_0(k)J_1(k\sqrt{n})\right) = 0.$$

Viewing d_0 as a function of $\sqrt{n}$ we compute

$$d_0'(k) = k\left(kJ_1(k)J_0'(k\sqrt{n}) - J_0(k)J_1(k\sqrt{n}) - k\sqrt{n}J_0(k)J_1'(k\sqrt{n})\right)$$

where differentiation is with respect to $\sqrt{n}$. Hence

$$d_0'(k)\big|_{\sqrt{n}=1} = k\left(kJ_1(k)J_0'(k) - J_0(k)J_1(k) - kJ_0(k)J_1'(k)\right).$$

But $J_0'(t) = -J_1(t)$ and $\dfrac{d}{dt}(tJ_1(t)) = tJ_0(t)$ and hence

$$d_0'(k)\big|_{\sqrt{n}=1} = -k^2\left(J_1^2(k) + J_0^2(k)\right) \tag{42}$$

that is,

$$f(k) = \lim_{\sqrt{n}\to 1^+} \frac{d_0(k)}{\sqrt{n}-1} = -k^2\left(J_1^2(k) + J_0^2(k)\right) \tag{43}$$

Since $J_1(k)$ and $J_0(k)$ do not have any common zeros, $f(k)$ is strictly negative for $k \neq 0$ real, that is, the only zeros of $f(k)$, $k \neq 0$, are complex. Furthermore, $f(k)$ is an even entire function of exponential type that is bounded on the real axis and hence by Hadamard's factorization theorem $f(k)$ has an infinite number of complex zeros. By Hurwitz's theorem in analytic function theory [Colton and Kress 1983, p. 213], we can now conclude that for n close enough to one $d_0(k) = 0$ has complex roots, thus establishing the existence of complex transmission eigenvalues for the unit disk and constant $n > 1$ sufficiently small (Note that by Montel's theorem [Colton and Kress 1983, p. 213] the convergence in (43) is uniform on compact subsets of the complex plane.)

A more comprehensive investigation of the existence of complex transmission eigenvalues for spherically stratified media in $\mathbb{R}^2$ and $\mathbb{R}^3$ has been recently initiated in [Leung and Colton 2012]. Based on tools of analytic function theory, the authors has shown that infinitely many complex transmission eigenvalues can exist. We state here their main results and refer the reader to the paper for the details of proofs.

Theorem 3.4 [Leung and Colton 2012]. *Consider the transmission eigenvalue problem* (24)–(27) *where* $B := \{x \in \mathbb{R}^d : |x| < 1\}$, $d = 2, 3$ *and* $n = n(r) > 0$ *is a positive constant. Then:*

(i) *In* $\mathbb{R}^2$, *if* $n \neq 1$ *then there exists an infinite number of complex eigenvalues.*

(ii) *In $\mathbb{R}^3$, if n is a positive integer not equal to one then all transmission eigenvalues corresponding to spherically symmetric eigenfunctions are real. On the other hand if n is a rational positive number $n = p/q$ such that either $q < p < 2q$ or $p < q < 2p$ then there exists an infinite number of complex eigenvalues.*

Note that complex transmission eigenvalues for n rational satisfying the assumptions of Theorem 3.4(ii) all must lie in a strip parallel to real axis. We remark that in [Leung and Colton 2012] the authors also show the existence of infinitely many transmission eigenvalues in $\mathbb{R}^3$ for some particular cases of inhomogeneous spherically stratified media $n(r)$. The existence of complex eigenvalues indicates that the transmission eigenvalue problem for spherically stratified media is nonselfadjoint. In the coming section we show that this is indeed the case in general.

We end this section by considering the transmission eigenvalue problem for absorbing media in $\mathbb{R}^3$ [Cakoni et al. 2012]. When both the scattering obstacle and the background medium are absorbing it is still possible to have real transmission eigenvalues which is easy to see in the case of a spherically stratified medium. In particular, let $B := \{x \in \mathbb{R}^3 : |x| < a\}$ and consider the interior transmission eigenvalue problem

$$\Delta_3 w + k^2 \left(\epsilon_1(r) + i \frac{\gamma_1(r)}{k} \right) w = 0 \quad \text{in } B, \tag{44}$$

$$\Delta_3 v + k^2 \left(\epsilon_0 + i \frac{\gamma_0}{k} \right) v = 0 \quad \text{in } B, \tag{45}$$

$$v = w \quad \text{on } \partial B, \tag{46}$$

$$\frac{\partial v}{\partial r} = \frac{\partial w}{\partial r} \quad \text{on } \partial B, \tag{47}$$

where $\epsilon_1(r)$ and $\gamma_1(r)$ are continuous functions of r in $\bar{B}$ such that $\epsilon_1(a) = \epsilon_0$ and ϵ_0 and γ_0 are positive constants. We look for a solution of (44)–(47) in the form

$$v(r) = c_1 j_0(k \tilde{n}_0 r) \quad \text{and} \quad w(r) = c_2 \frac{y(r)}{r} \tag{48}$$

where $\tilde{n}_0 := (\epsilon_0 + i \gamma_0/k)^{1/2}$ (where the branch cut is chosen such that $\tilde{n}_0$ has positive real part), j_0 is a spherical Bessel function of order zero, $y(r)$ is a solution of

$$y'' + k^2 \left(\epsilon_1(r) + i \frac{\gamma_1(r)}{k} \right) y = 0, \tag{49}$$

$$y(0) = 0, \quad y'(0) = 1 \tag{50}$$

for $0 < r < a$, and c_1 and c_2 are constants. Then there exist constants c_1 and c_2, not both zero, such that (48) will be a nontrivial solution of (44)–(47) provided that the corresponding $d_0(k)$ given by (35) satisfies $d_0(k) = 0$. We again derive an asymptotic expansion for $y(r)$ for large k to show that for appropriate choices of n_0 and γ_0 there exist an infinite set of positive values of k such that $d_0(k) = 0$ holds.

Following [Erdélyi 1956, pp. 84, 89], we see that (49) has a fundamental set of solutions $y_1(r)$ and $y_2(r)$ defined for $r \in [a, b]$ such that

$$y_j(r) = Y_j(r)\left[1 + O\left(\frac{1}{k}\right)\right] \tag{51}$$

as $k \to \infty$, uniformly for $0 \leq r \leq a$ where

$$Y_j(r) = \exp[\beta_{oj}k + \beta_{1j}],$$
$$(\beta'_{oj})^2 + \epsilon_1(r) = 0 \quad \text{and} \quad 2\beta'_{oj}\beta_{1j} + i\gamma_1(r) + \beta''_{oj} = 0. \tag{52}$$

From (3A) we see that, modulo arbitrary constants,

$$\beta_{0j} = \pm\int_0^r \sqrt{\epsilon_1(\rho)}\,d\rho \quad \text{and} \quad \beta_{ij} = \mp\frac{1}{2}\int_0^r \frac{\gamma_1(\rho)}{\sqrt{\epsilon_1(r)}}d\rho + \log[\epsilon_1(r)]^{-1/4}$$

where $j = 1$ corresponds to the upper sign and $j = 2$ corresponds to the lower sign. Substituting back into (51) and using the initial condition (50) we see that

$$y(r) =$$
$$\frac{1}{ik[\epsilon_1(0)\epsilon_1(r)]^{1/4}}\sinh\left[ik\int_0^r \sqrt{\epsilon_1(\rho)}\,d\rho - \frac{1}{2}\int_0^r \frac{\gamma_1(\rho)}{\sqrt{\epsilon_1(\rho)}}d\rho\right] + O\left(\frac{1}{k^2}\right) \tag{53}$$

as $k \to \infty$. Similarly,

$$j_0(k\tilde{n}_0 r) = \frac{1}{ik\sqrt{\epsilon_0}r}\sinh\left[ik\sqrt{\epsilon_0}r - \frac{1}{2}\frac{\gamma_0}{\sqrt{\epsilon_0}}r\right] + O\left(\frac{1}{k^2}\right) \tag{54}$$

as $k \to \infty$. Using (53), (54), and the fact that these expressions can be differentiated with respect to r, implies that, as $k \to \infty$,

$$d = \frac{1}{ika^2[\epsilon_1(0)\epsilon_0]^{1/4}}$$
$$\times \sinh\left[ik\sqrt{\epsilon_0}a - ik\int_0^a \sqrt{\epsilon_1(\rho)}\,d\rho - \frac{1}{2}\frac{\gamma_0 a}{\sqrt{\epsilon_0}} + \frac{1}{2}\int_0^a \frac{\gamma_1(\rho)}{\sqrt{\epsilon_1(\rho)}}d\rho\right]$$
$$+ O\left(\frac{1}{k^2}\right) \tag{55}$$

We now want to use (55) to deduce the existence of transmission eigenvalues. We first note that since j_0 is an even function of its argument, $j_0(k\tilde{n}_0 r)$ is an entire function of k of order one and finite type. By representing $y(r)$ in terms of j_0 via a transformation operator (29) it is seen that $y(r)$ also has this property and hence so does d. Furthermore, d is bounded as $k \to \infty$. For $k < 0$ d has the asymptotic behavior (55) with γ_0 replaced by $-\gamma_0$ and γ_1 replaced by $-\gamma_1$ and hence d is also bounded as $k \to -\infty$. By analyticity k is bounded on any compact subset of the real axis and therefore $d(k)$ is bounded on the real axis. Now assume that there are not an infinity number of (complex) zeros of $d(k)$. Then by Hadamard's factorization theorem $d(k)$ is of the form

$$d(k) = k^m e^{ak+b} \prod_{\ell=1}^{n} \left(1 - \frac{k}{k_\ell}\right) e^{k/k_\ell}$$

for integers m and n and constants a and b. But this contradicts the asymptotic behavior of $d(k)$. Hence $d(k)$ has an infinite number of (complex) zeros, that is, there exist an infinite number of transmission eigenvalues.

3B. *The existence and discreteness of real transmission eigenvalues, for real contrast of the same sign in D.*

We now turn our attention to the transmission eigenvalue problem (20)–(23). The main assumption in this section is that $\Im(n) = 0$ and that the contrast $n - 1$ does not change sign and is bounded away from zero inside D. Under this assumption it is now possible to write (20)–(23) as an equivalent eigenvalue problem for $u = w - v \in H_0^2(D)$ as solution of the fourth-order equation

$$(\Delta + k^2 n)\frac{1}{n-1}(\Delta + k^2)u = 0 \tag{56}$$

which in variational form, after integration by parts, is formulated as finding a function $u \in H_0^2(D)$ such that

$$\int_D \frac{1}{n-1}(\Delta u + k^2 u)(\Delta \bar{v} + k^2 n \bar{v})\, dx = 0 \quad \text{for all } v \in H_0^2(D). \tag{57}$$

The functions v and w are related to u through

$$v = -\frac{1}{k^2(n-1)}(\Delta u + k^2 u) \quad \text{and} \quad w = -\frac{1}{k^2(n-1)}(\Delta u + k^2 n u).$$

In our discussion we must distinguish between the two cases $n > 1$ and $n < 1$. To fix our ideas, we consider in details only the case where $n(x) - 1 \geq \delta > 0$ in D. (A similar analysis can be done for $1 - n(x) \geq \delta > 0$, see [Cakoni et al. 2010e; Cakoni and Haddar 2009]). Let us define

$$n_* = \inf_D(n) \quad \text{and} \quad n^* = \sup_D(n).$$

The following result was first obtained in [Colton et al. 2007] (see also [Cakoni et al. 2007]) and provides a Faber–Krahn type inequality for the first transmission eigenvalue.

Theorem 3.5. *Assume that* $1 < n_* \leq n(x) \leq n^* < \infty$. *Then*

$$k_1^2 > \frac{\lambda_1(D)}{n^*} \qquad (58)$$

where k_1^2 *is the smallest transmission eigenvalue and* $\lambda_1(D)$ *is the first Dirichlet eigenvalue of* $-\Delta$ *on* D.

Proof. Taking $v = u$ in (57) and using Green's theorem and the zero boundary value for u we obtain that

$$0 = \int_D \frac{1}{n-1}(\Delta u + k^2 u)(\Delta \bar{u} + k^2 n \bar{u})\, dx$$

$$= \int_D \frac{1}{n-1}|\Delta u + k^2 n u|^2\, dx + k^2 \int_D (|\nabla u|^2 - k^2 n |u|^2)\, dx. \quad (59)$$

Since $n - 1 \geq n_* - 1 > 0$, if

$$\int_D (|\nabla u|^2 - k^2 n |u|^2)\, dx \geq 0, \qquad (60)$$

then $\Delta u + k^2 n u = 0$ in D, which together with the fact $u \in H_0^2(D)$ implies that $u = 0$. Consequently we obtain that $w = v = 0$, whence k is not a transmission eigenvalue. But,

$$\inf_{u \in H_0^2(D)} \frac{(\nabla u, \nabla u)_{L^2(D)}}{(u,u)_{L^2(D)}} = \inf_{u \in H_0^1(D)} \frac{(\nabla u, \nabla u)_{L^2(D)}}{(u,u)_{L^2(D)}} = \lambda_1(D) \qquad (61)$$

where $(\cdot,\cdot)_{L^2(D)}$ denotes the L^2-inner product. Hence we have

$$\int_D (|\nabla u|^2 - k^2 n |u|^2)\, dx \geq \|u\|^2_{L^2(D)}(\lambda_1(D) - k^2 n^*).$$

Thus, (60) is satisfied whenever $k^2 \leq \lambda_1(D)/n^*$. Thus, we have shown that any transmission eigenvalue k (in particular the smallest transmission eigenvalue k_1), satisfies $k^2 > \lambda_1(D)/n^*$. $\square$

Remark 3.1. From Theorem 3.5 it follows that if $1 < n_* \leq n(x) \leq n^* < \infty$ in D and k_1 is the smallest transmission eigenvalue, then $n^* > \lambda_1(D)/k_1^2$ which provides a lower bound for $\sup_D(n)$.

To understand the structure of the interior transmission eigenvalue problem we first observe that, setting $k^2 := \tau$, (57) can be written as

$$\mathbb{T}u - \tau \mathbb{T}_1 u + \tau^2 \mathbb{T}_2 u = 0, \qquad (62)$$

where $\mathbb{T}: H_0^2(D) \to H_0^2(D)$ is the bounded, positive definite self-adjoint operator defined by means of the Riesz representation theorem:

$$(\mathbb{T}u, v)_{H^2(D)} = \int_D \frac{1}{n-1} \Delta u\, \Delta \bar{v}\, dx \quad \text{for all } u, v \in H_0^2(D)$$

(note that the $H^2(D)$ norm of a field with zero Cauchy data on ∂D is equivalent to the $L^2(D)$ norm of its Laplacian), $\mathbb{T}_1: H_0^2(D) \to H_0^2(D)$ is the bounded compact self-adjoint operator defined by means of the Riesz representation theorem:

$$\begin{aligned}(\mathbb{T}_1 u, v)_{H^2(D)} &= -\int_D \frac{1}{n-1}(\Delta u\, \bar{v} + u\, \Delta \bar{v})\, dx - \int_D \Delta u\, \bar{v}\, dx \\ &= -\int_D \frac{1}{n-1}(\Delta u\, \bar{v} + u\, \Delta \bar{v})\, dx + \int_D \nabla u \cdot \nabla \bar{v}\, dx \quad (63)\end{aligned}$$

for all $u, v \in H_0^2(D)$, and $\mathbb{T}_2: H_0^2(D) \to H_0^2(D)$ is the bounded compact nonnegative self-adjoint operator defined by mean of the Riesz representation theorem

$$(\mathbb{T}_2 u, v)_{H^2(D)} = \int_D \frac{n}{n-1} u\, \bar{v}\, dx \quad \text{for all } u, v \in H_0^2(D)$$

(compactness of $\mathbb{T}_1$ and $\mathbb{T}_2$ is a consequence of the compact embedding of $H_0^2(D)$ and $H_0^1(D)$ in $L^2(D)$). Since $\mathbb{T}^{-1}$ exists we have that (62) becomes

$$u - \tau \mathbb{K}_1 u + \tau^2 \mathbb{K}_2 u = 0, \quad (64)$$

where the compact self-adjoint operators

$$\mathbb{K}_1 : H_0^2(D) \to H_0^2(D) \quad \text{and} \quad \mathbb{K}_2 : H_0^2(D) \to H_0^2(D)$$

are given by

$$\mathbb{K}_1 = \mathbb{T}^{-1/2} \mathbb{T}_1 \mathbb{T}^{-1/2},$$
$$\mathbb{K}_2 = \mathbb{T}^{-1/2} \mathbb{T}_2 \mathbb{T}^{-1/2}.$$

(Note that if A is a bounded, positive and self-adjoint operator on a Hilbert space U, the operator $A^{1/2}$ is defined by $A^{1/2} = \int_0^\infty \lambda^{1/2} dE_\lambda$ where dE_λ is the spectral measure associated with A). Hence, setting

$$U := (u, \tau \mathbb{K}_2^{1/2} u),$$

the interior transmission eigenvalue problem becomes the eigenvalue problem

$$\left(\mathbb{K} - \frac{1}{\tau}I\right) U = 0, \quad U \in H_0^2(D) \times H_0^2(D)$$

for the compact nonselfadjoint operator $K : H_0^2(D) \times H_0^2(D) \to H_0^2(D) \times H_0^2(D)$ given by

$$K := \begin{pmatrix} \mathbb{K}_1 & -\mathbb{K}_2^{1/2} \\ \mathbb{K}_2^{1/2} & 0 \end{pmatrix}.$$

Note that although the operators in each term of the matrix are selfadjoint the matrix operator K is not. This expression for K clearly reveals that the transmission eigenvalue problem is nonselfadjoint. However, from the discussion above we obtain a simpler proof of the following result previously proved in [Colton et al. 1989; Colton and Päivärinta 2000; Rynne and Sleeman 1991] (see also [Colton and Kress 1998]) using analytic Fredholm theory.

Theorem 3.6. *The set of real transmission eigenvalues is at most discrete with $+\infty$ as the only (possible) accumulation point. Furthermore, the multiplicity of each transmission eigenvalue is finite.*

The nonselfadjointness nature of the interior transmission eigenvalue problem calls for new techniques to prove the existence of transmission eigenvalues. For this reason the existence of transmission eigenvalues remained an open problem until it was shown in [Päivärinta and Sylvester 2008] that for large enough index of refraction n there exits at least one transmission eigenvalue. The existence of transmission eigenvalues was completely resolved in [Cakoni et al. 2010e], where the existence of an infinite set of transmission eigenvalues was proven only under the assumption that $n > 1$ or $0 < n < 1$. We present the proof given there. To this end we return to the variational formulation (57). Using the Riesz representation theorem we now define the bounded linear operators $\mathbb{A}_\tau : H_0^2(D) \to H_0^2(D)$ and $\mathbb{B} : H_0^2(D) \to H_0^2(D)$ by

$$(\mathbb{A}_\tau u, v)_{H^2(D)} = \int_D \frac{1}{n-1}\big((\Delta u + \tau u)(\Delta \bar{v} + \tau \bar{v}) + \tau^2 u \bar{v}\big)\, dx \qquad (65)$$

and

$$(\mathbb{B} u, v)_{H^2(D)} = \int_D \nabla u \cdot \nabla \bar{v}\, dx. \qquad (66)$$

Obviously, both operators $\mathbb{A}_\tau$ and $\mathbb{B}$ are self-adjoint. Furthermore, since the sesquilinear form $\mathcal{A}_\tau$ is a coercive sesquilinear form on $H_0^2(D) \times H_0^2(D)$, the operator $\mathbb{A}_\tau$ is positive definite and hence invertible. Indeed, since

$$\frac{1}{n(x)-1} > \frac{1}{n^*-1} = \gamma > 0$$

almost everywhere in D, we have

$$\begin{aligned}(\mathbb{A}_\tau u, v)_{H^2(D)} &\geq \gamma\|\Delta u + \tau u\|_{L^2}^2 + \tau^2\|u\|_{L^2}^2\\
&\geq \gamma\|\Delta u\|_{L^2}^2 - 2\gamma\tau\|\Delta u\|_{L^2}\|u\|_{L^2} + (\gamma+1)\tau^2\|u\|_{L^2}^2\\
&= \epsilon\left(\tau\|u\|_{L^2} - \frac{\gamma}{\epsilon}\|\Delta u\|_{L^2(D)}\right)^2 + \left(\gamma - \frac{\gamma^2}{\epsilon}\right)\|\Delta u\|_{L^2(D)}^2\\
&\qquad\qquad\qquad\qquad\qquad\qquad + (1+\gamma-\epsilon)\tau^2\|u\|_{L^2}^2\\
&\geq \left(\gamma - \frac{\gamma^2}{\epsilon}\right)\|\Delta u\|_{L^2(D)}^2 + (1+\gamma-\epsilon)\tau^2\|u\|_{L^2}^2 \qquad(67)\end{aligned}$$

for some $\gamma < \epsilon < \gamma+1$. Furthermore, since $\nabla u \in H_0^1(D)^2$, using the Poincaré inequality we have that

$$\|\nabla u\|_{L^2(D)}^2 \leq \frac{1}{\lambda_1(D)}\|\Delta u\|_{L^2(D)}^2 \qquad(68)$$

where $\lambda_1(D)$ is the first Dirichlet eigenvalue of $-\Delta$ on D. Hence we can conclude that

$$(\mathbb{A}_\tau u, u)_{H^2(D)} \geq C_\tau \|u\|_{H^2(D)}^2$$

for some positive constant C_τ. We now consider the operator $\mathbb{B}$. By definition $\mathbb{B}$ is a nonnegative operator and furthermore, since $H_0^1(D)$ is compactly embedded in $L^2(D)$ and $\nabla u \in H_0^1(D)$, we can conclude that $\mathbb{B}: H_0^2(D) \to H_0^2(D)$ is a compact operator. Finally, it is obvious by definition that the mapping $\tau \to \mathbb{A}_\tau$ is continuous from $(0, +\infty)$ to the set of self-adjoint positive definite operators. In terms of the above operators we can rewrite (57) as

$$(\mathbb{A}_\tau u - \tau \mathbb{B} u, v)_{H^2(D)} = 0 \quad \text{for all } v \in H_0^2(D), \qquad(69)$$

which means that k is a transmission eigenvalue if and only if $\tau := k^2$ is such that the kernel of the operator $\mathbb{A}_\tau u - \tau \mathbb{B}$ is not trivial. In order to analyze the kernel of this operator we consider the auxiliary generalized eigenvalue problems

$$\mathbb{A}_\tau u - \lambda(\tau)\mathbb{B} u = 0 \quad u \in H_0^2(D). \qquad(70)$$

It is known [Cakoni and Haddar 2009] that for a fixed τ there exists an increasing sequence $\{\lambda_j(\tau)\}_{j=1}^\infty$ of positive eigenvalues of the generalized eigenvalue problem (70), such that $\lambda_j(\tau) \to +\infty$ as $j \to +\infty$. Furthermore, these eigenvalues satisfy the min-max principle

$$\lambda_j(\tau) = \min_{W \subset \mathcal{U}_j} \max_{u \in W\setminus\{0\}} \frac{(\mathbb{A}_\tau u, u)}{(\mathbb{B} u, u)} \qquad(71)$$

where $\mathcal{U}_j$ denotes the set of all j dimensional subspaces W of $H_0^2(D)$ such that $W \cap \ker(\mathbb{B}) = \{0\}$, which ensures that $\lambda_j(\tau)$ depends continuously on $\tau \in (0, \infty)$.

In particular, a transmission eigenvalue $k > 0$ is such that $\tau := k^2$ solves $\lambda(\tau) - \tau = 0$ where $\lambda(\tau)$ is an eigenvalue corresponding to (70). Thus, to prove that transmission eigenvalues exist we use the following theorem:

Theorem 3.7 [Cakoni and Haddar 2009]. *Let $\tau \mapsto \mathbb{A}_\tau$ be a continuous mapping from $]0, \infty[$ to the set of self-adjoint and positive definite bounded linear operators on a Hilbert space $H_0^2(D)$ and let $\mathbb{B}$ be a self-adjoint and non negative compact bounded linear operator on $H_0^2(D)$. We assume that there exists two positive constants $\tau_0 > 0$ and $\tau_1 > 0$ such that*

(1) $\mathbb{A}_{\tau_0} - \tau_0 \mathbb{B}$ *is positive on* $H_0^2(D)$,

(2) $\mathbb{A}_{\tau_1} - \tau_1 \mathbb{B}$ *is non positive on a m-dimensional subspace W_m of $H_0^2(D)$.*

Then each of the equations $\lambda_j(\tau) = \tau$ for $j = 1, \ldots, k$, has at least one solution in $[\tau_0, \tau_1]$ where $\lambda_j(\tau)$ is the j-th eigenvalue (counting multiplicity) of the generalized eigenvalue problem (70).

Now we are ready to prove the existence theorem.

Theorem 3.8. *Assume that $1 < n_* \leq n(x) \leq n^* < \infty$. There exists an infinite set of real transmission eigenvalues with $+\infty$ as the only accumulation point.*

Proof. First we recall that from Theorem 3.5 we have that as long as $0 < \tau_0 \leq \lambda_1(D)/n^*$ the operator $\mathbb{A}_{\tau_0} u - \tau_0 \mathbb{B}$ is positive on $H_0^2(D)$, whence the assumption 1. of Theorem 3.7 is satisfied for such τ_0. Next let k_{1,n_*} be the first transmission eigenvalue for the ball B_1 of radius one, that is, $B_1 := \{x \in \mathbb{R}^d : |x| < 1\}$, $d = 2, 3$, and constant index of refraction n_* (i.e., corresponding to (24)–(27) for $B := B_1$ and $n(r) := n_*$). This transmission eigenvalue is the first zero of

$$W(k) = \det \begin{pmatrix} j_0(k) & j_0(k\sqrt{n_*}) \\ -j_0'(k) & -\sqrt{n_*}\, j_0'(k\sqrt{n_*}) \end{pmatrix} = 0 \quad \text{in } \mathbb{R}^3 \qquad (72)$$

where j_0 is the spherical Bessel function of order zero, or

$$W(k) = \det \begin{pmatrix} J_0(k) & J_0(k\sqrt{n_*}) \\ -J_0'(k) & -\sqrt{n_*}\, J_0'(k\sqrt{n_*}) \end{pmatrix} = 0 \quad \text{in } \mathbb{R}^2 \qquad (73)$$

where J_0 is the Bessel function of order zero (if the first zero of the above determinant is not the first transmission eigenvalue, the latter will be a zero of a similar determinant corresponding to higher-order Bessel functions or spherical Bessel functions). By a scaling argument, it is obvious that $k_{\epsilon,n_*} := k_{1,n_*}/\epsilon$ is the first transmission eigenvalue corresponding to the ball of radius $\epsilon > 0$ with index of refraction n_*. Now take $\epsilon > 0$ small enough such that D contains

$m := m(\epsilon) \geq 1$ disjoint balls $B_\epsilon^1, B_\epsilon^2, \ldots, B_\epsilon^m$ of radius ϵ, that is, $\overline{B}_\epsilon^j \subset D$, $j = 1, \ldots, m$, and $\overline{B}_\epsilon^j \cap \overline{B}_\epsilon^i = \emptyset$ for $j \neq i$. Then $k_{\epsilon,n_*} := k_{1,n_*}/\epsilon$ is the first transmission eigenvalue for each of these balls with index of refraction n_* and let $u^{B_\epsilon^j, n_*} \in H_0^2(B_\epsilon^j)$, $j = 1, \ldots, m$ be the corresponding eigenfunctions. We have $u^{B_\epsilon^j, n_*} \in H_0^2(B_\epsilon^j)$ and

$$\int_{B_\epsilon^j} \frac{1}{n_* - 1} (\Delta u^{B_\epsilon^j, n_*} + k_{\epsilon,n_*}^2 u^{B_\epsilon^j, n_*})(\Delta \bar{u}^{B_\epsilon^j, n_*} + k_{\epsilon,n_*}^2 n_* \bar{u}^{B_\epsilon^j, n_*}) \, dx = 0. \quad (74)$$

The extension by zero $\tilde{u}^j$ of $u^{B_\epsilon^j, n_*}$ to the whole D is obviously in $H_0^2(D)$ due to the boundary conditions on $\partial B_{\epsilon,n_*}^j$. Furthermore, the vectors $\{\tilde{u}^1, \tilde{u}^2, \ldots, \tilde{u}^m\}$ are linearly independent and orthogonal in $H_0^2(D)$ since they have disjoint supports and from (74) we have that, for $j = 1, \ldots, m$,

$$0 = \int_D \frac{1}{n_* - 1} (\Delta \tilde{u}^j + k_{\epsilon,n_*}^2 \tilde{u}^j)(\Delta \bar{\tilde{u}}^j + k_{\epsilon,n_*}^2 n_* \bar{\tilde{u}}^j) \, dx \quad (75)$$

$$= \int_D \frac{1}{n_* - 1} |\Delta \tilde{u}^j + k_{\epsilon,n_*}^2 \tilde{u}^j|^2 \, dx + k_{\epsilon,n_*}^4 \int_D |\tilde{u}^j|^2 \, dx - k_{\epsilon,n_*}^2 \int_D |\nabla \tilde{u}^j|^2 \, dx.$$

Let W_m be the m-dimensional subspace of $H_0^2(D)$ spanned by $\{\tilde{u}^1, \tilde{u}^2, \ldots, \tilde{u}^m\}$. Since each $\tilde{u}^j$, $j = 1, \ldots, m$ satisfies (75) and they have disjoint supports, we have that for $\tau_1 := k_{\epsilon,n_*}^2$ and for every $\tilde{u} \in \mathcal{U}$

$$(\mathbb{A}_{\tau_1} \tilde{u} - \tau_1 \mathbb{B} \tilde{u}, \tilde{u})_{H_0^2(D)}$$

$$= \int_D \frac{1}{n-1} |\Delta \tilde{u} + \tau_1 \tilde{u}|^2 \, dx + \tau_1^2 \int_D |\tilde{u}|^2 \, dx - \tau_1 \int_D |\nabla \tilde{u}|^2 \, dx$$

$$\leq \int_D \frac{1}{n_* - 1} |\Delta \tilde{u} + \tau_1 \tilde{u}|^2 \, dx + \tau_1^2 \int_D |\tilde{u}|^2 \, dx - \tau_1 \int_D |\nabla \tilde{u}|^2 \, dx = 0. \quad (76)$$

Thus assumption (2) of Theorem 3.7 is also satisfied, so we can conclude that there are $m(\epsilon)$ transmission eigenvalues (counting multiplicity) inside $[\tau_0, k_{\epsilon,n_*}]$. Note that $m(\epsilon)$ and k_{ϵ,n_*} both go to $+\infty$ as $\epsilon \to 0$. Since the multiplicity of each eigenvalue is finite we have shown, by letting $\epsilon \to 0$, that there exists a countably infinite set of transmission eigenvalues that accumulate at ∞. $\square$

In a similar way it is possible to prove the following theorem.

Theorem 3.9 [Cakoni et al. 2010e]. *Assume that $0 < n_* \leq n(x) \leq n^* < 1$. There exists an infinite set of real transmission eigenvalues with $+\infty$ as the only accumulation point.*

The proof of the existence of transmission eigenvalues given above provides a framework to obtain lower and upper bounds for the first transmission eigenvalue. To this end denote by $k_1(n, D) > 0$ the first real transmission eigenvalue corresponding to n and D. From the proof of Theorem 3.8 it is easy to see the following monotonicity results for the first transmission eigenvalue (see [Cakoni et al. 2010e] for the details of the proof).

Theorem 3.10. *Let $n_* = \inf_D(n)$ and $n^* = \sup_D(n)$, and B_1 and B_2 be two balls such that $B_1 \subset D$ and $D \subset B_2$.*

(i) *If the index of refraction $n(x)$ satisfies $1 < n_* \leq n(x) \leq n^* < \infty$, then*

$$0 < k_1(n^*, B_2) \leq k_1(n^*, D) \leq k_1(n(x), D) \leq k_1(n_*, D) \leq k_1(n_*, B_1). \quad (77)$$

(ii) *If the index of refraction $n(x)$ satisfies $0 < n_* \leq n(x) \leq n^* < 1$, then*

$$0 < k_1(n_*, B_2) \leq k_1(n_*, D) \leq k_1(n(x), D) \leq k_1(n^*, D) \leq k_1(n^*, B_1). \quad (78)$$

We remark that from the proof of Theorem 3.10 it is easy to see that for a fixed D the monotonicity result

$$k_j(n^*, D) \leq k_j(n(x), D) \leq k_j(n_*, D)$$

holds for all transmission eigenvalues k_j such that $\tau := k_j^2$ is solution of any of $\lambda_j(\tau) - \tau = 0$. Theorem 3.10 shows in particular that for constant index of refraction the first transmission eigenvalue $k_1(n, D)$ as a function of n for D fixed is monotonically increasing if $n > 1$ and is monotonically decreasing if $0 < n < 1$. In fact in [Cakoni et al. 2010a] it is shown that this monotonicity is strict which leads to the following uniqueness result of the constant index of refraction in terms of the first transmission eigenvalue.

Theorem 3.11. *The constant index of refraction n is uniquely determined from a knowledge of the corresponding smallest transmission eigenvalue $k_1(n, D) > 0$ provided that it is known a priori that either $n > 1$ or $0 < n < 1$.*

Proof. Here, we show the proof for the case of $n > 1$ (see [Cakoni et al. 2010a] for the case of $0 < n < 1$). Assume two homogeneous media with constant index of refraction n_1 and n_2 such that $1 < n_1 < n_2$, and let $u_1 := w_1 - v_1$, where w_1, v_1 is the nonzero solution of (20)–(23) with $n(x) := n_1$ corresponding to the first transmission eigenvalue $k_1(n_1, D)$. Now, setting $\tau_1 = k_1(n_1, D)$ and after normalizing u_1 such that $\nabla u_1 = 1$, we have

$$\frac{1}{n_1-1}\|\Delta u_1 + \tau_1 u_1\|_{L^2(D)}^2 + \tau_1^2\|u_1\|_{L^2(D)}^2 = \tau_1 = \lambda(\tau_1, n_1)$$

Furthermore, we have

$$\frac{1}{n_2-1}\|\Delta u + \tau u\|_{L^2(D)}^2 + \tau^2\|u\|_{L^2(D)}^2 < \frac{1}{n_1-1}\|\Delta u + \tau u\|_D^2 + \tau^2\|u\|_{L^2(D)}^2$$

for all $u \in H_0^2(D)$ such that $\|\nabla u\|_D = 1$ and all $\tau > 0$. In particular, for $u = u_1$ and $\tau = \tau_1$,

$$\frac{1}{n_2-1}\|\Delta u_1 + \tau_1 u_1\|_{L^2(D)}^2 + \tau_1^2\|u_1\|_{L^2(D)}^2$$
$$< \frac{1}{n_1-1}\|\Delta u_1 + \tau_1 u_1\|_{L^2(D)}^2 + \tau_1^2\|u_1\|_{L^2(D)}^2 = \lambda(\tau_1, n_1).$$

But

$$\lambda(\tau_1, n_2) \leq \frac{1}{n_2-1}\|\Delta u_1 + \tau_1 u_1\|_{L^2(D)}^2 + \tau_1^2\|u_1\|_{L^2(D)}^2 < \lambda(\tau_1, n_1)$$

and hence for this τ_1 we have a strict inequality, that is,

$$\lambda(\tau_1, n_2) < \lambda(\tau_1, n_1). \tag{79}$$

Obviously (79) implies the first zero τ_2 of $\lambda(\tau, n_2) - \tau = 0$ is such that $\tau_2 < \tau_1$ and therefore we have that $k_1(n_2, D) < k_1(n_1, D)$ for the first transmission eigenvalues $k_1(n_1, D)$ and $k_1(n_2, D)$ corresponding to n_1 and n_2, respectively. Hence we have shown that if $n_1 > 1$ and $n_2 > 1$ are such $n_1 \neq n_2$ then $k_1(n_1, D) \neq k_1(n_2, D)$, which proves uniqueness. □

3C. *The case of inhomogeneous media with cavities.* Motivated by a recent application of transmission eigenvalues to detect cavities inside dielectric materials [Cakoni et al. 2008], we now discuss briefly the structure of transmission eigenvalues for the case of a nonabsorbing inhomogeneous medium with cavities, that is, inhomogeneous medium D with regions $D_0 \subset D$ where the index of refraction is the same as the background medium. The interior transmission problem for inhomogeneous medium with cavities is investigated in [Cakoni et al. 2010b; 2010e; Cossonnière and Haddar 2011], and is also the first attempt to relax the aforementioned assumptions on the contrast. More precisely, inside D we consider a region $D_0 \subset D$ which can possibly be multiply connected such that $\mathbb{R}^d \setminus \overline{D}_0$, $d = 2, 3$ is connected and assume that its boundary ∂D_0 is piece-wise smooth. Here ν denotes the unit outward normal to ∂D and ∂D_0. Now we consider the interior transmission eigenvalue problem (20)–(23) with $n \in L^\infty(D)$ a real valued function such that $n \geq c > 0$, $n = 1$ in D_0 and $n - 1 \geq \tilde{c} > 0$ or $1 - n \geq \tilde{c} > 0$ almost everywhere in $D \setminus \overline{D}_0$. In particular, $1/|n-1| \in L^\infty(D \setminus \overline{D}_0)$. Following the analytic framework developed in [Cakoni et al. 2010b], we introduce the Hilbert space

$$V_0(D, D_0, k) := \{u \in H_0^2(D) \text{ such that } \Delta u + k^2 u = 0 \text{ in } D_0\}$$

equipped with the $H^2(D)$ scalar product and look for the solution v and w both in $L^2(D)$ such that $u = w - v$ in $V_0(D, D_0, k)$. It is shown in [Cakoni et al.

2010b] that (20)–(23), with n satisfying the above assumptions, can be written in the variational form

$$\int_{D\setminus\bar{D}_0} \frac{1}{n-1}(\Delta+k^2)u\,(\Delta+k^2)\bar{\psi}\,dx + k^2\int_{D\setminus\bar{D}_0}(\Delta u+k^2 u)\,\bar{\psi}\,dx = 0 \quad (80)$$

for all $\psi \in V_0(D, D_0, k)$. Next let us define the following bounded sesquilinear forms on $V_0(D, D_0, k) \times V_0(D, D_0, k)$:

$$\mathcal{A}(u,\psi) = \pm\int_{D\setminus\bar{D}_0} \frac{1}{n-1}(\Delta u\,\Delta\bar\psi + \nabla u\cdot\nabla\bar\psi + u\,\bar\psi)\,dx$$
$$+ \int_{D_0}(\nabla u\cdot\nabla\bar\psi + u\,\bar\psi)\,dx \quad (81)$$

and

$$\mathcal{B}_k(u,\psi) = \pm k^2\int_{D\setminus\bar{D}_0}\frac{1}{n-1}(u(\Delta\bar\psi+k^2\bar\psi)+(\Delta u+k^2 nu)\bar\psi)dx$$
$$\mp \int_{D\setminus\bar{D}_0}\frac{1}{n-1}(\nabla u\cdot\nabla\bar\psi+u\,\bar\psi)\,dx - \int_{D_0}(\nabla u\cdot\nabla\bar\psi+u\,\bar\psi)\,dx, \quad (82)$$

where the upper sign corresponds to the case when $n-1 \geq \tilde{c} > 0$ and the lower sign corresponds to the case when $1-n \geq \tilde{c} > 0$ almost everywhere in $D\setminus\bar{D}_0$. Hence k is a transmission eigenvalue if and only if the homogeneous problem

$$\mathcal{A}(u_0,\psi) + \mathcal{B}_k(u_0,\psi) = 0 \quad \text{for all } \psi \in V_0(D, D_0, k) \quad (83)$$

has a nonzero solution. Let $A_k : V_0(D, D_0, k) \to V_0(D, D_0, k)$ and B_k be the self-adjoint operators associated with $\mathcal{A}$ and $\mathcal{B}_k$, respectively, by using the Riesz representation theorem. In [Cakoni et al. 2010b] it is shown that the operator $A_k : V_0(D, D_0, k) \to V_0(D, D_0, k)$ is positive definite, that is,

$$A_k^{-1} : V_0(D, D_0, k) \to V_0(D, D_0, k)$$

exists, and the operator $B_k : V_0(D, D_0, k) \to V_0(D, D_0, k)$ is compact. Hence we can define the operator $A_k^{-1/2}$ which is also bounded, positive definite and self-adjoint. Thus (83) is equivalent to finding $u \in V_0(D, D_0, k)$ such that

$$u + A_k^{-1/2} B_k A_k^{-1/2} u = 0. \quad (84)$$

In particular, it is obvious that k is a transmission eigenvalue if and only if the operator

$$I_k + A_k^{-1/2} B_k A_k^{-1/2} : V_0(D, D_0, k) \to V_0(D, D_0, k) \quad (85)$$

has a nontrivial kernel where I_k is the identity operator on $V_0(D, D_0, k)$. To avoid dealing with function spaces depending on k we introduce the orthogonal projection operator P_k from $H_0^2(D)$ onto $V_0(D, D_0, k)$ and the corresponding injection $R_k : V_0(D, D_0, k) \to H_0^2(D)$. Then one easily sees that $I_k + A_k^{-1/2} B_k A_k^{-1/2}$ is injective on $V_0(D, D_0, k)$ if and only if

$$I + R_k A_k^{-1/2} B_k A_k^{-1/2} P_k : H_0^2(D) \to H_0^2(D) \tag{86}$$

is injective. Furthermore, as discussed in [Cakoni et al. 2010b],

$$T_k := R_k A_k^{-1/2} B_k A_k^{-1/2} P_k : H_0^2(D) \to H_0^2(D)$$

is a compact operator and the mapping $k \to R_k A_k^{-1/2} B_k A_k^{-1/2} P_k$ is continuous. Therefore, from the max-min principle for the eigenvalues $\lambda(k)$ of the compact and self-adjoint operator $R_k A_k^{-1/2} B_k A_k^{-1/2} P_k$ we can conclude that $\lambda(k)$ is a continuous function of k. Finally, it is clear that the multiplicity of a transmission eigenvalue is finite since it corresponds to the multiplicity of the eigenvalue $\lambda(k) = -1$. Now the problem is brought into the right framework, similar to the one in Section 3B, to prove the discreteness and existence of transmission eigenvalues. Using the analytic Fredholm theory [Colton and Kress 1998], it is proven in [Cakoni et al. 2010b] that real transmission eigenvalues form at most a discrete set with $+\infty$ as the only possible accumulation point. Concerning the existence of transmission eigenvalues, it is now possible to apply a similar procedure as in Section 3B. In particular, we can use a slightly modified version of Theorem 3.7 (see also Theorem 4.7) to show that each equation $\lambda_j(k) + 1 = 0$ has at least one solution, which are transmission eigenvalues, where $\{\lambda_j(k)\}_{j=0}^\infty$ is the increasing sequence of eigenvalues of the auxiliary eigenvalue problem

$$(I - \lambda(k) R_k A_k^{-1/2} B_k A_k^{-1/2} P_k) u = 0.$$

In the next theorem, we set $n_* := \inf_{D \setminus \bar{D}_0}(n)$, $n^* := \sup_{D \setminus \bar{D}_0}(n)$ and recall that $\lambda_1(D)$ denotes the first Dirichlet eigenvalue for $-\Delta$ on D.

Theorem 3.12 [Cakoni et al. 2010b; 2010e]. *Let $n \in L^\infty(D)$, $n = 1$ in D_0 and assume that n satisfies either $1 < n_* \leq n(x) \leq n^* < \infty$ or $0 < n_* \leq n(x) \leq n^* < 1$ on $D \setminus \bar{D}_0$. Then the set of real transmission eigenvalues is discrete with no finite accumulation points, and there exist infinitely many transmission eigenvalues accumulating at $+\infty$.*

As byproduct of the proof of Theorem 3.12 it is possible to show the following monotonicity result for the first transmission eigenvalue. For a fixed D, denote by $k_1(D_0, n)$ the first transmission eigenvalue corresponding to the void D_0 and the index of refraction n.

Theorem 3.13 [Cossonnière and Haddar 2011, Theorem 2.10]. *If $D_0 \subseteq \tilde{D}_0$ and $n(x) \leq \tilde{n}(x)$ for almost every $x \in D$ then*

(i) $k_1(D_0, \tilde{n}) \leq k_1(\tilde{D}_0, n)$ *if* $n - 1 \geq \alpha > 0$ *and* $\tilde{n} - 1 \geq \tilde{\alpha} > 0$

(ii) $k_1(D_0, n) \leq k_1(\tilde{D}_0, \tilde{n})$ *if* $1 - n \geq \beta > 0$ *and* $1 - \tilde{n} \geq \tilde{\beta} > 0$.

These results are useful in nondestructive testing to detect voids inside inhomogeneous nonabsorbing media using transmission eigenvalues [Cakoni et al. 2008].

We end this section by remarking that the study of transmission eigenvalue problem in the general case of absorbing media and background has been initiated in [Cakoni et al. 2012] where it was proven that the set of transmission eigenvalues on the open right complex half-plane is at most discrete provided that the contrast in the real part of the index of refraction does not change sign in D. Furthermore using perturbation theory it is possible to show that if the absorption in the inhomogeneous medium and (possibly) in the background is small enough then there exist a finite number of complex transmission eigenvalues each near a real transmission eigenvalue associated with the corresponding nonabsorbing medium and background.

3D. Discussion.

The case of the contrast changing sign inside D. The crucial assumption in the above analysis is that the contrast does not change sign inside D, i.e., $n - 1$ is either positive or negative and bounded away from zero in D. Although using weighted Sobolev spaces it is possible to consider the case when $n - 1$ goes smoothly to zero at the boundary ∂D [Colton et al. 1989; Hickmann 2012; Serov and Sylvester 2012], the real interest is in investigating the case when $n - 1$ is allowed to change sign inside D. The question of discreteness of transmission eigenvalues in the latter case has been related to the uniqueness of the sound speed for the wave equation with arbitrary source, which is a question that arises in thermoacoustic imagining (Finch, personal communication). In the general case $n \geq c > 0$ with no assumptions on the sign of $n - 1$, the study of the transmission eigenvalue problem is completely open. However, recently in [Sylvester 2012] progress has been made in the study of discreteness of transmission eigenvalues under more relaxed assumptions on the contrast $n - 1$, namely requiring that $n - 1$ or $1 - n$ is positive only in a neighborhood of ∂D. More specifically:

Theorem 3.14 [Sylvester 2012]. *Suppose that there are real numbers*

$$m^* \geq m_* > 0$$

and a unit complex number $e^{i\theta}$ in the open right half-plane such that the following conditions are satisfied:

(1) $\Re(e^{i\theta}(n(x)-1)) > m_*$ in some neighborhood of ∂D or that $n(x)$ is real on all of D, and satisfies $n(x) - 1 \leq -m_*$ in some neighborhood of D.
(2) $|n(x) - 1| < m^*$ in all of D.
(3) $\Re(n(x)) \geq \delta > 0$ in all of D.

Then the spectrum of (20)–(23) (*i.e., the set of transmission eigenvalues*) *consists of a* (*possibly empty*) *discrete set of eigenvalues with finite dimensional generalized eigenspaces. Eigenspaces corresponding to different eigenvalues are linearly independent. The eigenvalues and the generalized eigenspaces depend continuously on n in the $L^\infty(D)$ topology.*

Sylvester uses the concept of upper triangular compact operator to prove the Fredholm property of the transmission eigenvalue problem and employs careful estimates to control solutions to the Helmholtz equation inside D by its values in a neighborhood of the boundary in order to show that the resolvent is not empty. The Fredholm property of the transmission eigenvalue problem can also be proven using an integral equation approach [Cossonnière 2011]. In Section 4B we present the proof of similar discreteness results for the transmission eigenvalue problems with $A \neq I$ based on a T-coercivity approach.

The location of transmission eigenvalues. Results concerning complex transmission eigenvalues for the problem (20)–(23) are limited to indicating eigenvalue free zones in the complex plane. A first attempt to localize transmission eigenvalues on the complex plane in done in [Cakoni et al. 2010a]. However to our knowledge the best result on location of transmission eigenvalues is given in [Hitrik et al. 2011a] where it is shown that almost all transmission eigenvalues k^2 are confined to a parabolic neighborhood of the positive real axis. More specifically:

Theorem 3.15 ([Hitrik et al. 2011a]). *Assume that D has C^∞ boundary, $n \in C^\infty(\overline{D})$ and $1 < \alpha \leq n \leq \beta$. Then there exists a $0 < \delta < 1$ and $C > 1$ both independent of n (but depending on α and β) such that all transmission eigenvalues $\tau := k^2 \in \mathbb{C}$ with $|\tau| > C$ satisfies $\Re(\tau) > 0$ and $\Im(\tau) \leq C|\tau|^{1-\delta}$.*

We do not include the proof (see the original paper) since it employs an approach quite different from the analytical framework developed in this article. Note that although the transmission eigenvalue problem (20)–(23) has the structure of quadratic pencils of operators (62), it appears that available results on quadratic pencils [Markus 1988] are not applicable to the transmission eigenvalue problem due to the incorrect signs of the involved operators. We also remark that some rough estimates on complex eigenvalues for the general case of absorbing media and background are obtained in [Cakoni et al. 2012].

We close the first part of this expose on the transmission eigenvalue problem by noting that in [Hitrik et al. 2010] the discreteness and existence of transmission eigenvalue are investigated for the case of (20)–(23) where the Laplace operator is replaced by a higher-order differential operator with constant coefficient of even order. Such a framework is applicable to the Dirac system and the plate equation.

4. The transmission eigenvalue problem for anisotropic media

We continue our discussion of the interior transmission problem by considering in this section the case where $A \neq I$. We recall that the transmission eigenvalue problem now has the form

$$\nabla \cdot A(x)\nabla w + k^2 n w = 0 \quad \text{in } D, \tag{87}$$

$$\Delta v + k^2 v = 0 \quad \text{in } D, \tag{88}$$

$$w = v \quad \text{on } \partial D, \tag{89}$$

$$\frac{\partial w}{\partial \nu_A} = \frac{\partial v}{\partial \nu} \quad \text{on } \partial D, \tag{90}$$

where we assume that

$$A_* := \inf_{x \in D} \inf_{\substack{\xi \in \mathbb{R}^3 \\ |\xi|=1}} (\xi \cdot A(x)\xi) > 0, \quad A^* := \sup_{x \in D} \sup_{\substack{\xi \in \mathbb{R}^3 \\ |\xi|=1}} (\xi \cdot A(x)\xi) < \infty,$$

$$n_* := \inf_{x \in D} n(x) > 0, \quad n^* := \sup_{x \in D} n(x) < \infty. \tag{91}$$

The analysis of transmission eigenvalues for this configuration uses different approaches depending on whether $n = 1$ or $n \neq 1$. In particular, the case where $n(x) \equiv 1$, can be brought into a similar form to the problem discuss in Section 3B but for vector fields. Hence we first proceed with this case.

4A. *The case* $n = 1$. When $n = 1$ after making an appropriate change of unknown functions, we can write (87)–(90) in a similar form as in the case of $A = I$ presented in Section 3B (we follow the approach developed in [Cakoni et al. 2009]). Letting $N := A^{-1}$, in terms of new vector valued functions

$$\boldsymbol{w} = A\nabla w, \quad \text{and} \quad \boldsymbol{v} = \nabla v,$$

the problem above can be written as

$$\nabla(\nabla \cdot \boldsymbol{w}) + k^2 N \boldsymbol{w} = 0 \quad \text{in } D, \tag{92}$$

$$\nabla(\nabla \cdot \boldsymbol{v}) + k^2 \boldsymbol{v} = 0 \quad \text{in } D, \tag{93}$$

$$\nu \cdot \boldsymbol{w} = \nu \cdot \boldsymbol{v} \quad \text{on } \partial D, \tag{94}$$

$$\nabla \cdot \boldsymbol{w} = \nabla \cdot \boldsymbol{v} \quad \text{on } \partial D. \tag{95}$$

Equations (92) and (93) are respectively obtained after taking the gradient of (87) and (88). The problem (92)–(95) has a similar structure to that of (20)–(23) in the sense that the main operators appearing in (92)–(93) are the same. We therefore can analyze this problem by reformulating it as an eigenvalue problem for the fourth-order partial differential equation assuming that $(N - I)^{-1} \in L^\infty(D)$, which is equivalent to assuming that $(I - A)^{-1} \in L^\infty(D)$ (given the initial hypothesis made on A and since $N - I = A^{-1}(I - A)$).

A suitable function space setting is based on

$$H(\mathrm{div}, D) := \{u \in (L^2(D))^d : \nabla \cdot u \in L^2(D)\}, \quad d = 2, 3,$$
$$H_0(\mathrm{div}, D) := \{u \in H(\mathrm{div}, D) : v \cdot u = 0 \text{ on } \partial D\},$$

and

$$\mathcal{H}(D) := \{u \in H(\mathrm{div}, D) : \nabla \cdot u \in H^1(D)\},$$
$$\mathcal{H}_0(D) := \{u \in H_0(\mathrm{div}, D) : \nabla \cdot u \in H^1_0(D)\},$$

equipped with the scalar product $(u, v)_{\mathcal{H}(D)} := (u, v)_{L^2(D)} + (\nabla \cdot u, \nabla \cdot v)_{H^1(D)}$ and corresponding norm $\|\cdot\|_\mathcal{H}$.

A solution w, v of the interior transmission eigenvalue problem (92)–(95) is defined as $u \in (L^2(D))^d$ and $v \in (L^2(D))^d$ satisfying (92)–(93) in the distributional sense and such that $w - v \in \mathcal{H}_0(D)$. We therefore consider the following definition.

Definition 4.1. Transmission eigenvalues corresponding to (92)–(95) are the values of $k > 0$ for which there exist nonzero solutions $w \in L^2(D)$ and $v \in L^2(D)$ such that $w - v$ is in $\mathcal{H}_0(D)$.

Setting $u := w - v$, we first observe that $u \in \mathcal{H}_0(D)$ and

$$(\nabla \nabla \cdot + k^2 N)(N - I)^{-1}(\nabla \nabla \cdot u + k^2 u) = 0 \quad \text{in } D. \tag{96}$$

The latter can be written in the variational form

$$\int_D (N - I)^{-1}(\nabla \nabla \cdot u + k^2 u) \cdot (\nabla \nabla \cdot \bar{v} + k^2 N \bar{v}) \, dx = 0$$
$$\text{for all } v \in \mathcal{H}_0(D). \tag{97}$$

Consequently, $k > 0$ is a transmission eigenvalue if and only if there exists a nontrivial solution $u \in \mathcal{H}_0(D)$ of (97). We now sketch the main steps of the proof of discreteness and existence of real transmission eigenvalues highlighting the new aspects of (97). To this end we see that (97) can be written as an operator equation

$$\mathbb{A}_\tau u - \tau \mathbb{B} u = 0 \quad \text{and} \quad \tilde{\mathbb{A}}_\tau u - \tau \mathbb{B} u = 0 \quad \text{for } u \in \mathcal{H}_0(D). \tag{98}$$

Here the bounded linear operators $\mathbb{A}_\tau : \mathcal{H}_0(D) \to \mathcal{H}_0(D)$, $\tilde{\mathbb{A}}_\tau : \mathcal{H}_0(D) \to \mathcal{H}_0(D)$ and $\mathbb{B} : \mathcal{H}_0(D) \to \mathcal{H}_0(D)$ are the operators defined using the Riesz representation theorem for the sesquilinear forms $\mathcal{A}_\tau$, $\tilde{\mathcal{A}}$ and $\mathcal{B}$ defined by

$$\mathcal{A}_\tau(u,v) := \left((N-I)^{-1}(\nabla\nabla\cdot u + \tau u), (\nabla\nabla\cdot v + \tau v)\right)_D + \tau^2(u,v)_D, \quad (99)$$

$$\tilde{\mathcal{A}}_\tau(u,v) := \left(N(I-N)^{-1}(\nabla\nabla\cdot u + \tau u), (\nabla\nabla\cdot v + \tau v)\right)_D$$
$$+ (\nabla\nabla\cdot u, \nabla\nabla\cdot v)_D, \quad (100)$$

$$\mathcal{B}(u,v) := (\nabla\cdot u, \nabla\cdot v)_D, \quad (101)$$

where $(\cdot,\cdot)_D$ denotes the $L^2(D)$-inner product. Then one can prove (see also (67)):

Lemma 4.1 ([Cakoni et al. 2009]). *The operators $\mathbb{A}_\tau : \mathcal{H}_0(D) \to \mathcal{H}_0(D)$, $\tilde{\mathbb{A}}_\tau : \mathcal{H}_0(D) \to \mathcal{H}_0(D)$, $\tau > 0$ and $\mathbb{B} : \mathcal{H}_0(D) \to \mathcal{H}_0(D)$ are self-adjoint. Furthermore, $\mathbb{B}$ is a positive compact operator.*

If $(I-A)^{-1}A$ is a bounded positive definite matrix function on D, then $\mathbb{A}_\tau$ is a positive definite operator and

$$(\mathbb{A}_\tau u - \tau \mathbb{B} u, u)_{\mathcal{H}_0(D)} \geq \alpha \|u\|^2_{\mathcal{H}_0(D)} > 0$$

for all $0 < \tau < \lambda_1(D)A_*$ and $u \in \mathcal{H}_0(D)$.

If $(A-I)^{-1}$ is a bounded positive definite matrix function on D, then $\tilde{\mathbb{A}}_\tau$ is a positive definite operator and

$$\left(\tilde{\mathbb{A}}_\tau u - \tau \mathbb{B} u, u\right)_{\mathcal{H}_0(D)} \geq \alpha \|u\|^2_{\mathcal{H}_0(D)} > 0$$

for all $0 < \tau < \lambda_1(D)$ and $u \in \mathcal{H}_0(D)$. □

Note that the kernel of $\mathbb{B} : \mathcal{H}_0(D) \to \mathcal{H}_0(D)$ is given by

$$\text{Kernel}(\mathbb{B}) = \{u \in \mathcal{H}_0(D) \quad \text{such that} \quad u := \operatorname{curl}\varphi, \; \varphi \in H(\operatorname{curl}, D)\}.$$

To carry over the approach of Section 3B to our eigenvalue problem (98), we also need to consider the corresponding transmission eigenvalue problems for a ball with constant index of refraction. To this end, we recall that it can be shown by separation of variables [Cakoni and Kirsch 2010], that

$$a_0 \Delta w + k^2 w = 0 \quad \text{in} \quad B, \quad (102)$$

$$\Delta v + k^2 v = 0 \quad \text{in} \quad B, \quad (103)$$

$$w = v \quad \text{on} \quad \partial B, \quad (104)$$

$$a_0 \frac{\partial w}{\partial \nu} = \frac{\partial v}{\partial \nu} \quad \text{on} \quad \partial B_R, \quad (105)$$

has a countable discrete set of eigenvalues, where $B := B_R \subset \mathbb{R}^d$ is the ball of radius R centered at the origin and $a_0 > 0$ a constant different from one. We now have all the ingredients to proceed with the approach of Section 3B. Following exactly the lines of the proof of Theorem 3.8 it is now possible to show the existence of infinitely many transmission eigenvalues accumulating at infinity. The discreteness of real transmission eigenvalue can be obtained by using the analytic Fredholm theory as was done in [Cakoni et al. 2009] or alternatively following the proof of Theorem 3.6. As a byproduct of the proof we can also obtain estimates for the first transmission eigenvalue corresponding to the anisotropic medium. Let us denote by $k_1(A_*, B)$ and $k_1(A^*, B)$ the first transmission eigenvalue of (102)–(105) with index of refraction $a_0 := A_*$ and $a_0 := 1/A^*$, respectively. Then the following theorem holds.

Theorem 4.1. *Assume that either $A^* < 1$ or $A_* > 1$. Then problem (92)–(95) has an infinite countable set of real transmission eigenvalues with $+\infty$ as the only accumulation point. Furthermore, let $k_1(A(x), D)$ be the first transmission eigenvalue for (92)–(95) and B_1 and B_2 be two balls such that $B_1 \subset D$ and $D \subset B_2$, Then*

$$0 < k_1(A^*, B_2) \leq k_1(A^*, D) \leq k_1(A(x), D) \leq k_1(A_*, D) \leq k_1(A_*, B_1) \quad \text{if } A^* < 1,$$
$$0 < k_1(A_*, B_2) \leq k_1(A_*, D) \leq k_1(A(x), D) \leq k_1(A^*, D) \leq k_1(A^*, B_1) \quad \text{if } A_* > 1.$$

Note that A_* is the infimum of the lowest eigenvalue of the matrix A and A^* is the largest eigenvalue of the matrix A. We end this section by noting that we also have the following Faber–Krahn inequality similar to Theorem 3.5:

$$k_1^2(A(x), D) \geq \lambda_1(D) A_* \quad \text{if } A^* < 1,$$
$$k_1^2(A(x), D) \geq \lambda_1(D) \quad \text{if } A_* > 1,$$

where again $\lambda_1(D)$ is the first Dirichlet eigenvalue of $-\Delta$ in D.

4B. *The case $n \neq 1$.* The case $n \neq 1$ is treated in a different way from the two previous cases for $n = 1$ since now it is not possible to obtain a fourth-order formulation. In particular in this case, as will be seen soon, the natural variational framework for (87)–(90) is $H^1(D) \times H^1(D)$. Here, we define transmission eigenvalues as follows:

Definition 4.2. Transmission eigenvalues corresponding to (87)–(90) are the values of $k \in \mathbb{C}$ for which there exist nonzero solutions $w \in H^1(D)$ and $v \in H^1(D)$, where the equations (87) and (88) are satisfied in the distributional sense whereas the boundary conditions (89) and (90) are satisfied in the sense of traces in $H^{1/2}(\partial D)$ and $H^{-1/2}(\partial D)$, respectively.

This case has been subject of several investigations [Bonnet-BenDhia et al. 2011; Cakoni et al. 2002; Cakoni and Kirsch 2010]. Here we present the latest results on existence and discreteness of transmission eigenvalues. In particular, the existence of real transmission eigenvalues is shown only in the cases where the contrasts $A - I$ and $n - 1$ do not change sign in D (see Section 4B), whereas the discreteness of the set of transmission eigenvalues is shown under less restrictive conditions on the sign of the contrasts using a relatively simple approach known as T-coercivity. The latter is the subject of the discussion in the next section which follows [Bonnet-BenDhia et al. 2011].

Discreteness of transmission eigenvalues. The goal of this section is to prove discreteness of transmission eigenvalues under sign assumptions on the contrasts that hold only in the neighborhood $\mathcal{V}$ of the boundary ∂D (a result of this type is also mentioned in Section 3D for the case of $A = I$). To this end we use the T-coercivity approach introduced in [Bonnet-Ben Dhia et al. 2010] and [Chesnel 2012]. Following [Bonnet-BenDhia et al. 2011], we first observe that $(w, v) \in H^1(D) \times H^1(D)$ satisfies (87)–(88) if and only if $(w, v) \in X(D)$ satisfies the (natural) variational problem

$$a_k((w, v), (w', v')) = 0 \quad \text{for all } (w', v') \in X(D), \tag{106}$$

where

$$a_k((w, v), (w', v')) := (A\nabla w, \nabla w')_D - (\nabla v, \nabla v')_D - k^2 \big((nw, w')_D - (v, v')_D\big)$$

for all (w, v) and (w, v) in $X(D)$ and

$$X(D) := \{(w, v) \in H^1(D) \times H^1(D) \mid w - v \in H_0^1(D)\}.$$

With the help of the Riesz representation theorem, we define the operator $\mathcal{A}_k$ from $X(D)$ to $X(D)$ such that

$$(\mathcal{A}_k(w, v), (w', v'))_{H^1(D) \times H^1(D)} = a_k((w, v), (w', v'))$$

for all $((w, v), (w', v')) \in X(D) \times X(D)$. It is clear that $\mathcal{A}_k$ depends analytically on $k \in \mathbb{C}$. Moreover from the compact embedding of $X(D)$ into $L^2(D) \times L^2(D)$ one easily observes that

$$\mathcal{A}_k - \mathcal{A}_{k'} : X(D) \to X(D)$$

is compact for all k, k' in $\mathbb{C}$. In order to prove discreteness of the set of transmission eigenvalues, one only needs to prove the invertibility of $\mathcal{A}_k$ for one k in $\mathbb{C}$. For the latter, it would have been sufficient to prove that a_k is coercive for some k in $\mathbb{C}$. Unfortunately this cannot be true in general, but we can show that a_k is T-coercive which turns out to be sufficient for our purpose. The idea behind the T-coercivity method is to consider an equivalent formulation of (106) where a_k is replaced by a_k^T defined by

$$a_k^T((w,v),(w',v')) := a_k((w,v), T(w',v')) \tag{107}$$

for all $(w,v), (w',v') \in X(D)$, with T being an ad hoc isomorphism of $X(D)$. Indeed, $(w,v) \in X(D)$ satisfies

$$a_k((w,v),(w',v')) = 0 \quad \text{for all } (w',v') \in X(D)$$

if and only if it satisfies $a_k^T((w,v),(w',v')) = 0$ for all $(w',v') \in X(D)$. Assume that T and k are chosen so that a_k^T is coercive. Then using the Lax–Milgram theorem and the fact that T is an isomorphism of $X(D)$, one deduces that $\mathcal{A}_k$ is an isomorphism on $X(D)$. We shall apply this technique to prove the following lemma where here and in the sequel $\mathcal{V}(\partial D)$ denotes a neighborhood of the boundary ∂D inside D. To this end, we set

$$A_\star := \inf_{\substack{x \in \mathcal{V}(\partial D)}} \inf_{\substack{\xi \in \mathbb{R}^3 \\ |\xi|=1}} (\xi \cdot A(x)\xi) > 0, \quad A^\star := \sup_{\substack{x \in \mathcal{V}(\partial D)}} \sup_{\substack{\xi \in \mathbb{R}^3 \\ |\xi|=1}} (\xi \cdot A(x)\xi) < \infty, \tag{108}$$

$$n_\star := \inf_{\mathcal{V}(\partial D)} n(x) > 0, \qquad n^\star := \sup_{\mathcal{V}(\partial D)} n(x) < \infty.$$

The difference between the $\ast$-constants in (91) and $\star$-constants in (108) is that, in the first set of constants, the infimum and supremum are taken over the entire D, whereas in the second they are taken only over the neighborhood $\mathcal{V}$ of ∂D.

Lemma 4.2. *Assume that either $A(x) \leq A^\star I < I$ and $n(x) \leq n^\star < 1$, or $A(x) \geq A_\star I > I$ and $n(x) \geq n_\star > 1$ almost everywhere on $\mathcal{V}(\partial D)$. Then there exists $k = i\kappa$, with $\kappa \in \mathbb{R}$, such that the operator $\mathcal{A}_k$ is an isomorphism on $X(D)$.*

Proof. We consider first the case when $A(x) \leq A^\star I < I$ and $n(x) \leq n^\star < 1$ almost everywhere on $\mathcal{V}(\partial D)$. Introduce $\chi \in \mathcal{C}^\infty(\bar{D})$ a cut off function equal to 1 in a neighborhood of ∂D, with support in $\mathcal{V}(\partial D) \cap D$ and such that $0 \leq \chi \leq 1$, and consider the isomorphism ($T^2 = I$) of $X(D)$ defined by $T(w,v) = (w - 2\chi v, -v)$. We will prove that $a_{i\kappa}^T$ defined in (107) is coercive for some $\kappa \in \mathbb{R}$. For all $(w,v) \in X(D)$ one has

$$|a_{i\kappa}^T((w,v),(w,v))| = |(A\nabla w, \nabla w)_D + (\nabla v, \nabla v)_D - 2(A\nabla w, \nabla(\chi v))_D$$
$$+ \kappa^2 ((nw,w)_D + (v,v)_D - 2(nw, \chi v)_D)|. \tag{109}$$

Using Young's inequality, one can write, for all $\alpha > 0, \beta > 0, \eta > 0$,

$$2|(A\nabla w, \nabla(\chi v))_D|$$
$$\leq 2|(\chi A\nabla w, \nabla v)_V| + 2|(A\nabla w, \nabla(\chi)v)_V|$$
$$\leq \eta(A\nabla w, \nabla w)_V + \eta^{-1}(A\nabla v, \nabla v)_V \qquad (110)$$
$$+ \alpha(A\nabla w, \nabla w)_V + \alpha^{-1}(A\nabla(\chi)v, \nabla(\chi)v)_V,$$

$$2|(nw, \chi v)_D| \leq \beta(nw, w)_V + \beta^{-1}(nv, v)_V,$$

where again $(\cdot, \cdot)_\mathbb{O}$ for a generic bounded region $\mathbb{O} \subset \mathbb{R}^d$, $d = 2, 3$, denotes the $L^2(\mathbb{O})$-inner product. Substituting (110) into (109), one obtains

$$\left| a_{ik}^T((w,v),(w,v)) \right|$$
$$\geq (A\nabla w, \nabla w)_{D\setminus \overline{V}} + (\nabla v, \nabla v)_{D\setminus \overline{V}} + \kappa^2\big((nw, w)_{D\setminus \overline{V}} + (v, v)_{D\setminus \overline{V}}\big)$$
$$+ \big((1-\eta-\alpha)A\nabla w, \nabla w\big)_V + \big((I - \eta^{-1}A)\nabla v, \nabla v\big)_V$$
$$+ \kappa^2\big((1-\beta)nw, w\big)_V + \big((\kappa^2(1-\beta^{-1}n) - \sup_V |\nabla \chi|^2 A^* \alpha^{-1})v, v\big)_V.$$

Taking η, α and β such that $A^* < \eta < 1$, $n^* < \beta < 1$ and $0 < \alpha < 1 - \eta$, we obtain the coercivity of a_{ik}^T for κ large enough. This gives the desired result for the first case.

The case $A(x) \geq A_\star I > I$ and $n(x) \geq n_\star > 1$ almost everywhere on $\mathcal{V}(\partial D)$ can be treated in a similar way by using $T(w, v) := (w, -v + 2\chi w)$. □

We therefore we have the following theorem.

Theorem 4.2. *Assume that either $A(x) \leq A^\star I < I$ and $n(x) \leq n^\star < 1$, or $A(x) \geq A_\star I > I$ and $n(x) \geq n_\star > 1$ almost everywhere on $\mathcal{V}(\partial D)$. Then the set of transmission eigenvalues is discrete in $\mathbb{C}$.*

As another direct consequence of Lemma 4.2 and the compact embedding of $X(D)$ into $L^2(D) \times L^2(D)$, we remark that the operator $\mathcal{A}_k : X(D) \to X(D)$ is Fredholm for all $k \in \mathbb{C}$ provided that only $A(x) \leq A^\star I < I$ or $A(x) \geq A_\star I > I$ almost everywhere in $\mathcal{V}(\partial D)$. Consequently, with a stronger assumption on A, namely assuming that $A - I$ is either positive definite or negative definite in D, one can relax the conditions on n in order to prove discreteness of transmission eigenvalues. To this end, taking $w' = v' = 1$ in (106), we first notice that the transmission eigenvectors (w, v) (i.e., the solution of (87)–(88) corresponding to an eigenvalue k) satisfy $k^2 \int_D (nw - v) dx = 0$. This leads us to introduce the subspace of eigenvectors

$$Y(D) := \left\{ (w, v) \in X(D) \mid \int_D (nw - v) dx = 0 \right\}.$$

Now, suppose $\int_D (n-1)dx \neq 0$. Arguing by contradiction, one can prove the existence of a Poincaré constant $C_P > 0$ (which depends on D and also on n through $Y(D)$) such that

$$\|w\|_D^2 + \|v\|_D^2 \leq C_P(\|\nabla w\|_D^2 + \|\nabla v\|_D^2) \quad \text{for all } (w, v) \in Y(D). \quad (111)$$

Moreover, one can check that $k \neq 0$ is a transmission eigenvalue if and only if there exists a non trivial element $(w, v) \in Y(D)$ such that

$$a_k((w, v), (w', v')) = 0 \quad \text{for all } (w', v') \in Y(D).$$

Using this new variational formulation and (111) we can now prove the following theorem.

Theorem 4.3. *Suppose $\int_D (n-1)dx \neq 0$ and $A^* < 1$ or $A_* > 1$. Then the set of transmission eigenvalues is discrete in $\mathbb{C}$. Moreover, the nonzero eigenvalue of smallest magnitude k_1 satisfies the Faber–Krahn-type estimate*

$$|k_1|^2 \geq \frac{A_*(1-\sqrt{A^*})}{C_P \max(n^*, 1)(1+\sqrt{n^*})} \quad \text{if } A^* < 1,$$

$$|k_1|^2 \geq \frac{1 - 1/\sqrt{A_*}}{C_P \max(n^*, 1)(1 + 1/\sqrt{n_*})} \quad \text{if } A_* > 1,$$

with C_P defined in (111).

Proof. We consider first the case $A^* < 1$. Set

$$\lambda(v) := 2 \frac{\int_D (n-1)v}{\int_D (n-1)}$$

and consider the isomorphism of $Y(D)$ defined by

$$T(w, v) := \big(w - 2v + \lambda(v), -v + \lambda(v)\big).$$

Notice that $\lambda(\lambda(v)) = 2\lambda(v)$ so that $T^2 = I$. For all $(w, v) \in Y(D)$, one has

$$\big|a_k^T((w, v), (w, v))\big|$$
$$= \big|(A\nabla w, \nabla w)_D + (\nabla v, \nabla v)_D - 2(A\nabla w, \nabla v)_D$$
$$\qquad - k^2\big((nw, w)_D + (v, v)_D - 2(nw, v)_D\big)\big|$$
$$\geq (A\nabla w, \nabla w)_D + (\nabla v, \nabla v)_D - 2|(A\nabla w, \nabla v)_D|$$
$$\qquad - |k|^2 \big((nw, w)_D + (v, v)_D + 2|(nw, v)_D|\big)$$
$$\geq (1 - \sqrt{A^*})\big((A\nabla w, \nabla w)_D + (\nabla v, \nabla v)_D\big)$$
$$\qquad - |k|^2 (1 + \sqrt{n^*})\big((nw, w)_D + (v, v)_D\big).$$

Consequently, for $k \in \mathbb{C}$ such that

$$|k|^2 < \frac{A_*(1-\sqrt{A^*})}{C_P \max(n^*,1)(1+\sqrt{n^*})},$$

a_k^T is coercive on $Y(D)$. The claim of the theorem follows from analytic Fredholm theory.

The case $A_* > 1$ can be treated in an analogous way by using the isomorphism T of $Y(D)$ defined by

$$T(w,v) := (w - \lambda(w), -v + 2w - \lambda(w)). \qquad \square$$

We remark that in particular, if $n^* < 1$ or if $1 < n_*$, then $\int_D (n-1)dx \neq 0$ and Theorem 4.3 proves that the set of interior transmission eigenvalues is discrete which recovers previously known results in [Cakoni et al. 2002; Cakoni and Kirsch 2010]. In those cases the Faber–Krahn type estimates can be made more explicit. For instance if $A^* < 1$ and $1 < n_*$, noticing that for $k^2 \in \mathbb{R}$,

$$\Re[a_k^T((w,v),(w,v))] = (A\nabla(w-v), \nabla(w-v))_D - k^2((n(w-v),(w-v))_D \\ + ((I-A)\nabla v, \nabla v)_D + ((1-n)v,v)_D),$$

where the isomorphism T is defined by $T(w,v) = (w - 2v, -v)$, one easily deduces that the first real transmission eigenvalue k_1 such that $k_1 \neq 0$ satisfies

$$k_1^2 \geq A_* \lambda_1(D)/n^*$$

where $\lambda_1(D)$ is the first Dirichlet eigenvalue of $-\Delta$ on D which is also proved in [Cakoni and Kirsch 2010] using a different technique.

We end this section with a result on the location of transmission eigenvalues, again requiring the sign assumption on the contrasts only on a neighborhood of the boundary ∂D.

Theorem 4.4. *Under the hypothesis of Theorem 4.2 there exist two positive constants ρ and δ such that if $k \in \mathbb{C}$ satisfies $|k| > \rho$ and $|\Re(k)| < \delta |\Im(k)|$, then k is not a transmission eigenvalue.*

Proof. Here we give the proof only in the case of $A(x) \leq A^* I < I$ and $n(x) \leq n^* < 1$ almost everywhere on $\mathcal{V}(\partial D)$. The case of $A(x) \geq A_* I > I$ and $n(x) \geq n_* > 1$ almost everywhere on $\mathcal{V}(\partial D)$ can be treated using similar adaptations as in the proof of Lemma 4.2.

Consider again the isomorphism T defined by $T(w,v) = (w - 2\chi v, -v)$ where χ is as in the proof of Lemma 4.2 where we already proved that for $\kappa \in \mathbb{R}$

with $|\kappa|$ large enough, the following coercivity property holds:

$$|a_{i\kappa}^T((w,v),(w,v))|$$
$$\geq C_1(\|w\|_{H^1(D)}^2 + \|v\|_{H^1(D)}^2) + C_2\kappa^2(\|w\|_D^2 + \|v\|_D^2), \quad (112)$$

where the constants $C_1, C_2 > 0$ are independent of κ. Take now $k = i\kappa e^{i\theta}$ with $\theta \in [-\pi/2; \pi/2k]$. One has

$$|a_k^T((w,v),(w,v)) - a_{i\kappa}^T((w,v),(w,v))|$$
$$\geq C_3|1 - e^{2i\theta}|\kappa^2(\|w\|_D^2 + \|v\|_D^2), \quad (113)$$

with $C_3 > 0$ independent of κ. Combining (112) and (113), one finds

$$|a_k^T((w,v),(w,v))|$$
$$\geq |a_{i\kappa}^T((w,v),(w,v))| - C_3\kappa^2|1 - e^{2i\theta}|(\|w\|_D^2 + \|v\|_D^2)$$
$$\geq C_1(\|w\|_{H^1(D)}^2 + \|v\|_{H^1(D)}^2) + (C_2 - C_3|1 - e^{2i\theta}|)\kappa^2(\|w\|_D^2 + \|v\|_D^2).$$

Choosing θ small enough, to have for example $C_3|1 - e^{2i\theta}| \leq C_2/2$, one obtains the desired result. □

As mentioned in Section 3D, Theorem 3.15, proven in [Hitrik et al. 2011a], provides a more precise location of transmission eigenvalues in the case when $A = I$. We also remark that related results on the discreteness of transmission eigenvalues are obtained in [Lakshtanov and Vainberg 2012].

Existence of transmission eigenvalues. We now turn our attention to the existence of real transmission eigenvalues which unfortunately can only be shown under restrictive assumptions on $A - I$ and $n - 1$. The proposed approach presented here follows the lines of [Cakoni and Kirsch 2010] which, inspired by the original existence proof in the case $A = I$ discussed in Section 3B, tries to formulate the transmission eigenvalue problem as a problem for the difference $u := w - v$. However, due to the lack of symmetry, the problem for u is no longer a quadratic eigenvalue problem but it takes the form of a more complicated nonlinear eigenvalue problem as is explained in the following.

Setting $\tau := k^2$, the transmission eigenvalue problem reads: Find

$$(w,v) \in H^1(D) \times H^1(D)$$

that satisfies

$$\nabla \cdot A\nabla w + \tau n w = 0 \quad \text{and} \quad \Delta v + \tau v = 0 \quad \text{in } D, \quad (114)$$
$$w = v \quad \text{and} \quad \nu \cdot A\nabla w = \nu \cdot \nabla v \quad \text{on } \partial D. \quad (115)$$

We first observe that if (w, v) satisfies (87)–(88), subtracting the second equation in (114) from the first we obtain

$$\nabla \cdot A\nabla u + \tau n u = \nabla \cdot (A - I)\nabla v + \tau(n-1)v \quad \text{in } D, \tag{116}$$

$$\nu \cdot A\nabla u = \nu \cdot (A - I)\nabla v \quad \text{on } \partial D, \tag{117}$$

where $u := w - v$, and in addition we have

$$\Delta v + \tau v = 0 \quad \text{in } D, \tag{118}$$

$$u = 0 \quad \text{on } \partial D. \tag{119}$$

It is easy to verify that (w, v) in $H^1(D) \times H^1(D)$ satisfies (6)–(7) if and only if (u, v) is in $H_0^1(D) \times H^1(D)$ and satisfies (116)–(118). The proof consists in expressing v in terms of u, using (116), and substituting the resulting expression into (118) in order to formulate the eigenvalue problem only in terms of u. In the case $A = I$, that is, $A - I = 0$, this substitution is simple and leads to an explicit expression for the equation satisfied by u. In the current case the substitution requires the inversion of the operator $\nabla \cdot [(A - I)\nabla \cdot] + \tau(n-1)$ with a Neumann boundary condition. It is then obvious that the case where $A - I$ and $n - 1$ have the same sign is more problematic since in that case the operator may not be invertible for special values of τ. This is why we only treat the simpler case of $A - I$ and $n - 1$ having opposite signs almost everywhere in D.

To this end we see that for given $u \in H_0^1(D)$, the problem (116) for $v \in H^1(D)$ is equivalent to the variational formulation

$$\int_D \left[(A-I)\nabla v \cdot \nabla \overline{\psi} - \tau(n-1)v\overline{\psi}\right] dx = \int_D \left[A\nabla u \cdot \nabla \overline{\psi} - \tau n u \overline{\psi}\right] dx \tag{120}$$

for all $\psi \in H^1(D)$. The following result concerning the invertibility of the operator associated with (120) can be proven in a standard way using the Lax–Milgram lemma.

Lemma 4.3. *Assume that either $(A_* - 1) > 0$ and $(n^* - 1) < 0$, or $(A^* - 1) < 0$ and $(n_* - 1) > 0$. Then there exists $\delta > 0$ such that for every $u \in H_0^1(D)$ and $\tau \in \mathbb{C}$ with $\Re \tau > -\delta$ there exists a unique solution $v := v_u \in H^1(D)$ of (120). The operator $A_\tau : H_0^1(D) \to H^1(D)$, defined by $u \mapsto v_u$, is bounded and depends analytically on $\tau \in \{z \in \mathbb{C} : \Re(z) > -\delta\}$.*

We now set $v_u := A_\tau u$ and denote by $\mathbb{L}_\tau u \in H_0^1(D)$ the unique Riesz representation of the bounded conjugate-linear functional

$$\psi \mapsto \int_D \left[\nabla v_u \cdot \nabla \overline{\psi} - \tau v_u \overline{\psi}\right] dx \quad \text{for } \psi \in H_0^1(D),$$

that is,

$$(\mathbb{L}_\tau u, \psi)_{H^1(D)} = \int_D \left[\nabla v_u \cdot \nabla \bar\psi - \tau\, v_u\, \bar\psi\right] dx \quad \text{for } \psi \in H_0^1(D). \tag{121}$$

Obviously, $\mathbb{L}_\tau$ also depends analytically on $\tau \in \{z \in \mathbb{C} : \Re z > -\delta\}$. Now we are able to connect a transmission eigenfunction, i.e., a nontrivial solution (w, v) of (6)–(7), to the kernel of the operator $\mathbb{L}_\tau$.

Theorem 4.5. (a) *Let $(w, v) \in H^1(D) \times H^1(D)$ be a transmission eigenfunction corresponding to some $\tau > 0$. Then $u = v - w \in H_0^1(D)$ satisfies $\mathbb{L}_\tau u = 0$.*

(b) *Let $u \in H_0^1(D)$ satisfy $\mathbb{L}_\tau u = 0$ for some $\tau > 0$. Furthermore, let $v = v_u = A_\tau u \in H^1(D)$ be as in Lemma 4.3, i.e., the solution of (120). Then $(w, v) \in H^1(D) \times H^1(D)$ is a transmission eigenfunction where $w = v - u$.*

The proof of this theorem is a simple consequence of the observation that the first equation in (118) is equivalent to

$$\int_D \left[\nabla v \cdot \nabla \bar\psi - \tau\, v\, \bar\psi\right] dx = 0 \quad \text{for all } \psi \in H_0^1(D). \tag{122}$$

The operator $\mathbb{L}_\tau$ plays a similar role as the operator $\mathbb{A}_\tau - \tau \mathbb{B}$ in (69) for the case of $A = I$. The following properties are the main ingredients needed in order to prove the existence of transmission eigenvalues.

Theorem 4.6. (a) *The operator $\mathbb{L}_\tau : H_0^1(D) \to H_0^1(D)$ is selfadjoint for all $\tau \in \mathbb{R}_{\geq 0}$.*

(b) *Let $\sigma = 1$ if $(A_* - 1) > 0$ and $(n^* - 1) < 0$, and $\sigma = -1$ if $(A^* - 1) < 0$ and $(n_* - 1) > 0$. Then $\sigma \mathbb{L}_0 : H_0^1(D) \to H_0^1(D)$ is coercive, that is, $(\sigma \mathbb{L}_0 u, u)_{H^1(D)} \geq c \|u\|_{H^1(D)}^2$ for all $u \in H_0^1(D)$ and $c > 0$ independent of u.*

(c) *$\mathbb{L}_\tau - \mathbb{L}_0$ is compact in $H_0^1(D)$.*

(d) *There exists at most a countable number of $\tau > 0$ for which $\mathbb{L}_\tau$ fails to be injective with infinity the only possible accumulation point.*

Proof. (a) First we show that $\mathbb{L}_\tau$ is selfadjoint for all $\tau \in \mathbb{R}_{\geq 0}$. To this end for every $u_1, u_2 \in H_0^1(D)$ let $v_1 := v_{u_1}$ and $v_2 := v_{u_2}$ be the corresponding solution of (120). Then

$$\begin{aligned}(\mathbb{L}_\tau u_1, u_2)_{H^1(D)} &= \int_D \left[\nabla v_1 \cdot \nabla \bar u_2 - \tau\, v_1 \bar u_2\right] dx \\ &= \int_D \left[A \nabla v_1 \cdot \nabla \bar u_2 - \tau n\, v_1\, \bar u_2\right] dx - \int_D \left[(A-I)\nabla v_1 \cdot \nabla \bar u_2 - \tau (n-1) v_1\, \bar u_2\right] dx.\end{aligned} \tag{123}$$

Using (120) twice, first for $u = u_2$ and the corresponding $v = v_2$ and $\psi = v_1$ and then for $u = u_1$ and the corresponding $v = v_1$ and $\psi = u_2$, yields

$$(\mathbb{L}_\tau u_1, u_2)_{H^1(D)} = \int_D \left[(A - I)\nabla v_1 \cdot \nabla \bar{v}_2 - \tau(n-1) v_1 \bar{v}_2 \right] dx$$
$$- \int_D \left[A \nabla u_1 \cdot \nabla \bar{u}_2 - \tau n u_1 \bar{u}_2 \right] dx \quad (124)$$

which is a selfadjoint expression for u_1 and u_2.

(b) Next we show that $\sigma \mathbb{L}_0 : H_0^1(D) \to H_0^1(D)$ is a coercive operator. Using the definition of $\mathbb{L}_0$ in (121) and the fact that $v = v_u = u + w$ we have

$$(\mathbb{L}_0 u, u)_{H^1(D)} = \int_D \nabla v \cdot \nabla \bar{u} \, dx = \int_D |\nabla u|^2 \, dx + \int_D \nabla w \cdot \nabla \bar{u} \, dx. \quad (125)$$

From (120) for $\tau = 0$ and $\psi = w$ we have

$$\int_D \nabla w \cdot \nabla \bar{u} \, dx = \int_D (A - I)\nabla w \cdot \nabla \bar{w} \, dx. \quad (126)$$

If $(A_* - 1) > 0$ then $\int_D (A-I)\nabla w \cdot \nabla \bar{w} \, dx \geq (A_* - 1)\|\nabla w\|_{L^2(D)}^2 \geq 0$; hence

$$(\mathbb{L}_0 u, u)_{H^1(D)} \geq \int_D |\nabla u|^2 \, dx.$$

From Poincaré's inequality in $H_0^1(D)$ we have that $\|\nabla u\|_{L^2(D)}$ is an equivalent norm in $H_0^1(D)$ and this proves the coercivity of $\mathbb{L}_0$. If $(A^* - 1) < 0$, from (124) with $u_1 = u_2 = u$ and $\tau = 0$ we have

$$-(\mathbb{L}_0 u, u)_{H^1(D)}$$
$$= -\int_D (A - I)\nabla v \cdot \nabla \bar{v} \, dx + \int_D A \nabla u \cdot \nabla \bar{u} \, dx \geq A_* \int_D |\nabla u|^2 \, dx,$$

which proves the coercivity of $-\mathbb{L}_0$ since $A_* > 0$.

(c) This now follows from the compact embedding of $H_0^1(D)$ into $L^2(D)$.

(d) Since $(\sigma \mathbb{L}_0)^{-1}$ exists and $\tau \mapsto \mathbb{L}_\tau$ is analytic on $\{z \in \mathbb{C} : \Re(z) > -\delta\}$, this follows directly from the analytic Fredholm theory. We remark that this part is also a consequence of the more general result of Theorem 4.3. $\square$

We are now in the position to establish the existence of infinitely many real transmission eigenvalues, i.e., the existence of a sequence of $\tau_j \in \mathbb{R}$, $j \in \mathbb{N}$, and corresponding $u_j \in H_0^1(D)$ such that $u_j \neq 0$ and $\mathbb{L}_{\tau_j} u_j = 0$. Obviously, these $\tau > 0$ are such that the kernel of $\mathbb{I} - \mathbb{T}_\tau$ is not trivial, where

$$-\sigma(\sigma \mathbb{L}_0)^{-1/2}(\mathbb{L}_\tau - \mathbb{L}_0)(\sigma \mathbb{L}_0)^{-1/2}$$

is compact, which corresponds to 1 being an eigenvalue of the compact self-adjoint operator $\mathbb{T}_\tau$. From the discussion above we conclude that transmission eigenvalues $k > 0$ have finite multiplicity and are such that $\tau := k^2$ are solutions to $\mu_j(\tau) = 1$ where $\{\mu_j(\tau)\}_1^{+\infty}$ is the increasing sequence of the eigenvalues of $\mathbb{T}_\tau$. Note that from max-min principle $\mu_j(\tau)$ depend continuously on τ which the core of the proof the following theorem (see e.g. [Päivärinta and Sylvester 2008] for the proof).

Theorem 4.7. *Assume that*

(1) *there is a $\tau_0 \geq 0$ such that $\sigma \mathbb{L}_{\tau_0}$ is positive on $H_0^1(D)$ and*

(2) *there is a $\tau_1 > \tau_0$ such that $\sigma \mathbb{L}_{\tau_1}$ is non positive on some m-dimensional subspace W_m of $H_0^1(D)$.*

Then there are m values of τ in $[\tau_0, \tau_1]$ counting their multiplicity for which $\mathbb{L}_\tau$ fails to be injective.

Using now Theorem 4.7 and adapting the ideas developed in Section 3B and Section 4A, we can prove the main theorem of this section.

Theorem 4.8. *Suppose that the matrix valued function A and the function n are such that either $(A_* - 1) > 0$ and $(n^* - 1) < 0$, or $(A^* - 1) < 0$ and $(n_* - 1) > 0$. Then there exists an infinite sequence of transmission eigenvalues $k_j > 0$ with $+\infty$ as their only accumulation point.*

Proof. We sketch the proof only for the case of $(A_*-1) > 0$ and $(n^*-1) < 0$ (i.e., $\sigma = 1$ in Theorem 4.7). First, we recall that the assumption *(1)* of Theorem 4.7 is satisfied with $\tau_0 = 0$ i.e., $(\mathbb{L}_0 u, u)_{H^1(D)} > 0$ for all $u \in H_0^1(D)$ with $u \neq 0$. Next, by the definition of $\mathbb{L}_\tau$ and the fact that $v = w + u$ have

$$(\mathbb{L}_\tau u, u)_{H^1(D)} = \int_D \left[\nabla v \cdot \nabla \bar{u} - \tau v \bar{u}\right] dx$$
$$= \int_D \left[\nabla w \cdot \nabla \bar{u} - \tau w \bar{u} + |\nabla u|^2 - \tau |u|^2\right] dx. \quad (127)$$

We also have that w satisfies

$$\int_D \left[(A - I)\nabla w \cdot \nabla \bar{\psi} - \tau (n-1) w \bar{\psi}\right] dx = \int_D \left[\nabla u \cdot \nabla \bar{\psi} - \tau u \bar{\psi}\right] dx \quad (128)$$

for all $\psi \in H^1(D)$. Now taking $\psi = w$ in (128) and substituting the result into (127) yields

$$(\mathbb{L}_\tau u, u)_{H^1(D)}$$
$$= \int_D \left[(A - I)\nabla w \cdot \nabla \bar{w} - \tau (n-1) |w|^2 + |\nabla u|^2 - \tau |u|^2\right] dx. \quad (129)$$

Let now $B_r \subset D$ be an arbitrary ball of radius r included in D and let
$$\hat{\tau} := k_1^2(A_*, n^*, B_r),$$
where $k_1(A_*, n^*, B_r)$ is the first transmission eigenvalue corresponding to the ball B_r with constant contrasts $A = A_* I$ and $n = n^*$ (we refer to [Cakoni and Kirsch 2010] for the existence of transmission eigenvalues in this case which is again proved by separation of variables and using the asymptotic behavior of Bessel functions). Let $\hat{v}$, $\hat{w}$ be the nonzero solutions to the corresponding homogeneous interior transmission problem, i.e., the solution of (87)–(90) with $D = B_r$, $A = A_* I$ and $n = n^*$ and set
$$\hat{u} := \hat{v} - \hat{w} \in H_0^1(B_r).$$
We denote the corresponding operator by $\hat{\mathbb{L}}_\tau$. Of course, by construction we have that (129) still holds, i.e., since $\hat{\mathbb{L}}_{\hat{\tau}} \hat{u} = 0$,
$$0 = (\hat{\mathbb{L}}_{\hat{\tau}} \hat{u}, \hat{u})_{H^1(B_r)}$$
$$= \int_{B_r} \left[(A_* - 1)|\nabla \hat{w}|^2 - \hat{\tau}(n^* - 1)|\hat{w}|^2 + |\nabla \hat{u}|^2 - \hat{\tau}|\hat{u}|^2 \right] dx. \quad (130)$$

Next we denote by $\tilde{u} \in H_0^1(D)$ the extension of $\hat{u} \in H_0^1(B_r)$ by zero to the whole of D and let $\tilde{v} := v_{\tilde{u}}$ be the corresponding solution to (120) and $\tilde{w} := \tilde{v} - \tilde{u}$. In particular $\tilde{w} \in H^1(D)$ satisfies

$$\int_D \left[(A - I) \nabla \tilde{w} \cdot \nabla \overline{\psi} - \hat{\tau} p \tilde{w} \overline{\psi} \right] dx$$
$$= \int_D \left[\nabla \tilde{u} \cdot \nabla \overline{\psi} - \hat{\tau} \tilde{u} \overline{\psi} \right] dx = \int_{B_r} \left[\nabla \hat{u} \cdot \nabla \overline{\psi} - \hat{\tau} \hat{u} \overline{\psi} \right] dx$$
$$= \int_{B_r} \left[(A_* - 1) \nabla \hat{w} \cdot \nabla \overline{\psi} - \hat{\tau}(n^* - 1) \hat{w} \overline{\psi} \right] dx \quad (131)$$

for all $\psi \in H^1(D)$. Therefore, for $\psi = \tilde{w}$ we have, by the Cauchy–Schwarz inequality,

$$\int_D (A - I) \nabla \tilde{w} \cdot \nabla \overline{\tilde{w}} - \hat{\tau}(n - 1)|\tilde{w}|^2 dx = \int_{B_r} (A_* - 1) \nabla \hat{w} \cdot \nabla \overline{\tilde{w}} + \hat{\tau}|n^* - 1| \hat{w} \overline{\tilde{w}} dx$$
$$\leq \left[\int_{B_r} (A_* - 1)|\nabla \hat{w}|^2 + \hat{\tau}|n^* - 1||\hat{w}|^2 dx \right]^{\frac{1}{2}} \left[\int_{B_r} (A_* - 1)|\nabla \tilde{w}|^2 + \hat{\tau}|n^* - 1||\tilde{w}|^2 dx \right]^{\frac{1}{2}}$$
$$\leq \left[\int_{B_r} (A_* - 1)|\nabla \hat{w}|^2 - \hat{\tau}(n^* - 1)|\hat{w}|^2 dx \right]^{\frac{1}{2}} \left[\int_D (A - I) \nabla \tilde{w} \cdot \nabla \overline{\tilde{w}} - \hat{\tau}(n - 1)|\tilde{w}|^2 dx \right]^{\frac{1}{2}},$$

since $|n-1| = 1-n > 1-n^* = |n^*-1|$ and thus

$$\int_D \left[(A-I)\nabla\tilde{w}\cdot\nabla\overline{\tilde{w}} - \hat{\tau}\,(n-1)\,|\tilde{w}|^2\right]dx$$
$$\leq \int_{B_r}\left[(A_*-1)\,|\nabla\hat{w}|^2 - \hat{\tau}\,(n^*-1)\,|\hat{w}|^2\right]dx.$$

Substituting this into (129) for $\tau = \hat{\tau}$ and $u = \tilde{u}$ yields

$$\left(\mathbb{L}_{\hat{\tau}}\tilde{u},\tilde{u}\right)_{H^1(D)}$$
$$= \int_D \left[(A-I)\nabla\tilde{w}\cdot\nabla\overline{\tilde{w}} - \hat{\tau}\,(n-1)\,|\tilde{w}|^2 + |\nabla\tilde{u}|^2 - \hat{\tau}\,|\tilde{u}|^2\right]dx$$
$$\leq \int_{B_r}\left[(A_*-1)|\nabla\hat{w}|^2 - \hat{\tau}\,(n^*-1)\,|\hat{w}|^2 + |\nabla\hat{u}|^2 - \hat{\tau}\,|\hat{u}|^2\right]dx$$
$$= 0, \tag{132}$$

by (130). Hence from Theorem 4.7 we have that there is a transmission eigenvalue $k > 0$, such that in $k^2 \in (0, \hat{\tau}]$. Finally, repeating this argument for balls of arbitrary small radius we can show the existence of infinitely many transmission eigenvalues exactly in the same way as in the proof Theorem 3.8. □

We can also obtain better bounds for the first transmission eigenvalue:

Theorem 4.9 ([Cakoni and Kirsch 2010]). *Let $B_R \subset D$ be the largest ball contained in D and $\lambda_1(D)$ the first Dirichlet eigenvalue of $-\Delta$ on D. Furthermore, let $k_1(A(x), n(x), D)$ be the first transmission eigenvalue corresponding to (87)–(90).*

(1) *If $(A_* - 1) > 0$ and $(n^* - 1) < 0$ then*

$$\lambda_1(D) \leq k_1^2(A(x), n(x), D) \leq k_1^2(A_*, n^*, B_R)$$

where $k_1(A_*, n^*, B_R)$ is the first transmission eigenvalue corresponding to the ball B_R with $A = A_* I$ and $n = n^*$.

(2) *If $(A^* - 1) < 0$ and $(n_* - 1) > 0$ then*

$$\frac{A_*}{n^*}\lambda_1(D) \leq k_1^2(A(x), n(x), D) \leq k_1^2(A^*, n_*, B_R)$$

where $k_1(A^*, n_*, B_R)$ is the first transmission eigenvalue corresponding to the ball B_R with $A = A^* I$ and $n = n_*$.

We end our discussion in this section by making a few comments on the case when $(A - I)$ and $(n - 1)$ have the same sign. As indicated above, if one follows a similar procedure, then one is faced with the problem that (120) is not solvable for all τ. Thus we are forced to put restrictions on τ, $(A - I)$ and $(n - 1)$ which only allow us to prove the existence of at least one transmission eigenvalue. In particular, skipping the details, we set

$$\hat{\tau}(r, A_*) := k_1^2(\frac{A_* + 1}{2}, 1, B_r)$$

(with the notation of Theorem 4.9 for the right-hand side), where the ball B_r of radius r is such that $B_r \subset D$. Then if $(n^* - 1) > 0$ is small enough such that

$$(n^* - 1) < \frac{\mu(D, n)}{2\hat{\tau}(r, A_*)} (A_* - 1) \tag{133}$$

where

$$\mu(D, n) := \inf_{\substack{\psi \in H^1(D) \\ \int_D (n-1)\psi\, dx = 0}} \frac{\|\nabla \psi\|_{L^2(D)}^2}{\|\psi\|_{L^2(D)}^2},$$

then there exists at least one real transmission eigenvalue in the interval

$$\left(0, k_1\left(\frac{A_* + 1}{2}, 1, B_r\right)\right]. \tag{134}$$

In fact, if $(n^* - 1)$ is small enough such that (133) is satisfied for an $r > 0$ such that in D we can fit m balls of radius r, then one can show [Cakoni and Kirsch 2010] that there are m real transmission eigenvalues in the interval (134) counting their multiplicity. It is still an open problem to prove the existence of infinitely many real transmission eigenvalues in this case.

5. Conclusions and open problems

In this survey we have presented a collection of results on the transmission eigenvalue problem corresponding to scattering by an inhomogeneous medium with emphasis on the derivation of the existence, discreteness and inequalities for transmission eigenvalues. Although we have focused on theoretical results, computational methods for transmission eigenvalues as well as their use in obtaining information on the material properties of inhomogeneous media from scattering data can be found in [Cakoni et al. 2009; 2007; 2010d; Colton et al. 2010; Cossonnière 2011; Giovanni and Haddar 2011; Sun 2011]. A similar analysis has been done in [Cakoni et al. 2012; Cossonnière 2011] for inhomogeneous media containing obstacles inside. The transmission eigenvalue problem has also been investigated for the case of Maxwell's equation where technical complications arise due to the structure of the spaces needed to study these equations (see [Cakoni et al. 2011; 2010d; 2010e; Cakoni and Kirsch 2010; Cakoni and Haddar 2009; Cossonnière and Haddar 2011; Haddar 2004; Kirsch 2009)]. The transmission eigenvalue problem associated with the scattering problem for anisotropic linear elasticity has been investigated in [Bellis et al. 2012; Bellis and Guzina 2010]. As previously mentioned, [Hitrik et al. 2010; 2011b] investigate transmission eigenvalues for higher-order operators with constant coefficients.

Despite extensive research and much recent progress on the transmission eigenvalue problem there are still many open questions that call for new ideas. In our opinion some important questions that impact both the theoretical understanding of the transmission eigenvalue problem as well as their application to inverse scattering theory are the following:

(1) Do complex transmission eigenvalues exists for general nonabsorbing media?

(2) Do real transmission eigenvalues exist for absorbing media and absorbing background?

(3) Can the existence of real transmission eigenvalues for nonabsorbing media be established if the assumptions on the sign of the contrast are weakened?

(4) What would the necessary conditions be on the contrasts that guaranty the discreteness of transmission eigenvalues?

(5) Can Faber–Krahn type inequalities be established for the higher eigenvalues?

(6) Can completeness results be established for transmission eigenfunctions, that is, nonzero solutions to transmission eigenvalue problem corresponding to transmission eigenvalues? (We remark that in [Hitrik et al. 2011b] the completeness question is positively answered for transmission eigenvalue problem for operators of order higher than 3. The proof breaks down for operators of order two which are the cases considered in this paper and are related to most of the practical problems in scattering theory.)

(7) Can an inverse spectral problem be developed for the general transmission eigenvalue problem? We also believe that a better understanding of the physical interpretation of transmission eigenvalues and their connection to the wave equation could provide an alternative way of determining transmission eigenvalues from the (possibly time-dependent) scattering data.

Acknowledgement

The authors thanks Professor David Colton for his valuable comments on spherically stratified media, in particular for the observation that resulted in an improved version of Theorem 3.1.

References

[Aktosun et al. 2011] T. Aktosun, D. Gintides, and V. G. Papanicolaou, "The uniqueness in the inverse problem for transmission eigenvalues for the spherically symmetric variable-speed wave equation", *Inverse Problems* **27**:11 (2011), 115004. MR 2851910 Zbl 1231.35295

[Arens 2004] T. Arens, "Why linear sampling works", *Inverse Problems* **20**:1 (2004), 163–173. MR 2005b:35036 Zbl 1055.35131

[Arens and Lechleiter 2009] T. Arens and A. Lechleiter, "The linear sampling method revisited", *J. Integral Equations Appl.* **21**:2 (2009), 179–202. MR 2011a:35564 Zbl 1237.65118

[Bellis and Guzina 2010] C. Bellis and B. B. Guzina, "On the existence and uniqueness of a solution to the interior transmission problem for piecewise-homogeneous solids", *J. Elasticity* **101**:1 (2010), 29–57. MR 2012d:74028 Zbl 1243.74038

[Bellis et al. 2012] C. Bellis, F. Cakoni, and B. Guzina, "Nature of the transmission eigenvalue spectrum for elastic bodies", *IMA J. Appl. Math.* (2012).

[Bonnet-Ben Dhia et al. 2010] A. S. Bonnet-Ben Dhia, P. Ciarlet, Jr., and C. M. Zwölf, "Time harmonic wave diffraction problems in materials with sign-shifting coefficients", *J. Comput. Appl. Math.* **234**:6 (2010), 1912–1919. MR 2011m:78009 Zbl 1202.78026

[Bonnet-BenDhia et al. 2011] A.-S. Bonnet-BenDhia, L. Chesnel, and H. Haddar, "On the use of T-coercivity to study the interior transmission eigenvalue problem", preprint, 2011, Available at http://goo.gl/G6fRw. to appear in *C. R. Acad. Sci., Ser. I*.

[Cakoni and Colton 2006] F. Cakoni and D. Colton, *Qualitative methods in inverse scattering theory: An introduction*, Interaction of Mechanics and Mathematics, Springer, Berlin, 2006. MR 2008c:35334 Zbl 1099.78008

[Cakoni and Haddar 2009] F. Cakoni and H. Haddar, "On the existence of transmission eigenvalues in an inhomogeneous medium", *Appl. Anal.* **88**:4 (2009), 475–493. MR 2010m:35557 Zbl 1168.35448

[Cakoni and Kirsch 2010] F. Cakoni and A. Kirsch, "On the interior transmission eigenvalue problem", *Int. J. Comput. Sci. Math.* **3**:1-2 (2010), 142–167. MR 2011f:78005 Zbl 1204.78008

[Cakoni et al. 2002] F. Cakoni, D. Colton, and H. Haddar, "The linear sampling method for anisotropic media", *J. Comput. Appl. Math.* **146**:2 (2002), 285–299. MR 2003i:35219

[Cakoni et al. 2007] F. Cakoni, D. Colton, and P. Monk, "On the use of transmission eigenvalues to estimate the index of refraction from far field data", *Inverse Problems* **23**:2 (2007), 507–522. MR 2008d:78016 Zbl 1115.78008

[Cakoni et al. 2008] F. Cakoni, M. Çayören, and D. Colton, "Transmission eigenvalues and the nondestructive testing of dielectrics", *Inverse Problems* **24**:6 (2008), 065016. MR 2010d:35040 Zbl 1157.35497

[Cakoni et al. 2009] F. Cakoni, D. Colton, and H. Haddar, "The computation of lower bounds for the norm of the index of refraction in an anisotropic media from far field data", *J. Integral Equations Appl.* **21**:2 (2009), 203–227. MR 2011b:35555 Zbl 1173.35722

[Cakoni et al. 2010a] F. Cakoni, D. Colton, and D. Gintides, "The interior transmission eigenvalue problem", *SIAM J. Math. Anal.* **42**:6 (2010), 2912–2921. MR 2012b:35364 Zbl 1219.35352

[Cakoni et al. 2010b] F. Cakoni, D. Colton, and H. Haddar, "The interior transmission problem for regions with cavities", *SIAM J. Math. Anal.* **42**:1 (2010), 145–162. MR 2011b:35556 Zbl 1209.35135

[Cakoni et al. 2010c] F. Cakoni, D. Colton, and H. Haddar, "On the determination of Dirichlet or transmission eigenvalues from far field data", *C. R. Math. Acad. Sci. Paris* **348**:7-8 (2010), 379–383. MR 2011b:35077 Zbl 1189.35204

[Cakoni et al. 2010d] F. Cakoni, D. Colton, P. Monk, and J. Sun, "The inverse electromagnetic scattering problem for anisotropic media", *Inverse Problems* **26**:7 (2010), 074004. MR 2011h:78015 Zbl 1197.35314

[Cakoni et al. 2010e] F. Cakoni, D. Gintides, and H. Haddar, "The existence of an infinite discrete set of transmission eigenvalues", *SIAM J. Math. Anal.* **42**:1 (2010), 237–255. MR 2011b:35557 Zbl 1210.35282

[Cakoni et al. 2011] F. Cakoni, D. Colton, and P. Monk, *The linear sampling method in inverse electromagnetic scattering*, CBMS-NSF Regional Conference Series in Applied Mathematics **80**, Society for Industrial and Applied Mathematics (SIAM), Philadelphia, PA, 2011. MR 2012e:78019 Zbl 1221.78001

[Cakoni et al. 2012] F. Cakoni, D. Colton, and H. Haddar, "The interior transmission eigenvalue problem for absorbing media", *Inverse Problems* **28**:4 (2012), 045005.

[Chesnel 2012] L. Chesnel, *Étude de quelques problèmes de transmission avec changement de signe. Application aux mĭamatrĭaux*, Ph.D. thesis, École Polytechnique, Paris, 2012.

[Colton 1979] D. Colton, "The construction of solutions to acoustic scattering problems in a spherically stratified medium. II", *Quart. J. Mech. Appl. Math.* **32**:1 (1979), 53–62. MR 58 #19822b Zbl 0431.73023

[Colton and Kress 1983] D. L. Colton and R. Kress, *Integral equation methods in scattering theory*, Pure and Applied Mathematics, Wiley, New York, 1983. MR 85d:35001 Zbl 0522.35001

[Colton and Kress 1998] D. Colton and R. Kress, *Inverse acoustic and electromagnetic scattering theory*, 2nd ed., Applied Mathematical Sciences **93**, Springer, Berlin, 1998. MR 99c:35181 Zbl 0893.35138

[Colton and Kress 2001] D. Colton and R. Kress, "On the denseness of Herglotz wave functions and electromagnetic Herglotz pairs in Sobolev spaces", *Math. Methods Appl. Sci.* **24**:16 (2001), 1289–1303. MR 2002k:78008 Zbl 0998.35034

[Colton and Monk 1988] D. Colton and P. Monk, "The inverse scattering problem for time-harmonic acoustic waves in an inhomogeneous medium", *Quart. J. Mech. Appl. Math.* **41**:1 (1988), 97–125. MR 89i:76080 Zbl 0637.73026

[Colton and Päivärinta 2000] D. Colton and L. Päivärinta, "Transmission eigenvalues and a problem of Hans Lewy", *J. Comput. Appl. Math.* **117**:2 (2000), 91–104. MR 2001c:35165 Zbl 0957.65093

[Colton and Sleeman 2001] D. Colton and B. D. Sleeman, "An approximation property of importance in inverse scattering theory", *Proc. Edinb. Math. Soc.* (2) **44**:3 (2001), 449–454. MR 2002j:35304 Zbl 0992.35115

[Colton et al. 1989] D. Colton, A. Kirsch, and L. Päivärinta, "Far-field patterns for acoustic waves in an inhomogeneous medium", *SIAM J. Math. Anal.* **20**:6 (1989), 1472–1483. MR 90i:35072 Zbl 0681.76084

[Colton et al. 2007] D. Colton, L. Päivärinta, and J. Sylvester, "The interior transmission problem", *Inverse Probl. Imaging* **1**:1 (2007), 13–28. MR 2008j:35027 Zbl 1130.35132

[Colton et al. 2010] D. Colton, P. Monk, and J. Sun, "Analytical and computational methods for transmission eigenvalues", *Inverse Problems* **26**:4 (2010), 045011. MR 2010m:35559 Zbl 1192.78024

[Cossonnière 2011] A. Cossonnière, *Valeurs propres de transmission et leur utilisation dans l'identification d'inclusions à partir de mesures eléctromagnétiques*, Ph.D. Thesis, Institut National des Sciences Appliquées, Toulouse, 2011, Available at http://eprint.insa-toulouse.fr/archive/00000417/.

[Cossonnière and Haddar 2011] A. Cossonnière and H. Haddar, "The electromagnetic interior transmission problem for regions with cavities", *SIAM J. Math. Anal.* **43**:4 (2011), 1698–1715. MR 2012h:35374 Zbl 1229.78014

[Davis 1963] P. J. Davis, *Interpolation and approximation*, Blaisdell Publishing Co., New York, 1963. MR 28 #393 Zbl 0111.06003

[Erdélyi 1956] A. Erdélyi, *Asymptotic expansions*, Dover, New York, 1956. MR 17,1202c Zbl 0070.29002

[Giovanni and Haddar 2011] G. Giovanni and H. Haddar, "Computing estimates on material properties from transmission eigenvalues", research report RR-7729, INRIA, 2011.

[Haddar 2004] H. Haddar, "The interior transmission problem for anisotropic Maxwell's equations and its applications to the inverse problem", *Math. Methods Appl. Sci.* **27**:18 (2004), 2111–2129. MR 2005i:35255 Zbl 1062.35168

[Hickmann 2012] K. Hickmann, "Interior transmission problem with refractive index having C^2-transition to the background medium", *Applicable Analysis* **91**:9 (2012), 1675–1690.

[Hitrik et al. 2010] M. Hitrik, K. Krupchyk, P. Ola, and L. Päivärinta, "Transmission eigenvalues for operators with constant coefficients", *SIAM J. Math. Anal.* **42**:6 (2010), 2965–2986. MR 2011j:35053 Zbl 05936776

[Hitrik et al. 2011a] M. Hitrik, K. Krupchyk, P. Ola, and L. Päivärinta, "The interior transmission problem and bounds on transmission eigenvalues", *Math. Res. Lett.* **18**:2 (2011), 279–293. MR 2012c:35079 Zbl 1241.47057

[Hitrik et al. 2011b] M. Hitrik, K. Krupchyk, P. Ola, and L. Päivärinta, "Transmission eigenvalues for elliptic operators", *SIAM J. Math. Anal.* **43**:6 (2011), 2630–2639. MR 2873234 Zbl 1233.35148

[Kirsch 1986] A. Kirsch, "The denseness of the far field patterns for the transmission problem", *IMA J. Appl. Math.* **37**:3 (1986), 213–225. MR 90b:35059 Zbl 0652.35104

[Kirsch 2009] A. Kirsch, "On the existence of transmission eigenvalues", *Inverse Probl. Imaging* **3**:2 (2009), 155–172. MR 2010m:35346 Zbl 1186.35122

[Kirsch and Grinberg 2008] A. Kirsch and N. Grinberg, *The factorization method for inverse problems*, Oxford Lecture Series in Mathematics and its Applications **36**, Oxford University Press, Oxford, 2008. MR 2009k:35322 Zbl 1222.35001

[Lakshtanov and Vainberg 2012] E. Lakshtanov and B. Vainberg, "Ellipticity in the interior transmission problem in anisotropic media", *SIAM J. Math. Anal.* **44**:2 (2012), 1165–1174. MR 2914264 Zbl 06070883

[Lax and Phillips 1967] P. D. Lax and R. S. Phillips, *Scattering theory*, Pure and Applied Mathematics **26**, Academic Press, New York, 1967. MR 36 #530 Zbl 0186.16301

[Leung and Colton 2012] Y.-J. Leung and D. Colton, "Complex transmission eigenvalues for spherically stratified media", *Inverse Problems* **28**:7 (2012), 075005.

[Markus 1988] A. S. Markus, *Introduction to the spectral theory of polynomial operator pencils*, Translations of Mathematical Monographs **71**, American Mathematical Society, Providence, RI, 1988. MR 89h:47023 Zbl 0678.47005

[McLaughlin and Polyakov 1994] J. R. McLaughlin and P. L. Polyakov, "On the uniqueness of a spherically symmetric speed of sound from transmission eigenvalues", *J. Differential Equations* **107**:2 (1994), 351–382. MR 94m:35311 Zbl 0803.35163

[McLaughlin et al. 1994] J. R. McLaughlin, P. L. Polyakov, and P. E. Sacks, "Reconstruction of a spherically symmetric speed of sound", *SIAM J. Appl. Math.* **54**:5 (1994), 1203–1223. MR 95f:34017 Zbl 0809.34024

[Päivärinta and Sylvester 2008] L. Päivärinta and J. Sylvester, "Transmission eigenvalues", *SIAM J. Math. Anal.* **40**:2 (2008), 738–753. MR 2009m:35358 Zbl 1159.81411

[Rynne and Sleeman 1991] B. P. Rynne and B. D. Sleeman, "The interior transmission problem and inverse scattering from inhomogeneous media", *SIAM J. Math. Anal.* **22**:6 (1991), 1755–1762. MR 93f:76088 Zbl 0733.76065

[Serov and Sylvester 2012] V. Serov and J. Sylvester, "Transmission eigenvalues for degenerate and singular cases", *Inverse Problems* **28**:6 (2012), 065004.

[Sun 2011] J. Sun, "Estimation of transmission eigenvalues and the index of refraction from Cauchy data", *Inverse Problems* **27**:1 (2011), 015009. MR 2012d:65261 Zbl 1208.35176

[Sylvester 2012] J. Sylvester, "Discreteness of transmission eigenvalues via upper triangular compact operators", *SIAM J. Math. Anal.* **44**:1 (2012), 341–354. MR 2888291 Zbl 1238.81172

[Weck 2004] N. Weck, "Approximation by Maxwell–Herglotz-fields", *Math. Methods Appl. Sci.* **27**:5 (2004), 603–621. MR 2005b:35271 Zbl 1044.35097

cakoni@math.udel.edu Department of Mathematical Sciences,
University of Delaware, Newark, DE 19716, United States

haddar@cmap.polytechnique.fr INRIA Saclay Ile de France/CMAP École Polytechnique,
Route de Saclay, 91128 Palaiseau Cedex, France